Coastal Risk Management in a Changing Climate

Coastal Risk Management in a Changing Climate

Edited by

Barbara Zanuttigh

Associate Professor, Department of Civil, Chemicals Environmental and Materials Engineering University of Bologna, Viale Risorgimento 2 40136 Bologna, Italy

Robert Nicholls

Professor, Faculty of Engineering and the Environment University of Southampton Highfield, Southampton, SO17 1BJ, UK

Jean Paul Vanderlinden

Professor, Cultures, Environments, Arctic, Representations Climate Research Center, Université de Versailles Saint-Quentin-en-Yvelines 11 boulevard d'Alembert, 78280 Guyancourt, France

Hans F. Burcharth

Professor, Department of Civil Engineering, Aalborg University Sofiendalsvej 9-11, DK-9200 Aalborg SV, Denmark

Richard C. Thompson

Professor, Marine Biology and Ecology Research Centre School of Marine Science and Engineering, Plymouth University Drake Circus, Plymouth, PL4 8AA, UK

AMSTERDAM • BOSTON • HEIDELBERG • LONDON
NEW YORK • OXFORD • PARIS • SAN DIEGO
SAN FRANCISCO • SINGAPORE • SYDNEY • TOKYO
Butterworth-Heinemann is an imprint of Elsevier

Butterworth-Heinemann is an imprint of Elsevier
The Boulevard, Langford Lane, Kidlington, Oxford OX5 1GB, UK
225 Wyman Street, Waltham, MA 02451, USA

Library of Congress Cataloging-in-Publication Data
Coastal risk management in a changing climate / edited by Barbara Zanuttigh.
pages cm
ISBN 978-0-12-397310-8 (paperback)
1. Coastal zone management–Case studies. 2. Coast changes–Risk assessment–Case studies. 3. Climatic changes–Risk assessment–Case studies. 4. Coastal engineering–Case studies. I. Zanuttigh, Barbara, editor of compilation.
HT391.C49816 2014
333.91'7–dc23

2014035128

British Library Cataloguing in Publication Data
A catalogue record for this book is available from the British Library

ISBN: 978-0-12-397310-8

For all information on all Butterworth-Heinemann publications
visit our Web site at http://store.elsevier.com/

Cover credit: Blankenberge, Belgium. By courtesy of the Flanders Marine Institute (www.vliz.be), Belgium.

Contents

Acknowledgments

The Editors wish to express a special thanks to Dr. Susan Hanson, University of Southampton, and Dr. Simon Hoggart, University of Plymouth, for their invaluable effort and kind assistance contributing to this book.

The whole Authors team—and especially the First Editor as Project Coordinator—would like to gratefully acknowledge the support of the European Commission through the Grant 244104, THESEUS "Innovative technologies for safer European coastal areas", www.theseusproject.eu, 2009–2013, funded within the FP7 theme 6 "Environment and climate change".

Grateful thanks to Simon Claus and Daphnis De Pooter of the Flanders Marine Institute, for the effort spent on increasing risk awareness and disseminating the results of THESEUS and of this book.

List of Contributors

Surname, Name	Title	Affiliation (Acronym as THESEUS partner)	Address / Contact
Airoldi, Laura	PhD, Associate Professor	Dipartimento di Scienze Biologiche, Geologiche ed Ambientali, University of Bologna (UniBo)	Via S. Alberto 163, 48123, Ravenna, Italy laura.airoldi@unibo.it
Andreeva, Nataliya	PhD	Institute of Oceanology, Bulgarian Academy of Sciences (IO-BAS)	PO box 152, Varna, Bulgaria n.andreeva@io-bas.bg
Androulidakis, Ioannis	PhD, Engineer	Laboratory of Maritime Engineering and Maritime Works, Department of Civil Engineering, Aristotle University of Thessaloniki (AUTh)	Aristotle University Campus, 54124 Thessaloniki, Greece iandroul@civil.auth.gr
Angelelli, Elisa	PhD	Department of Civil, Chemical, Environmental and Materials Engineering, University of Bologna (UniBo)	Viale Risorgimento 2 40136 Bologna, Italy elisa.angelelli4@unibo.it
Baztan, Juan	PhD	Marine Science for Society. Cultures, Environments, Arctic, Representations, Climate research Center, Université de Versailles Saint-Quentin-en-Yvelines (UVSQ)	11 boulevard d'Alembert, 78280 Guyancourt, France. jbaztan@marine-sciences-for-society.org
van Belzen, Jim	MSc	Royal Netherlands Institute for Sea Research, Spatial Ecology Department (NIOZ)	Korringaweg 7, 4401NT, Yerseke, The Netherlands Jim.van.Belzen@nioz.nl
Bi, Qilong	PhD Student	Hydraulics Section, Civil Engineering Department, KU Leuven (KUL)	Kasteelpark Arenberg 40, postbus 2448, 3001 Leuven Qilong.Bi@bwk.kuleuven.be

Bichot, Amandine	MsC	EID Méditerranée (EID)	165 avenue Paul Rimbaud 34184 Montpellier Cedex4 France abichot@eid-med.org
Bilton, David T	PhD, Associate Professor and Reader in Aquatic Biology	Marine Biology and Ecology Research Centre, School of Marine Science and Engineering, University of Plymouth (UoP)	Drake Circus, Plymouth, PL4 8AA, UK D.Bilton@plymouth.ac.uk
Bohn, Katrin	PhD, Post-doctoral Research Associate	1 - Ocean and Earth Science, National Oceanography Centre Southampton, University of Southampton (UoS) 2 - School of Ocean Sciences, Bangor University (BU)	1- Ocean and Earth Science, National Oceanography Centre Southampton, Waterfront Campus, University of Southampton (UoS), Southampton, SO14 3ZH, UK 2 - Askew Street, Menai Bridge, LL59 5AB, UK Katrin.Bohn@soton.ac.uk
Bouma, Tjeerd J	PhD	Royal Netherlands Institute for Sea Research, Spatial Ecology Department (NIOZ)	Korringaweg 7, 4401NT, Yerseke, The Netherlands Tjeerd.Bouma@nioz.nl
Bozzeda, Fabio	PhD	Dip. Scienze Biologiche, Geologiche Ambientali, University of Bologna – Ravenna Campus (UniBo)	Via S. Alberto, 161 48123-Ravenna fabio.bozzeda2@unibo.it
Brivois, Olivier		Bureau de Recherches Géologiques et minières (BRGM)	3 avenue Claude-Guillemin BP 36009 45060 Orléans Cedex 2, France o.brivois@brgm.fr
Burcharth, Hans F	Professor, Dr. Techn.	Department of Civil Engineering, Aalborg University (AAU)	Sofiendalsvej 9-11 DK-9200 Aalborg SV, Denmark hfb@civil.aau.dk
Chen, Zhongyuan	Professor	State Key Laboratory for Estuarine and Coastal Research, East China Normal University (SKLEC)	No. 3663, Zhongshan North Road Shanghai 200062, China z.chen@sklec.ecnu.edu.cn

(Continued)

Surname, Name	Title	Affiliation (Acronym as THESEUS partner)	Address / Contact
Coates, Tracey	PhD	Flood Hazard Research Centre, Middlesex University (MU) (currently: School of Geography, Geology and the Environment at Kingston University)	The Burroughs, Hendon London NW4 4BT (currently: Penryhn Road, Kingston upon Thames, UK, T.Coates@kingston.ac.uk)
Colangelo, Marina Antonia	PhD, Senior Researcher	Dip. Scienze Biologiche, Geologiche Ambientali, University of Bologna – Ravenna Campus (UniBo)	Via S. Alberto, 161 48123-Ravenna marina.colangelo@unibo.it
Le Cozannet, Goneri		Bureau de Recherches Géologiques et minières (BRGM)	3 avenue Claude-Guillemin BP 36009 45060 Orléans Cedex 2, France g.lecozannet@brgm.fr
Cui, Lifang		State Key Laboratory of Estuarine and Coastal Research, East China Normal University (SKLEC)	3663 N. Zhongshan Rd, 200062, Shanghai, P. R. China
Davies, Andrew J.	PhD	School of Ocean Sciences, Bangor University (BU)	Askew Street, Menai Bridge, LL59 5AB, UK andrew.j.davies@bangor.ac.uk
Díaz-Simal, Pedro	PhD, Associate Proffesor	Environmental Hydraulics Institute "IH Cantabria", Universidad de Cantabria (UC)	C/Isabel Torres n° 15, 39011, Santander, Spain pedro.diaz@unican.es
Doncheva, Valentina	PhD, Associate Professor	Institute of Oceanology, Bulgarian Academy of Sciences (IO-BAS)	PO box 152, Varna, Bulgaria doncheva@io-bas.bg
Durán, Gerardo	PhD Student	Instituto de Ingeniería, Universidad Nacional Autónoma de México (UNAM)	Circuito Escolar s/n, 04510 Coyoacán, DF. México GDuranV@iingen.unam.mx

Echavarri, Beatriz	PhD, Associate Researcher	Environmental Hydraulics Institute "IH Cantabria", Universidad de Cantabria (UC)	C/Isabel Torres n° 15, 39011, Santander, Spain beatriz.echavarri@unican.es
Eftimova, Petya	PhD Student	Institute of Oceanology, Bulgarian Academy of Sciences (IO-BAS)	PO box 152, Varna, Bulgaria eftimova@io-bas.bg
Escudero, Mireille del Carmen	PhD Student	Instituto de Ingeniería, Universidad Nacional Autónoma de México (UNAM)	Circuito Escolar s/n 04510 Coyoacán, DF. México MEscuderoC@iingen.unam.mx
Felts, Didier		Coordonnées des Centres d'études techniques de l'Equipement (CETMEF)	24, rue Carton BP 58 33019 Bordeaux Cedex, France didier.felts@cerema.fr
Ferrario, Filippo	MSc	Dipartimento di Scienze Biologiche, Geologiche ed Ambientali, University of Bologna (UniBo)	Via S. Alberto 163 48123, Ravenna, Italy filippo.ferrario@unibo.it
Firth, Louise B	PhD, University Fellow	Ryan Institute, National University of Ireland Galway; School of Ocean Sciences, Bangor University (BU)	Menai Bridge, Anglesey, LL59 5AB University Road, Galway, Ireland louise.firth@nuigalway.ie
Flipova-Marinova, Mariana	Professor in Botany	Department of Natural History, Varna Regional Museum of History (IO-BAS)	41 Maria Louisa Blvd., 9000 Varna, Bulgaria marianafilipova@yahoo.com
Formentin, Sara Mizar	PhD student, Engineer	Department of Civil, Chemical, Environmental and Materials Engineering, University of Bologna (UniBo)	Viale Risorgimento 2 40136 Bologna, Italy saramizar.formentin2@unibo.it
Fox, Andrew	Dr	COAST Research Group, School of Marine Science and Engineering, Plymouth University (UoP)	Drake Circus, Plymouth PL4 8AA, United Kingdom andrew.fox@plymouth.ac.uk

(Continued)

Surname, Name	Title	Affiliation (Acronym as THESEUS partner)	Address / Contact
Galiatsatou, Panagiota	PhD, Research Associate	Laboratory of Hydraulics and Hydraulic Works, Aristotle University of Thessaloniki (AUTh)	Aristotle University, Dept. of Civil Eng., Thessaloniki, 54124, Greece pgaliats@civil.auth.gr
Galván, Cristina	PhD Student	University of Cantabria (UC)	C/Isabel Torres n° 15, 39011, Santander, Spain galvanc@unican.es
Ge, Jianzhong	PhD	State Key Laboratory for Estuarine and Coastal Research, East China Normal University (SKLEC)	No. 3663, Zhongshan North Road Shanghai 200062, China gegsdw@hotmail.com
González Dávila, Osiel	Research Associate Dr.	Department of International and European Economic Studies, Athens University of Economics and Business (AUEB)	76, Patission Str. 10434, Athens, Greece osiel.davila@soas.ac.uk
Gonzalez, Raul	Dr	COAST Research Group, School of Marine Science and Engineering, Plymouth University, Drake Circus (UoP)	Plymouth PL4 8AA, United Kingdom raul.gonzalez@plymouth.ac.uk
Hanley, Michael	PhD	Marine Biology and Ecology Research Centre, School of Biological Sciences, University of Plymouth (UoP)	Drake Circus, Plymouth, PL4 8AA, UK Michael.Hanley@plymouth.ac.uk
Hanson, Susan	PhD	Faculty of Engineering and the Environment, University of Southampton (Soton)	Highfield, Southampton, SO17 1BJ, UK s.e.hanson@soton.ac.uk

Hawkins, Stephen J	PhD, DSc, FSB Dean of Natural and Environmental Sciences Professor of Natural Sciences	Ocean and Earth Science, National Oceanography Centre Southampton, University of Southampton (Soton)	Ocean and Earth Science, National Oceanography Centre Southampton, Waterfront Campus, University of Southampton (UoS), Southampton, SO14 3ZH, UK S.J.Hawkins@soton.ac.uk
Heurtefeux, Hugues	Civil Engineer	Entente Interdépartementale pour la démoustication du littoral méditerranéen (EID)	165 avenue Paul Rimbaud 34184 Montpellier Cedex4 France hheurtefeux@eid-med.org
Higuera, Pablo	PhD Student	Environmental Hydraulics Institute "IH Cantabria", Universidad de Cantabria (UC)	C/Isabel Torres n°15, Parque Cientifico y Tecnologico de Cantabria, 39011 Santander, Spain
Hissel, François		Centre d'Études Techniques Maritimes Et Fluviales (CETMEF)	134 rue de Beauvais - CS 60039 60280 Margny Lès Compiègne, France francois.hissel@cerema.fr
Hoggart, Simon	PhD, Post-doctoral Research Associate	Marine Biology and Ecology Research Centre, School of Marine Science and Engineering, University of Plymouth (UoP)	Drake Circus, Plymouth, PL4 8AA, UK Simon.Hoggart@plymouth.ac.uk
Horrillo-Caraballo, Jose M.	Dr	School of Marine Science and Engineering, University of Plymouth (UoP)	Drake Circus, Plymouth, Devon, PL4 8AA Jose.Horrillo-Caraballo@plymouth.ac.uk
Izaguirre Lasa, Cristina	Associate Researcher	Environmental Hydraulics Institute "IH Cantabria", Universidad de Cantabria (UC)	c/ Isabel Torres no15, Parque Científico y Tecnológico de Cantabria, 39011 Santander izaguirrec@unican.es

(Continued)

Surname, Name	Title	Affiliation (Acronym as THESEUS partner)	Address / Contact
Kane, Idrissa Oumar		Cultures, Environments, Arctic, Representations, Climate research Center, Université de Versailles Saint-Quentin-en-Yvelines (UVSQ)	11 boulevard d'Alembert, 78280 Guyancourt, France Idrissa-oumar.kane@reeds.uvsq.fr
Kappenberg, Jens	Dr.-Ing.	Helmholtz-Zentrum Geesthacht Centre for Materials and Coastal Research (HZG)	Max-Planck-Str. 1, 21502 Geesthacht, Germany jens.kappenberg@hzg.de
Karambas, Theofanis	Associate Professor	Laboratory of Hydraulics and Hydraulic Works, Aristotle University of Thessaloniki (AUTh)	Aristotle University, Dept. of Civil Eng., Thessaloniki, 54124, Greece karambas@civil.auth.gr
Keremedchiev, Stoyan	PhD, Associate Professor	Institute of Oceanology, Bulgarian Academy of Sciences (IO-BAS)	PO box 152, Varna, Bulgaria s.keremedchiev@io-bas.bg
Kirilova, Dimitrichka	Technician	Institute of Oceanology, Bulgarian Academy of Sciences (IO-BAS)	PO box 152, Varna, Bulgaria coastal@io-bas.bg
Koftis, Theoharris	PhD, Research Associate	Laboratory of Hydraulics and Hydraulic Works, Aristotle University of Thessaloniki (AUTh)	Hydraulics Laboratory, University Campus, AUTh, Thesaloniki 54124 thkoftis@civil.auth.gr
Kombiadou, Katerina	PhD	Laboratory of Maritime Engineering and Maritime Works, Department of Civil Engineering, Aristotle University of Thessaloniki (AUTh)	Aristotle University Campus, 54124 Thessaloniki, Greece kobiadou@civil.auth.gr

Konsulova, Tsenka	PhD, Associate Professor	Institute of Oceanology, Bulgarian Academy of Sciences (IO-BAS)	PO box 152, Varna, Bulgaria konsulova@io-bas.bg
Kotsev, Iliyan	PhD Student	Institute of Oceanology, Bulgarian Academy of Sciences (IO-BAS)	PO box 152, Varna, Bulgaria ikotsev@io-bas.bg
Koundouri, Phoebe	Associate Professor Dr.	Department of International and European Economic Studies, Athens University of Economics and Business (AUEB)	76, Patission Str. 10434, Athens, Greece pkoundouri@aueb.gr
Kowalska, Beata	MSc Eng. Senior Researcher	Institute of Meteorology and Water Management O/M Gdynia – National Research Institute (IMGW)	42 Waszyngtona, 81-342 Gdynia, Poland, beata.kowalska@imgw.pl
Krestenitis, Yannis	Professor	Laboratory of Maritime Engineering and Maritime Works, Department of Civil Engineering, Aristotle University of Thessaloniki (AUTh)	Aristotle University Campus, 54124 Thessaloniki, Greece ynkrest@civil.auth.gr
Kuznetsov, Sergey	Professor, PhD	Coast & Shelf Laboratory P.P.Shirshov Institute of Oceanology of Russian Academy of Sciences (IORAS)	Nakhimovskii prospect 36 Moscow 117997 Russia kuznetsov@ocean.ru
Kwan, Sum		Port Works Division, Civil Engineering and Development Department, HKSAR Government (UoP)	4/F, CEDD Building, 101, Princess Margaret Road, Kowloon, Hong Kong samuelshkwan@cedd.gov.hk
Laborie, Vanessya		Centre d'Études Techniques Maritimes Et Fluviales (CETMEF)	134 rue de Beauvais - CS 60039 60280 Margny Lès Compiègne, France vanessya.laborie@cerema.fr

(Continued)

Surname, Name	Title	Affiliation (Acronym as THESEUS partner)	Address / Contact
Lara, Javier L	PhD, Associate Professor	Environmental Hydraulics Institute "IH Cantabria", Universidad de Cantabria (UC)	C/Isabel Torres n° 15, 39011, Santander, Spain jav.lopez@unican.es
Lendzion, Jacek	Senior Researcher	Institute of Meteorology and Water Management O/M Gdynia – National Research Institute (IMGW)	42 Waszyngtona, 81-342 Gdynia, Poland, jaceklendzion@gmail.com
Leont'yev, Igor	DSc	P.P. Shirshov Institute of Oceanology of the Russian Academy of Sciences (IO RAS)	igor.leontiev@gmail.com
Li, Maotian	Associate Professor	State Key Laboratory for Estuarine and Coastal Research (SKLEC), East China Normal University	No. 3663, Zhongshan North Road Shanghai 200062, China mtli@sklec.ecnu.edu.cn
Losada Rodriguez, Iñigo	PhD, Full Professor	Environmental Hydraulics Institute "IH Cantabria", Universidad de Cantabria (UC)	C/Isabel Torres n° 15, 39011, Santander, Spain losadai@unican.es
Lykke Andersen, Thomas	PhD, Associate Professor	Department of Civil Engineering, Aalborg University (AAU)	Sofiendalsvej 9-11 DK-9200 Aalborg SV, Denmark tla@civil.aau.dk
Martinelli, Luca	PhD, Assistant Professor	Padova University, ICEA Dept. Part of the Consortium for coordination of research activities concerning the Venice lagoon system (CORILA)	v. Ognissanti, 39 35129, Padova, Italy, luca.martinelli@unipd.it

Maza, Maria	PhD Student	Environmental Hydraulics Institute "IH Cantabria", Universidad de Cantabria (UC)	C/Isabel Torres n° 15, 39011, Santander, Spain mazame@unican.es
McFadden, Loraine	PhD	Flood Hazard Research Centre Middlesex University (MU)	The Burroughs, Hendon London NW4 4BT, UK
Medina, Raul	PhD, Full Professor	Environmental Hydraulics Institute "IH Cantabria", Universidad de Cantabria (UC)	C/Isabel Torres n° 15, 39011, Santander, Spain raul.medina@unican.es
Mendez Incera, Fernando	Associate Professor	Environmental Hydraulics Institute "IH Cantabria", Universidad de Cantabria (UC)	c/ Isabel Torres no15, Parque Científico y Tecnológico de Cantabria, 39011 Santander mendezf@unican.es
Mendoza, Edgar	PhD, Assistant Professor	Instituto de Ingeniería, Universidad Nacional Autónoma de México (UNAM)	Circuito Escolar s/n 04510 Coyoacán, DF. México EMendozaB@iingen.unam.mx
Miętus, Mirosław	Professor	University of Gdańsk, Institute of Geography (IMGW)	4 Bażyńskiego, 80-952 Gdańsk, Poland, mietus@ug.edu.pl
Monbaliu, Jaak	Professor	Hydraulics Section, Civil Engineering Department , KU Leuven (KUL)	Kasteelpark Arenberg 40, postbus 2448, 3001 Leuven jaak.monbaliu@bwk.kuleuven.be
Narayan, Siddarth	PhD	Faculty of Engineering and the Environment, University of Southampton (Soton)	Highfield, Southampton, SO17 1BJ, UK sid.narayan@soton.ac.uk
Nguyen, Dan Kim	PhD, Senior Researcher	Institute of Marine and Inland Waterways (CETMEF)	CEREMA-DTEMF/EDF R&D, Laboratoire Saint-Venant, Bâtiment O 6, quai Watier B.P. 49 -78401 Chatou Cedex, France kimdan_nguyen@yahoo.fr

(Continued)

Surname, Name	Title	Affiliation (Acronym as THESEUS partner)	Address / Contact
Nicholls, Robert	Professor	Faculty of Engineering and the Environment, University of Southampton (Soton)	Highfield, Southampton, SO17 1BJ, UK r.j.nicholls@soton.ac.uk
Nikolaev, Rosen	Assoc. Prof. PhD	Department of Mathematical Sciences, University of Economics - Varna	77 "Kniaz Boris I" blvd., 9000 Varna, Bulgaria nikolaev_rosen@ue-varna.bg
Nørgaard, Jørgen Quvang Harck	PhD, Assistant Professor	Department of Civil Engineering, Aalborg University (AAU)	Department of Civil Engineering Sofiendalsvej 9-11 DK-9200 Aalborg SV, Denmark jhn@civil.aau.dk
Ohle, Nino	MSc	Hamburg Port Authority (HPA)	Neuer Wandrahm 4, 20457 Hamburg, Germany, nino.ohle@hpa.hamburg.de
Oliveros, Carlos		Bureau de Recherches Géologiques et minières (BRGM)	3 avenue Claude-Guillemin BP 36009 45060 Orléans Cedex 2, France c.oliveros@brgm.fr
Ondiviela Eizaguirre, Barbara	Associate Researcher	Environmental Hydraulics Institute "IH Cantabria", Universidad de Cantabria (UC)	c/ Isabel Torres no15, Parque Científico y Tecnológico de Cantabria, 39011 Santander ondiviela@unican.es
Ostrowski, Rafał	DSc, Associate Professor	Institute of Hydro-Engineering of the Polish Academy of Sciences (IBW PAN)	7 Kościerska, 80-328 Gdańsk, Poland, rafal.o@ibwpan.gda.pl

Pan, Shunqi	PhD	School of Marine Science and Engineering, Plymouth University (UoP) (currently: School of Engineering, Cardiff University)	Drake Circus, Plymouth, PL4 8AA, UK (currently: The Parade, Cardiff, CF24 3AA, UK) PanS2@cf.ac.uk
Parker, Dennis	Professor, PhD	Flood Hazard Research Centre Middlesex University (MU)	The Burroughs, Hendon London NW4 4BT denjpark@btinternet.com
Penning-Rowsell, Edmund	Professor, PhD	Flood Hazard Research Centre Middlesex University (MU)	The Burroughs Hendon London NW4 4BT, UK
Pescaroli, Gianluca	Research Fellow	University of Bologna (UniBo)	via Filippo Re, 6 - 40126 BOLOGNA, Italy
Petkov, Jordan	Professor, PhD	Department of Mathematical Sciences, University of Economics - Varna	77 "Kniaz Boris I" blvd., 9000 Varna, Bulgaria jr_petkov@ue-varna.bg
Pietrantoni, Luca	Associate Professor	University of Bologna (UniBo)	via Filippo Re, 6 - 40126 BOLOGNA, Italy luca.pietrantoni@unibo.it
Piotrowska, Danuta	Technician	Institute of Hydro-Engineering of the Polish Academy of Sciences (IBW PAN)	7 Kościerska, 80-328 Gdańsk, Poland, dapio@ibwpan.gda.pl
Polonsky, Alexander	Professor	Marine Hydrophysical Institute - Ukrainian National Academy of Scienses (MHI)	2 Kapitanskaya st. Sevastopol, Ukraine apolonsky5@mail.ru
Prati, Gabriele	PhD	University of Bologna (UniBo)	via Filippo Re, 6 - 40126 BOLOGNA, Italy gabriele.prati@unibo.it
Prévot, Guirec	Engineer	Centre d'Études Techniques Maritimes Et Fluviales (CETMEF)	134, rue de Beauvais, F-60280 Margny lès Compiègne, France guirec.prevot@developpement-durable.gouv.fr

(Continued)

Surname, Name	Title	Affiliation (Acronym as THESEUS partner)	Address / Contact
Prinos, Panayotis	Professor	Laboratory of Hydraulics and Hydraulic Works, Aristotle University of Thessaloniki (AUTh)	Aristotle University, Dept. of Civil Eng., Thessaloniki, 54124, Greece prinosp@civil.auth.gr
Pruszak, Zbigniew	Professor	Institute of Hydro-Engineering of the Polish Academy of Sciences (IBW PAN)	7 Kościerska, 80-328 Gdańsk, Poland, zbig@ibwpan.gda.pl
Puente, Araceli	PhD, Associate Professor	Environmental Hydraulics Institute "IH Cantabria", Universidad de Cantabria (UC)	C/Isabel Torres n° 15, 39011, Santander, Spain araceli.puente@unican.es
Reeve, Dominic E	Professor	School of Marine Science and Engineering, Plymouth University (UoP) (currently: Swn-U College of Engineering, Swansea University)	Drake Circus, Plymouth, PL4 8AA, UK (Singleton Park, Swansea, SA2 8PP, UK D.E.Reeve@swansea.ac.uk)
Rulleau, Bénédicte	Assistant Professor, PhD.	Cultures, Environments, Arctic, Representations, Climate Research Center, Université de Versailles Saint-Quentin-en-Yvelines (UVSQ)	11 boulevard d'Alembert, 78280 Guyancourt, France benedicte.rulleau@uvsq.fr
Rundle, Simon D	PhD	Marine Biology and Ecology Research Centre, School of Marine Science and Engineering, University of Plymouth (UoP)	Drake Circus, Plymouth, PL4 8AA, UK S.Rundle@plymouth.ac.uk
Ruol, Piero	Full Professor, PhD	Padova University, ICEA Dept., as part of the Consortium for coordination of research activities concerning the Venice lagoon system (CORILA)	Via Ognissanti, 39 35129 Padova – Italy piero.ruol@unipd.it

Saprykina, Yana	PhD	Coast & Shelf Laboratory P.P.Shirshov Institute of oceanology of Russian Academy of Sciences (IORAS)	Nakhimovskii prospect 36 Moscow 117997 Russia saprykina@ocean.ru
Schönhofer, Jan	MSc	Institute of Hydro-Engineering of the Polish Academy of Sciences (IBW PAN)	7 Kościerska, 80-328 Gdańsk, Poland, jan.schonhofer@ibwpan.gda.pl
Schuster, Dagmar	MSc	Hamburg Port Authority (HPA)	Neuer Wandrahm 4, 20457 Hamburg, Germany, dagmar.schuster@hpa.hamburg.de
Sergent, Philippe	PhD, Research Director	Centre d'Études Techniques Maritimes Et Fluviales (CETMEF)	134, rue de Beauvais, F-60280 Margny lès Compiègne, France philippe.sergent@developpement-durable.gouv.fr
Silva, Rodolfo	Professor	Instituto de Ingeniería, Universidad Nacional Autónoma de México (UNAM)	Circuito Escolar s/n, 04510 Coyoacán, DF. México RSilvaC@iingen.unam.mx
Simmonds, David	Associate Professor	School of Marine Science and Engineering (Faculty of Science & Environment), Plymouth University (UoP)	Drake Circus, Plymouth, PL4 8AA, UK D.Simmonds@plymouth.ac.uk
Skaja, Marek	BSC, Research Engineer	Institute of Hydro-Engineering of the Polish Academy of Sciences (IBW PAN)	7 Kościerska, 80-328 Gdańsk, Poland, mskaja@ibwpan.gda.pl
Skov, Martin W	PhD	School of Ocean Sciences, Bangor University (BU)	Askew Street, Menai Bridge, LL59 5AB, UK mwskov@bangor.ac.uk

(Continued)

Surname, Name	Title	Affiliation (Acronym as THESEUS partner)	Address / Contact
Smaoui, Hassan	PhD, Senior Researcher	Institute of Marine and Inland Waterways (CETMEF)	UTC-CEREMA/DTEM, Laboratoire Roberval, Bâtiment H Université de Technologie de Compiègne, CS 60319, 60203 Compiègne, France Hassan.Smaoui@developpement-durable.gouv.fr
Sothmann, Janina	MSc, PhD Student	Helmholtz-Zentrum Geesthacht Centre for Materials and Coastal Research (HZG)	Max-Planck-Str. 1, 21502 Geesthacht, Germany janina.sothmann@hzg.de
Stanisławczyk, Ida	MSc, Senior Researcher	Institute of Meteorology and Water Management O/M Gdynia – National Research Institute (IMGW)	42 Waszyngtona, 81-342 Gdynia, Poland, ida.stanislawczyk@imgw.pl
Steendam, Gosse Jan	Head of Department Flood Defence/Senior Consultant	INFRAM International BV (INFRAM)	P.O. Box 16, 8316 ZG Marknesse, The Netherlands
Stithou, Mavra	Research Associate, PhD	Department of International and European Economic Studies, Athens University of Economics and Business (AUEB)	Athens University of Economics and Business-Research Centre76, Patission Str. 10434, Athens, Greece stithoumavra@yahoo.com
Strain, Elisabeth M	PhD, Research Fellow	Dipartimento di Scienze Biologiche, Geologiche ed Ambientali, University of Bologna (UniBo)	Private Bag 129, Hobart, Tasmania, Australia 7001 estrain@utas.edu.au
Szmytkiewicz, Marek	DSc, Associate Professor	Institute of Hydro-Engineering of the Polish Academy of Sciences (IBW PAN)	7 Kościerska, 80-328 Gdańsk, Poland, mareksz@ibwpan.gda.pl

Szmytkiewicz, Piotr	PhD, Assistant Professor	Institute of Hydro-Engineering of the Polish Academy of Sciences (IBW PAN)	7 Kościerska, 80-328 Gdańsk, Poland, p.szmytkiewicz@ibwpan.gda.pl
Sztobryn, Marzenna	PhD, Assistant Professor	Institute of Meteorology and Water Management O/M Gdynia – National Research Institute (IMGW)	42 Waszyngtona, 81-342 Gdynia, Poland, marzenna.sztobryn@imgw.pl
Thierry, Blanche		Centre d'Études Techniques Maritimes Et Fluviales (CETMEF)	135 rue de Beauvais - CS 60039 60280 Margny Lès Compiègne, France
Thompson, Richard C	Professor	Marine Biology and Ecology Research Centre School of Marine Science and Engineering, Plymouth University (UoP)	Drake Circus, Plymouth, PL4 8AA, UK R.C.Thompson@plymouth.ac.uk
Tobio, Mirian Jimenez	PhD Student	Environmental Hydraulics Institute "IH Cantabria", Universidad de Cantabria (UC)	c/ Isabel Torres no15, Parque Científico y Tecnológico de Cantabria, 39011 Santander jimeneztm@unican.es
Todorova, Valentina	PhD, Associate Professor	Institute of Oceanology, Bulgarian Academy of Sciences (IO-BAS)	PO box 152, Varna, Bulgaria v.todorova@io-bas.bg
Toorman, Erik A.	Professor	Hydraulics Section, Civil Engineering Department, KU Leuven (KUL)	Kasteelpark Arenberg 40, postbus 2448, 3001 Leuven erik.toorman@bwk.kuleuven.be
Touili, Nabil		Cultures, Environments, Arctic, Representations, Climate (CEARC) Research Center, Université de Versailles Saint-Quentin-en-Yvelines (UVSQ)	11 boulevard d'Alembert, 78280 Guyancourt, France nabil.touili@uvsq.fr
Trifonova, Ekaterina	Engineer	Institute of Oceanology, Bulgarian Academy of Sciences (IO-BAS)	PO box 152, Varna, Bulgaria trifonova@io-bas.bg

(Continued)

Surname, Name	Title	Affiliation (Acronym as THESEUS partner)	Address / Contact
Valchev, Nikolay	PhD, Associate Professor	Institute of Oceanology, Bulgarian Academy of Sciences (IO-BAS)	PO box 152, Varna, Bulgaria valchev@io-bas.bg
Vanderlinden, Jean Paul	Professor	Cultures, Environments, Arctic, Representations, Climate Research Center, Université de Versailles Saint-Quentin-en-Yvelines (UVSQ)	11 boulevard d'Alembert, 78280 Guyancourt, France jean-paul.vanderlinden@uvsq.fr
Vergiev, Stoyan	PhD student	Department of Natural History, Varna Regional Museum of History (IO-BAS)	41 Maria Louisa Blvd., 9000 Varna, Bulgaria stvergiev@abv.bg
Villatoro, Monique Marie	Postdoctoral Student	Instituto de Ingeniería, Universidad Nacional Autónoma de México (UNAM)	Circuito Escolar s/n 04510 Coyoacán, DF. México MVillatoroL@iingen.unam.mx
de Vries, Wout		INFRAM International BV (INFRAM)	P.O. Box 16, 8316 ZG Marknesse, The Netherlands wout.devries@rws.nl
Weisse, Ralf	Dr. Senior Researcher	Helmholtz-Zentrum Geesthacht Centre for Materials and Coastal Research (HZG)	Max-Planck-Str. 1, D-21502 Geesthacht, Germany ralf.weisse@hzg.de
White, Anissia	BSc	Marine Biology and Ecology Research Centre, School of Biological Sciences, University of Plymouth (UoP)	Drake Circus, Plymouth, PL4 8AA, UK Anissia.White@plymouth.ac.uk

Willems, Patrick	Professor	Hydraulics Section, Civil Engineering Department, KU Leuven (KUL)	Kasteelpark Arenberg 40, postbus 2448, 3001 Leuven patrick.willems@bwk.kuleuven.be
Zagonari, Fabio	Associate Professor	Department of Economics, University of Bologna (UniBo)	Via Angherà 22 47921 Rimini – Italy fabio.zagonari@unibo.it
Zawadzka-Kahlau, Elżbieta	PhD, Senior Researcher	Institute of Meteorology and Water Management O/M Gdynia – National Research Institute (IMGW)	42 Waszyngtona, 81-342 Gdynia, Poland, elazawadzka46@gmail.com
Zhang, Liquan	Professor	State Key Laboratory of Estuarine and Coastal Research, East China Normal University (SKLEC)	3663 Zhongshan Road North, Shanghai 200062, China lqzhang@sklec.ecnu.edu.cn
Zhongyuan, Chen	Professor	State Key Laboratory of Estuarine and Coastal Research, East China Normal University (SKLEC)	3664 N. Zhongshan Rd, 200062, Shanghai, P. R. China z.chen@sklec.ecnu.edu.cn
Zhu, Zhenchang	MSc	1. State Key Laboratory of Estuarine and Coastal Research, East China Normal University (SKLEC) 2. Royal Netherlands Institute for Sea Research (NIOZ)	Korringaweg 7, 4400 AC Yerseke, the Netherlands Zhenchang.zhu@nioz.nl
Zanuttigh, Barbara	Associate Professor, PhD	Department of Civil, Chemicals, Environmental and Materials Engineering, University of Bologna (UniBo)	Viale Risorgimento 2 40136 Bologna, Italy barbara.zanuttigh@unibo.it

Chapter 1

Introduction

Barbara Zanuttigh[1], Robert Nicholls[2], Susan Hanson[2]
[1] *Department of Civil, Chemicals, Environmental and Materials Engineering, University of Bologna (UniBo), Bologna, Italy*
[2] *Faculty of Engineering and the Environment, University of Southampton (Soton), Southampton, BJ, UK*

1.1 Background

Coastal areas are important settlement zones and play a vital role in the wealth of many nations. Nearly 25% of the world's population lives within 62 miles (100 km) of a shoreline. Roughly 600 million people live in coastal regions that are less than 10 m above sea level, and this figure is likely to increase more rapidly (up to 50% over the next 25 years) than global population projections, reflecting coastal urbanization (Lichter, Vafeidis, Nicholls, & Kaiser, 2011; McGranahan, Balk, & Anderson, 2007). Many of the world's largest cities are located at the coast (e.g., Tokyo, Guangzhou, Bangkok, Istanbul, London, Lagos, New York, Buenos Aires) (Hanson et al., 2011), and of course include critical infrastructures such as energy facilities, ports, and airports. Coasts are also home to important and productive ecosystems that are increasingly valued by society.

Large coastal stretches are exposed to erosion and flooding worldwide (Hinkel et al., 2013; Hinkel, van Vuuren, Nicholls, & Klein, 2013). Coastal flooding has caused more than 3 million deaths since 1700, with a strong concentration around the Bay of Bengal (see Doocy, Daniels, Murray, & Kirsch, 2013; Nicholls, 2006). During

Coastal Risk Management in a Changing Climate. http://dx.doi.org/10.1016/B978-0-12-397310-8.00001-4

the worst sea surge recorded in modern European history, in the North Sea Surge in 1953, more than 2000 people lost their lives in England and the Netherlands, whereas 300 people died in Germany in 1962, which highlights the large areas in Europe that are also below normal high tide level. Most recent storms have been very economically disruptive because of flooding, for example, Hurricanes Katrina (2005) and Sandy (2012) in the United States and Typhoon Haiyan (2013) in the Philippines.

All the available assessments up to the Intergovernmental Panel on Climate Change Fourth Assessment show that world's coasts are threatened by sea-level rise and climate change in a variety of ways (Nicholls et al., 2007). Deltas, low-lying coastal plains, islands and barrier islands, beaches, coastal wetlands, and estuaries appear most affected by the acceleration in sea-level rise, although the local response will depend on the total sediment budget, for example, the Nile (Marriner, Flaux, Morhange, & Stanley, 2013) and in Bangladesh (Sarwar & Woodroffe, 2013). For erosion, this includes the indirect influence of sea-level rise on the beach sediment budget via infilling of coastal embayments or basins. As the sea level rises, estuaries and lagoons attempt to maintain equilibrium by raising their bed elevation and hence potentially act as a major sink of sand that is often derived from adjacent coasts, aggravating their erosion. For flooding, local subsidence has often been important, and this is likely to remain an issue in the future in susceptible areas, particularly for the extensive deltaic areas in Southern and Eastern Asia (e.g., Ericson, Vorosmarty, Dingman, Ward, & Meybeck, 2006; Syvitski et al., 2009).

These changes also have consequences for coastal habitats and ecosystems that have generated a societal response in the form of global, regional, and local conservation efforts (e.g., Ramsar Convention, European Directives). Climate change combines with and amplifies nonclimate stressors on coastal ecosystems (e.g., Jackson, Nordstrom, Feagin, & Smith, 2013; Jennerjahn & Mitchell, 2013). Intense development and overpopulation, poverty, internal conflict, fragmentation and loss of habitat, overfishing, pollution, and spread of invasive species will impair the resilience of ecosystems (i.e., the ability of the ecosystem to maintain its integrity and to continue to provide critical goods and services to coastal communities, such as fisheries, storm protection, erosion control, water storage, groundwater recharge, pollution abatement, and retention and cycling of nutrients and sediments) (e.g., van Slobbe et al., 2013; UK NEA, 2011). Healthy coastal habitats such as mangroves, coral reefs, estuaries, seagrass beds, and dune communities function as self-repairing "natural infrastructure" for flood mitigation, in contrast to human-built infrastructure, thus minimizing maintenance costs. When these critical resources are compromised, coastal ecosystems can deteriorate—and weakened, unhealthy coastal ecosystems are less resilient to climate change and variability and coastal flood risk may rise.

The application of risk assessment and management is one of the most important environmental policy developments of the past few decades. Modern societies recognize that their activities both depend upon and have consequences for the environment, and risk assessment can be used as a method for determining how and where to intervene for maximum benefit. Effects of human activity have an impact on sociopolitical institutions and environmentally dependent systems, such as the economy, human health, and natural ecosystems at different spatial and temporal scales. To be effective, flood risk management strategies therefore need to be developed with a multidisciplinary, long-term (many decades) perspective to include factors such as climate change, especially sea-level rise, coastal development pressures, and habitat implications. This is challenging because this is often beyond typical financial, political, and management decision timescales. The most comprehensive responses are seen in wealthy countries that have experienced significant coastal flooding (e.g., Japan, UK, The Netherlands).

Although in the media and the public mind, coastal floods and erosion are often seen as unnatural and nonallowable, they are natural occurrences, and no amount of investment can reduce risks to zero; the residual risk will always remain and society needs to define the acceptable level of this residual.

Traditional technical flood and erosion defenses have shown their limits. First, traditional coastal defenses have had an impact on their environment: in recent years, greater attention had been paid to the design and selection of coastal defense structures and technologies based on an integrated analysis of their performance and on their environmental and socioeconomic implications. Then, of course, they are facing their sensitivity to climate change: because of the expected sea-level rise and possible increase of storminess, thousands of kilometers of coastal dikes will be exposed to waves with heights exceeding the design value and will need to be improved or a higher risk accepted. Finally, what society expects from defenses is changing (e.g., widespread moves from hard to soft defenses to "hold the line" and maintain valuable breaches), and these needs will continue to evolve (e.g., the potential for widespread application of multipurpose structures such as surf reefs or marine renewable energy devices).

A more systems-based view of the coast, which is implied earlier in the chapter, suggests the possibility of enhancing overall coastal resilience rather than relying on single management measures. The management of risk levels also requires understanding the flood system as it responds to a range of internal and external drivers as well as the implementation of different risk mitigation options. In this way, more appropriate and locally acceptable decisions can be made that address local problems, while recognizing wider or more distant implications (Dawson et al., 2009; Jongman, Ward, & Aerts, 2012; Kok & Grossmann, 2010). Recent studies (e.g., Resilient Coasts Initiative, 2009) recommend six principles to enhance coastal resilience: (1) require risk-based land use planning; (2) design adaptable infrastructure and building code standards to meet future risk; (3) strengthen ecosystems as part of a risk mitigation strategy; (4) develop flexible adaptation plans;

(5) maintain a viable private property and casualty insurance market; and (6) integrate climate change impacts into due diligence for investment and lending. Although these ideas and concepts are well known, there is less practical experience of their application, especially applying a combination of different approaches together.

The effective management of coastal flooding therefore requires the active involvement and cooperation of multiple actors (national/regional governments, the private sector, research bodies, civil society, and community-based organizations and communities) playing differential but complementary well-defined roles across spatial and temporal scales. Learning from past and recent extreme events supports the development of a participatory framework in which all stakeholders have something to contribute to the process of risk governance and mutual communication. This, in turn, promotes societal acceptance of any proposed management approach.

1.2 THESEUS Project

Responding to coastal disasters and enhancing resilience to coastal erosion and flooding needs a holistic, participatory, and interdisciplinary approach in which science is embedded in the social, cultural, and economic context of coastal communities. It is in this spirit that the THESEUS project (www.theseusproject.eu, 2009–2013) aimed at delivering a safer (low-risk) coast for human use/development and healthy coastal habitats as sea levels rise and climate changes (and the economy continues to grow). The primary project objective was to develop an integrated methodology for planning sustainable defense strategies for the management of coastal erosion and flooding that addresses technical, social, economic, and environmental aspects.

The THESEUS consortium consisted of 31 partners; 25 from 12 EC Member States, four from International Cooperation Partners Countries (China, Mexico, Russia, Ukraine) and two from Third Countries (Taiwan and the United States). Among the Consortium were partners with expertise ranging from coastal and civil engineering; marine ecology; social sciences and economics; and meteorology and climate change to computer science and geographic information systems.

The most relevant issues addressed by THESEUS are synthesized as follows:

- A conceptual basis for flood assessment.
  The conceptual framework was based on the Source-Pathway-Receptor-Consequence model (FLOODsite Consortium, 2009; Narayan et al., 2014; Thorne, Evans, & Penning, 2007) and was aimed at providing a clear definition of the flood system and a conceptual map showing the inherent, causal

relationships and their interdependencies. Unlike many earlier flood assessments, THESEUS accounted for the consequences of flooding on both the human and natural systems and recognized that "risk" is a human or societal perception rather than an inherent characteristic of the natural system.

- Innovation and best practices in coastal engineering.
  THESEUS studied the resilience of existing management and explored new innovative approaches. For example, the resistance of grass covered sea dikes as well as methods for upgrading conventional rubble mound coastal structures, the use of artificial reefs as a wave dampening structure, and best practice for beach nourishment. THESEUS also systematically analyzed the use of floating wave energy converters for coastal protection purpose.
- Preserving and enhancing coastal ecosystems.
  The work in THESEUS shows how adopting a systems perspective allows the process understanding of habitats to be integrated with coastal engineering and social aspects of flooding to augment and increase the options available to flood risk managers. A better understanding of the habitat vulnerability to erosion and flooding process was also achieved and synthesized by means of threshold values and indicators.
- The governance dimension.
  THESEUS analyzed risk perception in terms of paradigmatic tensions associated with conflicting pertinence, normative, and evidence claims. As such, THESEUS allowed for the clear identification of norms as the main source of varying perceptions regarding coastal risks. THESEUS also analyzed a range of socially based mitigation options such as a simplified model of insurance, spatial planning, business recovery, and emergency planning that both reduce the consequences and improve resilience to flood events.
- Decision making for a sustainable coast.
  Within the large scope of THESEUS, one of the main objectives was to design a tool to help decision makers in defining optimal strategies to minimize risk in the short-, medium-, and long-term scenarios. The resulting software reproduces in a simplified way the most relevant physical processes (coastal erosion and flooding) induced by waves and sea levels, taking into account physical and nonphysical drivers, such as climate change, subsidence, population growth, and economic development.
- Science and policy.
  THESEUS strengthened the science and policy dialogue through three key measures: (1) learning from experience in study sites; (2) timely involvement of different actors (stakeholders, coastal managers, and policy makers); and (3) strategic dissemination, including products targeted to the audience, such as policy briefs, webinars, multimedia material, informative booklets, and of course the decision support system tool.

1.3 This Book

The book is composed of two main parts, providing the reader with up-to-date information and results based on relevant outcomes of the THESEUS project.

The first part (Chapters 2–6) presents a methodology for analyzing the coastal flood system and scoping the best combination of mitigation options including engineering, ecologically based, social, and economic measures. The second part (Chapter 7) presents the application of the developed methodology for risk assessment and mitigation to selected case studies.

More in details the book outline includes:

- a discussion of the coastal flood system in Chapter 2;
- risk-based conceptual design of innovative coastal structures (such as barriers for wave energy conversion, artificial reefs, and bottom vegetation) and technologies (nourishment and dredging, sand banks, sediment reservoirs, beach dewatering, and by-pass systems) in Chapter 3;
- evaluation of the impact "adaptation" technology (controlled seawater inundation) and estimation of the role of innovative mitigation measures based on the environment such as restoration of existing habitat or habitat creation in Chapter 4;
- setup of technologies and mitigation measures based on society and economy such as: spatial planning, controlled urbanization, flood resistant adaptations to buildings; innovative methods of siting, designing and managing business operations; evacuation plans; increase of social resilience; and adaptive capacity in Chapter 5;
- a holistic methodology for risk assessment and mitigation with specific reference to the Source Pathway Receptor and Consequence Model in Chapter 6;
- application to real worldwide cases in Chapter 7.

References

Dawson, R. J., Dickson, M. E., Nicholls, R. J., Hall, J. W., Walkden, M. J. A., Stansby, P. K., et al. (2009). Integrated analysis of risks of coastal flooding and cliff erosion under scenarios of long term change. *Climatic Change, 95*(1–2), 249–288. http://dx.doi.org/10.1007/s10584-008-9532-8.

Doocy, S., Daniels, A., Murray, S., & Kirsch, T. D. (2013). The human impact of floods: a historical review of events 1980–2009 and systematic literature review. *PLoS Currents, 5*. http://dx.doi.org/10.1371/currents.dis.f4deb457904936b07c09daa98ee8171a.

Ericson, J. P., Vorosmarty, C. J., Dingman, S. L., Ward, L. G., & Meybeck, M. (2006). Effective sea-level rise and deltas: causes of change and human dimension implications. *Global and Planetary Change, 50*(1–2), 63–82. http://dx.doi.org/10.1016/j.gloplacha.2005.07.004.

FLOODsite Consortium. (2009). *FLOODsite; integrated flood risk analysis and management methodologies*. FLOODsite Consortium http://www.floodsite.net/default.htm. Accessed 15.11.13.

Hanson, S., Nicholls, R., Ranger, N., Hallegatte, S., Corfee-Morlot, J., Herweijer, C., et al. (2011). A global ranking of port cities with high exposure to climate extremes. *Climatic Change, 104*(1), 89–112. http://dx.doi.org/10.1007/s10584-010-9977-4.

Hinkel, J., Nicholls, R. J., Tol, R. S. J., Wang, Z. B., Hamilton, J. M., Boot, G., et al. (2013). A global analysis of erosion of sandy beaches and sea-level rise: an application of DIVA. *Global and Planetary Change, 111*, 150–158. http://dx.doi.org/10.1016/j.gloplacha.2013.09.002.

Hinkel, J., van Vuuren, D. P., Nicholls, R. J., & Klein, R. J. T. (2013). The effects of adaptation and mitigation on coastal flood impacts during the 21st century. An application of the DIVA and IMAGE models. *Climatic Change, 117*(4), 783–794. http://dx.doi.org/10.1007/s10584-012-0564-8.

Jackson, N. L., Nordstrom, K. F., Feagin, R. A., & Smith, W. K. (2013). Coastal geomorphology and restoration. *Geomorphology, 199*, 1–7. http://dx.doi.org/10.1016/j.geomorph.2013.06.027.

Jennerjahn, T. C., & Mitchell, S. B. (2013). Pressures, stresses, shocks and trends in estuarine ecosystems an introduction and synthesis. *Estuarine Coastal and Shelf Science, 130*, 1–8. http://dx.doi.org/10.1016/j.ecss.2013.07.008.

Jongman, B., Ward, P. J., & Aerts, J. (2012). Global exposure to river and coastal flooding: long term trends and changes. *Global Environmental Change-Human and Policy Dimensions, 22*(4), 823–835. http://dx.doi.org/10.1016/j.gloenvcha.2012.07.004.

Kok, J.-L., & Grossmann, M. (2010). Large-scale assessment of flood risk and the effects of mitigation measures along the Elbe River. *Natural Hazards, 52*(1), 143–166. http://dx.doi.org/10.1007/s11069-009-9363-6.

Lichter, M., Vafeidis, A. T., Nicholls, R. J., & Kaiser, G. (2011). Exploring data-related uncertainties in analyses of land area and population in the "Low-Elevation Coastal Zone" (LECZ). *Journal of Coastal Research, 27*(4), 757–768. http://dx.doi.org/10.2112/jcoastres-d-10-00072.1.

Marriner, N., Flaux, C., Morhange, C., & Stanley, J. D. (2013). Tracking Nile Delta vulnerability to Holocene change. *Plos One, 8*(7). http://dx.doi.org/10.1371/journal.pone.0069195.

McGranahan, G., Balk, D., & Anderson, B. (2007). The rising tide: assessing the risks of climate change and human settlements in low elevation coastal zones. *Environment and Urbanization, 19*(1), 17–37.

Narayan, S., Nicholls, R. J., Clarke, D., Hanson, S., Reeve, D. E., Horillo-Caraballo, J., et al. (2014). The SPR systems model as the conceptual foundation for rapid integrated risk appraisals: lessons from Europe. *Coastal Engineering, 87*, 15–31. http://dx.doi.org/10.1016/j.coastaleng.2013.10.021.

Nicholls, R. J. (2006). Storm surges in coastal areas. In M. Arnold, R. S. Chen, U. Deichmann, M. Dilley, & A. L. Lerner-Lam (Eds.), *Natural disaster hotspots: Case studies (disaster risk management 6)*. Washington DC, USA: World Bank Publications.

Nicholls, R. J., Wong, P. P., Burkett, V. R., Codignotto, J. O., Hay, J. E., McLean, R. F., et al. (2007). Coastal systems and lowlying areas. In M. L. Parry, O. F. Canziani, J. P. Palutikof, P. J. van der Linden, & C. E. Hanson (Eds.), *Climate change 2007: Impacts, adaptation and vulnerability. Contribution of working group II to the fourth assessment report of the Intergovernmental Panel on Climate Change*. Cambridge, UK and New York, NY, USA: Cambridge University Press.

Resilient Coasts Initiative. (2009). *Resilient coasts: a blueprint for action*. The Heinz Center for Science, Economics and There Environment and CERES http://www.ceres.org/resources/reports/resilient-coasts-blueprint-for-action-2009/view. Accessed 13.11.13.

Sarwar, M. G. M., & Woodroffe, C. D. (2013). Rates of shoreline change along the coast of Bangladesh. *Journal of Coastal Conservation, 17*(3), 515–526. http://dx.doi.org/10.1007/s11852-013-0251-6.

van Slobbe, E., de Vriend, H. J., Aarninkhof, S., Lulofs, K., de Vries, M., & Dircke, P. (2013). Building with nature: In search of resilient storm surge protection strategies. *Natural Hazards, 65*(1), 947–966. http://dx.doi.org/10.1007/s11069-012-0342-y.

Syvitski, J. P. M., Kettner, A. J., Overeem, I., Hutton, E. W. H., Hannon, M. T., Brakenridge, G. R., et al. (2009). Sinking deltas due to human activities. *Nature Geoscience, 2*(10), 681–686. http://dx.doi.org/10.1038/ngeo629.

Thorne, C., Evans, E., & Penning-Rowsell, E. C. (Eds.). (2007). *Future flooding and coastal erosion risks*. London, UK: Thomas Telford.

UK NEA. (2011). *UK national ecosystem assessment: Understanding nature's value to society. Technical report*. Cambridge, UK: United Nations Environment Programme and World Conservation Monitoring Centre (UNEP-WCMC). http://uknea.unep-wcmc.org/Default.aspx. Accessed 13.11.13.

Developing a Holistic Approach to Assessing and Managing Coastal Flood Risk

Chapter 2

Robert Nicholls[5], Barbara Zanuttigh[1], Jean Paul Vanderlinden[2], Ralf Weisse[3], Rodolfo Silva[4], Susan Hanson[5], Siddarth Narayan[5], Simon Hoggart[6], Richard C. Thompson[6], Wout de Vries[7], Phoebe Koundouri[8]

[1] *Department of Civil, Chemicals, Environmental and Materials Engineering, University of Bologna (UniBo), Bologna, Italy*

[2] *Cultures, Environments, Arctic, Representations, Climate Research Center, Université de Versailles Saint-Quentin-en-Yvelines (UVSQ), Guyancourt, France*

[3] *Helmholtz-Zentrum Geesthacht Centre for Materials and Coastal Research (HZG), Geesthacht, Germany*

[4] *Instituto de Ingeniería, Universidad Nacional Autónoma de México (UNAM), Coyoacán, DF. México*

[5] *Faculty of Engineering and the Environment, University of Southampton (Soton), Southampton, UK*

[6] *Marine Biology and Ecology Research Centre, School of Marine Science and Engineering, University of Plymouth (UoP), Plymouth, UK*

[7] *INFRAM International BV (INFRAM), Marknesse, The Netherlands*

[8] *Department of International and European Economic Studies, Athens University of Economics and Business (AUEB), Athens, Greece*

Coastal Risk Management in a Changing Climate. http://dx.doi.org/10.1016/B978-0-12-397310-8.00002-6

2.1 Introduction

Coastal flood systems can be large and complex, and they change with time. These issues create several challenges with gaining a comprehensive understanding of these systems.

Hurricane Katrina in 2005 in New Orleans, USA was one of the costliest coastal flood disasters in history (Seed et al., 2008) and provided several lessons for flood risk management. Key was the lack of a systemic overview of the state of flood defenses prior to the event, mainly due to the size and complexity of both the New Orleans coastal defense system and the responsible organizations. Administrative delineations, especially where the flood system crosses political boundaries that do not recognize the full extent of the natural flood system, are significant constraints to a full understanding of the flood system. A similar lack of overview on emergency response measures and flood defenses led to aggravation of damage during the July 2007 floods in England (Pitt, 2008) and storm Xynthia in France in 2010 (Kolen et al., 2010; Lumbroso & Vinet, 2011). In addition, development contrary to spatial planning laws and a lack of knowledge of potential flood routes within the system caused authorities and inhabitants to be taken by surprise (Kolen et al., 2010).

Coastal flood systems often comprise many land-use types and an equal variety of stakeholders and experts. This is further complicated by interdependencies between the flood system elements. Understanding these interdependencies across the flood system and quantifying their effects on flooding due to changes in the states of particular elements is a significant challenge. A simplified model of the topological relationships allows users to understand effects on the system as particular elements change, or as new or improved information about these elements is obtained.

Changes over time of coastal flood systems include the following aspects that are nonlinearly correlated.

- Climate change
  Global climate change is a key influence on coastal flooding. Global sea levels have risen over the last 100 years and this is expected to continue and accelerate in the next century due to global warming and other possible changes in climate (Church et al., 2013, Church, Woodworth, Aarup, & Wilson, 2010; Nicholls et al., 2014). Any long-term rise in mean sea level will have an effect on extreme levels as it moves the entire distribution of sea levels toward higher values; i.e., it changes the baseline to which the other factors are added (Haigh, Nicholls, & Wells, 2009). Importantly, there is large uncertainty about the magnitude of future change, including other climate change (offshore waves, surges, etc.), although we are confident that sea levels will rise.

- Socioeconomic change
  Changes in coastal population and the coastal economy and resulting change in the use of the coastal zone for infrastructure are key issues for managing flood events. Globally, there has been massive expansion in people and assets at risk in the coastal zone, raising the consequences of flooding (McGranahan, Balk, & Anderson, 2007). This is expected to continue through the twenty-first century (Hanson et al., 2011; Nicholls et al., 2012a), and needs to be considered in risk management.
- Morphological and habitat change
  The coastal system is dynamic and can respond in a number of ways to any external perturbation. For instance, natural coastal habitats such as mangroves and salt-marshes provide protection during flood events, but these are themselves often affected by the same events. Habitat losses affect flooding during subsequent events. Similarly, erosion of beaches is widespread and influences flood risk (Dawson et al., 2009). The challenge here is to capture the flood system in its entirety, including key information on all potential flood routes, including those that may operate under the most extreme conditions over the time-frame of interest.

Historically, responses to reduce negative flood impacts have concentrated on the use of physical defenses to reduce the probability of flooding. More recently, a more holistic emphasis that includes the natural environment and nonstructural measures is considered a more prudent approach to risk management (Renn, 2008; Thorne, Evans, & Penning, 2007). Such an approach embraces the notion of flood resilience and vulnerability reduction as well as prevention (Figure 2.1). Consideration of the flood system beyond traditional flood modeling needs to include the natural environment, the concerns of stakeholders and public groups, information on local risk perception, and existing flood management governance (Figure 2.1, Aven & Renn, 2010, Chapter 5).

In order to overcome the challenges described, such an approach should also be able to integrate information on the different types of elements across the system and provide an overview of the relevant topological relationships and interdependencies. Ultimately this should inform more detailed models to provide a more complete picture of all relevant inputs, as well as the system itself. Establishing a baseline conceptual understanding of the existing flood system from these perspectives provides a logical basis for decision-making and encourages:

- Consideration of the social and political context in which flood events occur;
- The use of the information to develop policies and development controls around current and future development;
- The inclusion of the role of the natural environment both in flood management as well as its intrinsic value, as required in the Habitats Directive (European

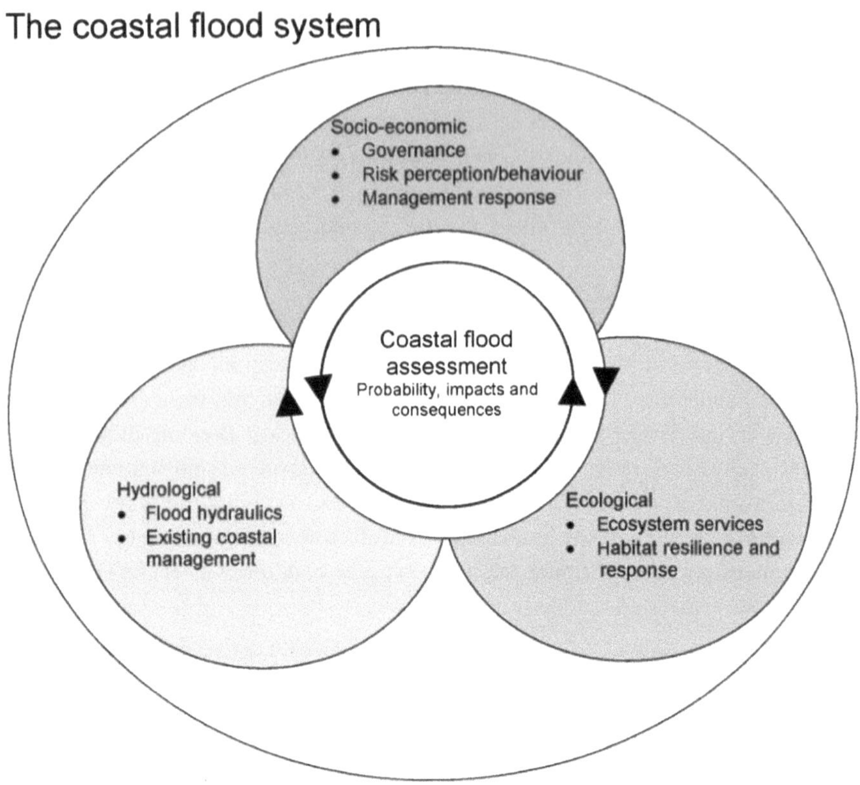

Figure 2.1 The coastal flood system: important aspects for flood assessment and mitigation selection.

Parliament and Council of the European Commission, 2007) and other habitat designations;

- Investigation of the full range of flood-mitigation options across the spectrum of engineering, ecological, and social measures (Chapter 6).

The present chapter looks at coastal floodplains in a systemic sense, including their social and ecological dimensions, and outlines the range of potential flood mitigation measures. This allows Chapters 3–5 to consider in detail a range of innovative engineering, ecological, and socioeconomic mitigation measures, respectively.

2.2 Flood Nomenclature: Vulnerability, Risk, and Resilience

Given the complexity of flooding and flood issues that has already been outlined, a range of concepts and associated terminology is used. In lay discussion these terms are often used interchangeably, providing potential for confusion and miscommunication. Here we lay out a common set of definitions for key terms to facilitate their use and discussion through the book. This draws on earlier work such as the FLOODSite project (http://www.floodsite.net/) and the Netherlands Knowledge for Climate Research Program (http://knowledgeforclimate.climateresearchnetherlands.nl/).

Floods are defined by physical characteristics such as flood depth, extent, and duration, which are fairly transparent. However, there are many systemic concepts such as flood risk, flood vulnerability, and flood resilience which are potentially ambiguous and require definition. All coastal floods begin with a hazardous event, often a storm that produces surges and waves (sources of flooding). However, a source of flooding does not automatically lead to a flood event, as this depends on the pathways and receptors. If the land is high enough, or if it is well protected by dikes and dunes, then the storm does not lead to a flood. However, if the seawater overtops the protection, or breaches the dikes or dunes, than the low-lying land behind it will be subject to flooding with consequences for the associated receptors.

The notion of *vulnerability* is the degree to which a system is susceptible to, and unable to cope with, adverse effects of the change agent, in this case floods. Flood vulnerability is a function of the character and magnitude of flooding and variation to which a system is exposed—the sensitivity and adaptive capacity of that system. A range of flood vulnerability indices have been developed to operationalize this concept (e.g., Balica, Douben, & Wright, 2009). Vulnerability assessment has been conducted in a range of contexts with a view to understand and reduce this vulnerability, including to floods.

The notion of *flood risk* is a combination of probability and consequences, often expressed as an annual mean damage (or consequence) (see Penning-Rowsell et al., 2013). Hence, risk can be expressed as a number, and the units of consequences may be related to flood victims and flood damage to homes, businesses, and nature. To explore the meaning of risk, the four quadrants of probability and consequences can be explored, considering high and low situations. If the probability of a major flood is high and the consequences are high, then this is a "high"-risk situation. On the other hand, if the probability of a flood is small and the consequences are small, then this is a "low"-risk situation. The situation where floods have a high probability and low consequences is also a "low"-risk situation, and with the regular experience of flooding, people and the environment often adjust to such floods, further reducing the consequences. Lastly, there is the situation with low probability of flooding, but with high (very serious) consequences, such as London, the Netherlands, Shanghai, and Tokyo. The challenge in these cases is to reduce the probability of flooding to acceptable levels, and reduce the consequences wherever possible. However, in these situations a failure can be catastrophic, as shown in New Orleans in 2005 (Kates, Colten, Laska, & Leatherman, 2006). Hence while you can calculate risk, the level of risk that we are willing to accept remains a political question.

The notion of *flood resilience* is related to vulnerability and describes the systemic ability to experience flooding with minimum damage and rapid recovery. It can be seen as a design approach that can reduce the damage that occurs due to flooding. For example, it could involve constructing a building in such a way that although floodwater may enter the building, its impact is minimized and recovery is rapid. Resilience operates at multiple scales from individual buildings to communities, towns, and cities. In this more aggregate sense, resilience can be provided by multiple measures that reduce damage and promote recovery, and hybrid approaches can be taken and need to be considered. This might include combinations of warnings, evacuation and emergency plans, land use planning, traditional hard and soft defenses, building construction approaches, provision of insurance, etc.

Reducing vulnerability and flood risk is a key goal of flood risk management, and this can be accomplished in many ways, as discussed later in this chapter and the book as a whole. Enhancing flood resilience is one method for achieving this goal.

2.3 Describing the Coastal Flood System: The Source-Pathway-Receptor-Consequence Model

Risk assessments are used to evaluate and support the selection of appropriate mitigation options for diverse environmental problems including the regulation of hazardous waste sites, industrial chemicals, and pesticides; or the management of ecosystems affected by multiple physical, chemical, or biological stressors (e.g., Arnoldi, 2009). Along Europe's coastlines, the EU Floods Directives requires member states to draw up regularly updated Preliminary Flood Risk Assessments considering potential impacts on human health and life, the environment, cultural heritage, and economic activity. These risk assessments can be used to identify areas at high risk of flooding. Following this, models can be used to produce flood and flood-risk maps for specified events. The development of flood-risk assessments has become an increasingly significant issue over recent years following storm events such as winter storm Xynthia in 2010 (flood damage in excess of €1.2 billion and at least 47 lives lost; Lumbroso & Vinet, 2011), and a recognition that impacts, while initially of local importance, can escalate to have consequences for the wider economy (e.g., Hallegatte, Green, Nicholls, & Corfee-Morlot, 2013). Important examples include the effect of Hurricane Katrina (2005) on global oil prices, and the Bangkok 2011 floods on the global availability of hard drives. As flood risk cannot be eliminated, but only reduced, the management of these risks requires understanding the flood system as it responds to a range of planned and unplanned interventions, (e.g., floodplain development, better defenses) as well as external changes (e.g., climate change). In this way, locally relevant decisions can be made while recognizing other wider or more distant implications.

The management of flood risk also requires fully engaging local communities within the risk-assessment process and making roles and responsibilities explicit, bringing together all the stakeholders in the coastal zone. In this process, it is essential to promote relevant, evidence- and knowledge-based contributions from a range of stakeholders, and to establish a shared understanding of the flood system (Charles, 2012).

This shared vision of the flood system should:

- Address the specific aims of flood management;
- Be accepted by all the disciplines contributing to the flood assessment to ensure integration and transferability of inputs/outputs;
- Illustrate where/how management options are influential in the system;
- Be understandable by stakeholders to enable clear communication of management options;
- Work across different temporal and spatial scales and levels of detail.

This shared view should enable the integration of methods and approaches, both qualitative and quantitative, to promote the understanding of flood events, their impacts, and the opportunities for mitigation measures. Establishing such a shared view before continuing with more detailed analysis is worthwhile, because the debate it engenders helps to develop a balanced view across the different perspectives, which often facilitates successful decisions made later. It also helps to create a common vocabulary, which is a fundamental necessity in integrated flood-risk assessments where the use and interpretation of words varies across scientific disciplines. This process creates a comprehensive picture of the areas of the flood system that are potentially affected by an event, providing useful information for higher levels of analysis. This requires a clear methodological approach, conceptual model, and analytical framework (Robinson, 2008).

Coastal flood risk studies—studies that focus on the evaluation of coastal flood impacts to human assets—conceptualize the coastal floodplain in terms of two components: (1) flood defenses that prevent or reduce the ingress of floodwater and (2) the floodplain behind the defenses comprising all the features considered to be at risk from flooding (Bakewell & Luff, 2008; Burzel et al., 2010; FLOODsite Consortium, 2009; NCDEM 2009). The quantitative evaluation of risk in these studies is usually performed using numerical hydraulic models. The modeling procedure parallels the conceptual description of the floodplains, starting from hydraulic boundary conditions and incorporating the influence of coastal flood defenses in order to evaluate the flood probabilities and damages to specific locations within the floodplain. Related floodplain systems such as habitats or physical coastal systems are often represented as external forces and pressures with little or no consideration of spatial and temporal feedbacks (Verwaest et al., 2008). Hence these treatments are often incomplete.

An alternative, more comprehensive way of visualizing the process of flood risk estimation and all its components is the Source-Pathway-Receptor-Consequence (SPRC) conceptual model (Gouldby & Samuels, 2005). The model was first used in the environmental sciences to describe the propagation of a pollutant from a source, through a conducting pathway, to a potential receptor (Holdgate, 1979). It was first adopted in coastal flooding in the UK by the Foresight: Future Flooding report (Evans et al., 2004). It has subsequently been used in several coastal flood risk studies (Burzel et al., 2010; FLOODsite Consortium, 2009; North Carolina Division of Emergency Management, 2009; THESEUS OD1.15, 2012), and is increasingly underpinning wider flood-risk management. Based on conventional approaches to flood-risk estimation, the SPRC model visualizes flood-risk estimation as a linear process involving a "Source" of flooding, flood "Pathways," and affected "Receptors" associated with different "Consequences" (Figure 2.2, Table 2.1).

The SPRC model recognizes the principle that the component parts of a system can best be understood in the context of relationships with each other (and with other systems), rather than in isolation. Consequently, it considers flood management within an overall system, highlighting where external drivers can be influential, and—importantly—where system vulnerability can be reduced or exacerbated following intervention. Fundamental to the approach is the defining of relationships between system components at a relevant scale to provide understanding and insight into the flood system under investigation. At its simplest, the concept is a linear representation of a flood event from the Source (of the floodwaters) through the Pathway (route of the floodwaters) to the Receptors (where the water culminates) and calculation of the effect of floodwater on the Receptors (Consequences); see Table 2.1.

The SPRC model presents a snapshot of the floodplain state. This is in turn is driven by boundary conditions operating at a range of spatial and time-scales such as offshore water levels and waves, climate change effects, and human influences such as coastal zone management decisions and actions. Therefore the SPRC model is usually nested within broader approaches such as the Driver-Pressure-State-Impact-Response (DPSIR) framework that conceptualize the influence of pressures and drivers external to the floodplain (e.g., Gregory, Atkins, Burdon, & Elliott, 2013; Kristensen, 2004; Lee, 2013; Zhang & Xue, 2013). The DPSIR assumes cause–effect relationships between interacting components of social, economic, and environmental systems (Carr et al., 2007). By identifying where external factors influence the flood system, the DPSIR framework helps identification of where management interventions (acting as Drivers) influence the Consequences of a flood event. It also illustrates the circular nature of flood management, with an intervention affecting consequences that will influence society's response, which in turn will determine future management interventions.

Figure 2.3 illustrates that the SPRC model can be divided into two components based on its nesting within the DPSIR: a floodplain state description (SPR) and a

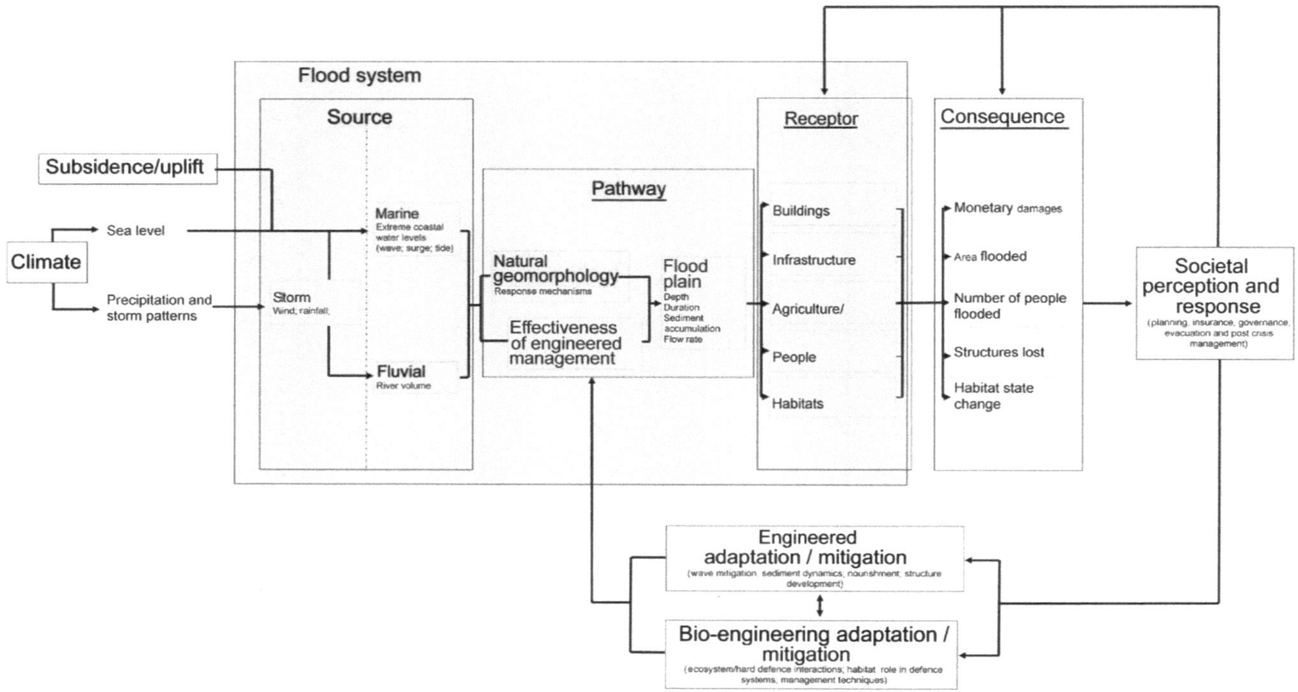

Figure 2.2 SPRC diagram showing where external Drivers can mitigate the Consequences of a flood event at the local scale. *(From Narayan et al., 2014.)*

TABLE 2.1 Definitions and Components of the SPRC Model

Category	Definition	Components
Source	Where the floodwaters originate	Sea: waves, surges, tides, mean sea level River: volume/flow
Pathway	The route for the Source to reach the Receptor	Various land uses seaward of any Receptor, including existing coastal management (e.g., built defenses, nourishment) and habitats
Receptor	Land use and buildings/structures in the floodplain	Urban areas, infrastructure, farmland, habitats, etc.
Consequence	Impact of flooding on the Receptor	Direct/indirect and tangible/intangible consequences for each Receptor (via various valuation methods)

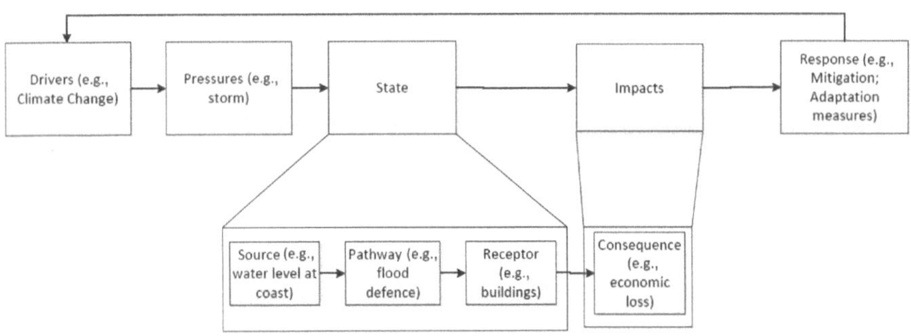

Figure 2.3 Nesting of the SPR-C model within the DSPIR framework. *(From Narayan et al., 2014.)*

description of the consequences to changes in this state (C). Flood risk assessments typically follow this division, using the SPR model to assess flood probabilities of elements within the floodplain and separate economic models to evaluate flood consequences.

The Source component of the SPR model usually describes the sources of flooding such as waves and water levels. The Pathway component generally refers to all floodplain elements that influence flood propagation within the floodplain. The Receptor component of the model is commonly used to describe the economic cost of a flood event estimated using existing observations and depth-damage relationships (Penning-Rowsell et al., 2013).

It is important to remember here that there may be several Pathways to the same Receptor, and it is useful to identify these in order to fully appreciate potential risk or damages. For example, a house sited in a floodplain directly behind a dyke may appear to be adequately protected, but if a neighboring defense is of a lesser standard (a "weak link") it may fail and the house still flood. Building on the underlying systems approach of the model, mapping of the Receptors and their Pathways encourages the exploration of the wider environmental setting, physical functioning of the site, and spatial variability within the system (Thorne, Evans & Penning-Rowsell, 2007). In this way, the SPRC model offers the opportunity to develop a more comprehensive representation of the flood system, acknowledging the complex network nature of the system (Narayan et al., 2012; Narayan et al., 2014). The mapping also shows that individual elements may be classified as either a Receptor or Pathway depending on the analysis being undertaken and its relative position within the floodplain. This is especially relevant for coastal habitats, which are often considered as Pathways but not as Receptors, which may change as a result of flood events or more continual processes (e.g., sea-level rise).

Though the conventional conceptual model visualizes a linear system of Source, Pathway, and Receptor, in practice a typical flood risk assessment uses a range of diverse models and inputs to describe and analyze the state of the coastal floodplain. Furthermore, the types and nature of models and inputs may differ depending on the scale and extent of detail of a particular assessment.

The key drivers (see Section 2.8.1) affecting the coastal floodplain are (1) climate change, which can affect Sources such as sea level, storm frequency and intensity, and rainfall patterns (increasing or decreasing the extreme water levels during a flood event), and in some cases a non-climate factor: subsidence; (2) sediment supply, which influences Pathways and ecological receptors, coastal geomorphology, and ecosystems; and (3) socioeconomic change, which can alter the type and extent of human receptors within the floodplain (e.g., Thorne, Evans & Penning-Rowsell, 2007). Once the relevant drivers have been determined in any floodplain, the relative importance of each driver can be evaluated based on expert judgment to assess potential impacts on future flood risk. This is based on a score for each driver impact according to its influence on flood risk (altering probability or consequences) under the given driver scenario and time slice (Evans et al., 2004; THESEUS OD1.10, 2012).

Depending on a number of factors, including the scope of the assessment, the relevant drivers and the availability of data and modeling methods, a wide range of numerical models may be used within a flood-risk assessment to describe the state of the coastal floodplain. Figure 2.4 illustrates the possible range and diversity across scales and levels of detail of typical flood-risk assessments, all of which use the linear SPR model described above to conceptualize the coastal floodplain. To ensure the development of a common language and shared understanding of the flood

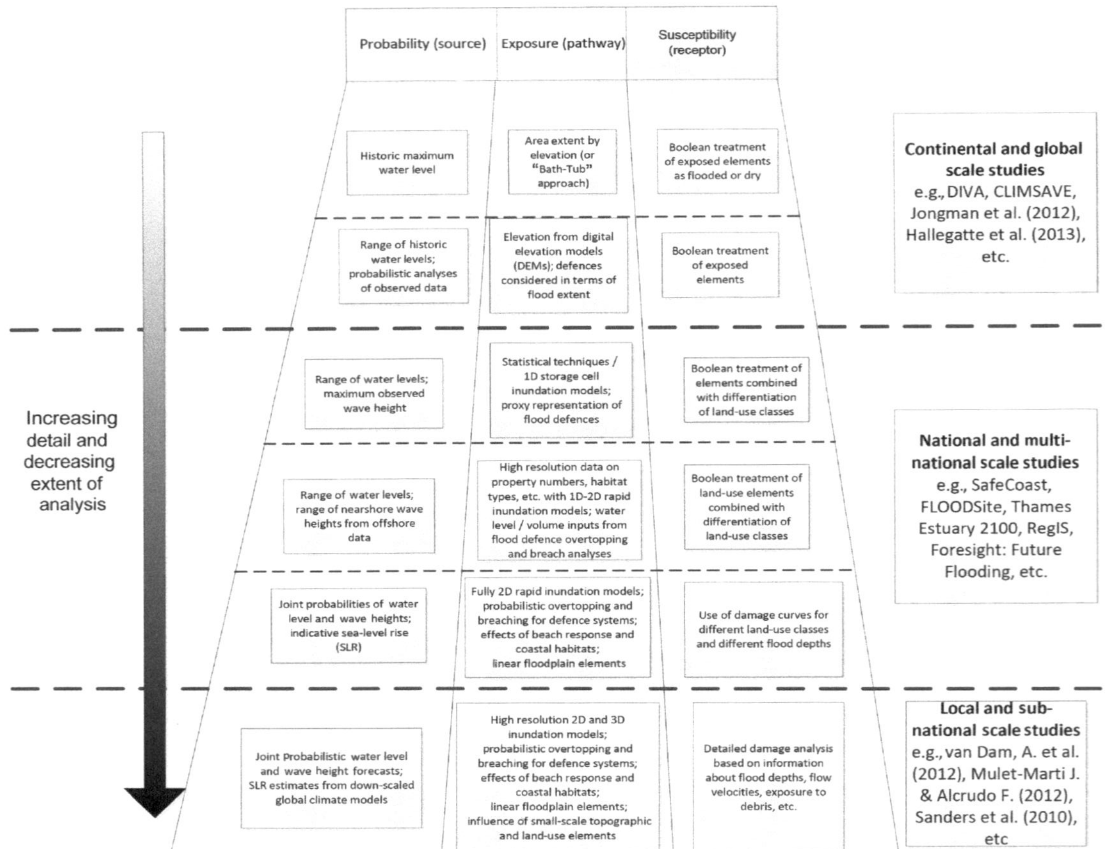

Figure 2.4 Types of flood-risk studies in terms of the SPR model. *(From Narayan et al., 2014.)*

system across all stakeholders before the use of these detailed models, a conceptual description of the flood system is developed via the quasi-2D SPR (Narayan et al., 2014) at the start of the flood risk study.

2.3.1 SPR Mapping

The quasi-2D SPR model is built in four steps.

Step 1: The landward boundaries of the coastal floodplain system are first decided using a planar water level model for the most extreme water level being considered. This is done under the assumption of a worst-case scenario where complete failure (or absence) of defenses is assumed. This assumption will indicate the full extent of the natural floodplain system and ensure that all system elements are included in subsequent analyses. The seaward boundary of the floodplain system is placed at Mean Low Water Neaps to ensure inclusion of all inter-tidal floodplain elements seaward of the coastline.

Step 2: Once the floodplain boundaries are defined, all floodplain elements, including flood defenses, are mapped as unique entities classified based on land use. This provides a platform for future integration of any analysis with the socioeconomic aspects of a flood event, such as economic consequences or land-use planning scenarios.

Step 3: Next the relationships between the identified elements are defined. A link is identified between any two elements if the elements share a geographical boundary, including engineered flood defenses. Flood compartments created by these defenses can therefore be studied as part of the larger natural floodplain system, rather than as isolated subsystems. The elements and links are then schematized to a systems map. The move from a geographical map to a systems map allows easy, rapid, and comprehensive analyses of the topological relationships between different elements regardless of their location or size.

Step 4: Once the system diagram is built, the sources of flooding are identified on all boundaries and, if necessary, within the system boundaries. These sources are also schematized, and all links between them and directly connected system elements are identified.

The mapping process and resultant SPR maps (see Chapter 7) represent the components and linkages within the flood system in a defined space, without scaling or topographical constraints. Spatial limits for the SPR mapping need to be clearly defined considering the geomorphological setting of the site and an upper elevation of the area should be defined, e.g., set above current flood levels (allowing for sea-level rise). This sets the boundary for any information that needs to be collected in the flood-risk assessment. This ensures that the SPR analysis extends beyond any expected flood extents within the mid- to long-term (e.g., 50–100 years). For sites that extend over large areas, a hierarchical SPR approach can be useful, allowing a detailed analysis to be carried out in areas of interest, while recognizing the wider context (Monbaliu et al., 2014; Villatoro et al., 2014, Chapter 7).

Once developed, the SPR maps provide a comprehensive characterization of the coastal flood system network that can be interpreted and analyzed in a number of ways. Key aspects include (1) the potential for multiple potential flood routes to particular Receptors (e.g., the Gironde estuary case-study; Monbaliu et al., 2014); (2) areas where reliance on existing coastal defense structures is notable; (3) the most beneficial locations for new or upgraded defenses; or (4) where land use may be modified. The identification of existing management can also identify routes for floodwaters that are not directly related to coastal mitigation measures but to urban management and flooding (such as storm drains, culverts, or sewers). The application of a system mapping process provides an inexpensive but rigorous and comprehensive model of the coastal flood system, essential not just for understanding these systems but also for defining the system components of analysis for the more detailed assessment required as part of flood-risk assessments.

Following the SPR mapping, an improved baseline suitable for the modeling of flood events is created. This commonly requires a base layer of digital topography and land use that can then be used to generate flood and erosion maps and estimate economic damages and potential human losses. Consequently, the use of a Geographic Information System (GIS) approach to capture, manipulate, process, and display spatial or geo-referenced data is beneficial (see Rodriguez, Montoya, Sanchez, & Carreno, 2009, and Chapter 6).

2.3.2 Defining the Sources in the Coastal System

Sources are essentially classified into three groups according to flood duration: short-term (storm surge, wind waves, tides, runoff due to downpours); seasonal (river high/low waters); and long-term processes (sea-level rise, local land surface vertical movement). Historical analysis (long and homogeneous time series of water levels or discharges) can be used to establish existing return periods for different extreme events. Extreme water levels from the sea are caused by a combination of several factors: (1) high astronomical tides due to the sun and the moon, (2) storm surges due to high winds and low atmospheric pressure, and (3) waves caused by local high winds or far-traveled swell from oceanic fetches. Hence, tropical or extra-tropical storms can both produce extreme sea levels and cause flooding. Changes in any of these factors may alter the characteristics of a flood event. Historically, the long-term change in mean sea level has contributed to changing extreme sea levels, and globally this is increasing the frequency of high sea levels (Menéndez & Woodworth, 2010).

Understanding extreme sea levels involves an analysis of the above components, including questions of statistical dependence and independence (e.g., Haigh,

Nicholls, & Wells, 2010). Often such time series are unavailable, are not of the required length or suffer from data inhomogeneity such as problems related to changing instrumental accuracy, observational practices, analysis routines, etc. (WASA-Group, 1998; Weisse & Von Storch, 2009). This may lead to incorrect inferences unless properly addressed. Alternatively, a number of re-analyses, hindcasts, and downscaling activities can be used, in particular for tide surge and wave conditions: e.g., PRUDENCE (Prediction of Regional scenarios and Uncertainties for Defining EuropeaN Climate change risks and Effects, http://prudence.dmi.dk/) or HIPOCAS (Hindcast of Dynamic Processes of the Ocean and Coastal Areas of Europe, http://www.mar.ist.utl.pt/hipocas/). The resultant local estimates of extreme sea levels and wave climate can be used to specify the environmental loading conditions on the coastal defenses.

2.3.3 Defining Pathways in the Coastal System

Pathways are the routes and processes that are active during a flood event and transport floodwaters from a source to a receptor. Thus, without a Pathway, a flood event cannot have any Consequences. On many occasions, an individual Pathway may have multiple Receptors, and individual Receptors may have multiple Pathways. As Pathways include the components of the flood system (identified in the SPR mapping) through or over which floodwaters flow, they can capture the different defense failure mechanisms, such as overtopping versus breaching, which can produce different flooding patterns. Pathways can be also receptors. Coastal structures (Chapter 3) affect the flood extent and they can stop flooding, fail or be degraded by the intensity of the Sources, and act as a Pathway (see also Section 6.2.1). Similarly, coastal habitats (see Chapter 4) may be regarded as Pathways as far as they offer some coastal protection in terms of wave energy reduction or increased beach stability (e.g., biogenic reefs, dunes), and of course they are also Receptors whose survival or modification depends on the Sources.

2.3.4 Defining Receptors in the Coastal System

Receptors are usually defined by either what can be found on, the use of, or the intrinsic value of the land that has the potential to flood. They are mainly, although not exclusively, found above the lowest water level for the site and can form part of either the human or natural system. The initial information for the identification of potential Receptors is generally land use, supplemented with more detailed habitat mapping and additional socioeconomic information. An example of a broad-scale receptor classification is shown in Table 2.2.

Consequences may be specific to the identified Receptor: e.g., for habitats, the area lost and species change due to flood duration; or more general: e.g., for

TABLE 2.2 Example Broad-Scale Classification of Receptors

System	Receptor Classification	Land Use Examples
Human	Buildings (residential)	Houses
	Buildings (nonresidential)	Factories, storage facilities
	Infrastructure	Roads, hospitals, airport
	Agriculture	Arable land, grazing
	Mariculture	Mussel farming, fish farms, oyster beds
Natural	Natural element	Beach, spit, salt marsh, mudflat
	Habitat	Dune, salt marsh, kelp beds

buildings/infrastructure, the damages based on depth-damage curves, number of people flooded, number of houses flooded, etc. Analysis of the Pathways and Receptors of a flood system, enshrined by existing flood management regimes, is discussed in Section 2.4.

2.3.5 Defining Consequences in the Coastal System

The development of the SPR mapping encourages the identification of direct Consequences of flooding related to the nature of the Receptor(s). The mapping of the consequences of a flood event is usually done after quantifying the flood probability of the different parts of the floodplain, as described in Figure 2.3. The process—and probability—of flooding is driven by the physical state of the flood pathways. However, the consequence of a flood event is felt only by an element that functions as a Receptor—even though this element may also function as a flood Pathway. For instance, the flooding of a beach, apart from acting as a flood Pathway, may result in tangible economic losses to the local tourist industry. Some floodplain elements may function primarily as Receptors. For instance, critical infrastructure such as hospital buildings are elements for which the consequence of flooding is of immediate concern. Consideration of the Pathway effect of the building will depend on the detail and sophistication of the data and numerical models used for later analysis (see Figure 2.4). On the other hand, the flooding of infrastructure such as a pumping station will be of relevance both for flood propagation (as a Pathway) as well as in terms of the direct economic costs of replacing damaged parts (Receptor–Consequence). A first overview of typical classes of Receptors and their associated consequences for a flood event is given in Table 2.3. Once the physical characteristics of a flood event (i.e., flood extents, depths, probabilities) are mapped onto to the floodplain system description, these can be combined with information on depth-damage curves and cost estimates for specific Receptor types (Zanuttigh et al., 2014; Sections 6.6 and 7.6) to obtain the

TABLE 2.3 Example of Direct Consequences of Flooding Associated with Receptors

Receptor	Example of Direct Consequences
ALL	Area permanently flooded (land loss) Area temporarily flooded/displaced
(Critical) infrastructure	Physical flood damage
Buildings—residential	People temporarily flooded Building/content damage
Buildings—commercial/industrial	Area temporarily flooded Building/content damage
Habitat	Habitat state change
Agriculture	Flood damage to crops Change of agricultural practices (e.g., crops to pastoral)
Recreation	Flood damage to recreational facilities

consequences of a flood event. Analysis of the consequences of a flood event is described further in Section 2.5.

2.4 Assessment of Existing Flood Management

Analysis of present conditions, including existing defenses, policies, regulations, and governance arrangements (Chapter 6), is an essential part of the flood risk assessment process (Penning-Rowsell et al., 2014). It provides the background against which any future management options will be taken and identifies those responsible for implementing such a strategy. Including those involved in policy development or decision-making also offers the opportunity to more fully integrate science into policy (de Vries et al., 2011).

Surveys can be used to characterize the risk governance in coastal floodplains based on five "building blocks":

- Administrative organization of coastal management;
- Legal system;
- Financing system;
- Economy of intervention measures;
- Participation level of stakeholders.

Many sites have complex institutional structures for flood management, with responsibilities found at local, regional, and national, as well as international, levels. This information will need to be collected in a systematic manner from local policymakers, managerial authorities, and administrators. Presenting the conceptual model of the flood system (Section 2.3) is often a beneficial aid to these discussions. Experience within the THESEUS project showed that institutional culture, traditions, and capabilities are of great significance to (innovation in) risk management, and could be of at least the same importance as technical issues in risk assessment and reduction choices (THESEUS OD4.8, 2013).

Existing management structures, flood policies, and defense design often reflect the relative importance and current understanding of coastal flooding and its consequences in any floodplain area (Aven & Renn, 2010, Chapter 5). Legal obligations, frequency of occurrence, economic value of the protected area, and previous experience with flood events are all influential. Where the flood risk is considered greater than other risks, in Europe there is often a strong tendency to focus on the reduction of the probability of flooding using physical coastal defenses and/or nourishment schemes (Chapter 3). Conversely, where flood risk is considered as one of many risks, the range of flood management options examined is often wider. In other regions, risk reduction may have other foci: for instance, in the USA there is focus on building standards and raising the levels of buildings above flood levels.

Stakeholder interviews are probably the most appropriate methods to identify the current governance structures (de Vries et al., 2011). Such interviews could be supported with the help of a structured or semi-structured questionnaire, which should be sent in advance to the interviewees. An additional benefit of undertaking group interviews is that they can bring together, sometimes for the first time, stakeholders with management responsibilities in the coastal zone. Possible feedback to participants of the resulting report is essential, particularly where there may be ethical issues or wider implications in the accumulation of the information.

The experience in the THESEUS study sites across Europe (Penning-Rowsell et al., 2014, Chapter 7) shows that the institutional arrangements in many coastal situations are complicated, almost invariably multilevel, and potentially confusing to the public. Central government is almost always involved, because of the large investment required for engineering mitigation works to reduce risks from flooding and their involvement in spatial planning legislation at the coast. It is recognized that these investments and powers cannot simply come from the communities at risk, but need support from the general taxpayers and national or regional-level legislators. Further, in most of the sites there is a provision for sustainable coastal zone management, within the existing legislation. However, not all laws and regulations are properly enforced. Outside Europe, these factors will require further assessment, as this model may not apply.

2.5 Flood Damage

Flood damage is defined as all the varieties of harm provoked by flooding. It includes all detrimental effects on people, their health, and properties; on public and private infrastructure, ecological systems, cultural heritage, and economic activities (Messner & Meyer, 2006). Understanding the nature of flood damage is important in assessing flood risk. For most people, the benefits of flood risk reduction at the coast is the direct flood damage on property and economic activity avoided as a result of schemes to reduce either the frequency or impact of flooding (Penning-Rowsell et al., 2013). However, the consequences of flooding for people are more complex. Following Smith and Ward (1998), we can classify flood losses into direct and indirect losses. Direct losses are caused by the physical contact of the floodwater with humans, property, or other objects, and the location of the flood will indirectly affect networks and social activities, causing indirect losses (e.g., disruptions of traffic, trade, and public services). Further, we can distinguish between immediate or long-term consequences and tangible or intangible consequences. Such consequences depend on the land uses found within the floodplain. Immediate impacts of flooding can include loss of human life, damage to property and infrastructure, and destruction of crops and livestock. Examples of long-term impacts include the interruption to communication networks and critical infrastructure (such as power plants, roads, hospitals, etc.) that can have significant impacts on social and economic activities. More difficult to assess are the intangible impacts; for example, the psychological effects of loss of life, displacement, and property damage can be long-lasting (see Table 2.4). Methods of assessing these impacts are equally varied. They range from quantitative (financial or economic) to more qualitative approaches (see Chapter 6).

A key concept in flood-loss estimation is the concept of damage functions or loss functions. They relate damage for a specific element at risk to the features of the flooding. These functions are similar to dose–response functions or fragility curves in other fields (Merz Kreibich, Schwarze, & Thieken, 2010; Penning-Rowsell et al., 2013). Flood damage losses are a function of the nature and extent of the flooding, including its duration, velocity, and the contamination of the floodwaters by sewage and other pollutants. It is important to ensure that for the purposes of flood-risk management there is consistency in the assessment of damages: this often means that only the national economic losses caused by floods and coastal erosion are assessed, rather than the financial losses to individuals and organizations affected, severe though those may be.

Protecting property from flooding is considered in investment decision-making through approaches such as cost–benefit tests that, for example, the UK Treasury uses, and which are becoming more commonly applied throughout the world. Also environments are often now protected—sometimes irrespective of cost—courtesy of national and European legislation (creating Ramsar sites, Special Protection Areas, Sites of Special Scientific Interest, etc.). Nevertheless, the "social" effects of floods need to be considered: those caused by the disruption of people and communities that do not or cannot carry a monetary price tag. Floods can cause health impacts that are enduring, including the stress and trauma created months or years afterward

TABLE 2.4 A Typology of Flood Losses with Examples

		Measurement	
		Tangible	**Intangible**
Forms of flood losses	Direct	■ Damage to private buildings and contents ■ Destruction of infrastructure such as roads and railroads ■ Erosion of agricultural soil, destruction of harvest ■ Damage to livestock ■ Evacuation and rescue measures ■ Business interruption inside the flooded area ■ Clean-up costs	■ Loss of life, injuries, loss of memorabilia ■ Psychological distress, damage to cultural heritage ■ Negative effects on habitats/ecosystems
	Indirect	■ Disruption of public services outside the flooded area ■ Induced production losses to companies outside the flooded area (e.g., suppliers of flooded companies) ■ Cost of traffic disruption ■ Loss of tax revenue due to migration of companies in the aftermath of floods	■ Inconvenience of post-flood recovery ■ Trauma ■ Loss of trust in authorities

Source: Adapted from Merz, Kreibich, Schwarze, and Thieken (2010) whenever floods threaten to reoccur. Loss of treasured possessions in floods can be heartbreaking, and much more significant than financial losses, which are now commonly recovered through government compensation schemes or household insurance policies. These impacts are seen as the net effect of the threat, the mediating influences (e.g., flood defenses) that moderate that threat for the affected population, and the support capacity in households, communities, and indeed the nation that helps to promote resilience in that population and the capacity to recover from the threat, the event, and its effects. In this respect, the health and mental health effects of flooding need to be considered, so that these can be accounted when evaluating policy options at the coast.

Natural disasters such as flooding can affect people's health in a number of ways (Tapsell, Penning-Rowsell, Tunstall, & Wilson, 2002); good health being defined as complete physical, mental, and social well-being. Many impacts are associated with the trauma of flooding and living subsequently for long periods in damp and dirty conditions. The close proximity of people living in cramped conditions in their homes following flooding mean that some of these adverse health effects can be passed from person to person within the household, particularly where pre-existing health issues are present. Hence, the effects of flooding on people's health and general well-being can continue for many months after the actual flood event. People suffer from psychological health impacts from the stress of the flooding (Tapsell, Penning-Rowsell, Tunstall, & Wilson, 2002). Stress arises from the difference between the perceived demand the event places upon the individual and the resources the individual can draw upon to adapt to that demand. The severity of the impact represents the degree to which coping and support capacity are insufficient to cope with the challenge and costs of responding.

The conclusion is that the impacts of flooding on people are more extensive and complex than have hitherto been appreciated. Hence, assessments of the effect of flood risk reduction measures on these more intangible impacts are flawed and incomplete if only monetary losses are used within the necessary project-appraisal and option analysis methods.

2.6 The Social Context of Flooding at the Coast

The social context of floods is a critical dimension of any system-based analysis of floods. All human groups are not equal when facing floods, and within coastal communities parts of the population may be more vulnerable to floods and their consequences. A review of social vulnerability analysis to floods indicates that the following key dimensions must be taken into account: demographics (age, population density, migratory status), wealth (absolute and its distribution), health status, and mobility. McElwee (2010), Baum, Horton, & Choy (2008) and Coninx & Bachus

TABLE 2.5 Benefits and Limitations of Qualitative Assessment Methods

Techniques	Structured Interviews, Focus Groups, Surveys, Questionnaires
Benefits	Engage stakeholders in the flood-management process Provide depth, detail, and context for more quantitative approaches Ensure identification and focus on relevant issues for stakeholders Identify people's individual experiences, building up a picture of the diversity of stakeholders' views and why these exist Attempt to avoid pre-judgments, identifies trends and emergent themes Can be cyclical, with analysis informing subsequent data collection and further analysis Focus groups promote openness by allowing different views to be expressed
Limitations	Identification of relevant individuals Time consuming; available time may dictate number of participants, length of interviews, and analysis Not easy to generalize or systematically compare a small number of interviews Highly dependent on skills of the interviewer

(2007) provide detailed examples for Vietnam, the Gold Coast (Australia), and climate change, respectively. Social vulnerability is a complex phenomenon, and no single measure comprehensively includes all aspects of vulnerability (Adger, Hughes, Folke, Carpenter, & Rockstrom, 2005). Factors such as those listed above can all be considered, but vulnerability is site-specific and some relationships between social characteristics and vulnerability are unlikely to be linear or readily transferable. While there seems to be a consensus on the dimensions to be taken into account, their local articulation varies because of local variation in governance, cultures, and perceptions, and this requires evaluation in any assessment.

A review of governance structures and perceptions should thus take place at the beginning of any flood risk assessment, and the stakeholders contacted should be encouraged to participate throughout the assessment process (Section 2.4, Chapter 5). Information is generally collected from stakeholders using qualitative methodologies: individual interviews, semi-structured interviews, and focus groups. These are time-consuming processes to apply, with distinct benefits and limitations (Table 2.5). Ultimately, a focus on the participation of local communities and authorities has two major benefits:

- Optimal use is made of the know-how and skills of local communities, taking into account their wishes and needs;
- The involvement and shared responsibility of local parties in coastal risk assessment will guarantee a sound community basis for the development of management plans (Chapter 6).

Recently, the Social Vulnerability Index has been suggested as a comparative spatial assessment of human-induced vulnerability to environmental hazards (Cutter

Boruff, & Shirley, 2003; Wisner et al., 2004). This index is based on a large set of measurable variables that can be grouped into main common factors such as population structure, gender, income, socioeconomic status, and renters (www.csc.noaa.gov/slr). Analysis and mapping of social vulnerability should also consider identifying critical facilities or resources to help prioritize potential hazard mitigation.

2.7 Coastal Habitats Within the Flood System

Healthy and productive environments are the basis of sustainable development and human welfare, providing both direct resources and key ecosystem services such as coastal defense (Millenium Ecosystem Assessment, 2005). Over the last few decades there has been a growing recognition of the importance of coastal ecosystems for their biodiversity and aesthetic value, and regulations are increasingly being implemented to promote their conservation across the world (e.g., European Commission, 1992; Nicholls et al., 2012a). However, one aspect of the flood system that is not always fully recognized by coastal managers and stakeholders is the role of coastal habitats in determining the impacts of flood events (Jones et al., 2011, pp. 411–458).

In general, coastal habitats can tolerate a degree of flooding by sea water, and are able to adjust in accordance with both short- and long-term processes if there is the required space and sedimentary material to do so. Short-term and seasonal processes are part of the dynamic coastal environment. For example, a salt marsh requires flooding on a daily basis, while the erosion and deposition of sediment driven by coastal processes is part of a dynamic sand dune system. However, coastal grasslands are not tolerant to flooding, but are able to regenerate after temporary flooding through the soil seed bank (Kalamees & Zobel, 2002). If coastal flooding becomes more regular, there is likely to be a change in the species composition of coastal grasslands. If water inundation occurs as a result of long-term processes (e.g., sea level rise), it is assumed that this is permanent. Unless defended, this will result in the loss of terrestrial habitat areas, although it is important to recognize that intertidal/subtidal habitats will be gained. If habitats have the ability to "retreat" (the affected terrestrial habitats can move landward), these newly occupied territories may be considered as additional coastal habitat. Alternatively, where there is no possibility for habitat retreat because of natural or anthropogenic barriers (e.g., cliffs or sea walls/dikes), areas of lost habitat cannot be compensated. Overall there is a need for understanding of the possible effects of changing inundation duration on any coastal habitat.

Within ecology, vulnerability is considered to relate to the inherent properties of an ecosystem that determine resistance and resilience to any perturbation. An ecosystem can be defined as resistant if it has a high ability to withstand disturbance events. For example, resistance can be measured in a salt marsh as the biomass lost or anticipated to be lost as a consequence of equal disturbance events. However, once a system is damaged by a flood or storm event, the system is more vulnerable to greater damage by a second flooding event even if this second event is of lower magnitude. Resilience, on the other hand, is the time the ecosystem needs to recover to the state from before the disturbance event took place. Resilience can be measured as the recovery time. Hence, more vulnerable ecosystems are less resistant against a disturbance event, or resilient with longer recovery times. Therefore, the most vulnerable ecosystems have both low resistance and resilience, and the persistence of such systems is unlikely.

2.8 The Physical Context of Flooding at the Coast

2.8.1 Extreme Water Levels

As explained in Section 2.3.2, extreme water levels from the sea are caused by a combination of several factors: high astronomical tides, storm surges, and waves. Changes in any of these factors may change the magnitude and impact of a flood event. Long-term changes in relative mean sea level have also contributed to changing extreme sea levels: globally this is the most relevant factor for changing extreme water levels (Menéndez & Woodworth, 2010). For some coastal areas (deltas, estuaries, bays, and inlets), river input can also influence flooding, particularly when high flows coincide with extreme coastal conditions (e.g., Gilbert & Horner, 1984, p. 216; Monbaliu et al., 2014). Local effects from water works, land subsidence, gas extraction, etc. may also be substantial, and where this is the case should be considered. This is particularly a problem on recent alluvial sediments and on deltas, such as widely found in South, South-East and East Asia (Nicholls et al., 2014). Consequently any coastal hazard assessment should first aim at identifying processes potentially relevant at the local scale and subsequently at estimating their variability and their past and potential future contributions to extreme water-level changes. In many cases, the most important factors comprise hydrology; effects from water works and engineering; coastal ocean conditions such as mean sea level, waves, and storm surges; and river flow and subsidence. A combination of statistical analyses of existing data, together with modeling studies, represents the most commonly used approach.

2.8.2 Coastal Morphology

Long term morphological transformations are to be expected in coastal areas as a result of gradients in long-term sediment transport, sea level rise, and changes of the environmental dynamics (Cowell et al., 2003; Nicholls et al., 2012b, 15 pp; Stive,

Cowell, & Nicholls, 2009, pp. 158–179). How the coastline will respond in the future depends not only on the complex physical processes involved in the hydrodynamics and morphodynamics of coastal regions but also on the interactions between the physical environment and anthropogenic actions. Furthering our understanding of morphological change will improve the management of coastal flooding in the future, as we anticipate future adverse conditions.

Morphological change can be rapid, being induced by tsunamis (e.g., Japan, Tanaka et al., 2012, and Chile, Morton, Gelfenbaum, Buckley, & Richmond, 2011), rainfall (e.g., landslides in Canada, Guthrie, Mitchell, Lanquaye-Opoku, & Evans, 2010), and storms (e.g., Hurricane Wilma in Cancun, Mexico, Silva et al., 2012), Hurricane Sandy in eastern USA (Trembanis et al., 2013); or can be caused by gradual stresses, generated by a deficit in sediment supply, either due to long-term reductions or temporary inequalities (e.g., Varna, Bulgaria, Villatoro et al., 2014), seasonal changes in wave climate (e.g., modifications of the profile and plan shape in beaches in Wales, Thomas, et al., 2010), drift inequalities (e.g., in the South of Spain, Moreno et al., 2010), and, finally, by longer-term changes in climate (sea level change and human activities).

According to Stive et al. (2002), Cooper (2009, pp. 129–152), and Oumeraci (2004), coastline changes take place at a variety of temporal and spatial scales, and proper understanding of coastal behavior requires that short-term patterns of change be placed in a longer-term perspective. The cumulative effects of a change at any time-scale are normally nonlinear, dynamic, complex, and sensitive to anthropogenic interference and climate variability, and hence are subject to large uncertainties. The impact of high-magnitude, low-frequency events, such as those mentioned above, may also be important. In terms of coastal morphology, coastal managers need to be aware of obvious, visible features such as foreshore slope, the width and height of the berm, and the backshore width. Other features, controlled by the geological setting or underlying topography of the area, must also be considered. In addition, a checklist for the factors to be taken into consideration should include sediment type (grain diameter, form, sorting, mineral composition, moisture content, stratification (e.g., Camenen, 2007; Alcerreca et al., 2013), sediment volume, and supply. Climatic and oceanographic settings also form part of the analysis (waves: height, period, and angle; tides: range, diurnal pattern, and stage; currents: velocity and direction; and winds: velocity and direction). Given the inherently low predictability level and the broad range of time and space scales of these processes, monitoring is strongly recommended (Nicholls, Townend, Bradbury, Ramsbottom, & Day, 2013).

The key processes involved in coastal morphological changes are a loop of erosion, transport, storage, and deposition of sediment. Coastal erosion, which can be understood as land loss, landward movement of the shoreline, or a reduction in beach volume, is most likely to cause coastal flooding due to climate change. In some areas, in addition to the effects of climate change on coastal morphology, local erosion processes, and coastal squeeze must also be considered. The term "coastal squeeze" is used to describe the retreat of the shoreline in areas where landward retreat of the geomorphic feature is not possible, "squeezing" the width of the coast and its protective function. The urbanization of coastal areas has turned coastal erosion from a natural phenomenon into an urgent problem of growing intensity in many areas around the world. Dynamic ecosystems and their undeveloped coastal landscapes are gradually disappearing, due to a lack of space and sediment, (Cooper, 2009, pp. 129–152).

While natural erosion may have a short-term (temporary) or long-term (chronic) character, man-made erosion generally has a chronic character. For this reason, before undertaking any action to protect an area from the effects of erosion, coastal managers must first identify and understand which type of erosion (temporary or chronic) occurs (Eurosion, 2004). Some examples are presented in Chapter 7 of this book. A systematic cataloging is therefore needed which not only accounts for the effects of erosion but also for the main causes, for the relevant time and space scale, and for the order of magnitude of recession at the site under consideration. This is a prerequisite to deciding what action to take: protection against wave attack, increasing dune sediment storage, or fixing the shoreline and/or infrastructure protection (e.g., Mendoza et al., 2014 and Chapter 3 show innovative cases). In the case of chronic erosion, improving the protective role of natural structures may also be implemented.

Presently, numerical modeling is the means most often used to predict long-term morpho-dynamical changes. The available numerical models are not completely satisfactory and are subject to uncertainties from diverse sources. Therefore, we need to account for the inherent stochastic variability of the influencing parameters, modeling errors, and also human and organizational errors (Oumeraci, 2004), as well as focusing attention on new methods as they emerge.

Most numerical models are integrated by hydrodynamic, sediment transport, and morphological modules. It is necessary to have field measurements and observations to calibrate and to validate the models. Since the models do not directly consider the interaction between hydrodynamics and sediment transport, an update loop is required to account for the influence of morphological change on the hydrodynamics and sediment transport. Specific modeling methods (see Chapter 7 of this book) are needed to predict the response of the coastline at different scales. These methods come with varying levels of reliability, accuracy, and required expertise, and may be seen basically as process-based numerical models, behavior-based numerical models, statistical analyses, geomorphological analyses, parametric equilibrium models, and other, emerging techniques (Sutherland, 2007).

The future geomorphic evolution of coastal areas is heavily dependent on many factors that are difficult to consider, such as the relation between the physical and biological characteristics of ecosystems and their resilience, among others. Therefore, the use and protection of coastal zones will always be subject to conflict and compromise. We need to take into account the evolving socioeconomic demands, impacts of human interventions, and morphological changes. Hence, engineering solutions have to deal with each case individually.

2.8.3 Defense Failure

On defended coasts, flooding arises when defenses fail in some way. Three separate mechanisms of such failure can be considered: overflow, overtopping, and breaching. In these mechanisms, still water levels can flow over (overflow), wave action propels water over the crest (overtopping), or the defense is lowered (breaching). Few defenses remain undamaged under heavy overtopping, and this can initiate a breaching response—hence one type of failure can lead progressively to another. Failure to close flood gates, mobile barriers, and sluices can also allow flooding: these are mainly caused by errors in human operation. As already noted, natural geomorphic and habitat features may directly or indirectly provide flood protection. For instance, beaches and dunes can attenuate wave affects prior to contact with structural defenses, or act as a line of defense in their own right; while offshore barriers, breakwaters, reefs, salt marshes, and mudflats may dissipate wave energy further away from the floodplain. In general terms, we wish to minimise defense failures and avoid breaching as the volume of water that passes on to the floodplain can increase by several orders of magnitude, greatly increasing the consequences of any flood. This topic is considered in more detail by Kortenhaus (2012). Of course defense failure should be accounted for when planning coastal management and assessing its performance (see Section 6.2.1).

2.8.4 Modeling Flood Events

There are numerous hydrodynamic models that can be used to simulate the propagation of floodwater across floodplain areas. These models generally solve a form of the two-dimensional shallow water equations and range in complexity from raster-based approaches (Bates & De Roo, 2000; Bates et al., 2005, 2010; Bradbrook Lane, Waller, & Bates, 2004; Dottori and Todini, 2011) to more complex finite volume approaches (Lane & Richards, 1998) that solve the full two-dimensional equations.

The accurate representation of the complex dynamics of sea–river interaction and/or beach reshaping and run-up requires a three-dimensional solution of the three-dimensional Reynolds-averaged Navier Stokes equations (RANS equations). This in turn necessitates of an approximate numerical technique such as finite differences, finite elements, or finite volumes. A number of codes are available for local predictions of three-dimensional velocity fields in main channels and floodplains, such as MATO-3D (Posada, Silva, & de Brye, 2008) and FLUENT. However, these approaches are computationally expensive (run time of several days) and thus far have only been applied to channels of a limited domain size and regular geometry (Woodhead, 2007).

Two-dimensional approaches typically use depth-averaged velocity obtained by integrating the Reynolds averaged Navier–Stokes equations over the flow depth. Examples are the St. Venant equations, which assume a hydrostatic pressure distribution, or the Boussinesq equations (Woodhead, 2007). The 2D class of models includes full solutions of the two-dimensional St. Venant or shallow water equations and simplified shallow water models where certain terms, such as inertia, are omitted from the controlling equations. Models of this class are MATO (Posada, Silva, Simmonds, & Pedrozo, 2007, pp. 205–241), TUFLOW, Mike 21, TELEMAC, LISFLOOD-FP, and Delf-FLS (e.g., Neelz & Pender, 2009). The computational time is from hours to days. One of the first numerical models developed for wave overtopping-induced erosion of the inner slope of grassed sea-dikes is the BREaching of Inhomogeneous sea Dikes (BREID), developed by Tuan and Oumeraci (2010). On the other hand, 2D models are commonly applied to a broad variety of problems including urban inundation. Simpler 1D methods (Wadey, 2013) such as Mike 11, HEC-RAS, and Infoworks RS (ISIS) (Neelz & Pender, 2009), with computation time in the order of minutes to hours, represent the flooding process under the assumption that the floodplain flow is equal to the channel flow.

Where a broad-scale assessment of extents and depths of flooding is required, GIS-based flood inundation or flood-spreading models (Brown, 2006; Poulter & Halpin, 2008) can be a cost-effective solution. These models do not solve hydraulic equations but perform flood mapping through the spreading of water levels or volumes across a Digital Elevation Model (DEM) by using several techniques (Chen, Hill, & Urbano, 2009; Gouldby, Sayers, Mulet-Marti, Hassan, & Benwell, 2008; Wang, Wan, & Palmer, 2010; Zerger, Smith, Hunter, & Jones, 2002). The computational time ranges from seconds to a few minutes, depending on modifications introduced in the algorithms; therefore these approaches can be easily implemented in Decision Support Systems (see Section 6.6). However, to provide the user with sufficient accuracy they require high-resolution topographic data and are less suited to application in flat areas.

Complementary methodologies to simulate extreme long-term flooding water-level scenarios, based on the application of combinations of different types of hydrodynamic models and probabilistic distributions of the atmospheric and river discharges forcing, are presented in Chapter 7.

2.9 Handling Uncertainty

Uncertainty permeates the whole process of risk assessment and is often ignored. There are two main causes: (1) lack of knowledge either about relevant data or about whether a particular effect will occur; and (2) as a result of the random nature of the events, which itself depends on natural circumstances and their timespan. These random events can include:

1. Errors in the probabilities of flood events: e.g., through the extrapolation of short time series to flood discharges.
2. Inundation area and depth: imprecision due to generalized digital terrain models or because of difficulties in estimating failure probabilities of flood defenses.
3. Type and location of elements at risk: inaccuracies because of generalizations in spatial resolution and categorization of land use data.
4. Value of elements at risk: values are often approximations or have to be disaggregated or have to cope with nonmarketable elements such as valuable habitats.
5. Susceptibility of elements at risk: damage functions are often derived from poor empirical data.

Hazard forecasting and risk assessment systems traditionally concentrated on separately modeling single phenomenon such as sea level, rainfall, waves, river discharges, flash flooding, etc. Each forecasting system comprises a linear flow of data and a combination of different models; for instance: starting with a meteorological model, which provides forcing to a shallow water flow model for forecasting tides and surges; a wave model for forecasting waves nearshore and further models to predict subsequent overtopping and/or flooding; and river flow models for forecasting pluvial and fluvial flooding. The weaknesses (or limitations) of these flood modeling systems include:

- Lack of inter-operability between model components;
- A tendency to consider only a single source of hazard;
- Lack of ensemble or data-assimilation techniques;
- Absence of tracking of estimation errors for uncertainty analysis;
- The need to constrain uncertainties and narrow prediction bounds with model refinement;
- Assessment of the potential associated risk is often limited or even absent with respect to vulnerability and resilience;
- Assumption of historic/static data on the condition of geomorphic landforms and defenses.

Cascading forecast uncertainty in coupled models is an important step to improve the quality of hydrological forecasts (Cloke & Pappenberger, 2009). However, the best methodology to quantify the total predictive uncertainty in hydrology is still debated (Beven, Smith, & Freer, 2008). Sources of uncertainty in the hydrometeorological forecast chain are numerous and include the meteorological forcing, corrections, and downscaling procedure of the meteorological predictions; antecedent conditions of the system (meteorological and hydrological) observation networks; methods of data assimilation (discharge, soil moisture); geometry of the system (including flood defense structures); possibility of infrastructure failure (dykes or backing-up of drains); characteristics of the system (in the form of model parameters); and limitations of the hydrological model to fully represent processes (for example, surface and subsurface flow processes in the flood generation and routing). The importance of the individual components varies in time, depending on the dominant flow regime, and in space, as each catchment is unique (Beven, 2000). It also depends on the interactions between the space-time scales of the predicted event, the main catchment characteristics (area and response time), and the resolution of the meteorological forcing data (Thirel, Rousset-Regimbeau, Martin, & Habets, 2008). A full uncertainty analysis can track all sources of uncertainty and estimate both their relative importance in the system and the total uncertainty from the combination of each component (Pappenberger et al., 2005). The total magnitude of the uncertainty influences the quality of the predictions, the interpretation of model output forecasts, and ultimately its use in decision-making (Ramos, Mathevet, Thielenc, & Pappenberger, 2010).

Many of the issues of projecting future change are addressed by presenting risk as a range of values rather than a single number. This provides an envelope within which the actual future is expected to occur. There are two main approaches: the use of scenarios and probabilistic approaches.

The use of scenarios in risk assessments recognizes that the future is unknowable. For example, knowledge about future socioeconomic developments is limited. In turn, this leads to uncertainties in future greenhouse gas emissions. Further, when subjected to the same emission scenario, different climate models will show different responses, reflecting both imperfect knowledge of the underlying physical mechanisms and internal (natural) climate variability.

A number of different scenarios should be used that sample the underlying assumptions that appear plausible. Commonly an ensemble of climate change simulations obtained from different models and scenarios is used. Scenarios cannot be associated with a likelihood of occurrence and represent "plausible futures" rather than probable outcomes (Von Storch & Zwiers 2013). Hence, scenarios generally address questions of the type "What may happen if …?". The benefit of using scenarios is that decision-makers consider a range of views of what may unfold and understand broad sensitivities of the flood system. Hence, they can develop suitable policies/management. A focus is on options that are robust to the range of existing uncertainty and flexible; that is, they may be adopted in the course of time when expected changes manifest and uncertainty becomes smaller, raising the approach of defining and selecting adaptive pathways (Ranger, Reeder, & Lowe, 2013; Tarrant & Sawyers, 2012). Hence, there can be benefits in considering scenarios that have a low chance of occurring (Randall & Ertel, 2005), to test for the long-term robustness and feasibility of different adaptation approaches over time and the range of scenarios.

In the context of historical changes and present conditions, probabilistic or statistical approaches can be used. For example, the definition of return periods and their uncertainties has become more common with the increase in data availability and computing power. However, this still depends on the availability of data. Extreme events pose a particular set of challenges for implementing probabilistic approaches, because their relative infrequency makes it difficult to obtain adequate data for estimating the probabilities, and this gets worse as return periods increase (Milly, Wetherald, Dunne, & Delworth, 2002).

Communication of the uncertainty within a flood assessment is good scientific practice, maximizing credibility and minimizing misinterpretation, bias, and different interpretations (Kloprogge, van der Sluijs, & Wardekker, 2007). Ineffective communication of scientific research to decision-makers and the public has often proved a barrier to uptake of knowledge by stakeholders. Uncertainty information concerning probabilities is particularly prone to biases, as the concepts themselves are not easy to understand; risk experts separate the probability and magnitude components of a risk, but for nonscientific audiences the perception of risk is often directly linked to consequences and specifically to consequence experienced by the users involved in the assessment. This can lead to an under-appreciation of low-probability, high-impact events (Kloprogge, van der Sluijs, & Wardekker, 2007).

2.10 Capturing Future Changes

Timing and time-scales are important cross-cutting themes that need more attention when dealing with the identification and management of extreme climate and weather events, disasters, and adaptation strategies. The first key issue when dealing with timing and time-scales is the fact that different hazards and their recurrence intervals might fundamentally change with time. This implies that the identification and assessment of risk, exposure, and vulnerability also needs to address multiple time-scales. At present, most of the climate change scenarios focus on climatic change up to the year 2100, while projections of vulnerability often just use present socioeconomic data. However, a key challenge for enhancing knowledge of exposure and vulnerability as key determinants of risk requires improved data and methods to project and identify directions and different development pathways in demographic, socioeconomic, and political trends that can illustrate potential increases or decreases in vulnerability with the same time horizon as the changes in the climate system related to physical-biogeochemical projections (Birkmann, Garschagen, Kraas, & Quang, 2010). This is challenging, as future socioeconomic conditions are more uncertain than biophysical conditions, and for example, a maximum of time frames of 25–30 years is normal in government. Furthermore, the time dependency of risk analysis, particularly if the analysis is conducted at a specific point in time, has been shown to be critical (e.g., Setiadi, 2011). These types of issues should also be considered, but the details of how and to what degree will vary from study to study.

2.10.1 Use of Drivers and Scenario Analysis

As the SPRC model describes the flood system at a single moment in time, the conceptual system needs to sit within a wider analytical framework that allows for time and external and internal changes as a result of different Drivers. Including Drivers is essential when looking at the evolution of the flood system (and flood risk) over time, and requires clarity early in the flood risk assessment (Millner, 2012). This effectively addresses the uncertainties faced when looking at future situations and can range from uncertainties inherent in the modeling process (including scientific understanding of the system) to the range of possible socioeconomic futures and projections of climate change that can affect flooding. Participatory approaches including stakeholder engagement are good practice, maximizing credibility and minimizing misinterpretation, bias, and differences by readers and users (Kloprogge, van der Sluijs, & Wardekker, 2007).

Many of the challenges of communicating possible change are addressed by presenting risk as a range of values rather than a single number. Scenarios (storylines) are often used to illustrate different plausible relationships between cause and outcome, illustrating how current and alternative development paths might affect the future (Moss et al., 2010; Nakićenović et al., 2000; Nicholls et al., 2012a). Hence, scenarios can have multiple dimensions depending on the question being posed. In addition to considering the Drivers in isolation, one approach is to use a range of scenarios which vary the underlying assumptions: at the minimum, estimations can reflect where everything works to expectations—a best-case scenario—and where nothing does—a worst-case scenario; the difference between the best-case and worst-case values can then be used as a measure of the range of risk. There can also be benefits to considering scenarios that have a low probability of occurring (Nicholls et al., 2014; Randall & Ertel, 2005).

How individual parameters within the scenario are represented also needs to be decided (see Table 2.6). For the quantitative components of the system, such as water levels and number of people, future projections commonly draw on global or national level data and are downscaled using statistical methods. For example, with the increase in data availability and computing power, methods such as standard deviation and probabilities have become more common, particularly for the translation of climate model outputs for detailed flood modeling. For some parameters, however, the use of such data to represent local changes could raise the question of plausibility, as a different pattern of change could be experienced: for example, a city may increase in population despite regional or country projections of population decline.

TABLE 2.6 Examples of Representative Scenarios and Data for the Different Aspects of the Flood System

Data Type	Data Source	Social Aspects	Ecological Aspects	Hydrological Aspects
Qualitative	Global	SRES or SSP scenarios		SRES or RCP scenarios
(Semi) qualitative		Human typologies	Vulnerability/resilience assessment (expert opinion)	
Quantitative	Global - national	Downscaled existing population and GDP projections	Designated areas	Water levels and discharge modeled from global climate models (long-term)
	Local	Local data on population and GDP, census data, land-use maps, habitat maps, development plans, buildings database	Changes in specific indicator parameters (e.g., species diversity, salinity, area)	Projections based on 30+ years of historical data—can be short-term only (10 years)

For the more qualitative aspects of the flood system, such as public perception and human behavior, deciding how (or even whether) to incorporate them is a challenge for assessments largely based on quantitative flood modeling. This represents a key research challenge.

2.10.2 Climate Change

Linking changes in coastal flooding and erosion risk with climate change requires the combination of data, numerical and statistical models, and empirical approaches. For coastal flooding, analyses of wind, wave, tide, surge conditions, and changes in mean sea level are required. In some cases, contributions from river floods need to be considered. It is often found that mean sea level changes appears to be the most relevant climatic factor contributing to observed and potential future changes in extreme sea levels, while changes in other components (such as the wave or storm surge climate) are small and/or highly uncertain (Weisse, von Storch, Niemeyer, & Knaack, 2012). Determining future scenarios of change can be demanding, so it is important to use methods appropriate to available resources (e.g., Nicholls et al., 2014).

2.10.2.1 Mean Sea Level

Over the twentieth century global mean sea level increased on average 1–2 mm/year (Bindoff et al., 2007; Church et al., 2013; Jevrejeva, Moore, Grinsted, & Woodworth, 2008). For the satellite era from 1993 onwards, somewhat higher estimates of about 3 mm/year are provided that are broadly consistent with that derived from tide-gauge data only (Church et al., 2008; Church et al., 2013). For the future, a further increase in global mean sea level is expected primarily as a result of two processes: ocean thermal expansion in consequence of rising temperatures and freshwater input as a result of melting glaciers and ice sheets. Contributions from melting ice sheets are thought to be minor for the twentieth century, but are potentially the largest contributors in the future (Church et al., 2008; Church et al., 2013).

Regional changes in mean sea level can deviate substantially from the global mean. Regional factors therefore need to be considered. Ideally such assessments are based on global projections taking into account regional effects such as vertical land movement or effects from changing ocean circulation (e.g., Katsman et al. 2008, Katsman et al., 2011; Lowe et al., 2009; Slangen, Katsman, van de Wal, Vermeersen, & Riva, 2012). If such regional projections are unavailable, a pragmatic two-way approach can be adopted: extrapolation of observed regional decadal trends for short and medium time scales may be used as long as the decadal variability in the observed records is larger than the changes projected by the IPCC (e.g., Weisse et al., 2014).

For the long time scales (2080s), global projections based on IPCC scenarios may be used if no other regional information is available.

2.10.2.2 Wind Fields

Surges and waves and their long-term changes are intimately connected with statistics of storms, changes in wind speed, direction, and/or frequency. Wave run-up and setup are relevant processes to consider that may influence coastal flooding. Future changes in extreme wave climate depend on corresponding changes in the atmospheric wind fields that are highly uncertain (Christensen et al., 2007). An ensemble approach taking a number of different climate change scenarios and projections from different models into account is therefore applicable for both surges and wave climate. In this way, a range of possible future developments is obtained, allowing for a better quantification of uncertainties. To do so, a simple descriptive approach rather that formal significance tests is proposed for characterizing the information in an ensemble of scenarios (Von Storch & Zwiers, 2013). It should be noted that the statistics of present winds, surges, and waves is quite uncertain, so determining changes will be difficult.

2.10.2.3 Extreme Water Levels

Extreme sea levels may show pronounced fluctuations on seasonal, inter-annual, and decadal time scales. The latter may be associated, e.g., with the nodal cycle (e.g., Méndez, Menéndez, & Losada, 2007; Menéndez et al., 2009). Such fluctuations may be important for any risk assessment related to extreme sea levels. A time-dependent extreme value analysis is therefore proposed to assess the magnitude and phase of such fluctuations. An approach to do so is presented in Méndez, Menendez, & Losada, (2007) and Menéndez et al. 2009. By taking the different time-scales of sea level variability into account, the approach is able to produce short-term forecasts of the probability of exceeding a certain extreme sea level, allowing for the definition of a time-varying flood risk.

2.10.3 Population and Economy Change

For a long-term flood-risk assessment, potential changes in population and asset value should be considered. Specific knowledge may be available at local level and the short term (e.g., development plans), but over longer periods appropriate socioeconomic scenarios need to be created. In particular, population, gross domestic product (GDP), and other scenarios relevant at the scale of the study sites are required. These localized scenarios need to represent coherent, internally consistent, and plausible descriptions of possible trajectories of future conditions based on self-consistent storylines or images of the future. They also need to agree with relevant stakeholders for credibility purposes. The high level of indeterminacy of these factors should be conveyed to local and national stakeholders: these scenarios must be presented as food for thought and action, rather than robust projections of the future.

These social and economic scenario will also need to consider cross-scale interactions (Turner et al., 2003a,b). However, the practical application and analysis of these interacting influences on vulnerability from different spatial scales is a major challenge and in most cases not sufficiently understood. Furthermore, vulnerability analysis, particularly linked to the identification of institutional vulnerability, must consider the various functional scales of climate change, natural hazards, vulnerability, and administrative systems. In most cases, current disaster management instruments and measures of urban or spatial planning as well as water management tools operate on different functional scales compared to climate change. For example, policy setting and management of climate change and of disaster risk reduction are usually the responsibility of different institutions or departments; thus it is a challenge to develop a coherent and integrated strategy (Birkmann & von Teichman, 2010). Consequently, functional and spatial scale mismatches might even be part of institutional vulnerabilities that limit the ability of governance system to adequately respond to hazards and changes induced by climate change. This illustrates the potential complexity of this aspect of flood-risk assessment and the need for clarity on the questions being asked.

2.10.4 Habitat Loss or Change

Coastal habitats can tolerate a degree of flooding by sea water, but are vulnerable to changes in flooding regime, and permanent changes can occur. Including quantitative modeling of the full range of potential impacts on all coastal habitats may not be possible within many coastal flood assessments due to our limited detailed understanding. Hence, an alternative, robust method is needed. Coastal habitats (including seagrass meadows, biogenic reefs, rocky shores, and salt marshes) are generally not affected by individual flood events, but over time, persistent changes in Drivers will have an effect; for example, negative secondary impacts from flooding such as sedimentation over biogenic reefs. Terrestrial habitats (including coastal grasslands, grazing marshes, and sand dunes) can be damaged by seawater flooding. Estuarine transition zones can be regarded as further sources of floodwater particularly when the freshwater level is high. These habitats also act as efficient pathways for transferring floodwaters. As these habitats regularly experience changes in salinity over varying time periods, consequences from one-off flood events are negligible.

However, salinization is likely to produce negative impacts on the species found there; although these habitats will be able to recover, permanent changes to the saline regime potentially though changes in sea level will result in permanent changes to the communities found. These changes can be considered as changes in Pathways for the landward Receptors, but these habitats also represent Receptors in terms of habitat designations, etc.

2.11 Conclusions

This chapter has shown that a holistic approach to coastal flood risk management has significant benefits. Changes in the physical, ecological, and socioeconomic dimensions identified within the SPRC concept all have the potential to change levels of risk. Hence if they can be managed appropriately, this has the potential to regulate levels of risk. By considering these three main aspects of the flood system, a more resilient approach can be adopted that is capable of dealing with flood events better than more traditional approaches (Wardekker, de Jong, Knoop, & van der Sluijs, 2010). Chapters 3, 4, and 5 deal in detail with the different kind of mitigation measures addressing specifically engineering, ecological, and socioeconomic components of the coastal system, respectively. It is important to remember that there are potential interactions between the different classes of techniques—for example, beach nourishment (soft engineering) may have ecological consequences, and the effects of ecological management may have influences on the design of artificial structures. In fact, some of these interactions may be planned and intended. The considerations when selecting individual or combinations of techniques are discussed in Chapter 6, while Chapter 7 looks at examples.

References

Adger, W. N., Hughes, T. P., Folke, C., Carpenter, S. R., & Rockstrom, J. (2005). Social-ecological resilience to coastal disasters. *Science, 309*(5737), 1036–1039. http://dx.doi.org/10.1126/science.1112122.

Alcérreca-Huerta, J. C., Silva-Casarín, R., & Mendoza Baldwin, E. (2013). Simple settling velocity formula for calcareous sand. *Journal of Hydraulic, Research, 51*(2), 215–219. http://dx.doi.org/10.1080/00221686.2012.753645.

Arnoldi, J. (2009). *Risk: an introduction*. Cambridge, UK: Polity.

Aven, T., & Renn, O. (2010). *Risk appraisal*. In *Risk management and governance*. Berlin Heidelberg: Springer, 131.

Bakewell, I., & Luff, S. (2008). *North London strategic flood risk assessment*. Final Report for the North London Waste Plan. London, UK, Mouchel. http://www.nlwp.net/downloads/north_london_sfra_final_august_08.pdf Accessed September 2014

Balica, S. F., Douben, N., & Wright, N. G. (2009). Flood vulnerability indices at varying spatial scales. *Water Science & Technology, 60*, 2571–2580.

Bates, P. D., Dawson, R. J., Hall, J. W., Matthew, S. H. F., Nicholls, R. J., Wicks, J., et al. (2005). Simplified two-dimensional numerical modelling of coastal flooding and example applications. *Coastal Engineering, 52*, 793–810. http://dx.doi.org/10.1016/j.coastaleng.2005.06.001.

Bates, P. D., Horritt, M. S., & Fewtrell, T. J. (2010). A simple inertial formulation of the shallow water equations for efficient two-dimensional flood inundation modelling. *Journal of Hydrology, 387*, 33–45.

Bates, P. D., & De Roo, A. P. J. (2000). A simple raster-based model for flood inundation simulation. *Journal of Hydrology, 236*(1–2), 54–77. http://dx.doi.org/10.1016/s0022-1694(00)00278-x.

Baum, S., Horton, S., & Choy, D. L. (2008). Local urban communities and extreme weather events: mapping social vulnerability to flood. *Australasian Journal of Regional Studies, 14*(3), 251–273.

Beven, K. J., Smith, P. J., & Freer, J. E. (2008). So just why would a modeler choose to be incoherent? *Journal of Hydrology, 354*(1–4), 15–32.

Beven, K. J. (2000). Uniqueness of place and process representations in hydrological modelling. *Hydrology and Earth System Sciences, 4*(2), 203–213.

Bindoff, N., Willebrand, J., Artale, V., Cazenave, A., Gregory, J., Gulev, S., et al. (2007). Observations: oceanic climate change and sea level. In S. Solomon, D. Qin, & M. Manning (Eds.), *Climate change 2007: The physical science basis. Contribution of working group I to the fourth assessment report of the intergovernmental panel on climate change*. Cambridge, UK: Cambridge University Press.

Birkmann, J., & von Teichman, K. (2010). Integrating disaster risk reduction and climate change adaptation: key challenges – scales, knowledge, and norms. *Sustainability Science, 5*(2), 171–184.

Birkmann, J., Garschagen, M., Kraas, F., & Quang, N. (2010). Adaptive urban governance: new challenges for the second generation of urban adaptation strategies to climate change. *Sustainability Science, 5*(2), 185–206.

Bradbrook, K. F., Lane, S. N., Waller, S. G., & Bates, P. D. (2004). Two dimensional diffusion wave modelling of flood inundation using a simplified channel representation. *International Journal of River Basin Management, 2*(3), 211–223. http://dx.doi.org/10.1080/15715124.2004.9635233.

Brown, I. (2006). Modelling future landscape change on coastal floodplains using a rule-based GIS. *Environmental Modelling & Software, 21*, 1479–1490.

Burzel, A., Dassanayake, D. R., Naulin, M., Kortenhaus, A., Oumeraci, H., Wahl, et al.. (2010). Integrated flood risk analysis for extreme storm surges (XTREMRISK). In J. McKee Smith, & P. Lynett (Eds.), *Proceedings of the 32nd Conference on Coastal Engineering*. Shanghai, China: ASCE. http://dx.doi.org/10.9753/icce.v32.management.9, https://journals.tdl.org/icce/index.php/icce/article/view/1435. Accessed September 2014.

Camenen, B. (2007). Simple and general formula for the settling velocity of particles. *Journal of Hydraulic Engineering, 133*(2), 229–233.

Carr, E. R., Wingard, P. M., Yorty, S. C., Thompson, M. C., Jensen, N. K., & Roberson, J. (2007). Applying DPSIR to sustainable development. *International Journal of Sustainable Development and World Ecology, 14*, 543–555.

Charles, A. (2012). People, oceans and scale: governance, livelihoods and climate change adaptation in marine social–ecological systems. *Current Opinion in Environmental Sustainability, 4*(3), 351–357. http://dx.doi.org/10.1016/j.cosust.2012.05.011.

Chen, J., Hill, A. A., & Urbano, L. D. (2009). A GIS-based model for urban flood inundation. *Journal of Hydrology, 373*(1–2), 19–184.

Christensen, J. H., Hewitson, B., Busuioc, A., Chen, A., Gao, X., Held, I., et al. (2007). Regional climate projections. In S. Solomon, D. Qin, & M. Manning (Eds.), *Climate change 2007: The physical science basis. Contribution of working group I to the fourth assessment report of the intergovernmental panel on climate change*. Cambridge, UK: Cambridge University Press.

Church, J. A., Clark, P. U., Cazenave, A., Gregory, J. M., Jevrejeva (UK), S., Levermann, A., et al. (2013). *Sea level change. Climate Change 2013: The Physical Science Basis*. Contribution of Working Group I to the Fifth Assessment Report of the Intergovernmental Panel on Climate Change. Intergovernmental Panel on Climate Change, Geneva, Switzerland.

Church, J. A., White, N. J., Aarup, T., Wilson, W. S., Woodworth, P. L., Domingues, C. M., et al. (2008). Understanding global sea levels: past, present and future. *Sustainability Science, 3*(1), 9–22. http://dx.doi.org/10.1007/s11625-008-0042-4.

Church, J., Woodworth, P. L., Aarup, T., & Wilson, S. (Eds.). (2010). *Understanding sea-level rise and variability.* Chichester, UK: John Wiley & Sons.

Cloke, H., & Pappenberger, F. (2009). Ensemble flood forecasting: a review. *Journal of Hydrology, 375*(3–4), 613–626.

Coninx, I., & Bachus, K. (2007). Integrating social vulnerability to floods in a climate change context. In *Proceedings of the international conference on adaptative and integrated water management, coping with complexity and uncertainty, Basel, Switzerland.*

Cooper, A. (2009). Decadal to centennial shoreline change illustrated by case studies from Ireland and South Africa. In K. Pye, & B. Ritchie (Eds.), *The measurement of coastal change* (pp. 129–152). University of Aberdeen.

Cowell, P. J., Stive, M. J. F., Niedoroda, A. W., De Vriend, H. J., Swift, D. J. P., Kaminsky, G. M., et al. (2003). The coastal tract. Part 1: a conceptual approach to aggregated modelling of low-order coastal change. *Journal of Coastal Research, 19*(4), 812–827.

Cutter, S. L., Boruff, B. J., & Shirley, W. L. (2003). Social vulnerability to environmental hazards. *Social Science Quarterly, 84*(2), 242–261.

Dawson, R. J., Dickson, M., Nicholls, R. J., Hall, J., Walkden, M. J. A., Stansby, P. K., et al. (2009). Integrated analysis of risks of coastal flooding and cliff erosion under scenarios of long term change. *Climatic Change, 95*, 249–288. http://dx.doi.org/10.1007/s10584-008-9532-8.

Dottori, F., & Todini, E. (2011). Developments of a flood inundation model based on the cellular automata approach: Testing different methods to improve model performance. *Physics and Chemistry of the Earth, 36*(7–8), 266–280. http://dx.doi.org/10.1016/j.pce.2011.02.004.

EA. (2011). *Flood and coastal risk management – An introduction to delivering the water framework directive. Operational Instruction 871_11.* Bristol, UK: Environment Agency.

European Commission. (1992). *Habitats directive council Directive 92/43/EEC on the conservation of natural habitats and of wild fauna and flora.* Brussels, Belgium: European Commission. http://ec.europa.eu/environment/nature/legislation/habitatsdirective/index_en.htm Accessed December 2012.

European Parliament and Council of the European Commission. (2007). *Directive 2007/60/EC on the assessment and management of flood risks (October 23, 2007).* Brussels, Belgium: European Union.

Eurosion. (2004). *Shoreline management guide. Living with coastal erosion in Europe: Sediment and space for sustainability. A guide to coastal erosion management practices in Europe.* Final version – 30 June 2004. Directorate General Environment European Commission Accessed January 2014 http://www.eurosion.org/reports-online/reports.html.

Evans, E., Ashley, R., Hall, J., Penning-Rowsell, E. C., Saul, A., Sayers, P., et al. (2004). *Foresight. Future flooding. Scientific summary.* London, UK: Office of Science and Technology (OST).

FLOODsite Consortium. (2009). *FLOODsite; Integrated flood risk analysis and management methodologies.* FLOODsite Consortium. http://www.floodsite.net/default.htm. Accessed 15.11.13.

Gilbert, S., & Horner, R. (1984). *The Thames barrier.* London: Thomas Telford Limited.

Gouldby, B., & Samuels, P. (2005). *Language of risk—project definitions. Integrating flood risk analysis and management methodologies.* Floodsite Project Report T32-04-0. FLOODsite www.floodsite.net/html/partner_area/project_docs/FLOODsite_Language_of_Risk_v4_0_P1(1).pdf Accessed September 2014.

Gouldby, B., Sayers, P., Mulet-Marti, J., Hassan, M., & Benwell, D. (2008). A methodology for regional-scale flood risk assessment. *Proceedings of the Institution of Civil Engineers – Water Management, 161*(3), 169–182.

Gregory, A. J., Atkins, J. P., Burdon, D., & Elliott, M. (2013). A problem structuring method for ecosystem-based management: the DPSIR modelling process. *European Journal of Operational Research, 227*(3), 558–569. http://dx.doi.org/10.1016/j.ejor.2012.11.020.

Guthrie, R. H., Mitchell, S. J., Lanquaye-Opoku, N., & Evans, S. G. (2010). Extreme weather and landslide initiation in coastal British Columbia. *Quarterly Journal of Engineering Geology and Hydrogeology, 43*(4), 417–428.

Haigh, I., Nicholls, R. J., & Wells, N. (2009). Twentieth-century changes in extreme still sea levels in the English Channel. *Continental Shelf Research, 29*, 2083–2098.

Haigh, I., Nicholls, R. J., & Wells, N. (2010). A comparison of the main methods for estimating probabilities of extreme still water levels. *Coastal Engineering, 57*(9), 838–849.

Hallegatte, S. (2009). Strategies to adapt to an uncertain climate change. *Global Environmental Change – Human and Policy Dimensions, 19*(2), 240–247. http://dx.doi.org/10.1016/j.gloenvcha.2008.12.003.

Hallegatte, S., Green, C., Nicholls, R. J., & Corfee-Morlot, J. (2013). Future flood losses in major coastal cities. *Nature Climate Change, 3*(9), 802–806. http://dx.doi.org/10.1038/nclimate1979, http://www.nature.com/nclimate/journal/v3/n9/abs/nclimate1979.html#supplementary-information.

Hanson, S., Nicholls, R. J., Patmore, N., Hallegatte, S., Corfee-Morlot, J., Herweijer, C., et al. (2011). A global ranking of port cities with high exposure to climate extremes. *Climatic Change, 140*(1), 89–111.

Holdgate, M. W. (1979). *A perspective of environmental pollution.* UK: Cambridge University Press Cambridge.

Jevrejeva, S., Moore, J. C., Grinsted, A., & Woodworth, P. L. (2008). Recent global sea level acceleration started over 200 years ago? *Geophysical Research Letters, 35*(8). doi L0871510.1029/2008gl033611.

Jones, L., Angus, S., Cooper, A., Doody, P., Everard, M., Garbutt, A., et al. (2011). Coastal margins. In *UK national ecosystem assessment: Understanding nature's value to society.* Cambridge, UK: United Nations Environment Programme and World Conservation Monitoring Centre (UNEP-WCMC). http://uknea.unep-wcmc.org/Resources/tabid/82/Default.aspx.

Kalamees, R., & Zobel, M. (2002). The role of the seed bank in gap regeneration in a calcareous grassland community. *Ecology, 83*(4), 1017–1025. http://dx.doi.org/10.2307/3071910.

Kates, R. W., Colten, C. E., Laska, S., & Leatherman, S. P. (2006). Reconstruction of New Orleans after Hurricane Katrina: a research perspective. *Proceedings of the National Academy of Sciences of the United States of America, 103*, 14653–14660. http://dx.doi.org/10.1073/pnas.0605726103.

Katsman, C. A., Hazeleger, W., Drijfhout, S. S., van Oldenborgh, G. J., & Burgers, G. J. H. (2008). Climate scenarios of sea level rise for the northeast Atlantic Ocean: a study including the effects of ocean dynamics and gravity changes induced by ice melt. *Climatic Change, 91*(3–4), 351–374.

Katsman, C. A., Sterl, A., Beersma, J. J., van den Brink, H. W., Church, J. A., Hazeleger, W., et al. (2011). Exploring high-end scenarios for local sea level rise to develop flood protection strategies for a low-lying delta – The Netherlands as an example. *Climatic Change.* http://dx.doi.org/10.1007/s10584-011-0037-5.

Kloprogge, P., van der Sluijs, J., & Wardekker, A. (2007). *Uncertainty communication: Issues and good practice.* Utrecht, The Netherlands: Copernicus Institute for Sustainable Development and Innovation, Utrecht University. http://www.nusap.net/downloads/reports/uncertainty_communication.pdf Accessed November 2012.

Kolen, B., Slomp, R., van Balen, W., Terpstra, T., Bottema, M., & Nieuwenhuis, S. (2010). *Learning from French experiences with storm Xynthia: Damages after a flood.* Lelystad, the Netherlands: HKV 'Lijn in Water' and Rijkswaterstaat, Waterdienst.

Kortenhaus, A. (2012). Performance of coastal flood defences. In P. Sayers (Ed.), *Flood risk: Planning, design and management of flood defence infrastructure* (pp. 103–123). London: Ice Publishing.

Kristensen, P. (2004). *The DPSIR Framework, Workshop presentation on comprehensive assessment of the vulnerability of water resources to environmental change in Africa using River Basin Approach.* Nairobi, Kenya. http://greenresistance.files.wordpress.com/2008/10/dpsir-1.pdf Accessed September 2014

Lane, S. N., & Richards, K. S. (1998). High resolution, two-dimensional spatial modelling of flow processes in a multi-thread channel. *Hydrological Processes, 12*(8), 1279–1298. http://dx.doi.org/10.1002/(sici)1099-1085(19980630)12, 8<1279::aid-hyp615>3.3.co;2–5.

Lee, H. C. (2013). DPSIR and disaster risk analysis. *Disaster Advances, 6*(1), 48–53.

Lowe, J. A., Howard, T. P., Pardaens, A., Tinker, J., Holt, J., Wakelin, S., et al. (2009). *UK climate projections science report: Marine and coastal projections.* Exeter, UK: Met Office Hadley Centre.

Lumbroso, D. M., & Vinet, F. (2011). A comparison of the causes, effects and aftermaths of the coastal flooding of England in 1953 and France in 2010. *Natural Hazards and Earth System Sciences, 11*, 2321–2333.

McElwee, P. (2010). *The social dimension of adaptation to climate change in Vietnam.* World Bank discussion paper. Washington DC: World Bank.

McGranahan, G., Balk, D., & Anderson, B. (2007). The rising tide: assessing the risks of climate change and human settlements in low elevation coastal zones. *Environment & Urbanization, 19*(1), 17–37.

Méndez, F., Menéndez, M. A. L., & Losada, I. (2007). Analyzing monthly extreme sea levels with a time-dependent GEV model. *Jounal of Atmospheric and Oceanic Technology, 24*, 894–911.

Mendoza, E., Silva, R., Zanuttigh, B., Angelelli, E., Andersen, T. L., Martinelli, L., et al. (2014). Beach response to wave energy converter farms acting as coastal defence. *Coastal Engineering, 87*, 97–111. http://dx.doi.org/10.1016/j.coastaleng.2013.10.018.

Menéndez, M., & Woodworth, P. L. (2010). Changes in extreme high water levels based on a quasi-global tide-gauge data set. *Journal of Geophysical Research – Oceans, 115*, 15. http://dx.doi.org/C10011 10.1029/2009jc005997.

Menéndez, M., Méndez, F. J., Izaguirre, C., Luceño, A., & Losada, I. J. (2009). The influence of seasonality on estimating return values of significant wave height. *Coastal Engineering, 2009.* http://dx.doi.org/10.1016/j.coastaleng.2008.07.004.

Merz, B., Kreibich, H., Schwarze, R., & Thieken, A. (2010). Assessment of economic flood damage. *Natural Hazards and Earth System Sciences, 10*, 1697–1724. www.nat-hazards-earth-syst-sci.net/10/1697/2010/.

Messner, F., & Meyer, V. (2006). Flood damage, vulnerability and risk perception - challenges for flood damage research. In Jochen Schanze, Evzen Zeman, & Jiri Marsalek (Eds.), *Flood risk management - Hazards, vulnerability and mitigation measures, Nato Science Series* (pp. 149–167). Springer Publisher.

Millenium Ecosystem Assessment. (2005). *Ecosystems and human well-being: A framework for assessment.* Washington DC, USA: Island Press. http://www.maweb.org/en/Index.aspx Accessed November 2012.

Millner, A. (2012). Climate prediction for adaptation: who needs what? *Climatic Change, 110*(1–2), 143–167. http://dx.doi.org/10.1007/s10584-011-0073-1.

Milly, P. C. D., Wetherald, R. T., Dunne, K. A., & Delworth, T. L. (2002). Increasing risk of great floods in a change climate. *Nature, 415*, 514–517.

Monbaliu, J., Chen, Z., Felts, D., Ge, J., Hissel, F., Kappenberg, J., et al. (2014). Risk assessment of estuaries under climate change: lessons from Western Europe. *Coastal Engineering, 87*, 32–49. http://dx.doi.org/10.1016/j.coastaleng.2014.01.001.

Moreno, I. M., Ávila, A., & Losada, M.Á (2010). Morphodynamics of intermittent coastal lagoons in Southern Spain: Zahara de los Atunes. *Geomorphology, 121*(3–4), 305–316. http://dx.doi.org/http://dx.doi.org/10.1016/j.geomorph.2010.04.028.

Morton, R. A., Gelfenbaum, G., Buckley, M. L., & Richmond, B. M. (2011). Geological effects and implications of the 2010 tsunami along the central coast of Chile. *Sedimentary Geology, 242*(1), 34–51.

Moss, R. H., Edmonds, J. A., Hibbard, K. A., Manning, M. R., Rose, S. K., van Vuuren, D. P., et al. (2010). The next generation of scenarios for climate change research and assessment. *Nature, 463*(7282), 747–756. http://dx.doi.org/10.1038/nature08823.

Nakićenović, N., Alcamo, J., Davis, G., Devries, B., Fenhann, J., Gaffin, S., et al. (2000). *Special report on emissions scenarios*. Cambridge, UK: Cambridge University Press.

Narayan, S., Hanson, S., Nicholls, R. J., Clarke, D., Willems, P., Ntegeka, V., et al. (2012). A holistic model for coastal flooding using system diagrams and the source-pathway-receptor (SPR) concept. *Natural Hazards and Earth System Sciences, 12*(5), 1431–1439. http://dx.doi.org/10.5194/nhess-12-1431-2012.

Narayan, S., Nicholls, R. J., Clarke, D., Hanson, S., Reeve, D., Horrillo-Caraballo, J. M., et al. (2014). The 2D Source – pathway – Receptor model: a participative approach for coastal flood risk assessments. *Coastal Engineering, 87*, 15–31. http://dx.doi.org/10.1016/j.coastaleng.2013.10.021.

NCDEM. (2009). *North Carolina sea level rise risk management study – potential impacts, risk assessments and management strategies*. Raleigh, NC, USA: North Carolina Division of Emergency Management (NCDEM), Office of Geospatial and Technology Management.

Neelz, S., & Pender, G. (2009). *Desktop review of 2D hydraulic modelling packages*. Science Report: SC080035, Bristol, England.

Nicholls, R. J., Townend, I. H., Bradbury, A. P., Ramsbottom, D., & Day, S. A. (2013). Planning for long-term coastal change: experiences from England and Wales. *Ocean Engineering, 71*, 3–16.

Nicholls, R. J., Bradbury, A., Burningham, H., Dix, J. K., Ellis, M., French, J., et al. (2012b). iCOASST – integrating coastal sediment systems. In *ICCE2012 international conference on coastal engineering, Santander, ES, July 01–06, 2012* (p. 15).

Nicholls, R. J., Hanson, S. E., Lowe, J. A., Warrick, R. A., Lu, X., & Long, A. J. (2014). Sea-level scenarios for evaluating coastal impacts. *Wiley Interdisciplinary Reviews: Climate Change, 5*(1), 129–150. http://dx.doi.org/10.1002/wcc.253.

Nicholls, R. J., Woodroffe, C. D., Burkett, V., Hay, J., Wong, P. P., & Nurse, L. (2012a). Scenarios for coastal vulnerability assessment. In E. Wolanski, & D. S. McLusky (Eds.), *Ecological economics of estuaries and coasts: Vol. 12. Treatise on Estuarine and coastal science* (pp. 289–303). Maryland Heights, US: Elsevier.

Oumeraci, H. (2004). *Sustainable coastal flood defences: scientific and modelling challenges towards an integrated risk-based design concept*. In *Proc. first IMA international conference on flood risk assessment, IMA-institute of mathematics and its applications, session* (Vol. 1); (pp. 9–24).

Pappenberger, F., Beven, K. J., Hunter, M. N., Bates, P. D., Gouweleeuw, B. T., Thielen, J., et al. (2005). Cascading model uncertainty from medium range weather forecasts (10 days) through a rainfall-runoff model to flood inundation predictions within the European flood forecasting system (EFFS). *Hydrology and Earth System Sciences, 9*(4), 381–393.

Penning-Rowsell, E., Parker, D. J., de Vries, W. S., Zanuttigh, B., Simmonds, D., Trifonova, E., et al. (2014). Innovation in coastal risk management: an exploratory analysis of risk governance issues at eight THESEUS study sites. *Coastal Engineering, 87*, 210–217. http://dx.doi.org/10.1016/j.coastaleng.2013.12.005.

Penning-Rowsell, E. C., Priest, S., Parker, D., Morris, J., Tunstall, S., Viavattene, C., et al. (2013). *Flood and coastal erosion risk management: A manual for economic appraisal*. London: Routledge.

Pitt, M. (2008). *Learning lessons from the 2007 floods. Pitt review*. London, UK: Cabinet Office. http://webarchive.nationalarchives.gov.uk/20080906001345/http://cabinetoffice.gov.uk/thepittreview/final_report.aspx Accessed 28.11.13.

Posada, G., Silva, R., & de Brye, S. (2008). Three dimensional hydrodynamic model with multiquadtree meshes. *American Journal of Environmental Sciences, 4*(3), 209–222. http://dx.doi.org/10.3844/ajessp.2008.245.258.

Posada, G., Silva, R., Simmonds, D., & Pedrozo, A. (2007). *Chapter VIII. A 2D hydrodynamic model with multi-quadtree mesh. "Ocean engineering research advances."* New York: Nova Science Publishers, Inc.

Poulter, B., & Halpin, P. N. (2008). Raster modelling of coastal flooding from sea-level rise. *International Journal of Geographical Information Science, 22*(2), 167–182.

Ramos, M. H., Mathevet, T., Thielenc, J., & Pappenberger, F. (2010). Communicating uncertainty in hydro-meteorological forecasts: mission impossible? *Meteorological Applications, 17*, 223–235. http://dx.doi.org/10.1002/met.202.

Randall, D., & Ertel, C. (2005). *Moving beyond the future*. Special report: Managing risk. London, UK: Financial Times http://gbn.com/consulting/article_details.php?id=22. Accessed November 2012.

Ranger, N., Reeder, T., & Lowe, J. (2013). Addressing 'deep' uncertainty over long-term climate in major infrastructure projects: four innovations of the Thames Estuary 2100 Project. *EURO Journal on Decision Processes, 1*(3–4), 233–262. http://dx.doi.org/10.1007/s40070-013-0014-5.

Renn, O. (2008). *Risk governance: Coping with uncertainty in a complex world*. London, UK: Earthscan.

Robinson, S. (2008). Conceptual modelling for simulation part II: a framework for conceptual modelling. *Journal of the Operational Research Society, 59*(3), 291–304. http://dx.doi.org/10.1057/palgrave.jors.2602369.

Rodriguez, I., Montoya, I., Sanchez, M. J., & Carreno, F. (2009). Geographic information systems applied to integrated coastal zone management. *Geomorphology, 107*(1–2), 100–105. http://dx.doi.org/10.1016/j.geomorph.2007.05.023.

Seed, R. B., Bea, R. G., Abdelmalak, R. I., Athanasopoulos-Zekkos, A., Boutwell, G. P., Briaud, J. L., et al. (2008). New Orleans and Hurricane Katrina. I: Introduction, overview, and the east flank. *Journal of Geotechnical and Geoenvironmental Engineering, 134*(5), 701–717. http://dx.doi.org/10.1061/(asce)1090-0241(2008)134:5(701).

Setiadi, N. (2011). Daily mobility – excursus Padang, Indonesia. In D. Chang Seng, & J. Birkmann (Eds.), *Migration and global environmental change, foresight SR4b, vulnerability assessment, risk management and adaptive planning sectionEarly warning in the context of environmental shocks: Demographic change, dynamic exposure to hazards, and the role of EWS in migration flows and human displacement* (pp. 41–45). United Nations University for Environment and Human Security, Government Office for Science.

Silva, R., Ruiz, G., Mariño-Tapia, I., Posada, G., Mendoza, E., & Escalante, E. (2012). Manmade vulnerability of the Cancun beach system: the case of hurricane Wilma. *CLEAN–Soil, Air, Water, 40*(9), 911–919.

Slangen, A. B. A., Katsman, C. A., van de Wal, R. S. W., Vermeersen, L. L. A., & Riva, R. E. M. (2012). Towards regional projections of twenty-first century sea-level change based on IPCC SRES scenarios. *Climate Dynamics*. http://dx.doi.org/10.1007/s00382-011-1057-6.

Smith, K., & Ward, R. (1998). *Floods: Physical process and human impacts*. Chichester: John Wiley & Sons Ltd, 376 p.

Stive, M. J. F., Aarninkhof, S. G. J., Hamm, L., Hanson, H., Larson, M., Wijnberg, K. M., et al. (2002). Variability of shore and shoreline evolution. *Coastal Engineering, 47*(2), 211–235. http://dx.doi.org/10.1016/S0378-3839(02)00126-6.

Stive, M. J. F., Cowell, P., & Nicholls, R. J. (2009). Beaches, cliffs and deltas. In *Geomorphology and global environmental change*. Cambridge, UK: Cambridge University Press.

Sutherland, J. (2007). *Model types. From coastal Wiki*. http://www.coastalwiki.org/wiki/Model_types. Accessed January 2014.

Tanaka, H., Tinh, N. X., Umeda, M., Hirao, R., Pradjoko, E., Mano, A., et al. (2012). Coastal and estuarine morphology changes induced by the 2011 Great East Japan Earthquake Tsunami. *Coastal Engineering Journal, 54*(01).

Tapsell, S. M., Penning-Rowsell, E. C., Tunstall, S. M., & Wilson, T. L. (2002). Vulnerability to flooding: health and social dimensions. *Philosophical Transactions of the Royal Society London A, 360*, 1511–1525.

Tarrant, O., & Sayers, P. (2012). Managing flood risk in the Thames Estuary - The development of a long term robust and flexible strategy. In P. Sayers (Ed.), *Flood risk management - The design and management of supporting infrastructure*. London, UK: Thomas Telford.

THESEUS OD1.10. (2012). *Report on quantified scenarios analysis of drivers and impacts of changing flood risk.* http://www.theseusproject.eu/resources/documents/Deliverables/Official-Deliverables-(OD)/OD1.10-(M18)-Report-on-quantified-scenarios-analysis-of-drivers-and-impacts-of-changing-flood-risk/.

THESEUS OD1.15. (2012). *Integrated report on risk assessment in study sites.* http://www.theseusproject.eu/resources/documents/Deliverables/Official-Deliverables-(OD)/OD1.15-(M24)-Integrated-report-on-risk-assessment-in-study-sites.

THESEUS OD4.8. (2013). *Structured portfolio of tested operational innovative tools and rotocols for policy and management purposes of coastal flooding risks.* http://www.theseusproject.eu/resources/documents/Deliverables/Official-Deliverables-(OD)/OD4.8-(M36)-Structured-portfolio-of-tested-operationalinnovative-tools-and-protocols-for-policy-and-management-purposes-of-coastal-flooding-risks./.

Thirel, G., Rousset-Regimbeau, F., Martin, E., & Habets, F. (2008). On the impacts of short-range meteorological forecasts for ensemble stream flow predictions. *Journal of Hydrolometerology, 9*(6), 1301–1317.

Thomas, T., Phillips, M. R., & Williams, A. T. (2010). Mesoscale evolution of a headland bay: Beach rotation processes. *Geomorphology, 123*(1–2), 129–141.

Thorne, C., Evans, E., & Penning-Rowsell, E. C. (Eds.). (2007). *Future flooding and coastal erosion risks*. London, UK: Thomas Telford.

Trembanis, A., DuVal, C., Beaudoin, J., Schmidt, V., Miller, D., & Mayer, L. (2013). A detailed seabed signature from Hurricane Sandy revealed in bedforms and scour. *Geochemistry, Geophysics, Geosystems, 14*(10), 4334–4340.

Tuan, T. Q., & Oumeraci, H. (2010). A numerical model of wave overtopping on seadikes. *Coastal Engineering, 57*, 757–772.

Turner, B. L., II, Matson, P. A., McCarthy, J. J., Corell, R. W., Christensen, L., Eckley, N., Hovelsrud-Broda, G. K., Kasperson, J. X., Kasperson, R. E., Luers, A., Martello, M. L., Mathiesen, S., Naylor, R., Polsky, C., Pulsipher, A., Schiller, A., Selin, H., & Tyler, N. (2003b). Illustrating the coupled human-environment system for vulnerability analysis: Three case studies. *Proceedings of the National Academy of Sciences, 100*(14), 8080–8085.

Turner, B. L., Kasperson, R. E., Matson, P. A., McCarthy, J. J., Corell, R. W., Christensen, L., Eckley, N., Kasperson, J. X., Luers, A., Martello, M. L., Polsky, C., Pulsipher, A., & Schiller, A. (2003a). A framework for vulnerability analysis in sustainability science. *Proceedings of the National Academy of Sciences, 100*(14), 8074–8079.

Verwaest, T., Van der Biest, K., Vanpoucke, P., Reyns, J., Vanderkimpen, P., De Vos, et al.. (2008). Coastal flooding risk calculations for the Belgian coast. In J. McKee Smith, & P. Lynett (Eds.), *Proceedings of the 31st International Conference on Coastal Engineering* (pp. 4193–4203). Hamburg, Germany: ASCE.

Villatoro, M., Silva, R., Méndez, F. J., Zanuttigh, B., Pan, S., Trifonova, E., et al. (2014). An approach to assess flooding and erosion risk for open beaches in a changing climate. *Coastal Engineering, 87*, 50–76. http://dx.doi.org/10.1016/j.coastaleng.2013.11.009.

Von Storch, H., & Zwiers, F. (2013). Testing ensembles of climate change scenarios for "statistical significance". *Climatic Change, 117*(1–2), 1–9. http://dx.doi.org/10.1007/s10584-012-0551-0.

de Vries, W. S., Zanuttigh, B., Steendam, G. J., Kloosterboer, H., Van der Nat, A., & Graaff, H. (2011). Integrating science and policy for creating tools for safer European coasts in a changing climate. *Irrigation and Drainage, 60*, 77–83. http://dx.doi.org/10.1002/ird.670.

Wadey, M. P. (2013). *Understanding defence failures and coastal flood events: a case study approach* (Doctoral dissertation). University of Southampton.

Wang, C., Wan, T., & Palmer, I. (2010). Urban flood risk analysis for determining optimal flood protection levels based on digital terrain model and flood spreading model. *The Visual Computer*, 1–13.

Wardekker, J. A., de Jong, A., Knoop, J. M., & van der Sluijs, J. P. (2010). Operationalizing a resilience approach to adapting an urban delta to uncertain climate changes. *Technological Forecasting and Social Change*, 987–998.

WASA-Group. (1998). Changing waves and storms in the Northeast Atlantic? *Bulletin of the American Meteorological Society, 79*, 741–760.

Weisse, R., & Von Storch, H. (2009). *Marine climate and climate change: Storms, wind waves and storm surges*. Berlin, Germany: Springer Praxis.

Weisse, R., von Storch, H., Niemeyer, H. D., & Knaack, H. (2012). Changing North Sea storm surge climate: an increasing hazard? *Ocean & Coastal Management, 68*, 58–68. http://dx.doi.org/10.1016/j.ocecoaman.2011.09.005.

Weisse, R., Bellafiore, D., Menendez, M., Mendez, F., Nicholls, R., Umgiesser, G., et al. (2014). Changing extreme sea levels along European Coasts. *Coastal Engineering, 87*, 4–14. http://dx.doi.org/10.1016/j.coastaleng.2013.10.017.

Wisner, B., Blaikie, P., Cannon, T., & Davis, I. (2004). *At Risk: Natural Hazards, People's Vulnerability and Disasters* (2nd ed.). London, UK: Routledge.

Woodhead, S. (2007). *Evaluation of inundation models – limits and capabilities of models. FLOODsite, April 2007*. Report T08-07-01, 18–19.

Zanuttigh, B., Simcic, D., Bagli, S., Bozzeda, F., Pietrantoni, L., Zagonari, F., et al. (2014). THESEUS decision support system for coastal risk management. *Coastal Engineering*. http://dx.doi.org/10.1016/j.coastaleng.2013.11.013.

Zanuttigh, B. (2011). Coastal flood protection: what perspective in a changing climate? the THESEUS approach. *Environmental Science and Policy, 14*, 845–863. Elsevier.

Zerger, A., Smith, D. I., Hunter, G. J., & Jones, S. D. (2002). Riding the storm: a comparison of uncertainty modelling techniques for storm surge risk management. *Applied Geography, 22*(3), 307–330.

Zhang, X., & Xue, X. (2013). Analysis of marine environmental problems in a rapidly urbanising coastal area using the DPSIR framework: a case study in Xiamen, China. *Journal of Environmental Planning and Management, 56*(5), 720–742. http://dx.doi.org/10.1080/09640568.2012.698985.

Chapter 3

Innovative Engineering Solutions and Best Practices to Mitigate Coastal Risk

Hans F. Burcharth[1], Barbara Zanuttigh[18], Thomas Lykke Andersen[1], Javier L. Lara[2], Gosse Jan Steendam[23], Piero Ruol[3], Philippe Sergent[4], Rafał Ostrowski[5], Rodolfo Silva[6], Luca Martinelli[7], Jørgen Quvang Harck Nørgaard[1], Edgar Mendoza[6], David Simmonds[8], Nino Ohle[9], Jens Kappenberg[10], Shunqi Pan[11], Dan Kim Nguyen[12], Erik A. Toorman[13], Panayotis Prinos[14], Simon Hoggart[15], Zhongyuan Chen[16], Danuta Piotrowska[5], Zbigniew Pruszak[5], Jan Schönhofer[5], Marek Skaja[5], Piotr Szmytkiewicz[5], Marek Szmytkiewicz[5], Igor Leont'yev[17], Elisa Angelelli[18], Sara Mizar Formentin[18], Hassan Smaoui[12], Qilong Bi[19], Janina Sothmann[10], Dagmar Schuster[9], Maotian Li[20], Jianzhong Ge[20], Jacek Lendzion[21], Theoharris Koftis[14], Sergey Kuznetsov[22], Araceli Puente[2], Beatriz Echavarri[2], Raul Medina[2], Pedro Díaz-Simal[2], Iñigo Losada Rodriguez[2], Maria Maza[2], Pablo Higuera[2]

[1] *Department of Civil Engineering, Aalborg University (AAU), Aalborg SV, Denmark*

[2] *Environmental Hydraulics Institute "IH Cantabria", Universidad de Cantabria (UC), Santander, Spain*

[3] *Padova University, ICEA Department, as part of the Consortium for Coordination of Research Activities Concerning the Venice Lagoon System (CORILA), Padova, Italy*

[4] *Centre d'Ütudes Techniques Maritimes Et Fluviales (CETMEF), Margny-lès-Compiègne, France*

Coastal Risk Management in a Changing Climate. http://dx.doi.org/10.1016/B978-0-12-397310-8.00003-8

[5] *Institute of Hydro-Engineering of the Polish Academy of Sciences (IBW PAN), Gdańsk, Poland*
[6] *Instituto de Ingeniería, Universidad Nacional Autónoma de México (UNAM), Coyoacán, DF. México*
[7] *Padova University, ICEA Department Part of the Consortium for Coordination of Research Activities Concerning the Venice Lagoon System (CORILA), Padova, Italy*
[8] *School of Marine Science and Engineering (Faculty of Science & Environment), Plymouth University (UoP), Plymouth, UK*
[9] *Hamburg Port Authority (HPA), Hamburg, Germany*
[10] *Helmholtz-Zentrum Geesthacht Centre for Materials and Coastal Research (HZG), Germany, Geesthacht, Germany*
[11] *School of Marine Science and Engineering, Plymouth University (UoP) (currently: School of Engineering, Cardiff University), Plymouth, UK*
[12] *Institute of Marine and Inland Waterways (CETMEF), Compiègne, France*
[13] *Hydraulics Section, Civil Engineering Department, Katholic University of Leuven (KUL), Leuven*
[14] *Laboratory of Hydraulics and Hydraulic Works, Aristotle University of Thessaloniki (AUTh), Thessaloniki, Greece*
[15] *Marine Biology and Ecology Research Centre, School of Marine Science and Engineering, University of Plymouth (UoP), UK*
[16] *State Key Laboratory of Estuarine and Coastal Research, East China Normal University (SKLEC), Shanghai, China*
[17] *P. P. Shirshov Institute of Oceanology of the Russian Academy of Sciences (IO RAS), Moscow, Russia*
[18] *Department of Civil, Chemical, Environmental and Materials Engineering, University of Bologna (UniBo), Bologna, Italy*
[19] *Hydraulics Section, Civil Engineering Department, KU Leuven (KUL), Leuven*
[20] *State Key Laboratory for Estuarine and Coastal Research (SKLEC), East China Normal University, Shanghai, China*
[21] *Institute of Meteorology and Water Management O/M Gdynia – National Research Institute (IMGW), Gdynia, Poland*
[22] *Coast & Shelf Laboratory, P. P. Shirshov Institute of Oceanology of Russian Academy of Sciences (IORAS), Moscow, Russia*
[23] *INFRAM International BV (INFRAM), Marknesse, The Netherlands*

3.1 Introduction

Engineering solutions are widely used for the mitigation of flood and erosion risks and have new challenges because of the expected effects induced by climate change; in particular, sea-level rise and increased storminess.

This chapter describes both active methods of mitigation based on the reduction of the incident wave energy, such as the use of wave energy converters, floating breakwaters and artificial reefs, and passive methods, consisting of increase in overtopping resistance of dikes, improvement of resilience of breakwaters against failures, and the use of beach nourishment as well as tailored dredging operations.

Section 3.2 investigates the wave transmission offered by floating breakwaters (FBs) or arrays of floating wave energy converters (F-WECs) located in the near-shore area. In particular, three kinds of FBs and four different promising F-WEC concepts are described. The analyzed structures are characterized by a modest impact on the shoreline (in terms of wave energy attenuation), independent from tide or sea level rise. However, high costs and insufficient reliability in case of extreme waves have been a hindrance so far to their systematic exploitation.

Section 3.3 discusses the use of various forms of submerged barriers aimed at affording a degree of protection to an exposed coastline by reducing the incident wave energy and at the same time assuring a lower environmental impact than traditional breakwaters. This is achieved through a manipulation of the key hydro-dynamic processes that creates a shielding of the coastline and often entails a corresponding modification of the sediment dynamics around these structures and also on the adjacent coastal stretches.

Sections 3.4 and 3.5 examine the threats for coastal dikes from sea-level rise and, in particular when built in shallow waters where the maximum wave height is depth limited, also because of an increase in wave heights. This causes both larger wave impacts on the structures and more wave overtopping.

Traditionally, the dikes would be heightened but other measures may be less costly and also less space consuming.

Section 3.4 examines the solution to allow substantial more water to overtop grass covered dikes, a solution that might be feasible as long as it does not cause the dike to breach and the amount of overtopping water does not cause too much inconvenience or damage to economic or ecological values behind the dike. Some guidelines to assess and increase the overtopping resistance of dikes to cope with the mentioned effects of climate change are provided.

Section 3.5 is devoted to reinforcement of rubble mound coastal protection structures that can compensate the effects of sea-level rise in terms of wave over-topping, armor stability, and crown wall stability. The options consist, for instance, in limiting overtopping by modifying the crown wall, in improving armor stability by adding an armor layer or by using milder armor slope. An option is also to reduce the incident wave energy by building a detached low-crested breakwater or by sand nourishment. In this section the design steps of structure upgrading to resist the effect of climate change are presented.

Section 3.6 focuses on the nourishment strategy and on the management of sediment stocks by providing methodologies and practical tools for analyzing nourishment efficiency (in terms of wave energy reduction and of nourishment stability) when the nourishment is performed on the emerged and submerged part of the beach. Specifically, simplified methods for estimating the resilience of coastal profiles characterized by nourishments and/or dunes for different combinations of nourishment position and extension, and dune widths and heights are given. The process of silting up of underwater excavations and the sediment distribution during dredging operations are also treated. Finally, the efficiency of artificial sandbanks to mitigate the effects of storm surges in estuaries is discussed.

Overall remarks and guidance about engineering techniques, their environmental impact and their sensitivity to climate change are finally drawn in Section 3.7.

3.2 Floating Breakwaters and Wave Energy Converters

3.2.1 Introduction

The aim of this section is to examine the use of FBs and F-WECs to reduce the inshore wave height for coastal protection purposes.

The search for innovative approaches aiming at flood risk mitigation in coastal areas, and particularly deltas and estuaries of large rivers, has drawn attention to the potentialities of floating structures. Floating structure performance is almost independent of relatively small variation of the water level such as those induced by tide, surge, or sea-level rise, and therefore these structures may be particularly well suited in view of the expected climate change.

FBs and F-WECs are analyzed together because these structures approximately have the same effect on wave transmission. FBs are only used to decrease the wave energy content in a specific area (e.g., to protect a small craft harbor or a sketch of eroding coast), whereas F-WECs also aim at converting (i.e., dissipating) some of the wave energy into electricity.

The wave transmitted behind a floating device is reduced because part of the incident energy is reflected and part is dissipated (or harvested, in the case of WECs). The wave is transmitted either directly, below or above the structure, or indirectly, as a consequence of the device movements. Direct energy transmission is affected by the intrinsic nature of the waves (irregular, oblique, and short-crested) and by the overtopping process. Energy transmission related to the device movements is the result of the interaction of the incident wave and the structure dynamics. A dominant role is played by the mooring forces and the reactions because of connections between device modules.

It should be stressed that the application of FBs here proposed is inherently different from the traditional one, which aims at sheltering marinas. For example, in coastal areas, there is a considerable benefit if a 2 m high incident wave is reduced by

20%, because this may prevent flooding to occur (or may drive a significantly lower sediment transport), whereas such high waves cannot be tolerated in marinas. A general discussion on the traditional FB function may be found for instance in Tsinker (1994) or Headland (1995) .

Similarly, the primary goal of WECs is to harvest energy, so that the layout design (discussed for instance by Babarit (2013)) is not optimized for coastal defense. Nevertheless, in some cases, it may be worthy to modify the design and narrow the gap between devices to lower the wave transmission.

The first studies on the effects of floating bodies, and WECs in particular, on wave propagation are relatively recent. Budal (1977), Evans (1979), and Falnes (1980) first investigated the diffraction and radiation from the heaving oscillation under regular and unidirectional waves for regular bathymetries.

Little attention has been paid to the response of the coastline in the presence of parks of WECs. To the authors' knowledge, Millar, Smith, and Reeve (2007) first studied the change of wave height at the shoreline induced by a generic wave farm. They showed that when the WEC park, actually only represented as a generic transmitting obstacle (4 km wide), is far from the coast (20 km), the wave disturbance is negligible. Several studies followed, focusing on ways to describe more realistically the transmission characteristics of WECs (e.g., Angelelli & Zanuttigh, 2012; Carballo & Iglesias, 2013; Palha, Mendes, Conceiç, Brito-Melo, & Sarmento, 2010;. The first papers that focused on the morphologic response are apparently those by Ruol, Zanuttigh, Kofoed, Martinelli, and Frigaard (2010) and Zanuttigh et al. (2010), which briefly analyzed the effects on the long-shore sediment transport in a specific installation located in the near-shore area. Up to date, the only extensive work focusing the changes of the shoreline induces by wave farms is apparently the one by Mendoza et al. (2014).

This section will first present the different types of FBs, with special reference to the rectangular, π-shaped, and permeable ones, for which wave transmission characteristics as well as construction, installation, and cost issues are given. Then, after a brief description of the WECs principle of operation, four different F-WECs suited to near-shore deployment are examined. These F-WECs include an overtopping device (Wave Dragon, www.wavedragon.net), a multichamber oscillating water column (OWC) device (Sea Breath, www.seabreath.it), a wave activated body (DEXA, www.dexawave.com), and a new concept device (Blow-Jet). For these devices, the ability to reduce the incident wave height is described on the basis of experimental investigations (Martinelli, Pezzutto, & Ruol, 2013; Nørgaard & Lykke Andersen, 2011; Zanuttigh, Angelelli, & Kofoed, 2013). Special attention is then dedicated to the layout of the array, based on a set of numerical tools (Angelelli & Zanuttigh, 2012; Nørgaard & Lykke Andersen, 2012). The importance of the mooring system design is also discussed, introducing different design alternatives and critical issues. Finally, some considerations about costs will be tentatively given also for these new types of structures.

3.2.2 Floating Breakwaters

Apart from very few cases (namely very large and costly structures, such as the breakwater in the Principality of Monaco; Ortega & Floriano, 2009), FBs are used today to protect small harbors or marinas where the wave lengths and heights are not too large. In case of long waves, the incident wave is fully transmitted, because of the FB "rides" the waves. Furthermore, high waves can induce large forces on the structure and on the mooring system, so that it is generally not suited to high energetic environments as open seas.

3.2.2.1 Classification and Description of Selected Devices

FBs can be effective in coastal areas with mild wave environment conditions. Among the conditions that favor FB installation there are:

- poor foundation: floating breakwaters might be a proper solution where poor foundations possibilities prohibit the application of traditional breakwaters;
- deep water: in water depths greater than 6 m, traditional breakwaters are often more expensive than floating ones.

Other advantages are related to:

- water quality: FBs do not substantially interact with water circulation and fish migration;
- visual impact: FBs have a minimum intrusion on the horizon, independently from the tidal range; and
- layout relocation: FBs can usually be rearranged into a new layout with minimum effort.

FBs are commonly divided into four general categories: box, pontoon, mat, and tethered float. Some floating breakwaters in each category are shown in Figure 3.1. The "box" and "pontoon" categories have much larger model and prototype experience.

Box type breakwaters are composed of reinforced concrete modules with a rectangular cross-section. A typical module is 15–20 m long, 2.5–5 m wide, and 1–2 m high. They are either empty inside or, more often, filled with a polystyrene or similar materials. In the former case, the risk of structure sinking is not negligible. Connections are either flexible, allowing preferably only the roll along the FB axis, or pre- or posttensioned, to make them act as a single unit. In the latter case, the efficiency is higher, but the risk of damage is also larger. Interconnection between adjacent modules and mooring system are primary points of concern for this kind of structures.

Barge-type FBs are sometimes built with used barges and ballasted to the desired draft with sand or rock.

Pontoon-type FBs include several models such as the catamaran type, the Alaska type, and the A-frame type. These shapes aim at increasing the ratio between width and the incident wavelength, and therefore the ability to attenuate the waves, in an economic way.

The mat and tethered float types are wide, on the order of 10–20 m, but draft is quite small (less than 1.0 m). Within the mat category, most are made with tires. They are subjected to lower anchor loads, reflect less, and dissipate relatively more than previous FB types. Although the effectiveness is low, they also have a low cost and they can be easily removed and constructed with unskilled labor and minimal equipment. Clearly, the large environmental impact is a strong limit to its applicability.

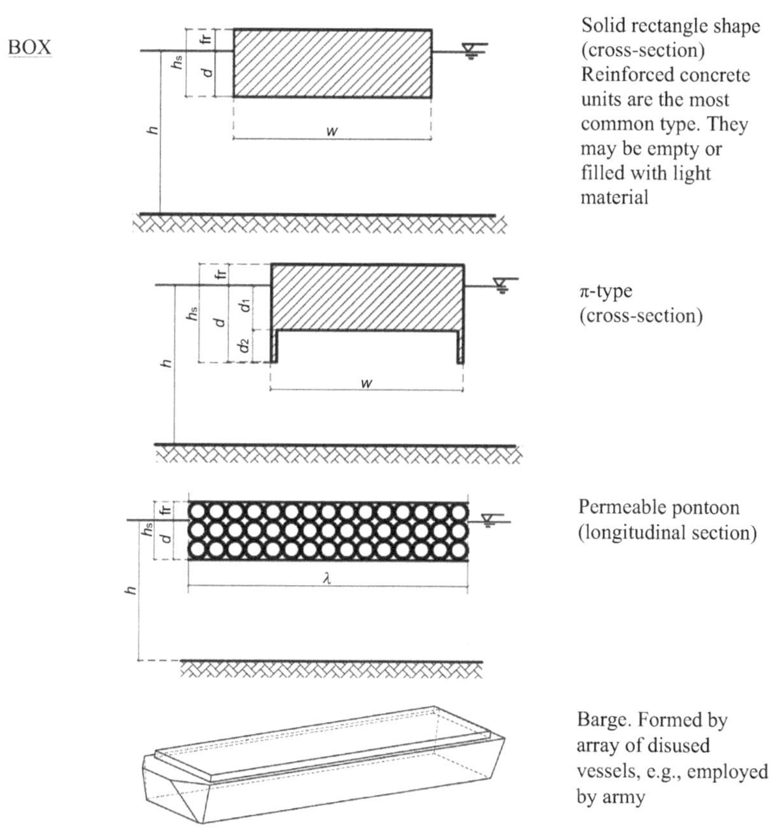

Figure 3.1 Various Types of Breakwaters. In the schemes: h = water depth, W = width of the pontoon, fr = freeboard of the FB, h_s = height of the pontoon, d = draught of the pontoon, and λ = length of the single module forming the FB. *From McCartney, J. Waterway Port Coastal and Ocean Eng., 1985, 111, 2, 304–318, modified.*

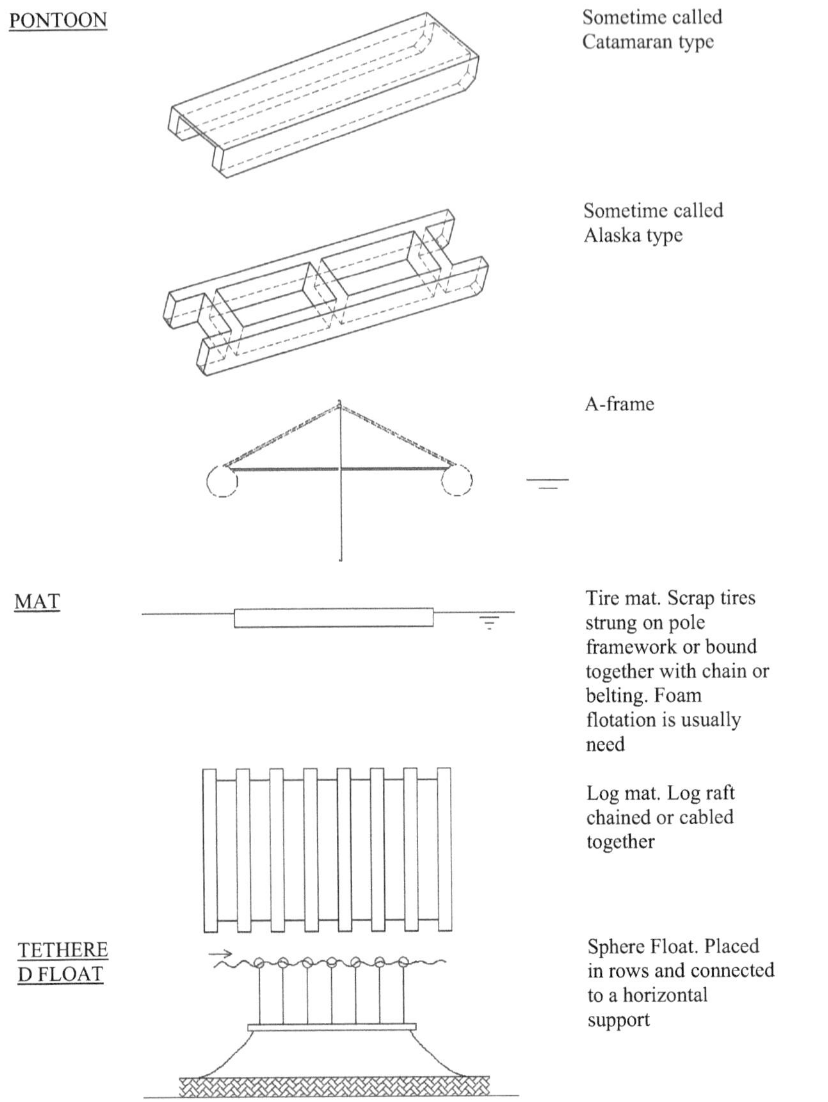

Figure 3.1 Continued

A growing number of companies provide prefabricated floating elongated modules, having two vertical plates protruding downwards from the sides. The FBs are formed by arrays of individually chain-moored modules, connected more or less rigidly (Figure 3.2). Because their cross-section resembles a Greek π, they are referred to as π-type FBs Table 3.1. The description of these devices is given hereafter. This type is believed to be more economical compared with other with

Figure 3.2 Prototype of π-type floating breakwater.

TABLE 3.1 Range of Applicability of $K_{T,D}$ for Box-type FBs

Reference	Wave Conditions	Fixed Structure	Moving Structure	h/L	w/h	d/h
Williams (1988)	Regular	X	X	0.21–0.60	0.80	0.10–0.40
Tolba (1988)	Regular	X	X	0.09–0.52	0.50	0.20–0.25
Koutandos et al. (2005)	Reg/Irr	X	X	0.05–0.32	1.00	0.20–0.33
Rahman et al. (2006)	Regular		X	0.15–0.61	0.50	0.22
Koftis-Prinos (2006)	Regular	X		0.15	1.00	0.23–0.46
Cox et al. (2007)	Reg/Irr		X	0.15–0.67	0.57	0.40
Martinelli et al. (2008)	Irregular		X	0.40–0.98	0.40	0.07
Uzaki et al. (2010)	Regular		X	0.14–1.22	0.53	0.10

TABLE 3.2 Range of Applicability of $k_{T,D}$ for π-type FBs

Wave Conditions	H_s (cm)	T_p (s)	h/L	w/h	d/h	$w/(h-d)$
Irregular	1.5–8.5	0.45–1.12	0.27–1.58	0.50–1.00	0.20–0.48	0.62–1.91

Figure 3.3 Prototype of permeable floating breakwater pontoon.

different geometries, such as the simple rectangular shape, and the possible application as barriers for coastal protection is examined.

The permeable pontoon (Figure 3.3) refers to another device suitable to protect an area from short-period waves (e.g., boats generated ones) and consists of hollow pipes connected to each other in a few rows, with the tube axis perpendicular to protected area. The prototypes are investigated for different pipe diameter and for various draughts.

3.2.2.2 Wave Transmission

The wave transmitted inshore a generic floating body has usually a height directly proportional to the incident wave one. Wave transmission is therefore usually described by a coefficient, K_T, defined as the ratio between transmitted and incident significant wave heights ($H_{m0,t}/H_{m0,i}$).

The consolidated formula given by Macagno (1954) for the transmission coefficient $K_{T,M}$, is:

$$K_{tM} = \frac{1}{\sqrt{1 + \left(kw \frac{\sinh(kh)}{2\cosh(kh - kd)}\right)^2}} \tag{3.1}$$

where k is wave number, w the width, d the draft, and h is the water depth. The formula is suited to a box-type FB, subject to regular waves incident perpendicular to the FB, and assumes that the body does not move under the incident wave (i.e., it remains essentially fixed at the floating position under the wave action), that the body is neither too thin (say, $W/d > 1$) nor too close to the bed (say, $d/h < 0.8$), that it is infinitely long.

In case of real FBs in which the modules are separated by gaps, the wave transmission is partly due to the transmission of energy past the device ($K_{T,D}$) and partly due to the diffraction at the sides and through the gaps.

$K_{T,D}$ can more reliably be obtained from physical model tests and is used to define the wave transmission that would occur for a theoretical infinitely long device (no gap between them). This means that the energy flux dissipated, absorbed or reflected by the floating body ($P_{a,r}$) may be evaluated based on the transmitted one:

$$P_{a,r}Q\left(\left(1 - K_{T,D}\right)H_{m0,i}\right)^2 b_d \tag{3.2}$$

where b_d is the width of the device and $H_{m0,i}$ is the incident wave height.

Based on the expression for $K_{T,D}$, the total transmission coefficient inshore one row of devices ($K_{T,1-row}$) can be calculated based on b_d and the gap width (s), by assuming that through the gaps the energy flux is fully transmitted:

$$K_{T,1-row} = \sqrt{\frac{K_{T,D}^2\, b_d + s}{b_d + s}} \tag{3.3}$$

This approach was proposed by Nørgaard and Lykke Andersen (2012) with specific reference to WECs, but may be used for FBs as well.

Note that it is generally found that K_T decreases with increasing wave obliquity (Abul-Azm & Gesraha, 2000; Martinelli, Ruol, & Zanuttigh, 2008).

A more accurate evaluation of $K_{T,D}$ specific for three different kinds of FBs is given in the following section.

Transmission Coefficients for Box-Type FB

A new method for the prediction of the wave transmission for box-type FBs is proposed based on a parametric analysis of the results derived from various experimental studies: Williams (1988), Tolba (1988), Koutandos, Prinos, and Gironella (2005), Rahman, Mizutani, and Kawasaki (2006), Koftis and Prinos (2006), Cox, Coghlan, and Kerry (2007), Martinelli, Ruol, and Zanuttigh (2008), Uzaki, Ikehata, Matsunaga, and Finkl (2010). These studies addressed two main FB categories (i.e., the fixed structure and the FBs having various degrees of freedom such as free heave, free rolling structure, structure with mooring chains, or pile supported).

Two formulae were developed by introducing a new parameter, $(h/L)k_{T=0.5}$, which is defined as the h/L (L being the regular or mean wavelength) in correspondence of which the structure provides $K_{T,D} = 0.50$. This parameter $(h/L)k_{T=0.5}$ is calculated by using respectively Eqn (3.4) for the fixed structure and Eqn (3.5) for the moving structure. The experimental data from the previously mentioned studies are plotted against the normalized $(h/L)/(h/L)k_{T=0.5}$, in Figure 3.4(a) and B for the fixed

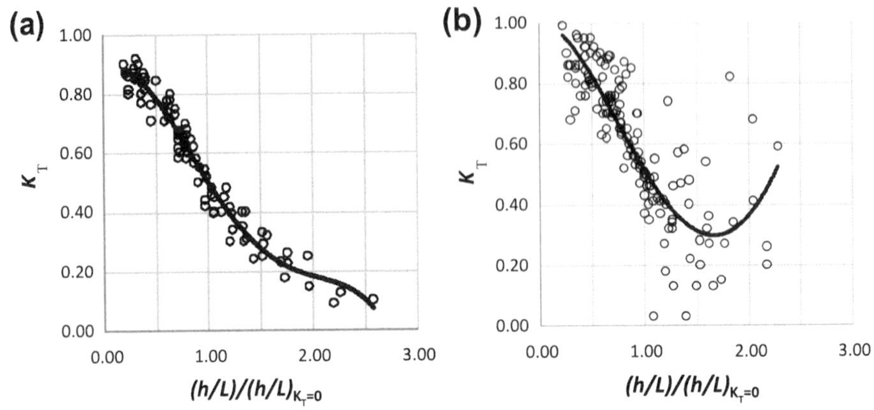

Figure 3.4 Variation of K_T for fixed (a) and moving FB (b).

and moving structure respectively, together with an eye-fitting curve. $K_{T,D}$ is obtained from the eye-fitting curve for the calculated $(h/L)/(h/L)k_{T=0.5}$. Larger dispersion in the results for the floating structure is shown when compared to the fixed structure because of the additional parameters (mooring lines, stiffness, etc.) that affect the FB performance. The proposed formulae (3.4) and (3.5) are valid for $0.25 < (h/L)/(h/L)k_{T=0.5} < 2.50$. More generally, Table 3.1 summarizes the studies considered in the fitting and the basic parameter range for each study, which are essentially the limits of application of the formulae.

For fixed structures

$$\left(\frac{h}{L}\right)_{K_T=0.5} = 0.42\left(\frac{w}{h}\right)^{-0.57} \exp\left\{-2.61\frac{d}{h}\right\} \tag{3.4}$$

For moving structures

$$\left(\frac{h}{L}\right)_{K_T=0.5} = 0.65\left(\frac{w}{h}\right)^{-0.51} \exp\left\{-2.60\frac{d}{h}\right\} \tag{3.5}$$

Transmission Coefficients for π-type FB

The formula for the transmission coefficient $K_{t,D}$ of a π-type FB is based on an experimental study performed by Ruol, Martinelli, and Pezzutto (2012, 2013). The efficiency of various π-type pontoons (different draught and width) for different mooring system (chains, piles, or tethered) has been evaluated for irregular wave conditions. Based on the results, $K_{T,D}$ can be evaluated as:

$$K_{T,D} = \beta(\chi)K_{t,M} \tag{3.6}$$

where $K_{T,M}$ is the Macagno's formula shown in Eqn (3.1) and $\beta(\chi)$ is a correction factor given in Eqn (3.7).

Macagno's formula was validated for regular waves. For irregular waves, the wavenumber k in Eqn (3.1) should be calculated with an average wave period, in general lower than the peak. In the absence of other information, it is proposed to use as equivalent regular period $T = T_p/1.1$.

The expression for $\beta(\chi)$ is

$$\beta = \frac{1}{1 + \left(\frac{\chi - \chi_o}{\sigma}\right) e^{-\left(\frac{\chi - \chi_o}{\sigma}\right)^2}} \tag{3.7}$$

where: $\chi_o = 0.7919$ (with 95% confidence interval 0.7801–0.8037), $\sigma = 0.1922$ (0.1741–0.2103), and χ = scaling parameter given by:

$$\chi = \frac{T_p}{2\pi} \sqrt{\frac{g}{d + 0.35w}} \tag{3.8}$$

where: T_p = equivalent peak wave period and g = acceleration of gravity. For perpendicular wave attacks, T_p coincides with the peak wave period. In the presence of oblique waves, T_p derives from an oblique wavelength $L_p' = L_p \cos(\theta)$ where θ is the wave obliquity and L_p is the peak wavelength. T_p in Eqn (8) is then easily obtained from the dispersion relationship, using local water depth h and apparent wavelength L_p.

Formula (3.6) is valid for $0.20 < d/h < 0.6$, $w/(h-d) < 2$ and for $0.50 < \chi < 1.50$. Table 3.3 summarizes the tested conditions.

Transmission Coefficients for Permeable FB

The formula of $K_{T,D}$ for a permeable FB is based on the results from flume experiments in scale 1:10. The permeable FB is placed parallel to wave fronts. The anchor system just fixes the average position of the FBs, but it does not significantly restrict its movements at the time scale of the wave period. Based on the results, $K_{T,D}$ can be evaluated by:

$$K_{T,D} = 0.92 - X_{bl} w/L \tag{3.9}$$

TABLE 3.3 Range of Applicability of $k_{T,D}$ for the Analyzed Porous FBs

Wave Conditions	H (cm)	T (s)	h/L	w/h	d/h	Porosity
Regular	5.89–17.04	0.72–1.65	0.17–0.74	0.40–0.67	0.12–0.39	0.75–-0.87

where X_{bl} is the part of wave energy flux blocked by FB, derived from linear wave theory as follows:

$$X_{bl} = \frac{\sinh(2kh) - \sinh(2k(h-d)) + 2kd}{\sinh(2kh) + 2kh} \quad (3.10)$$

Equation (3.9) is valid for $0.1 < X_{bl}\, w/L < 0.55$ and for a porosity of the breakwater (volume of holes divided by total) varying from 0.75 to 0.87.

Comparison of FBs Performance

Figure 3.5 examines, based on the proposed empirical formulas, the variation of the $K_{T,D}$ for different cross-section types (π-type, fixed, and moving box type and the permeable FB) all characterized by the same dimensions, i.e., $w = 4.0$ m, $d = 1.5$ m, and all installed in a water depth of $h = 5.0$ m, for varying periods of the regular and perpendicular incident wave T.

It is obvious that the increased stiffness of the fixed structure makes the FB more efficient in wave attenuation; actually, the fixed structure represents the limit state for the floating structure performance. The performance of the moving box-type is comparable to the π-type and has the same form for such specific configuration. As it was expected, the permeable FB is less efficient in terms of wave attenuation.

3.2.2.3 Construction and Installation

Box, π-type, and Pontoon units are usually cast in tailor designed formworks either in the factory or less frequently on site. Units are shipped to the installation site (by

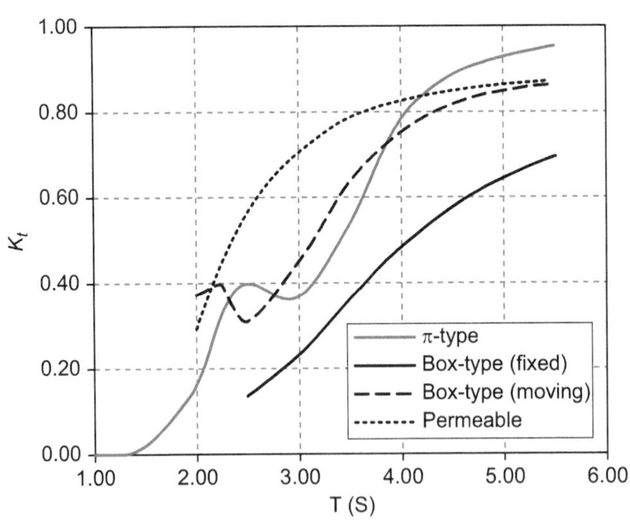

Figure 3.5 Variation of *KT* with T for various types of FB pontoon with $w = 4$ m, $d = 1.5$ m placed in water depth $h = 5$ m.

road or by sea) where they are launched usually with the help of mobile cranes. Once in the water, the units are joined in groups of two or three elements near the quay and then towed in position where each group is temporarily anchored to the anchor blocks by means of ropes. The groups are joined together in pairs and their transversal joints are tightened.

The permanent anchoring system is then installed. This is usually done by means of chains (less frequently by high-capacity ropes). The main technical challenges are to keep the offset to a minimum and possibly allow berthing during calm seas, and yet resist to extreme loads. The mooring systems are discussed more in detail in Section 3.1.3.4.

A final snag check is performed; and if needed, joints are retightened and anchoring lines retensioned. The trim of the system is eventually corrected by means of additional buoyancy blocks.

3.2.2.4 Costs

There are several variables that influence the costs of an FB, such as the cost of the structures itself (different companies in the market propose different prices, based on commercial strategies), the location, the type of installation, etc., so that a factor 2 uncertainty is unavoidable at this stage.

With reference to recent experience (2005–2010) specifically related to π-type FBs, the cost of the modules is generally in the range of 0.5–1.0 €/kg. For example, a 3-m-wide FB at the factory may cost around 2000 €/m, provided that the breakwater is sufficiently long, and cost of installation may be an additional 50% of the cost of the structure.

3.2.3 Wave Energy Converters

Several concepts have been investigated during the past three decades for harvesting energy. This has, over recent years, resulted in more than 1000 patents.

3.2.3.1 Classification and Description of Selected Devices

WECs can be classified in several ways: according to the device location, to its position in relation with the incoming wave direction, or to the conversion principle (Falcão & Rodrigues, 2002).

The "historical" classification is based on the device location:

1. Shoreline (first generation)
   For practical and economic reasons, WEC units were first directly installed at the shoreline, therefore it is possible to call them "first-generation" devices. In general, these devices cannot be classified as floating because they have their own

foundation on the coast or in breakwater structures for harbor protection. Main advantages are:

a. unnecessary design of mooring systems;
b. absence of long submerged electrical cables; and
c. easy installation and maintenance.

On the other hand, these devices can exploit a milder wave climate with respect to the offshore available wave power and therefore to be effective the installation site should be a hot spot. However, locations that would be suitable for these installations are often protected areas. Specific design issues include:

- sedimentation inside the chambers (in the case of OWC devices, see the last type of classification) or in the storage basins (in case of overtopped devices) in case of sandy coastlines, a problem that has still not been analyzed in depth; and
- high wave reflection and eventual scour at the foundation to be prevented by an appropriate protection layer.

2. Near-shore (second generation)
These devices are generally floating and are typically installed at water depths of 20–30 m and at a distance of the order of 500 m from the shore. In this area, the wave power is higher than at the shore. The near-shore devices require high costs of installation and maintenance with respect to the onshore devices, but at the same time their visual and environmental impact is generally low.

3. Offshore (third generation)
As a consequence of the higher available wave energy density and of the continuous technology development (taking advantage of the knowledge developed for offshore oil platforms), third-generation devices are installed at the open sea even if the installation costs rise. These devices are typically installed in deep water (i.e., water depth greater than 40 m) and consist of moored floating units, usually employing principles and ideas from the first- and second-generation devices.

A second classification is based on the WEC ability to intercept and attenuate waves.

1. Point absorber
A point absorber is a device with a small size compared with the wave length. It is able to capture energy, regardless the incoming wave direction, from a wave front potentially much larger than its physical dimension. These devices normally involve a floating structure composed of a first relatively immobile component and a second moving component, which follows the wave motion (e.g., a floating buoy which slides outside a fixed cylinder).

2. Attenuator
An attenuator has its principal axis parallel to the incoming wave direction, it is usually a floating device composed by several segments (for a total device length

typically greater than the wave length), and the mooring system should allow the unit to weathervane into the predominant wave direction.

3. Terminator
   A terminator has its principal axis parallel to the incoming wave front (i.e., orthogonal to the incoming wave direction) and acts as a significant obstacle to wave propagation. Typical installations were onshore or near-shore; however, nowadays there are also offshore floating offshore terminators such as the Wave Dragon, see Figure 3.6.

The last classification is based on the three main operating principles:

1. Oscillating water column (OWC)
   This OWC-WEC usually consists of a partially submerged structure with a chamber at the bottom (i.e., below the seawater line) from which waves enter the structure. Wave crests cause the water column to rise and compress the air in the chamber, whereas the wave troughs cause the water column to fall and the air pressure to drop. The power takeoff system is typically based on air turbines, which are activated by the flux induced by the alternate compression/decompression of the air in the chamber. To this purpose, the Wells self-rectifying air turbines (i.e., able to rotate regardless to the airflow direction) are commonly used.
2. Overtopping device (OTD)
   The power production concept is based on the conversion of the potential energy: waves run up along a ramp, overtop in a storage reservoir above sea level, and the power is then obtained by exploiting the difference of water level between the reservoir and the sea (usually by low head Kaplan turbines), similarly to a hydroelectric plant. This technology has been later adapted to offshore use by means of floating ramps.

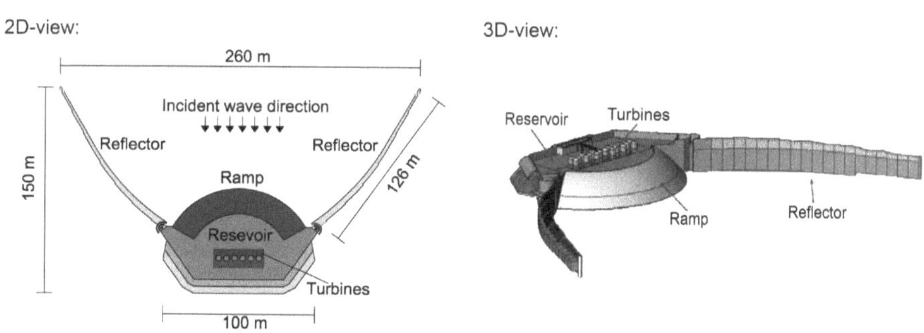

Figure 3.6 (left) Illustration and measures of considered prototype WD model (Nørgaard, Lykke Andersen, & Kofoed, 2011). (right) Three-dimensional illustration of the floating WD (Tedd, 2007).

3. Wave-activated bodies (WAB)
   These devices are usually composed of two or more parts that interact because of the progressive wave action along the device. In fact, waves activate the oscillatory relative motions of parts of the device or of one part with respect to a fixed reference (that may also be a gyro).

The categories described previously include most, but not all, of the technologies being developed today. For example, there are devices that exploit the wave energy and water particles movement at the surface (oscillating wave surge converters). A more detailed collection of the WEC concepts is summarized in Clément et al. (2002) as well as in Ingram, Smith, Bittencourt-Ferreira, and Smith (2011).

The F-WECs investigated in the following are: the Wave Dragon, the DEXA, the SeaBreath, and the Blow Jet.

The dimensions, such as length or depth of installation, need to be based on the actual location, and for the moment there are only hypothetical designs or visions of the inventors. There is no real typical area of installation because they have not yet been installed at full scale (for commercial purposes).

The Wave Dragon device (WD, www.wavedragon.net) is an OTD type of WEC. The hypothetic prototype of the WD has dimensions $\lambda \times W = 260 \times 150$ m, and it is schematically shown in Figure 3.6. Two reflectors focus the incoming waves toward a ramp on the main body. Electricity is produced from a set of hydroturbines when the water drains through the turbines. The device is moored and is able to turn and face the mean incident wave direction. Air chambers and an advanced pneumatic system are used to adjust the floating level of the WD. The harvesting capability in the North Sea is estimated to be 24 kW/m.

The Blow Jet device, illustrated in Figure 3.7, reproduces the hydraulic behavior of the well-known naturally occurring blowhole structure. The device ($\lambda \times W = 6 \times 28$ m) consists of a tapered pipe with an internal profile similar to that of a brass

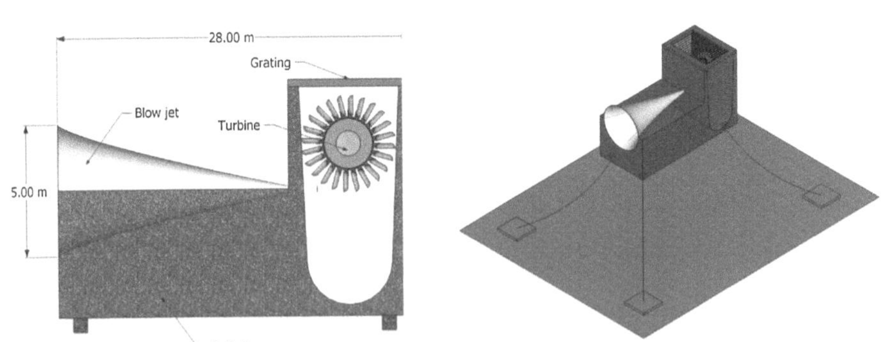

Figure 3.7 Illustration of the blow-jet.

tuba. Its hydraulic operation goes from the wave steeping and concentration inside the device to the expulsion of an intermittent jet that can be directed toward a turbine.

The SeaBreath device (www.seabreath.it; Martinelli et al., 2013; Martinelli, Ruol, & Cortellazzo, 2012) is an OWC elongated structure formed by a series of aligned rectangular chambers with open bottom. The device is aligned perpendicularly to the incident wave crests, i.e., it is of the attenuator type of WECs. The device is illustrated in Figure 3.8 ($\lambda \times W = 30$ m $\times$ 150 m). When the air in each chamber is compressed, it is directed by nonreturn valves toward a longitudinal high pressure duct; when the air is decompressed, the low-pressure air is directed toward a low pressure duct. A unidirectional impulse turbine placed between the high- and low-pressure ducts is used to generate energy. In practice, as the incident wave propagates through the different chambers, a "wave of air" is pushed through the turbine. The induced flow does not oscillate and the produced power is more steady compared with a single OWC chamber, with consequent benefits for the turbine efficiency.

The DEXA device (www.dexawave.com) belongs to the WAB type. It consists of two rigid pontoons with a hinge in between, which allows each pontoon to pivot in relation to the other (Figure 3.9). The draft is such that at rest the free water surface passes in correspondence of the axis of the four buoyant cylinders. The Power Take-Off (PTO) system consists of a low-pressure power transmission technology and is placed close to the center of the system to maximize the stabilization force.

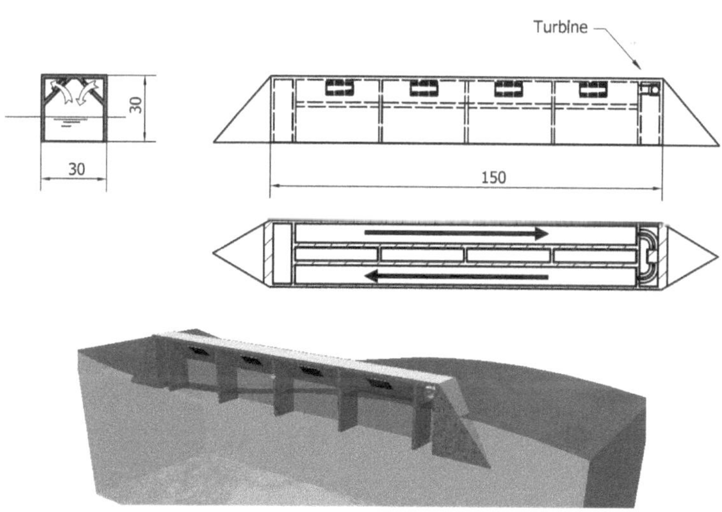

Figure 3.8 Sections of the SeaBreath.

Figure 3.9 Image of the full scale DEXA prototype.

3.2.3.2 Wave Transmission

Basically, the Macagno formula (Eqn (3.1)) may be used to predict the order of magnitude of the transmission coefficient also for nonrectangular shapes, provided that "equivalent" draft and width are used: essentially, d and W should be evaluated considering an equivalent volume of water reasonably constrained by the body during the wave propagation.

Section 3.2.2 presents the definition of $K_{\mathrm{T,1-row}}$ (Eqn (3.3)), that describes the transmission coefficient for a row of aligned FBs, accounts for the possible presence of gaps. The same equation is suited to describe the effect of a row of terminator types of WECs, always separated by gaps.

A different behavior is expected for the point absorber and the attenuator types. These devices absorb the energy also from the sides. It would therefore be inappropriate to use Eqn (3.3), because it is based on the assumption that the energy flowing through the gap is not absorbed. Rather, the "*apparent transmission width*" b_{a} concept may be used, i.e., the width of an apparent sponge that absorbs all the incoming energy (no transmission for a width b_{a}). Note that this is not equivalent to the "*capture width*," which defines the width of an apparent device that converts all the incoming energy.

Knowing (typically thanks to experimental investigations) the transmission coefficient of an array of devices displaced in one row ($K_{\mathrm{T,1-row}}$) for a specific case (i.e., with known gap spacing s and device width b_{d}), the apparent transmission width b_{a} can be calculated based assuming the energy flux is null through b_{a} and fully transmitted elsewhere:

$$b_{\mathrm{a}} = \left(1 - K_{\mathrm{T,1-row}}^2\right)(b_{\mathrm{d}} + s) \tag{3.11}$$

Then, for a generic gap width, the transmission behind a one-row array of point absorber or attenuator types of WECS may be evaluated using Eqn (3.3) and assuming that the device width is b_a instead of the actual one b_d.

For complex arrays, such as in case of: (1) more than one row; (2) significant diffraction effect from a finite length of the park of structures, or (3) wave-reforming by winds, the overall transmission coefficient has to be based on numerical models (see Section 2.1.5).

3.2.3.2.1 Transmission Coefficient for the Wave Dragon

The transmission behind a single WD was measured in physical model tests by investigating the influence of the wave period for application in high energetic seas with long waves. It was found that the wave transmission characteristics of a single device can be approximated with relatively good accuracy using a simple computation model based on the vertical distribution of the wave energy flux from linear theory (Nørgaard & Lykke Andersen, 2012), assuming that the energy flowing below the structure is fully transmitted, and that part of the energy incident the reflectors is focused toward the central body, in a proportion that is given by physical model experiments. Figure 3.10 shows the overall wave transmission coefficient $K_{T,D}$ as function of the WD width relative to the peak wave length, W/L_p and the WD width relative to the water depth. The influence of the wave length is very important, whereas the influence of the water depth seems to be negligible.

3.2.3.2.2 Transmission Coefficient for the Blow-Jet

The most relevant characteristic of the Blow-Jet device is the amplification of the orbital wave velocity, and the consequent production of a water jet suitable for energy generation.

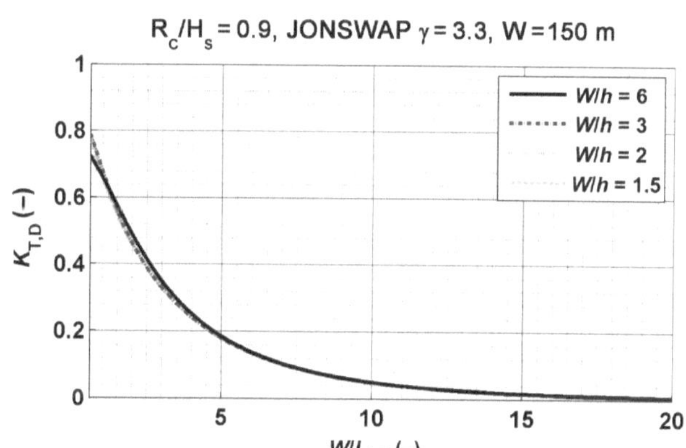

Figure 3.10 Estimated overall wave transmission coefficient from a single 24 kW/m WD device as function of the WD width to peak wave length ratio. The method is only validated in the range: $W/L_p = 0.905$ to $W/L_p = 1.232$.

For the Blow-Jet, tests with regular waves were carried out for a fixed structure considering:

- relative submergence of the device equal to 1/2 or 1/3 of the entrance diameter;
- symmetry axis inclination 0° or 10° with respect to the horizon.

The empirical formula for $K_{T,D}$ obtained for the fixed structure by fitting to the model test data is given by

$$K_{T,D} = 0.084(W/L_0 - 0.0049)^{-0.55} \tag{3.12}$$

where W is the draft at the entrance of the Blow-Jet and L_0 the deep water wave length. Figure 3.11 shows that the influence of h is insignificant, whereas L_0 is the governing parameter. Equation (3.12) is valid for $0.02 < W/L_0 < 0.22$.

3.2.3.2.3 Transmission Coefficient for the DEXA

Wave attenuation behind the DEXA device was studied under a variety of wave attacks in case of a single WEC in 1:30 scale and of an array of three WECs in 1:60 scale (Zanuttigh & Angelelli, 2013; Zanuttigh et al., 2013). In both conditions, the devices were moored with a realistic CALM (catenary anchor leg mooring) or spread system composed of chains. Based on these experiments, the following guidelines have been drawn.

The device design should consider that:

- the device length l (in cross-shore direction) has to be "tuned" on the basis of the local peak wave length (i.e., the more frequent wave length within the typical yearly climate); specifically, the condition $l/L_p \approx 1.0$ has to be assured for most of the year because it leads to the best compromise between wave transmission and power production performance;

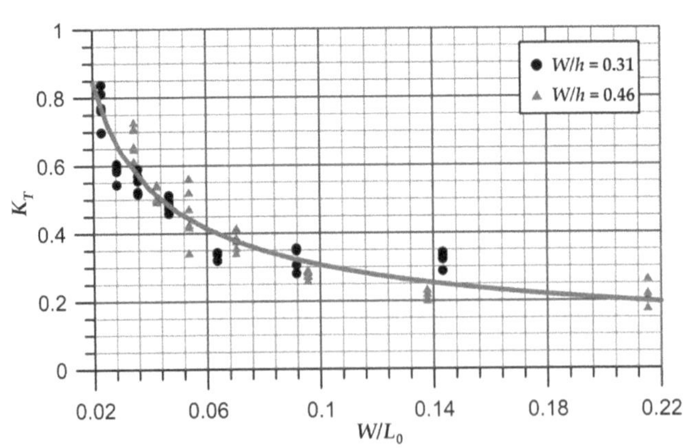

Figure 3.11 Best fit curve for the transmission coefficient from a single blow-jet.

- if other design features are kept constant, a heavier system gives better result for reducing the incident wave energy.

The measured values of the efficiency η can be approximated by the curve (Figure 3.12):

$$\eta = -0.5154 \cdot (1/L_p)^2 + 1.0901 \cdot (1/L_p) - 0.3417 \quad (3.13)$$

The measured values of the wave transmission coefficients for a single device $K_{T,D}$ are fitted by

$$K_{T,D} = -0.276 \cdot (1/L_p)^2 + 0.4304 \cdot (1/L_p) + 0.6781 \quad (3.14)$$

The performance of the device was obtained based on physical model tests in scale 1:30 (i.e., in case of a single device carrying the PTO system), in the range $0.49 < l/L_p < 1.26$ and for perpendicular waves. Water level was varied ± 1 m at prototype scale, and $K_{T,D}$ showed a negligible dependence on h.

3.2.3.2.4 Transmission Coefficient for the Seabreath

The Seabreath was experimentally tested in wave flume under regular and irregular waves to evaluate the hydrodynamics and energy conversion capabilities of the device (Martinelli et al., 2013).

The transmission characteristics are derived by two series of tests. In one configuration, including both fixed and floating conditions, this attenuator type

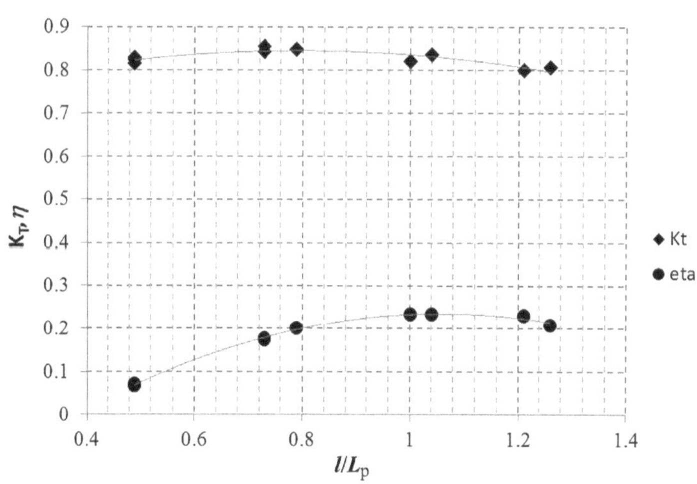

Figure 3.12 DEXA wave transmission coefficient and energy conversion efficiency as a function of the device cross-shore length divided by the peak wave length. Spread mooring systems, perpendicular waves. The fitting curves represent Eqns (3.13) and (3.14).

device was indeed placed with its axis parallel to the flume axis. The device width was only a fraction of the flume width aiming at representing the behavior of an array of structures under perpendicular waves, with a certain gap between the elements. The investigated porosity p was 70% (defined as $p = s/(s + b_d) = 70\%$).

In the second configuration, the model was fixed, with its axis rotated 25°, obtaining a structure that occupies the whole flume width. This case roughly represents an array of devices under waves with $\theta = 25°$ obliquity, with a length of each device equal to $(s + b_d)/\sin(\theta)$ to obtain a complete "overlapping" (the resultant apparent gap is zero).

Figure 3.13 (left) shows the experimental results for the oblique configuration. Figure 3.13 (right) compares the data for ($\theta = 25°$) with the results for perpendicular structures ($\theta = 0°$) using fixed and floating structures. The floating structure allows much more transmission when the peak wave period is close to the resonance one ($k^2WP \cong 2$).

The transmission coefficient from a Seabreath array of devices can be estimated as

$$K_{T,1-\text{row}} = [1.67 - \log 10(k^2WP/\cos\theta]/2.3 \quad \text{for} \quad 0.3 < k^2WP/\cos(\theta) < 20 \tag{3.15}$$

where k is the wavenumber (regular or spectral average $T_{m-1,0}$), W is the device width, θ is the wave obliquity, and P is the draft. In Eqn (3.14) $K_{T,\ 1-\text{row}}$ is upper limited by 1 and lower limited by 0.

3.2.3.3 Arrays of WECs

Few studies are available in the literature on the hydrodynamic performance of wave farms (Angelelli & Zanuttigh, 2012, 2013). Although the concept of

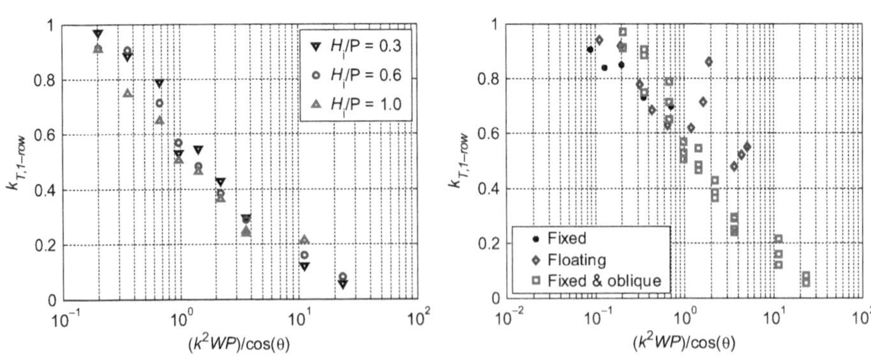

Figure 3.13 Wave transmission coefficient for a one-row array of Seabreath with 25° wave obliquity (left panel), and comparison with perpendicular waves (right).

park effect, i.e., the interference on the WEC performance induced by the other WECs, is well-known (e.g., Falnes, 1980), there are not many indications about optimal layout and interdistances among WECs. Furthermore, contributions on the installation of WEC in a wave farm are essentially based on numerical simulations, and mainly related to point absorbers (see the reviews by Babarit, 2010, 2013).

The numerical methods adopted so far to model the wave field around WECs are essentially based on three different approaches.

The first approach is the traditional simplified modeling adopted for vessels and floating breakwaters, which is relative to the two-dimensional (2D) case (indefinitely long structure with perpendicular wave attack) and is based on the hypothesis of irrotational flow. Available commercial codes as WAMIT (www.wamit.com) and AQWA (www.ansys.com) are based on the Boundary Element Method. These models are typically implemented for horizontal bottom, linear waves and do not account for viscous dissipation, hypotheses that become less accurate if referred to shallow water conditions (Cruz, Sykes, Siddorn, & Eatock Taylor, 2009).

The second approach consists of 2D or three-dimensional (3D) models typically developed to assess the impact on the littoral of wind farms and then applied also to wave farms where the piles or devices are represented through an equivalent bottom friction or percentage of wave energy absorption respectively (Palha et al., 2010). These simulations are usually done for environmental impact assessment purposes and do not consider model calibration.

The third approach uses 2DV and 3D RANS-volume of fluid (VOF) models to represent velocities and pressure around floating bodies. Simulations are usually always calibrated against experimental data in wave flume or tank. The first applications were performed on single and multiple floating breakwaters but fixed; more advanced research proposed 3D models combined with the external solution of the motion equation and iteratively solved. In this way, it is possible to reproduce the presence of PTO or moorings by means of the external forces applied to the WEC (Agamloh, Wallace, & Von Jouanne, 2008).

Recent works applied the 2D Boussinesq model MIKE 21 to reproduce the hydrodynamics around a single WD (Beels, Troch, De Visch, Kofoed, & De Backer, 2010; Nørgaard et al., 2011) and around arrays of WD and DEXA devices (Nørgaard & Lykke Andersen, 2012; Angelelli & Zanuttigh, 2012). The devices were reproduced as porous bodies whose characteristics were calibrated based on the wave reflection and transmission coefficients derived from physical tests. Figure 3.13 shows the field of the wave disturbance coefficients K_d (local wave height nondimensionalized with the incident wave height at the wave maker) around two rows of DEXA devices. Measured and modeled K_d are compared in Figure 3.14 for a given mooring stiffness, which was found to significantly reduce the wave transmission through the device (see also Figure 3.15).

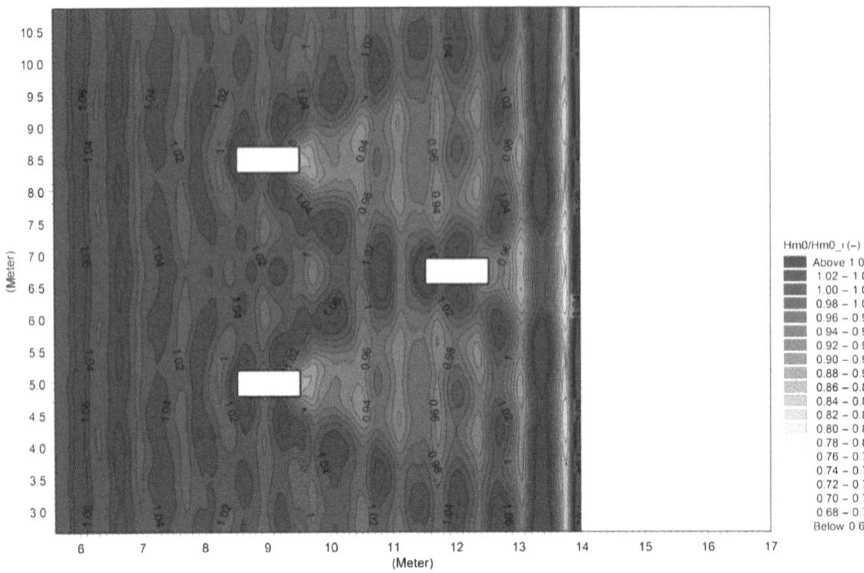

Figure 3.14 Wave disturbance coefficients K_d (local wave height nondimensionalized with the incident wave height at the wave maker) around an array of DEXA devices.

Physical and numerical experiments carried out on arrays of WABs (Angelelli & Zanuttigh, 2012, 2013) showed that:

- the optimal long-shore gap width is function of the radiated waves (i.e., the extension of the wake effects) generated by the device motions; it should be considered also that the distance among the devices affects the PTO system;

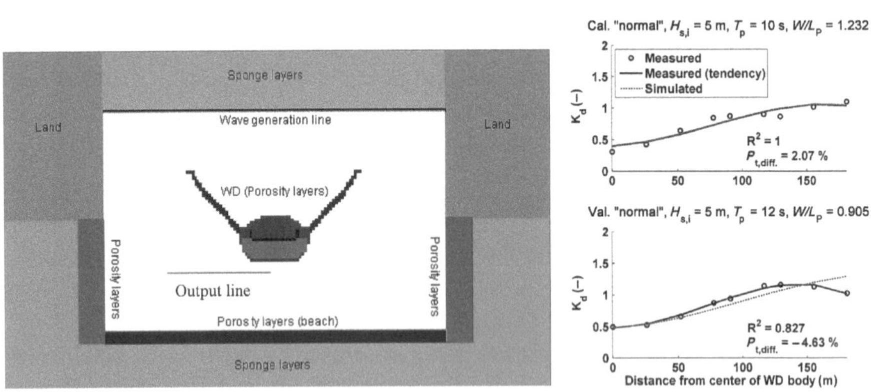

Figure 3.15 Wave Dragon implemented in MIKE21 BW using so-called porosity layers *(left)* modeled and measured wave disturbance behind Wave Dragon *(right)* (Nørgaard & Lykke Andersen, 2012).

- the optimal cross-shore gap width has to be greater or equal to the distance at which waves regenerate behind the device, accounting also for its surge motions;
- a staggered configuration optimizes the device density and therefore the power production of the installation, the reduction of the wave height inshore the array and the residual wave height behind each line; and
- the design of the layout is strictly related to the space required by the mooring systems, to let the devices and the mooring lines to freely move without interacting; wave diffraction effects however are more significant than the pre-tension level of the mooring lines.

The calibrated MIKE21 BW model was used to study the wake behind a farm of staggered WDs in one of the study sites to be presented in Chapter 7, Section 7.9, Santander Bay, Spain. The wave field in Santander Bay is illustrated in Figure 3.16 in absence and presence of the WDs using realistic mooring stiffness. When using the farm layout that provides the best compromise between electricity production and coastal protection, the wave power is reduced by approximately 50% at the shoreline in Santander Bay. It should be mentioned though that this calculation is performed without taking into account wave reformation because of wind.

More recently, a new method was proposed (Babarit, Folley, Charrayre, Peyrard, & Benoit, 2013) that couples the Boundary Element Model (BEM) approach and phase resolved or phase averaged approaches. The idea is first to use the wave models to calculate the unperturbed wave resource at the WEC location. Then, the perturbation can be calculated using BEM solvers. The perturbation potential is composed of two parts: a near field (which decays rapidly with the distance from the WEC) and a far field (which decays much slower than the near field). For phase-resolved models, the method requires using domain decomposition and

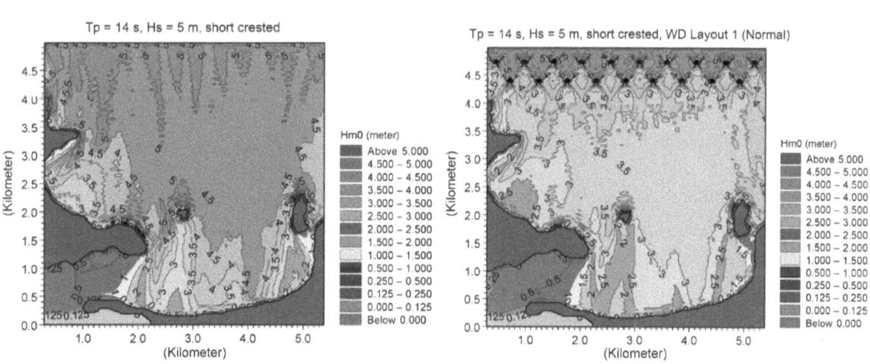

Figure 3.16 The wave field in Santander bay in absence *(left)* and presence *(right)* of a staggered array of wave dragons.

introducing a wave generation line in the wave model. For phase average models, far-field coefficients provide an alternative to the empirical approach for the calibration of source and sink coefficients employed until now.

3.2.3.4 Mooring System

The installation of F-WECs requires a suitable design of the mooring system, that shall account for high waves in relatively shallow water depths (i.e., in a very nonlinear context). Although to a smaller extent, this issue is also applicable to the mooring of FBs.

The design process of the mooring system concerns the configuration of the individual lines, their interaction with the floating body (fairleads) and the bottom (through drag embedded, pile, gravity or plate anchors), and the overall layout (i.e., the disposition of the many lines forming the mooring system). The design should focus on mechanical resistance, durability, reliability, deployment and dismissal phases, and environmental impact. Environmental factors to be considered include damage to the local environment, visual impact, and any effect on the local ecosystem.

The main technical challenges are to allow for weathervaning (a correct orientation relatively to the predominant waves), and in some cases to favor energy harvesting by constructive interaction with the system dynamics.

The following requirements are common to most floating structures (including FBs):

- to maintain the floating structure on station within specified tolerances under normal operating load and extreme storm load conditions;
- to avoid impact between lines, or with other structures;
- to ensure the mooring is sufficiently compliant to accommodate the tidal range at the installation location, and sufficiently stiff to allow berthing for inspection and maintenance purposes; and
- to make sure the setup allows removal of mooring lines for inspection and maintenance.

Specific requirements for the moorings of F-WECs are:

- the movements of the device must not permit tension loads in the electrical transmission cable(s) and should allow for suitable specified clearance distances between devices in multiple installations; contact between mooring lines, or contact with the device itself must be avoided;
- the system must be sufficiently compliant to the environmental loading to reduce the forces acting on anchors, mooring lines, and the device itself to a minimum (unless the stiffness of the mooring itself is an active element in the wave energy conversion principle used);

- the removal of single devices and should not affect the mooring of adjacent devices in the array; and
- the system as a whole should last for 20 years or more, with replacement of particular components at no less than 5 years.

Mooring systems may be classified on the basis of the layout configuration: most FBs are anchored with spread moorings, whereas WECs are moored in several ways, and in particular spread, turret, and single (Fitzgerald, 2009; Harris, Johanning, & Wolfram, 2004; Martinelli et al., 2012). Spread moorings are common but they are not always suitable when the device is requested to weathervane into the incident waves, because spread mooring tend to constrain the heading angle. To enable a vessel to rotate, single-point attachment from the vessel to a fixed or floating structure/buoy/turret is used, hence the term single-point mooring. Active mooring or dynamic positioning (propulsion) could be an expensive station keeping option.

Individual lines are frequently a combination of chains, wire ropes, and synthetic line, such as polypropylene (and derivate products with high resistance and small elasticity) or nylon, depending on the type of mooring and the environment in which it is deployed, and involve several additional hardware components (e.g., shackles, links).

The most economic option for the lines in a spread mooring system is reasonably the free-hanging catenary configuration (i.e., a set of diverging chains). Unfortunately, this may not be able to allow for a sufficient extension without excessive loads when the tidal range is large. Another disadvantage of such a configuration could be the restraining stiffness affecting the WEC dynamic. Especially in case of WAB devices, the use of these multicatenary mooring configurations could provide some flexibility in particular modes of motion and improve the energy extraction.

Some tethered mooring systems with substantial preload tend to fully constrain the floating body movements, and in this case, ideal for submerged applications, the mooring is designed to directly absorb the applied load.

The most common condition, however, is associated with a different conceptual approach that takes advantage of the oscillatory nature of the waves. The mooring force balances the average external load, rather than the maximum. Therefore, the floating body is forced to move, and to activate its inertia in order to respond to the instantaneous unbalance between the external loads, mainly from wind, currents, and waves, and the reaction loads from hydrostatic pressure and the mooring system constraint.

Such an approach is generally quite convenient. Nevertheless, there are three critical cases: (1) when the movements of F-WECs exceed the limits of the station-keeping footprint area; (2) when the movements are not congruent with the mooring system, and the reaction force exceeds the design one; and (3) when resonance conditions are reached, so that the dynamic reaction exceeds the quasi-static equivalent load.

Unfortunately, these three conditions are likely to occur simultaneously: the nonlinear response to a real wave has a load component at low frequency that induces resonant conditions and low-frequency drift-like movements of large amplitude. Although the vertical oscillations are easily predictable, being basically that of the surface elevation (i.e., of the order of the sum of tide, surge, and wave amplitude), the evaluation of horizontal movements requires a detailed investigation, and the low-frequency oscillations may be quite larger than the wave amplitude. Furthermore, because of the nonlinear response of the mooring system, the large movements are associated to extreme reactions that are likely to break the chains or permanently displace the anchors.

Critical design issues are related to reliability and durability. The wear at the bottom (i.e., where the mooring line touches the bed) is one of the most frequent limits to chain durability.

For FBs anchored in deep waters, yet subject to waves of limited height, the severe station keeping targets may require that the mooring lines are designed to withstand snapping loads (Martinelli et al., 2008). This is a particular condition that requires special attention (i.e., accurate laboratory testing).

On the contrary, mooring of WECs, which are generally characterized by a larger total mass and are subject to larger waves, must be designed to avoid snapping. A critical design issue is associated to the underestimation of the floating body drift caused by resonance between the mooring system and the long period drift induced by load second-order wave interactions (Fitzgerald, 2009). The occasional large resonance–induced drift may expose the lines to snapping and, consequently, to failure, whereas in the absence of resonance, the applied load are considerably lower, giving a false image of security. In fact, even prototype-scale testing may be misleading and should be numerically interpreted in view of possible resonance conditions.

3.2.3.5 Costs Analysis

In projects related to power generation, the cost is typically measured by the cost of energy production and it is expressed in €/kWh. Because marine energy conversion is a relatively recent technology field and is mainly at its research stage, there is a lack of application of economic assessment methods and procedures.

A WEC installation includes the following cost items:

- costs associated with the device, i.e., costs determined by its components (body/ies and PTO), its functions (for example, the energy production), and the design choices made;
- mooring system costs;

- costs of all the cables, including cable among devices in the wave farm installation and between the installation site and the shoreline; and
- costs of the network transmission.

Furthermore, if one considers the entire WEC lifecycle, also the costs of decommissioning should be accounted for and depend on several factors; for example, the number of WECs to be removed and the selected decommissioning type (for instance, the removal of the WEC and then its demolition or the WEC disposal at sea, such as in the form of an artificial reef).

Figure 3.17 proposes the breakdown of the capital costs for an array of WECs (Carbon Trust, 2006).

The WEC annual operational expenditure is primarily associated with:

- ordinary maintenance: these costs are defined as the costs for the achievement of the design power production. These costs also include items such as the cost of the monitoring, insurance, vessel and equipment transportation, labor, and consumable parts; and
- restoration (i.e., extraordinary maintenance): it is strictly associated to the "mean failure time," which depends on many factors (e.g., the possible failure modes and the installation environment).

The simplest and most common approach is to assume that there is no connection between the two categories of maintenance: scheduled and unplanned, even if some scheduled maintenances are generally dependent on the reparation costs. For WECs, the maintenance is likely to be not infrequent and therefore accessibility of the site and wave climate become key parameters for planning the device location.

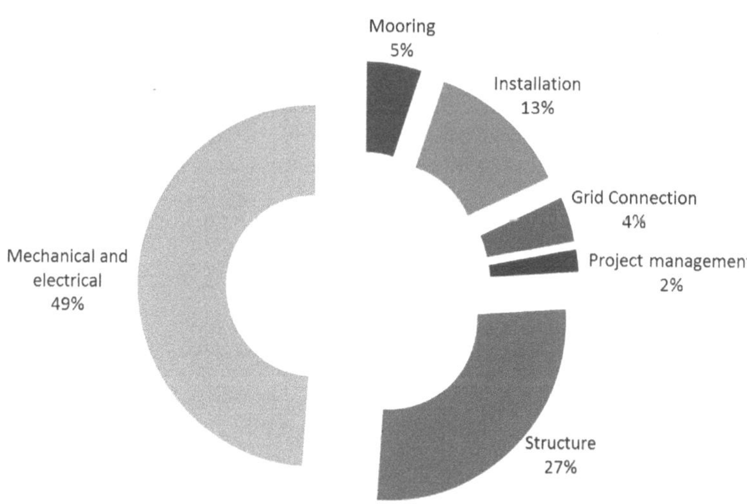

Figure 3.17 Influence rate of the capital costs, proposed by Carbon Trust (2006).

The amount of produced energy determines the revenue of the project. The performance of a WEC is generally derived from the sea trials assuming that the device will not be interrupted because of maintenance or failures. Therefore the energy output of a WEC farm is function only of the available wave power and of the number of installed devices. A revenue analysis, however, should specify the uncertainty of the power production because of the uncertainty of the device performance and the expected downtime based on the persistence of high energetic waves. Furthermore, the electricity prices can be very volatile depending on the market, adding another source of uncertainty.

3.2.4 Final Remarks

Differently from fixed traditional breakwaters parallel to the coast, near-shore floating structures aim at lowering the incident wave heights without inducing the breaking phenomenon.

The selection of the device typology may be assessed (Table 3.4) based on the following few engineering variables: wave height, water depth, wave length, and tidal range.

FBs are usually placed where the wave period is shorter than 4 s and waves are mild ($H < 1.5$ m). To address higher and longer waves, structures need to be very large, moorings need to be carefully designed, and F-WECs may be more suited than FBs.

F-WECs are designed for highly energetic seas, with average energy flux per unit span not smaller than 10 kW/m. F-WECs are generally much larger than FBs, and are therefore placed much more distant from the shore. Actually, the distance from the shore is a quite important variable to be designed. In fact, when placing F-WECs for coastal protection beside energy production, the configuration must necessarily result as a compromise between both purposes. A smaller distance from the shore decreases the available power (e.g., because bottom friction). On the other hand, when F-WECs are placed too far from the shore, the waves regenerate between the devices and the coast, so that sheltering effects are not observed. Moreover, the greater the distance from the shore, the higher the costs of maintenance and operation.

TABLE 3.4 Range of Applicability of FBs and WECs

Device	FB	WEC: OVT Type	WEC: OWC or WAB Type
Operation conditions	$T_p < 4$ s $H_s < 1.5$ m	Available power >10 kW/m $T_p < 15$ s	Available power >5 kW/m $T_p < 10$ s
Extreme conditions	$H_s < 2.5$ m	Device dependent	Device dependent

A single device reduces the wave height only locally because of diffraction and directional spreading. To protect the coastline, an array of WEC devices is needed. The distances among devices in an array depend on the device type and on the mooring system configuration. The change of wave height behind an array can be approximately calculated with numerical wave propagation models by implementing each device and its transmission properties. A feasible reduction of wave height is on the order of 20% of the incident wave.

When selecting the mutual distance between devices in a row or in an array, the following considerations should be made.

- Interaction between the devices can lead to unexpected resonance (requires physical modeling investigation);
- A minimum gap width is required to avoid collisions and to let the devices and the moorings freely moving allowing for operation and maintenance;
- A gap width as small as possible leads to an economic optimization of the farm layout, being less marine space required to deploy the devices. In addition, lower cable costs result for the same target energy production and at the same time lower transmission is obtained, thus improving coastal protection;
- If the array configuration is chosen wisely (staggered grid) then the second row of devices addresses almost the same amount of available power as the front row and leads to a more compact layout.

Some considerations are specific of the nature of the WEC operational conditions. In the presence of very small or extreme waves, WECs do not harvest energy and therefore are less efficient in terms of reducing wave transmission. For this reason, it should be noted that OTDs have generally a lower operational threshold in terms of wave height. In the presence of extreme waves, many F-WECs do not operate because they are put in "safe mode" to preserve the structure and the mechanical parts, and this is a disadvantage in terms of coastal protection. The floating OWCs and WABs are usually more effective in a narrow range of wave frequencies, whereas OTDs are effective for a larger range of wave lengths.

3.3 Innovative Submerged Structures

3.3.1 Introduction

Among all the alternatives developed worldwide for coastal defense, those that have found best acceptance in recent years are the more environmentally friendly designs. Moreover, if the aesthetic impact is also considered (e.g., touristic industry), submerged structures, also known as reef breakwaters, seem to be a very profitable option regarding overall costs and construction time (Burcharth, Hawkins,

Zanuttigh, & Lamberti, 2007; Ranasinghe, Larson, & Savioli, 2010). The construction of this type of structure effects a reduction of onshore wave energy to landward. This may be desirable to prevent coastal erosion, to protect land and infrastructure from flooding, to protect from damage or remedy the poor design of existing coastal structures. They may also be used to increase the stability of beaches, possibly in combination with sand nourishments.

The use of submerged structures has been a common practice in the past two decades, especially in Europe, and more particularly along the Mediterranean coast. The term low-crested structures (LCS) traditionally refers to rubble-mound breakwaters whose crown elevation is near the still-water level. LCS are usually constructed as "detached" structures, i.e., detached from the shoreline, and are aligned approximately parallel to the local shoreline in sites where partial protection from wave attack is required (Burcharth et al., 2007). These structures may be submerged, emerged, or both alternately, and are characterized by frequent and significant levels of wave overtopping. The main aim of LCS is to reduce the wave loads on the coast through a series of wave transformation processes occurring at, on or around the structure: namely wave reflection and energy dissipation because of both wave breaking on the crest and flows inside the breakwater body.

In the past decade, an interesting trend has been to explore the use of more innovative submerged structures that employ either natural features of the coastal zone, or some artificial representation of them. Examples of the features that might be considered suitable for their coastal protection function include seagrass and oyster beds. The goal of these structures is the same as the LCS, but they are designed to reduce the environmental and economic impact at any specific location. These structures may or may be not considered as permanent solutions to beach erosion problems.

Generally speaking, permanent solutions are the more expensive; in fact they are sometimes around 10 times higher in cost than other solutions, although they may offer similar coastal protection, i.e., Woolmer, Syvret, and FitzGerald (2011). Wave climate, sediment properties, and social and economic aspects need to be taken into account when selecting a solution (Baine, 2001), e.g., permanent solutions may be considered as optimal in places where intense storms are expected, but if the estimated damage to the area and reconstruction costs are not very high, a cheaper solution may be the best option as well as offering a more ecological service.

This section will consider the use of innovative coastal structures as effective alternatives to the more traditional approaches employed in coastal protection and is organized as follows. First, a short description of the more relevant physical processes when waves meet submerged structures is presented. Then, guidance is provided about design, cost, and the morphological changes induced by these structures.

3.3.2 Physical Processes

Submerged strcutures (SS) reduce the incoming wave energy across the structure by partially reflecting the waves at the toe, by triggering wave breaking at and on the structure, and by dissipation related to the wave-induced porous flow through the structure. The main objective is reducing wave height at the coast and the long-shore sediment transport. No significant differences can be found in the use of innovative submerged structures compared with traditional Low crested structures (LCS) in terms of wave transformation and wave-induced hydrodynamics at the coastline, although some processes are enhanced, mainly with the aim of increasing wave damping. The degree of protection generated by both LCS and SS depend on their geometries and maritime climate conditions, Pilarczyk (2003).

3.3.2.1 Wave Hydrodynamics

When the incident wave train impinges on a submerged structure, part of the energy is reflected back to the sea and part is transmitted in the leeside zone, Silva, Salles, and Palacio (2002). Reflection is an important characteristic of submerged structures because this type of structure is generally narrow and built with relatively steep slopes, Losada, Silva, and Losada (1996a). The interferences between the incident and reflected waves give rise to standing wave patterns that affect the near-field flow and subsequently the stability of the structure, Lara, Garcia, and Losada (2006).

Besides, as waves shoal on the rising front of the structure, significant nonlinear effects occur, resulting in the amplification of bound waves (phase-locked with the primary wave), which may result in an amplification of the surf beat at the beach. Most of the incident wave energy is lost on the crest of the structure, essentially by breaking, Gourlay (1996).

Part of the energy is also dissipated by air entrainment and friction at the solid skeleton interface and within the gaps created between the individual rocks, for rocky traditional LCS, or within the artificial holes created for that purpose, for ecologically compatible elements. For nonbreaking waves, the flow resistance in the porous media is the main dissipation mechanism.

Significant nonlinear interactions occur in the zone of the structure crest between the various wave phases and some energy is transferred from the fundamental wave frequency to higher harmonics (Losada et al., 1996a,b). In the deeper waters of the leeside zone, the wave field is characterized by waves of complex form, with a lower height and a lower mean period. In addition, wave interaction with submerged structures also gives rise to a series of 3D phenomena, such as diffraction, 3D wave breaking and currents system generation, i.e., de Alegría-Arzaburu, Mariño-Tapia, Enriquez, Silva, and González-Leija (2013) and Silva et al. (2013) .

3.3.2.2 *Currents and Tide Effects*

Submerged structures with a small freeboard typically induce wave breaking on their crest or at their offshore or inshore slope leading to the generation of new current patterns (Johnson et al., 2005).

In the case of detached breakwater systems, or potentially any series of structures that provide shelter, wave breaking and transmission can drive current circulation in the shadows of the structures and also induce fast-flowing currents around the breakwater heads (ends) where there is potential for significant scour and a need for careful design (Johnson et al., 2005; Villani, Bosboom, Zijlema, & Stive, 2012).

Tidal setting and sea-level rise, in different time scales, are also important when considering the design and performance of structures near the coastline. Tides and other water level variations determine the depth of water over or around a structure, thus influencing how strongly surface water waves and currents interact with the structure (Dean, Chen, & Browder, 1997).

3.3.2.3 *Morphological Changes*

It is generally understood that waves stir up sediment and currents transport this mobilized sediment. Thus, the deployment of a coastal structure or some other form of barrier will not only change existing currents and regions of wave breaking, it will also affect the sediment pathways and the areas of erosion and deposition (Stauble, 2008). Where a barrier produces a wave shadow, the wave height is lower and deposition is likely to occur (Fairley, Davidson, & Kingston, 2009). Waves can diffract around such barriers, approaching the shoreline obliquely, where they break and push sediment into the lee. Refraction over the deposited sediment in the lee reinforces this pattern of deposition.

Thus a barrier of finite extent will impact upon the shoreline by initiating the growth of features referred to as a salient, double salients, or eventually a tidal tombolo (Ranasinghe et al., 2010). These are tongues of sand that grow from the coast in the wave shadow, out toward the lee of the barrier. Tombolos, which resemble a permanent isthmus of sediment, will interrupt any littoral currents. This can result in updrift accretion and downdrift starvation of sediment, which may be undesirable in certain locations.

Careful design guidance is needed to assist in the planning of any each type of structure or barrier. In the United Kingdom, guidance for the placement of submerged and surface piercing breakwaters in tidal sandy environments is available (Environment Agency, 2010). Complementary information can be found also in the Coastal Engineering Manual, U.S. Army Corps of Engineers (2002), and the Japanese artificial reef design, Ministry of Construction (1992).

3.3.2.4 *Additional Issues*

Local scour is induced by the interaction of the wave and the elements that constitute the structure, particularly on the seaward side, but it may also be important in the leeside for plunging breakers (Sumer et al., 2005). One way to avoid this is by the installation of a geotextile carpet, with a toe of rocks on the seaward side.

Submerged coastal structures can induce downdrift erosion and updrift accretion on coastlines with a net direction of sediment transport, but to a lesser degree than with surface piercing structures with crowns above the mean sea level.

Most submerged structures give their best performance when placed over sand, where they can be partially buried and friction can enhance their stability. If they are placed over rock, they can eventually be displaced by waves.

Land subsidence can cause structure settlement and therefore decrease the degree of coastal protection offered, similar to the effect of sea-level rise. The expected sea-level rise should be considered for structures with a useful life of 25 years or more. Indeed in the United Kingdom, for example, new schemes are appraised in terms of their performance and maintenance with consideration of the effects of subsidence and climate change over the next 100 years.

Submerged breakwaters may be effective tsunami control structures, Silva, Losada, and Losada (2000), thanks to the turbulence generated in the porous medium. Impermeable structures induce less energy dissipation and eventually the waves only shoal.

3.3.3 Examples of Submerged Structures

Traditional hard-engineered structures might comprise water surface-piercing, low-crested structures, submerged reef structures, and other forms of breakwater. "Softer" technologies that can also be considered are the use of submerged geotextile structures, or the installation of artificial dissipative blocks on the seabed to increase wave damping and to promote new habitats (Sundar & Sannasiraj, 2013). This section focuses on the hydrodynamic and morphological changes induced by LCS and rocky reefs, geobags, and artificial reefs made of natural rock or geometric concrete units. Specific issues related to design and costs are pointed out for applicability and performance purposes.

Despite the extended use of submerged structures, certain hydraulic conditions must be satisfied for a submerged structure to operate adequately. First, the protected beach must be capable of dealing with the transmitted waves. Because of the submerged nature of the barrier an important percentage of the waves will pass beyond the structure, which means that some agitation will be present. So, if the area intended to be protected is vulnerable even to these small waves, a submerged structure is not an option. Furthermore, the geometry and location of the structure must produce wave breaking for the mean wave regime and for a significant number of storms. If too many waves pass over the structure without breaking, the energy dissipation is nullified and the beach response may be even worse than it would be without such a structure. Another obvious factor to consider is the danger to navigation that submerged structures will cause.

3.3.3.1 Low-Crested Structures and Rocky Reefs

There are several reasons why LCS and rocky reefs are chosen for beach protection. Most importantly, their reduced visual intrusion makes them valuable, particularly in zones with important aesthetic constraints. Where economic constraints are foremost, low-level structures, especially submerged ones, are generally more stable and cheaper in that they use less material (d'Angremond, Van Der Meer, & De Jong, 1996). Similarly, a further economic benefit is the possibility of rehabilitating other, existing structures just by reducing the incident wave conditions with a low-crested breakwater. Finally, because of the high level of transmission, LCS induce in-shore water renovation, which is desirable, particularly for recreational waters as well as improving water oxygenation, which is valuable for animals and plants living in the protected area leeward of the breakwaters.

Burcharth et al. (2007) presented very comprehensive design guidelines for standard LCS, providing methodological tools for the engineering design and for the prediction of performance and environmental impacts of this kind of structure, pointing out the importance in the selection of the material to be used as well as the selection of the shape and location of the LCS.

Recently, V-shaped breakwaters have been suggested as an option, (Briggs, 2001). Geometrically similar to fishtail breakwaters, Figure 3.18, they protect the

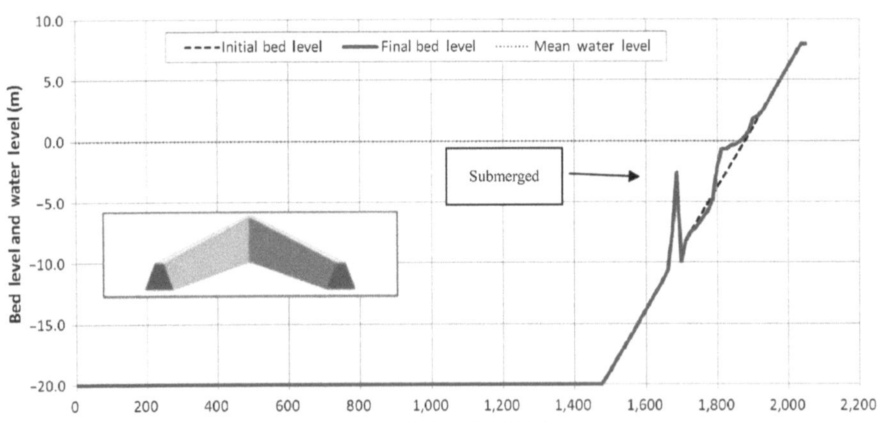

Figure 3.18 Beach profile (central section) and location of V-shaped submerged breakwater (insert: breakwater form).

beach from a wider range of incident waves than parallel ones. Although their performance on wave transformation and their impact on beach morphology are yet to be fully understood, a consideration of their performance can be modeled using an appropriate process-based modeling approach (see 3.3.4.1).

3.3.3.2 Geobags

Submerged or semisubmerged sand- or gravel-filled geotextile bags (geobags) are semirigid structures that represent a reversible option that can be implemented in a few weeks (Figure 3.19). The main advantage of these structures is that they are relatively inexpensive to install. Geobags have become increasingly popular as an alternative to conventional hard structures, especially in those situations where rapid implementation of stabilization measures are required.

The geotextile must be able to withstand the extreme environmental conditions where it is installed, including marine salinity, temperature fluctuations, and exposure to sunlight. The dimensions of the structure required influence the size of the geobags used. As a rule of thumb, the length of each geobag unit should be five to 25 times its height, and the larger the bag, the greater the mechanical stability offered against failures because of mechanisms such as sliding, overturning, etc. However if a large unit breaks, the area of coast left unprotected will be larger. The degree of wave energy dissipation is controlled by the relative crown width; the greater the width, the greater the degree of protection.

The hydrodynamic effects (e.g., reflection, transmission, diffraction, shoaling, breaking) induced by structures made of geobags are very similar to those induced by impermeable structures made of concrete. Although the geobags can reflect a large amount of incident wave energy, where the relative freeboard is not large enough, waves are steepened and the breaking process is more intense than without the structure and therefore sediment transport is increased. For this reason, it is

Figure 3.19 Submerged geobag structure for temporary shore protection in the Caribbean Sea.

recommended that the maximum freeboard must not exceed half of the annual mean significant wave height at the mean sea water level, see Silva, Mendoza, and Chávez (2012).

Normally, scour protection must be considered at the toe of the structure on the seaward side to prevent the geobags sliding seaward as a result of the increment of wave energy from the reflection effect. With the presence of sand bars, it is advisable to locate these structures seaward of the bars (Short, 1992).

Local, natural (native) material is recommended for filling the geobags to reduce costs and contamination hazards in case of a break in the bags. The filling process must be completed in phases to prevent breakage of the geotextile and allow water drainage. When very fine material (clay and/or silt) is used as a filling, the weave of the geotextiles is a key parameter that should be selected to minimize the chance of leakage.

3.3.3.3 *Artificial Reefs*

This relatively new approach involves the emplacement of submerged permeable structures parallel to the coast and crowned at the mean sea level, to provide wave dissipation, mimicking the effects of a natural reef. These can be constructed from natural rock or from artificial f concrete units with specific functional geometries (e.g., WAD – Wave Attenuation Device – units or Reef Balls).

When designing these artificial reefs, the following aspects must be addressed. First, in storm conditions with associated storm surge, the effectiveness of an artificial reef decreases considerably. Stability (sliding and overturning), settlement and scour should be carefully considered during the design phase. Occasionally, sliding or overturning of the units occurs from intense wave and current effects or settling deficiencies; in this case, the units can strike one another and will eventually break. Therefore, to improve the stability of the pieces, the possibility of adding more weight to the bases of the pieces should be considered where local conditions call for this. For temporary deployment, these protective structures have the advantage that they can be relocated with little technical effort.

Where natural rocks are not used, special geometrical structural elements can be manufactured nearby and installed on site using a small barge crane. Examples of these include WADs (http://coastlinesolution.com/), ERCON (http://www.uni-due.de/nomatec/themen_ercon_inhalt_en.html), Florida Special-Artificial Reef (http://floridasportfishing.com/magazine/nautical-art-jewelry/reefmaker-artificial-reefs), and Reef Balls (http://www.reefball.org/index.html). However, extra care is necessary during installation, because some designs, though effective when deployed, are susceptible to damage if carelessly handled. Thus construction with these elements is straightforward provided that operations are carried out in calm conditions (see, for example, Figure 3.20).

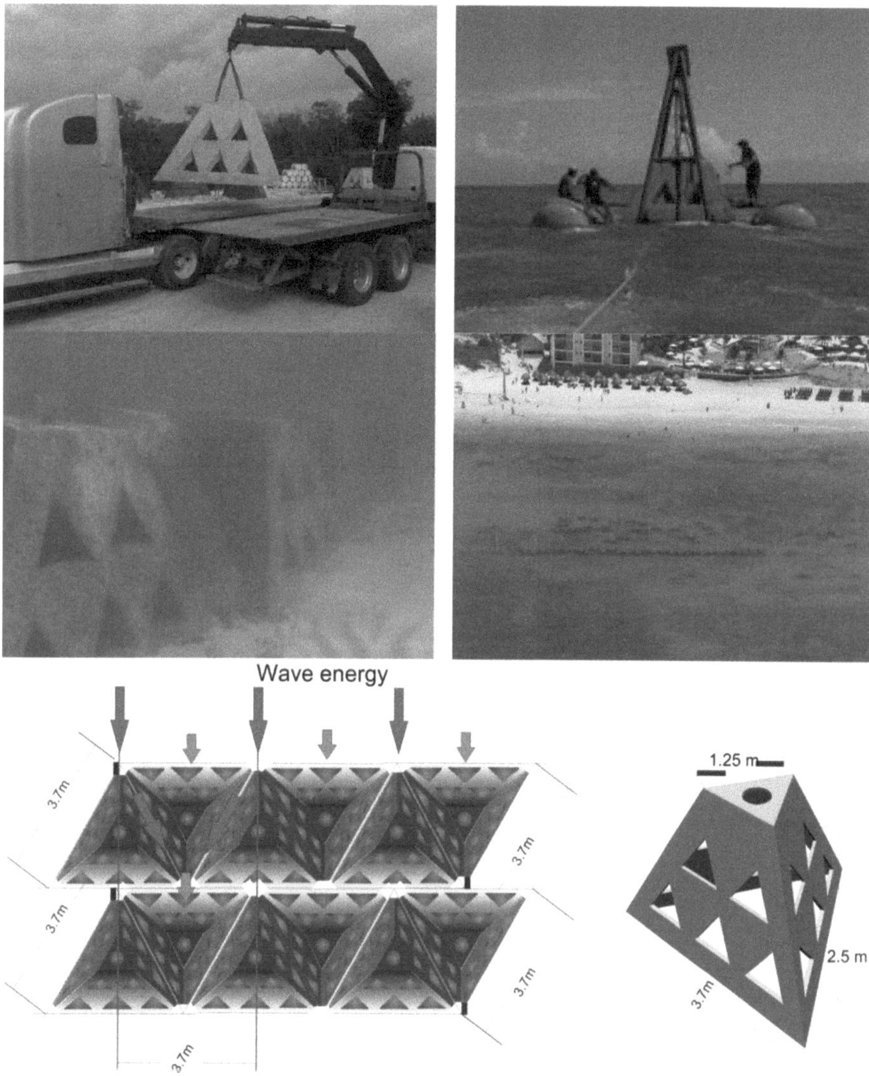

Figure 3.20 An example of an artificial reef made of WADs in the Caribbean Sea.

3.3.4 Further Design Considerations

3.3.4.1 Impact of Structures on Coastline

To predict the effect a structure may have on the adjacent coastline, the effective modification to the hydrodynamic and sediment dynamic processes needs to be considered. This is often achieved through numerical or physical modeling approaches, through which designs can be appraised or optimized to achieve the desired level of protection and an acceptable level of morphodynamic adjustment.

For instance, the design of a scheme of V-shaped breakwaters on an open tidal coast has been considered by Pan, Horrillo-Caraballo, Reeve, and Simmonds (2013). This required the use of a numerical model that incorporated the relevant processes: wave transformation, refraction, diffraction, wave breaking, and the resulting current generation and sediment transport formulae for the sediment size in addition to a tidal modulation of the water level. Both sea-facing (VS) and land-facing (VL) configurations of the breakwaters were considered (Figures 3.21(a) and (b)). The geometry of the breakwater designs in these simulations is as follows: the arm of the V-shaped breakwater is about 100 m long and both ends are 50 m from its baseline, so that the angle between the two arms is 120°. The breakwaters are sitting on a plane beach with a 1:20 bed slope, with breakwaters were located at approximately 10 m water depth (i.e., approximately 200 m from the shoreline at the mean tidal level) and 200 m apart from each other.

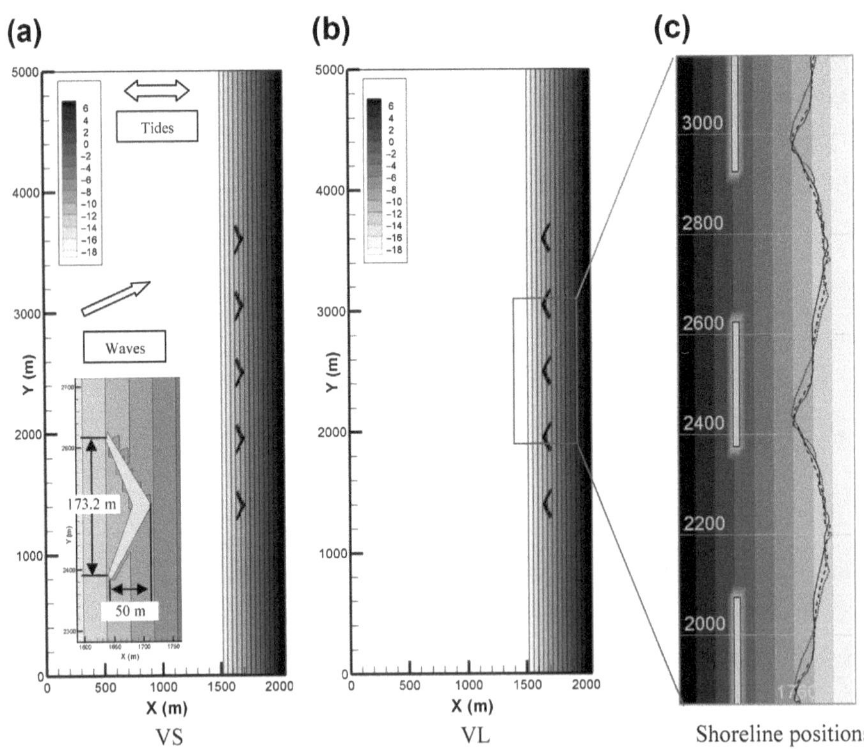

Figure 3.21 Computational domain and V-shaped breakwaters. (a) Sea-facing (VS); (b) land-facing (VL); (c) shoreline position after 120 h for VS (dotted), VL (dashed), and shore parallel (solid) breakwaters.

The results presented in Figure 3.21(c) show the effect of the V-shaped breakwaters in comparison to the typical impact of constructing shore parallel segmented breakwaters. That is, the coastline responds with the formation of salient features behind the breakwaters, generated by the altered breaking patterns and currents in the lee of the structures and the ensuing pattern of sediment deposition.

This modeling has been used to show that the use of V-shaped breakwaters has a similar protective effect to that of shore parallel segmented breakwaters in a tidal setting but may be more cost-effective in terms of the amount of material required in their construction and in terms of the wider range of angles of wave attack for which they are effective.

3.3.4.2 Influence of Choice of Armor Unit Designs

Having decided on the design of a submerged structure, the materials and choice of armor units from which the structure form is assembled needs careful consideration. This may be driven by the economics of the situation or may be informed by physical testing of candidate structures. An example of such tests is proposed by Mendoza, Silva, Enriquez-Ortiz, Díaz-Hernández, and Lara (2011). These tests evaluated the efficiency of a submerged dike design (Figure 3.22), constructed from either rock, concrete cubes, WADs, or geobags (Figure 3.23).

The submerged structure was placed in front of an eroded beach profile (i.e., a profile that had been allowed to freely distort by a storm or similar event). The laboratory work included testing a nonprotected profile that served as a control. A small group of significant wave heights and peak periods were selected to test the structures, that is 10 and 15 cm and 1 and 2 s (i.e., mean and storm wave conditions). All the tests were carried out using JONSWAP spectra with Hasselmann et al., (1973) $\gamma = 3.3$. The dike position was set to have the shallowest freeboard possible while still being considered as submerged, all in front of the same initial profile. A more detailed description of the experimental setup and characteristics can be found in Mendoza et al. (2011).

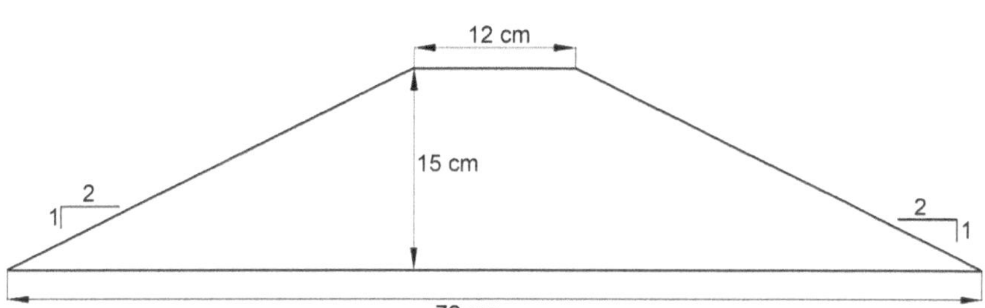

Figure 3.22 General submerged structure cross-section.

Figure 3.23 Submerged structures tested (from left to right cubes, geobags, WADs, and rocks).

For the purpose of comparison, the ratio of the volume of sand removed, Vol_{rem}, above the still water level was compared to the initial volume of sand, Vol_{ini}, above the still water level in the non-protected case and Vol_{rem}/Vol_{ini}, was considered for each type of structure. The removed volume of sand above the still-water level was evaluated as the difference between total volumes of sand before and after the tests.

From the experimental data, the equation obtained to estimate the ratio Vol_{rem}/Vol_{ini}, is:

$$\frac{Vol_{rem}}{Vol_{ini}} = a\frac{H}{L} + b \tag{3.16}$$

where L is the local wave length, H the wave height, hs the height of the structure, and h the still water level and the coefficients a and b can be computed from Table 3.5. Figure 3.24 shows the values of Vol_{rem}/Vol_{ini} for the mean wave climate and for storm conditions.

From Figure 3.24, it can be seen that in absence of structures, the beach profile continues to erode. An impermeable structure is efficient only for small wave lengths (i.e., under breaking wave conditions). The structure with highest porosity, the WAD, provided the best performance, proving the relevance of dissipation from interstitial friction. The structure made of rounded rocks, presented reduced friction that results in poor performance; it is thus recommended to use more angular shaped units. The structure constructed of conventional concrete cubes gives better beach protection than a submerged breakwater built of rocks or geobags.

TABLE 3.5 Values of the Arameters *a* and *b* for Different Cases

	No Structure	WADs	Cubes	Rocks	Geobags
a	$-1180.3\frac{hs}{h}+1217.9$	$-3105.7\frac{hs}{h}+3060.8$	$-1021.8\frac{hs}{h}+1050.3$	$-1677.3\frac{hs}{h}+1646.6$	$-1644.6\frac{hs}{h}+1633.1$
b	$27.7\frac{hs}{h}-27.9$	$192.9\frac{hs}{h}-187.3$	$-68.6\frac{hs}{h}+65.1$	$52.6\frac{hs}{h}-54$	$57.1\frac{hs}{h}-65.3$

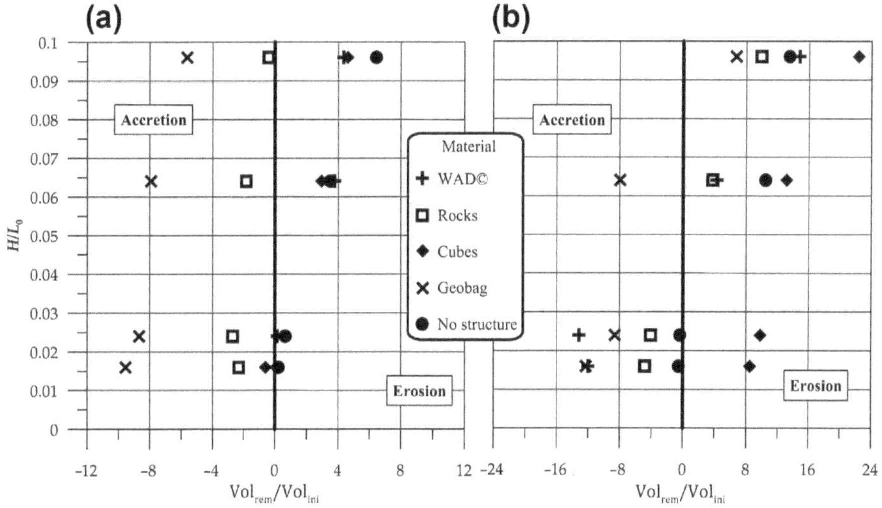

Figure 3.24 Relative volumes removed from the beach profile Vol_{rem}/Vol_{ini}. (a) For mean wave climate and (b) for storm conditions.

3.3.4.3 Cost and Lifetime

When appraising competing coastal protection scheme designs, the lifetime cost of schemes, including construction and maintenance is often the main factor in addition to issues of buildability, and impact on environment and stakeholders. For instance, geobags are an important alternative approach to the construction of hard defense measures and particularly attractive because of their lower cost. In general terms, the cost of a structure made of geobags involves engineering design, materials, conveying or pumping of the filling material into the bags, installation, and maintenance. As a reference, the total cost for 100 m of Geocylinder of 4.5 m width and 3.5 m height is around €100,000 to €150,000. Depending on the properties of the geotextile and the environmental conditions, geobags have a lifetime of between 3 and 5 years without maintenance, which could be extended to 30–50 years, with the replacement of damaged sections. It should be noted, however, that geobags are more vulnerable to vandalism and accidental damage than "harder" solutions.

The cost of artificial reefs involves engineering design, manufacturing, installation, monitoring, and payment of patent rights. As an example, a submerged reef 100 m long, formed by two rows of WADs of 2 m height and 1 m width, with each WAD weighing 3.4 tons would cost about €100,000. Other types of elements, such as the Reef Ball have similar costs to the WAD elements. The lifetime of artificial reefs depends greatly on its performance in the first few years, but it may be up to 50 years, depending on the quality of the concrete (Metha, 2001). With time, as they accumulate biomass, they are more stable, given that they gain weight and settle better.

3.3.5 Final Remarks

Submerged structures and LCS are increasingly used to protect coastlines around the world given their aesthetic value and cost. LCS and submerged structures offer a valuable, increasingly popular solution at locations where complete protection from waves is not required and a moderate degree of energy transmission is even desirable. They can be an efficient protection for beach stability when used in combination with nourishment interventions and they are also used to protect harbor entrances and reduce siltation in entrance channels. The structural design of LCS and other, more innovative types of breakwaters should take various factors into consideration. The structure freeboard must be selected according to the amount of protection required and the tidal range. Then, the fact that the smallest wave heights have the highest transmission coefficients must be considered, remembering that higher wave heights have smaller transmission coefficients. Under extreme meteorological conditions, it is possible that the effect of storm surge affects the functioning of the breakwater. These structures are usually considered as lying parallel to the coast, though the distance from the shore is relative to the currents generated by the waves. The material chosen for the construction of the breakwater also influences its degree of success. Cost factors and visual impact as well as environmental issues must all be weighed in selecting an appropriate type of structure at any given location.

3.4 Overtopping Resistance of Grass-covered Landward Slopes of Dikes

3.4.1 Introduction

Both breakwaters and dikes are subject to climate change as water levels rise and wave climate may worsen. This section only handles measures on grass covered crests and landward slopes of dikes. The protection of the outer or seaward slope is not dealt with in this section.

Given the anticipated sea-level rise, thousands of kilometers of coastal dikes in Europe will be exposed to waves with heights exceeding the original design heights This is in particular true for all structures built in shallow waters, where the limited water depth imposes a maximum wave height because of wave breaking. In addition, sea-level rise results in higher waves, which will lead to larger overtopping volumes. Sea-level rise will therefore provoke problems in terms of function, mechanical stability and safety of goods and persons.

To be able to comply with the safety levels under changing circumstances resulting from climate change, significant stretches of coastal dikes have to be reinforced. Traditionally, reinforcement usually results in higher and wider dikes.

Given the increasing economic and social pressure on the coastal stretch, space is too limited for simply allowing for such traditional reinforcement. In response, the concept of overtopping resistant dikes (i.e., resistant against the impact of wave overtopping) is being developed. Instead of allowing no or a very limited volume of wave overtopping under design conditions, this innovative approach allows for significant larger (but still limited) amount of water overtopping a dike, without causing damage to the dike and its subsequent failure.

Before measures to improve the overtopping resistance are taken it is necessary to know the actual overtopping resistance of existing structures. Most dikes and levees have a grass cover on the crest and landward slope. Until a few years ago, there was hardly any knowledge on the strength of such a grass cover in relation to overtopping waves. To gain this knowledge, a new device was invented, i.e., the Wave Overtopping Simulator (WOS) (Van der Meer, Bernardini, Snijders, & Regeling, 2006), with which the actual resistance of a grass cover can be measured in situ (Figure 3.25). Results of the tests with the WOS will be described in this section (see also Van der Meer et al., 2006; Van der Meer, Bernardini, Steendam, Akkerman, & Hoffmans, 2007; Van der Meer, Steendam, de Raat, & Bernardini, 2008; Van der Meer et al., 2009; Van der Meer, Hardeman, Steendam, Schttrumpf, & Verheij, 2010; Van der Meer, Thornton, & Hughes, 2011; Steendam et al., 2008; Steendam, van der Meer, Hardeman, & van Hoven, 2010; Steendam, Peeters, van der Meer, Van Doorslaer, & Trouw, 2011; 2012; Akkerman et al., 2007).

Figure 3.25 Test with wave overtopping simulator.

3.4.2 Overtopping Resistance

3.4.2.1 Existing Structures

A great deal of artificial flood defenses along coasts and rivers are conventional dikes and levees. The primary function of these dikes and levees is to keep water out of a potential flood area. These structures are designed in such a way that they are geotechnical stable under both normal and extreme conditions.

A typical geometry of dikes and levees can be described by an outer or seaward slope, a crest, and an inner or landward slope (Figure 3.27). The slope angles as well as the width of the crest may vary. In the past, dikes were mainly constructed from clay because of its high erosion resistance and its very small permeability. Nowadays, because of limited availability of clay, the main body of dikes is constructed of sand and covered by a substantial thick layer of clay. To further enhance the surface resistance to erosion, the upper layer is usually covered by grass (see Figure 3.26).

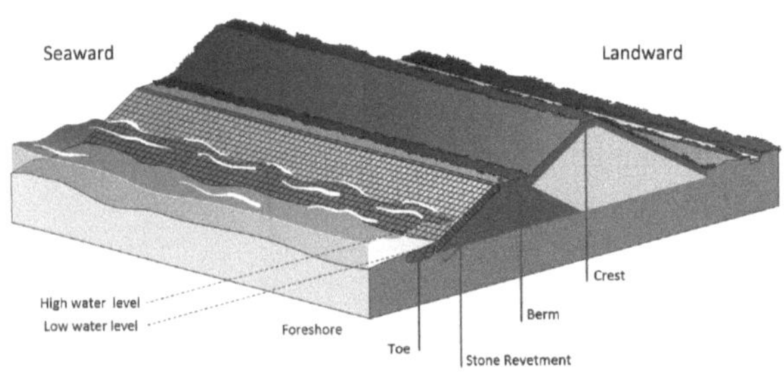

Figure 3.26 Typical cross-section of a dike.

Figure 3.27 View of a typical Dutch dike.

The function of the landward slopes is the stability of the dike structure and the guidance of overtopping waves or water flowing over the crest to the inlands. The landward slopes must be able to resist loads induced by design conditions. The slope angles as well as the width of the crest vary.

Sea dikes have a typical landward slope 1:3. River dikes are mostly steeper, up to 1:2. The height of the dikes also varies a lot, depending on the local hydraulic conditions and on the norm frequency of water level for which the dike is designed. This varies from 1/10,000 per year for sea defenses to 1:1250 years for dikes along rivers.

The landward slope of dikes proved to be a weak spot during the flooding of 1953 in the Netherlands, when more than 1800 people drowned. Back then, the dikes were much lower than today. This event led to the Delta Law stating that the dikes should be heightened. It was also recommended to decrease the landward slopes (from typically 1:2) to 1:3 or even milder. If the slopes were milder, the erosive impact of overtopping waves was expected to be less, and therefore now the dikes are heightened and the landward slopes are milder.

Experience elsewhere shows the failure of the landward slopes from overtopping waves. For example, New Orleans, during Hurricane Katrina, experienced failures of the landward slopes before the water level reached the crest of the levees. Damages from Denmark and Germany are also known and documented.

3.4.2.2 Safety Assessment and Design for Dikes

The general design practice over the past 50 years and more is that the required crest height is determined by almost no or little wave overtopping. First this was taken as the 2% run-up level, which was later translated to tolerable average overtopping discharges around 0.1–1 l/s per m.

There are reasons to expect that extra strength or safety in the mechanism of erosion by wave overtopping is present. But it does not mean that the design philosophy should be changed, because infiltration and sliding may give failure for fairly low wave overtopping and it is also required to have a safety margin in a proper design.

The design practice of allowing hardly any overtopping under very extreme storms (return periods up to 10,000 years) has led to strong and high dikes. There is pressure to become even more safe for flooding and the assumed faster sea-level rise may give more severe storm conditions. Continuation of the design policy will mean that almost all dikes have to be raised and improved again in the next 50 years. The safety assessment procedure in the Netherlands also considers a tolerable overtopping discharge of 1 l/s per m (in some situations more).

For the assessment of erosion of inner slopes by wave overtopping, there might be a good reason to decide not to improve the height of the dike, if overtopping discharges are found to be a little larger than 1 l/s per m. But what would be the effect of a larger tolerable overtopping discharge on the crest height? The answer is of course dependent on the actual situation, such as wave conditions considered and geometry of the dike. By assuming a significant wave height of 2 m, wave steepness of 0.04 and an outer slope of 1:4, the result is given in Figure 3.28.

A tolerable overtopping discharge of 1 l/s per m leads to a required crest height of 3.84 m. If this discharge is increased to 10 l/s per m, the required crest height reduces by 1.23 m, which is about a 30% reduction! If we want another 30% or 1.23 m reduction (a required crest height close to 1.4 m), this means that the tolerable overtopping discharge becomes 100 l/s per m. That would be totally unacceptable for a grass slope. Because of the shape of the curve, the effect is largest if small allowable overtopping discharges are increased. The effect is relatively smaller for increasing even more. In other words, it seems attractive, for the situation in the Netherlands, to increase the tolerable overtopping for safety assessment to 5–10 l/s per m; it becomes less attractive to increase it further.

3.4.2.3 Research on Actual Resistance Against Overtopping

As mentioned, the effect of overtopping water passing the crest of dikes is not known. To gain this knowledge, a large program was initiated by the Dutch government (Ministry of Transport, Public Works and Water Management) called Strength and Loads Water Defenses. Part of this program is the project for research on the strength of grass-covered landward slopes of dikes under loads induced by

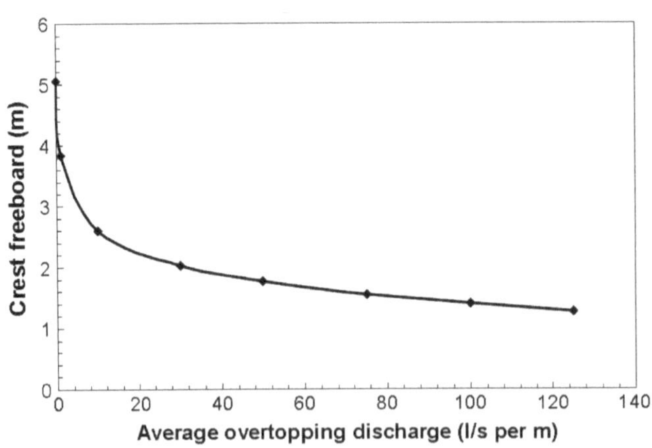

Figure 3.28 Relationship between overtopping discharge and required crest freeboard. The plot refers to the conditions: $H_s = 2$ m; $s_{op} = 0.04$ and $\cot\alpha = 4$, where: H_s = the significant wave height at the seaward toe of the dike, S_{op} = wave steepness = H_s/L_{op}, L_{op} = wave length offshore, and α = angle of the dike seaward slope.

waves overtopping. This research is performed with help of the WOS (see Van der Meer et al., 2007; Steendam et al., 2010).

A certain amount of conservatism was expected in the current Dutch design methods. Among others, the resistance of grass-covered landward slopes against erosion caused by overtopping waves was expected to be much better than assumed in the design methods. First results confirm the expectation of larger strength of these slopes, but how strong are they exactly? To find out how strong a grass-covered slope is, research was performed on the failure mechanisms induced by overtopping waves. With the test by means of the WOS together with the model tests by Schüttrumpf (2001), Van Gent (2002), and Nørgaard, Lykke Andersen, & Burcharth (2013a), now some knowledge exists of the hydraulics on the slope, specifically the thickness of the water tongue and the velocities over the slope. However, the hydraulic loads on the slope itself are still not known well.

The research with the WOS resulted in a new Dutch Technical Report Deltares (2012). A major observation is that a closed grass sod proved to be very resilient against the erosive forces of massive wave overtopping volumes. On the other hand, rough herbal growth and open patches in the sod can make it vulnerable to erosion. From 2007 to 2012, the research with the WOS was also conclusive on the effect of objects and transitions from slope to berm and transitions from a grass cover to other (hard) revetment types. These objects and transitions proved to be weak spots. Therefore management and maintenance of the grass cover on dikes is of great importance. Management and maintenance should focus on developing good grass coverage, preventing and repairing damage, and creating a smooth surface with as few irregularities as possible. Special attention should be given to discontinuities such as transitions to other (hard) cover materials, objects, and transitions from slope to horizontal berms or hinterland. Furthermore, the research mentioned provided in-depth insight on the hydraulics of overtopping waves. The research showed that the tolerable amount of overtopping may be increased from the current (Dutch) design level of 1 l/s per to 5 l/s per m if specific restrictions are met. These specific restrictions include the presence of a well-covered slope, no damages, only few irregularities and absence of objects. Guidance on assessment of strength of grass covered landward slopes of dikes is given in Deltares (2012).

3.4.2.4 Description of Resistance

Current design criteria for landward slopes are based on the CIRIA curves (as described in Technical Note 71) and Hewlett, Boorman, and Bramley (1985). In Figure 3.29, design limit/safety assessment curves are given. Basically, these curves are determined on the basis of overflow experiments (water level higher than crest level) instead of overtopping. Overtopping water has other characteristics than

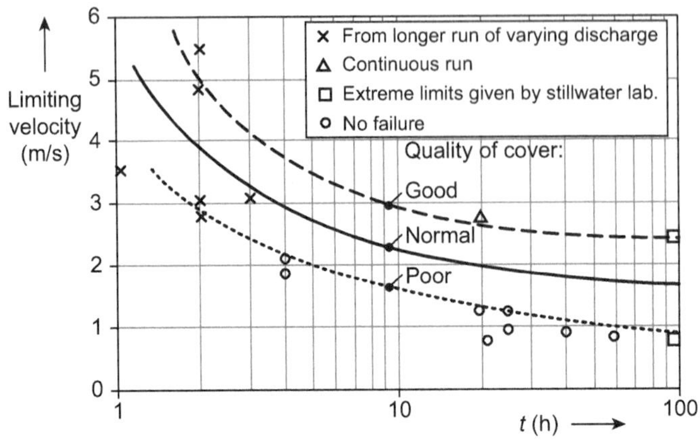

Figure 3.29 Resistance of grass covers according to CIRIA technical note 71.

overflow of water. The velocities of overtopping waves may be much larger and there is a cumulative effect.

Wave run-up and overtopping discharges of various types of coastal structures is well described in the EurOtop Manual (2007), including the distribution of overtopping wave volumes (see also www.overtopping-manual.com. A calculation tool is also given on this Website.). In the next section, the main formulae for estimating overtopping discharge are presented.

For breaking waves ($\gamma_b\, \xi_{m-1,0} < 2$):

$$\frac{q}{\sqrt{gH_{m0}^3}} = \frac{0.067}{\sqrt{\tan\alpha}} \gamma_b\, \xi_{m-1,0} \exp\left(-4.3 \frac{R_c}{H_{m0}} \frac{1}{\xi_{m-1,0}\, \gamma_b\, \gamma_f\, \gamma_\beta}\right) \tag{3.17}$$

with a maximum (for nonbreaking waves generally reached when γ_b $\xi_{m-1,0} > \approx 2$):

$$\frac{q}{\sqrt{gH_{m0}^3}} = 0.20 \exp\left(-2.30 \frac{R_c}{H_{m0}} \frac{1}{\gamma_f\, \gamma_\beta}\right) \tag{3.18}$$

With

q = mean overtopping discharge (m^3/s per m)
g = gravitational acceleration (m/s^2)
γ_b = berm reduction coefficient (–)
γ_f = roughness reduction coefficient (–)
γ_β = angular wave attack reduction coefficient (–)

$$\gamma_\beta = 1 - 0.0033|\beta| \quad for: \quad 0° \leq \beta \leq 80°$$
$$\gamma_\beta = 0.736\,|\beta| \quad for: \quad \beta > 80°$$

R_c = crest freeboard (m)
H_{m0} = spectral significant wave height (m)
α = slope (°)
$\xi_{m-1,0}$ = surf parameter calculated with the H_{m0} and $T_{m-1,0}$: energetic period (s):

$$\xi_{m-1,0} = \frac{\tan\alpha}{\sqrt{\frac{H_{m0}}{L_{m-1,0}}}} \quad \text{with } L_{m-1,0} = \frac{g}{2\pi} T^2_{m-1,0}$$

$L_{m-1,0}$: energetic wave length (m)

The overtopping discharge, q, is the total volume of overtopped water (per unit length) in a certain duration, and is divided by this duration. There will be a certain number of overtopping waves that produce a distribution of overtopping wave volumes. The distribution can be characterized by many small overtopping waves and a few significant larger ones (see also the EurOtop Manual, 2007).

The distribution can be described by:

$$P_V = P\left(\underline{V} \leq V\right) = 1 - \exp\left[-\left(\frac{V}{a}\right)\right]^b \quad (3.19)$$

$$a = 0.84 T_m \frac{q}{P_{OV}} = (0.84\, T_m\, q \cdot N_w)/N_{OW} = 0.84\, q \cdot t / N_{OW} \;\; ; b = 0.75 \quad (3.20)$$

with: P_V = probability of the overtopping volume V being smaller than V, V = overtopping wave volume (m3/m), T_m = mean wave period (s), q = mean overtopping discharge (m^3/s per m width), N_w = number of incident waves, N_{ow} = number of overtopping waves, and t = duration of test or storm (s).

In reality, the overtopping wave volumes occur randomly. The shape factor b was considered for rarely overtopped structures to constant and equal to 0.75. Over the past two years, the theory has been adjusted and the factor b has been made dependent on the relative crest freeboard by Hughes, Thornton, van der Meer, and Scholl (2012):

$$b = \left[\exp\left(-0.6 \frac{R_c}{H_{m0}}\right)\right]^{1.8} + 0.64 \quad (3.21)$$

For values $R_c/H_{m0} > 1.5$, the value of b is indeed approximately 0.75. However, with smaller crest freeboards, the b-value may increase significantly, leading to a gentler distribution of overtopping volumes. Equation (3.19) was developed in case of smooth impermeable dikes.

If b varies, the scale factor a will vary too. In Eqn (3.20), b was assumed to be 0.75. However, to predict the value of a, use can be made of:

$$a = \left[\left(\frac{1}{\Gamma\left(1+\frac{1}{b}\right)} + 1.6\,N_{\mathrm{ow}}^{-0.75} - 0.009\right)\right]\left(\frac{qT_{\mathrm{m}}}{P_{\mathrm{ov}}}\right) \tag{3.22}$$

from Hughes et al. (2012).

Commonly, the mean discharge per meter of width, q ($m^3/s/m$), is used as criterion for wave overtopping. However, in reality, there is no constant discharge over the crest of a dike during overtopping. In designing grass-covered dikes in the Netherlands, the limit is set to a mean overtopping discharge of 0.1 or 1 l/s/m depending on the wave climate. In special circumstances, even 10 l/s/m is allowed if it can be proven that this amount can be coped with by the slope of the dike and can be dealt with in the hinterland. In addition, the accessibility to the dike is hindered with high overtopping rates. In EurOtop (2007) it is shown what impact the amount of overtopping has on, for example, pedestrians and cars.

The mean overtopping discharge is not directly related to the expected damage caused by overtopping. Therefore, from the tests with the overtopping simulator a new method to estimate time of failure of a specific dike under attack of wave overtopping is developed (Van der Meer et al., 2010). This new method is based on hydraulic overload caused by overtopping. It was found that small overtopping volumes do not contribute to damage, whereas large waves do. The larger the volume (correlated to the front velocity), the larger the impact. The cumulative effect of the impacts generates damage. Only if a critical velocity is exceeded on the specific dike section, it contributes to the generation or development of damage. The general formula is given by:

$$\sum_{i=0}^{n}\left(\alpha_{\mathrm{M}}u_{\mathrm{i}}^{2} - u_{\mathrm{c}}^{2}\right) = \mathrm{D} \tag{3.23}$$

where:

D = damage number (m^2/s^2)

U_{i} = maximum front velocity in overtopping wave (m/s)

α_{M} = amplification factor

u_{c} = critical (depth averaged) velocity representing the strength of grass on clay (m/s)

Σ = summation over all overtopping waves with $u > u_{\mathrm{c}}$

Note that the actual velocity can be corrected with the amplification factor α_{M}, for example, to include the effect caused by an external object or by the transition zone between inner slope and too.

To keep the crest and the landward slope from becoming instable during overtopping, the surface should be sufficiently erosion resistant. At overtopping tests performed with the WOS, mean overtopping discharges were simulated up to 125 l/s per m with maximum overtopping volumes of 5500 l/m. Corresponding velocities found on a 1:3 slope were approximately 10 m/s. In the case of open asphalt concrete and Elastocoast (and in some cases also on grass), no damage was found at these velocities.

3.4.2.5 Design Steps for an Overtopping Resistant Dike

To design an overtopping resistant dike the following (technical), design steps have to be taken:

1. evaluate if overtopping resistant dikes are an option (physically in the hinterland, socially acceptable, effect on economics, etc.). If yes, proceed with the following steps;
2. evaluate (design) wave climate ($H_{s,toe}$, $T_{m-1,0}$, mean water level (MWL));
3. establish the configuration of existing or designed dike (slope angles, crest level, cover layer materials (roughness), berm elevations, berm width, etc.);
4. compute estimated mean overtopping discharge with results of steps 2 and 3 (methods and equations in 3.3.2.4 and EurOtop Manual, 2007); and
5. evaluate discharge to design criterion. The design criterion is a conservative indication of what resistance against erosive impact of water overtopping the dike. In the Netherlands, dikes are designed on mean overtopping discharge less than 1 l/s per m. But this does not mean the dike will breach if the mean overtopping discharge is larger. In a safety assessment, the impact of larger mean discharges can be evaluated, for example, with the curve mentioned in Section 3.4.2.4. If the calculated estimation of mean overtopping discharge is larger than the allowable mean discharge according to the curve, the slope has to be reinforced. As mentioned, instead of evaluating the impact of overtopping water with the curve also a more sophisticated method can be used by means of evaluating the cumulating the individual overtopping wave flow velocities on the crest and landward slope to the criteria of damage as stated in Van der Meer et al. (2010). The velocities themselves can be evaluated by means of the formula given in Section 3.4.2.7. If the cumulative hydraulic load exceeds the criterion a reinforcement of the slope is in order.

As an example, a conceptual design is given which is applied at several sites in the Netherlands. An open stone asphalt layer is put on a geotextile on the crest and landward slope of a dike.

In the case of an open stone asphalt, the thickness of the layer is commonly given by a minimum thickness that can be placed in situ. Sometimes, for environmental or aesthetic reasons, the asphalt is covered by a soil layer covered with grass. From an erosion resistance point of view, this additional grass cover gives no additional strength because these two layers do not sufficiently connect. It may even be negative because the roots reaching the asphalt may cause disintegration. In such cases, an additional 0.1 m of asphalt to the layer thickness is advised.

The characteristics of open stone asphalt are:

- estimated lifetime of about 30 years;
- maintenance consists mostly of filling of cracks and surface treatment; and
- typical cross-section of landward slope: see Figure 3.30.

If the asphalt has to be located below the wave run-up zone, impacts of waves should also be taken into account. Specific design criteria are given in Pilarczyk (1998) , whereas a design graph for asphalt under wave attack is provided in STOWA (2010), see Figure 3.31.

3.4.2.6 Costs

Costs of making dikes overtopping resistant differ a lot per type of measure, project, and country. As an example, costs in a project in the Netherlands of applying an asphalt layer at the landward slope as an overtopping resistant dike were calculated to be 75 €/m^2. This includes excavation of the top layer, supplying and applying the geotextile, and asphalt. Also engineering and tale costs are included. No value-added tax or other taxes are included. Below is an example of the calculation of costs is given in detail (Table 3.6). The example cost estimation is from an early phase (preliminary design) in the project.

The example is given for open stone asphalt. Use of other materials is also possible. For example, reinforcement of grass covered with a geogrid is widely

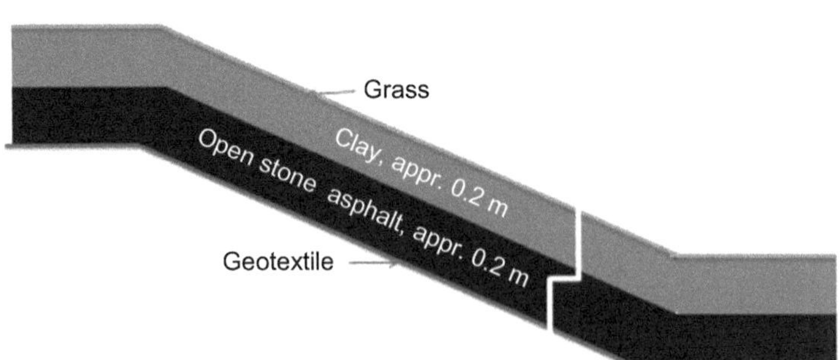

Figure 3.30 Typical cross-section of the landward slope of a dike with open stone asphalt.

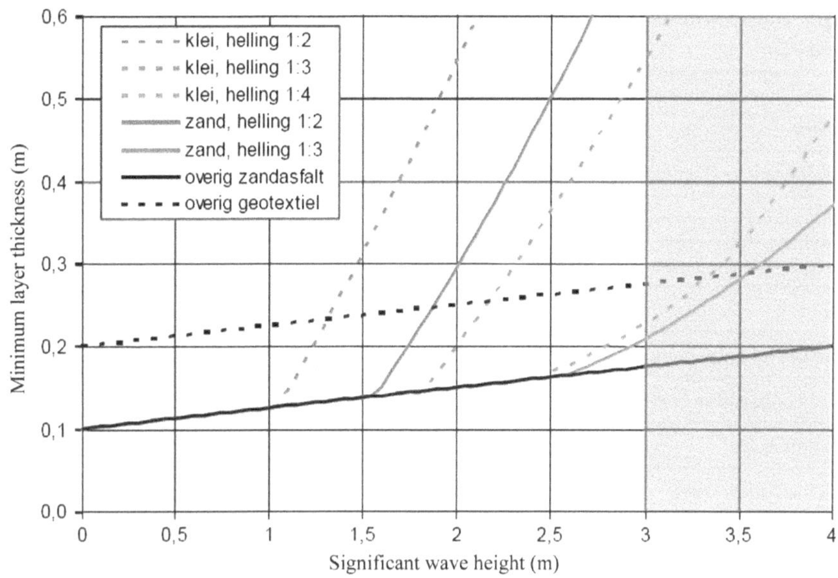

Figure 3.31 Design graph open stone asphalt (STOWA, 2010).

available. Extensive tests have been and are being performed by, for example, Deltares/Infram (Steendam et al., 2008) and Colorado State University (Hughes et al., 2013). Also, other hard (permeable) materials may be suitable for making landward slopes of dikes resistant against overtopping. The costs of the alternatives may differ a lot from the example given previously.

3.4.2.7 Modeling the Characteristics of the Overtopping Flow along Dike Slopes

Equations for flow depth and velocity along the dike slopes are based on physical model investigations, such as Schüttrumpf (2001), Schüttrumpf and Oumeraci (2005), Van Gent (2002), Schüttrumpf and Van Gent (2003) and Nørgaard et al. (2013a). The EurOtop Manual (2007) also presents such equations. By assuming a Rayleigh distribution for the flow velocity and flow depth, the squared velocity and flow depth can be calculated for each overtopping wave volume with a certain probability of exceedance. Such calculations can lead to graphs of flow velocity or flow depth versus overtopping wave volume. It was, however, observed that these curves deviate from each other. For the same overtopping wave volumes, lower flow velocities and flow depths were found if the overtopping discharges increased. The same happened for the flow duration. However, this is physically not possible because a decrease in flow velocity should result in an increase in flow depth or flow

TABLE 3.6 Example Cost Calculation Overtopping Resistance with Open Stone Asphalt

Activity	%	Estimation €/m^2	(Sub)Total €
Excavation of top layer landward slope		2.50	
Supply and apply open stone asphalt		37.50	
Supply and place geotextile		4.00	
Total			44.00
Construction costs to be detailed	15	6.60	
Direct construction costs			50.60
One-time costs	1	0.44	
Implementation costs	10	4.40	
General costs	7	3.08	
Profit and/or costs for calculated risks	4	1.76	
Indirect construction costs	22	9.68	
Planned construction costs			60.28
Object risk construction costs	15	9.04	
Engineering	6	3.62	
Additional costs	3	1.81	
Grand total			**74.75**

duration (mass balance). Therefore, it must be concluded that present knowledge and prediction formulae for flow velocity, flow depth, and flow duration do not yet give consistent answers. Probably the flow depth and velocity must become more dependent on the wave period.

Some tests were performed with the WOS with the purpose to measure solely hydraulic parameters, such as flow depth, velocity, and overtopping duration. These tests were carried out on a sandy river dike (Steendam et al., 2010) by repeating the same overtopping wave volumes three times, which increased in time from 200 l/m to 5′500 l/m (the maximum capacity of the WOS). The measurements carried out along the landward slope with five surf boards to measure flow depths (Van der Meer et al., 2009) and paddle wheels mounted in the surfboards to measure the flow velocity, are well fitted by the following relation:

$$u = 5.0V^{0.34} \tag{3.24}$$

where the flow velocity u is expressed in m/s and the related wave overtopping volume V is in m^3/m. The influence from oblique and short-crested waves on overtopping flow velocities and overtopping flow depths was investigated in model tests (Nørgaard et al. 2013a). Based on these model tests, the following extensions were made to the existing 2D formulae (see Section 3.4.2.4) for the estimation of the influence of obliqueness of waves on the mean overtopping discharge.

$$\text{long}-\text{crested waves}: \gamma_\beta = \begin{cases} 1.0 & for \quad 0^\circ \le \beta \le 10^\circ \\ cos\,(\beta - 10^\circ) & for \quad 10^\circ \le \beta \le 45^\circ \end{cases} \tag{3.25}$$

$$\text{short}-\text{crested waves}: \gamma_\beta = 0.95 - 0.0035\,\beta \tag{3.26}$$

where γ_β is the reduction factor used in the wave overtopping discharge formula and β is the incident wave direction (being 0° perpendicular to the dike).

Moreover, it was found that flow depths and velocities over the crest are Rayleigh distributed when incident wave heights are also Rayleigh distributed and that the flow directions on the dike crest are similar to the incident wave direction.

By using the extended formulae (3.25) and (3.26), it was found that at large relative crest ratios R_c/H_{m0}, especially flow velocities, are significantly reduced by oblique waves compared with perpendicular waves. In addition, the overtopping flow parameters are more reduced in short crested waves compared with long crested waves.

In recent years, a great effort has been made in improving numerical models to study wave–structure interaction problems. Models based on the use of the complete Navier–Stokes equations have shown (i.e., Losada et al., 2008; Lara et al., 2008; Guanche et al., 2009) that they can overcome the inherent limitations presented in Nonlinear Shallow Water and Boussinesq equation models related mainly to wave dispersion and breaking, vertical flow characterization, nonhydrostatic pressure fields, and flows inside porous coastal structures. Navier–Stokes models can provide a refined description of the wave, flow, and pressure fields around the structure. Magnitudes from wave overtopping discharges or pressure field distributions (both instantaneous and averaged) can be obtained from the numerical model predictions.

Among these models, the 2DV IH-2VOF model (Lara et al., 2011) was tested (Raosa, Zanuttigh, Lara, & Hughes, 2012) against experiments on emerging and overwashed dikes (Hughes & Nadal, 2009) and reproduced the wave overtopping volumes on the dike crest well (see Figure 3.32) when forced with the same water levels measured in front of the smooth impermeable structure. The numerical flow depths and velocities over the emerged and zero freeboard dike fit accurately to the theoretical approach by Schuttrumpf and Oumeraci (2005), see Figure 3.33. The IH-2VOF model allows gathering of the velocity profiles over the flow depth in time, which is usually not measured in the laboratory or at prototype scale. Peak velocities along the profile differ in some cases with a factor 2 from the average velocity of the

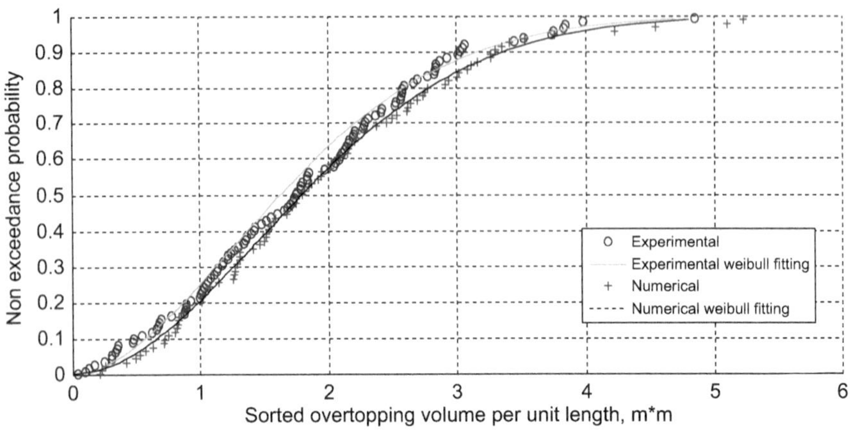

Figure 3.32 Comparison of computed and measured overtopping volumes at the dike crest offshore edge (Raosa et al., 2012).

2% of the highest overtopping waves $u_{2\%}$ at the measurement point (Raosa et al., 2012). This result is particularly relevant for design purposes, considering the dependence of the formation of dike erosion holes on flow velocity derived from previous studies (Van der Meer et al., 2010).

Recently, del Jesus, Lara, and Losada (2012) and Lara et al. (2012) presented a 3D numerical model called IH-3VOF, which solves the 3D VARANS equations with the help of two volume-averaged turbulence models, a k–ε and an SST model. Free

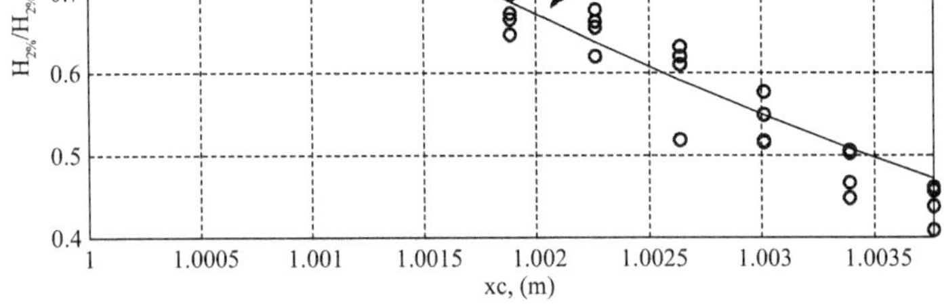

Figure 3.33 Evolution of the flow depth and of the overtopping velocities on emerging dike crests: the points are extracted from numerical simulations and compared with the theoretical curve (Raosa et al., 2012).

surface tracking is carried out with the VOF technique. The model was used to study the influence of wave obliquity on wave overtopping. Different snapshots show waves overtopping along the structure following oblique wave incidence angles are shown in Figure 3.34. The model shows a good performance when comparing wave induced overtopping discharge for oblique waves (Nørgaard et al., 2013a).

3.4.3 Final Remarks

As sea level rises, thousands of kilometers of coastal dikes in Europe will be exposed to waves with heights exceeding the original design level, which will lead to larger overtopping volumes. These larger volumes are not anticipated for in the design and therefore significant stretches of coastal dikes have to be reinforced, usually resulting in higher and wider dikes. Given the increasing economic and social pressure on the coastal stretch, space is too limited for simply allowing for such traditional reinforcement. In response, the concept of overtopping resistant dikes (i.e., resistant against the impact of wave overtopping) is being developed.

The design criterion, determining the necessary height of the dike, is described as the wave overtopping with a mean discharge per meter of width, q ($m^3/s/m$).

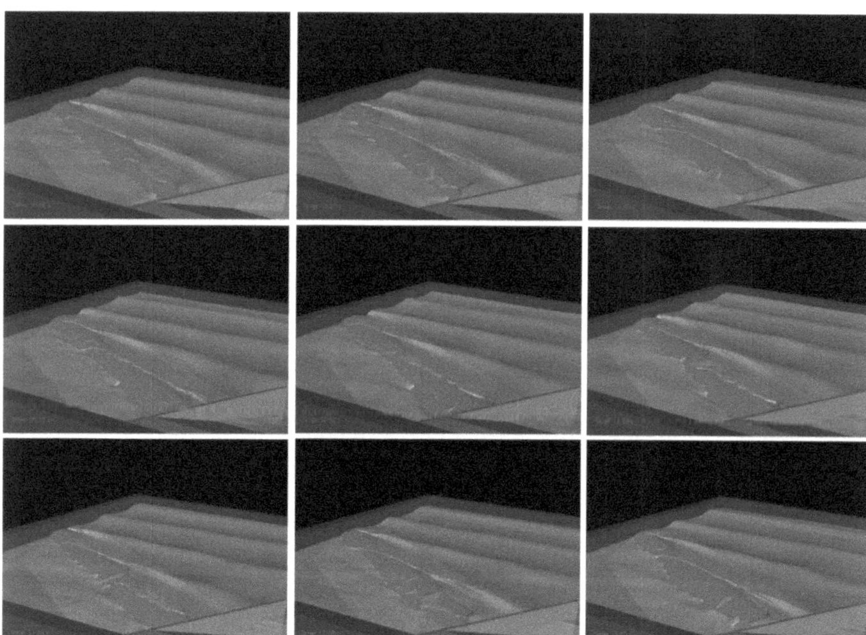

Figure 3.34 Overtopping over the breakwater (running from left to right) obtained by calibrating the IH-3VOF model against laboratory measurements (Nørgaard et al., 2013a).

However, in reality there is no constant discharge over the crest of a dike during storm conditions. In designing grass-covered dikes, mostly the limit is set to a mean overtopping discharge of 0.1 or 1 l/s/m depending on the wave climate. Heightening this criterion may be profitable. An increase in tolerable overtopping discharge by a factor 10 may lead to a 30% reduction in needed crest freeboard. Therefore, if the dike could cope with more water overtopping, a significant reduction of crest level may be achieved, leading to lower investments.

Overtopping water has other characteristics than overflow of water. The velocities of overtopping waves may be much larger and there is a cumulative effect. From tests with the WOS, a new method to assess the resistance against erosion from water overtopping the dike was developed. This new method is based on hydraulic overload caused by overtopping waves. From recent research, it was also found that some extensions can be made to the existing overtopping formulae from the EurOtop Manual. By using the extended formulae, it was found that at large relative crest ratios R_c/H_{m0}, especially flow velocities, are significantly reduced by oblique waves compared with perpendicular waves. In addition, the overtopping flow parameters are more reduced in short-crested waves compared with long-crested waves.

Existing equations for flow depth and velocity along the dike slopes are based on physical model investigations; therefore, it must be concluded that present knowledge and prediction formulae for flow velocity, flow depth, and flow duration do not yet give consistent answers. Probably the flow depth and velocity must become more dependent on wave period. To gain knowledge on this matter, some tests with the WOS were performed to measure solely hydraulic parameters, such as flow depth, velocity, and overtopping duration. From these tests, the following relation was found:

$$u = 5.0V^{0.34}$$

In recent years, a great effort has been made in improving numerical models to study wave–structure interaction problems. Among these models, the 2DV IH-2VOF model was tested against experiments on emerging and overwashed dikes and reproduced the wave overtopping volumes on the dike crest well when forced with the same water levels measured in front of the smooth impermeable structure. The model allows also gathering of velocity profiles over the flow depth in time. The results may be relevant for design purposes considering the dependence of the formation of dike erosion holes on flow velocities.

To prevent slopes from failing as large volumes of water overtop the dike, reinforcements can be made. For example an open stone asphalt layer can be put on a geotextile on the crest and landward slope of a dike but also other hard (permeable) materials may be suitable for making landward slopes of dikes resistant against overtopping. The costs of making dikes overtopping-resistant differ a lot per type of measure, project, and country.

3.5 Upgrade of Conventional Rubble Mound Breakwaters and Revetments

3.5.1 Introduction

As already anticipated in Section 3.4, with the MWL rise, coastal defense structures will be exposed to larger wave heights than the design values, in particular all the structures that are built in shallow waters where the depth imposes the maximal wave height because of bathymetric breaking. If MWL rise is 1 m, the crest of these structures will have to be raised between 2 and 3 m to keep the same overtopping volumes. Moreover, the structures will observe more frequent and more severe damage unless the armor layers are reinforced. Schematically, with the increase of damages, the stakeholder will adopt one of the following scenarios according to the severity of changes: (1) repair the structures as it is, (2) reinforce the structures, (3) demolish and redesign the structures, or (4) accept coastal realignment.

This section is devoted to reinforcement of the structures. Options of reinforcement can compensate MWL rise in terms of wave overtopping, armor stability, and crown wall stability. These options consists in limiting overtopping by modifying for example the crown wall, improving armor stability by adding an armor layer or by using milder armor slope and reducing the incident wave energy by building a detached low-crested breakwater or by sand nourishment.

In this section, hydraulic performance of existing and reinforced structure is assessed and the design steps for reinforcement with climate change are presented.

3.5.2 Existing Structures

Maritime rubble mound breakwaters and revetments (see Figure 3.35) are structures built with sloping seaward front armored with quarry rock or concrete armor units. These type of structures are generally used in shallow water. Rubble mound

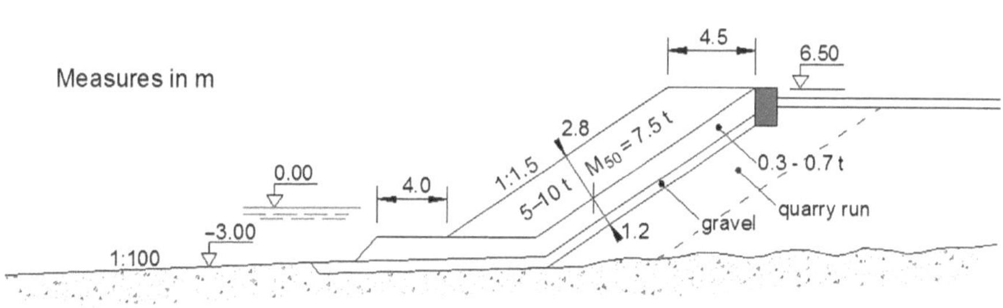

Figure 3.35 Example of a rubble mound revetment (Burcharth et al., in press).

structures are used to reduce coastal erosion and to stabilize the coastline and in harbors to protect access channels and reduce dredging costs.

3.5.3 Increase in Hydraulic Conditions

Climate changes might lead to increased mean water level and increase of extreme wind speeds and thus resulting in larger waves and larger wind generated setup.

To assess the impact of an increased water level and an increase in offshore significant wave height on the response of rubble mound structures, the significant waves at the toe of the structure are needed that, for example, can be assessed by the analytical method of Goda (2000) for a uniform seabed slope or alternatively based on numerical models.

The maximum wave at the toe can also be given by Goda or alternatively by a shallow-water wave-height distribution such as the Battjes and Groenendijk (2000). Figure 3.36 presents the effect of MWL rise in front of the structure for unchanged offshore waves. The MWL rise can be represented as a movement of the structure offshore with a distance equal to the bottom slope multiplied by the water level rise.

Unless the offshore waves change significantly, the structures presently in the shoaling zone will not suffer from a significant change of the intensity in wave attacks. The structures presently in the breaking zone (i.e., in shallow water) will be impacted by larger wave heights as

$$\frac{\Delta H_{m0}}{H_{m0}} \approx \frac{\Delta h}{h}.$$

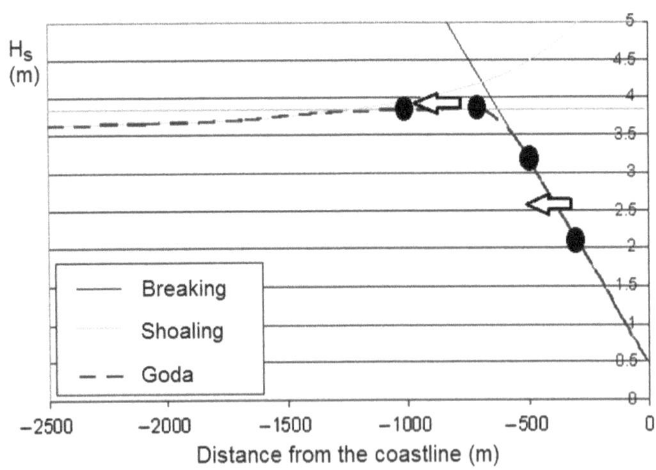

Figure 3.36 Example of wave propagation assessed by the analytical method of Goda (2000).

3.5.4 Reinforcement Options

To face climate change, several concepts of reinforcement of structures were proposed and analyzed (Burcharth et al., in press) through physical model tests in the THESEUS project, see Figure 3.37. The physical model tests aimed at verifying in particular that these concepts were able to compensate MWL rise in terms of wave overtopping, armor stability, and crown wall stability.

3.5.5 Analysis of Hydraulic Performance of Existing and Reinforced Structure

In general, evaluation of hydraulic performance of rubble mound structures is evaluated based on physical model tests or empirical formulae fitted to such test

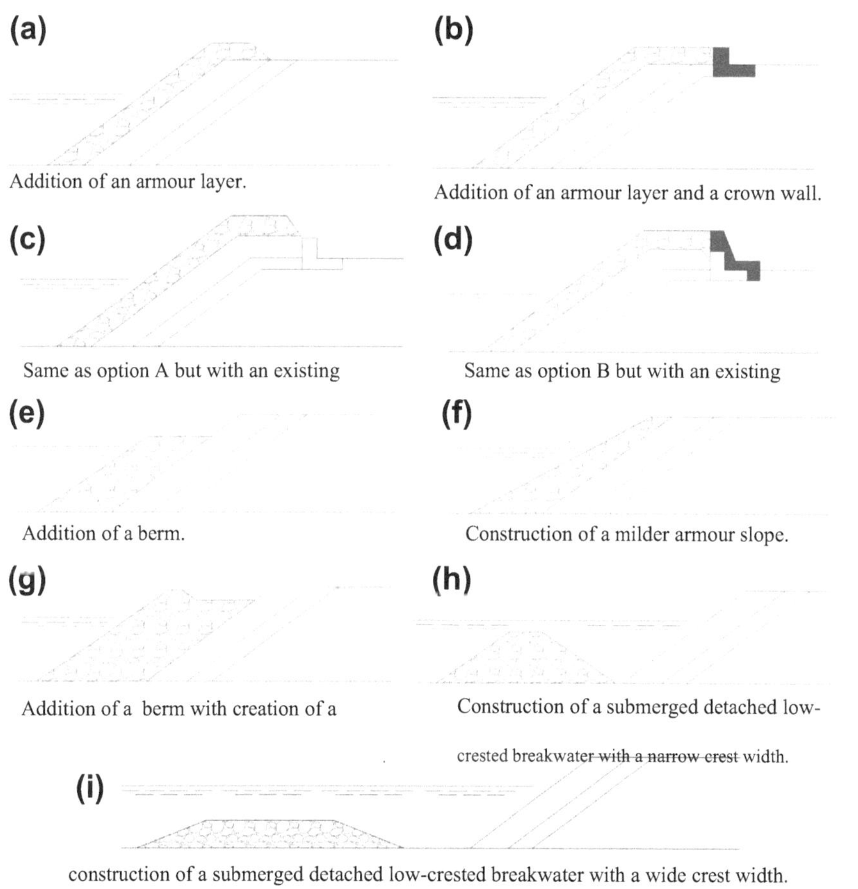

Figure 3.37 Reinforcement options. *After Burcharth et al. in press.*

results. Numerical models might also be used to extend the database if the model is properly calibrated and validated on model test data.

Climate changes are most critical for structures located in the breaking zone. However, most empirical formulae are based on waves that are not breaking on the foreshore ($H_{m0}/h < 0.2$). The consequence is typically that the formulae will give quite conservative results when they were applied for structures in the breaking zone. If the existing structure is designed based on deep water empirical formulae, the design thus include some extra safety.

Empirical deep water formulae have been recently updated to correctly assess the structural response in shallow water. An example is the formulae by Nørgaard, Lykke Andersen, and Burcharth (2013b) for loads on crown walls. This formula is based on Pedersen (1996) but updated to correctly take into account the effect of non-Rayleigh–distributed wave heights in shallow water. Another example is the formula by Nørgaard et al. (2014) for distribution of the individual wave overtopping volumes to cover shallow water effects.

Nørgaard (2013) evaluated the performance of an existing structure exposed to climate changes and found that crown wall stability is extremely sensitive to climate changes. Moreover, he found the increase in mean and maximum overtopping volumes also to be very important, whereas armor stability typically is less critical although more frequent repairs might be needed. However, for structures with wet rear side the rear side armor stability might though be very critical because of the increased overtopping.

Burcharth et al., in press argue that an important issue for erodible seabeds might be the changed morphology from climate changes, which can cause a steepening of the profile. He analyzed upgraded solutions for no steepening of a 1:100 sea bed profile and for a hypothetical steepening to 1:30. The steeper profile leads to significantly larger design waves at the toe, and the upgrading costs were also significantly larger.

Different design formulae related to the options presented in Figure 3.37 are listed in Table 3.7.

3.5.6 Design Steps

Ten design steps are proposed to select the preferred design among the alternatives considering total costs in service lifetime.

1. Define service lifetime for the structure (N years).
2. Define geometrical and environmental restrictions/limitations for the upgraded structure, for example, increase in crest level not allowed, concrete armor units not allowed, etc.
3. Define the new scenario design sea states in terms of combined long-term statistics of offshore wave height, wave length, water depth, and angle of incidence.

TABLE 3.7 List of Options and Main Design Formulae

	Mean Overtopping Discharge	Armour Stability Crown Wall Stability
Initial design	TAW (2002) EurOtop (2007) CLASH neural network	Hudson (1953) Van der Meer (1988) Van Gent (2003) Nørgaard et al. (2013)
Plain or parapet crown wall	Reduction factor γ_v for crown wall Tuan, Cat, and Trung (2009) CLASH neural network	Armour stability is reduced by the raise of the crown wall Crown wall stability is reduced by the raise of the crown wall
Berm with creation of a front reservoir (Figure 3.37(g))	*Laboratory tests have to be done*	
Berm (Figure 3.37(e))	Reduction factor γ_b for berm Lykke Andersen (2006) CLASH neural network	If fully or partly reshapable: Lykke Andersen and Burcharth (2010) If hardly reshaping: Lykke Andersen, Van der Meer, Burcharth, and Sigurdarson (2012)
Milder armor slope (Figure 3.37(f))	Existing tools on the safe side because of higher dissipation in the thick armor layer. *Laboratory tests have to be done*	Hudson (1953), van der Meer (1988), van Gent (2003)
Addition of an armor layer (Figure 3.37(a) and (c))	Existing tools on the safe side because of higher dissipation in the thick armor layer. *Laboratory tests have to be done*	Stability is increased by the additional armor layer *Laboratory tests have to be done*
Simultaneous addition of an armor layer and a crown wall (Figure 3.37(b) and (d))	Overtopping is reduced by additional dissipation and by parapet *Laboratory tests have to be done*	Stability is reduced by the presence of parapet and increased by the additional armor layer *Laboratory tests have to be done*
Submerged detached low-crested breakwater (Figure 3.37(h) and (i))	Modification of incident waves Pina and Valdez (1990), Burcharth et al. (2007)	Modification of incident waves Pina and Valdez (1990), Burcharth et al. (2007)

a. Define the present long-term statistics.

Long-term statistics represent statistics of data at high tide (offshore wave heights and sea levels). Long-term statistics enable the analysis of exceedance probability of hazards: offshore wave height at high tide (more precisely the maximal wave height between two successive high tides) and wind setup at high tide (the difference between the maximal observed level and the predicted level around high tide). A joint probability of offshore wave heights and sea levels can also be obtained.

These long-term statistics are assessed to give the exceedance probability of impacts (overtopping discharge and structure stability) in relation with following steps 5 and 6.

Nearshore wave heights at high tide are used directly if data are available.

Table with data (offshore wave heights and sea levels) is then built:

- from measurements or from hindcasts
- from extrapolated joint probability of wave heights and sea levels (Hawkes, Gouldby, & Tawn, 2002)
- variability of wave period or wave steepness can be included
- variability of direction can be included

a. Define climate change (that can be reduced to MWL rise Δh). In the latter case, the wave height at the toe of breakwater is also modified because of MWL rise. Generally speaking, climate change include MWL rise but also modification of storm intensity and frequency. Modification of storms leads to changes of wind set up, wave height, wave period, and wave direction.

4. From offshore wave height H_0' determine H_s or H_{m0} in front of the structure by using for example Goda (2000) method or numerical models.

5. Define performance criteria for the upgraded structure.

a. Define for each type of impact (overtopping, armor damage, etc.), the acceptable occurrence probability p_f (design probability) over the structure lifetime N (in years), and calculate the corresponding return periods M (in years) by the formula (for example, $N = 100$ and $M = 500$):

$$p_{\mathrm{f}} = 1 - \left(1 - \frac{1}{M}\right)^N \tag{3.27}$$

A limit state is a condition of a structure beyond which it no longer fulfills the relevant design criteria. The ISO Standard 2394 (1998) on "Reliability of Structures" prescribes a format for safety implementation in which safety classification is based on the importance of the structure and the consequences of malfunction, and for design both a "Serviceability Limit State" (SLS) and an "Ultimate Limit State" (ULS) must at minimum be considered with damage criteria assigned to these limit states. Repairable limit state might be applied as well. Note that the return period M will be different for impacts corresponding to SLS and ULS because most probably a smaller p_{f} is selected for ULS than for SLS.

b. Impact type 1: Overtopping discharges and related exceedance probabilities for several design scenarios representing SLS and ULS.

c. Impact type 2: Structure stability. Types of damages, damage levels and related exceedance probabilities for SLS and ULS:
   - Toe, front, crest, and rear side (if relevant) armor unit displacements
   - Crown wall breakage, sliding, geotechnical slip failures
   - Geotechnical overall stability, overall, and differential settlements with preliminary soil investigation (De Rouck, Van Doorslaer, Goemaere, & Verhaeghe, 2010)

6. Explore weaknesses in the performance of existing structures with respect to overtopping, structural, and geotechnical stability given the actual (today's) design sea states and the new scenario design sea states.
7. Design upgrading of structure by use of formulae of Table 3.7 in particular, Neural networks like CLASH for wave overtopping (crest level assessment of coastal structures by full-scale monitoring, neural network prediction, and hazard analysis on permissible wave overtopping) and computational models. Several conceptional alternatives must be designed. All needed formulae have to be taken from the literature and presented in particular in Table 3.7.

   The total cost of the structure is estimated as a sum of the costs of construction and the expected costs of repairs and failure.
8. For each alternative, calculate the construction cost of the upgrading (i.e., calculate volumes and define unit costs for the various materials). (Note that unit costs vary tremendously with volume and site.)

   The direct cost of construction is written as a function of the design variables (i.e., the volumes of the different materials and their cost per volume).
9. For each alternative, define repair policy and subsequent estimation (by simulation) of the service lifetime repair costs.

   The expected costs of failure are calculated for the total lifetime of the breakwater (e.g., 100 years) and represent economic damage in case of failure as well as costs of repair. The formula to extract the costs of failure is:

$$I_{\text{failure}} = \sum_{i=1}^{N} \left(\frac{C_{\text{ULS}} P_{\text{f,ULS}}(z)}{(1+r)^i} + \frac{C_{\text{SLS}} P_{\text{f,SLS}}(z)}{(1+r)^i} \right) \tag{3.28}$$

where C_{ULS} and C_{SLS} = damage costs and $P_{\text{f,ULS}}$ and $P_{\text{f,SLS}}$ = probabilities of failure in case of ULS and SLS, respectively; r = the interest rate; and N = the reference period for ULS and SLS failure (e.g., lifetime). The probabilities of failure, $P_{\text{f,ULS}}$, $P_{\text{f,SLS}}$, are expressed in an annual scale. Costs of maintenance (repairs) can also be added to the construction and expected failure costs.

Four levels of damage of armor layers might be considered: initial damage; intermediate damage; important damage; and failure. Each of these levels

TABLE 3.8 Example of Repair Strategy for Rock Armored Breakwater

Limit State Damage Levels	Relative Number of Displaced Units	Repair Policy
Initial	0–5%	No repair
SLS serviceability (Minor damage, only to armor)	5%	Repair of armor layer
Repairable limit state repairable (Major damage to armor and filter layer)	15%	Repair of armor and filter layer
ULS ultimate (Failure)	30%	Repair of armor, filter layer(s) and core

corresponds to a stability coefficient which can be related to a percentage of moved armor units. A repair strategy might be as shown in Table 3.8:

10. Select the preferred design among the alternatives considering total costs in service lifetime.

3.5.7 Final Remarks

To compensate for MWL rise, the reinforcement of the structures is often the most appropriate strategy. The solution strategy will clearly depend on if increase in crest level is allowed or not.

Various reinforcement options were recently tested and some empirical formulae were updated (Burcharth et al., in press; Nørgaard et al., 2014) to correctly assess the structural response in shallow water when originally the formula was only based on deep water data. Particular attention must be paid to the crown wall stability, the wave overtopping volumes, and also rear side armor stability as a consequence of wave overtopping.

The preferred design among all the alternatives is the design that minimizes the total costs in service lifetime.

For erodible seabeds, the variation of the bottom from climate change in front of the structures is an important issue. A steeper profile will lead also to larger upgrading costs.

3.6 Management of Sediment Resources

3.6.1 Introduction

Beach nourishment comprises the placement of sediment in the shoreface to advance the shoreline and it is nowadays the preferred technique to face naturally or anthropogenically induced beach erosion.

Feasibility of the artificial shore nourishment strongly depends on availability of appropriate sediment, as well as possibility of sediment delivery and placement. Borrow areas must satisfy a few criteria, comprising grain size distributions, accessibility (convenient water depth, small distance from a nourished shore, navigable ease), and legal requirements.

The artificial shore nourishment ventures are almost always accompanied (preceded) by dredging operations. The latter works can affect the environment equally or more significantly than the shore nourishment itself. Reliable prediction of dynamics of sediment plumes induced by dredging operations is crucial in the assessment of relevant environmental impacts.

Long-term shore evolution of the artificial nourishment, with focus on the lifetime and the optimization of the nourished volumes is discussed in Section 3.5.1.1. A simplified methodology is proposed in Section 3.5.1.2 to determine resilience of coastal profiles by identifying the key profile parameters contributing to the beach hydraulic vulnerability. The optimized placement of sediments on the submerged versus emerged beach is discussed in Section 3.5.1.3. This aspect is enriched in Section 3.5.1.4 by formulation of the dimensionless bed shear stresses to optimize the location of near-shore borrow areas. Section 3.5.1.5 addresses the issue of protected and reinforced nourishment intervention, for instance through sand accumulation fences and groins. Information on environmental impacts of dredging and nourishing operations is given in Section 3.5.1.6.

The management of sediment stocks (dredging and nourishment) in transitional waters, such as river mouths, estuaries, and tidal flats, requires a specific approach and dedicated tools. The assessment of the estuarine morphological changes is provided in Section 3.6.2.1. The influence of dredging activities on tidal processes is described in Section 3.6.2.2. An innovative physically based bottom roughness predictor is proposed in Section 3.6.2.3 for fully developed turbulent flow down to the bottom into the transient and viscous boundary layers. It accounts both for the physical roughness (skin friction and form roughness) as well as the energy dissipation due to high suspended loads, including bed load transport and sheet flow conditions (drag enhancement) or high-concentrated mud flow (drag reduction). Another issue concerning a tidal estuary is related to the flood threat in an upstream city. The feasibility of mitigating this hazard by construction of artificial sandbanks in the river mouth is discussed in Section 3.6.2.4.

3.6.2 Nourishment

3.6.2.1 Evolution of the Nourished Shore

The prediction of the shoreline evolution in different time scales is very important for making decisions on shore nourishment interventions and requires both the assessment of the sedimentary budget (Leont'yev, 2008) and the prediction of sea-level changes (Hunter, 2010; CSIRO, 2010).

Short-term seabed deformations are usually predicted by means of process-based models (Leont'yev, 2003; Van Rijn, Tonnon, & Walstra, 2011). For the description of the long-term evolution (on "engineering" and "geological" time scales), only the aggregated effect of primary mechanisms is considered in the so called behavior-oriented models (Cowell, Roy, & Jones, 1995; Stive & de Vriend, 1995).

The SPELT model (Leont'yev, 2012) is a synthesis of both of the previously mentioned approaches. It takes into account the changes in coastal profile from a sediment budget imbalance and so it provides more realistic description of the long-term coastal evolution.

The SPELT model is based on the mass conservation equation that is expressed in the following form

$$\frac{\partial h}{\partial t} = Er - Ac + w, \tag{3.29}$$

where h is the water depth, t is the time, w is the rate of sea level change, and Er and Ac are, respectively, the erosion and accumulation rates depending on water depth, sea bed slope ($\partial h/\partial x$), and distance from the coast (x):

$$Er = \frac{1}{\alpha_{\mathrm{e}}}\frac{Q_E}{h_*}\left(1 - \frac{h}{h_*}\right)^{\mathrm{m}}\frac{\partial h}{\partial x}, \quad Ac = \frac{1}{\alpha_{\mathrm{a}}}\frac{Q_{\mathrm{E}} + B}{l_*}\left(1 - \frac{x - x_0}{l_*}\right)^{\mathrm{n}}. \tag{3.30}$$

Here, Q_{E} is the potential annual volume of erosion depending on annual wave energy flux to the coast F_Σ:

$$Q_{\mathrm{E}} = c_{\mathrm{e}}\alpha_{\mathrm{e}}\frac{F_\Sigma}{\rho g l_*}, \quad \alpha_{\mathrm{e}} = \frac{1}{m+1} + \beta_0\frac{l_{\mathrm{b}}}{h_*}, \quad \alpha_{\mathrm{a}} = \frac{1}{n+1} + \frac{l_{\mathrm{b}}}{l_*}, \tag{3.31}$$

where c_e is a calibration coefficient of order 0.1, h^* is the closure depth, l^* is the cross-shore distance from the shoreline position x_0 to the point where $h=h^*$, l_b is the beach width, $\beta_0=(\partial h/\partial x)_0$ is the sea bed slope at the shoreline, ρ is the water density, g is the acceleration from gravity, and B is the sediment budget imbalance.

The equilibrium of the coastal profile is derived from Eqns (3.30) and (3.31) when $Er = Ac$ and $B = 0$

$$\frac{h}{h_*} = 1 - \left(1 - \frac{x}{l_*}\right)^p, \quad p = \frac{n+1}{m+1}. \tag{3.32}$$

According to Eqn (3.32), $\beta_0=p(h^*/l^*)$.

Figure 3.38 shows a comparison between calculated and observed coastal profiles at the sites of Lubiatowo (the Baltic Sea), Anapskaya peresyp (the Black Sea), and Terschelling (the North Sea), where the bed is composed of medium-fine sand and

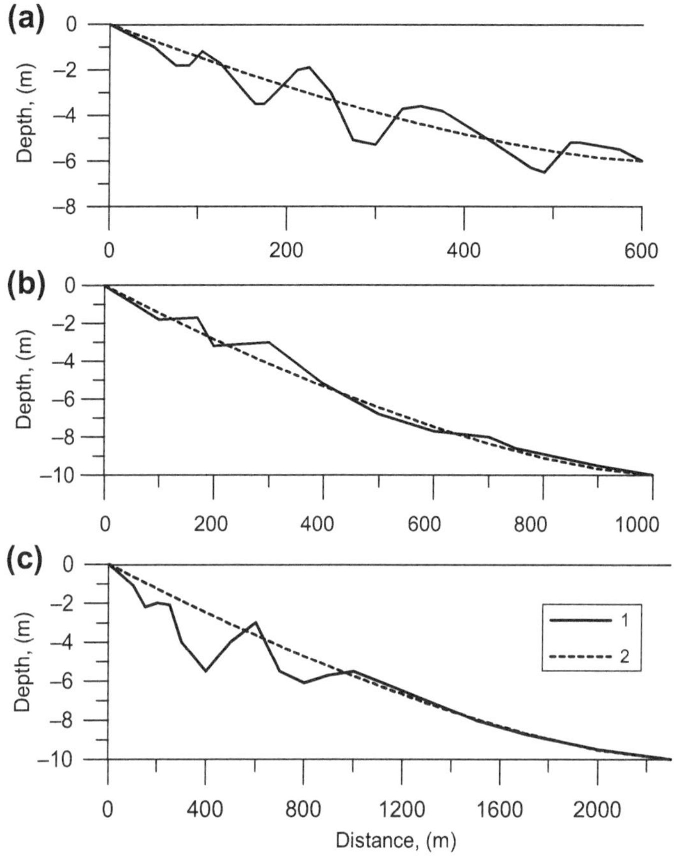

Figure 3.38 Comparison of the observed (1) and theoretical (2) profiles of submerged coastal slopes: (a) Lubiatowo (the Baltic Sea), (b) Anapskaya peresyp (the Black Sea), and (c) Terschelling (the North Sea). Mean sand grain size is 0.22, 0.25, and 0.18 mm, respectively. The exponent $p = 1.5$ was used for the calculations in Eqn (3.32).

appreciably complicated by submarine bars. The relationship (3.32) quite satisfactorily reproduces the average patterns of the profiles. Based on the comparison of the analytical results and of the field data, the parameters in Eqn (3.32) are estimated to be $m = 2$, $n = 3.5$, $p = 1.5$.

The integration of Eqn (3.29) over the profile length l_* leads to the following equations for the coast line displacement rate $\partial x_0/\partial t$, resulting from sea-level change

$$\frac{\partial x_0}{\partial t} = -\frac{w(l_* + l_b)}{h_* + z_b}, \tag{3.33}$$

and from a sediment budget imbalance B

$$\frac{\partial x_0}{\partial t} = \frac{B}{\frac{p}{p+1} h_* + z_b}, \tag{3.34}$$

where z_b is the beach height (elevation of berm or dune crest). The sediment budget imbalance results in changes of length and slope of the profile. In case of excessive sediment feed ($B > 0$), the coastline tends to be shifted seaward, whereas sediment deficiency ($B < 0$) results in coastline recession. Because the lowest point of the profile corresponding to $h*$ is fixed, its average slope increases in the former case and decreases in the latter one. It follows from Eqns (3.33) and (3.34) that variations of $l*$ from B have an effect on the profile erosion or accretion under sea-level changes.

The integration of Eqns (3.33) and (3.34) and their combination provide the shoreline position in time and the profile evolution.

To assess the value of B, the longshore and cross-shore sediment fluxes in a given coast segment should be determined. For example, Figure 3.39 shows the distribution of longshore sediment fluxes in the Gulf of Gdańsk (the Baltic Sea). Calculations led to the conclusion that the erosion or accretion caused by longshore fluxes is in most compensated by accretion or erosion from cross-shore sediment transport. In other words, the sediment budget is in balance over a greater part of coast (with the exception of end parts of Vistula Spit and Hel Peninsula), and so the shoreline position may be influenced essentially by the changes in sea level. The calculations by the SPELT model predict a gradual retreat of the shore from sea-level rise of about 0.3–0.4 m/year for both Vistula Spit and Hel Peninsula (see details regarding the Hel Peninsula site in Chapter 7, Section 7.8). For the maintenance of the present coastline

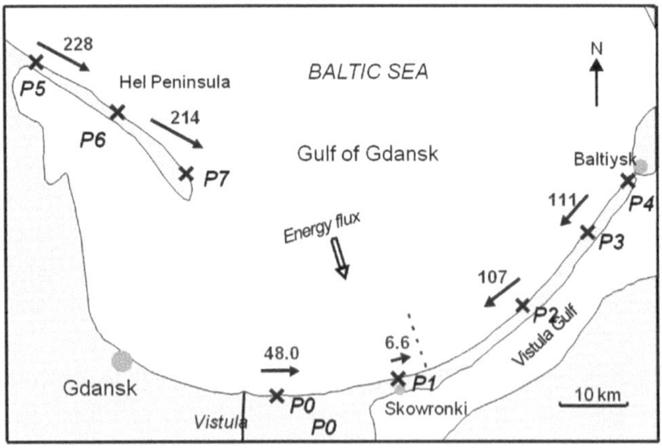

Figure 3.39 Coastlines of the Vistula Spit and Hel Peninsula. The sites under consideration are marked with the crosses. Arrows and nearby numbers show the direction and the value of the alongshore sediment flux in 10^3 m^3/y.

at the sites representing special value, it is possible to recommend adding a material in volume of 3 m^3/m/year.

3.6.2.2 Resilient Coastal Profiles

Resilience of coastal profiles is particularly relevant for hazard management and planning (Bates et al., 2005; Purvis, Bates, & Hayes, 2008). For this purpose, a one-dimensional conceptual model was developed to identify the key profile parameters contributing to the beach hydraulic vulnerability and it has been validated through application to wide stretches of the Emilia Romagna coastline (Martinelli, Zanuttigh, & Corbau, 2010).

The model requires:

- for the topography: collection of cross-shore profiles of the area, including both emerged and submerged beach; sediment characterization; selection of representative profiles; description of the profiles by means of a common simplified scheme, for instance a multiline using few key variables such as foreshore and beach slope, dune height, beach width, and closure depth; and
- for the climate: collection of meteomarine data including historical series of waves, storm surges, sea levels; statistical analysis; and reconstruction of climate scenarios.

All the relevant topographical and climate variables are statistically characterized by means of an appropriate distribution function and modeled by using STRUREL software (http://www.strurel.de) or equivalent.

To define in a simple and quantitative way the flooding process, the failure mechanism is given by

$$(Z_m + R_{u2\%}) - (R_{c0} - S_{ubs}) >= 0 \tag{3.35}$$

where Z_m is the average water depth derived from tide and climate change effects; S_{ubs} is the subsidence; $R_{u2\%}$ is wave run-up corresponding to the characteristic value of 2% exceeding probability; and R_{c0} is thc crest height of the sea bank (equal to the beach height plus the dune height, if applicable).

In Eqn (3.35), beach erosion is not physically described as beach reshaping during storms. Beach morphological changes are considered by means of the statistical description of beach profiles based on available surveys.

Further, the failure function Eqn (3.35) does not account for the magnitude of inland flooding (e.g., the effect of few overtopping waves is not distinguished from the effect of a complete flooding), thus overestimating the damage consequences. In practice, because the limit state causing failure in Eqn (3.35) is represented by $R_{u2\%}$, it can be considered that failure is caused by some degree of flooding which is perceived by the local inhabitants.

Wave run-up $R_{u2\%}$ is estimated based on CIRIA/CUR (1991):

$$R_s/H_s = 1.35 \cdot \xi_p \quad \text{if } \xi_p \leq 2; \quad R_s/H_s = 3.00 - 0.15 \cdot \xi_p \quad \text{if } \xi_p \geq 2 \qquad (3.36)$$

where ξ_p is the Iribarren parameter, the subscript "p" denotes the peak wave period, R_s is significant run-up, related to $R_{u2\%}$ by the Rayleigh distribution, and the number of waves and H_s is the significant wave height accounting for the sheltering effect of barriers for coastal protection.

The Iribarren parameter ξ_p in Eqn (3.36) is intended for a straight slope, whereas profiles of the emerged beach in presence of dunes or sea banks show complex geometries. So far there are no methods suited for the level II probabilistic framework to evaluate wave run-up on variable slopes. The following procedure is proposed:

- the profile is divided into reaches with constant slope;
- wave run-up R_s is calculated over each slope, in the order (i.e., from offshore to inshore) based on Eqn (3.36);
- over each slope there is a dissipation of wave energy, so that the reduced wave height incident over the next slope is derived from an energy balance. Dissipated energy is evaluated as the one associated to the wave for which R_s equals the crest of each reach; and
- reduction of incident wave height due to coastal structures is represented only when the structures are parallel to the coast and is evaluated by means of simple formulas (Van der Meer, Briganti, Zanuttigh, & Wang, 2005).

Meteomarine and geometric variables included in the analysis are beach elevation, dune height and slope, tide, subsidence, distance from the shoreline of isobath −5 m, offshore significant wave height and steepness, and transmission coefficient.

For each variable introduced in the model, the output consists of the partial safety factors, i.e., the most probable value that each variable would assume at failure, and of the sensitivities, i.e., the contribution of each variable to the overall failure probability. The variable sensitivities are very useful where model uncertainty is high.

The developed procedure allows thus coastal managers:

- to obtain the flooding probability along a wide coastline; and
- to prioritize and set up proper defense planning strategies based on the contribution of each variable to the failure; for instance, dune reinforcement prioritized with respect of building a barrier or nourish the beach when a threshold beach width is reached to avoid more severe consequences.

The methodology however is strongly affected by some of the simplifying assumptions, and specifically: by the method used for the evaluation of run-up in presence of the dune (i.e., with variable foreshore slope) and by the missing representation of the breaching mechanism.

3.6.2.3 Emerged and Submerged Nourishment

Nourishments can be performed on the emerged or on the submerged beach. The latter case is usually less expensive if the sand is dredged offshore or transported by sea and it is expected to have a lower lifetime, but also a lower environmental impact because it does not alter significantly the beach slope, which is a key parameter for the survival of benthic communities, see Section 3.6.2.6. Moreover, the nourishment on the submerged beach allows for the use of finer sand, which can be more frequently found in offshore submarine deposits. The chance to use alternative sources of sand is particularly relevant if one considers the huge amount of material already needed to preserve beaches, the increased need of the sand resource to compensate the loss induced by sea-level rise, and the impact produced by quarries.

Sand volumes for nourishments on the emerged beach are usually estimated based on historical data of sediment transport in the area. Alternatively, the interventions are planned based on the results from long term runs of one-dimensional models or based on the combination of the desired beach width and simple theoretical considerations. Because the experience of nourishments on the submerged beach is rather limited, indications for the volumes and for the placement depth can be derived from numerical simulations. Within the THESEUS project (THESEUS OD2.7, 2013) morphological and morphodynamic simulations were carried out with the MIKE 21 FM numerical suite to estimate the performance of nourishment interventions on the submerged beach depending on the degree of protection and on the nourishment depth. The following schematic configurations (cases, in the following) were examined (see Figure 3.40):

1. native profile (i.e., Dean equilibrium profile);
2. nourishment placed at −4 m depth;
3. nourishment placed at −3 m depth; and
4. nourishment at a −3 m depth protected by a submerged barrier.

The bathymetries for cases 2–4 were derived under the assumption that the beach profile corresponds to the equilibrium profile by Dean with the specific following parameters

- nourishment volume, $V = 300\ m^3/m$;
- sediment size of the nourishment material, $d_{50} = 0.1$ mm;
- sediment size of the native beach, $d_{50} = 0.12$ mm; and
- emerged beach slope equal to 1/20.

The data used for the profiles are typical of the Emilia Romagna coast and specifically of the Cesenatico (see details regarding the site in Chapter 7, Section 7.6).

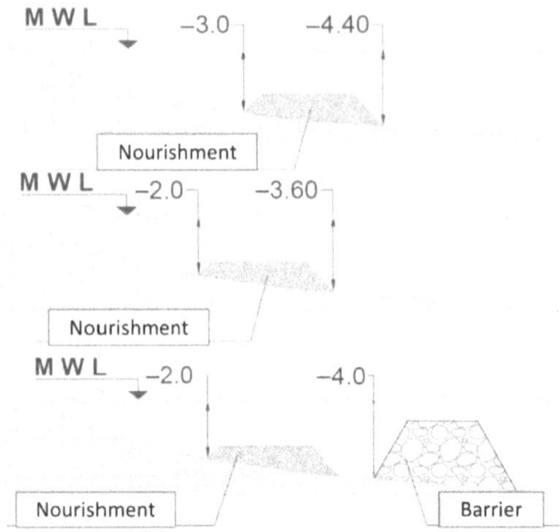

Figure 3.40 Schemes of the submerged nourished profiles analyzed with morphological and morphodynamic modeling. From top to bottom: cases 2, 3, and 4 as they are referred to in the text. The nourishment volume is kept constant. The profile adopted is a Dean profile based on the characteristics of the Emilia Romagna coastline. Case 1, not reported here, corresponds to the nonnourished and unprotected beach profile common to all schemes.

The climate conditions reproduced the typical yearly wave climate in morphological simulations (Table 3.9.) and a typical 1-year return period storm in morphodynamic simulations (Figure 3.41), being the storm energy around the 6% of the total annual wave energy.

Cases 2 and 3 do correspond to different breaking conditions: in case 2, the waves propagate across the nourishment berm without much breaking, whereas in case 3, there is a relevant energy dissipation focused on the berm.

The main results of the morphological and morphodynamic simulations are reported in Table 3.9. The unprotected nourishment schemes induce modest and similar effects, with lower induced current in case 3 than in case 2.

Beach reshaping at the end of the storm shows the development of a bar in correspondence of the breaking line (Figure 3.42). From the bar onshore, some accretion is observed, whereas almost all the nourished sand offshore the bar is eroded. Consequently, the interaction with the initial profile sometimes leads to accretion and sometime to erosion. Case 3 (i.e., the placement of the nourished material so as to induce breaking) results in lower erosion (about the 68% after the storm) than case 2.

The protected nourishment scheme (i.e., case 4) leads to the best performance also during the storm, suggesting the effectiveness of combining hard and soft defense interventions.

TABLE 3.9 Tested Wave Conditions (Significant wave height H_s, Off-shore peak wave steepness s_{op}) in Morphological Simulations and Main Results in Terms of Wave Transmission Coefficient K_t and Long-shore Current Speed V. The Values of K_t and V are Estimated for each Wave Attack as Average Values along the Isobath −1.25 m (i.e., Inshore all the Interventions). These Values are then Averaged Over the Year by Using as Weight the Probability of Occurrence of P each Wave Condition. Sand Losses/Accretions during the Storm used in Morphodynamic Modeling (Figure 3.41) in the Area of the Intervention (Between the Isobaths −5 m and +2 m) are also Compared

				Dean profile		Submerged −3 m		Submerged −2 m		Protected −2 m	
H_s(m)	Direction, °	s_{op}	P (%)	K_t	V (m/s)	K_t	V (m/s)	K_t	V (m/s)	K_t	V (m/s)
0.30	50	0.03	0.265	0.88	0.026	0.88	0.024	1.01	0.009	0.24	0.011
1.33	50	0.04	0.057	0.53	0.238	0.47	0.185	0.38	0.093	0.45	0.096
1.74	50	0.04	0.016	0.44	0.241	0.39	0.164	0.31	0.087	0.49	0.104
0.79	80	0.04	0.242	0.79	0.152	0.75	0.141	0.64	0.097	0.44	0.066
1.86	80	0.04	0.007	0.46	0.193	0.41	0.248	0.34	0.13	0.57	0.095
0.29	120	0.03	0.387	0.85	0.046	0.82	0.043	0.84	0.042	0.24	0.040
1.22	120	0.04	0.027	0.61	0.329	0.57	0.290	0.48	0.232	0.55	0.231
K_t, V – average values				**0.81**	**0.089**	**0.78**	**0.080**	**0.79**	**0.056**	**0.32**	**0.048**
Storm: sand loss (−)/accretion				**−2039 m^3**		**−21,208 m^3**		**−14,511 m^3**		**14,981 m^3**	

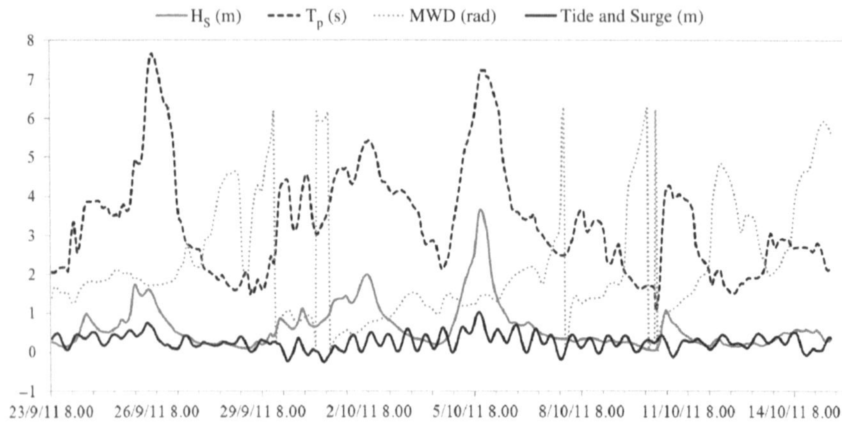

Figure 3.41 Significant wave height H_s, peak wave period T_p, mean wave direction MWD, storm and tidal level Z during the 21 days of the typical storm with return period 1 year in Cesenatico.

3.6.2.4 Optimized Location of Near-shore Borrow Areas

In a number of cases, suitable borrow areas can be found near the site to be nourished at depths of several meters and therefore some hundreds of meters seaward of the shoreline. These areas can be dredged only if the excavations do not negatively affect

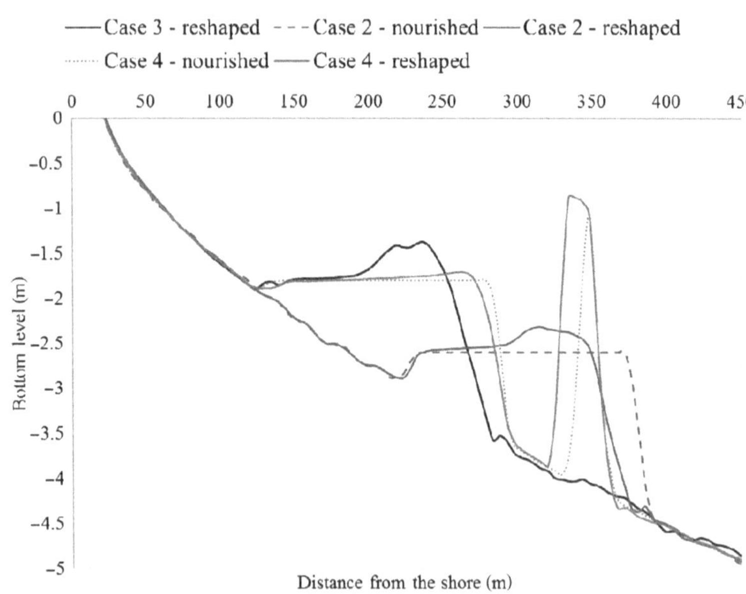

Figure 3.42 Beach reshaping after the typical storm (whose characteristics are summarized in Figure 3.41) for cases 2, 3, and 4.

the near-shore hydrodynamic and lithodynamic processes. In particular, the near-shore excavations must not increase the wave energy close to the nourished beach and coastal or harbor structures.

Dredging in near-shore regions with direct placement of the sediment on the nearby shore was successfully applied in a few sites worldwide, also on exposed coasts (Dean, 2002).

A simple one-dimensional approach is here proposed to assess the onshore effects of near-shore borrow areas. The input deep water wave rays are assumed perpendicular to the shore, so such effects as wave diffraction and refraction are not taken into account. On the other hand, wave energy dissipation from breaking is considered as the most important process, disturbance of which can cause intensification of shore erosion. Moreover, waves approaching the shore perpendicularly constitute favorable conditions for the most severe erosion.

The wave transformation process can be modeled by any reliable theoretical approach, e.g., Battjes and Janssen (1978). These calculations are next supplemented by the simplified bed shear stress concept.

The dimensionless bed shear stresses $\theta_{2.5}$ at various locations in the nearshore zone are determined using the formula proposed by Nielsen (1992):

$$\theta_{2.5} = \frac{1}{2} f_{2.5} \frac{(a_{1m}\omega)^2}{(s-1)gd} \tag{3.37}$$

with:

$$f_{2.5} = \exp\left[5.213\left(\frac{2.5d}{a_{1m}}\right)^{0.194} - 5.977\right] \tag{3.38}$$

where ω is the angular frequency, d is the mean grain diameter, s is the relative density ($s = \rho_s/\rho \approx 2.65$), $a_{1m} = U_{1m}/\omega$ is the amplitude of the near-bed oscillatory flow, and U_{1m} is the maximum near-bed (free stream) oscillatory velocity, i.e., for sinusoidal waves $U(\omega t) = U_{1m}\sin(\omega t)$.

The condition $\theta_{2.5} \approx 0.05$ is assumed for the incipient motion of sand grains. For $\theta_{2.5} = 0.2$–0.3, one can expect a slightly more intensive sediment transport and the development of sea bed ripples. Under severe hydrodynamic conditions, with $\theta_{2.5} = 0.8$–1.0, the ripple marks disappear and the mass transport of sand (sheet flow) starts.

The wave transformation process was modeled with the Battjes and Janssen (1978) approach, adapted for an arbitrary (also multibar) cross-shore profile. Calculations were carried out for the representative cross-shore profile of Hel Peninsula, Gulf of Gdańsk, Poland (see details regarding the site in Chapter 7, Section 7.8). The profile is given in Figure 3.43. The deep water wave conditions were assumed corresponding to the return period of 10 years, with the significant wave height

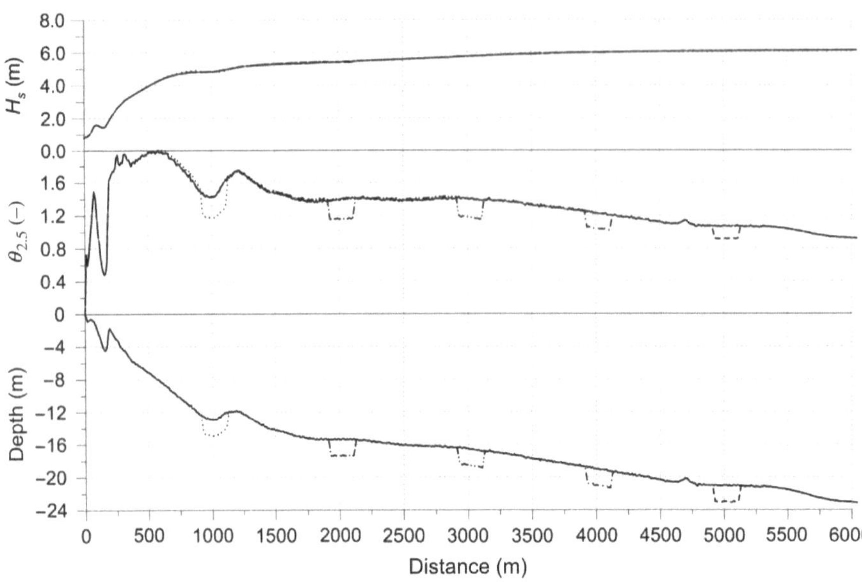

Figure 3.43 Calculated distributions of significant wave height H_s and dimensionless bed shear stress $\theta_{2.5}$ on the typical cross-shore profile of Hel Peninsula for 10-year storm conditions: natural profile (solid line) and with underwater excavations (dashed and dotted lines).

$H_s = 6.27$ m and the peak wave period $T_p = 12$ s. The waves are assumed to approach the shore perpendicularly and the tidal effect is neglected. According to the sediment samples available in the area, the sediment mean grain diameter was assumed to be $d = 0.2$ mm. Different locations of the borrow areas (all being 2 m deep and 200 m wide) were examined, see Figure 3.43.

In Figure 3.43, values of the dimensionless shear stress $\theta_{2.5}$ on the entire cross-shore profile (with the exception of two locations in the region 0–180 m from the shoreline) are high enough to cause the sand transport (sheet flow). Figure 3.43 shows, however, that the presence of the underwater excavation at any of the considered location does not affect the wave transformation pattern. The bed shear stress is affected only close to the excavations.

Further calculations were performed by assuming that the sandy material is dredged from the natural offshore bar (1000–1600 m offshore, at depths of 12.0–14.5 m) and is placed close to the shoreline, in the bar trough (0–185 m from the shoreline, at depths 0.0–4.5 m). Figure 3.44 shows that the wave height slightly increases in the region of the dredged offshore bar and remarkably decreases close to the shoreline. The induced wave energy dissipation farther from the shoreline is convenient in terms of the sediment stability on the cross-shore profile.

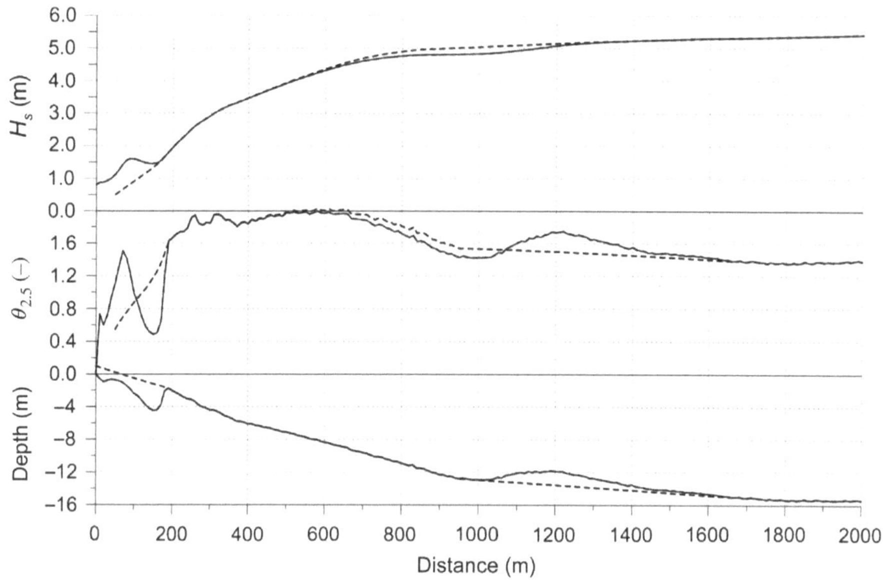

Figure 3.44 Calculated distributions of significant wave height H_s and dimensionless bed shear stress $\theta_{2.5}$ on the cross-shore profile of Hel Peninsula for 10-year storm conditions: natural profile (solid line) and with beach nourished by sediment taken from the offshore bar (dashed line).

The change of the bed shear stresses in close vicinity of the shoreline seems also convenient. The dimensionless shear stress $\theta_{2.5}$ amounts to about 1.4 at location 2000 m offshore. The maxima attain values of about 2.0 at distances 300–600 m from the shoreline. In the presence of the artificial beach, $\theta_{2.5}$ starts to decrease rapidly from the value of 1.4 at a distance of about 170 m from the shoreline to about 0.6 near the shoreline.

The computational results presented in Figures 3.43 and 3.44 concern the extreme conditions, occurring during the 10-year storm. According to the depth of closure, the representative hydrodynamic impact on the shore is associated with the extreme 1-year storm. Taking into account milder wave conditions, locations of the borrow areas closer to the shore at depths smaller than 12 m, might be considered.

The results of computations for the natural profile and the profile with the excavation 200 m wide and 2 m deep (600–800 m from the shoreline, at depths of 8–11 m) for 1-year storm conditions ($H_s = 3.13$ m, $T_p = 8$ s) are shown in Figure 3.45. As before, the most unfavorable situation was assumed, in which waves approach the shore perpendicularly.

The results of computations presented in Figure 3.45 indicate the lack of effect of the borrow hole on the wave energy dissipation process (wave height H_s) and the sand motion driving force (shear stress $\theta_{2.5}$) in the nearshore region between the

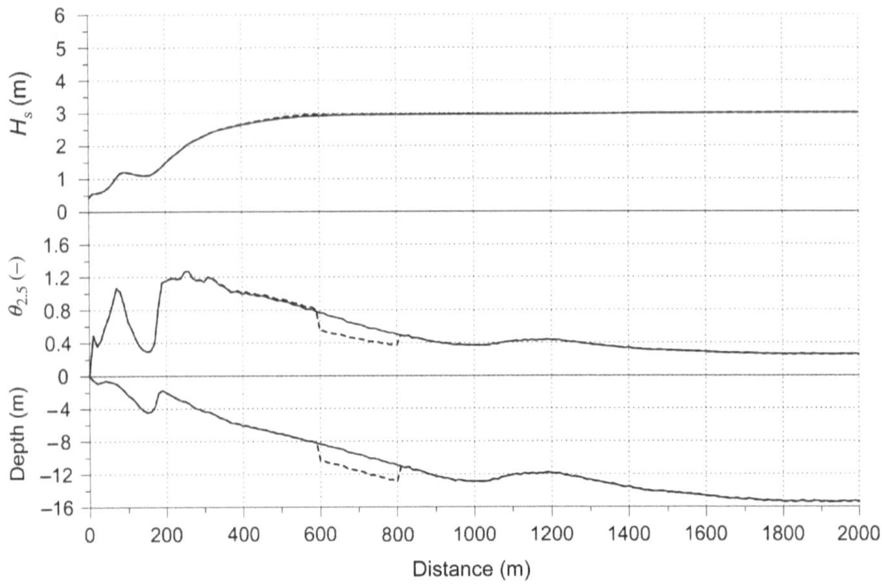

Figure 3.45 Calculated distributions of significant wave height H_s and dimensionless bed shear stress $\theta_{2.5}$ on the typical cross-shore profile of Hel Peninsula for 1-year storm conditions: natural profile (solid line) and with underwater excavation (dashed line) 600–800 m from the shoreline, at depths of 8–11 m.

excavation and the shoreline. It should be pointed out that seawards of the excavation $\theta_{2.5}$ values imply a moderately intensive sediment motion (the seabed ripple flow regime). Thus, the borrow area in the considered case is located at smaller depth than the depth of closure without negative consequences to the nearshore seabed and the shore stability.

The presented generic approach can be helpful in indication of the places where the lithodynamic processes may negatively evolve after the appearance of the borrow holes. This approach may also be used to identify whether and how quickly the borrow holes will be silted up (e.g., by the long-shore sediment transport), yielding the opportunity for the renourishment operations within the same or similar scheme. Certainly, while carrying out such studies, local site-specific conditions ought to be considered, such as the offshore wave climate, the cross-shore profile shape and sand grain diameters as well as the nourishment/dredging needs and planned dimensions of the excavations.

3.6.2.5 *Supplementary Measures: Fences and Groins*

Sand accumulation fences can play an important role in shore protection against erosion and flooding on many European coasts, e.g., the Baltic and the

Mediterranean (THESEUS OD2.1, 2010). The fences, constructed as rows of wooden sticks at the dune toe, trap wind-transported sand, and accelerate accretion of the dune at its seaward side. Because of the use of natural and cheap material, such a human influence on dune evolution is highly recommended. Similarly, a timber pile groin system can be an important supplementary measure with respect to artificial beach nourishment. The presence of groins can considerably improve the nourishment efficiency if the groins are undamaged. Various kinds of groin malfunctioning can affect the coexistence of nourishment and groins in different ways (Ostrowski, Pruszak, Schönhofer, Szmytkiewicz, & Szmytkiewicz, 2012; THESEUS OD2.7, 2013).

Field observations showed (THESEUS OD2.1, 2010) that the use of fences can be recommended for the shores in microtidal regions. It can be supposed that the application of fences can also be profitable in some cases of mesotidal coastal areas, characterized by moderate wave climate and mildly inclined near-shore sea bottom. Specific criteria regarding the geometry of the (natural or artificially nourished) emerged beach have to be satisfied.

The following input data are necessary for the profitability check (see Figure 3.46):

w Emerged beach width above the long-term mean sea level at high tide, excluding storm surge;

h Height of the dune toe above the long-term mean sea level at high tide, excluding storm surge;

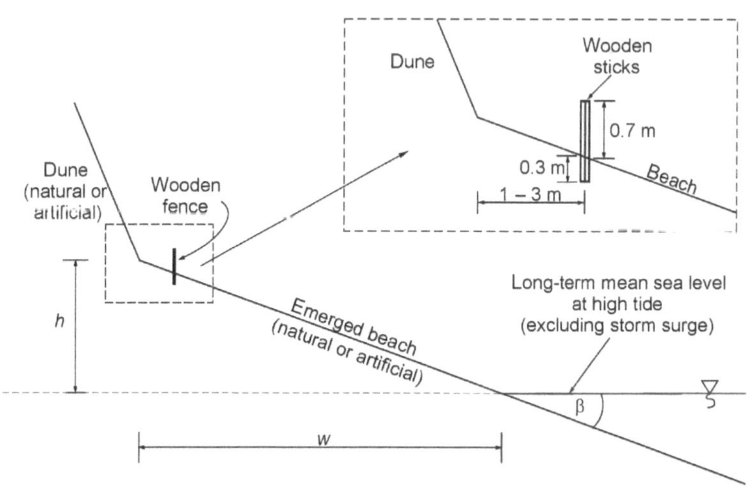

Figure 3.46 Definition sketch: applicability criteria and design of sand accumulation fences.

β Average seabed slope angle (with respect to horizontal) in the nearshore zone (i.e., seabed slope average over a 500 m long distance from the shoreline);
H_b Representative (typical significant) breaking wave height;
T Mean wave period.

The profitability check is performed through the following steps.
The surf scale parameter ε is calculated as

$$\varepsilon = \frac{2\pi^2 H_b}{gT^2 \tan^2 \beta} \tag{3.39}$$

Then the following criteria have to be checked:

1. $w > 35$ m
2. $h >$ 100-year return period storm surge
3. $\varepsilon > 150$

If conditions (1) and (2) are satisfied, there is a wide beach available all the time as a source of material for wind-induced accumulation of sand at the dune toe. Criterion (3) defines the seashore dissipating most of wave energy in front of the shoreline (Ostrowski et al., 2010). In such conditions, erosive effect of wave run-up is reduced.

The construction of the fascine fences is recommended when criteria (1), (2), and (3) are satisfied.

The fence should be placed as locally straight line segments at a distance of 1–3 m from the dune seaward toe. The fascine sticks about 1.0 m long ought to be buried 0.3 m deep in the beach sand, so that the sticks will jut out 0.7 m on the beach surface (Figure 3.47).

The implementation of combinations of artificial shore nourishment and coastal structures may decrease the need of maintenance (THESEUS OD2.1, 2010; THESEUS OD2.7, 2013).

A recent numerical study was carried out to assess the effects induced by groins on nourishment efficiency (Ostrowski et al., 2012; THESEUS OD2.7, 2013).

The following schematic configurations were simulated with the Delft 3D modeling suite:

- no groins;
- groins in good conditions;
- missing piles in the root part of the groin;
- missing piles in the central part of the groin; and
- missing piles at the groin end.

In case of damages, the length of the damaged area was assumed to be 1/3 of the whole groin length.

Figure 3.47 Construction of the sand accumulation fence.

The computational layout was based on the existing system of timber palisade groins stretching along the Hel Peninsula shore (Gulf of Gdańsk, Poland; for details of this site refer to Chapter 7, Section 7.8). The modeling area stretched 500 m in the long-shore direction and 1200 m in the cross-shore direction. Bearing in mind the structural features and permeability of the groins (palisades with narrow gaps between the piles), the values of both the reflection coefficient and the transmission coefficient were assumed to be equal to 0.5. The other conditions were assumed as follows:

- groins length equal to 60 m (counted seawards from the shoreline);
- span between neighboring groins equal to 90 m;
- bathymetric layout based on a representative cross-shore profile after nourishment (Figure 3.48).

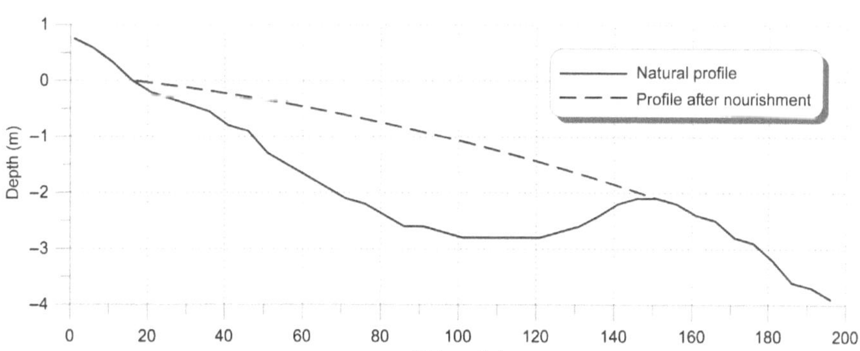

Figure 3.48 Cross-shore profile assumed in modeling of erosion of nourished shore with groins and without groins.

Numerical simulations were carried out for a few frequent typical storms of the South Baltic, characterized by significant wave heights $H_s = 1.0, 1.5, 2.0$ m, and peak wave periods $T_p = 4.0, 5.5$, and 7.0 s, respectively. In all these cases, the wave to shoreline angle was $\alpha = 45°$. These parameters were assumed at the offshore location, namely 1200 m from the shoreline over a 14-m depth.

The computational procedure consisted of the following steps:

1. modeling of sediment transport and nearshore morphodynamics of the scheme with groins, within an area including three groins (10,800 m^2);
2. quantification of the volume of sediment washed away from the considered area for the scheme with groins, V_{groins};
3. modeling of sediment transport and nearshore morphodynamics of the natural shore (without groins) within an area of the same extension;
4. quantification of the volume of sediment washed away from the natural area, $V_{natural}$; and
5. calculation of the coefficient W:

$$W = \left(1 - \frac{V_{groins}}{V_{natural}}\right) \cdot 100\% \quad (3.40)$$

The presence of a groin system reduces flow velocities significantly as well as the rates of sediment transport induced by the wave–current interactions. This in turn preserves the sediment to be washed out from the fields between groins.

The results of the computations of W for complete (undamaged) groins are depicted in Figure 3.49. It can be seen from Figure 3.49 that coexistence of complete groins with artificial shore nourishment causes that less sediment is washed away from the near-shore zone than for beach fill without groins. In the considered cases, the presence of groins reduces the sediment erosion of about 30–45% after 6 days.

The computational results, together with field observations, have also revealed that a lack of single piles in the palisade groin does not affect its efficiency, whereas wider breaches, comprising a few piles, result in growth of flow velocities through the groin, which in turn causes intensive sea bed erosion at the breaches. Groins with breaches near the shoreline are much more harmful to the shore than the total lack of groins.

3.6.2.6 Environmental, Social, and Economic Implications

Environmental impacts of nourishment and dredging include:

- morphological and bathymetric changes of the sea bottom and shoreline;
- changes in geotechnical and textural characteristics of superficial sediments;
- input of significant quantities of suspended fine sediment leading to an increase in turbidity and light attenuation;

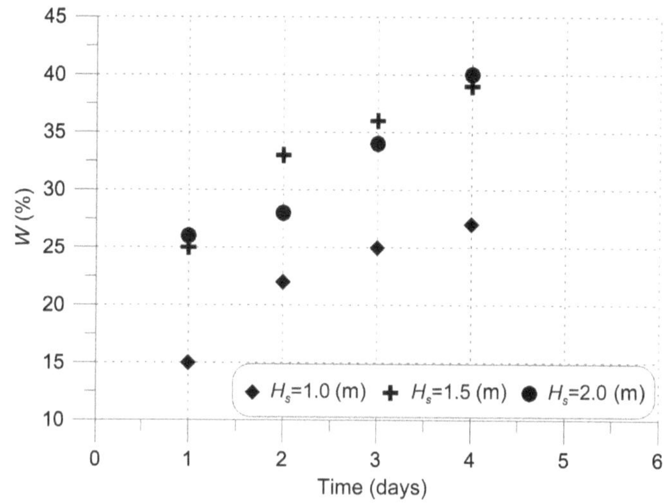

Figure 3.49 Coefficient W as function of time. The value $W=0$ means that no supporting role of groins can be expected, whereas $W=100\%$ denotes the fully effective support of groins.

- changes in benthic organisms as a result of removal of organisms and burying, or indirectly as consequence of the modification of the beach slope, grain size, and type.

The environmental impacts associated with these activities occur at different spatial and temporal scales and relate to (1) dredging to obtain the required material, (2) transport of the sediments from dredging zone to the nourishment area, and (3) placement of the sediments on the shore.

The excavation and extraction of material to nourish beaches can come from a variety of sources including: river beds, river months, navigable channels, generic coastal environment, marine deposits, and relict sand deposits. In the case of marine deposits, the dredging may cause the following physical alterations:

- modification of wave and current patterns; this is dependent on the position of the pit with respect to the active beach (Uda, Agemori, & Chujo, 1986; Kojima, Ijima, & Nakamuta, 1987; Bender & Dean, 2003). It is recommended to avoid near-shore borrow areas and to dredge offshore;
- changes in the morphological and bathymetric features of the seafloor including alterations to grain size, compaction, oxygenation, and organic matter content that are characteristic of superficial sediments;
- changes in the chemical and physical characteristics of the water column and in the concentrations of suspended particles. The level of change will depend on the hydrodynamics (i.e., waves, currents, and tides), which also impacts plume dispersion during dredging operations (Nicoletti, Paganelli, & Gabellini, 2006).

Beach nourishment can cause ecological damage to sandy beach habitats (Blott & Pye, 2004) and biota (Nelson, 1988; Peterson & Bishop, 2005; Speybroeck et al., 2006). Nourishment can impact individual organisms or entire ecosystems; nourishment and dredging may also produce indirect impacts at a larger scale because of plume dispersion, sediment resuspension, and changes to light properties in the water column.

After the sediment has been dredged for use in beach nourishment, the material has to be transported to the beach; the transport itself can generate disturbance. Special attention should be paid when the transport area or its immediate surroundings include sensitive habitats, such as seagrass meadows, coral reefs, or if the area is a known crossing path for marine mammals. Also, any recreational (e.g., bathing sailing) use of the site should be considered because nourishment can negatively affect the site.

The material used for shore nourishment should be physically and chemically compatible with the native sand, if sediment with a different grain size is used to replenish beaches, severe ecological impacts may occur and recovery of the benthic community can be increased (Peterson, Hickerson, & Johnson, 2000; Peterson, Bishop, Johnson, D'Anna, & Manning, 2006). The impact of differences between nourished and native sand composition can be substantial. Differences in the characteristics of the sediment can result in modifications of the beach slope and changes to sediment transport regime. If nourished sediment is finer than the native sediment particle resuspension will increase resulting in an increase in turbidity. If the nourished sediment is courser than the native sediment beach slope will increase and sediment transport will decrease. The placement of nourished material on the emerged beach can affect sediment transport, mainly if there are changes in the grain size and slope. On the other hand, submerged beach nourishment may induce waves breaking on the nourishment berm resulting in deposition of finer sediments on the beach during storms. Even from a tourism perspective, nourished sand that visibly differs from the native one in color and dimensions may affect tourist and recreational activities from visual impact and use.

The environmental impacts of dredging for the purpose of beach nourishment can have significant social and economic implications, as synthesized in Figure 3.50.

Nourishment may be applied alone or in combination with preexisting hard structures (Colosio, Abbiati, & Airoldi, 2007). If nourishment is applied in combination with breakwaters, it may have only a limited effect on nearshore habitats and species assemblages because the sediment is relatively stable; however, the minor direct effect of nourishment is largely canceled out by the direct impact of the breakwaters (Colosio et al., 2007).

Artificially flattened and extended sandy beaches can be colonized by rapidly moving opportunistic macrofauna; under these conditions, few species dominate and biodiversity is reduced (Peterson & Bishop, 2005). The process of beach bulldozing

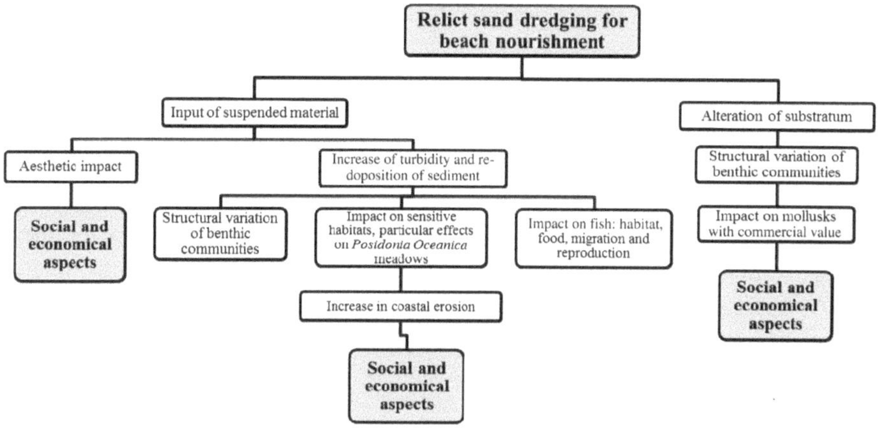

Figure 3.50 Environmental disturbance and economic effects related to dredging for beach nourishment. *Based on Nicoletti et al. (2006).*

or scraping consists of mechanically redistributing beach sand from the littoral zone to the upper beach. This process can damage intertidal fauna, with the degree of damage is related to season. The heaviest impact occurs in summer because this is when reproduction and settlement for many benthic organisms takes place. Furthermore operations such as grooming, linked to the tourist season, produce additional impacts to the benthic system by reworking the sand that had been settled. Benthic organisms are an important food sources for fish and birds and a reduction in their number will have a negative effect on feeding populations.

Recovery of benthic organisms however can occur relatively fast within a few months rather than years of renourishment. This is because sandy–beach species are naturally adapted to severe physical disturbance. However, if the profile of the nourished beach and the imported sediments do not match the original conditions, recovery of benthos is unlikely (Peterson et al., 2000, 2006). Understanding this process can help inform beach nourishment and lessen the ecological impact. From an ecological perspective, these issues have to be considered to minimize impacts of beach nourishment projects.

- the project scale, duration, timing, zone, fill method, and sediment match are all important in determining the level of impact of beach nourishment on benthic communities. The extent of impact may be mitigated if timed appropriately, specifically avoiding the summer months. For that reason, it is fundamental to know the lifecycle of species present at the site to allow the recovery of populations;
- many of the activities linked to beach management are repeated in time and space, and even if individually small, may have negative ecological consequences. Recovery may be delayed if repeated nourishment occurs in the same area (Dolan & Stewart, 2006);

- interventions should be decided in agreement with local, regional, and national planners along with ecologists.

Knowledge of reproductive biology, lifecycles, and species interactions can guide the timing of interventions and may help to indicate likely rates of recovery. Therefore baseline data should be gathered before any coastal protection schemes are carried out; this can be used to help identify appropriate interventions and to distinguish impacts from seasonal variability. Long-term studies focusing on identifying temporal impact trajectories of nourishment on sediments and macrofauna under a variety of environmental conditions are necessary. Further detail on the management of sandy habitats from an ecological perspective is given in Section 4.3.

3.6.3 Estuarine Morphology

3.6.3.1 Assessment of Estuarine Erodibility

Over time, sediment transport in estuarine river channels and on tidal flats has caused an extensive topographic change in mostly all estuarine environments worldwide and is therefore closely linked to the efficient estuarine navigation and dredging maintenance (Dredging Operations Technical Support, 1994; Stive, Ranasinghe, & Colwell, 2010). For the assessment of the estuarine erodibility, especially under the projection of future sea level/climate change, the analysis of topographical change will be critical for a better estuarine management. By quantifying the sediment transport budget both in the river channel and on the tidal flat, it is possible to establish the relationship of estuarine erodibility with morphodynamic processes.

The proposed methodology consists of simple tools that are based on long-term observation and practice. Specifically, the methodology is here applied to test the Elbe estuarine and coastal topographic erodibility (details about the Elbe estuary can be found in Chapter 7, Section 7.5). The tools proposed by this study essentially include: (1) hydromorphodynamic correlation and (2) coefficient of erodibility, on the basis of numerous fieldwork and laboratory tests carried out in the Elbe estuary (Kappenberg & Grabemann, 2001; Ge et al., 2013).

The method was applied to three representative cross-sections of the outer Elbe estuary (see Figure 3.51), which were derived from annual sea charts (soundings). Along the cross-sections, the average water depth and width of the river channel (H_c, < 0 m; W_c), the average elevation and width of the tidal flat (H_t, > 0 m; W_t) were calculated.

The following first-order relationship between cross-section areas and ebb flow discharge is assumed:

$$S_i = a + b \cdot Q_i, \quad i = 1 \cdots 3 \tag{3.41}$$

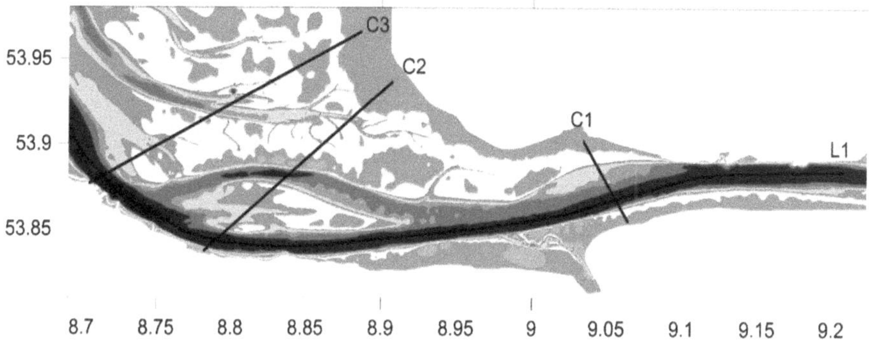

Figure 3.51 Three representative cross-sections in the Elbe estuary can be selected: C1, single channel; C2, anabranch river channel + tidal flat; and C3, multichannel + low tidal flat.

where:

S_i = cross-section area of C_i,
Q_i = maximum ebb flow discharge through cross-section.

Using the FVCOM model (Ge, Chen, Qi, Ding, & Beardsley, 2012), the hydrodynamic conditions in the mouth of the Elbe estuary were simulated from 1927 to 2006 (Chapter 7, Section 7.5) to determine the annual values S_i and Q_i. The results for cross-section C1 are displayed as an example in Figure 3.52. For C1 the hydromorphodynamic correlation was found to be:

$$S_1 = 0.53 + 0.75Q_1, R^2 = 0.66 \tag{3.42}$$

As a result, the estuarine erodibility at the cross-section C1 can be derived by substituting the values of Q_i in formula (3.42).

The coefficient of erodibility P can be used to quantitatively describe the morphodynamic development of the cross-sections. P is defined as

$$P = W/H$$

where W is the river cross-section width and H is the average water depth at the given cross-section.

P can be further divided into the coefficient of river channel erodibility P_c and the coefficient of tidal flat erodibility P_t that are similarly defined as:

$$P_c = W_c/H_c, \ P_t = W_t/H_t \tag{3.43}$$

where W_c and W_t are, respectively, the cross-section widths of the river channel and of the tidal flat; H_c is the average water depth (<0 m msl) in the river channel cross-sections; and H_t is the average elevation (>0 m msl) of tidal flat cross-sections.

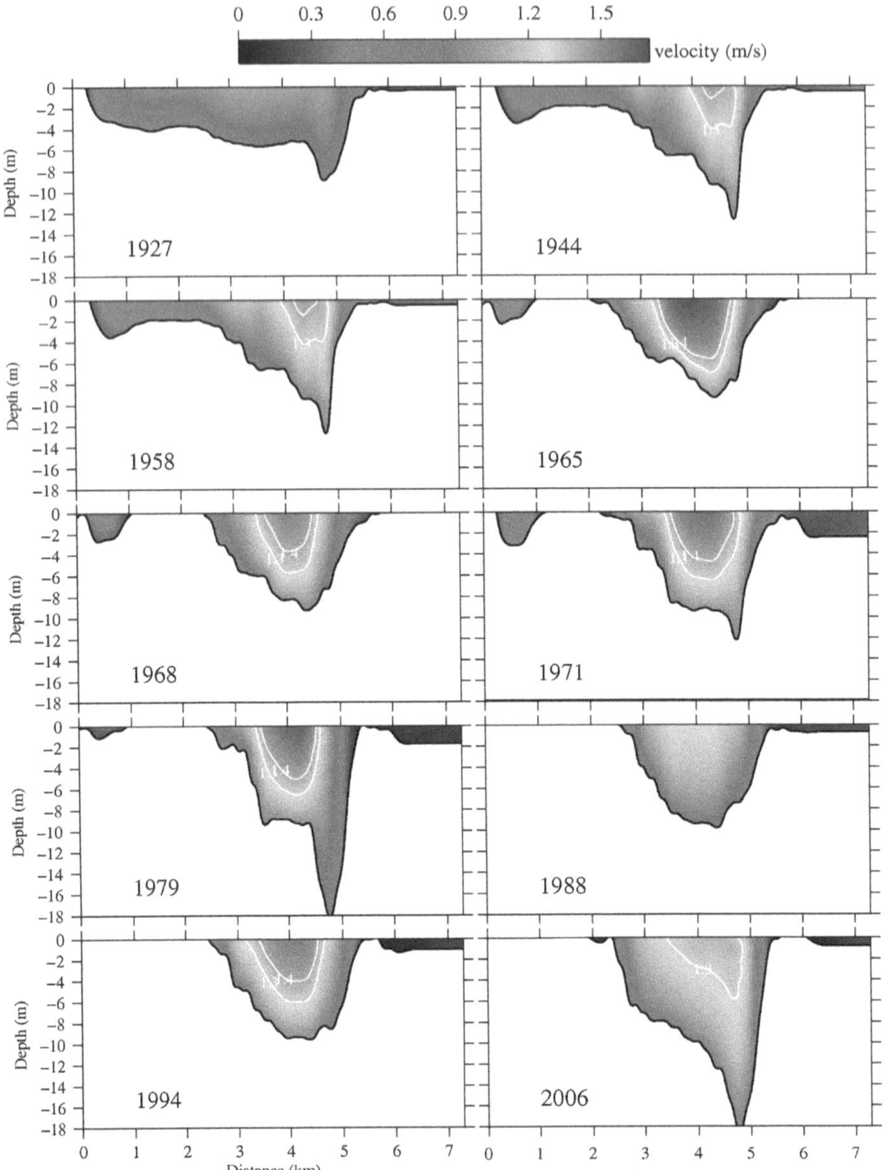

Figure 3.52 Cross-section distribution of ebb maximum at cross-section C1.

Because the ratio *W*/*H* is negatively correlated to flow velocity, if *P* decreases, the river channel is deepened (eroded), and vice versa. By determining the average P_c and P_t for each year, the annual change of erodibility can be obtained. Figure 3.53 shows the time-series of the erodibility in the channels and on the tidal flats for the three different cross-sections in Figure 3.51.

The combination of these tools was also been applied in the Chinese Yellow river estuary (Shen, Zhang, Yan, & Wang, 2009; Sun, Feng, & Zhang, 2009), showing successful achievements in assessing the estuarine morphodynamic processes. This is particularly effective for coastal management, in terms of necessary dredging, future flood prevention, and mitigation of hazardous conditions.

3.6.3.2 Influence of Turbidity on Tidal Properties

In hypersynchronous estuaries, where tidal damping by friction is less than the effect of the upstream convergence, the tidal amplitude and power dissipation can reach a maximum within the estuary. Therefore, these large tidal ranges may continue to affect areas upstream of the estuary that are generally intensively urbanized. In addition, in the hypersynchronous estuaries, tidal forces push salt water upstream of the estuary to form an area called "salt wedge." Turbulence from tidal forces induced resuspension of sediment available on the river bed. Concurrently with these processes, the phenomenon of flocculation of dissolved material in the river water was observed when it comes into contact with the salt wedge. The area combining the two processes is characterized by a high concentration of matter (generally silty

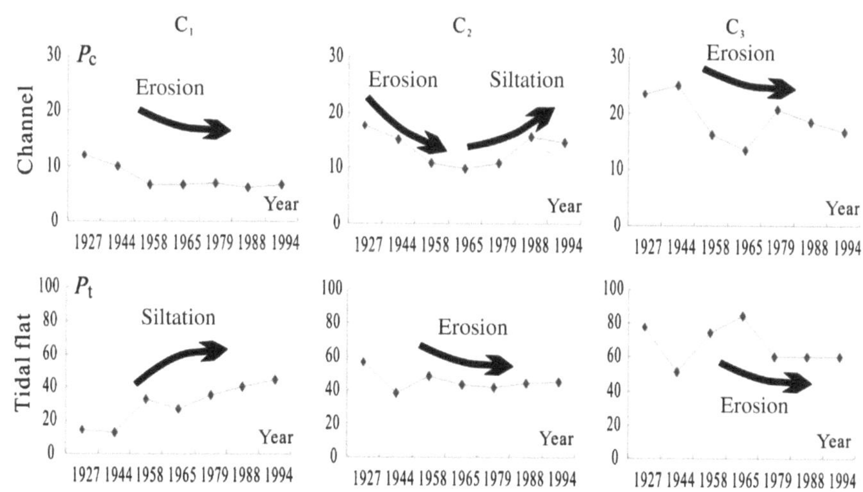

Figure 3.53 Siltation and erosion calculated by the coefficient of the erodibility (*P*); cross-sections C1, C2, and C3 refer to Figure 3.51; decreasing P_c implies erosion in river channel, and vice versa; and increasing P_t implies siltation in tidal flat.

or muddy material) in the water column and on the bed of the estuary. This phenomenon is called the estuarine turbidity maximum (ETM). The ETM moves with the tide depending on the tidal forces and on the river discharge.

When the bathymetry measurement undertaken by regular navigation highlights the accumulation of ETM that can interfere with navigation, it usually requires dredging operations either to maintain navigation channels or for ecological considerations. To reduce the increase of the tidal amplitude that can cause flooding, the dredging of sediments at the navigation channels and at the other deposit locations combined with artificial nourishment intervention is recommended. Considering the "flooding risk/dredging disadvantages" ratio, this estuary management option has the advantage of adding value to the dredged sediments and of preventing and limiting flood disasters caused by the combination of sea-level rise, high tide, and storm surge.

The presence of ETM significantly influences the hydrodynamic and hydro-sedimentary processes in the estuary, in terms of viscosity, density, and height of the water flow. To investigate these influences, we adapted the model ECOMOD-3D model (Estuary COastal MODel-3D), developed and validated (by comparison with field data) by Nguyen (1988). This model can simulate the hydrodynamics, saltwater intrusion, and suspended noncohesive sediment transport in coastal and estuarine environment. In this model, the turbulent effects are taken into account by a two-equations closure model of type "$k-l$." The adaptation of the ECOMOD-3D model consisted to implement an ETM dynamics module (cohesive sediment transport equations); it includes both sediment settling and consolidation at the bottom.

By using the rheology of the studied cohesive sediment, both the density and the viscosity are now computed as a function of the estimated SPM (suspended particulate matter) by this module. Therefore the presence of ETM modifies the hydrodynamics of the estuary by means of modified density and viscosity laws.

The ETM is the high accumulation of the cohesive sediment in some areas of the estuary. Its movement depends strongly on the bottom topography, hydrodynamic, and atmospheric conditions of the estuary. This accumulation requires dredging to maintain the channel in the standards of the navigation and the safety. To reduce the costs of these operations, it is necessary to locate the positions of ETM before proceeding to the dredging. For this purpose, the adapted version of the ECOMOD-3D model is applied to the Gironde estuary to predict the ETM movement for an optimized and effective dredging operation.

Figure 3.54 shows the time evolution of the ETM (i.e., the 3D concentration of the SPM). This figure illustrates that the ETM moves according to tidal movement. This information may be exploited by teams of dredging for example to limit their intervention during the slack water where ETM is stationary. Consequently the

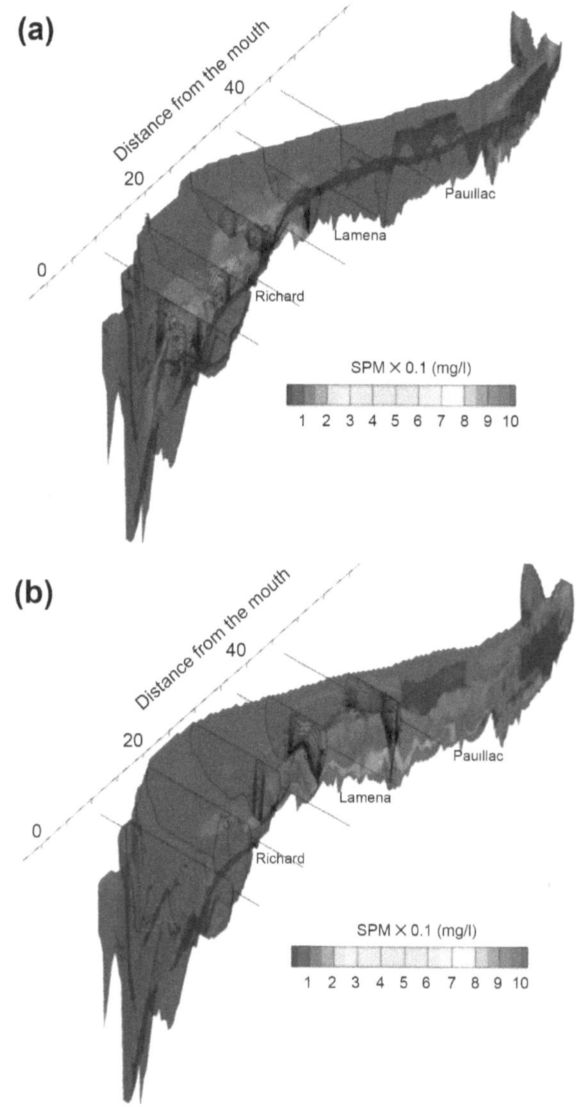

Figure 3.54 Location of the ETM region: (a) ETM distribution at low water, (b) ETM distribution at high water. (Mean river flow rate 700 m^3/s, tidal coefficient 58–79).

location of the ETM is particularly relevant and can be regarded as a strategy contributing to the optimization of the dredging.

The effects of dredging operations on the tidal amplitude can be better assessed if also the ETM is taken into account. Figure 3.55 shows the results from numerical

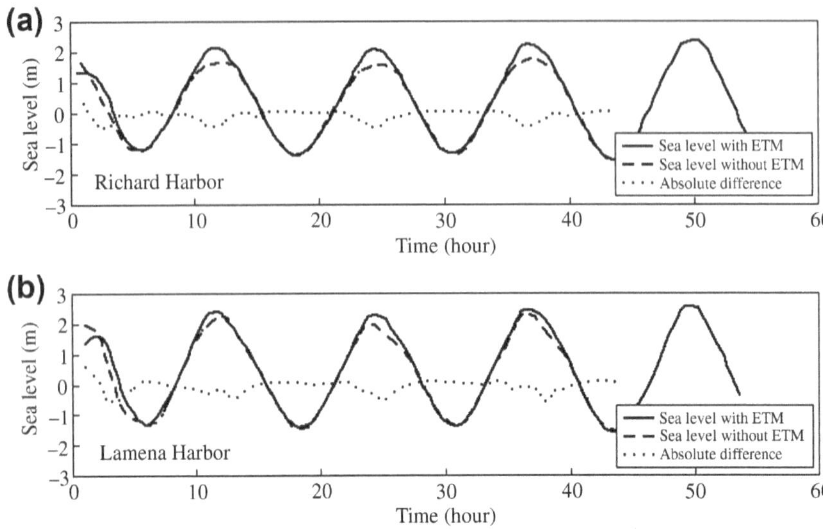

Figure 3.55 Influence of the ETM on the tide propagation. (Mean river flow rate 700 m^3/s, tidal coefficient 58–79).

modeling of the tidal wave amplitude at the same locations in time with and without ETM effects. These results were derived (Smaoui, Nguyen, & Sergent, 2012) from the application of ECOMOD-3D to the Gironde estuary (details in Chapter 7, Section 7.4).

The localized presence of an ETM in estuary area results in reduction in the total water depth. During the passage of the tidal wave in this area, the amplitude of the wave increases to conserve the total mass. This amplitude increase may be the cause of coastal flooding of the estuary. To avoid such a disastrous situation, the dredging operation can be effective. Indeed, the extraction of aggregates by this operation induced increasing the total water depth and consequently reduces the amplitude of the tidal wave in the dredged area (conservation of mass).

Figure 3.55 shows the time evolution of the water level for instance at Richard Harbor and Lamena Harbor (north of Saint Estèphe) during 60 h (five tidal cycles). These harbors were selected to analyze the simulation results because it has been observed that the ETM is present quasipermanently in these ports.

3.6.3.3 Modeling Near-bottom Sediment Transport

Wave and current energy is partially dissipated by bottom friction, which becomes more and more important with decreasing water depth. In theoretical and numerical models, this energy loss is described by an empirical or semiempirical roughness closure formula (Van Rijn, 1993). In practice, a single roughness parameter (sometimes related to the grain size of the bottom sediment or to bed form dimensions) is introduced and is calibrated by comparison of predicted and measured water levels.

Laboratory and numerical experiments demonstrated that in presence of suspended sediments the traditional approach is not able to predict the correct velocity fields, especially near the bottom (Toorman & Bi, 2011; 2013). This has serious implications for flows over shallow areas (i.e., near-shore and intertidal areas) and for the estimation of sediment budgets in particular.

For the purpose of overcoming these implications, a new modeling strategy was developed by (Toorman & Bi, 2011; 2012; 2013), and was implemented in a model for analyzing the Western Scheldt estuary (details about this site can be found in Chapter 7, Section 7.1). It consists of a new generic friction model that accounts not only for the energy dissipation caused by the flow over the bottom roughness structures, but also includes the dissipation induced by the inertia of the suspended particles (Toorman, 2011). The latter is no longer negligible above the bed where high concentrations of suspended matter are encountered. This process explains the drag modulation by suspended matter reported in the literature (Toorman & Bi, 2013). Two versions of the bottom-friction predictor included one for the implementation in 2DH (two-dimensional horizontal) models and one for 3D models.

The friction model for a 2DH model directly computes the bed shear stress as follows (Toorman & Bi, 2012):

$$\begin{aligned}\tau_b/\rho &= u_*^2 = f_A u_{*\text{turb}}^2 + u_{*\text{lam}}^2 \\ &= f_A\left(\frac{\kappa U}{\ln(h/z_0) - 1 + z_0/h}\right)^2 \\ &\quad + \left(\sqrt{\left((z_0+\beta\phi h)\frac{U}{h}\right)^2 + 3\frac{U}{h}\nu(\phi)} - (z_0+\beta\phi h)\frac{U}{h}\right)^2\end{aligned} \tag{3.44}$$

where: τ_b = the bed shear stress, ρ = the density of water, u^* — the shear velocity, u^*_{turb} = the shear velocity for fully developed turbulent open-channel flow, u^*_{turb} = the shear velocity for laminar open-channel flow $f_A = (1-\exp(-h_+/A_+))^2$, the turbulence damping factor (with the water depth nondimensionalized as $h_+ = \sqrt{3Uh/\nu}$, and A_+ an empirical value to be calibrated), κ = the von Karman coefficient (which decreases from the clear water value 0.41 to lower values depending on the sediment load), U = the local depth-averaged flow velocity, h = the local water depth, $z_0 = k_s/30$, the effective roughness length scale (with k_s the equivalent Nikuradse roughness), and ν = the suspension viscosity (function of the concentration), ϕ = the volumetric suspended particle concentration, β = an empirical parameter (to be calibrated).

The dissipative effect of suspended sediment is incorporated into the closures for the effective roughness (which actually is a length scale related to the turbulent eddies generated by vortex shedding over roughness elements and in the wake of particles) and the suspension viscosity (e.g., including steric hindrance and granular friction of dense sand suspension or non-Newtonian behavior of fluid mud).

Unlike in traditional hydrodynamic models (e.g., Hervouet, 2007; Amoudry, 2008), this new bottom-friction model accounts for the water depth and remains valid in turbulent, transient, and laminar flows until an intertidal area falls dry.

Moreover, the new friction law avoids the numerical instabilities typical of traditional models, in which the classical friction law requires a threshold minimum value of water level to be imposed. In the improved model, the roughness increases with decreasing water depth and the friction tends to infinity as soon as the inundation threshold, taken equal to the equivalent roughness height z_0, is reached, preventing flow. The model allows therefore a more accurate prediction of hydrodynamics over intertidal flats because no inundation threshold needs to be specified any longer. The different model behavior is clearly demonstrated in Figure 3.56, where the bed shear stresses at peak flood in the Scheldt estuary are represented (see Chapter 7, Section 7.1).

The 3D version of this improved model requires additional modifications to the distribution along the water depth of the $k-\varepsilon$ turbulence model, where k denotes the turbulent kinetic energy and ε its dissipation rate. Also the bottom boundary conditions have to be changed because the traditional ones are not valid for sediment-laden flows. Because of the energy dissipation by suspended particles, the available energy to produce turbulence is reduced as is the distance from the bed where the turbulence generation achieves its maximum. Therefore, a low-Reynolds version of the k-ε model and corresponding new boundary conditions for the near-bed values of the variables k and ε were developed based on the new mixing-length model (Toorman, 2011).

This promising methodology still needs further testing and tuning because of a lack of sufficient sediment transport data close to the bed. It is also worth mentioning that it does not include wave effects; therefore, further research should be performed to improve the prediction of beach morphodynamics and the generation and flow of gravity currents caused by storms.

3.6.3.4 Management of Sediment Stocks: Artificial Sandbanks

An unconventional use of sediment resources is the construction of emerged and submerged sandbanks to dampen the tidal wave energy with the deposits at the mouth of tidal estuaries. The objective of this intervention is to reduce high water levels and current velocities upstream of the artificial sandbanks in the inner estuary and along its pathway. This measure can be applied to funnel shaped

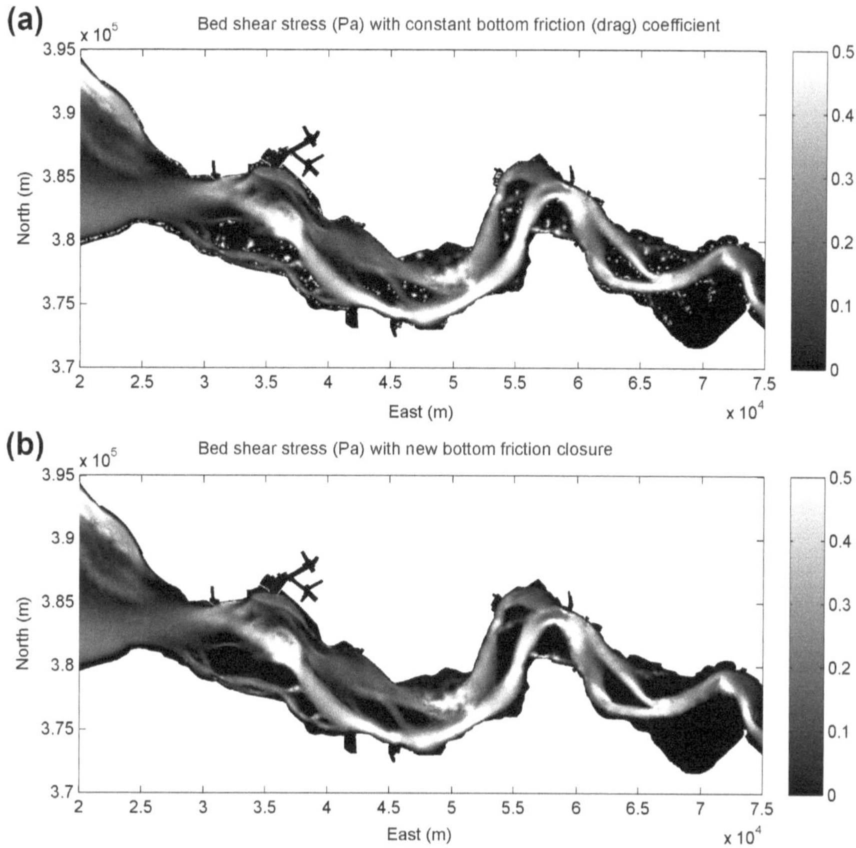

Figure 3.56 Comparison of the bed shear stress distribution at peak flood in the Western Scheldt estuary for a traditional constant roughness parameter model (a) and the new friction law (b), in the case of a very small inundation threshold of 1 mm. Without the improved roughness model excessive erosion is predicted in the intertidal areas (white spots on the dark tidal flats in (a), because of the incorrect prediction of the bed shear stress.

estuaries, where stocks of sand are available and the wave climate is sufficiently mild, so that the sandbanks can be easily sustained (Glindemann, Thode, Dücker, & Witte, 2006).

This idea was tested in the German Elbe estuary and in the inner part of the German Bight (see Figure 3.57) by means of the numerical Tidal, Residual, Indertidal Mudflat, Nested, Parallelized (TRIMNP) hydrodynamic model (Sothmann, Schuster, Kappenberg, & Ohle, 2011) in 2DH mode (details can be found in Chapter 7, Section 7.5). The 2DH mode was selected because of the main emphasis of this work on water surface elevations.

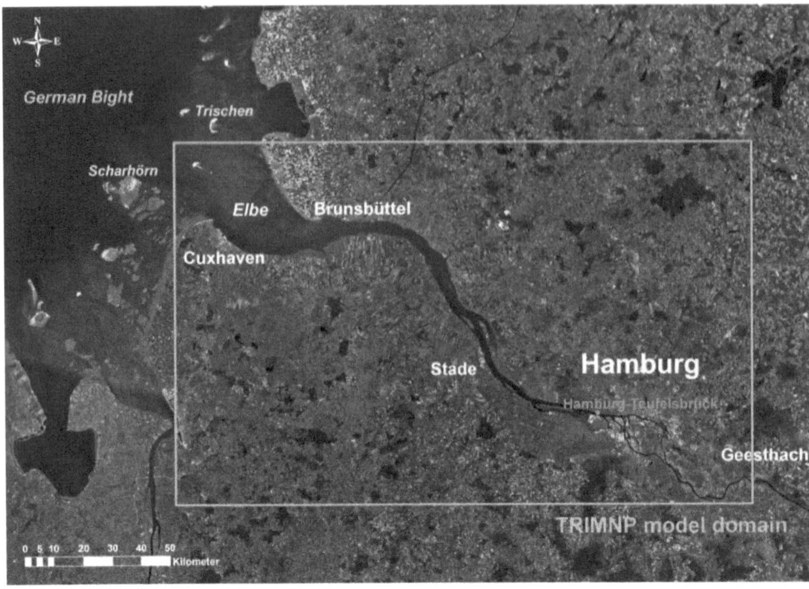

Figure 3.57 Investigation area of the Elbe estuary with indication of the TRIMNP model domain.

The following specific issues were examined:

- the effectiveness and efficiency of artificial sandbanks in the mouth of the estuary for mitigating the effects of storm surges regarding local effects and effects along the estuary and in the region of Hamburg;
- the sensitivity and the robustness of sandbanks against wave and current attack; and
- design aspects and implementation strategy of the artificial sandbanks.

Figure 3.58 gives an overview of the several sandbank scenarios investigated in the mouth of the Elbe Estuary. Both emerged and submerged sandbank designs were studied.

The effects of the seven tested different sandbanks scenarios (year 2006 condition) in the receptor area of Hamburg are summarized in Table 3.10 in terms of changes of high water (HW) levels and maximum current velocities U_{max}.

An important aspect in securing the sand islands is the prevention of erosion processes in the area, and specifically the

- erosion in the intertidal zone due to waves and currents and the;
- erosion in the supratidal zone due to wind and storm surges.

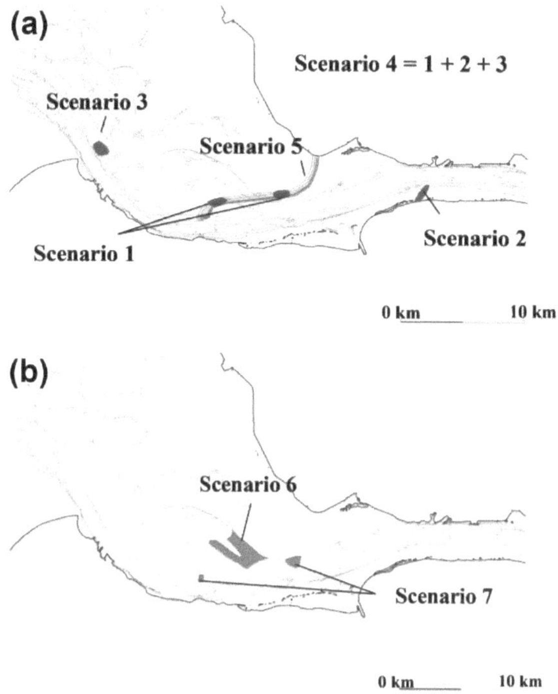

Figure 3.58 Location of emerged sandbanks (a) and submerged sandbanks (b) in the mouth of the Elbe estuary.

To determine the factors governing the erosion processes the wave model SWAN (Simulating WAves Nearshore) (Ris, 1997) was applied in the investigation area. Currents were taken from the output of the hydrodynamic model. The results showed that rubble mound revetments are necessary in the intertidal zone to ensure the sustainability of the layout (see Chapter 7, Section 7.5). Erosion in the supratidal zone occurs more frequently directly after construction because in this stage there is no protection offered by vegetation. The transplanting of vegetation after construction would decrease the erosion processes and would also be beneficial for the generation of estuarine habitats.

The artificial sandbanks in the mouth of the Elbe estuary may dampen the tidal energy to some extent. Their efficiency in this respect does not depend so much on their geographical position, but on their elevation. For the investigated sandbank scenarios, the reduction in HW levels in Hamburg is lower than 0.03 m compared with the mean annual tidal range of 3.6 m. In general, the dissipation or the redirection of the incoming tidal energy at the river mouth is not achieved to an extent that would be relevant for flood protection in Hamburg. There are local modifications of the current fields and the water elevations in the vicinity of the sandbanks, but the

TABLE 3.10 Summary of the Effects of the Various Sandbank Scenarios on the Annual Average Tidal High Water (HW) Levels and Maximum Tidal Current Velocity U_{max} at the Station Hamburg-Teufelsbrück under Current Conditions (Reference Year 2006). The Effects are Reported as Decreases of the Mean Values μ and Standard Deviations σ with Respect to the Values in Absence of Sand Banks

Scenario No.: Year	Decrease in HW Level (in m) (Annual Mean μ)	Change in HW Level (in m) (SD σ)	Decrease in Maximum Current Velocity U_{max} (in m/s) (Annual Mean μ)	Change in Maximum Current Velocity U_{max} (in m/s) (SD σ)
1: 2006	0.00	0.02	0.00	0.01
2: 2006	0.00	0.02	0.00	0.01
3: 2006	0.01	0.03	0.00	0.01
4: 2006	0.03	0.03	0.01	0.01
5: 2006	0.13	0.04	<0.01	0.01
6: 2006	0.01	0.02	0.00	0.01
7: 2006	0.02	0.03	0.01	0.01

tidal wave in the estuary as a whole is very little affected. Only the more or less hypothetical dam scenario 5 (see Figure 3.58 and Table 3.10) produces the desired reflection and damping of the tidal wave and a 10% decrease of HWL in Hamburg.

Keeping in mind the nautical and environmental constraints, it must be stated that there are narrow margins for constructions of sandbanks in the mouth of the Elbe to influence the HWLs in and close to Hamburg.

As a general conclusion, the efficiency of the introduced sandbanks increases with the reduction of the hydraulic radius at the respective estuarine cross-section. In other estuaries with other nautical and environmental constraints and under different hydro- and morphodynamic conditions, the efficiency of artificial sandbanks may be higher than in the Elbe case study site. Other possible beneficial effects of the sandbanks might be the influence on sediment transport and the creation of habitats.

3.6.4 Final Remarks

Nourishment interventions are widespread worldwide to face beach erosion for their low environmental impact and for their social and economic immediate benefits. However, the problem to find the required amount of compatible sand for beach maintenance is already an issue in some countries and will worsen in time because of the expected effect of climate change.

It is therefore crucial to set up long-term management plans of combined dredging and nourishments and to explore the potential use of near-shore and offshore borrow areas.

Nourishment efficiency can be significantly improved by means of additional protections, such as breakwaters, groins, fences, and dune maintenance. The structures have to be monitored and maintained, because malfunctioning groins are much more harmful to beach stability than the absence of protection.

Short- and long-term prediction of morphological evolution of coastal stretches requires the run of many numerical simulations after a detailed model calibration with field data.

Simplified methodologies are proposed here

- to optimize the volume and the placement of the nourished sand with the purpose of increasing the resilience of coastal profiles to wave run-up and flood risk; this approach is a statistically based approach that allows to identify the sensitivity of flood risk to selected meteomarine and geometrical variables but does not account for beach reshaping during storms; and
- to reproduce the effects of near-shore borrow areas and nourishments performed on the submerged beach; these traditional approaches model the effects of selected storms on different schematic configurations through calibrated 2D numerical tools.

An upgraded modeling tool proved to more accurately represent current velocities in presence of sediment suspension in near-shore and intertidal areas. Specifically, it is relevant to correctly reproduce the presence of ETM in estuaries to plan for dredging operations because the localized presence of an ETM results in reduced water depth and therefore increased tidal wave amplitude that may cause flooding at the estuary.

An unconventional example of integrated management of sandy stocks is the construction of sandbanks at the mouth of funnel-shaped estuaries to dampen the tidal wave energy with the locally deposited material. The objective of this intervention is to reduce high-water levels and current velocities upstream of the artificial sandbanks in the inner estuary and along its pathway. In the case here presented, the Elbe estuary, the sandbanks have insufficient efficiency to justify their construction. The efficiency can be increased with the reduction of the hydraulic radius at the respective estuarine cross-section.

3.7 Conclusions

Although maintaining their primary role of defending the coastal areas from flooding and the beaches from erosion, coastal structures have nowadays to face new challenges: the sensitivity to sea-level rise, the resilience to more frequent extreme events and larger overtopping volumes, the integrated design to cope with multiple use and growing development of coastal areas.

Adaptation to sea-level rise can be achieved by means of floating structures, such as floating breakwaters and floating wave energy converters, which can at the same time reduce incident wave heights, the former in milder climates or inside estuaries and the latter in more energetic environments. Wave energy converters are also a particularly attractive because they may combine wave energy production and with coastal protection purposes. However, recent research suggests that the wave attenuation is modest and that wave diffraction effects have to be carefully evaluated while designing near-shore installations.

To face sea-level rise, existing revetments, breakwaters, and dikes should be appropriately reinforced, paying particular attention to the crown wall stability, the wave overtopping volumes, and also the rear side armor stability as a consequence of wave overtopping. However, an attractive alternative can be the construction of overtopping resistant dikes. These structures are based on the innovative concept of tolerating a higher (but still limited) overtopping discharge, reducing the costs of the structure and creating new buffer areas that can be beneficial for environmental and societal purposes.

Other defenses, such as submerged or low-crested structures are already built to offer a moderate degree of wave protection that will decrease in time from sea-level rise. These structures can be particularly suited to enhance beach stabilization in combination with nourishment interventions and to protect harbor entrances and reduce siltation in entrance channels. Moreover, they can be designed by using ecocompatible or ecologically friendly material or specific layout to promote biodiversity enhancement, attract valuable species, and promote also recreational activities such as diving and surfing.

Besides hard defenses, nourishment interventions are of course particularly sensitive to the effects of climate change that would lead to a significant increase of the nourishment material required for traditional beach maintenance and pose the challenge of finding alternative sources of sands (i.e., near-shore and offshore borrow areas, dredged sediments from rivers or harbor entrances with periodic and systematic planning). Adopting alternative nourishment concepts, such as the well-known "sand motor" methodology developed by Stive (http://www.dezandmotor.nl/en-GB/the-sand-motor/) might be a solution.

Similarly, estuaries and tidal flats might be significantly affected by sea-level rise and increase in tidal wave amplitude. Therefore timely dredging operations are essential for protection against flooding enhanced by the propagating waves on the reduced water depth that usually occur at the estuary mouth from sand deposition. In the case of funnel-shaped estuaries, dredged materials may be used to artificially build sandbanks to reduce high water levels and current velocities in the inner estuary and along its pathway.

Any coastal defense structure or intervention will alter the natural environment, resulting in a wide range of ecological impacts that need to be considered by the coastal manager.

Impacts may induce the complete loss of a habitat, for instance if a structure replaces a natural habitat or if a sea wall disconnects an intertidal area from the sea. Habitats may also be directly modified through the process of beach nourishment and dredging (see Section 3.5.1.6) or indirectly modified through the alteration of coastal processes.

Coastal defense is the primary role for any structure and although it is not possible to improve on nature, it is possible to alter the ecological outcomes of a structure to achieve different ecological endpoints. Structures can be modified to alter biodiversity value; for example, the addition of rock pools and increased surface heterogeneity can encourage the establishment of a wider range of organisms, including commercially important species (e.g., see Martins et al., (2010); details of these modifications are given in Section 4.6)

An alternative to constructing hard defense structures is to ensure that the coastal defense properties of natural habitats are fully utilized. Natural habitats such as saltmarshes (Section 4.2), biogenic reefs (Section 4.4), and seagrass meadows (Section 4.5) can be used to attenuate waves, whereas sand dunes can be an important feature in preventing coastal erosion and flooding (Section 4.3). Overall, the key to successful coastal flood protection is to ensure that the appropriate mitigation option is chosen after a full consideration of the engineering, ecological, and socioeconomic implications (see Chapter 6).

References

Abul-Azm, A. G., & Gesraha, M. R. (2000). Approximation to the hydrodynamics of floating pontoons under oblique waves. *Ocean Engineering, 27*, 365–384.

Agamloh, E., Wallace, A., & Von Jouanne, A. (2008). Application of fluid-structure interaction simulation of an ocean wave energy extraction device. *Renewable Energy, 33*(4), 748–757.

Akkerman, G. J., Bernardini, P., van der Meer, J. W., Verheij, H., & van Hoven, A. (2007). Field tests on sea defences subject to wave overtopping. *Proceedings of Coastal Structures Conference*. Venice.

de Alegría-Arzaburu, A. R., Mariño-Tapia, I., Enriquez, C., Silva, R., & González-Leija, M. (2013). The role of Fringing coral reefs on Beach morphodynamics. *Geomorphology*.

Amoudry, L. (2008). *A review on coastal sediment transport modeling*. Report POL_ID 189. Liverpool (UK): Proudman Oceanographic Laboratory.

Angelelli, E., & Zanuttigh, B. (2012). A farm of wave activated bodies for coastal protection purposes. In *Proc. 33rd international conference on coastal engineering, July 1–6 2012, Santander, ES*.

Angelelli, E., & Zanuttigh, B. (2013). Wave disturbance induced by a one-line array of floating wave energy converters. In *Proc 10th EWTEC 2013, Aalborg*.

d'Angremond, K., Van Der Meer, J. W., & De Jong, R. J. (1996). Wave transmission at low-crested structures. *Coastal Engineering Proceedings, 1*(25).

Babarit, A. (2010). Impact of long separating distances on the energy production of two interacting wave energy converters. *Ocean Engineering, 37*, 718–729.

Babarit, A. (2013). On the park effect in arrays of oscillating wave energy converters. *Renewable Energy, 58*, 68–78. October.

Babarit, A., Folley, M., Charrayre, F., Peyrard, C., & Benoit, M. (2013). On the modelling of WECs in wave models using far field coefficients. In *Proc. EWTEC, Aalborg, Sep 2013*.

Baine, M. (2001). Artificial reefs: a review of their design, application, management and performance. *Ocean and Coastal Management, 44*(2001), 241–259.

Bates, P. D., Dawson, R. J., Hall, J. W., Horritt, M. S., Nicholls, R. J., Wicks, J., et al. (2005). Simplified two-dimensional numerical modelling of coastal flooding and example applications. *Coastal Engineering, 52*, 793–810.

Battjes, J. A., & Janssen, J. P. F. M. (1978). *Energy loss and set-up due to breaking of random waves*. In *Proceedings of 16th international conference on coastal engineering* (Vol. I), 1978.

Battjes, J. A., & Groenendijk, H. W. (2000). Wave height distributions on shallow foreshores. *Coastal Engineering, 40*, 161–182.

Beels, C., Troch, P., De Visch, K., Kofoed, J. P., & De Backer, G. (2010). Application of the time-dependent mild-slope equations for the simulation of wake effects in the lee of a farm of wave dragon wave energy converters. *Renewable Energy, 35*, 1644–1661.

Bender, C. J., & Dean, R. J. (2003). Wave field modification by bathymetric anomalies and resulting shoreline changes: a review with recent results. *Coastal Engineering, 49*, 125–153.

Blott, S. J., & Pye, K. (2004). Morphological and sedimentological changes on an artificially nourished beach, Lincolnshire, UK. *Journal of Coastal Research, 20*, 214–233.

Briggs, M. J. (2001). *Performance characteristics of a rapidly installed breakwater system*. Vicksburg: US Army Corps of Engineers, Engineer Research and Development Center, 238 pp. 01–13. http://www.worldcat.org/search?qt=hotseries&q=se%3A%22ERDC%2FCHL+TR%22.

Budal, K. (1977). "Theory of absorption of wave power by a system of interacting bodies". *Journal of Ship Research, 21*, 248–253.

Burchart, H. F., Hawkins, S. J., Zanuttigh, B., & Lamberti, A. (Eds.). (2007). *Environmental design guidelines for low crested coastal structures*. Elsevier.

Burcharth, H., Hawkins, S., Zanuttigh, B., & Lamberti, A. (2007). *Environmental design guidelines for low crested coastal structures*. The Netherlands: Elsevier (400 pp.).

Burcharth. H.F. Lykke Andersen. T. and Lopez Lara. J. (2013). "Upgrading of coastal defence structures against increased loadings caused by climate change". *Coastal Engineering* (in press).

Carballo, R., & Iglesias, G. (2013). Wave farm impact based on realistic wave-WEC interaction. *Energy, 51*(1), 216–229.

Carbon Trust. (2006). *Future marine energy. Findings of the marine energy challenge: Cost competitiveness and growth of wave and tidal stream energy*. www.thecarbontrust.co.uk.

CIRIA/CUR. *Manual on the use of rock in coastal and shoreline engineering* (1991). London: Construction Industry Research and Information Association. Special publ. 83.

Clément, A., McCullen, P., Falcão, A., Fiorentino, A., Gardner, F., Hammarlund, K., et al. (2002). *Wave energy in Europe: current status and perspectives, 6*(5), 405–431.

Colosio, F., Abbiati, M., & Airoldi, L. (2007). Effects of beach nourishment on sediments and benthic assemblages. *Marine Pollution Bulletin, 54*, 1197–1206.

Cowell, P. J., Roy, P. S., & Jones, R. A. (1995). Simulation of large-scale coastal change using a morphological behaviour model. *Marine Geology, 126*, 45–61.

Cox, R., Coghlan, I., & Kerry, C. (2007). *Floating breakwater performance in irregular waves with particular emphasis on wave transmission and reflection, energy dissipation, motion and restraining forces*. In *Proc. coastal structures '07, Venice, Italy, 1*, (pp. 351–362).

Cruz, J., Sykes, R., Siddorn, P., & Eatock Taylor, R. (2009). Wave farm design: preliminary studies on the influences of wave climate, array layout and farm control. In *Proc. EWTEC 2009, Uppsala,*

Sweden. http://www.see.ed.ac.uk/~shs/Wave%20Energy/EWTEC%202009/EWTEC%202009%20(D)/papers/218.pdf.

CSIRO. (2010). *Sea level projections.* http://www.cmar.csiro.au/sealevel/sl_proj_21st.html.

Evans, D. V. (1979). Some theoretical aspects of three dimensional wave energy absorbers. In *Proc. Of the 1st Symposium on Wave Energy Utilization.* Sweden: Gothenburg.

De Rouck, J., Van Doorslaer, K., Goemaere, J., & Verhaeghe, H. (2010). Geotechnical design of breakwaters in Ostend on very soft soil. In *ASCE, proc. ICCE 2010, Shanghai, China.*

Dean, R. G. (2002). *Beach nourishment. Theory and practice. Advanced series on ocean engineering* (Vol. 18). World Scientific Publishing Co. Pte. Ltd.

Dean, R. G., Chen, R., & Browder, A. E. (1997). Full scale monitoring study of a submerged breakwater, Palm Beach, Florida, USA. *Coastal Engineering, 29*(3), 291–315.

Deltares. (2012). *Technisch Rapport Toetsen Grasbekledingen op Dijken* (only in Dutch).

Dolan, R., & Stewart, D. (2006). A concept for reducing ecological impacts of beach nourishment and tidal inlet bypassing. *Shore and Beach, 74*(1), 28–31.

Dredging Operations Technical Support. (1994). *Water wave attenuation over nearshore underwater mudbanks and mud berms.* Technical Note DRP-1–14. Vicksburg (MS): U.S. Army Engineer Waterways Experiment Station, 16 pp.

Environment Agency. (2010). *Guidance for outline design of nearshore detached breakwaters on sandy macro-tidal coasts,* ISBN 978-1-84911-178-2. Project: SC060026/R1. Environment Agency report SCHO0210BRYO-E-P.

EurOtop. (2007). In T. Pullen, N. W. H. Allsop, T. Bruce, A. Kortenhaus, H. Schüttrumpf, & J. W. Van der Meer (Eds.), *European manual for the assessment of wave overtopping.* Available from www.overtopping-manual.com.

Evans, D. V. (1979). Some theoretical aspects of three dimensional wave energy absorbers. *In Proc. the 1st Symposium on Wave Energy Utilization,* Gothenburg, Sweden.

Fairley, I., Davidson, M., & Kingston, K. (2009). The morpho-dynamics of a beach protected by detached breakwaters in a high energy tidal environment. *Journal of Coastal Research, 56,* 607–611.

Falcão, A., & Rodrigues, R. (2002). Stochastic modelling of OWC wave power plant performance. *Applied Ocean Research, 24*(2), 59–71.

Falnes, J. (1980). Radiation impedance matrix and optimum power absorption for interacting oscillators in surface waves. *Applied Ocean Research, 2,* 75–80.

Fitzgerald. J. (2009). *"Position mooring of wave energy converters"* (Ph.D. thesis). Goteborg, Sweden: at the Chalmers University of Technology.

Ge, J., Chen, C., Qi, J., Ding, P., & Beardsley, R. C. (2012). A dike–groyne algorithm in a terrain-following coordinate ocean model (FVCOM): Development, validation and application. *Ocean Modelling, 47*(C), 26–40. http://dx.doi.org/10.1016/j.ocemod.2012.01.006.

Ge, J., Much, D., Kappenberg, J., Nino, O., Ding, P., & Chen, J. (2013). Simulating storm flooding maps over Hafen City and sea level rise scenarios. *Journal of Flood Risk Management.* http://dx.doi.org/10.1111/jfr3.12054 (published online).

Glindemann, H., Thode, K., Dücker, H. P., & Witte, H. H. (2006). *Concept for a sustainable development of the Tidal Elbe River as an artery of the metropolitan region Hamburg and beyond. A contribution for discussion by Hamburg Port Authority and the Federal Administration for Waterways and Navigation.* http://www.tideelbe.de/files/strategiepapier_tideelbe_komplett_english_final_020408.pdf.

Goda, Y. (2000). *Random seas and design of maritime structure.* In P. L.-F. Liu (Ed.), *Advanced series on ocean engineering* (Vol. 15). Singapore: World Scientific, 444 page(s).

Gourlay, M. R. (1996). Wave set-up on coral reefs. 1. Set-up and wave-generated flow on an idealised two dimensional horizontal reef. *Coastal Engineering, 27*(3), 161–193.

Guanche, R., Losada, I. J., & Lara, J. L. (2009). Numerical modelling of coastal structures stability. *Coastal Engineering, 56*(5–6), 543–558. ELSEVIER.

Harris, R. E., Johanning, L., & Wolfram, J. (2004). *Mooring systems for wave energy converters: A review of design issues and choices*. Edinburgh, UK: Heriot-Watt University.

Hasselmann, K., Barnett, T. P., Bouws, E., Carlson, H., Cartwright, D. E., Enke, K., Ewing, J. A., Gienapp, H., Hasselmann, D. E., Kruseman, P., Meerburg, A., Mller, P., Olbers, D. J., Richter, K., Sell, W., & Walden, H. (1973). Measurements of wind-wave growth and swell decay during the Joint North Sea Wave Project (JONSWAP)′ Ergnzungsheft zur Deutschen Hydrographischen Zeitschrift Reihe. A(8) (Nr. 12). p. 95.

Hawkes, P. J., Gouldby, B. P., & Tawn, J. A. (2002). The joint probability of waves and water levels in coastal engineering design. *Journal of Hydraulic Research, 40*(3), 241–251.

Headland, J. R. (1995). Floating breakwaters. In G. P. Tsinker (Ed.), *Marine Structures Engineering: specialised applications* (pp. 367–411). Chapman & Hall.

Hervouet, J. M. (2007). *Hydrodynamics of free surface Flows: Modelling with the finite element method*. Chichester (UK): John Wiley & Sons.

Hewlett, H. W. M., Boorman, L. A., & Bramley, M. E. (1985). *Reinforcement of steep grassed waterways*. CIRIA report 166, London.

Hudson, R. Y. (1953). Wave forces on breakwaters. *Transactions of the American Society of Civil Engineers, 118*, 653–674.

Hughes, S., Thornton, C., Scholl, B., Youngblood, N., Beasley, J., Tucker, R., et al. (2013). *Wave overtopping Resiliency of grass and Turf reinforcement mats on Sandy soils*. Edinburgh: ICE.

Hughes, S., Thornton, C., van der Meer, J. W., & Scholl, B. (2012). Improvements in describing wave overtopping processes. In *ASCE, proc. ICCE 2012, santander, Spain*.

Hughes, S. A., & Nadal, N. C. (2009). Laboratory study of combined wave overtopping and storm surge overflow of a levee. *Coastal Engineering, 56*, 244–259.

Hunter, J. R. (2010). Estimating sea-level extremes under conditions of uncertain sea-level rise. *Climate Change, 99*, 331–350. http://dx.doi.org/10.1007/s 10584-009-9671-6.

Ingram, D., Smith, G., Bittencourt-Ferreira, C., & Smith, H. (2011). Protocols for the equitable assessment of marine energy converters. In *European ocean energy association*.

del Jesus, M., Lara, J. L., & Losada, I. J. (2012). Three-dimensional interaction of waves and porous coastal structures. *Part I: Numerical Model Formulation. Coastal Engineering, 64*, 57–72.

Johnson, H. K., Karambas, T. V., Avgeris, I., Zanuttigh, B., Gonzalez-Marco, D., et al. (2005). Modelling of waves and currents around submerged breakwaters. *Coastal Engineering, 52*(10), 949–969.

Kappenberg, J., & Grabemann, I. (2001). Variability of mixing zones and estuarine turbidity in the Elbe and Wesser estuaries. *Estuary, 24*(5), 699–706.

Koftis, T., & Prinos, P. (2006). Experimental study on wave overtopping of floating breakwaters. In *Proc. of coastlab06, Porto, Portugal* (pp. 351–362).

Kojima, H., Ijima, T., & Nakamuta, T. (1987). *Impact of offshore dredging on beaches along the Genkai Sea. Japan*. In *Proc. 20th coastal engineering conference, 2*; (pp. 1281–1295).

Koutandos, E., Prinos, P., & Gironella, X. (2005). Floating breakwaters under regular and irregular wave forcing-reflection and transmission characteristics. *Journal of Hydraulic Research, 43*(2), 174–188.

Lara, J. L., Garcia, N., & Losada, I. J. (2006). RANS modelling applied to random wave interaction with submerged permeable structures. *Coastal Engineering, 53*(5), 395–417.

Lara, J. L., Losada, I. J., & Guanche, R. (2008). Wave interaction with low mound breakwaters using a RANS model. *Ocean Engineering, 35*(7), 1388–1400. ELSEVIER.

Lara, J. L., Ruju, A., & Losada, I. J. (2011). RANS modelling of long waves induced by a transient wave group on a beach. Proc. of the Royal Society A-Mathematical. *Physical and Engineering Sciences, 467*(2129), 1215–1242.

Lara, J., del Jesus, M., & Losada, I. J. (2012). Three dimensional interaction of waves and porous structures. Part II: Experimental validation. *Coastal Engineering, 64*, 26–42. ELSEVIER.

Leont'yev, I. O. (2003). Modeling the morphological response in a coastal zone for different temporal scales. In V. C. Lakhan (Ed.), *Advances in coastal modeling* (pp. 299–335). Amsterdam, The Netherlands: Elsevier Science Publishers.

Leont'yev, I. O. (2008). Budget of sediments and forecast of long-term coastal changes. *Oceanology, 48*(3), 467–476.

Leont'yev, I. O. (2012). Modeling beach profile evolution at centennial to millennial scales. *Oceanology, 52*(4), 588–599.

Losada, I. J., Silva, R., & Losada, M. A. (1996a). 3-D non-breaking regular wave interaction with submerged breakwaters. *Coastal Engineering, 28*, 229–248.

Losada, I. J., Silva, R., & Losada, M. A. (1996b). Interaction of non- breaking directional random waves with submerged break- waters. *Coastal Engineering, 28*, 249–266.

Losada, I. J., Lara, J. L., Guanche, R., & Gonzalez-Ondina, J. M. (2008). Numerical analysis of wave overtopping of rubble mound breakwaters. *Coastal Engineering, 55*(1), 47–62. ELSEVIER.

Lykke Andersen. T. (2006). Hydraulic response of rubble mound breakwaters. Scale effects – berm breakwaters (Ph.D. thesis). Denmark: Aalborg University Department of Civil engineering.

Lykke Andersen, T., & Burcharth, H. F. (2010). A new formula for front slope recession of berm breakwaters. *Coastal Engineering, 57*, 359–374.

Lykke Andersen, T., Van der Meer, J. W., Burcharth, H. F., & Sigurdarson, S. (2012). Stability of hardly reshaping berm breakwaters. In *Proc. 33rd int. conf. on coastal eng., Santander, Spain.*

Macagno, E. O. (1954). Houles dans un Can Presentent un Passage en Charge. *La Houille Blanche, 9*(1), 10–37.

Martinelli, L., Pezzutto, P., & Ruol, P. (2013). Experimentally based model to size the geometry of a new OWC device, with reference to the Mediterranean Sea wave climate. *Energies, 6*(9), 4696–4720.

Martinelli, L., Ruol, P., & Cortellazzo, G. (2012). *On mooring design of wave energy converters: the seabreath application.* In *ICCE 2012* (Vol. 32). ISSN: http://journals.tdl.org/icce/index.php/icce.

Martinelli, L., Ruol, P., & Zanuttigh, B. (2008). Wave basin experiments on floating breakwaters with different layouts. *Applied Ocean Research, 30*(3), 199–207.

Martinelli, L., Zanuttigh, B., & Corbau, C. (2010). Assessment of coastal flooding risk along the Emilia Romagna littoral, Italy. *Coastal Engineering, 57*(11–12), 1042–1058.

Mehta, P. K. (2001). Reducing the environmental impact of concrete. *Concrete International, 23*(10), 61–66.

Mendoza, E., Silva, R., Enriquez-Ortiz, C., Díaz-Hernández, G., & Lara, J. L. (2011). An Experimental Evaluation of Wave Energy dissipation Due to Submerged Structures. In *Coastal structures 2011. Japan.*

Mendoza, E., Silva, R., Zanuttigh, B., Angelelli, E., Andersen, T. L., Martinelli, L., & Ruol, P. (2014). Beach Response to Wave Energy Converter Farms Acting as Coastal Defence. *Coastal Engineering, 87*(Special issue), 97–111, 10.1016/j.coastaleng.2013.10.018.

Millar, D. L., Smith, H. C. M., & Reeve, D. E. (2007). Modelling analysis of the sensitivity of shoreline change to a wave farm. *Ocean Engineering, 34*(5–6), 884–901.

Ministry of Construction. (1992). *Artificial reef design manual.* Coastal Branch, River Division, Tokyo, Japan (in Japanese).

Nelson, W. G. (1988). An overview of the effects of beach nourishment on the sand beach fauna. In L. S. Tait (Ed.), *Problems and advancements in beach nourishment: 88. Beach preservation technology.* Tallahassee, FL: Florida Shore and Beach Preservation Association Inc.

Nguyen K. D., 1988. Modélisation numérique 2-D et 3-D de la circulation générale en milieux estuariens. Application a la Gironde. Thèse de 3eme cycle, Université Paris VI, 153p.

Nicoletti, L., Paganelli, D., & Gabellini, M. (2006). *Environmental aspects of relict sand dredging for beach nourishment: proposal for a monitoring protocol* (I Quaderni ICRAM, in Italian).

Nielsen, P. (2009). *Coastal and Estuarine Processes, Advanced Series on Ocean Engineering –volume 29.* World Scientific Publishing Co. Pte. Ltd, 343pp.

Nørgaard. J. Q. H. (2013). Upgrade and design of coastal structures exposed to climate changes. Aalborg: Aalborg University. Department of Civil Engineering. (DCE thesis; No. 51).

Nørgaard, J. H., & Lykke Andersen, T. (2012). Experimental and numerical investigation of wave transmission from a floating wave dragon wave energy converter. In *International ocean and polar engineering conference, Rhodes 2012 (ISOPE).*

Nørgaard, J. H., Andersen, T. L., & Burcharth, H. F. (2013). Wave Loads on Rubble Mound Breakwater Crown Walls in Deep and Shallow Water Wave Conditions. *Coastal Engineering, 80*(October), 137–147, 10.1016/j.coastaleng.2013.06.003.

Nørgaard, J. Q. H., Lykke Andersen, T., Burcharth, H. F., & Steendam, G. J. (2013). Analysis of overtopping flow on sea dikes in oblique and short-crested waves. *Coastal Engineering, 76.* June 2013, Pages 43–54, ISSN 0378-3839, 10.1016/j.coastaleng.2013.01.012.

Nørgaard, J. H., Lykke Andersen, T., & Kofoed, J. P. (2011). *Wave dragon wave energy converters used as coastal protection: a physical model test study.* Coastal Structures 2011. International Conference September 6–8, 2001. Yokohama, Japan., The Coasts, Oceans, Ports, and Rivers Institute (COPRI): The American Society of Civil Engineers.

Nørgaard, J. Q. H., Andersen, T. L., & Burcharth, H. F. (2014). Distribution of Individual Overtopping Wave Volumes in Deep and Shallow Water Wave Conditions. *Coastal Engineering, 83*, 15–23.

Ortega, L. M., & Floriano, M. A. (2009). *Condamine floating dock, Monaco. Encyclopedia of structural health monitoring.* Wiley Online Library.

Ostrowski, R., Pruszak, Z., Schönhofer, J., Szmytkiewicz, M., & Szmytkiewicz, P. (2012). Influence of damaged groins on nourished seashore. In *Proceedings of 33rd international conference on coastal engineering (Santander 2012).*

Ostrowski, R., Pruszak, Z., Skaja, M., Szmytkiewicz, M., Trifonova, E., Keremedchiev, S., et al. (2010). Hydrodynamics and lithodynamics of dissipative and reflective shores in view of field investigations. *Archives of Hydro-Engineering & Environmental Mechanics, IBW PAN, Gdańsk, 57*(3–4), 219–241.

Palha, A., Mendes, L., Conceiç, J. F., Brito-Melo, A., & Sarmento, A. (2010). The impact of wave energy farms in the shoreline wave climate: Portuguese pilot zone case study using Pelamis energy wave devices. *Renewable Energy, 35*(1), 62–77.

Pan, S., Horrillo-Caraballo, J. M., Reeve, D. E., & Simmonds, D. (2013). *Morphological modelling of V-shaped submerged breakwaters. Coasts, marine structures and breakwaters.* Edinburgh, UK: ICE.

Pedersen. J. (1996). *"Wave Forces and Overtopping on Crown Walls of Rubble Mound Breakwaters"* (Ph.D. thesis). Series paper 12, ISBN 0909–4296, Hydraulics & Coastal Engineering Lab., Dept. of Civil Engineering, Denmark: Aalborg University.

Peterson, C. H., & Bishop, M. J. (2005). Assessing the environmental impacts of beach nourishment. *Bioscience, 55*, 887–896.

Peterson, C. H., Bishop, M. J., Johnson, G. A., D'Anna, L. M., & Manning, L. M. (2006). Exploiting beach filling as an unaffordable experiment: benthic intertidal impacts propagating upwards to shore birds. *Journal of Experimental Marine Biology and Ecology, 338*, 205–221.

Peterson, C. H., Hickerson, D. H. M., & Johnson, G. G. (2000). Short-term consequences of nourishment and bulldozing on the dominant large invertebrates of a sandy beach. *Journal of Coastal Research, 16*, 368–378.

Pilarczyk, K. W. (1998). Dikes and Revetments: Design, Maintenance and Safety Assessment, published by A.A. Balkema. *ISBN*, 9054104554, 9789054104551, 562 pages.

Pilarczyk, K. W. (2003). Design of low-crested (submerged) structures–an overview. In *6th international conference on coastal and port engineering in developing countries, Colombo, Sri Lanka.*

Pina, G., & Valdez, J. M. (1990). Experiments on coastal protection submerged breakwaters: a way to look at the results. *Coastal Engineering*, 1592–1605.

Purvis, M. J., Bates, P. D., & Hayes, C. M. (2008). A probabilistic methodology to estimate future coastal flood risk due to sea level rise. *Coastal Engineering, 55*, 1062–1073.

Rahman, Md A., Mizutani, N., & Kawasaki, K. (2006). Numerical modeling of dynamic responses and mooring forces of submerged floating breakwater. *Coastal Engineering, 53*, 799–815.

Ranasinghe, R., Larson, M., & Savioli, J. (2010). Shoreline response to a single shore-parallel submerged breakwater. *Coastal Engineering, 57*, 1006–1017.

Raosa, A. N., Zanuttigh, B., Lara, J. L., & Hughes, S. (2012). 2DV RANS-VOF modelling of depths and velocities of overtopping waves at overwashed dikes. In *ICCE 2012, Santander.*

Ris, R. C. (1997). *Spectral modelling of wind waves in coastal areas*. report no. 97–4, TU Delft. Communications on Hydraulic and Geotechnical Engineering.

Roelvink, D., Reniers, A., Van Dongeren, A., Van Thiel de Vries, J., McCall, R., & Lescinski, J. (2009). Modelling storm impacts on beaches, dunes and barrier islands. *Coastal Engineering, 56*, 1133–1152.

Ruol, P., Martinelli, L., & Pezzutto, P. (2012). *Limits of the new transmission formula for Π-type floating breakwaters*. In *ICCE 2012* (Vol. 32). ISSN: 2156–1028. http://journals.tdl.org/icce/index.php/icce.

Ruol, P., Martinelli, L., & Pezzutto, P. (2013). Formula to Predict Transmission for Π-type Floating Breakwaters. *Journal of Waterway, Port, Coastal and Ocean Engineering, 139*(1), 1–8.

Ruol, P., Zanuttigh, B., Kofoed, J. P., Martinelli, L., & Frigaard, P. (2010). Near-shore floating wave Energy converters: benefits for coastal protection. In *Proc. ICCE No. 32(2010), Shanghai, China.* Paper #: structures 6.1 http://journals.tdl.org/ICCE/.

Schüttrumpf. H.F.R. (2001). *Wellenüberlaufströmung bei See-deichen* (Ph.D. thesis). Technical University Braunschweig.

Schüttrumpf, H., & Oumeraci, H. (2005). Layer thicknesses and velocities of wave overtopping flow at sea dikes. *Coastal Engineering, 52*(6), 473–495.

Schüttrumpf, H., & Van Gent, M. R. A. (2003). Wave overtopping at seadikes. In *ASCE, proc. coastal structures 2003* (pp. 431–443).

Shen, H. B., Zhang, X. F., Yan, J., & Wang, Y. H. (2009). Effect of ratio of width to depth of river channel on sediment transport capacity in Lower Yellow River. *Engineering Journal of Wuhan University, 43*(3), 273–278 (in Chinese).

Short, A. D. (1992). Beach systems of the central Netherlands coast: processes, morphology and structural impacts in a storm driven multi-bar system. *Marine Geology, 107*(1), 103–137.

Silva, R., Losada, I. J., & Losada, M. A. (2000). Reflection and transmission of tsunami waves by coastal structures. *Applied Ocean Research, 22*(4), 215–223.

Silva, R., Losada, M., & Salles, P. (2006). Modelling linear wave transformation induced by dissipative structures—Random waves. *Ocean Eng., 33*(16), 2174–2194.

Silva, R., Mendoza, E., & Chávez, X. (2012). *Diseño de prototipo de rompeolas sumergido para disipación de oleaje y recuperación de playas, mediante el modelado en laboratorio*. Technical report elaborated for AXIS Ingeniería, Mexico. 46 pp. (In Spanish).

Silva, R., Salles, P., & Palacio, A. (2002). Linear waves propagating over a rapidly varying finite porous bed. *Coastal Engineering, 44*(3), 239–260.

Smaoui, H., Nguyen, K. D., & Sergent, P. (2012). *Numerical modeling of the turbidity maximum in the Gironde estuary and its impact on tide propagation*. Theseus Project Report WP 2.6, 30p.

Sothmann, J., Schuster, D., Kappenberg, J., & Ohle, N. (2011). Efficiency of artificial sandbanks in the mouth of the Elbe Estuary for damping the incoming tidal energy. In *Proceedings of the 5th international short conference on applied coastal research (Aachen 2011).*

Speybroeck, J., Bonte, D., Courtens, W., Gheskiere, T., Grootaert, P., Maelfait, J. P., et al. (2006). Beach nourishment: an ecologically sound coastal defence alternative? A review. *Aquatic Conservation – Marine and Freshwater Ecosystems, 16*, 419–435.

Stauble, D. K. (2008). Coastal engineering project impact assessment using long-term morphodynamic change analysis. In *FSBPA conference proceedings*. In *http://www.fsbpa.com/08Proceedings/02Stauble2008.pdf.*

Steendam, G. J., van der Meer, J. W., Hardeman, B., & van Hoven, A. (2010). Destructive wave overtopping tests on grass covered landward slopes of dikes and transitions to berms. In *ASCE, proc. ICCE 2010, Shanghai.*

Steendam, G. J., Peeters, P., van der Meer, J. W., Van Doorslaer, K., & Trouw, K. (2011). Destructive wave overtopping tests on Flemish dikes. In *ASCE, proc. Coastal structures 2011, Yokohama, Japan.*

Steendam, G. J., de Vries, W., van der Meer, J. W., van Hoven, A., de Raat, G., & Frissel, J. Y. (2008). Influence of management and maintenance on erosive impact of wave overtopping on grass covered slopes of dikes; Tests. In Samuels, et al. (Eds.), *Proc. FloodRisk* (pp. 523–533). Oxford: UK. Flood Risk Management: Research and Practice, ISBN 978-0-415-48507-4.

Stive, M. J. F., Ranasinghe, R., & Colwell, P. J. (2010). Sea level rise and coastal erosion. Ch. 37. In Y. C. Kim (Ed.), *Handbook of coastal and ocean engineering* (pp. 1023–1037). New Jersey: World Scientific.

Stive, M. J. F., & de Vriend, H. J. (1995). Modelling shoreface profile evolution. *Marine Geology, 126*, 235–248.

STOWA, (2010). State of the art asfaltdijkbekledingen, Stichting Toegepast Onderzoek WAterbeheer. *Amersfoort, rapport STOWA*, 2010. W6(only in Dutch).

Sumer, B. M., Fredsøe, J., Lamberti, A., Zanuttigh, B., Dixen, M., Gislason, K., et al. (2005). Local scour at roundhead and along the trunk of low crested structures. *Coastal Engineering, 52*(10), 995–1025.

Sun, D. M., Feng, P., & Zhang, M. J. (2009). Numerical model for predicting the air losses in the operation process of air-cushioned surge chamber. *Hydraulic Engineering, 40*(10), 1227–1233 (in Chinese).

Sundar, V., & Sannasiraj, S. A. (2013). Artificial reefs: a review. *The International Journal of Ocean and Climate Systems, 4*(2), 117–124.

TAW. (2002). Technisch rapport golfoploop en golfoverslag bij dijken (Technicalreport on wave run-up and wave overtopping at dikes - in Dutch). In *Technical advisory committee on water defences, delft* (p. 50).

Tedd. J. (2007). *"Testing, analysis and control of wave dragon, wave energy converter"* (Ph.D. thesis). Defended in public at Aalborg University.

THESEUS. (2010). *Integrated inventory of data and prototype experience on coastal defences and technologies*. Official Deliverable 2.1 http://www.theseusproject.eu/index.php.

THESEUS. (2013). Integrated report on design of innovative coastal structures and best practices for coastal defence. *Results from Numerical, Experimental and Prototype Testing*. Official Deliverable 2.7 http://www.theseusproject.eu/index.php.

Tolba. E.R.A.S. (1998). *Behavior of floating breakwaters under wave action* (Ph.D. thesis). Suez Canal University.

Toorman. E.A. (2011). Low-reynolds modelling of high-concentrated near-bottom suspended sediment transport. Symposium on two-phase modelling for sediment dynamics in geophysical flows (thesis-2011, Paris, April 26–28, 2011), Abstract, 4 pp.

Toorman, E. A., & Bi, Q. (2011). Dealing with benthic high-concentrated layers and fluid mud in cohesive sediment transport modeling. 11th Int. Conf. on Cohesive Sediment Transport Processes (INTERCOH'11, Shanghai, 18–21 October 2011). Book of Abstracts, pp. 85–86.

Toorman, E. A., & Bi, Q. (2012). Section 2.4, EU FP7 FIELD_AC Report D3.3. In M. Liste-Muñoz, & J. Monbaliu (Eds.), *Improvements to sea-bed boundary conditions* (pp. 19–31).

Toorman, E. A., & Bi, Q. (2013). A practical model for drag modulation by suspended sediment, with application to the Scheldt estuary. 12th Int. Conf. on Cohesive Sediment Transport (INTERCOH2013, 20–24 October 2013, UFL, Gainesville, FL), Book of Abstracts, pp. 7–8.

Tuan, T. Q., Cat, V. M., & Trung, L. H. (2009). Experimental study on wave overtopping at seadikes with vertical crown-wall. In *Proceedings of the 5th international conference on APAC 2009.*

Tsinker, G. (1994). *Marine structure engineering: specialized application.* Chapman & Hall, International Thomson Publishing Inc.

U.S. Army Corps of Engineers. (2002). *Coastal engineering manual. Engineer manual 1110-2-1100.* Washington, D.C: U.S. Army Corps of Engineers (in 6 volumes).

Uda, T., Agemori, C., & Chujo, N. (1986). Beach changes caused by offshore dredging. *Coastal Engineering in Japan, 29*, 215–226.

Uzaki, K., Ikehata, Y., Matsunaga, N., & Finkl, C. (2010). Performance of the wave energy dissipation of a floating breakwater with truss structures and the quantification of transmission coefficients. *Journal of Coastal Research, 27*, 687–697.

Van der Meer, J. W. (1988). Rock slopes and gravel beaches under wave attack. PhD thesis, Delft University of Tehcnology, also Delft publication n. 36. http://www.vandermeerconsulting.nl/downloads/stability_b/1988_vandermeer_doctoralthesis.pdf

Van der Meer, J. W. (2011). *The Wave Run-up Simulator. Idea, necessity, theoretical background and design.* Van der Meer Consulting. Report vdm11355.

Van der Meer, J. W., Briganti, R., Zanuttigh, B., & Wang, B. (2005). Wave transmission and reflection at low crested structures: design formulae oblique wave attack and spectral change. *Coastal Engineering, 52*(10–11), 915–929.

Van der Meer, J. W., & Janssen, J. P. F. M. (1994). *Wave run-up and wave overtopping at dikes and revetments.* Delft Hydraulics.

Van der Meer, J. W., Hardeman, B., Steendam, G. J., Schttrumpf, H., & Verheij, H. (2010). Flow depths and velocities at crest and inner slope of a dike, in theory and with the wave overtopping simulator. In *ASCE, proc. ICCE 2010, Shanghai.*

Van der Meer, J. W., Thornton, C., & Hughes, S. (2011). Design and operation of the US Wave Overtopping Simulator. In *ASCE, proc. Coastal structures 2011, Yokohama, Japan.*

Van der Meer, J. W., Steendam, G. J., de Raat, G., & Bernardini, P. (2008). Further developments on the wave overtopping simulator. In *ASCE, proc. ICCE 2008, Hamburg* (pp. 2957–2969).

Van der Meer, J. W., Bernardini, P., Steendam, G. J., Akkerman, G. J., & Hoffmans, G. J. C. M. (2007). The wave overtopping simulator in action. In *Proc. coastal structures, Venice, Italy.*

Van der Meer, J. W., Bernardini, P., Snijders, W., & Regeling, H. J. (2006). The wave overtopping simulator. In *ASCE, ICCE 2006, San Diego* (pp. 4654–4666).

Van der Meer, J. W., Schrijver, R., Hardeman, B., van Hoven, A., Verheij, H., & Steendam, G. J. (2009). Guidance on erosion resistance of inner slopes of dikes from three years of testing with the Wave Overtopping Simulator. In *Proc. ICE, coasts, Marine structures and breakwaters 2009, Edinburgh, UK.*

Van Gent, M. R. A., Smale, A. J., & Kuiper, C. (2003). Stability of rock slopes with shallow foreshores. In Ja Melby (Ed.), *Proc 4th int coastal structures conf. Portland, 26-30 August 2003, ASCE Reston* (pp. 100–112).

Van Gent, M. R. A. (2002). "Wave overtopping events at dikes". *World Scientific, Proc. ICCE 2002, Cardif, 2*, 2203–2215.

Van Rijn, L. (1993). Effective bed roughness. Chapter 6. In *Principles of sediment transport in rivers, estuaries and coastal seas, Part 1.* Blokzijl (NL): Aqua Publications.

Van Rijn, L. C., Tonnon, P. K., & Walstra, D. J. R. (2011). Numerical modeling of erosion and accretion of plane sloping beaches at different scales. *Coastal Engineering, 58*, 637–655.

Victor, L., van der Meer, J. W., & Troch, P. (2012). Probability distribution of individual wave overtopping volumes for smooth impermeable steep slopes with low crest freeboards. *Coastal Engineering, 64*. June 2012, Pages 87–101, ISSN 0378-3839, 10.1016.

Villani, M., Bosboom, J., Zijlema, M., & Stive, M. (2012). Circulation patterns and shoreline response induced by submerged breakwaters. *Coastal Engineering Proceedings, 1*(33). http://dx.doi.org/10.9753/icce.v33.structures.25. structures.25.

WaveNet Project. www.waveenergy.net/Library/WaveNet%20Full%20Report(11.1).pdf

Williams, K. J. (1988). An experimental study of wave-obstacle interaction in a two-dimensional domain. *Journal of Hydraulic Research, 26*, 463–482.

Woolmer, A. P., Syvret, M., & FitzGerald, A. (2011). Restoration of Native Oyster, *Ostrea edulis*. In *South Wales: Options and approaches* (p. 93). CCW Contract Science Report No: 960.

Zanuttigh, B., & Angelelli, E. (2013). Experimental investigation of wave energy converters for coastal protection purpose. *Coastal Engineering, 80*, 148–159.

Zanuttigh, B., Angelelli, E., & Kofoed, J. P. (2013). Effects of mooring systems on the performance of a wave activated body energy converter. *Renewable Energy, 57*(9), 422–431.

Zanuttigh, B., Martinelli, L., Castagnetti, M., Ruol, P., Kofoed, J. P., & Frigaard, P. (2010). Integration of wave energy converters into coastal protection schemes. In *3rd international conference on ocean energy (ICOE), Bilbao*.

Chapter 4

Ecological Approaches to Coastal Risk Mitigation

Simon Hoggart[1], Stephen J. Hawkins[2], Katrin Bohn[2,3], Laura Airoldi[4], Jim van Belzen[5], Amandine Bichot[6], David T. Bilton[1], Tjeerd J. Bouma[5], Marina Antonia Colangelo[7], Andrew J. Davies[3], Filippo Ferrario[4], Louise B. Firth[8], Cristina Galván[9], Michael Hanley[10], Hugues Heurtefeux[11], Javier L. Lara[12], Iñigo Losada Rodriguez[12], Maria Maza[12], Barbara Ondiviela Eizaguirre[12], Simon D. Rundle[1], Martin W. Skov[3], Elisabeth M. Strain[13], Anissia White[10], Liquan Zhang[14], Zhenchang Zhu[14,15], Richard C. Thompson[1]

[1] *Marine Biology and Ecology Research Centre, School of Marine Science and Engineering, University of Plymouth (UoP), Plymouth, UK*

[2] *Ocean and Earth Science, National Oceanography Centre Southampton, University of Southampton (Soton), Southampton, UK*

[3] *School of Ocean Sciences, Bangor University (BU), Bangor, UK*

[4] *Dipartimento di Scienze Biologiche, Geologiche ed Ambientali, University of Bologna (UniBo) Ravenna, Italy*

[5] *Royal Netherlands Institute for Sea Research, Spatial Ecology Department (NIOZ), Yerseke, The Netherlands*

[6] *EID Méditerranée (EID), Montpellier, France*

[7] *Dip. Scienze Biologiche, Geologiche Ambientali, University of Bologna—Ravenna Campus (UniBo), Ravenna, Italy*

[8] *Ryan Institute, National University of Ireland, Galway, Ireland; School of Ocean Sciences, Bangor University (BU)*

[9] *University of Cantabria (UC), Santander, Spain*

Coastal Risk Management in a Changing Climate. http://dx.doi.org/10.1016/B978-0-12-397310-8.00004-X

[10] *Marine Biology and Ecology Research Centre, School of Biological Sciences, University of Plymouth (UoP), Plymouth, UK*
[11] *Entente Interdépartementale pour la Démoustication du Littoral Méditerranéen (EID), Montpellier, France*
[12] *Environmental Hydraulics Institute "IH Cantabria", Universidad de Cantabria (UC), Santander, Spain*
[13] *Dipartimento di Scienze Biologiche, Geologiche ed Ambientali, University of Bologna (UniBo), Tas, Australia*
[14] *State Key Laboratory of Estuarine and Coastal Research, East China Normal University (SKLEC), Shanghai, PR China*
[15] *Royal Netherlands Institute for Sea Research (NIOZ), Yerseke, The Netherlands*

4.1 Introduction

Coastal ecosystems provide a wide range of ecosystem goods and services that are of value to society (MEA, 2005). These include regulating services such as flood protection, supporting services such as nutrient cycling and water filtration, food provision, habitat provision including nursery habitats for juveniles some of which are commercially important species, and key cultural and recreational services. Costanza et al. (1997) estimated the global contribution of coastal habitats to ecosystem services to be in the order of $12.3 trillion per year.

It is clear that in addition to the more traditionally viewed impacts on intrinsic biodiversity value, the possible loss of key ecosystem services through poor management could have significant economic consequences. These valuable ecosystem services are compromised by development for residential, agricultural and industrial activities together with tourism and transport infrastructure such as roads, railways, and ports. In addition maritime fringing habitats and the intertidal zone are often degraded or lost because of land reclamation. Valuable economic activities along our coasts are often at risk from erosion and flooding, which can be exacerbated by rising and stormier seas as a result of climate change.

In response, coastal defense structures have been built to protect economic and societal assets. With rising sea levels hard engineered coastlines result in coastal squeeze (Doody, 2004) where maritime habitats such as sand dunes and saltmarshes are unable to migrate landward. Other pressures on coastal ecosystems include eutrophication at regional and local scales and pollution associated with our industrial society such as persistent organic compounds, pesticides from agriculture, antifouling paints and heavy metals together with hydrocarbons from oil spills and road run off.

Overexploitation of living resources also has a major impact on the maritime environment with both direct and indirect effects. These various impacts interact at a variety of scales, and are exacerbated by the pervasive effects of global environmental change such as rising temperatures, ocean acidification, and invasions by non-native species.

The focus of this chapter is the role of natural habitats in providing a key service of coastal protection. In particular, the emphasis is on how to manage coastal systems in face of multiple threats and stressors. The habitats that are considered in this chapter are sand dune and associated sandy beach systems, saltmarshes, seagrass meadows, and biogenic reefs (including mussel beds, oyster beds, and *Sabellaria* reefs). Because coastal ecosystems are found at the interface between the marine and terrestrial biome, their role within flood management is clear. Saltmarshes and dunes are obvious examples of effective barriers between the land and sea, whereas other coastal habitats can also play an important role in modifying coastal processes thereby reducing flood risk, especially shallow water habitats of the sub- and intertidal zones such as biogenic reefs (Bouma et al., 2014) and seagrass meadows (Ondiviela et al., 2014). Flood protection is provided directly by coastal habitats through attenuation of waves by the canopy structure of the saltmarsh plants or seagrass leaves or by acting as a physical barrier as the case with sand dunes. In general, the role of coastal habitats for direct attenuation of storm-induced impact diminishes with water depth; therefore, the most important habitats are in the high intertidal zone such as saltmarshes.

Coastal habitats also contribute to coastal protection indirectly by trapping and stabilizing sediment. As a consequence of sediment stabilization, the spatial pattern of wave energy dissipation along the cross-shore profile is different to a shore lacking the presence of saltmarsh plants (Knutson, 1988, pp. 161–174). Loss of the stabilizing species can therefore lead to a shift in the pattern of wave attenuation toward the shoreline, with higher rates of erosion as a direct consequence. Similar effects happen with other sediment-stabilizing species that construct biogenic reef structures, including mussels, oysters, or honeycomb worms (*Sabellaria* spp.). Therefore, insights in the key ecological processes and important stressors acting on these ecosystems are needed. Many of these habitats have considerable natural seasonal and interannual variability, and predictions about their relative abundance and response to climate change can be difficult to make.

In this chapter, we discuss the ecological approaches to coastal risk management focusing on two areas. We consider the use of natural habitats as a coastal defense option, discussing management and restoration techniques including both methods for seaward extension and managed realignment. We also review progress on incorporating ecological principles and criteria in the design and maintenance of hard coastal defense structures to achieve desired outcomes. Thus, we give a comparison of design configurations for "soft" and "hard" engineering techniques for coastal defense with a focus on the ecological implications of these approaches. Soft engineering solutions in general involve some form of habitat restoration, rehabilitation, or creation; as such, there is a spectrum of approaches ranging from complete restoration creating a "natural habitat" through to the creation of an artificial habitat that will require long term management. By comparing soft and hard engineering techniques, we explore the strategy of holding the line with managed retreat/realignment and the ecological implications of various strategies. Furthermore, examples are given of techniques for ecologically enhancing hard engineering structures including offshore breakwaters, shore-perpendicular structures such as groins and jetties, sea walls (Sections 3.2 and 3.4), and, lastly, wave energy converters (Section 3.1). The ecological consequences of building hard structures is discussed and focuses on the impact of the additional hard substrate facilitating the spread of invasive species and the impact of these structures on surrounding soft sedimentary habitats.

Options are considered within a framework of coastal management strategies that can be split into four options: (1) no active intervention, (2) managed realignment, (3) holding the line, and (4) advancing the line. Hold the line is defined by Defra (2006) as maintaining or upgrading coastal defense structures to improve or maintain the standard of protection provided by the existing defense line. Advance the existing defense line is defined as building new defenses on the seaward side of the original defenses (Defra, 2006). Both holding the line and advancing the line can be achieved by the creation of intertidal ecosystems for coastal defense on the seaward side of the land–sea transition. Managed realignment is defined as the removal or retreat of the primary coastal defense line to achieve sustainable coastal defense and/or to (re) create intertidal habitats though the accommodation of natural processes. Lastly, considerations of potential positive and negative interactions with other anthropogenic or natural stressors to all coastal habitats are addressed. Figure 4.1 shows how coastal management options link to the different schemes for coastal defense that are considered in this chapter.

For the management and restoration of soft habitats (Sections 4.2–4.5), saltmarshes are considered first of all because they are high in the intertidal zones and thus have greatest potential to attenuate waves and offer coastal protection. Dunes are considered next because they play an important role as a barrier. Seagrass meadows and biogenic reefs are discussed last because they may have a lesser direct effect in coastal protection, whereas their indirect effects by foreshore stabilization can be significant. The issues threatening each habitat and the solutions that can be used to overcome these problems are discussed. For each of the habitats, the practices of management and restoration are addressed, focusing on specific techniques that can be implemented. Finally, socioeconomic implications are discussed, including, where possible, an indication of the costs associated with each intervention. For the ecological enhancement of hard coastal defense structures (Section 4.6), we first discuss issues and guidelines for each type of hard defense structure. We then consider the ecological impacts of the structures and the methods to mitigate or enhance the structures from an ecological perspective. The socioeconomic

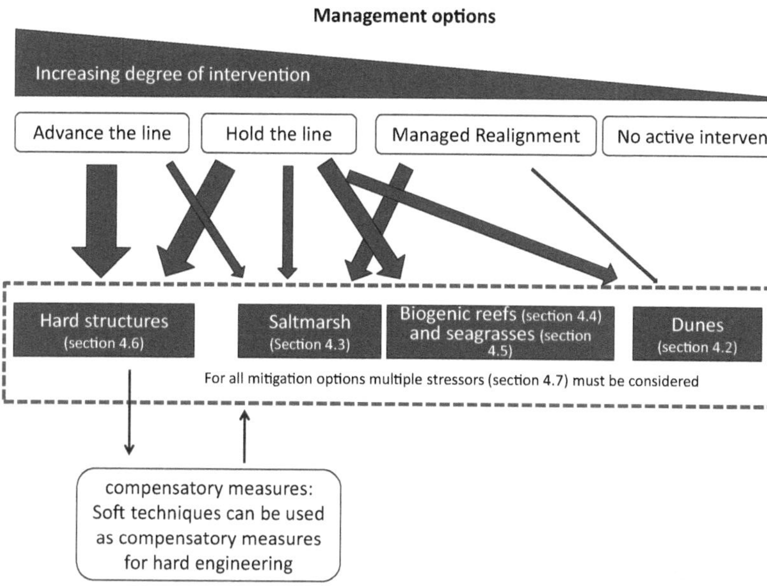

Figure 4.1 Diagram linking coastal management policies (white boxes) with the different management options (black boxes). Width of the arrows linking the policies with options represents the relative importance of that option to an active management strategy. Details are given to the sections in this chapter where the options are discussed. No active intervention is included here as it is an important arrangement consideration that can be applied to all habitants, however it is not linked by an arrow as it is not an intervention.

implications of these enhancements are considered and where possible, costs are given. Ecological considerations for managed realignment are discussed in Section 4.3.3 and interactions with other anthropogenic stressors are discussed in Section 4.7.

4.2 Management of Sandy Habitats: Sandy Beaches to Sand Dunes

4.2.1 Background

The beach dune habitat provides many ecosystem services, but is vulnerable to the effects of climate change human impacts (Schlacher et al., 2012). Currently, the majority of beaches around the world are classified as eroding (Bird, 2000) and sea-level rise and increased intensity of storms from climate change will accelerate beach erosion and retreat inland (Nordstrom, 2000). Sandy shores are particularly vulnerable to climate driven changes, particularly sea-level rise, changes in wave regime and storm intensity and frequency, and alterations to the sediment budget. They are among the most intensively used coastal ecosystems and this will likely increase in the future.

Figure 4.2 Aber Saltmarsh near Bangor at the Menai Strait, North Wales, UK: J. van Belzen.

Sandy coastlines are composed of three interrelated compartments: the subtidal beach, the emerged beach, and the dune. Sand dunes (Figure 4.2) form a barrier between the sea and land and therefore can play an important role in coastal defense. Dunes are important for many reasons, including acting as a barrier to storm surges and flooding and protecting landward development as well as other natural habitats (Paskoff, 1998). They absorb the impact of high-energy storms, preventing or delaying intrusions of water into inland areas while also acting as a source of sand to the beach when it is exposed to erosion (Gómez-Pina et al., 2002). Dunes are also home to unique flora and fauna and are extensively used for tourism and recreation. Dunes therefore provide an important resource as well as ecosystem services such as protecting infrastructure, farmland, property, and other inland natural habitats, particularly during severe winter storms.

Sand dunes are a sensitive habitat requiring management for their protection and restoration, particularly given predictions of rising and stormier seas increasing flood risks. Coastal dunes are found in highly dynamic environments because they are strongly influenced by both wind and sea. Sediments are exchanged by marine processes and wind, thus dunes can at times act as a sand stock. The variability of wind, both in direction and magnitude, can play an important role in the mobilization and landward sediment transport that controls dune formation. These spatiotemporal dynamics of dunes are a result of local feedback together with seasonal and other naturally occurring perturbations. Coastal areas affected by short-term perturbations can return to their predisturbance morphology and reach a dynamic equilibrium (Taramelli, Valentini, Dejana, Zucca, & Mandrone, 2011). Sand availability is necessary for dune formation and dune maintenance. Several factors affect the availability of sediment for dune formation, including changes in sea level, changes in sediment transport from off shore, the presence of vegetation, and impacts from human activities.

In front of sand dunes, sandy beaches are important for coastal defense because they provide the zone where waves break ahead of the transition to sand dunes found at higher elevations. The intertidal areas of beaches provide habitat for a diversity of fauna. These organisms are operationally classified as microbenthos (<38 μm), meiobenthos (38–500 μm), and macrobenthos (>500 μm). They play important roles such as filtration and organic matter deposition and are considered indicators of natural and anthropogenic changes in the environment (McIntyre, 1977). The benthic macrofauna of exposed sandy beaches is generally dominated by crustaceans, molluscs, and polychaetes while the meiofauna is typically dominated by nematodes, copepods, platyhelminths, polychaetes, and ostracods (McIntyre, 1977). Important biotic components of the beach ecosystem are mobile epibenthic predators, such as shrimps, crabs, fish that forage when the tide is high, and shorebirds that forage when the tide is out. Benthic organisms can reach high abundance and biomass (McLachlan & Brown, 2006). In general, shore morphodynamics influence the biota of beaches, ultimately shaping assemblage composition, community structure, zonation, and ecosystem functioning (reviewed in Defeo & McLachlan, 2005).

Exposed sandy beaches provide a range of ecosystem services for coastal protection. Key services include sediment storage and transport, coastal defense through wave dissipation and associated buffering against storms and offer a dynamic response to sea-level rise. Other ecosystem services provided by sandy beaches are the breakdown of organic material, mineralization and recycling of nutrients, water filtration, feeding resources for birds, nursery grounds for commercial fish, and recreation.

4.2.2 Management Techniques for Sandy Habitats

Sediment stabilization is necessary to preserve and restore natural sand dunes. Vegetation plays an important role in the stabilization of dunes because it traps sediment that is transported by wind, slows wind speed, and reduces wind erosion of sediment (Tsoar, 2005). The impact of vegetation on wind and dune evolution depends on plant density, morphology, distribution, and height (van Dijk, Arens, & van Boxel, 1999). In addition, the presence of vegetation on dunes is limited by the availability of moisture and high sediment supply preventing the establishment of vegetation through burial. Unprotected and unvegetated dunes are unable to withstand stressors including wind, wave erosion, and trampling, for example, by people. Because the natural dune rebuilding process can take several years, it may be desirable to artificially rebuild a dune (O'Connell, 2008). Ecological engineering techniques in combination with natural processes can be used to help protect and restore degraded dune systems and improve the resilience of dunes to withstand erosion. These techniques are designed to reduce the erosive action of wind and the impact of human pressure, in particular the effect of trampling.

Several techniques can be used to restore dunes. Sand can be added to the dune to reshape it to an appropriate height for coastal defense. Dunes can sometimes be of insufficient height to be effective against coastal flooding. They can also suffer morphological damage such as breaches within them. A second technique is to use sand catcher fences; these trap and stabilize sand, allowing the dune to grow in volume and height. The fences work by slowing down the wind speed, thus reducing the wind's potential to erode the sand and resulting in the trapping of sand. This technique is appropriate for degraded or newly created dunes, and the efficiency of the technique depends on the available sediment and beach width. Fences must be a suitable distance from the shoreline and must be constructed perpendicular to the wind direction. Once dunes are of an appropriate height, soil can be added to further protect the dunes and to encourage recovery and growth of vegetation. The ultimate aim is to provide protection from erosion by wind and rainfall, retain moisture at ground level during the summer, protect plants against frost, supply organic matter and minerals during decomposition, and avoid the development of unwanted, weedy vegetation. Addition of soil is suitable in areas with morphological damage and in areas undergoing revegetation operations to prepare substrate, especially in steeply sloping areas. A mulch of cut reed, biodegradable geotextile, or branches can be added on to dunes to conserve moisture, improve the fertility and health of the added soil, reduce weed growth, and enhance the visual appearance of the area.

Overall, vegetation cover is the best way to protect and stabilize sand. Dune vegetation, through its root systems of varying depths, increases stability by protecting the dune from wind erosion. Moreover, vegetation also traps sand, leads to accretion, and enhances local microclimatic conditions and so facilitates further vegetation growth. Revegetation is appropriate in areas that have undergone degradation. The planted species must be chosen according to the dune habitat that is to be restored. The success of the technique depends on biological traits and physiological adaptations of the planted species (e.g., the ability to tolerate salt water or burial by sand). Planting is most successful when carried out during spring. The slope of the dune is also important in determining the success of revegetation, with steeply sloping areas being more problematic. The reintroduction of all potential species is, however, usually not feasible for both technical as well as economic reasons. Certain key species such as *Ammophila* spp. and *Elymus farctus* tend to fulfill the essential function of promoting sand stabilization and accumulation, whereas others have only minimal effect. (Ley de la Vega, Favennec, Gallego-Fernández, & et Pascual Vidal, 2012, 124 pp.) It may therefore be advisable to reintroduce a limited number of key species. On European coasts, two grass species that are commonly used are *Ammophila* spp. (marram grass) and *E. farctus*. Both species are considered important for structuring vegetation and building dunes (García Mora, Gallego-Fernández, & García Novo, 2000).

Exclusion of recreational activity is necessary for any of these sand dune management techniques to be effective because dune vegetation can be fragile and sensitive to trampling. Exclosures can be used to protect the dunes from trampling and protect materials used in restoration such as geotextile or sand catcher fences. However, when exclosures are established, access to the beach must be maintained for recreational users, especially in the summer. This is often achieved via corridors with boardwalks to restrict and channel trampling during access.

Hard engineering structures such as seawalls, groins, and breakwaters are frequently constructed within beach environments and are the primary techniques used to control erosion (see Sections 3.2, 3.5, and 4.4). Alternatively, soft engineering options may be used, the most frequently used technique being beach nourishment (Section 3.5). Nourishment is preferred on both economic and conservation grounds (Finkl & Walker, 2004) and involves importing sand to replenish, restore, or nourish the beach. Beach nourishment is used primarily for the benefit of the tourism industry and can have significant negative ecological implications (Colosio, Abbiati, & Airoldi, 2007; Dugan, Hubbard, Rodil, Revell, & Schroeter, 2008).

4.2.3 Related Considerations for Sandy Habitats

When active revegetation is needed, it is necessary to determine the species and seed sources that will establish and perform well on the site without impeding natural community recovery and succession or compromising the diversity, genetic integrity, and long-term viability of resident wild populations. (Erickson, Wood, & Riley, 2003, 18 pp.) Linhart (1995, pp. 271–287) showed that plants or their progeny must come from nearby sites to maintain the ecological and genetic integrity of the biota. Also, local populations have adaptive advantages to local conditions and are therefore less vulnerable to abrupt environmental changes (Mortlock, 2000). The use of local seed sources in revegetation also helps avoid problems of genetic contamination of native populations. Introduction of nonlocal plants may lead to failure of the revegetation project and endanger the long-term biological health of local populations. In this instance the term "local" refers to a distance of ~100 m for herbs and ~1 km for woody plants. Nevertheless, mixing of seed sources, particularly in fragmented landscapes, may result in greater long-term viability of reseeded populations (Waters, Young, & Crosthwaite, 2007, pp. 85–93), reducing the potential for genetic impoverishment as a consequence of inbreeding. A mixture of seed sources increases the genetic diversity, ensuring remnant populations have sufficient genetic diversity and adaptive capacity to survive predicted climate change scenarios, thus improving resilience of the ecosystem change. For severe and large-scale disturbances, a mixture of genotypes or seed sources from ecologically different populations has therefore been suggested as a strategy for maximizing genetic variation and persistence in stressful environments. (Erickson et al., 2003, 18 pp.)

Paths across the dunes should be used to access the beach for pedestrians and authorized vehicles, protecting both the dune and its vegetation. Beach access should be constructed where users usually cross the dunes to facilitate public acceptance of these structures and their efficacy. Discussions should therefore be held with the different beach users, such as nautical clubs and fishermen. Paths must not be located in protected or sensitive areas.

Signage on beaches has become a common feature of the coastline. It aids awareness of the rehabilitation project, to beach-dune ecology, and to the importance of not trampling dune vegetation.

4.3 Management of Saltmarsh Habitats

4.3.1 Background

Saltmarshes are vegetated areas in the upper part of the intertidal zone at the interface between the marine and terrestrial biome (Figure 4.3, Allen, 2000). They consist of a

Figure 4.3 Sand dune at Walcheren, Scheldt Estuary, The Netherlands. *Photograph taken by J. van Belzen.*

limited range of salt and flood tolerant vegetation species. Those at the seaward margin are most tolerant of physiological stress from inundation with saline water and disturbance from waves and tidal currents. In the low marsh areas, positive interactions ("facilitation") within and among species are predominant (Bertness & Ellison, 1987). These interactions often ameliorate physical–chemical stresses, thereby allowing species from the higher marsh to extend downward. Higher in the marsh competition among species can be more important, with stress-tolerant, but slow growing species replaced by faster growing, but less tolerant species. In mid- to high-marsh areas, grazing from wild (e.g., geese, hare, deer) and farm (e.g., cows, sheep) animals often occurs and can have a major effect on the composition and size of saltmarsh vegetation (Bakker, 1985).

Pioneer species such as *Salicornia europaea* (annual glasswort) and *Spartina anglica* (common cordgrass) in sheltered conditions are usually the first species to colonize bare tidal flats (Garbutt & Wolters, 2008; Mossman, Brown, Davy, & Grant, 2010; Reed, Davidson-Arnott, & Perillo, 2000; Wolters, Garbutt, Bekker, Bakker, & Carey, 2008) because of their tolerance of the harsh abiotic conditions. These species have been shown to catalyze a saltmarsh development once established (Bertness & Ellison, 1987; Reed, Spencer, Murray, French, & Leonard, 1999). Moreover, these early successional species can in the longer term facilitate the growth of less robust species such as sea lavender (*Limonium vulgare*), greater sea-spurrey (*Spergularia media*), and common scurvy-grass (*Cochlearia officinalis*), which are more specifically adapted to higher saltmarsh elevations (Townend, Fletcher, Knappen, & Rossington, 2011) by stimulating marsh elevation and providing shelter. In addition, these pioneer plants can have positive effects on sediment aeration through evapotranspiration (Ursino, Silvestri, & Marani, 2004), which can prevent the development of anoxic conditions. Saltmarsh pioneers have frequently been labeled as "ecological engineers" (in the sense of Jones, Lawton, & Shachak, 1994) because, without their modifying effects on abiotic conditions, the colonization of higher plants may be limited by abiotic conditions at lower elevations (Allen, 1990; Langlois, Bonis, & Bouzille, 2003).

The ecosystem engineering effects of saltmarshes make them dynamic biogeomorphological ecosystems, with their development resulting from interactions among hydrodynamic processes, sedimentation, and vegetation. These interactions are fundamental in shaping and maintaining the structure of these systems (Allen, 1990; Temmerman et al., 2007). The canopy of a saltmarsh is able to attenuate tidal flow and waves, enhancing sediment accretion within the vegetation (e.g., see Bouma et al., 2007; Temmerman et al., 2007). Vegetation also diverts tidal currents, thereby accelerating flows some distance away from its canopy, which can result in increased erosion rates or scouring away from the saltmarsh (Bouma et al., 2009; Van Hulzen, Van Soelen, & Bouma, 2007; van Wesenbeeck, van de Koppel, Herman, & Bouma, 2008), creating patterns of gullies and channel networks characteristic of saltmarshes that are important for their form and function (Temmerman et al., 2007). These flow-deflecting effects are species-specific, indicating that marsh development differs depending on the pioneer species that come first (Bouma et al., 2013c). To benefit from saltmarshes from a coastal defiance perspective requires fundamental understanding of these key biogeomorphic processes and mechanisms, which ultimately drive long-term marsh dynamics, stability, and resilience (Bouma et al., 2014).

Unlike engineered structures, saltmarshes have the ability to grow and adapt to changing environmental conditions such as sea-level rise, provided that key physiological, hydrodynamic, and sedimentary conditions are met (Kirwan & Temmerman, 2009). Elevation adjustments are dependent on the indirect and direct mechanisms provided by plant growth (Reed et al., 1999). For marsh elevation to rise in relation to the tidal frame, sedimentation must exceed the rate of total subsidence (Cahoon & Reed, 1995), which occurs from compaction after settlement of suspended particles (Kaye & Barghoorn, 1964) or sea-level rise. The above-ground biomass of plants contributes to elevation rise by actively altering the hydrodynamics and sedimentary conditions (Bouma et al., 2009; Van Hulzen et al., 2007), e.g., by reducing current velocities (Koch et al., 2009; Leonard & Luther, 1995) and attenuating wave energy (Möller, Spencer, French, Leggett, & Dixon, 1999; van Proosdij, Davidson-Arnott, & Ollerhead, 2006; Yang et al., 2008). Such attenuation of hydrodynamic energy is species-specific, and depends upon characteristics such as shoot stiffness and standing biomass (Bouma et al., 2005; Bouma, De Vries, & Herman, 2010; Peralta, van Duren, Morris, & Bouma, 2008). Plant biomass also contributes organic matter directly to the soil surface, which can be a significant constituent of the net sediment input (Belluco et al., 2006; Cahoon & Reed, 1995; Day et al., 1999; Santín et al., 2009). Plant roots can help to bind accreted sediment and aboveground biomass provides shelter, therefore inhibiting resuspension of sediment and aiding marsh elevation rise (Friedrichs & Perry, 2001).

Elevation changes invoked by saltmarsh vegetation can lead to positive feedback mechanisms. Changes in elevation shorten inundation periods, improving growth conditions for vegetation (van Wesenbeeck, van de Koppel, Herman, Bertness et al., 2008). In turn, this leads to improvements of the geomorphological feedbacks, which can alter the morphology of saltmarshes (Temmerman et al., 2007). These feedbacks improve the stability and resilience of the saltmarsh to changes in environmental conditions. A consequence of these positive feedbacks is that establishment and restoration of saltmarshes might be hampered because of size thresholds, as the positive feedbacks can only occur once the vegetation has reached a sufficient size (Bouma et al., 2009). Vegetation transplantation is used as a technique in saltmarsh restoration, but for plants to establish they need to be of sufficient size (van Wesenbeeck, van de Koppel, Herman, Bertness et al., 2008), and be planted when stressful environmental conditions are reduced thereby providing windows of opportunity to successfully establish (Balke et al., 2013).

Because of their height in the intertidal zone, saltmarshes can be found in approximately the upper 30% of the intertidal zone and they can play an important role in coastal protection (Bouma et al., 2013a). When fully submerged, the water depths over saltmarshes are relatively shallow and waves are therefore easily attenuated (e.g., Bouma et al., 2005; Yang, Shi, Bouma, Ysebaert, & Luo, 2012; Ysebaert et al., 2011). When saltmarshes are not entirely flooded, they can also serve as an effective natural breakwater.

Saltmarshes also provide a myriad other ecosystem services. They play an important role in maintaining biodiversity, providing nursery habitats, and spawning grounds for commercially important fish and crustaceans as well as nesting and foraging grounds for birds. Furthermore, they can work as effective biological filters, improving water quality by removing excess nutrients and pollutants. Thus, managing saltmarshes as part of coastal defense an add significant additional benefits and functionality over hard engineering solutions, because (1) they are "living structures" that can grow and adapt to changing environmental conditions and (2) they can add functions and services beyond those offered by hard coastal defense.

Coastal squeeze is an increasing threat to saltmarshes worldwide because of the proliferation of man-made structures in the coastal zone (Wolters, Bakker, Bertness, Jefferies, & Moller, 2005). Coastal squeeze has severely reduced the extent of intertidal habitat along coastlines. Land reclamation converts saltmarsh areas into arable, industrial, or urban areas (Adam, 2002). The problem of coastal squeeze has been exacerbated by historic land reclamation from the sea mostly for agriculture (King & Lester, 1995). Because structures such as seawalls or embankments in urban areas have fixed the upper limits of marsh saltmarshes get "squeezed" between the upward moving lower marsh limit and the higher fixed marsh limit. As a consequence, the upper boundary of the marsh is not able to move freely with changes in sea level.

Sea-level rise as a result of climate change is a potential threat to the survival of saltmarshes worldwide. Accelerated levels of sea-level rise, which might outpace the rate at which saltmarshes can adjust to changing conditions, are of particular concern. Human disturbances can influence and interfere with the ability of saltmarshes to keep pace with sea-level rise and/or have a huge impact on the resilience of saltmarshes. Trampling can hinder vegetation growth and negatively affect recovery. Grazing by livestock can significantly affect the composition and structure of vegetation. Pollution and eutrophication may weaken the ability of vegetation to bind sediment and increase erosion (Deegan et al., 2012). Waves from ships can have a similar effect on a saltmarsh as an increase in storminess from global change (Airoldi & Beck, 2007). Nevertheless, various studies indicate that most saltmarshes will be able to keep up with sea-level rise if the ecogeomorphological feedbacks are not undermined and if sediment supply is sufficient (Kirwan et al., 2010; Kirwan & Temmerman, 2009). That is, if there is enough sediment available vegetation growth is not hampered or squeezed, the saltmarshes will maintain their relative height in the intertidal zone.

Saltmarshes typically exhibit a cyclic development, as the spatial extent is variable over time because of processes that result in expansion and contraction of the saltmarsh range (van de Koppel, van der Wal, Bakker, & Herman, 2005). These cyclic dynamics may be altered because saltmarshes are likely to experience increased levels of wave exposure in the future because storminess is predicted to increase from climate change (Young, Zieger, & Babanin, 2011). As waves are an important forcing factor of saltmarsh dynamics (Callaghan et al., 2010), this will have important consequences for the long-term stability of saltmarshes (Bouma et al., 2013a). The dynamics involved in the range expansion and contraction may be particularly affected. Higher energy waves are likely to increase scarp erosion rates and inhibit recolonization of the mudflat once eroded. This will lead to a net upward shift in the lower limit of the saltmarsh and may increase spatiotemporal variability of saltmarsh range dynamics (Bouma et al., 2013a; Taramelli et al., 2014). Thus, increased storminess may result in an additional contribution of a range shift further exacerbating coastal squeeze.

Overall, mechanistic insights in the processes driving cycles of biogeomorphical development are of vital importance for managing saltmarshes in a sustainable way, thereby integrating them into coastal defense strategies (Bouma et al., 2014).

4.3.2 Management Techniques for Maintaining and Extending Saltmarsh Habitats

Historically, hard coastal engineering has been regarded as the solution to protect coastal areas from flooding and erosion. Under increasing threats of sea-level rise, policymakers have to assess physical and economic limits on the height and size of hard defenses (King & Lester, 1995; Nicholson-Cole & Riordan, 2009). In response to rising sea levels and economic costs associated with building hard defense structures, existing defenses may in some cases be economically less favorable than soft defense. As a result, the use and creation of saltmarshes have recently come to the forefront of coastal defence and policy as a "soft," sustainable, and therefore cheaper alternative to defending coast lines with hard structures (French, 2006). Therefore, saltmarshes are increasingly managed or even created to protect the foreshore, with the marsh serving as a natural buffer to reduce the forces imposed on and erosion of the seawall or other defense structure, thereby decreasing maintenance and construction costs (King & Lester, 1995). Thus, from a coastal defense perspective, the accepted benefits of created or restored saltmarshes are increased wave attenuation and a localized reduction in sea-level rise from an increase in tidal storage. Creating saltmarshes can be cheaper than upgrading or repairing existing defenses; in addition to the economic benefits provided by coastal wetlands, their biodiversity and functional value is now recognized in law under the European Union

Habitats Directive (Council Directive 92/43/EEC). The directive seeks to maintain no-net-loss in total habitat area.

Using salt marshes for coastal defense can be realized in different ways as summarized in Table 4.1. If a marsh is already present, management should focus on maintaining and protecting the existing marsh. Alternatively, a marsh can be created where there is no marsh. This can be either on the seaward side of the existing coastline by creating conditions that allow marsh establishment on tidal areas or on the landward side by managed realignment (Table 4.1). The choice for either of these options depends on many different aspects, with social, economic, and ecological aspects being highly important (see Chapters 2 and 6). For example, in the United Kingdom, managed realignment (Section 4.3.4) is often done because: (1) maintenance or improvements of the existing sea defense structure are no longer economically or societally advantageous or (2) there is a need to create natural habitats or compensate for saltmarsh loss elsewhere as saltmarsh is a protected habitat under the Habitats Directive (Council Directive 92/43/EEC), in large deltas located below sea level, such

TABLE 4.1 Overview of Issues and Solutions Related with Different Approaches to Saltmarsh Management. X Marks Which Issues Are Appropriate to Each Approach to Saltmarsh Management.

	Approaches to Use Marshes for Coastal Defense			
		Create New Marsh		
Issues	Maintain Existing Marshes	Seaward Extension	Managed Realignment	Solutions
Biophysical boundary conditions (Section 4.3.2)				
1. Physiological tolerance to flooding		x	x	Adjust inundation period by modifying elevation
2. Hydrodynamic driven sediment dynamics	x	x		Reduce wave energy by hard or soft engineering solutions or stabilize sediment
3. Sediment supply	x	x	x	Enhance supply by piping/dredging
Socioecological constraints (Section 4.3.5)				
1. Ecosystem lost/ replaced		x		
2. Sociological acceptance			x	

an approach is not possible without making major costs in reallocating dikes. In such situations, seaward marsh creation in front of the existing dikes may be more favorable.

If saltmarshes are to be used as part of coastal defense plans, there is a need for accurate marsh restoration and marsh creation methods, particularly in light of sea-level rise and coastal squeeze. To use marshes as a reliable coastal defense option it is particularly important to manage the key processes that determine long-term marsh stability (Bouma et al., 2013a; Bouma et al., 2014; Taramelli et al., 2014). When saltmarshes are not present in areas where they may be desirable for coastal defense, options for marsh creation need to be explored based on mechanistic insights; meaning there must be a focus on the boundary conditions set by: (1) plant-physiological constraints on growth, (2) hydrodynamic boundaries on marsh development, and (3) sedimentary supply requirements to enable the saltmarsh to keep up with sea-level rise. The relative importance of each of these aspects will of course depend on the approach used to maintain/generate a marsh (Table 4.1). For example, in most cases, managed realignment sites are relatively sheltered, so that hydrodynamic driven sediment dynamics will often to be of minor important. Obtaining the appropriate elevation for plants to establish at a suitable inundation regime often imposes a major challenge to managed realignment sites. In the case of creating a marsh on the seaward side, attenuating hydrodynamic energy will often be necessary, whereas the elevational height does not always impose a problem.

In summary, when managing or creating saltmarsh ecosystems, the following three key processes should be taken into account in a site specific way, to actively create and/or promote development.

4.3.2.1 Plant-Physiological Constraints

Hydrodynamic forces and anoxic sedimentary conditions should not exceed the physiological limits of the vegetation (Bouma, Koutstaal, van Dongen, & Nielsen, 2001; Bouma et al., 2005). Inundation period of imposes dominant control on saltmarsh vegetation, and with it soil and water condition parameters co-vary. Inundation period, therefore, should not be too extensive so that vegetation can survive and grow fast enough to compensate for disturbance.

4.3.2.2 Constraints by Hydrodynamic Driven Sediment Dynamics

There are physical limits of the vegetation and the vegetation–sediment matrix created in relation to the wave forces it is able to resist (Bouma et al., 2009; Friess et al., 2012). Furthermore, if current velocities are too high, sediment can be eroded from the vegetation canopy because of turbulence (Balke et al., 2012). However, in most cases, an important constraint to the establishment of saltmarsh vegetation is the height of the bed-level change, which is driven by the bed shear stress induced by water flow (Bouma et al., 2013a). In many saltmarshes, the sediment dynamics are driven by the wave-forcing (Callaghan et al., 2010), which means that the wave-forcing regime sets the boundary conditions for vegetation establishment and survival in marshes. Fluctuation in sediment levels on the tidal flat adjacent to the marsh must remain small enough to reduce the risk for lateral erosion and allow vegetation to reestablish if saltmarsh is lost (Bouma et al., 2013a).

4.3.2.3 Requirements on Sediment Supply to Follow Sea-Level Rise

Key to sustainable management of saltmarsh ecosystems is appropriate management of sediments as long as boundary conditions, such as water quality and the inundation period are met. This means that sediment concentrations in the water column should be sufficiently high for saltmarshes to accrete enough to keep pace with sea-level rise. In Section 4.3.3, we discuss potential management approaches and solutions to meet the requirements imposed by these three key processes.

4.3.3 Solutions for Maintaining and/or Extending Marshes

4.3.3.1 Adjust Inundation Period by Changing Bed-Level Elevation

A correct (i.e., not too long) inundation period is key to the establishment and survival of saltmarshes, regardless of the management techniques applied (Table 4.1). Adjusting the elevation of the sediment bed has proven to be important in cases of managed realignment. Hence this issue is elaborated in more detail in Section 4.3.4. However, it is obvious that using sediment (e.g., as available from dredging) can facilitate marsh establishment, both seaward or via managed realignment techniques. It is important, however, that the sediments remain stable enough for seedlings to establish. This may be most important at wave-exposed sites, as may be the case for seaward marsh creation/extension. This aspect is discussed in more detail in the next section (Section 4.3.3.2).

4.3.3.2 Managing Hydrodynamic-Driven Sediment Dynamics

Low mobilization or deposition of sediments is necessary for seedling establishment of saltmarsh species (Balke et al., 2013; Bouma et al., 2014). Next to the management of the sediment supply, improving the long-term saltmarsh stability requires a reduction in sediment dynamics, defined as the temporal changes in the bed level due to erosion or accretion processes (Bouma et al., 2013a; Bouma et al., 2014). This can be achieved by either reducing the external forcing of the saltmarsh site or by increasing sediment stability. Waves and currents are important external forcing conditions that can determine the local sediment dynamics (Callaghan et al., 2010;

Shi, Yang, Wang, Bouma, & Zhu, 2012). A range of management options exist that allow the hydrodynamic forcing of a saltmarsh to be reduced. These include regulating shipping to reduce waves, adjusting bathymetry of adjacent mudflats using dredged materials (i.e. adding sediment) and controlling in-site conditions by stabilizing sediments.

Regulating shipping is a simple and effective method to reduce erosion of saltmarshes. Various measures are possible and include: (1) limiting the size of the ships passing the marsh; (2) limiting the speed of the ships that are passing the marsh; and (3) time limits relative to the tidal height when ships can pass, as if ships only pass when the saltmarsh and the pioneer vegetation is still emerged the shipping waves cannot affect the seedlings thereby promoting higher survival probability.

Hard structures such as breakwaters and groins (see Section 4.6) can be used to reduce wave forcing. However, they often have negative consequences for the sediment balance and hamper natural ecosystem development. Soft structures, such as mudflats, oyster reefs, mussel beds (Section 4.4), or seagrass meadows (Section 4.5), can be used to attenuate waves instead of hard structures (Beck et al., 2011; Piazza, Banks, & La Peyre, 2005).

The bathymetry of adjacent mudflats can be adjusted by adding dredged sediment in order to change the bathymetric profile and reduce the external forcing conditions that drive local sediment dynamics so that vegetation can establish. The simplest example is to increase the intertidal height. Also, creating an artificial mudflat (potentially with a biogenic reef or sediment stabilizing seagrass meadow; see Sections 4.4 and 4.5) in front of a saltmarsh to reduce wave forcing can be an effective way to reduce local sediment dynamics.

In-site conditions can be controlled by stabilizing sediments. Within the target saltmarsh site, stabilization of sediment can be used to aid the establishment and restoration success. Techniques to do this include the use of geotextiles or the temporarily reduction of benthic fauna.

Degradable geotextiles can be used to stabilize sediments in situ. This method is now actively used for dune revegetation (see Section 4.2), although its application in saltmarsh restoration is still experimental. The geotextile protects the sediment and provides a relatively stable matrix for vegetation. However, the geotextile must be securely fixed to the sediment, especially in the dynamic tidal environment. If the edges of the geotextile are detached because of wave action, it is possible that the whole geotextile can be eroded from the mudflat. A second disadvantage to using geotextiles in an environment of cohesive sediments is that it can cause or increase anoxic conditions. Geotextiles must be degradable to enable application in natural environments. The degradation of the organic material of the geotextile reduces oxygenation of the sediments and potentially leads to accumulation of toxic compounds from anoxic processes that may hamper the growth of saltmarsh vegetation. However, some saltmarsh species are known for their particular tolerance to such conditions. The pioneering grass species *Spartina anglica* is likely to be an especially good species to support the revegetation of mudflat using geotextiles to stabilize the sediments.

Studies on saltmarshes (van Wesenbeeck, van de Koppel, Herman, Bakker, & Bouma, 2007) and recent work on seagrasses (Suykerbuyk et al., 2012) have shown that benthic infauna can be an important agent that destabilizes the sediments and thereby hamper reestablishment of vegetation on the mudflat. Different methods can be used to control such destabilizing effects and their consequences for vegetation establishment and survival, allowing vegetation to successfully establish. Temporary sealing of a mudflat with plastics creates an anoxic environment locally, which kills most of the important sediment-destabilizing infauna. This could create a window of opportunity for seedlings to establish. However, this method is only feasible on a small scale, but may be a good option when manual revegetation is considered. Layers of seashells can be used to reduce the abundance of infaunal organisms that destabilizes the sediment (Suykerbuyk et al., 2012). The shell layers prevent bioturbating organisms disturbing restored vegetation. The benefits of his method are that it is an efficient way to modify the sediments over a long period and that it is relatively low cost (Suykerbuyk et al., 2012). Temporary sandbag walls can be used to increase the stability of the sediment by reducing the wave forcing (Christianen et al., 2013), facilitating the establishment of vegetation. Lastly, pesticides can be used to eradicate benthos, giving vegetation the time and opportunity to reestablish. However, the use of toxicants is not recommended as many other unwanted side effects can arise.

4.3.3.3 Managing Sediment Supply to Follow Sea-Level Rise

Management of sediment, specifically sediment supply, takes place on multiple scales, requiring different levels of management and spatial planning. Landscape scale management is needed to cover the whole watershed or catchment. Dikes and dams or other alterations of the watershed have in several cases reduced the amount of suspended solids that reach the saltmarsh surface. To sustainably manage saltmarshes, it is often necessary to restore suspended solids concentrations. These issues are now addressed in Europe through the Water Framework Directive (Directive 2000/60/EC).

Sediment concentrations can also be managed on a more local scale. This can be achieved using one of the following methods: (1) piping of sediment slurry or (2) dredge spoil. Piping of sediment slurry can be used in saltmarsh restoration or creation through sediment slurry amendments and is an increasingly common technique to increase the relative height of the saltmarsh. Although information is limited on its effectiveness and consequences, the few existing applications have demonstrated that if applied appropriately, sediment slurry can restore saltmarshes affected by submergence caused by subsidence or sea-level rise. In the 1990s, the hydraulic application of sediment slurry was employed to restore brown marsh areas, in the Mississippi Delta Plain, Louisiana (USA), which had subsided because of soil consolidation following a vegetation dieback event (Mendelssohn, Materne, & Schrift, 2006). Low-pressure hydraulic dredging was used to disperse sediments over the marsh surface (Cheramie, Francis, & Kilgen, 1995; Slocum, Mendelssohn, & Kuhn, 2005). Studies on the effectiveness and consequences of this application showed that when sediment slurry was applied at the correct elevations, the macro-invertebrate community, as well as aboveground marsh vegetation, recovered to reference conditions. In contrast, too much sediment can hamper recovery (Cornell, Craft, & Megonigal, 2007; Croft, Leonard, Alphin, Cahoon, & Posey, 2006; Schrift, Mendelssohn, & Materne, 2008). Addition of slurry can be a sustainable restoration technique when performed in the proper way, as illustrated by a study on the long-term effects of sediment subsidies on resilience and stability of saltmarshes (Stagg & Mendelssohn, 2011). Large-scale application is still limited and as such there is uncertainty regarding site-specific factors such as how much sediment should be applied to ensure vegetation recovery and long-term sustainability. Sediment quantity is difficult to judge because once the slurry is applied the bed level will lower as the sediments lose water and become compacted. Consideration must be given to establishing suitable postconstruction marsh elevations to ensure successful ecosystem restoration.

Dredge spoil can also be used in saltmarsh restoration or creation. Dredging is essential for the maintenance and development of ports, harbors, and waterways. In past decades, widening and deepening of ports and waterways across the world has been widespread and frequent to satisfy the growth of maritime trade and larger ships. Projects, such as deepening of the Mississippi estuary, Rotterdam Port, New York Harbor, and Panama Canal, can produce huge quantities of dredged materials exceeding 100 million cubic meters (Wang, Xie, & Wu, 2004). The potential use of dredged material for several aspects of marsh restoration process include increasing the sediment budget at low elevations, accelerating restoration trajectories, improving the geomorphology of the marsh platform, providing high marsh refuge areas for species that depend on these habitats, restoring upland dike elevations for off-site protection of people and property, and stabilizing shorelines to reduce erosion rates. Dredged material can be successfully used to enhance or create new mudflats (Bolam & Whomersley, 2005; Evans, Ward, Bone, & Leakey, 1998; Ray, 2000; Ray, Clarke, Wilber, & Fredette, 1994; Schratzberger, Bolam, Whomersley, & Warr, 2006) and saltmarshes (Lasalle, Landin, & Sims, 1991; Posey, Alphin, & Powell, 1997), which eventually function in a similar manner to natural systems. This way, salt marshes can be created at the seaward side to hold or extend the line.

The abundance of dredged material from channel-deepening projects, the maintenance dredging of major ports, on-site construction, and other projects provides opportunities to combine dredging needs with coastal marsh rehabilitation and restoration. However, the use of dredged material can be a contentious because some sediments are contaminated. Thus the use of clean dredged material is critical for its application on coastal marsh restoration and creation.

4.3.4 Considerations for Saltmarsh Management Using Managed Realignment as an Example

Managed realignment is currently defined as the process of allowing the shoreline to move backwards or forwards with management or control to limit movement (Defra, 2006, p. 19). In principle, the option is technically simple: a seawall is shifted inland and the old sea defense is (actively or passively) removed, allowing seawater to flood the area. Alternatively, a gate system can be introduced to allow the exchange of seawater. There are many drivers for managed realignment, which can be categorized into two broad categories: habitat creation or improvement of flood risk management. A significant benefit of managed realignment is that it can alleviate the problem of coastal squeeze. In this section, we specifically focus on the ecological considerations to achieve successful managed realignment.

The establishment of saltmarsh after managed realignment is not guaranteed (French, 1997). Managed realignment can be problematic if a motivation for the scheme is habitat creation, especially if there is a target ecosystem that managers would like to develop (e.g., saltmarsh). There is the risk that the scheme can be perceived as a failure if the desired habitat is not formed and a different type of habitat (e.g., tidal flat) develops. A problem with managed realignment occurs if there has been subsidence of the land to be realigned. Agricultural land that has been closed off from tidal flooding will not have received any sediment input. In combination with groundwater use (e.g., for irrigation of crops), this often results in subsidence of the land. Too strong a subsidence may hamper saltmarsh regeneration when reexposed to tides, so that the realignment will result in mudflat creation instead of the often-desired saltmarsh. Another challenge occurs if sites are disconnected from seed and propagule supply required for the establishment of saltmarsh plants. It should be noted that, even if the saltmarsh creation by managed realignment is successful, ecosystem functioning in most managed realignment sites is often less than that of natural saltmarshes (Mossman, Davy, & Grant, 2012). The knowledge needed to assess if a site is likely to develop into a marsh or not is similar to that needed for maintaining or extending existing marshes.

Politics and socioeconomics are important factors to take into account when deciding on a managed realignment scheme. Managed realignment can be a realistic option in locations where space is not a limitation, such as in parts of the United Kingdom. However, in large estuaries, where the majority of the land is located below sea level, such as, for example, the Netherlands, managed realignment may be difficult because it invokes the displacement of dikes to prevent large areas becoming flooded. Also, in densely populated countries, where space is limited, the local inhabitants can perceive managed realignment schemes as unacceptable due to the percived lose of land. Public opinion on these matters may gradually change based on information comparing the long-term effect of different coastal defense options: hold the line versus managed realignment. Quite abrupt changes in the mindset of people may also follow after extreme events such as storms lead to catastrophic breaches and flooding of urban areas, potentially resulting in the loss of lives (see Chapter 5).

4.3.4.1 The Practice of Managed Realignment

Managing the realignment process is also not straightforward because of the many interrelated biotic and abiotic factors controlling saltmarsh development (Townend et al., 2011). Because managed realignment is a comparatively new approach to coastal management (French, 2006), there is a lack of long-term evidence covering the extent to which restored saltmarshes function as natural saltmarshes (i.e., reference marshes nearby). As such, judging the long-term performance of managed realignment in the context of a changing climate and rising sea levels is still problematic. However, some evidence has been gained by studying historical storm-breached sites. Vegetation establishment in many of these studies, both in long-term storm breached/abandoned defenses and in recent managed realignment projects, has been shown to be slow and often not matching the composition of natural local reference marshes. In one UK estuary, French et al. (2000) concluded that the marshes still showed no sign of reestablishment, even 55 years after abandonment. Comparing both modern managed realignment and older storm-breached marshes with neighboring reference marshes in another estuary, Garbutt and Wolters (2008) found that 13 of the 18 realigned sites examined had lower plant species diversity than the neighboring reference marshes. On the other hand, Wolters et al. (2008) showed that species diversity in some managed realignment sites was comparable to local reference marshes within a 5-year period. Given the site-specific nature of managed realignment schemes, it is difficult to draw firm conclusions as to why some schemes had positive ecological outcomes and others less so.

A range of management techniques can be employed to accelerate the development of saltmarsh after realignment. For example, at places where the managed realignment area is too low in the tidal frame for saltmarshes to establish, engineering techniques, such as regulated tidal exchange, can be used to control saltmarsh development. The latter requires a solid background understanding of interrelating abiotic and biotic processes (French, 2006; Marani et al., 2006). In places where seed supply is limiting marsh development (Diggory & Parker, 2009) or sediments appears to be too unstable for seedling establishment (Reed et al., 1999), the deliberate planting of vegetation plugs into newly flooded areas can enhance saltmarsh development (Craft, Broome, & Campbell, 2002). However, these techniques have often been used with little consideration for which species are most efficient at attenuating waves and accreting sediment, and are also most appropriate in terms of biodiversity enhancement (Leonard & Luther, 1995). In addition, the abiotic/biotic prerequisites necessary for the optimum establishment of individual species are often neglected in the planning process of vegetation planting (Diggory & Parker, 2009). This aspect may thus be optimized to maximize specific aims, such as, for example, creating a high coastal defense value.

The long-term success of saltmarsh vegetation after the planting of ecosystem engineer species is not guaranteed, even if elevation rise has been initiated. Diggory and Parker (2009) state that if saltmarsh species are absent from the seed bank, salt-marsh development will be limited. Because the majority of saltmarsh seeds are transported short distances by water flow (Huiskes et al., 1995), it is essential to help ensure that a well-developed natural saltmarsh exists adjacent to the realignment site to provide a future seed supply (Erfanzadeh, Garbutt, Pétillon, Maelfait, & Hoffmann, 2010; Garbutt & Wolters, 2008; Wolters et al., 2008). Otherwise, deliberate seed sow may need to be implemented at an appropriate estuary location and time of year (Huiskes et al., 1995; Wolters et al., 2008).

4.3.5 Related Considerations for Saltmarsh Management

Using saltmarshes for coastal defense is effective (Shepard, Crain, & Beck, 2011) but requires the creation of a new marsh, either toward the sea or by managed realignment; a main drawback is that it requires a significant amount of space that was previously used for other purposes, whether intertidal ecosystems (in case of seawards extension) or agricultural land (managed realignment). Such space is not always available even when neighboring land may be historically reclaimed marshes, so that marsh creation is not always a viable option for coastal defense. From an ecological perspective, two main questions arise when using space to create saltmarshes for coastal defense: (1) which habitat will be lost and (2) how does this compare to the ecological value of the newly created marsh?

In cases where managed realignment is not an option (either through political or economic circumstances), creating or extending an existing marsh can be considered as a useful way to hold or extend the line and vice versa. In many cases, extension of marshes may be the better option from a management perspective, especially where erosion of the marsh occurs as a retreating cliff and/or if pioneer vegetation is in decline. A negative consequence of extending saltmarshes can be that (at least parts of) existing mudflats have to be sacrificed in the process, possibly leading to a loss of foraging grounds for birds and other wild life. Therefore, to evaluate the ecological impacts of this approach requires evaluation of both the saltmarsh and the ecosystem it replaces.

Extending existing marshes will always be at the expense of terrestrial (in case of managed realignment) or intertidal (in case of seaward extension) habitats. The areas that will be lost may represent important ecological values and may potentially outweigh the ecological value of the marsh that is created. For example, if little intertidal area remains, this may hamper the feeding of protected bird species on benthic communities or decrease the productivity and extent of nursery grounds.

With respect to the restoration, creation, or expansion of marshes, it is important to try to define which criteria must be met for those ecosystems to be of comparable value to natural marshes. Parameters such as species diversity and ecosystem functions might be the most obvious ones. Unfortunately, several studies indicate that apparently attractive "man-made" marshes may score lower on these quantitative measures than a naturally developed ecosystem (Moreno-Mateos, Power, Comin, & Yockteng, 2012; Mossman et al., 2012).

Socioeconomic implications and management costs will be widely variable between locations, depending on societal values and physical setting. The yearly costs of the proposed management solutions are highly location specific and, as such, they are not been considered here. With respect to the socioeconomic implications, there are major differences between regions within Europe. For example, in the province of Zeeland, the Netherlands, the long history of claiming land from the sea makes managed realignment very difficult to accept by society and it is an extremely sensitive political topic. This goes far beyond the scarcity of space, which is important throughout the Netherlands, given the high population density. In contrast, the same topic seems less sensitive just across the border in the Flemish part of Belgium. In the United Kingdom, on the other hand, managed realignment appears to be widely accepted at locations where it is no longer economically feasible to hold the line. Each management solution discussed must be carefully evaluated against the social setting and acceptances, including all stakeholder groups, to successfully enable the application of marshes for coastal protection.

In wave-exposed areas, *Spartina anglica* has been shown to attenuate flows very effectively. There is concern, however, over the invasive nature of *Spartina* species because they are very efficient competitors for resources (Yang et al., 2008). Liao, Luo, Fang, Chen, and Li (2008) believe that ecophysiological properties, such as a greater leaf area index and epiphytic nitrogen fixation in dead shoots of the introduced species *Spartina alterniflora*, have given the species a competitive advantage over native species in many marshes in China. Therefore, if allowed to spread, *Spartina* species may prevent the establishment of a more diverse array of native species. From this, a dilemma arises over the ultimate aims of site-specific restoration: should the saltmarsh be managed as a coastal defense or to recreate a lost habitat? It is important to note that in all natural communities (including saltmarshes), higher biodiversity tends to increase resilience to disturbance and therefore help the marsh become sustainable in the long term (Folke et al., 2004; Loreau et al., 2001; Tilman et al., 1997).

The example of salt marsh creation by managed realignment demonstrates that creation of saltmarshes involves many aspects. These vary from ecological considerations, based on species requirements in relation to innundation, as well as engineering aspects, taking appropriate measures to enable the system to develop in a desirable way. As indicated in Table 4.1, the relative importance of the biophysical boundary conditions and socioecological constraints will differ between the approaches chosen. Most importantly, if the desire is there to use and create salt marshes for coastal defense, this can often be achieved. In some extreme cases, the required technical measures may make it an unattractive option however, but in many cases marshes offer a valuable method worth exploring into depth = Much more is possible than is currently done, and a major limitation in some locations is the societal appetite for these kinds of solutions to coastal defense.

4.4 Management of Biogenic Reefs

4.4.1 Background

Biogenic reefs are created by living organisms when their shells either concreate together (e.g., oysters and serpulid worms), bind together by byssus threads (e.g., mussels), or when worms creating tubes with sand grains that form colonies (e.g., *Sabellaria* spp.), or by calcareous encrusting algae. These reef-building species can be considered bioengineering species or ecosystem engineers (Padilla, 2010). Generally, reefs are resilient and, in the case of *Sabellaria*, mussel or oyster reefs they can occur in naturally disturbed areas at the interface of sandy beaches and rocky reefs. The major reef-forming species in the Northeast Atlantic and Mediterranean Sea are the native oyster *Ostrea edulis*, the non-native oyster *Crassostrea gigas* (Figure 4.4), and the honeycomb worm *Sabellaria alveolata* (Figure 4.5, Table 4.2). Mussels (*Mytilus* spp.), see Figure 4.6, can form reefs in sedimentary habitats, for example at the entrance to the Exe in the United Kingdom and in the Wadden Sea, but more often form extensive sheets on rocky reefs. Other reef-forming species such as the mussel *Modiolus modiolus* and the polychaete worms *Sabellaria spinulosa* and *Serpula vermicularis* occur in deeper or more sheltered water and thus play a less significant role in coastal protection (Table 4.2). Calcareous algae can also form reef-like structures, for example, in the Mediterranean or in tropical waters as part of coral reefs (Jones et al., 1994).

Figure 4.4 Oyster bed in the Eastern Scheldt, The Netherlands. *By J. van Belzen.*

Figure 4.5 *Sabellaria alveolata* hummocks at Dubmill Point in the Solway Firth, UK. *By L. Firth.*

TABLE 4.2 List of Common Biogenic Reef-Forming Organisms Detailing the Species and Habitats Found. Figures 4.4–4.6 Show the Major Reef-Forming Species

	Species	Habitat
Oyster reefs	European oyster *Ostrea edulis* (native)	Low intertidal to shallow subtidal
	Pacific oyster *Crassostrea gigas* (non-native)	Low intertidal to shallow subtidal
Mussel beds	Edible mussel *Mytilus edulis*	High intertidal to shallow subtidal
	Horse mussel *Modiolus modiolus*	Subtidal
Reef-building polychaete worms	Honeycomb worm *Sabellaria alveolata*	Low intertidal
	Ross worm *Sabellaria spinulosa*	Subtidal
	Fan worm *Serpula vermicularis*	Subtidal

Figure 4.6 Mussel bed in the Wadden Sea, The Netherlands. *By J. van Belzen.*

Biogenic reefs provide a range of ecosystem services pertinent to coastal protection. These include services such as the regulation of disturbance through breaking waves, storms, and flooding as well as erosion control and sediment retention. Most oyster, mussel, and *Sabellaria* reefs occur intertidally and in shallow coastal waters, thus modifying water movements and cementing together semi-mobile cobbles and boulders (e.g., Naylor, 2005). In soft-sediment environments, such as the Wadden Sea, oysters and mussels can have a major impact on the composition and stability of the sediment. Sediment becomes more muddy and less erodible because of excretion of pseudo fecal matter and hydrodynamic alterations (van Leeuwen et al., 2010; Widdows & Brinsley, 2002). Reef-forming species thus can provide an important ecosystem service by potentially increasing shore stability and decreasing wave action incident on the shore (Hendrick, Foster-Smith, & Davies,

2011). Water movements measured over *Sabellaria* reefs (which are the most extensive intertidal reef building structures on European rocky shores) have shown that flow is modified; however, the reduction is fairly modest. Generally, waves above reefs are more turbulent than waves off reefs because of the breaking of waves above the reef structure. This is also likely to be the case over mussel sheets because of their low profile (A.J. Davies, unpublished data).

Another important effect of reef-building organisms in the intertidal is likely to be the stabilization of potentially semimobile cobbles and boulders. This occurs via cementing by byssus threads of mussels or via the tubes of *Sabellaria* (Meadows, Meadows, West, Shand, & Shaikh, 1998, pp. 331–347; Widdows & Brinsley, 2002). Sometimes both species cooccur and help fix boulders in combination. *Crassostrea* reefs can have a similar effect, but generally occur in areas sheltered from wave action, but with considerable currents such as at the entrances to lagoons and estuaries where cobbles are smaller and where reefs can form over coarse sedimentary systems. Although it is reducing water flow, this biogenic ecosystem engineering will halt erosion by stabilizing cobble and boulder fields (Hendrick et al., 2011). Boulders move less within biogenic reefs compared to outside such reefs.

Biogenic reefs are facing several threats. A pervasive issue is loss of habitat from coastal development and the exploitation of living resources. The latter especially affects mussel and oyster reefs. Beds of the native oyster *O. edulis* are now much diminished to the point of functional extinction in Europe because of a long history of exploitation for food, exacerbated by disease and an increase invasive species over the past 50–100 years (Beck et al., 2011; Reise, 1998). The non-native oyster *Crassostrea gigas* is, however, now forms extensive beds and may reach high densities (Airoldi & Beck, 2007). When cemented together, *C. gigas* can also provide a coastal protection function primarily by stabilizing the foreshore. It should be kept in mind, however, that they are displacing both native oysters and economically important mussel beds, as has been found in case of the Dutch Wadden Sea (Reise, 1998).

Loss of sediment is a major threat to *Sabellaria* reefs and aggregate dredging can remove sediment that is necessary for reef formation. Suspended particulate matter in a consistent and dependable supply is essential for the formation, development, and longevity of *S. alveolata* reefs (Holt, Rees, Hawkins, & Seed, 1998). Without it, reefs will gradually erode and eventually disappear. Coastal defenses such as groins can reduce sediment supply and have a negative impact on reef formation. At both Lyme Regis (Dorset) and on the Wirral in Northwest England, however, *Sabellaria* grows well on shore parallel and perpendicular breakwaters and groins (Figure 4.7). On the Wirral, it fills in gaps on cast concrete structures (Frost et al., 2004).

Climatic warming, on the other hand, is likely to increase the abundance of *S. alveolata*. *Sabellaria* reefs were found to respond with an expansion of their northern limit and are now more extensive than they were in the colder 1960s (Cunningham, Hawkins, Jones, & Burrows, 1984; Frost et al., 2004). The past few

Figure 4.7 *Sabellaria alveolata* growing between groin boulders at the Wirral (UK). *Photograph by Louise Firth.*

years have shown a return to colder winters (Wethey et al., 2011), which will lead to a contraction in range and reduction in abundance.

The abundance of reef-forming organisms is strongly determined by larval supply. Changes in hydrodynamic conditions may thus drastically impact the natural maintenance of reefs when supply of larvae is reduced. This applies to all reef-forming species. Extensive artificial defense systems, however, may expand the population of fouling organisms such as mussels.

4.4.2 Management Techniques for Biogenic Reefs

Only with concerted restoration activities and reduction of other impacts (e.g., dredging) will any coastal protection service be put back in place. Generic management options for existing biogenic reefs involve protection of the reef from disturbances such as trampling and coastal development. Restriction of access to the reef and establishing protected sites are likely to be successful methods for restoration and conservation. Some reef-forming species like mussels and oysters are of commercial interests. Their management should be based not only on considerations of their exploitation potential, but also considering the value of all the other ecosystem services they provide as habitats, first of all coastal protection (Beck et al., 2011). In the case of *Sabellaria*, aggregate extraction could be minimized in the area to ensure sediment supply.

Most reef-forming species depend on the sufficient supply of larvae and in case of *Sabellaria*, the availability of large amounts of sediments. Both will be largely determined by regional and local hydrodynamic conditions. Sea defenses modify these (see Section 4.6) and may thus interrupt the supply of larvae and sediment to the reefs. It is thus advisable that the construction of hard defense structures occurs at a suitable distance from existing reefs.

Besides maintaining existing biogenic reefs by applying the previously mentioned ecological management options, restoration of biogenic reefs is another option that can be followed to restore the coastal protection service provided by biogenic reefs. Unfortunately, methods for reef restoration are in their early development. Restoration of mussel and oyster reefs are more common and more easily achieved by encouraging the settlement of juvenile oyster spat on the desired substrate. The possibility of using geotextiles provides an option to enhance mussel spat settlement as widely used in aquaculture.

The management and restoration of biogenic reefs is unlikely to have any negative ecological impacts, provided that the restoration results in the proliferation of the targeted reef-forming species and not in the establishment or increase in abundance of unwanted species such as non-natives. Transplantation of living laboratory-settled *Sabellaria* is a potential mitigation option against the establishment of nontarget species and could seed further settlement because of gregarious settlement behavior (Davies, unpublished data).

It is likely that biogenic reefs will benefit the ecological status of an area (Scyphers, Powers, Heck, & Byron, 2011). The previously mentioned coastal protection service will reduce effects of flooding and soil erosion on the resident, terrestrial biological communities. Also, because reef-forming organisms typically are ecosystem engineers, they create habitats for a range of other organisms and increase local biodiversity (e.g., Dias & Paula, 2001; Dubois, Retiere, & Olivier, 2002). Many reef-building organisms are filter feeders, so they also provide important benefits in terms of maintenance of water quality or reduce the likelihood of harmful algal blooms, with associated negative ecological and economic consequences (Allen, Wilkinson, & Hawkins, 1995; Beck et al., 2011; Newell, 2004). Oysters and mussels can also help to remove excess nutrients from coastal bays by facilitating denitrification in surrounding sediments, which has tremendous economic value in areas where nutrient removal is a high priority for coastal policymakers (Newell, Fisher, Holyoke, & Cornwell, 2005).

4.4.3 Related Considerations for Biogenic Reefs

Soft engineering options such as the restoration of biogenic reefs can be considered as relatively cost-effective coastal protection services, especially when compared with the costs incurred during construction and maintenance of hard defense structures and considering the value of the ecological services gained (Grabowski et al., 2012). The implementation of full-protection management schemes for biogenic reefs may result in the loss of value for tourism of an area because of limitations in amenity use. Partial protection such as prohibition of harvesting of mussels and oysters but allowing some limited beach access may be an acceptable solution as this would restrict impacts through physical disturbance. Other positive socioeconomic impacts, especially of mussel beds, include the improvement of water quality and reductions in turbidity through biofiltration (Allen et al., 1995; Wilkinson et al., 1996), which is thought to be beneficial for the reintroduction of seagrass meadows in the Wadden Sea (van Katwijk, 2003, pp.173–195). Cost evaluations of restoration and management measures for biogenic reefs are not yet available as these associated coastal engineering methods are still being developed. However, Grabowski et al. (2012) estimate that the economic value of oyster reef services, excluding oyster harvesting, is between $5500 and $99,000 per hectare per year and that reefs recover their median restoration costs in 2–14 years.

4.5 Management of Seagrass Meadows

4.5.1 Background

Seagrasses (Figure 4.8) are marine flowering plants distributed along temperate and tropical coastlines, where they have the potential to provide coastal protection (Short et al., 2011). Because of this ecosystem service, the development and reinforcement of seagrass habitats can be considered an adaptation measure against climate change (Borsje et al., 2011; Zanuttigh, 2011). From a coastal defense perspective, seagrass meadows can act as a buffer from the impacts of climate change because of their capacity to attenuate waves, reduce current velocities, and promote sediment stabilization (Koch et al., 2009; Orth et al., 2006; Whanpetch et al., 2010). The coastal protection provided by seagrass meadows through the attenuation of waves is from a combination of the ability of the plants to break waves, separate the flow of water between individual stems, and increase bed roughness. This has a secondary benefit as it reduces near-bed flow velocity allowing sediment accretion. Mechanical properties of the plants are also important in the energy dissipation processes, as the stem density and flexibility of seagrass leaves is important in controlling the dissipation of wave energy.

Management strategies to restore (and even increase) the coastal protection function of seagrass meadows against climate change are influenced by the expected damage over habitats or ecosystems caused by increased sea level, increase in wave height, rates of erosion, changes in currents, and local environmental conditions such as nutrients, temperature, and light levels. Consequently, the severity of the damage will be determined by the resistance and resilience shown by natural habitats by these changes and local environmental variables regulating seagrasses presence and

Figure 4.8 (a) Above water photograph of a seagrass meadow in Salcombe Bay Devon. *(Photography by E. Jackson, CQ University, Australia.)* (b) Photograph of the seagrass meadow in Salcombe Bay below water. *Photograph by R. Ticehurst (University of Plymouth).*

distribution. Management and restoration strategies required to preserve and improve the coastal protection functions provided by the four European seagrass species: *Cymodosea nodosa, Posidonia oceanica, Zostera marina,* and *Zostera noltii* (Table 4.3). The restoration techniques for seagrass meadows are suitable for areas where the seagrasses are present or have been found recently but have been lost. Seagrass meadows provide a number of ecosystems services as well as flood protection; they act as nursery and feeding habitats for commercial and recreational important fish and shellfish species (Short & Neckles, 1999; Unsworth et al., 2008; Watson, Coles, & Lee Long, 1993). They also improve water transparency and quality through trapping and storing solids particles and dissolved nutrients as part of the

TABLE 4.3 Geographical, Maximum Depth Distribution, and Life Span of European Seagrass Species

	Cymodocea Nodosa	*Posidonia Oceanica*	*Zostera Marina*	*Zostera Noltii*
European distribution	Atlantic and Mediterranean	Mediterranean	Atlantic (except Canary Islands) and Mediterranean	Atlantic, Mediterranean, and Black seas
Maximum depth	Up to 30 m	Up to 40 m	Up to 10 m	Up to 10 m
Shoot life span	Years	Decades	Years	Months to years

nutrient cycling of coastal waters (Costanza et al., 1997; Orth et al., 2006). Likewise, they are considered a significant sink for carbon in the ocean and represent an important part of the global carbon cycle (Duarte, Middelburg, & Caraco, 2005; Smith, 1981). Seagrass meadows also support high levels of biodiversity. However, it must be noted that seagrasses are only found low in the intertidal and that therefore their role in wave attenuation is less than habitats such as saltmarshes that are found higher in the intertidal zone.

4.5.2 Management Techniques for Seagrass Meadows

Healthy seagrasses will be better able to adapt to global changes. (Björk, Short, Mcleod, & Beer, 2008, 56 pp.) Consequently, effective management is the basis for enhancing seagrass resilience and ensuring habitat survival. The first management strategy is to regulate activities that physically damage seagrass meadows. (Björk et al., 2008, 56 pp.) Outfalls, submarine cables, coastal constructions (e.g., harbors, marinas, breakwaters), oil exploration, dredging, disposal of soft-sediment material, filling works, and beach nourishment are all examples of activities that can damage seagrass meadows through direct physical disturbance. Physical damage can also be caused by boat propellers, moorings, anchors, certain mariculture practices, and destructive fishing practices such as bottom trawling or dragging (Duarte, Marbà, & Santos, 2004), which may damage seagrass communities by directly cutting the leaves (Short & Wyllie-Echeverria, 1996).

Water quality and clarity are critical issues to maintaining healthy seagrass meadows (Björk et al., 2008, 56 pp.) Changes in water quality and clarity are related to the release of nutrients, pollutants, or suspended material and dredging activities. Major causes of nutrient inputs can be related to both point and diffuse discharges from tertiary water-treatment plants and urban sewage (Borum, Duarte, Krause-Jensen, & Greve, 2004; Luque & Templado, 2004), industrial release, dredging and resuspension activities (Erftemeijer & Lewis, 2006), aquaculture (Borum et al., 2004), and inputs from streams and watersheds. Poor land use practices (e.g., removal of terrestrial vegetation) result in increased soil erosion and the delivery of substantial quantities of sediment through rivers and streams to estuaries and coastal waters, where suspended particles significantly increase the water turbidity (Borum et al., 2004). Most of the land-source pollution can be considerably reduced by adequate land-use practices to decrease nutrient and sediment runoff, reducing the use of fertilizers and pesticides and increasing filtration of effluent discarded into the water. (Björk et al., 2008, 56 pp.) River and coastal buffer zones (such as saltmarshes discussed in Section 4.3) are also highly effective in filtering nutrients, pollutants, and sediment runoff from water before it reaches seagrass meadows hence careful management of surrounding habitats is important for healthy seagrass meadows. Shellfish also remove suspended solids from surrounding waters, thereby increasing water clarity, which can help facilitate seagrass growth (Newell & Koch, 2004).

When seagrass meadows become damaged, die-off of plants can be rapid, whereas the natural recovery of disturbed seagrass habitats is comparatively slow (Lee & Park, 2008). As a result, numerous restoration projects transplanting seagrasses have been attempted in past decades (Bastyan & Cambridge, 2008; Fonseca, Kenworthy, Courtney, & Hall, 1994; Lee & Park, 2008; Short, Davis, Kopp, Short, & Burdick, 2002; Thorhaug, 1987; West, Jacobs, & Roberts, 1990; Zimmerman, Reguzzoni, & Alberte, 1995). The cost of restoration programs and the uncertainties associated with the success of the final results imply that careful analysis of the feasibility of these schemes is needed. Worldwide, the success of seagrass transplantation and restoration is around 30% (Fonseca, Kenworthy, & Thayer, 1998), although in some regions higher success rates have been reported (Orth et al., 2006). Restoration projects have been attempted or are being planned mostly at small scales (<1 ha) using a variety of techniques with both adult plants and seeds (Orth et al., 2006). Although interest in larger scale transplant programs is growing, some species are so difficult to transplant that restoration is not logistically or economically feasible, and longer-term studies are still needed. For example, *Posidonia oceanica* grows slowly and has only sparse sexual reproduction. Transplanting in intertidal areas is more straightforward in comparison to subtidal areas.

Species selection for transplantation depends on the geographical distribution, the tidal level, the life span (Table 4.3), and growth rates of the particular species. Knowledge on seagrass growth rates and success of reproductive effort is crucial to manage seagrass beds, particularly to derive expectations on the recolonization time required for seagrass meadows to recover (Marba & Duarte, 1994). Colonization is typically most successful with small species such as *Zostera noltii*. Transplanted patches of *Z. noltii* require a few months to a year to become an established meadow. Conversely, the recovery of transplanted patches of larger species such as *Posidonia oceanica*, which is a long-living, slow-growing species with limited spread of clones requires centuries to form a meadow (Marba & Duarte, 1994). Given the time it takes for *P. oceanica* to establish, efforts should be focused on the conservation and protection of existing beds. Meanwhile, *Zostera marina,* and *Cymodocea nodosa* have shorter life spans and rates of growth, making them more suitable for transplantation in enclosed bays and inlets and they can also facilitate the return of later successional species such as *Posidonia*.

Survival of seagrass transplants is strongly related to the local environmental conditions at the restoration site. The success of any transplantation depends on the selection of suitable areas for the settlement of specific species. Consequently, habitat suitability maps for the species selected are useful to identify sites with favorable conditions for settlement. This information should be considered alongside predictions about future climate change scenarios. Important considerations include information on characteristics of vegetated and unvegetated areas (e.g., current and historical seagrass distribution), proximity to natural beds, and key variables defining the ecological niche of the species such as sediment type, wave exposure, current velocities, depth, or water quality. This information may allow identification of the best possible locations for transplantation.

Many planting methods for seagrasses have been trialed including transplanting shoots, seedlings, and mature plants. Seeds have been developed to restore damaged seagrass beds as well as to create new ones (Lee & Park, 2008). In both cases, local genetic diversity must be maintained in the restored area. Results of studies performed to date show that the selection of the appropriate technique depends on a combination of factors including, but not restricted to, the species, transplanting site, and logistics. (Bortone, 2000, 318 pp.) It should be noted, however, that independently of the technique applied, restoration efforts should always be balanced against the capacity of seagrasses to recover naturally.

Seeding techniques are advisable when the seeds can settle and germinate, such as in areas with low seed herbivory and little physical disturbance (Lee & Park, 2008). This technique has the advantage that: (1) the impact on the donor population is minimized; (2) sowing seeds is a relatively cheap method for large-scale restoration (Orth, Luckenbach, & Moore, 1994, Orth et al., 2000; Harwell & Orth, 1999); and (3) working with seeds may also be less laborious and time consuming in comparison to transplanting cuttings or mature plants (Christensen, Díaz, & Diekmann, 2004). Seeds can be collected directly from mature reproductive shoots from established beds during seed release. Large quantities of seeds can be deposited on breaches and seeds can also be collected.

Transplantation techniques have focused on moving vegetative shoots from donor beds to restoration sites. Several techniques for transplanting vegetative shoots have been trialed and shown to be successful; however, most techniques are labor intensive and expensive (Lee & Park, 2008). Common approaches include: transplanting plugs of seagrass with the associated sediment that are harvested using a core tube; taking sediment blocks of seagrasses and then putting them into peat pots, which can then be transplanted into the receptor bed; and removing seagrass shoots along with bare roots and rhizomes from donor beds and transplanting the shoots. In general, the recommended time for seagrass transplanting is spring. For all plans, it is important to consider the structural design of meadows with a focus on improving the connectivity between meadows. There is also potential for the creation of new large meadows by connecting together smaller meadows to form a continuous areas (Alexandra, Duarte, & Krause-Jensen, 2004).

4.5.3 Related Considerations for Seagrass Meadows

Coastal defense services offered by seagrasses must be considered alongside the ecological interactions with other habitats and ecosystems also contributing to the coastal protection such as oyster reefs. Well-defined strategies related with the maintenance and restoration of the connectivity between these habitats and ecosystems are essential.

In any transplantation, donor beds may suffer in some way and care should be taken to not damage the donor beds. Therefore, passive recovery, if feasible, is the preferred strategy. Gaps within seagrass meadows formed by natural events are primarily recolonized by the clonal extension of the rhizomes of the plants located at the periphery of the gaps. Horizontal rhizome extension may be highly effective to recolonize small gaps (1 m^2), which can be closed within a few years. Gaps greater than 1 m^2 may require a decade or longer to recover and must involve sexual reproduction to initiate new patches within the denuded area (Alexandra et al., 2004). These horizontal extension rates are fast for *Z. noltii* and *C. nodosa* but slow for *P. oceanica* and *Z. marina* (0.1–0.9 m/y) (Marba & Duarte, 1998). Thus, depending on the size of the altered area, passive recovery could be the best management strategy to recover the coastal protection functions.

Given the wide variety of seagrass growth strategies and environmental settings under which seagrass beds occur, the cost of transplanting is not easy to determine. Costs of successfully establishing seagrass beds depend on site selection, experience of workers and managers, extent and rate of subsequent disturbance, water clarity and depth (e.g., light availability and quality), temperature, extent of monitoring following planting, and numerous other factors (Fonseca & Bell, 1998). Costs for planting ranged from $25,000 to $50,000 per hectare, with an average value of about $37,000 (Fonseca et al., 1998).

4.6 Ecological Design Considerations for Hard Structures

This section considers how the ecological outcomes of various coastal defense structures including offshore breakwaters, shore-perpendicular structures including groins and seawalls, and to a lesser extent wave energy converters (WECs) can be modified via alterations to design to achieve desired ecological outcomes. Potential modifications are described at a range of scales from small changes in surface rugosity at a scale of millimeters to centimeters or the addition of pools at a scale of 10s of cm to meters, to the design of whole structures or wider schemes on scales of 10–1000s of meters. Evidence of changes in the diversity and abundance of species in relation to various levels of habitat complexity are discussed along with the logistical considerations in the deployment/incorporation of habitat features within engineering schemes. Potential impacts of hard structures on natural assemblages of both soft and hard bottom habitats are considered together with the potential for these structures to facilitate the spread of invasive species (discussed in Section 4.7). Suggestions on design optimization, optimal placement, and efficiency from an ecological perspective are outlined. For each of the hard structures considered (including offshore breakwaters, shore-perpendicular structures including groins and jetties, seawalls, and WECs), the issues and practice of customizing these structures and their ecological impacts are considered along with ways of mitigating these impacts. Last, socioeconomic implications and costs are considered.

4.6.1 Background

Coastal development is proliferating in response to burgeoning human populations (Firth & Hawkins, 2011). The threats of sea-level rise, flooding, and erosion have led to a growing need to defend our coasts and in heavily urbanized areas, efforts largely comprise hard-substrate defense structures such as seawalls, breakwaters, and groins (Chapman & Bulleri, 2003; Chapman & Underwood, 2011; Dugan, Airoldi, Chapman, Walker, & Schlacher, 2011).

Shore-parallel, off-shore breakwaters reduce the intensity of wave action, thus reducing coastal erosion and the risks of flooding. Most commonly, they are constructed of natural hard stone such as granite, limestone, or of concrete, an artificial material. Seawalls are typically vertical or steeply sloping structures forming a boundary between soft sediment environment and urban infrastructure such as roads, railways, and buildings. They can be constructed of a variety of materials ranging from concrete, sheet steel, large quarried blocks, stone-filled gabions, and/or artificial blocks fronted by a decorative façade and are usually positioned parallel to the shore line. Groins are usually constructed perpendicular to the shoreline, although other configurations are built such as T- or Y-shaped groins. Their main purposes are prevention of sediment erosion as well as stabilization and protection of beaches and adjacent land, thereby averting flooding. They are built from concrete, stone, wood, gabions, or sandbags ("geotextiles") and usually extend from the backshore into the littoral zone. Several of these barrier-like structures may be constructed along the beach, forming groin fields between which sediment will be retained through the modification of long shore movement of sediments. Although more commonly shore-parallel when used for the construction of seawalls, gabions consisting of stone-filled baskets or wickerwork can also be used in the construction of groins perpendicular to the shore. Other shore-perpendicular structures such as jetties may serve multiple purposes, for example mooring boats, harbor protection, or directing river and current flow. Jetties are usually made of stone, earth, concrete, or wood.

WECs to fulfill the role of capturing wave energy for generating electricity but also reduce incident wave energy on the coast (Section 3.1). There are numerous WEC designs; some use hydraulic systems to convert wave motion to pump high pressure oil through motors that drive electrical generators and produce electricity, others rely on forcing water or air through turbines. In the lee of WECs, wave energy will be reduced but with potential consequences on water flow and sediment regimes.

4.6.2 Ecological Impacts of Artificial Hard Structures

The impacts of the construction of hard structures such as breakwaters, seawalls, shore-perpendicular structures, and WECs on the natural environment may originate at a local scale from: (1) the introduction of an artificial hard-substrata habitat that replaces previously soft-sedimentary environments and that differs in its properties from natural rocky shores; (2) changes in the hydrodynamic regime of the soft-sedimentary environment with resultant shifts in grain size and species composition adjacent to the structure, particularly on the landward side; (3) the accumulation of detrital materials produced by the fouling organisms (seaweeds and sedentary animals such as mussels) growing on the hard structures and/or redirection of detrital material produced elsewhere; and at a coastal landscape scale (4) the facilitation of the introduction and spread of non-native or other pest species as well as (5) the loss of genetic diversity and homogenization. Additionally, local interventions can scale up along whole coastlines as occurs in the Northern Adriatic region.

Introduced, artificial hard structures usually differ from natural rocky habitats in terms of substrate type, orientation, surface topography, and extent affecting habitat availability complexity on various spatial scales from the microhabitat (cm) to the macrohabitat (1–100 m) scale. They typically comprise fewer microhabitats such as crevices and rock pools than natural rocky habitats. The introduction of seawalls alters the slope of the shore in extreme scenarios from near horizontal to near vertical, substantially reducing the extent of the habitat available to intertidal organisms. The communities associated with artificial structures therefore tend to differ from the naturally occurring rocky habitats and they are often less diverse than natural rocky shores (e.g., Bulleri and Chapman, 2006; Firth et al., 2013; Gacia, Satta, & Martin, 2007; Moschella et al., 2005; Vaselli, Bulleri, & Benedetti-Cecchi, 2008) because of lower habitat diversity. Mobile species tend to be rarer because of fewer refuges, and genetic diversity of the colonizing organisms can be lower (Chapman, 2003; Fauvelot, Costantini, Virgilio, & Abbiati, 2012; Fauvelot, Bertozzi, Costantini, Airoldi, & Abbiati, 2009; Pister, 2009). Other studies reported reduced reproductive output and smaller individual size on artificial structures compared with natural rocky shores (Moreira, Chapman, & Underwood, 2006; Moreira, Chapman, & Underwood, 2007).

Shifts in local hydrodynamics may also negatively impact local biodiversity. Shore-parallel structures such as breakwaters and T-shaped groins are designed to decrease the intensity of wave action incident to the shore. This will ultimately lead to a shift in local hydrodynamics, reducing current flow at the landward side and increasing current flow at the seaward side. Changes in grain size composition and the associated infaunal communities are common, with coarser sediments accumulating at the seaward side and finer sediments at the landward side (Martin et al., 2005). The establishment of shore-perpendicular structures aims to alter long-shore currents (groins) or directing current flow (jetties). Changes in shore morphodynamics are thus likely to occur as beach maintenance through changes in sediment dynamics is often the initial aim. This, too, causes changes in grain size distribution and the associated soft-sedimentary communities. Several studies reported that current velocities increased around groins (Lamberti & Zanuttigh, 2005; Munari, Corbau, Simeoni, & Mistri, 2011), creating a mosaic of different soft sedimentary communities as a likely consequence. The side of the groin exposed to net water movements (upstream) is likely to suffer from considerable scouring which will lead to coarser sediments, high disturbance, and thus an impoverished community. On the downstream side, colonization is likely to be greater. Macrobenthic communities near groins may therefore be highly disturbed, most likely because of changes to beach hydrodynamics resulting in reduced oxygen penetration into the sediment and lower oxygen supply to the interstitial fauna with opportunist species being more common at the impacted site compared with undisturbed sites.

The creation of hard bottom habitats will also cause the deposition of detrital material produced by the fouling communities that colonize the structure. The accumulation of the detritus on the soft sediment around the hard structures may have profound impacts on composition, abundance, and community structure of adjacent infaunal communities (Franzitta, 2013, p. 94). Breakwaters such as those at Elmer, UK, can channel detrital pathways, leading to greater accumulation of seaweeds in strandlines and around the structures themselves (Jolly, 2008). These accumulations are unsightly and rot, smell, and can lead to large numbers of flies, negatively impacting the amenity value of the adjacent shoreline.

Artificial structures are frequently colonized by non-native species (Bulleri & Airoldi, 2005; Glasby, Connell, Holloway, & Hewitt, 2007; Moschella et al., 2005; Tyrrell & Byers, 2007; Vaselli et al., 2008). This is probably because coastal defense structures are often found in highly disturbed environments, favoring the establishment of opportunistic species such as non-native species (Airoldi et al. 2005a).

Also, defense structures themselves are often exposed to high levels of disturbance during construction and maintenance, further increasing the likelihood of colonization by opportunistic species (Airoldi & Bulleri, 2011). Hard structures may also function as stepping stones between unfavorable habitat patches and thus facilitate the spread and establishment of non-native species in an area. For example, the non-native algae *Codium fragile* ssp. *tomentosoides* and *Caulerpa racemosa* were reported to be among the dominant space-colonizing species on breakwaters along the Italian coastline, especially on the landward side where conditions are more sheltered compared with surrounding environments (Bulleri & Airoldi, 2005; Bulleri & Chapman, 2004; Vaselli et al., 2008).

Some of these non-native and opportunistic species can have substantial negative economic consequences. For example, it has recently been suggested that the proliferation of artificial coastal structures may be an important driver of the global increase in jellyfish blooms (Duarte et al., 2013). Support is provided by observations and experimental evidence demonstrating that jellyfish larvae settle in large numbers on artificial structures in coastal waters and develop into dense concentrations of jellyfish-producing polyps.

It should be noted, however, that some species may also benefit from the introduction of artificial habitats. In the Wirral, UK, for example, it was observed that several bird species, in particular oystercatchers, dunlins, and plovers, used groins for roosting and feeding (Firth, Hawkins and Sharps, unpublished observations). A greater variety of birds were found on and around the structure sites compared with that of control sites, although occupancy can be patchy. It would appear that groins modify potential feeding grounds, with little use of finer sediments landward of structures while feeding aggregations were observed to the seaward side of structures. Fish, mussel, or crab aggregations can also be found around structures, making them popular venues for recreational fishing (Airoldi et al., 2005b).

4.6.3 Minimizing Ecological Impact

It is well documented that the availability of heterogeneous surfaces facilitates the establishment of diverse biological communities on rocky substrata because of the provision of various microhabitat structures (Firth & Crowe, 2008; Firth & Crowe, 2010; Huston, 1979; Johnson, Frost, Mosley, Roberts, & Hawkins, 2003; Therriault & Kolasa, 2000; Thompson, Wilson, Tobin, Hill, & Hawkins, 1996). Pinn, Mitchell, and Corkill (2005) found that microhabitat diversity also facilitates the establishment of a more diverse community on artificial structures, for example, rock pools support higher diversity than free-draining rock. Unfortunately, pools tend not to occur on boulder mound breakwaters, groins, or seawalls other than at the base. The addition of variously sized pits, crevices, and rock pools into the defense structure can thus result in its colonization by a variety of species by providing refuges from predation and desiccation (Browne & Chapman, 2011; Martins, Thompson, Neto, Hawkins, & Jenkins, 2010; Moreira, Chapman, & Underwood, 2007). This may also enhance the colonization of native species and thus reduce the risk of biotic invasion of the assemblages of these structures. These features may also reduce the accumulation of detrital material: small pools, pits, and crevices provide nursery areas for grazing molluscs that help to rapidly reduce algal cover, decreasing production of detritus. This results in other secondary positive effects because it makes structures less slippery and therefore safer for amenity use. Depending on local environmental conditions and regional species pool, the provision of microhabitats may also be beneficial for commercially important species. Martins et al. (2010) for example found that the addition of pits resulted in an increase in the abundance of the highly valued, commercially harvested limpet *Patella candei* on hard structures around the Azores.

A number of measures can be used to mitigate some of the ecological impacts of the construction of hard structures many of which are relatively low cost. A key consideration of any modification is that it does not compromise the primary coastal defense function of the structure. Table 4.3 summarizes the interventions considered here. Microhabitat features can be easily incorporated into shore-perpendicular as well as shore-parallel defense structures during or after their construction. For example, surface complexity was added to the side of the concrete wave breaker armor units at Plymouth Breakwater by drilling holes during the construction process (Figure 4.9, Table 4.4). Similarly, rock pools were created as part of the construction

Figure 4.9 (a) Precast units placed on the seaward side of Plymouth Breakwater, Devon, England, for additional protection. (b) Depressions around the lifting loops create artificial rock pools that support a great diversity of organisms. (c) As part of the THESEUS project, holes have been drilled onto the side of the blocks to increase habitat heterogeneity. *All photographs by L. Firth.*

TABLE 4.4 Summary of the Potential Effect on Performance and Cost Implications of Interventions Described in this Chapter and Trailed During THESEUS Adapted from Firth et al. (2014)

Structure	Effect on Performance	Cost Implications
Drilled holes in wave-breaker blocks on Plymouth breakwater, England	Drilling 100 pits on the breakwater block equated to removing 0.00025% of the total volume, and as such can be regarded as having a negligible impact on the strength of the structure. However, pits may accelerate the rate of fracturing.	In freshly cured concrete, 100 holes can be drilled in 1 h, making this a low-cost option.
Drilled rock pools on Tywyn breakwater, Wales	Drilling 18 rock pools on the breakwater equated to removing 1% of the total volume, as such can be regarded as have a negligible impact on the strength of the structure. However, holes may accelerate the rate of fracturing.	The drilling of these rocks pools required two men for 2 days and cost £860. This is a low-cost option.
Modified seawall at Shaldon, England	The sea wall at Shaldon was constructed in two parts; there is an inner core that provides the strength of the structure and an outer facing added to improve the aesthetics of the structure. The modifications were made to the rock facing and as such will not affect the strength of the wall.	The total project cost £6.5 million and the estimated cost of the trail was £20K representing 0.3% of the coast of the entire project Naylor, Coombes, Venn, Roast, and Thompson (2012). The enhancements were a requirement for planning permission.
Gabions filled with different sized rocks	It is not anticipated that the size of the rock used in the gabions will have any effect on the performance of the gabions.	The approximate cost of the gabion baskets and quarried rock was £500. Gabions are a cheap option of coastal defense.
Precast concrete BIOBLOCK in groin at Colwyn Bay, Wales	The primary function of the BIOBLOCK is to dissipate wave energy. The BIOBLOCK was cast to weigh 5.4 t. This is on the heavy side of the types of boulders that are typically used in coastal defense structures in the United Kingdom (3–6 t). The positioning of a BIOBLOCK in the place of a boulder is considered to have a negligible effect on the structure. However, pits and pools may accelerate the rate of fracturing.	Only 1 BIOBLOCK was deployed at Colwyn bay as a demonstration project. The mold, concrete, and delivery cost approximately £2000. If BIOBLOCKs were available as a product, it is anticipated that they will not cost much more than a typical quarried boulder.
Gardening of *Cystoseira* spp. onto subtidal breakwaters in Italy	The gardening of canopy-forming algae is not expected to have any negative effect on the performance of the structure.	At present the technique has only been tested in small-scale experiments. In our experimental tests, the costs for a small patch of 15 × 15 cm were about 6 euros. We are now testing methodologies to expand the tests at larger scales and reduce the associated costs.

process when large concrete blocks were cast and deployed either side of the metal eye used in placement (Figure 4.10). These rock pools support diverse biological assemblages (Griffin et al., 2010). Small rock pools can be created postconstruction by drilling or coring holes in the breakwater or groin unit (Figure 4.11, Table 4.4). On seawalls, rock pools can be constructed either during construction by incorporating water-retaining ledges (Chapman & Blockley, 2009) or retrospectively by affixing water-retaining units such as plant pots to vertical walls (Browne & Chapman, 2011). Crevices can be artificially created on seawalls during construction by incorporating gaps between interlocking blocks. Pits can be created in seawalls either during construction by manipulating the mortar between bricks (Firth et al., 2014, Table 4.4) or retrospectively again by drilling into the substrata (Martins et al., 2010; Witt et al., 2012). A precast BIOBLOCK unit with multiple habitat types (Figure 4.10) including pits, crevices, and rock pools was developed and was found to be successfully colonized by a variety of species (Firth et al., 2014). This unit with its microhabitat features could easily be incorporated into shore-perpendicular or parallel coastal defense structures, both during and after construction. Such blocks could

Figure 4.10 BIOBLOCK deployed at Colwyn Bay, Wales, UK. (a) The unit being lifted into place; (b) top-down view of unit showing scale and the three faces with multiple habitat types; (c) areas (25 × 25 cm) with pits that are either shallow (2 cm, top) or deep (5 cm, bottom); and (d) sidelong view showing longitudinal horizontal crevices. *All photographs by D. Roberts (Bangor University).*

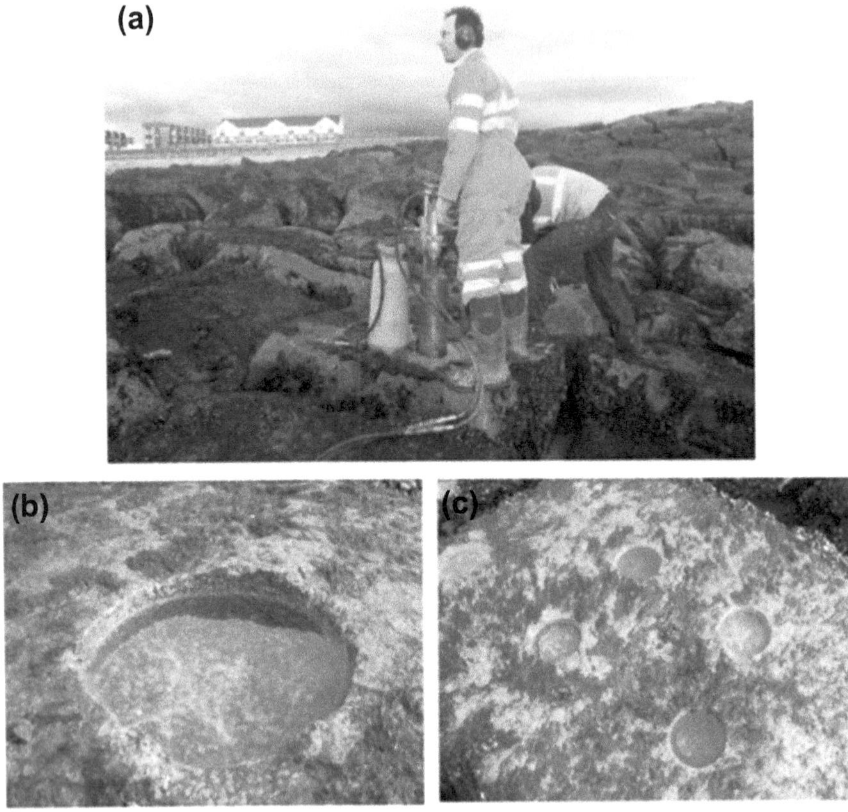

Figure 4.11 Drill-cored rock pools at Tywyn, Wales. (a) Contractors drilling the pools in November 2011. (b) Shallow (5 cm deep) pool. (c) Four pools on the surface of a boulder. *All photographs by L. Firth.*

be deployed among traditionally constructed breakwaters to locally enhance biodiversity.

The transplantation of desired species onto the breakwaters is another possibility to optimize the value of artificial coastal defense structures as substratum for possible forestations of vulnerable species where their habitats are under threat (Perkol-Finkel, Ferrario, Nicotera, Airoldi, & Cadotte, 2012), without compromising their original purpose of maintaining beaches for tourism. For example, in the Mediterranean Sea, experimental work is in progress to develop effective transplantation of the threatened canopy-forming, fucoid algae of the genus *Cystoseira* on a variety of coastal defense structures (Table 4.3) (Gianni et al., 2013; Perkol-Finkel & Airoldi, 2010; Perkol-Finkel et al., 2012). *Cystoseira* species are some of the most important marine ecosystem engineers, forming extended canopies comparable to forests on land that provide habitat, refuge, and food for many invertebrates and fishes at different life history stages (Benedetti-Cecchi et al., 2001; Thompson et al., 1996). *Cystoseira* forests, hence, play an important functional role in Mediterranean coastal ecosystems, sustaining complex food webs and maintaining a high biodiversity. Gardening of native habitat-forming species can also be a valuable means to prevent the establishment of non-native species, as space on the structure would be preempted by targeted species.

Care in the selection of building materials and consideration of the regional environmental context is recommended when aiming at incorporating habitat features into hard coastal defense structures. Many species colonize rough surfaces more easily compared to smooth surfaces. This was found at the Elmer defense structures (UK), where barnacles were more abundant on rough surfaces and in small cracks and crevices than on smooth surfaces (Moschella et al., 2005). Natural materials with a high degree of surface rugosity should thus be the preferred choice of material for the construction of hard defense structures whenever consistent with the regional environmental context. Also, particularly hard types of stone such as granite are often used for construction because they are less susceptible to weathering, thus minimizing damage to the defense structure. Granite tends to have a smooth surface, limiting the diversity of organisms that can colonize these structures. Softer rocks such as limestone are naturally rougher and weather more rapidly increasing spatial heterogeneity. This is reinforced by bioerosion by microscopic cyanobacteria as well as macrofauna further enhancing surface complexity and providing refuges for settlement. In those regions where naturally calcareous rocks prevail, it is thus advisable to use materials that incorporate as much such softer rock as possible to recreate the natural variability of rock hardness and weathering (Moschella et al., 2005). For example, the Plymouth Breakwater (UK) was constructed using a mixture of granite and limestone in the early part of the nineteenth century. This has resulted in the creation of rock pools where the limestone has eroded more easily than the adjacent granite.

Stone-filled cages known as gabions are an alternative to concrete in more sheltered areas. It was found that gabions constructed of stones of varying sizes offer refuges for mobile animals and support a higher biodiversity compared with gabions with only small or only large stones. This, too, can be easily and cost-effectively incorporated during the design phase and deployed on whole structures or part of structures to provide suitable habitats for living resources that might for example be the focus for recreational (such as angling, snorkeling) or educational (such as appreciation of marine life, rock pooling, or ornithology) activities (Table 4.4).

In the long term, ecological considerations in the design of marine infrastructure need to be integrated by careful planning of repair and maintenance during the lifetime of the structure. Airoldi and Bulleri (2011) found that maintenance caused a marked decrease in the cover of dominant space occupiers, such as mussels and oysters, and a significant increase in opportunistic and invasive forms, such as microbial films and weedy macroalgae. They also found that if interventions were made at certain times of the year the system recovered to the original state more rapidly compared with when interventions were done in other periods of the year, reflecting the reproductive biology windows of the dominant species. This is a type of design consideration that is extremely cost-effective because some impacts could be reduced by just planning the timing of the interventions more carefully.

It is also important to consider the level of maintenance works that may be required at a given site before other types of manipulations are undertaken (e.g., addition of pits, crevices or other surface features) to assess if they are worthwhile or whether the level of maintenance will mask the effects of the manipulations. Enhanced colonization by marine species facilitated by ecological engineering can sometimes elevate the strength and longevity of infrastructure thus reducing the frequency and magnitude of maintenance work. For example, biogenic buildup of organisms such as oysters, serpulid or sabellariid worms, barnacles, and corals can contribute to the stability and robustness of concrete-made infrastructure.

Pinn et al. (2005) found that older groins tend to support more species than younger groins as succession progressed. Thus biodiversity is likely to increase with time. Maintenance activities that will reset succession should thus be avoided as far as possible. Reducing disturbance because of poorly planned or unnecessary maintenance activities would also reduce the likelihood of colonization by non-native species. More research is needed addressing the practical implications of this particular issue.

Increasing the porosity of groins and breakwaters increases the wave energy transmitted through the structure thus lowering impacts on the soft sediment communities on the shoreward side. This approach must be optimized because increasing the porosity of the structure will decrease the effectiveness of the structure to attenuate wave energy. Impacts of defense structures depend on the type of management strategies deployed. A combination of "soft" and "hard" engineering solutions may be the preferred approach in a particular coastal defense scheme, reducing impacts on soft sedimentary habitats.

Also, Marine Renewable Energy Installations (MREIs) have the potential to attract mobile organisms and fouling species. Similarly to coastal defense structures, however, the habitats provided by these artificial installations differ from natural habitats and their expansion could lead to genetic diversity loss at regional scales, even if the underlying mechanisms are not yet fully clear (Fauvelot, Costantini, Virgilio, & Abbiati, 2012). The capacity of MREIs to provide suitable habitat for desired marine life could be increased through the addition of novel structural modifications (Inger et al., 2009; Wilson & Elliott, 2009; Witt et al., 2012). It may be possible to translate the outcomes of recent work examining habitat enhancement on other structures such as coastal defenses (Borsje et al., 2011; Martins et al., 2010;

Witt et al., 2012). The extent to which these structures attract marine life and the nature of the species attracted will largely be shaped by the location of the MREI and the design of its components (Burcharth, Hawkins, Zanuttigh, & Lamberti, 2007; Inger et al., 2009), with structural complexity of exposed surfaces being a key driver of the extent of colonization (Coombes et al., 2011; Petersen & Malm, 2006). A wider consideration is that MREIs may help to restore areas of seabed that have been lost through aggregate and dredging activities and commercial fishing (Thurstan, Roberts, Hawkins, & Neves, 2009; Witt et al., 2012).

4.6.4 Socioeconomic Implications

Although artificial structures increase the economic value of an area by offering protection from flooding and erosion, the economic costs potentially related to some of their negative impacts (such as the spread of invasive pests that could damage aquaculture or other nuisance species such as jellyfish that could damage tourism) are rarely considered. If structures are deemed necessary, it is therefore fundamental to incorporate as much ecological knowledge as possible to enhance their ecological value as habitat and mitigate some of the impacts.

If colonization is by habitat-forming species such as mussels and certain macroalgal species such as *Cystoseira barbata*, this may result in the establishment of biological communities more similar to native communities. This is also likely to prevent the establishment of non-native species (Arenas et al., 2006). Oysters or mussels often colonize defense structures and may be an important food source, in particular on a subsistence basis in less wealthy countries, thus providing multiple beneficial outcomes of defense structures. In more wealthy countries, collection of shellfish may become a recreational activity. The incorporation of certain design features such as pits and crevices into artificial structures may even result in an increase in abundance of overexploited, commercially important species (Martins et al., 2010), even more so increasing the economic value of the structure. Sea defense structures also attract fish which provide opportunities for recreational snorkeling, spear-fishing, and angling.

Negative socioeconomic effects, however, may result from the impacts associated with the defense structures if the native community changes through the introduction of new habitats, the loss of natural habitats, the introduction of non-native and pest species, and changes in existing communities and habitats. Groins in particular may impact on the economic value for tourism of a beach because they can be unsightly. Also, the enhancement of jellyfish in coastal areas that has been related to the proliferation of artificial structures (Duarte et al., 2013) could be very damaging to tourism. Depending on the location of the structure, its appearance can be important. In case of a new seawall at Shaldon, for example, it was important that the construction was visually attractive, resulting in the use of local stone.

MREIs also have the potential to function as fish-aggregating devices (Fayram & de Risi, 2007; Wilhelmsson, Malm, & Ohman, 2006). Many species of fish aggregate and can be found around floating objects, a phenomenon that fishermen have exploited for centuries (Castro, Santiago, & Santana–Ortega, 2001). Species richness and assemblage size are positively correlated with fish-aggregating device size (Nelson, 2003); as such, larger MREIs may have greater diversity of fish species. The attraction of fish will have ecological consequences and likely to attract larger predatory birds, fish, and marine mammals that feed on the concentrated prey resources.

4.7 Interaction of Multiple Stressors in the Coastal Zone

A significant proportion of the seashore has been altered for human use; thus, many native shallow habitats have been turned into artificial ones or are impacted by development and changes in sediment dynamics that increase erosion (Dugan et al., 2011). When coupled with climatic-driven changes, these cumulative anthropogenic perturbations will increasingly affect the stability, resilience, and productivity of ecosystems (Claudet & Fraschetti, 2010). Local habitat modification (10–100 m) and loss can also scale up to have broader scale implications on the ecological landscape (km) in extreme cases such as the Northern Adriatic to the regional scale (100–1000 km). Such changes in regime and extent of natural habitats compromise the provision of fundamental ecosystem services, including coastal protection. So far most management processes, including approaches to coastal risk management, have been developed in an impact-by-impact framework (Crowder et al., 2006), often delivered at a local scale. It is thus important to develop a management framework that incorporates knowledge of how natural systems are affected by multiple stressors (Crain, Kroeker, & Halpern, 2008). These concepts inform management of ecosystem resilience (Lotze, Coll, Magera, Ward-Paige, & Airoldi, 2011). If the resilience is progressively "eroded" by cumulative impacts, the system becomes vulnerable to regime shifts; these are critical transitions that are characterized by a different set of structures and processes (Scheffer, Carpenter, Foley, Folke, & Walker, 2001).

4.7.1 Basic Concepts and Definitions

There are three common models to describe the outcome of multiple stressors on biological systems: (1) the additive model, in which the response is equal to the sum of the effects of the individual stressors; (2) the antagonistic model, when the combined effect is less than the sum of the effects of the individual stressors; and (3) the synergistic model, in which the response to exposure to multiple stressors is greater than the sum of the effects of the individual stressors (Crain et al., 2008). Multiple stressors acting simultaneously also have the potential to interact, driving changes that are not at all predictable from the effects of single stressors ("ecological surprises" in the sense of Paine, Tegner, & Johnson, 1998). Such uncertainty reduces the predictive value of models of biodiversity change (Darling & Côté, 2008).

In most marine and coastal systems, the cumulative impact from several stressors tends to result in synergistic and antagonistic effects rather that additive ones, and synergies can double whenever more than two stressors were considered (Crain et al., 2008). From a management perspective, this has important implications: as the synergistic interactions between multiple stressors can exacerbate nonlinear responses of ecosystems to human impacts and limit their recovery capacity for recovery (Halpern, McLeod, Rosenberg, & Crowder, 2008). For additive effects, reducing the magnitude of any stressor should lead to a corresponding increase in the response of interest allowing for straightforward expectations of response to management and conservation actions. In contrast, mitigation of stressors involved in synergistic or antagonistic interactions will lead to either greater than or less than predicted results compared with additive models (Brown, Saunders, Possingham, & Richardson, 2013). Of particular concern are cooccurring stressors in which multiple stressors reinforce the direction of change and reduce the likelihood that a system can retrace the same trajectory during restoration as during degradation leading to hysteresis. Active restoration can reduce the divergence between the trajectories of environmental degradation and recovery, thereby reducing hysteresis. Unfortunately, at the moment, there is insufficient understanding of the effects of multiple stressors for most coastal systems or understanding of the scale of operation of the underlying ecological processes.

4.7.2 Key Cumulative Stressors That Can Impair Ecological Approaches to Coastal Risk Mitigation

Here attention is drawn to the cumulative stressors that are likely to act on European coastal marine ecosystems in parallel with environmental pressures arising from flooding and increased storminess. Stressors are defined as factors that interfere with the normal functioning of the system, referring to current use of the term in relevant policies (i.e., the Marine Strategy Directive 2008/56/EC). For each target habitat (saltmarshes, seagrass meadows, biogenic reefs, and sandy habitats including beaches and dunes), combinations of factors were identified that could cumulatively increase the vulnerability of the natural systems to phase shift, therefore negatively affecting the capability of the system to provide effective coastal protection. Table 4.5 synthesizes the major factors that can potentially increase or reduce the risk of step changes or in the worst case regime shifts in communities and ecosystems, thereby influencing the ecological sustainability of defense measures.

TABLE 4.5 List of Most Important Cooccurring Stressors that Could Increase the Vulnerability of the Natural Systems to Phase Shift at the THESEUS Study Sites (See Chapter 7)

Habitat	Key Stressors
Saltmarshes	Nutrient loading, pollution, dredging, land reclamation and use, changes in sedimentation, changes in hydrodynamics, shipping, tourism (including trampling) and water sports, destructive fisheries, grazing, invasive species, rising temperatures, changes in precipitation, increased storminess, sea-level rise
Sand dunes	Urbanization, land use, trampling and driving by vehicles, invasive species, sea-level rise, increased storminess, and erosion
Seagrass meadows	Nutrient loading, destructive fisheries, dredging and water turbidity, changes in sedimentation, coastal development, aquaculture, land reclamation, changes in hydrodynamics, invasive species, rising sea surface temperatures, increased storminess, changes in salinity, sea-level rise
Biogenic reefs (including *Sabellaria* reefs and *Ostrea edulis* beds)	Pollution, destructive fisheries, sedimentation, harvesting, anoxia, changes in hydrodynamics, invasive species, toxic algal blooms, increasing temperatures, ocean acidification, changes in salinity
Freshwater systems	Nutrient loading, pollution, urbanization, and channelization, changes in sedimentation, invasive species, rising water temperatures, reduction in freshwater input, increased frequency of storms

Pollution, eutrophication, drainage, urban development, conversion to agriculture and aquaculture, changes in sedimentation and hydraulic regimes, fishery overexploitation and disruptive fishing techniques, shipping, tourism and water sports, and invasive species were generally found to the most serious threats to most coastal ecosystems (see also Airoldi & Beck, 2007 for a review on past and current threats to European coastal habitats). Pollution, while being more regulated, especially from point sources, is pervasive (Thompson, Crowe, & Hawkins, 2002), and plastic litter is a global problem (Thompson, Moore, vom Saal, & Swan, 2009). The coasts are becoming increasingly developed, leading to habitat loss at local and regional scales (Airoldi & Beck, 2007; Lotze et al., 2006). Stressors acting on coastal habitats must be considered within the context of the site and the scale at which they act. Some stressors can be addressed more easily than others. For example, stressors that are local or regional in extent, such as eutrophication of coastal waters, can be addressed through better treatment of waste water entering coastal areas and improving land management by reduction fertilizer runoff. In Europe, the Water Framework Directive provides both the legislative framework and has prompted tools to enable monitoring and management. Other pollution impacts such as the effects of pesticides or antifoulants can be controlled by national or European legislation and international conventions to restrict pathways. Inshore fisheries can be controlled by European, national, and local regulations (e.g., in the United Kingdom by local bylaws). In contrast, there are stressors operating on global or regional scales which are not controllable. These include anthropogenic climate change driven stressors such as global warming and stormier and rising seas. The interactions of these unmanageable global stressors with those more tractable regional and local impacts can, however, be addressed. Awareness of these interactions can be built into management plans for stewardship. For natural coastal defenses such as sand dunes and saltmarshes, reducing trampling and grazing impacts would increase resilience to lower return time of extreme storm events in a warming world. For built hard structures in nutrient-rich areas, increasing topographic complexity to encourage greater colonization of grazing limpets and snails would reduce the likelihood of greater ephemeral algal cover, which could compromise recreational use of structures because of extensive coverage of slippery algae. Awareness of potential interactions can also be incorporated at the planning phase to enable a more precautionary approach by anticipating global environmental change. Consideration of all likely impacts and their interaction with potential benefits or impacts is essential before management intervention or structure placement in the coastal zone. Postimplementation monitoring and assessment of efficacy and any expected or unanticipated impacts is essential to enable ongoing adaptive management. Additionally, well-documented case studies provide interesting research opportunities as well as informing emerging best practice.

4.8 Conclusions

Ecological approaches to coastal risk mitigation enable natural coastal habitats including saltmarshes, sand dunes, biogenic reefs, and seagrass meadows to be managed and restored. For these systems to provide coastal defense services, they must be healthy and multiple interacting natural and anthropogenic stressors must also be considered and managed where possible.

Four options are available as coastal management strategies: (1) no active intervention, (2) managed realignment, (3) hold the line, and (4) advance the line. This chapter has shown how natural habitats can be used as a tool in each of these management options. As discussed in Section 4.3, saltmarshes can be used to reclaim land through techniques to extend the marsh seawards. Saltmarshes are also the common target habitat in the majority of managed realignment projects (see Section 4.3.2). Healthy saltmarsh and their associated mudflats along with the other intertidal habitats including, biogenic reefs (Section 4.4) and seagrass meadows (Section 4.5), can naturally contribute to holding the line, principally because of the wave attenuating properties of all these habitats. Thus restoration of degraded coastal habitats can play an important role in holding the line. Where natural habitats are bordered by an impermeable barrier such as a seawall, dike, road, or railway embankment, then the process of coastal squeeze through sea-level rise could eventually reduce their size and effectiveness as a natural coastal defense. In the longer term, some form of managed realignment would allow these habitats to continue to function as a viable part of coastal defense by providing space for landward expansion with rising sea levels, potentially leading to some form of managed realignment. Healthy sand dunes (Section 4.2) of sufficient size also provide a suitable barrier to hold the line. If sand dunes are to be used successfully to replace a hard defense structure through a program of habitat creation or regeneration, the dune systems need space to evolve and move inland as sea levels rise and climatic conditions change. Such a strategy is more akin to managed realignment than "hold the line" and should be implemented accordingly.

Managed realignment can be a valuable tool in coastal management and can offer many advantages. For example, it has been shown that if used in conjunction with hard defenses such as sea walls, it can reduce the cost and increase the life of sea walls. Newly created saltmarsh through managed realignment can provide wave attenuation, shoreline stabilization, and floodwater attenuation (Shepard et al., 2011). It is anticipated that the new saltmarsh should be able to keep pace with rising sea levels through sediment accretion as long as there is enough space for the marsh to grow and evolve. A carefully planned, managed realignment scheme should therefore consider this issue in order to anticipate such problems. There is still insufficient knowledge on the best way to implement managed realignment schemes, particularly because the success of current schemes has been variable and context dependent. For a managed realignment scheme to be successful, all aspects of marsh function need to be considered, including geomorphological, hydrological, and ecological processes.

Habitat restoration can be used to advance the line. The only viable example is saltmarsh extension in front of a seawall or dike, which can reduce the costs of building or maintaining such a structure. Sea-level rise could compromise this option in areas with low sediment supply. With sufficient resources and know-how, advancing the line is technically feasible. Doing nothing is, however, still a viable option in some rural settings such as low-value agricultural or amenity use land. This relies on natural coastal processes to stabilize the coastline. Such an approach is being used by the landowning, natural heritage charity National Trust in the United Kingdom (The National Trust, 2008) and is UK government policy for implementation in some rural areas. Reducing risks to lives and property is achieved by stricter planning regimes in areas prone to flooding and coastal erosion. Although doing nothing is an inexpensive option with little ecological impact, rising prices for agricultural land and concerns for food security may influence the tradeoffs between defense and unmanaged or managed retreat. Doing nothing is not an option in many parts of Europe. In the Netherlands, there is little available land and a strong cultural tradition of restling land from the sea.

Maintaining and restoring healthy natural habitats or managed realignment, however, are just a part of a range of coastal defense tools. There will always be a need to construct hard coastal defenses to guarantee a certain level of protection of coastal property and infrastructure. There is still scope to enhance the ecological potential of these structures through simple measures, preferably at the design stage, but also after construction. They do not need to be objects devoid of life; simple techniques such as increasing the surface heterogeneity or adding water-retaining features can make a significant difference to the biodiversity inhabiting these structures. Planners must be aware that new structures and modifications to existing structures could facilitate the spread of non-native and potentially invasive species. It is particularly important to consider the effects of hard structures in terms of habitat replacement and in modifying surrounding habitats.

The exact approach adopted for coastal management will ultimately depend on context. Geographic setting, economic imperatives, cultural values, and the interaction of complex ecological processes all need to be considered to make informed choices to reduce flood risk and sustainably develop our valuable coastlines.

References

Adam, P. (2002). Saltmarshes in a time of change. *Environmental Conservation, 29*, 39–61.

Airoldi, L., Abbiati, M., Beck, M. W., Hawkins, S. J., Johnson, P. R., Martin, D., et al. (2005a). An ecological perspective on the deployment and design of lowcrested and other hard coastal defence structures. *Coastal Engineering, 52*, 1073–1087.

Airoldi, L., Bacchiocchi, F., Cagliola, C., Bulleri, F., & Abbiati, M. (2005b). Impact of recreational harvesting on assemblages in artificial rocky habitats. *Marine Ecology Progress Series, 299*, 55–66.

Airoldi, L., & Beck, M. W. (2007). Loss, status and trends for coastal marine habitats of Europe. In R. N. Gibson, R. J. A. Atkinson, & J. D. M. Gordon (Eds.), *Oceanography and marine biology: An annual review*. Boca Raton: CRC Press-Taylor & Francis Group.

Airoldi, L., & Bulleri, F. (2011). Anthropogenic disturbance can determine the magnitude of opportunistic species responses on marine urban infrastructures. *Plos One, 6*, e22985.

Alexandra, H., Duarte, C. M., & Krause-Jensen, D. (2004). How long time does it take to recolonize seagrass beds? In J. Borum, C. M. Duarte, D. Krause-Jensen, & T. M. Greve (Eds.), *European seagrasses: An introduction to monitoring and management. The M&MS project. Monitoring and managing of european seagrasses EVK3-CT-2000–00044*.

Allen, J. (1990). Salt-marsh growth and stratification: a numerical model with special reference to the Severn Estuary, southwest Britain. *Marine Geology, 95*, 77–96.

Allen, J. R., Wilkinson, S. B., & Hawkins, S. J. (1995). Redeveloped docks as artificial lagoons: the development of brackish-water communities and potential for conservation of lagoonal species. *Aquatic Conservation: Marine and Freshwater Ecosystems, 5*, 299–309.

Allen, J. R. L. (2000). Morphodynamics of Holocene salt marshes: a review sketch from the Atlantic and Southern North Sea coasts of Europe. *Quaternary Science Reviews, 19*, 1155–1231.

Arenas, F., Bishop, J. D. D., Carlton, J. T., Dyrynda, P. J., Farnham, W. F., Gonzalez, D. J., & Wood, C. A. (2006). Alien species and other notable records from a rapid assessment survey of marinas on the south coast of England. *Journal of the Marine Biological Association of the United Kingdom, 86*(06), 1329–1337.

Bakker, J. P. (1985). The impact of grazing on plant communities, plant populations and soil conditions on salt marshes. *Vegetation, 62*, 391–398.

Balke, T., Klaassen, P. C., Garbutt, A., van der Wal, D., Herman, P. M. J., & Bouma, T. J. (2012). Conditional outcome of ecosystem engineering: a case study on tussocks of the salt marsh pioneer *Spartina anglica. Geomorphology, 153*, 232–238. http://dx.doi.org/10.1016/j.geomorph.2012.03.002.

Balke, T., Webb, E. L., van den Elzen, E., Galli, D., Herman, P. M. J., & Bouma, T. J. (2013). Seedling establishment in a dynamic sedimentary environment: a conceptual framework using mangroves. *Journal of Applied Ecology, 50*, 740–747. http://dx.doi.org/10.1111/1365-2664.12067.

Bertness, M. D., & Ellison, A. M. (1987). Determinants of pattern in a New England salt marsh plant community. *Ecological Monographs, 57*, 129–147.

Bastyan, G. R., & Cambridge, M. L. (2008). Transplantation as a method for restoring the seagrass *Posidonia australis. Estuarine, Coastal and Shelf Science, 79*, 289–299.

Beck, M. W., Brumbaugh, R. D., Airoldi, L., Carranza, A., Coen, L. D., Crawford, C., et al. (2011). Oyster reefs at risk and recommendations for conservation, restoration, and management. *BioScience, 61*, 107–116.

Belluco, E., Camuffo, M., Ferrari, S., Modenese, L., Silvestri, S., Marani, A., et al. (2006). Mapping salt-marsh vegetation by multispectral and hyperspectral remote sensing. *Remote Sensing of Environment, 105*, 54–67.

Benedetti-Cecchi, L., Pannacciulli, F., Bulleri, F., Moschella, P. S., Airoldi, L., Relini, G., et al. (2001). Predicting the consequences of anthropogenic disturbance: large-scale effects of loss of canopy algae on rocky shores. *Marine Ecology Progress Series, 214*, 137–150.

Bird, E. C. F. (2000). *Coastal geomorphology: An introduction*. Chichester: John Wiley.

Björk, M., Short, F., Mcleod, E., & Beer, S. (Eds.). (2008). *Managing seagrasses for resilience to climate change*. Gland, Switzerland: IUCN.

Bolam, S. G., & Whomersley, P. (2005). Development of macrofaunal communities on dredged material used for mudflat enhancement: a comparison of three beneficial use schemes after one year. *Marine Pollution Bulletin, 50*, 40–47. http://dx.doi.org/10.1016/j.marpolbul.2004.08.006.

Borsje, B. W., et al. (2011). How ecological engineering can serve in coastal protection. *Ecological Engineering, 37*, 113–122.

Bortone, S. A. (2000). *Seagrasses: Monitoring, ecology, physiology, and management*. In *CRC marine science series* (Vol. 16). Boca Raton: CRC Press, ISBN 0-8493-2045-3.

Borum, J., Duarte, C. M., Krause-Jensen, D., & Greve, T. M. (2004). *European seagrasses: An introduction to monitoring and management.*

Bouma, T. J., Koutstaal, B. P., van Dongen, M., & Nielsen, K. L. (2001). Coping with low nutrient availability and inundation: root growth responses of three halophytic grass species from different elevations along a flooding gradient. *Oecologia, 126*, 472–481. http://dx.doi.org/10.1007/s004420000545.

Bouma, T. J., De Vries, M. B., Low, E., Peralta, G., Tanczos, C., Van de Koppel, J., et al. (2005). Trade-offs related to ecosystem engineering: a case study on stiffness of emerging macrophytes. *Ecology, 86*, 2187–2199. http://dx.doi.org/10.1890/04-1588.

Bouma, T. J., van Duren, L. A., Temmerman, S., Claverie, T., Blanco-Garcia, A., Ysebaert, T., et al. (2007). Spatial patterns in flow- and sedimentation within vegetation patches: comparing field, flume and hydrodynamic modelling experiments. *Continental Shelf Research, 27*, 1020–1045.

Bouma, T. J., Friedrichs, M., van Wesenbeeck, B. K., Temmerman, S., Graf, G., & Herman, P. M. J. (2009). Density-dependent linkage of scale-dependent feedbacks: a flume study on the intertidal macrophyte *Spartina anglica. Oikos, 118*, 260–268. http://dx.doi.org/10.1111/j.1600-0706.2008.16892.x.

Bouma, T. J., De Vries, M. B., & Herman, P. M. J. (2010). Comparing ecosystem engineering efficiency of 2 plant species with contrasting growth strategies. *Ecology, 91*, 2696–2704. http://dx.doi.org/10.1890/09-0690.

Bouma, T. J., Temmerman, S., van Duren, L. A., Martini, E., vandenburwaene, W., callaghan, D. P., et al. (2013c). Organism traits determine the strength of scale-dependent bio-geomorphic feedbacks: a flume study on these intertidal plant species. *Geomorphology, 180–181*, 57–65.

Bouma, T. J., van Belzen, J., Balke, T., Zhu, Z., Airoldi, L., Blight, A. J., et al. (2014). Identifying knowledge gaps hampering application of intertidal habitats in coastal protection: Opportunities & steps to take. *Coastal Engineering, 87*, 147–157. http://dx.doi.org/10.1016/j.coastaleng.2013.11.014.

Brown, C. J., Saunders, M. I., Possingham, H. P., & Richardson, A. J. (2013). Managing for interactions between local and global stressors of ecosystems. *Plos One, 8*.

Bulleri, F., Abbiati, M., & Airoldi, L. (2006). The colonisation of human-made structures by the invasive alga codium fragile ssp. tomentosoides in the North Adriatic Sea (NE Mediterranean). *Hydrobiologia, 555*, 263–269.

Bulleri, F., & Airoldi, L. (2005). Artificial marine structures facilitate the spread of a non-indigenous green alga, *Codium fragile* ssp tomentosoides, in the north Adriatic Sea. *Journal of Applied Ecology, 42*, 1063–1072. http://dx.doi.org/10.1111/j.1365-2664.2005.01096.x.

Bulleri, F., & Chapman, M. G. (2004). Intertidal assemblages on artificial and natural habitats in marinas on the north-west coast of Italy. *Marine Biology, 145*, 381–391. http://dx.doi.org/10.1007/s00227-004-1316-8.

Browne, M. A., & Chapman, M. G. (2011). Ecologically informed engineering reduces loss of intertidal biodiversity on artificial shorelines. *Environmental Science and Technology, 45*, 8204–8207.

Burcharth, H. F., Hawkins, S. J., Zanuttigh, B., & Lamberti, A. (2007). *Environmental design guidelines for low crested coastal structures*. Amsterdam: Elsevier.

Cahoon, D. R., & Reed, D. J. (1995). Relationships among marsh surface topography, hydroperiod, and soil accretion in a deteriorating Louisiana salt marsh. *Journal of Coastal Research, 11*, 357–369.

Callaghan, D. P., Bouma, T. J., Klaassen, P., van der Wal, D., Stive, M. J. F., & Herman, P. M. J. (2010). Hydrodynamic forcing on salt-marsh development: Distinguishing the relative importance of waves and tidal flows. *Estuarine Coastal and Shelf Science, 89*, 73–88. http://dx.doi.org/10.1016/j.ecss.2010.05.013.

Castro, J. J., Santiago, J. A., & Santana-Ortega, A. T. (2001). A general theory on fish aggregation to floating objects: an alternative to the meeting point hypothesis. *Reviews in Fish Biology and Fisheries, 11*, 255–277.

Chapman, M. G. (2003). Paucity of mobile species on constructed seawalls: effects of urbanization on biodiversity. *Marine Ecology Progress Series, 264*, 21–29. http://dx.doi.org/10.3354/meps264021.

Chapman, M. G., & Blockley, D. (2009). Engineering novel habitats on urban infrastructure to increase intertidal biodiversity. *Oecologia, 161*, 625–635.

Chapman, M. G., & Bulleri, F. (2003). Intertidal seawalls—new features of landscape in intertidal environments. *Landscape and Urban Planning, 62*, 159–172. http://dx.doi.org/10.1016/s0169-2046(02) 00148–2.

Chapman, M. G., & Underwood, A. J. (2011). Evaluation of ecological engineering of "armoured" shorelines to improve their value as habitat. *Journal of Experimental Marine Biology and Ecology, 400*, 302–313. http://dx.doi.org/10.1016/j.jembe.2011.02.025.

Cheramie, K., Francis, R., & Kilgen, K. (1995). Thin layer spoil deposition: an alternative restorative tool for coastal marshes. In *Coastal zone: Proceedings of coastal zone '95, the ninth symposium on coastal and ocean management. 16–21 July*. Tampa, Florida, USA: American Society of Civil Engineers.

Christensen, P. B., Díaz, E., & Diekmann, O. (2004). Can transplanting accelerate the recovery of sea grasses? In J. Borum, C. M. Duarte, D. Krause-Jensen, & T. M. Greve (Eds.), *European seagrasses: An introduction to monitoring and management. The M&MS project. Monitoring and managing of european seagrasses EVK3-CT-2000–00044*.

Christianen, M. J. A., van Belzen, J., Herman, P. M. J., van Katwijk, M. M., Lamers, L. P. M., van Leent, P. J. M., et al. (2013). Low-canopy seagrass beds still provide important coastal protection services. *PloS One, 8*, e62413. http://dx.doi.org/10.1371/journal.pone.0062413.

Claudet, J., & Fraschetti, S. (2010). Human-driven impacts on marine habitats: a regional meta-analysis in the Mediterranean Sea. *Biological Conservation, 143*, 2195–2206. http://dx.doi.org/10.1016/j.biocon.2010.06.004.

Colosio, F., Abbiati, M., & Airoldi, L. (2007). Effects of beach nourishment on sediments and benthic assemblages. *Marine Pollution Bulletin, 54*, 1197–1206.

Coombes, M. A., Naylor, L. A., Thompson, R. C., Roast, S. D., Gomez-Pujol, L., & Fairhurst, R. J. (2011). Colonization and weathering of engineering materials by marine microorganisms: an SEM study. *Earth Surface Processes Landforms, 36*, 582–593.

Council Directive 92/43/EEC of 21 May 1992 on the conservation of natural habitats and of wild fauna and flora.

Cornell, J. A., Craft, C. B., & Megonigal, J. P. (2007). Ecosystem gas exchange across a created salt marsh chronosequence. *Wetlands, 27*, 240–250. http://dx.doi.org/10.1672/0277-5212(2007)27[240:egeaac]2.0.co;2.

Costanza, R., d'Arge, R., Groot, R. d., Farber, S., Grasso, M., Hannon, B., et al. (1997). The value of the World's ecosystem services and natural capital. *Nature, 387*, 253–260.

Craft, C., Broome, S., & Campbell, C. (2002). Fifteen years of vegetation and soil development after Brackish-water marsh creation. *Restoration Ecology, 10*, 248–258.

Crain, C. M., Kroeker, K., & Halpern, B. S. (2008). Interactive and cumulative effects of multiple human stressors in marine systems. *Ecology Letters, 11*, 1304–1315.

Croft, A. L., Leonard, L. A., Alphin, T. D., Cahoon, L. B., & Posey, M. H. (2006). The effects of thin layer sand renourishment on tidal marsh processes: Masonboro Island, North Carolina. *Estuaries and Coasts, 29*, 737–750.

Crowder, L. B., Osherenko, G., Young, O. R., Airame, S., Norse, E. A., Baron, N., et al. (2006). Sustainability—resolving mismatches in US ocean governance. *Science, 313*, 617–618. http://dx.doi.org/10.1126/science.1129706.

Cunningham, P. N., Hawkins, S. J., Jones, H. D., & Burrows, M. T. (1984). *The geographical distribution of* Sabellaria alveolata *(L.) in England, Wales and Scotland, with investigations into the community structure of, and the effects of trampling on* Sabellaria alveolata *colonies*. Report to the Nature Conservancy Council from the Department of Zoology. Manchester: Manchester University.

Darling, E. S., & Côté, I. M. (2008). Quantifying the evidence for ecological synergies. *Ecology Letters, 11*, 1278–1286. http://dx.doi.org/10.1111/j.1461-0248.2008.01243.x.

Day, J., Rybczyk, J., Scarton, F., Rismondo, A., Are, D., & Cecconi, G. (1999). Soil accretionary dynamics, sea-level rise and the survival of wetlands in Venice Lagoon: a field and modelling approach. *Estuarine, Coastal and Shelf Science, 49*, 607–628.

Deegan, L. A., et al. (2012). Coastal eutrophication as a driver of salt marsh loss. *Nature, 490*(7420), 388–392. http://dx.doi.org/10.1038/nature11533.

Defeo, O., & McLachlan, A. (2005). Patterns, processes and regulatory mechanisms in sandy beach macrofauna: a multi-scale analysis. *Marine Ecology Progress Series, 295*, 1–20.

Defra. (2006). *Shoreline management plan guidance. Volume 1: Aims and requirements. Volume 2: Procedures (plus 10 appendices)*. London: Department of Environment, Food and Rural Affairs.

Dias, A. S., & Paula, J. (2001). Associated fauna of *Sabellaria alveolata* colonies on the central coast of Portugal. *Journal of Marine Biological Association of United Kingdom, 80*(2001), 169–170.

Diggory, Z. E., & Parker, V. T. (2009). Seed supply and revegetation dynamics at restored tidal marshes, Napa river, California. *Restoration Ecology, 19*, 121–130.

Doody, J. P. (2004). "Coastal squeeze"—an historical perspective. *Journal of Coastal Conservation, 10*(1), 129–138.

Duarte, C. M., Middelburg, J., & Caraco, N. (2005). Major role of marine vegetation on the oceanic carbon cycle. *Biogeosciences, 2*, 1–8.

Duarte, C. M., Marbà, N., & Santos, R. (2004). What may cause loss of seagrasses? In J. Borum, C. M. Duarte, D. Krause-Jensen, & T. M. Greeve (Eds.), *European seagrasses: An introduction to monitoring and management. The M&MS project. Monitoring and managing of european seagrasses EVK3-CT-2000-00044* (pp. 24–32).

Duarte, C. M., Pitt, K. A., Lucas, C. H., Purcell, J. E., Uye, S., Robinson, K., et al. (2013). Is global ocean sprawl a cause of jellyfish blooms? *Frontiers in Ecology and the Environment, 11*, 91–97.

Dubois, S., Retiere, C., & Olivier, F. (2002). Biodiversity associated with *Sabellaria alveolata* (Polychaeta: Sabellariidae) reefs: effects of human disturbances. *Journal of the Marine Biological Association of the United Kingdom, 82*, 817–826.

Dugan, J. E., Airoldi, L., Chapman, M. G., Walker, S. J., & Schlacher, T. (2011). 8.02-Estuarine and coastal structures: environmental effects, a focus on shore and nearshore structures. In E. Wolanski, & D. McLusky (Eds.), *Treatise on estuarine and coastal science*. Waltham, UK: Academic Press.

Dugan, J. E., Hubbard, D. M., Rodil, I. F., Revell, D. L., & Schroeter, S. (2008). Ecological effects of coastal armoring on sandy beaches. *Marine Ecology-An Evolutionary Perspective, 29*, 160–170.

Erfanzadeh, R., Garbutt, A., Pétillon, J., Maelfait, J. P., & Hoffmann, M. (2010). Factors affecting the success of early salt-marsh colonizers: seed availability rather than site suitability and dispersal traits. *Plant Ecology, 206*, 335–347.

Erftemeijer, P., & Lewis, R. (2006). Environmental impacts of dredging on seagrasses: a review. *Marine Pollution Bulletin, 52*, 1553–1572.

Erickson, V. J., Wood, J., & Riley, S. A. (2003). *Guidelines for revegetation of invasive weed sites and other disturbed areas on national forests and grasslands in the Pacific Northwest.*

Evans, P. R., Ward, R. M., Bone, M., & Leakey, M. (1998). Creation of temperate-climate intertidal mudflats: factors affecting colonization and use by benthic invertebrates and their bird predators. *Marine Pollution Bulletin, 37*, 535–545.

Fauvelot, C., Bertozzi, F., Costantini, F., Airoldi, L., & Abbiati, M. (2009). Lower genetic diversity in the limpet *Patella caerulea* on urban coastal structures compared to natural rocky habitats. *Marine Biology, 156*, 2313–2323. http://dx.doi.org/10.1007/s00227-009-1259-1.

Fauvelot, C., Costantini, F., Virgilio, M., & Abbiati, M. (2012). Do artificial structures alter marine invertebrate genetic makeup? *Marine Biology, 159*, 2797–2807.

Fayram, A. H., & de Risi, A. (2007). The potential compatibility of offshore wind power and fisheries: an example using bluefin tuna in the Adriatic Sea. *Ocean and Coastal Management, 50*, 597–605.

Finkl, C. W., & Walker, H. J. (2004). Beach nourishment. In M. Schwartz (Ed.), *The encyclopedia of coastal science* (pp. 37–54). Dordrecht, The Netherlands: Kluwer Academic.

Firth, L. B., & Crowe, T. P. (2008). Large-scale coexistence and small-scale segregation of key species on rocky shores. *Hydrobiologia, 614*, 233–241. http://dx.doi.org/10.1007/s10750-008-9509-7.

Firth, L. B., & Crowe, T. P. (2010). Competition and habitat suitability: small-scale segregation underpins large-scale coexistence of key species on temperate rocky shores. *Oecologia, 162*, 163–174. http://dx.doi.org/10.1007/s00442-009-1441-7.

Firth, L. B., & Hawkins, S. J. (2011). Introductory comments - global change in marine ecosystems: patterns, processes and interactions with regional and local scale impacts. *Journal of Experimental Marine Biology and Ecology, 400*, 1–6. http://dx.doi.org/10.1016/j.jembe.2011.02.001.

Firth, L. B., Thompson, R. C., Abbiati, M., Airoldi, L., Bohn, K., Bouma, T. J., Bozzeda, F., Ceccherelli, V. U., Colangelo, M. A., Evans, A., Ferrario, F., Hanley, M. E., Hinz, H., Hoggart, S. P. G., Jackson, J., Moore, P., Morgan, E. H., Perkol-Finkel, S., Skov, M. W., Strain, E. M., van Belzen, J., & Hawkins, S. J. (2014). Between a rock and a hard place: environmental and engineering considerations when designing flood defence structures. *Coastal Engineering*. http://dx.doi.org/10.1016/j.coastaleng.2013.10.015.

Firth, L. B., Thompson, R. C., White, F. J., Schofield, M., Skov, M. W., Hoggart, S. P. G., et al. (2013). The importance of water-retaining features for biodiversity on artificial intertidal coastal defence structures. *Diversity and Distributions*. ppn/a-n/a doi 10.1111/ddi.12079.

Folke, C., Carpenter, S., Walker, B., Scheffer, M., Elmqvist, T., Gunderson, L., et al. (2004). Regime shifts, resilience, and biodiversity in ecosystem management. *Annual Review of Ecology, Evolution, and Systematics, 35*, 557–581.

Fonseca, M. F., & Bell, S. (1998). Influence of physical setting on seagrass landscapes near Beaufort, North Carolina, USA. *Marine Ecology Progress Series, 171*, 109–121.

Fonseca, M. S., Kenworthy, W. J., Courtney, F. X., & Hall, M. O. (1994). Seagrass planting in the southeastern United States: methods for accelerating habitat development. *Restoration Ecology, 2*, 198–212.

Fonseca, M. S., Kenworthy, W. J., & Thayer, G. W. (1998). *Guidelines for the conservation and restoration of seagrasses in the United States and adjacent waters*. In *NOAA coastal ocean program decision analyses series* (Vol. 12). Washington, DC: NOAA.

Franzitta (2013). *Effects of artificial defences and flooding on coastal habitats and assemblages* (Ph.D. thesis). Bologna: University of Bologna, pp. 94.

French, C. E., French, J. R., Clifford, N. J., & Watson, C. J. (2000). Sedimentation-erosion dynamics of abandoned reclamations: the role of waves and tides. *Continental Shelf Research, 20*, 1711–1733.

French, P. W. (1997). *Coastal and estuarine management*. London: Routledge.

French, P. W. (2006). Managed realignment—the developing story of a comparatively new approach to soft engineering. *Esturine and Coastal Shelf Science, 67*, 409–423.

Friedrichs, C. T., & Perry, J. E. (2001). Tidal salt marsh morphodynamics: a synthesis. *Journal of Coastal Research, 27*, 7–37.

Friess, D. A., Krauss, K. W., Horstman, E. M., Balke, T., Bouma, T. J., Galli, D., et al. (2012). Are all intertidal wetlands naturally created equal? Bottlenecks, thresholds and knowledge gaps to mangrove and saltmarsh ecosystems. *Biological Reviews, 87*, 346–366. http://dx.doi.org/10.1111/j.1469-185X.2011.00198.x.

Frost, M. T., Leaper, R., Mieszowska, N., Moschella, P., Murua, J., Smyth, C., et al. (2004). Recovery of a biodiversity action plan species in Northwest England: possible role of climate change, artificial habitat and water quality amelioration. Report No. 16. *Marine Biological Association of the United Kingdom*.

Gacia, E., Satta, M. P., & Martin, D. (2007). Low crested coastal defence structures on the Catalan coast of the Mediterranean Sea: how they compare with natural rocky shores. *Scientia Marina, 71*, 259–267.

García Mora, M. R., Gallego-Fernández, J. B., & García Novo, F. (2000). Plant diversity as a suitable tool for coastal dune monitoring. *Journal of Coastal Research, 16*, 990–995.

Garbutt, A., & Wolters, M. (2008). The natural regeneration of salt marsh on formerly reclaimed land. *Applied Vegetation Science, 11*, 335–344.

Gianni, F., Bartolini, F., Airoldi, L., Ballesteros, E., Francour, P., Guidetti., P...., & Mangialajo, L. (2013). Conservation and restoration of marine forests in the Mediterranean Sea and the potential role of Marine Protected Areas. *Advances in Oceanography and Limnology, 4*(2), 83–101.

Glasby, T. M., Connell, S. D., Holloway, M. G., & Hewitt, C. L. (2007). Nonindigenous biota on artificial structures: could habitat creation facilitate biological invasions? *Marine Biology, 151*, 887–895. http://dx.doi.org/10.1007/s00227-006-0552-5.

Gómez-Pina, G., Muñoz-Pérez, J. J., Ramírez, J. L., & Ley, C. (2002). Sand dune management problems and techniques, Spain. *Journal of Coastal Research, 36*(36), 325–332.

Grabowski, J. H., Brumbaugh, R. D., Conrad, R. F., Keeler, A. G., Opaluch, J. J., Peterson, C. H., et al. (2012). Economic valuation of ecosystem services provided by oyster reefs. *BioScience, 62*, 900–909.

Griffin, J., Noël, L. M. L. J., Crowe, T., Burrows, M. T., Hawkins, S. J., Thompson, R. C., et al. (2010). Consumer effects on ecosystem functioning in rock pools: roles of species richness and composition. *Marine Ecology Progress Series, 420*, 45–56. http://dx.doi.org/10.3354/meps08844.

Halpern, B. S., McLeod, K. L., Rosenberg, A. A., & Crowder, L. B. (2008). Managing for cumulative impacts in ecosystem-based management through ocean zoning. *Ocean and Coastal Management, 51*, 203–211.

Harwell, M. C., & Orth, R. J. (1999). Eelgrass (*Zostera marina* L.) seed protection for field experiments and implications for large scale restoration. *Aquatic Botany, 64*, 51–61.

Hendrick, V. J., Foster-Smith, R. L., & Davies, A. J. (2011). Biogenic reefs and the marine aggregate industry. *Marine ALSF Science Monograph Series No. 3. MEPF 1-/P149*, 60.

Holt, R. J., Rees, E. I., Hawkins, S. J., & Seed, R. (1998). *Biogenic reefs. An overview of dynamic and sensitivity characteristics for conservation management of marine SACs*. Report prepared by the Scottish Association for Marine Sciences (SAMS) for the UK Marine SACs Project, Task Manager, A.M.W. Wilson. SAMS. Available at http://www.ukmarinesac.org.uk/pdfs/biogreef.pdf.

Huiskes, A., Koutstaal, B., Herman, P., Beeftink, W., Markusse, M., & De Munck, W. (1995). Seed dispersal of halophytes in tidal salt marshes. *Journal of Ecology, 83*, 559–567.

Huston, M. (1979). General hypothesis of species diversity. *American Naturalist, 113*, 81–101. http://dx.doi.org/10.1086/283366.

Inger, R., Attrill, M. J., Bearhop, S., Broderick, A. C., Grecian, W. J., Hodgson, D. J., et al. (2009). Marine renewable energy: potential benefits to biodiversity? An urgent call for research. *Journal of Applied Ecology, 46*, 1145–1153.

Johnson, M. P., Frost, N. J., Mosley, M. W. J., Roberts, M. F., & Hawkins, S. J. (2003). The area-independent effects of habitat complexity on biodiversity vary between regions. *Ecology Letters, 6*, 126–132. http://dx.doi.org/10.1046/j.1461-0248.2003.00404.x.

Jolly, E. C. (2008). *The role of coastal defence structures in channeling production in coastal ecosystems* (Ph.D. thesis). University of Southampton.

Jones, C. G., Lawton, J. H., & Shachak, M. (1994). Organisms as ecosystem engineers. *Oikos, 69*, 373–386.

Kaye, C. A., & Barghoorn, E. S. (1964). Late quaternary sea-level change and crustal rise at Boston, Massachusetts, with notes on the autocompaction of peat. *Geological Society of America Bulletin, 75*, 63–80.

King, S. E., & Lester, J. N. (1995). The value of salt marsh as a sea defence. *Marine Pollution Bulletin, 30*, 180–189.

Kirwan, M., & Temmerman, S. (2009). Coastal marsh response to historical and future sea-level acceleration. *Quaternary Science Reviews, 28*, 1801–1808. http://dx.doi.org/10.1016/j.quascirev.2009.02.022.

Kirwan, M. L., Guntenspergen, G. R., D'Alpaos, A., Morris, J. T., Mudd, S. M., & Temmerman, S. (2010). Limits on the adaptability of coastal marshes to rising sea level. *Geophysical Research Letters, 37*. http://dx.doi.org/10.1029/2010gl045489.

Koch, E. W., Barbier, E. B., Silliman, B. R., Reed, D. J., Perillo, G. M. E., Hacker, S. D., et al. (2009). Non-linearity in ecosystem services: temporal and spatial variability in coastal protection. *Frontiers in Ecology and the Environment, 7*, 29–37.

Knutson, P. L. (1988). Role of coastal marshes in energy dissipation and shore protection. In D. D. Hook (Ed.), *Ecology of wetlands: Vol. 1. The ecology and management of wetlands*. Portland: Timber Press.

Lamberti, A., & Zanuttigh, B. (2005). An integrated approach to beach management in Lido di Dante, Italy. *Estuarine Coastal and Shelf Science, 62*, 441–451.

Langlois, E., Bonis, A., & Bouzille, J. (2003). Sediment and plant dynamics in saltmarshes pioneer zone: *Puccinellia maritima* as a key species? *Estuarine, Coastal and Shelf Science, 56*, 239–249.

Lasalle, M. W., Landin, M. C., & Sims, J. G. (1991). Evaluation of the flora and fauna of a *Spartina alternifolia* marsh establishes on dredged material in Winyah Bay, South Carolina. *Wetlands, 11*, 191–208.

Ley de la Vega, C., Favennec, J., Gallego-Fernández, J., & et Pascual Vidal, C. (Eds.). (2012). *Conservation des dunes côtières. Restauration et gestion durables en Méditerranée occidentale*. Gland, Suisse et Malaga, Espagne: UICN.

Lee, K.-S., & Park, J.-I. (2008). An effective transplanting technique using shells for restoration of *Zostera marina* habitats. *Marine Pollution Bulletin, 56*, 1015–1021.

Leonard, L. A., & Luther, M. E. (1995). Flow hydrodynamics in tidal marsh canopies. *Limnology and Oceanography, 40*, 1474–1484.

Liao, C. Z., Luo, Y. Q., Fang, C. M., Chen, J. K., & Li, B. (2008). Litter pool sizes, decomposition, and nitrogen dynamics in *Spartina alterniflora*-invaded and native coastal marshlands of the Yangtze Estuary. *Oecologia, 156*, 589–600.

Linhart, Y. B. (1995). Restoration, revegetation, and the importance of genetic and evolutionary perspectives. In B. A. Roundy, E. D. McArthur, J. S. Haley, & D. K. Mann (Eds.), *Proceedings: Wildland scrubs and arid land restoration symposium; October 19–21, 1993*. Las Vegas, NV.

Loreau, M., Naeem, S., Inchausti, P., Bengtsson, J., Grime, J., Hector, A., et al. (2001). Biodiversity and ecosystem functioning: current knowledge and future challenges. *Science, 294*, 804.

Lotze, H. K., Lenihan, H. S., Bourque, B. J., Bradbury, R. H., Cooke, R. G., Kay, M. C., et al. (2006). Depletion, degradation, and recovery potential of estuaries and coastal seas. *Science, 312*, 1806–1809. http://dx.doi.org/10.1126/science.1128035.

Lotze, H. K., Coll, M., Magera, A. M., Ward-Paige, C., & Airoldi, L. (2011). Recovery of marine animal populations and ecosystems. *Trends in Ecology and Evolution, 26*, 595–605.

Luque, A. A., & Templado, J. (2004). *Praderas y bosques marinos de Andalucía*. Consejería de Medio Ambiente. Junta de Andalucía & Universidad Autónoma de Madrid.

Marani, M., Belluco, E., Ferrari, S., Silvestri, S., D'Alpaos, A., Lanzoni, S., et al. (2006). Analysis, synthesis and modelling of high-resolution observations of salt-marsh eco-geomorphological patterns in the Venice lagoon. *Estuarine, Coastal and Shelf Science, 69*, 414–426.

Marba, N., & Duarte, C. M. (1994). Growth response of the seagrass *Cymodocea nodosa* to experimental burial and erosion. *Marine Ecology Progress Series, 107*, 307–311.

Marba, N., & Duarte, C. M. (1998). Rhizome elongation and seagrass clonal growth. *Marine Ecology Progress Series, 174*, 269–280.

Martin, D., Bertasi, F., Colangelo, M. A., de Vries, M., Frost, M., Hawkins, S. J., et al. (2005). Ecological impact of coastal defence structures on sediment and mobile fauna: evaluating and forecasting consequences of unavoidable modifications of native habitats. *Coastal Engineering, 52*, 1027–1051. http://dx.doi.org/10.1016/j.coastaleng.2005.09.006.

Martins, G. M., Thompson, R. C., Neto, A. I., Hawkins, S. J., & Jenkins, S. R. (2010). Enhancing stocks of the exploited limpet *Patella candei* d'Orbigny via modifications in coastal engineering. *Biological Conservation, 143*, 203–211. http://dx.doi.org/10.1016/j.biocon.2009.10.004.

McIntyre, A. D. (1977). Effects of pollution on inshore benthos. *Ecology of marine benthos*, 301–318.

McLachlan, A., & Brown, A. C. (2006). *The ecology of Sandy shores*. Burlington, MA, USA: Academic Press.

Meadows, P. S., Meadows, A., West, F. J. C., Shand, P. S., & Shaikh, M. A. (1998). *Mussels and mussel beds (*Mytilus edulis*) as stabilizers of sedimentary environments in the intertidal zone, 139.* Geological Society, London: Special Publications.

Mendelssohn, I., Materne, M., & Schrift, A. (2006). *Louisiana marsh dieback—factors controlling the restoration of brown marsh sites with small dredge sediment enrichment task 5.4*. Baton Rouge, LA, USA: Louisiana Department of Natural Resources. http://brownmarsh.com/data/V-4/Brown.Marsh.Task.5.4.Final.Report.pdf.

Millennium Ecosystem Assessment (MEA). (2005). *Ecosystems and human well-being: Synthesis*. Washington: Island Press. Available at: http://www.millenniumassessment.org/documents/document.356.aspx.pdf.

Möller, I., Spencer, T., French, J., Leggett, D., & Dixon, M. (1999). Wave transformation over salt marshes: a field and numerical modelling study from North Norfolk, England. *Estuarine, Coastal and Shelf Science, 49*, 411–426.

Moreira, J., Chapman, M. G., & Underwood, A. J. (2006). Seawalls do not sustain viable populations of limpets. *Marine Ecology Progress Series, 322*, 179–188. http://dx.doi.org/10.3354/meps322179.

Moreira, J., Chapman, M. G., & Underwood, A. J. (2007). Maintenance of chitons on seawalls using crevices on sandstone blocks as habitat in Sydney Harbour, Australia. *Journal of Experimental Marine Biology and Ecology, 347*, 134–143. http://dx.doi.org/10.1016/j.jembe.2007.04.001.

Moreno-Mateos, D., Power, M. E., Comin, F. A., & Yockteng, R. (2012). Structural and functional loss in restored wetland ecosystems. *PLoS Biology, 10*, e1001247. http://dx.doi.org/10.1371/journal.pbio.1001247.

Mortlock, W. (2000). Local seed for revegetation. *Ecological Management and Restoration, 1*(2), 93–101.

Moschella, P. S., Abbiati, M., Aberg, P., Airoldi, L., Anderson, J. M., Bacchiocchi, F., et al. (2005). Low-crested coastal defence structures as artificial habitats for marine life: using ecological criteria in design. *Coastal Engineering, 52*, 1053–1071. http://dx.doi.org/10.1016/j.coastaleng.2005.09.014.

Mossman, H. L., Brown, M. J. H., Davy, A. J., & Grant, A. (2010). Constraints on salt marsh development following managed coastal realignment: dispersal limitation or environmental tolerance? *Restoration Ecology, 20*, 65–75.

Mossman, H. L., Davy, A. J., & Grant, A. (2012). Does managed coastal realignment create saltmarshes with 'equivalent biological characteristics' to natural reference sites? *Journal of Applied Ecology, 49*, 1446–1456. http://dx.doi.org/10.1111/j.1365-2664.2012.02198.x.

Munari, C., Corbau, C., Simeoni, U., & Mistri, M. (2011). Coastal defence through low crested breakwater structures: jumping out of the frying pan into the fire? *Marine Pollution Bulletin, 62*, 1641–1651.

Naylor, L. A., Coombes, M. A., Venn, O., Roast, S. D., & Thompson, R. C. (2012). Facilitating ecological enhancement of coastal infrastructure: the role of policy, people and planning. *Environmental Science and Policy, 22*, 36–46.

Naylor, L. A. (2005). The contributions of biogeomorphology to the emerging field of geobiology. *Palaeogeography, Palaeoclimatology, Palaeoecology, 219*, 35–51.

Nelson, P. A. (2003). Marine fish assemblages associated with fish aggregating devices (FADs): effects of fish removal, FAD size, fouling communities, and prior recruits. *Fisheries Bulletin, 101*, 835–850.

Newell, R. I. E. (2004). Ecosystem influences of natural and cultivated populations of suspension-feeding bivalve mollusks: a review. *Journal of Shellfish Research, 23*, 51–61.

Newell, R. I. E., & Koch, E. W. (2004). Modeling seagrass density and distribution in response to changes in turbidity stemming from bivalve filtration and seagrass sediment stabilization. *Estuaries, 27*, 793–806.

Newell, R. I. E., Fisher, T. R., Holyoke, R. R., & Cornwell, J. C. (2005). Influence of eastern oysters on nitrogen and phosphorus regeneration in Chesapeake Bay, USA. In R. F. Dame, & S. Olenin (Eds.), *The comparative roles of suspension feeders in ecosystems* (pp. 93–120). Springer.

Nicholson-Cole, S., & Riordan, T. O. (2009). Adaptive governance for a changing coastline: science, policy and the public in search of a sustainable future. In N. W. Adger, I. Lorenzone, & K. L. O'Brien (Eds.), *Adapting to climate change thresholds values governance*. Cambridge, UK: Cambridge University Press.

Nordstrom, K. F. (2000). *Beaches and dunes on developed coasts*. Cambridge, UK: Cambridge University Press.

O'Connell, J. (December 2008). Coastal dune protection and restoration. Using "Cape" American beachgrass & fencing. *Marine Extension Bulletin*, 20.

Ondiviela, B., Losada, I. J., Lara, J. L., Maza, M., Galván, C., Bouma, T. J., & van Belzen, J. (2014). The role of seagrasses in coastal protection in a changing climate. *Coastal Engineering, 87*, 158–168.

Orth, R. J., Carruthers, T. J. B., Dennison, W. C., Duarte, C. M., Fourqurean, J. W., Heck, K. L., et al. (2006). A global crisis for seagrass ecosystems. *Bioscience, 56*, 987–996.

Orth, R. J., Harwell, M. C., Bailey, E. M., Bartholomew, A., Jawad, J. T., Lombana, A. V., et al. (2000). A review of issues in seagrass seed dormancy and germination: Implications for conservation and restoration. *Marine Ecology Progress Series, 200*, 277–288.

Orth, R. J., Luckenbach, M., & Moore, K. A. (1994). Seed dispersal in a marine macrophyte: implications for colonization and restoration. *Ecology, 75*, 1927–1939.

Paine, R. T., Tegner, M. J., & Johnson, E. A. (1998). Compounded perturbations yield ecological surprises. *Ecosystems, 1*, 535–545. http://dx.doi.org/10.1007/s100219900049.

Paskoff, R. (1998). *Les littoraux. Impact des aménagements sur leur évolution.* Armand Colin ed.

Petersen, J. K., & Malm, T. (2006). Offshore windmill farms: threats to or possibilities for the marine environment. *AMBIO: A Journal of the Human Environment, 35*(2), 75–80.

Peralta, G., van Duren, L. A., Morris, E. P., & Bouma, T. J. (2008). Consequences of shoot density and stiffness for ecosystem engineering by benthic macrophytes in flow dominated areas: a hydrodynamic flume study. *Marine Ecology Progress Series, 368*, 103–115.

Perkol-Finkel, S., & Airoldi, L. (2010). Loss and recovery potential of marine habitats: an experimental study of factors maintaining resilience in subtidal algal forests at the Adriatic Sea. *Plos One, 5*, e10791.

Perkol-Finkel, S., Ferrario, F., Nicotera, V., Airoldi, L., & Cadotte, M. (2012). Conservation challenges in urban seascapes: promoting the growth of threatened species on coastal infrastructures. *Journal of Applied Ecology, 49*, 1457–1466.

Piazza, B. P., Banks, P. D., & La Peyre, M. K. (2005). The potential for created oyster shell reefs as a sustainable shoreline protection strategy in Louisiana. *Restoration Ecology, 13*, 499–506.

Pinn, E. H., Mitchell, K., & Corkill, J. (2005). The assemblages of groynes in relation to substratum age, aspect and microhabitat. *Estuarine Coastal and Shelf Science, 62*, 271–282.

Pister, B. (2009). Urban marine ecology in southern California: the ability of riprap structures to serve as rocky intertidal habitat. *Marine Biology, 156*, 861–873. http://dx.doi.org/10.1007/s00227-009-1130-4.

Posey, M. H., Alphin, T. D., & Powell, C. M. (1997). Plant and infaunal communities associated with a created marsh. *Estuaries, 20*, 42–47. http://dx.doi.org/10.2307/1352718.

Ray, G. L. (2000). Infaunal assemblages on constructed intertidal mudflats at Jonesport, Maine (USA). *Marine Pollution Bulletin, 40*, 1186–1200. http://dx.doi.org/10.1016/s0025-326x(00)00083-7.

Ray, G. L., Clarke, D. G., Wilber, P., & Fredette, T. J. (1994). *Ecological evaluation of mudflat habitats on the coast of Maine constructed of dredged material. Environmental effects of dredging D-93-3.* Vicksburg. Mississippi, USA: United States Army Engineer Waterways Experiment Station.

Reed, D. J., Davidson-Arnott, R., & Perillo, G. M. E. (2000). 5 Estuaries, coastal marshes, tidal flats and coastal dunes. In J. E. Hobbie (Ed.), *Estuarine science: A synthetic approach to research and practice*. Washington, D.C: Island Press.

Reed, D. J., Spencer, T., Murray, A. L., French, J. R., & Leonard, L. (1999). Marsh surface sediment deposition and the role of tidal creeks: implications for created and managed coastal marshes. *Journal of Coastal Conservation, 5*, 81–90.

Reise, K. (1998). Pacific oysters invade mussel beds in the European Wadden Sea. *Senckenbergiana Maritima, 28*, 167–175.

Santín, C., De La Rosa, J., Knicker, H., Otero, X., àlvarez, M.à., & González-Vila, F. (2009). Effects of reclamation and regeneration processes on organic matter from estuarine soils and sediments. *Organic Geochemistry, 40*, 931–941.

Scheffer, M., Carpenter, S., Foley, J. A., Folke, C., & Walker, B. (2001). Catastrophic shifts in ecosystems. *Nature, 413*, 591–596. http://dx.doi.org/10.1038/35098000.

Schlacher, T. A, Noriega, R., Jones, A., & Dye, T. (2012). The effects of beach nourishment on benthic invertebrates in eastern Australia: impacts and variable recovery. *Science of The Total Environment*, 435–436, 411–417. http://dx.doi.org/10.1016/j.scitotenv.2012.06.071

Schratzberger, M., Bolam, S., Whomersley, P., & Warr, K. (2006). Differential response of nematode colonist communities to the intertidal placement of dredged material. *Journal of Experimental Marine Biology and Ecology, 334*, 244–255. http://dx.doi.org/10.1016/j.jembe.2006.02.003.

Schrift, A. M., Mendelssohn, I. A., & Materne, M. D. (2008). Salt marsh restoration with sediment-slurry amendments following a drought-induced large-scale disturbance. *Wetlands, 28*, 1071–1085. http://dx.doi.org/10.1672/07-78.1.

Scyphers, S. B., Powers, S. P., Heck, K. L., & Byron, D. (2011). Oyster reefs as natural breakwaters mitigate shoreline loss and facilitate fisheries. *Plos One, 6*.

Shepard, C. C., Crain, C. M., & Beck, M. W. (2011). The protective role of coastal marshes: a systematic review and meta-analysis. *Plos One, 6*.

Shi, B. W., Yang, S. L., Wang, Y. P., Bouma, T. J., & Zhu, Q. (2012). Relating accretion and erosion at an exposed tidal wetland to the bottom shear stress of combined current-wave action. *Geomorphology, 138*, 380–389. http://dx.doi.org/10.1016/j.geomorph.2011.10.004.

Short, F. T., Davis, R. C., Kopp, B. S., Short, C. A., & Burdick, D. M. (2002). Site-selection model for optimal transplantation of eelgrass *Zostera marina* in the northeastern US. *Marine Ecology Progress Series, 227*, 253–267.

Short, F. T., & Neckles, H. A. (1999). Review. The effects of global climate change on seagrasses. *Aquatic Botany, 63*, 169–196.

Short, F. T., Polidoro, B., Livingstone, S. R., Carpenter, K. E., Bandeira, S., Bujang, J. S., et al. (2011). Extinction risk assessment of the world's seagrass species. *Biological Conservation, 144*, 1961–1971.

Short, F. T., & Wyllie-Echeverria, S. (1996). Natural and human-induced disturbance of seagrasses. *Environmental Conservation, 23*, 17–27.

Slocum, M. G., Mendelssohn, I. A., & Kuhn, N. L. (2005). Effects of sediment slurry enrichment on salt marsh rehabilitation: plant and soil responses over seven years. *Estuaries, 28*, 519–528. http://dx.doi.org/10.1007/bf02696063.

Smith, S. V. (1981). Marine macrophytes as a global carbon sink. *Science, 211*, 838–840.

Stagg, C. L., & Mendelssohn, I. A. (2011). Controls on resilience and stability in a sediment-subsidized salt marsh. *Ecological Applications, 21*, 1731–1744.

Suykerbuyk, W., Bouma, T. J., van der Heide, T., Faust, C., Govers, L. L., Giesen, W. B. J. T., et al. (2012). Suppressing antagonistic bioengineering feedbacks doubles restoration success. *Ecological Applications, 22*, 1224–1231.

Taramelli, A., Valentini, E., Dejana, M., Zucca, F., & Mandrone, S. (2011). Modelling coastal processes by means of innovative integration of remote sensing time series analysis. *Geoscience and Remote Sensing Symposium, 2011 IEEE International*, 1547–1550.

Taramelli, A., Valentini, E., Cornacchia, L., Mandrone, S., Monbaliu, J., Hoggart, S. P. G., Thompson, R. C., & Zanuttigh, B. (2014). Modeling uncertainty in estuarine system by means of combined approach of optical and radar remote sensing. *Coastal Engineering, 87*, 77–96. http://dx.doi.org/10.1016/j.coastaleng.2013.11.001.

Temmerman, S., Bouma, T. J., Van de Koppel, J., Van der Wal, D., De Vries, M. B., & Herman, P. M. J. (2007). Vegetation causes channel erosion in a tidal landscape. *Geology, 35*, 631–634. http://dx.doi.org/10.1130/g23502a.1.

Therriault, T. W., & Kolasa, J. (2000). Explicit links among physical stress, habitat heterogeneity and biodiversity. *Oikos, 89*, 387–391. http://dx.doi.org/10.1034/j.1600-0706.2000.890220.x.

The National Trust. (2008). Shifting shores in the south west. *Living with a Changing Coastline*.

Thompson, R. C., Crowe, T. P., & Hawkins, S. J. (2002). Rocky intertidal communities: past environmental changes, present status and predictions for the next 25 years. *Environmental Conservation, 29*, 168–191. http://dx.doi.org/10.1017/s0376892902000115.

Thompson, R. C., Moore, C. J., vom Saal, F. S., & Swan, S. H. (2009). Plastics, the environment and human health: current consensus and future trends. *Philosophical Transactions of the Royal Society B-biological Sciences, 364*, 2153–2166. http://dx.doi.org/10.1098/rstb.2009.0053.

Thompson, R. C., Wilson, B. J., Tobin, M. L., Hill, A. S., & Hawkins, S. J. (1996). Biologically generated habitat provision and diversity of rocky shore organisms at a hierarchy of spatial scales. *Journal of Experimental Marine Biology and Ecology, 202*, 73–84. http://dx.doi.org/10.1016/0022-0981(96) 00032–9.

Thorhaug, A. (1987). Large-scale seagrass restoration in a damaged estuary. *Marine Pollution Bulletin, 18*, 442–446.

Thurstan, R., Roberts, C., Hawkins, J., & Neves, L. (2009). *Highly protected marine conservation zones: Defining damaging and disturbing activities*. CCW Policy Research Report No. 09/01.

Tilman, D., Knops, J., Wedin, D., Reich, P., Ritchie, M., & Siemann, E. (1997). The influence of functional diversity and composition on ecosystem processes. *Science, 277*, 1300–1302.

Townend, I., Fletcher, C., Knappen, M., & Rossington, K. (2011). A review of salt marsh dynamics. *Water and Environment Journal, 25*, 477–488. http://dx.doi.org/10.1111/j.1747-6593.2010.00243.x.

Tsoar, H. (2005). Sand dunes mobility and stability in relation to climate. *Physica A, 357*, 50–56.

Tyrrell, M. C., & Byers, J. E. (2007). Do artificial substrates favor nonindigenous fouling species over native species? *Journal of Experimental Marine Biology and Ecology, 342*, 54–60. http://dx.doi.org/10.1016/j.jembe.2006.10.014.

Unsworth, R. K. F., De Leon, P. S., Garrard, S. L., Jompa, J., Smith, D. J., & Bell, J. J. (2008). High connectivity of Indo-Pacific seagrass fish assemblages with mangrove and coral reef habitats. *Marine Ecology Progress Series, 353*, 213–224.

Ursino, N., Silvestri, S., & Marani, M. (2004). Subsurface flow and vegetation patterns in tidal environments. *Water Resources Research, 40*, W05115.

van Dijk, P. M., Arens, S. M., & van Boxel, J. H. (1999). Aeolian processes across transverse dunes. ii: modelling the sediment transport and profile development. *Earth Surface Processes and Landforms, 24*, 319–333.

van de Koppel, J., van der Wal, D., Bakker, J. P., & Herman, P. M. J. (2005). Self-organization and vegetation collapse in salt marsh ecosystems. *American Naturalist, 165*, E1–E12.

Van Hulzen, J. B., Van Soelen, J., & Bouma, T. J. (2007). Morphological variation and habitat modification are strongly correlated for the autogenic ecosystem engineer *Spartina anglica* (Common cordgrass). *Estuaries and Coasts, 30*, 3–11.

van Katwijk, M. M. (2003). Reintroduction of eelgrass (*Zostera marina* L.) in the Dutch Wadden Sea: a research overview and management vision. In W. J. Wolff, K. Essink, A. Kellermann, & M. A. van Leeuwe (Eds.), *Challenges to the Wadden Sea area. Proceedings of the 10th international scientific Wadden Sea symposium, Groningen, The Netherlands, October 31–November 3, 2000.*

Van Leeuwen, B., Augustijn, D. C. M., van Wesenbeeck, B. K., Hulscher, S. J. M. H., & de Vries, M. B. (2010). Modeling the influence of a young mussel bed on fine sediment dynamics on an intertidal flat in the Wadden Sea. *Ecological Engineering, 36*(2), 145–153.

van Proosdij, D., Davidson-Arnott, R. G. D., & Ollerhead, J. (2006). Controls on spatial patterns of sediment deposition across a macro-tidal salt marsh surface over single tidal cycles. *Estuarine, Coastal and Shelf Science, 69*, 64–86.

van Wesenbeeck, B. K., van de Koppel, J., Herman, P. M. J., Bakker, J. P., & Bouma, T. J. (2007). Biomechanical warfare in ecology; negative interactions between species by habitat modification. *OIKOS, 116*, 742–750.

van Wesenbeeck, B. K., van de Koppel, J., Herman, P. M. J., & Bouma, T. J. (2008a). Does scale-dependent feedback explain spatial complexity in salt-marsh ecosystems. *OIKOS, 117*, 152–159.

van Wesenbeeck, B. K., van de Koppel, J., Herman, P. M. J., Bertness, M. D., van der Wal, D., Bakker, J. P., et al. (2008). Potential for sudden shifts in transient systems: distinguishing between local and landscape-scale processes. *Ecosystems, 11*, 1133–1141. http://dx.doi.org/10.1007/s10021-008-9184-6.

Vaselli, S., Bulleri, F., & Benedetti-Cecchi, L. (2008). Hard coastal-defence structures as habitats for native and exotic rocky-bottom species. *Marine Environmental Research, 66*, 395–403. http://dx.doi.org/10.1016/j.marenvres.2008.06.002.

Wang, Y., Xie, H., & Wu. (2004). Present state and development of domestic and international dredging markets. *Journal of SSSR, 27.*

Waters, C. M., Young, A. G., & Crosthwaite, J. (2007). Genetic integrity as a target for natural capital restoration: weighing the costs and benefits. In J. Aronson, S. J. Milton, & J. N. Blignaut (Eds.), *Restoring natural capital: Science, business and practice*. Washington: Island Press.

Watson, R. A., Coles, R. G., & Lee Long, W. J. (1993). Simulation estimates of annual yield and landed values for commercial penaeid prawns from tropical seagrass habitat, Northern Queensland, Australia. *Australian Journal of Marine and Freshwater Research, 44*, 211–219.

West, R. J., Jacobs, N. E., & Roberts, D. E. (1990). Experimental transplanting of seagrasses in Botany Bay, Australia. *Marine Pollution Bulletin, 21*, 197–203.

Wethey, D. S., Woodin, S. A., Hilbish, T. J., Jones, S. J., Lima, F. P., & Brannock, P. M. (2011). Response of intertidal populations to climate: effects of extreme events versus long term change. *Journal of Experimental Marine Biology and Ecology, 400*, 132–144.

Whanpetch, N., Nakaoka, M., Mukai, H., Suzuki, T., Nojima, S., Kawai, T., et al. (2010). Temporal changes in benthic communities of seagrass beds impacted by a tsunami n the Andaman Sea, Thailand. *Estuarine, Coastal and Shelf Science, 87*, 246–252.

Widdows, J., & Brinsley, M. (2002). Impact of biotic and abiotic processes on sediment dynamics and the consequences to the structure and functioning of the intertidal zone. *Journal of Sea Research, 48*, 143–156.

Wilhelmsson, D., Malm, T., & Ohman, M. C. (2006). The influence of offshore windpower on demersal fish. *ICES Journal of Marine Science, 63*, 775–784.

Wilkinson, S. B., Zheng, W. Z., Allen, J. R., Fielding, N. J., Wanstall, V. C. R. G., & Hawkins, S. J. (1996). Water quality improvements in Liverpool docks: the role of filter feeders in algal and nutrient dynamics. *Marine Ecology, 17*, 197–211.

Wilson, J. C., & Elliott, M. (2009). The habitat—creation potential of offshore wind farms. *Wind Energy, 12*, 203–212.

Witt, M. J., Sheehan, E. V., Bearhop, S., Broderick, A. C., Conley, D. C., Cotterell, S. P., et al. (2012). Assessing wave energy effects on biodiversity: the wave hub experience. *Philosophical Transactions of the Royal Society a, 370*, 502–529.

Wolters, M., Bakker, J. P., Bertness, M. D., Jefferies, R. L., & Moller, I. (2005). Saltmarsh erosion and restoration in south-east England: squeezing the evidence requires realignment. *Journal of Applied Ecology, 42*, 844–851. http://dx.doi.org/10.1111/j.1365-2664.2005.01080.x.

Wolters, M., Garbutt, A., Bekker, R. M., Bakker, J. P., & Carey, P. D. (2008). Restoration of salt-marsh vegetation in relation to site suitability, species pool and dispersal traits. *Journal of Applied Ecology, 45*, 904–912.

Yang, S., Li, H., Ysebaert, T., Bouma, T., Zhang, W., Wang, Y., et al. (2008). Spatial and temporal variations in sediment grain size in tidal wetlands, Yangtze Delta: on the role of physical and biotic controls. *Estuarine, Coastal and Shelf Science, 77*, 657–671.

Yang, S. L., Shi, B. W., Bouma, T. J., Ysebaert, T., & Luo, X. X. (2012). Wave attenuation at a salt marsh margin: a case study of an exposed coast on the Yangtze Estuary. *Estuaries and Coasts, 35*, 169–182. http://dx.doi.org/10.1007/s12237-011-9424-4.

Young, I. R., Zieger, S., & Babanin, A. V. (2011). Global trends in wind speed and wave height. *Science, 332*, 451–455. http://dx.doi.org/10.1126/science.1197219.

Ysebaert, T., Yang, S.-L., Zhang, L., He, Q., Bouma, T. J., & Herman, P. M. J. (2011). Wave attenuation by two contrasting ecosystem engineering salt marsh macrophytes in the intertidal pioneer zone. *Wetlands, 31*, 1043–1054. http://dx.doi.org/10.1007/s13157-011-0240-1.

Zanuttigh, B. (2011). Coastal flood protection: what perspective in a changing climate? The THESEUS approach. *Environmental Science and Policy, 14*, 845–863.

Zimmerman, R. C., Reguzzoni, J. L., & Alberte, R. S. (1995). Eelgrass (*Zostera marina* L.) transplants in San Francisco Bay: role of light availability on metabolism, growth and survival. *Aquatic Botany, 51*, 67–86.

Nonstructural Approaches to Coastal Risk Mitigations

Chapter 5

Jean Paul Vanderlinden[1], Juan Baztan[2], Tracey Coates[3], Osiel González Dávila[4], François Hissel[5], Idrissa Oumar Kane[1], Phoebe Koundouri[4], Loraine McFadden[6], Dennis Parker[6], Edmund Penning-Rowsell[6], Luca Pietrantoni[7], Gianluca Pescaroli[7], Gabriele Prati[7], Bénédicte Rulleau[1], Mavra Stithou[4], Nabil Touili[1]

[1] *Cultures, Environments, Arctic, Representations, Climate Research Center, Université de Versailles Saint-Quentin-en-Yvelines (UVSQ), Guyancourt, France*

[2] *Marine Science for Society, Cultures, Environments, Arctic, Representations, Climate Research Center, Université de Versailles Saint-Quentin-en-Yvelines (UVSQ), Guyancourt, France*

[3] *Flood Hazard Research Centre, Middlesex University (MU) (currently: School of Geography, Geology and the Environment at Kingston University), Hendon, London*

[4] *Department of International and European Economic Studies, Athens University of Economics and Business (AUEB), Athens, Greece*

[5] *Centre d'Études Techniques Maritimes et Fluviales (CETMEF), Margny Lès Compiègne, France*

[6] *Flood Hazard Research Centre, Middlesex University (MU), London, UK*

[7] *University of Bologna (UniBo), Bologna, Italy*

Coastal Risk Management in a Changing Climate. http://dx.doi.org/10.1016/B978-0-12-397310-8.00005-1

5.1 Introduction

The purpose of this chapter is to present measures that deal with the governance of coastal risks associated with climate change. The focus is on measures entailing collective action at the local, regional, or national levels, and involving the private sector, the public sector, and/or civil society. Key risk mitigation options that are the product of collective action include insurance schemes, spatial planning, business continuity planning, postflood recovery, risk communication, and evacuation planning.

Although there has been much analysis in the past regarding these mitigation options as standalone measures, in this chapter they are presented as part of a potential portfolio in which their combination transcends their sum. Furthermore, these approaches to risk mitigation are dual in the sense they may impact both vulnerability (Adger, 2006) and resilience (Berkes, Colding, & Folke, 2003) simultaneously.

Although the mitigation options presented here are the product of collective action, these options are part of a broader process: risk governance (Renn, 2008). Risk governance, "structures and processes for collective decision making" (Renn, 2008), is the application of governance principles to risk and risk-related activities, and as such may involve the state, economic interests, civil society, and academia, and at various institutional scales. At the core of the principle of governance lie the concepts of horizontal integration (i.e., among state, economic interests, civil society, and academia) and vertical integration (i.e., among local, regional, national, and global scales). Risk governance is composed of five elements: preassessment, appraisal, characterization/evaluation, management, and communication (Renn, 2008). The first four elements are sequential and may be iterative, whereas the fifth element, communication, interacts with all the others.

In this chapter, the overarching goal of "safer coasts" can generally be framed as a contribution to flooding and erosion risk governance principles. Because the focus is on risk mitigation, this chapter looks at the "management" and "communication" phases of risk governance. Risk management consists of the major components of decision making and implementation; risk communication crosses all phases of risk governance. A key element that guided the analysis and the associated recommendation is the dual nature of nonstructural risk mitigation. Mitigating risk simultaneously aims to reduce vulnerability and increase resilience.

Reducing vulnerability is associated with a paradigmatic stance regarding the function of the risk system. To reduce vulnerability a (quasi) deterministic representation of the risk system must be available (see Chapter 2, the Source Pathway Receptor Consequence model). In such a representation, a causal chain, often linear, connects the hazard to its associated consequences. Vulnerability reduction is associated with key points in such a causal chain, in which it is possible, through mitigation, to reduce or break this connection to reduce the consequences. Although this approach to risk management can be quite powerful, it does not take into account the nondeterministic characteristics of the complex coastal risk system.

Increasing resilience is associated with a different way of thinking. Resilience is associated with nondeterministic complex system representations whereas a linear causal chain is not applicable (see Table 5.1).

TABLE 5.1 Resilience Principles of a Nondeterministic Complex System (After Wardekker, de Jong, Knoop, & van der Sluijs, 2010)

Resilience Principle	Description	Examples
Homeostasis	Multiple feedback loops within the coastal system counteract disturbances and stabilize the system.	Improving clarity of responsibilities, early warning systems, spatial planning strategies, flexible structures, infrastructures, and flood defenses.
Omnivory	The capacity to recover is increased by diversification of resources and means that may be mobilized in the event of a shock.	Diversifying and distributing power generation, diversifying transportation options, creating multifunctional spaces and buildings, and multiskilled planning teams.
High flux	A fast rate of movement of resources through the system ensures fast mobilization of these resources to cope with perturbations.	Reducing planning horizons, planning easy-to-modify land uses.
Flatness	The hierarchical levels relative to the base should not be top heavy, because overly hierarchical systems with no local formal competence to act are too inflexible and too slow (1) to cope with surprise and (2) to implement nonstandard highly local responses rapidly.	Enabling local populations to self-respond to disturbances (self-sufficiency, self-regulation, and self-organization), increasing public participation.
Buffering	Essential capacities of the system are overdimensioned such that critical thresholds in capacities are less likely to be crossed.	Creating low-elevation spaces (e.g., squares and parks) and ground floors, with nonessential functions that provide water storage capacity.
Redundancy	The system contains overlapping functions; if one fails, others can take over.	Creating multiple power supply routes and interlinked transportation, building access at multiple levels.

Another way to address this duality (i.e., reducing vulnerability and increasing resilience), which is also followed in this chapter, is to consider constantly that, ultimately, a coastal system is a complex, nondeterministic system, but some parts of the system may be treated as deterministic. This way of approaching risk governance allows for a clear dialogue between vulnerability reduction measures (such as those discussed in Chapters 3 and 4—limited in scope but part of well-established causal descriptions) and resilience enhancement measures (broad in scope, but not amenable to a simplified causal representation).

The socially based mitigation options presented here cover six main options—insurance, spatial planning, business continuity planning, postflood recovery, risk communication, and evacuation—and are discussed in terms of both vulnerability reduction and resilience enhancement. The chapter concludes by underscoring the key overarching characteristics of nonstructural mitigation options.

5.2 Insurance

5.2.1 Reducing Vulnerability through Insurance

The objective of this section is to provide guidance to policymakers in the implementation of efficient insurance programs. Insurance can offer the potential of an ex ante mitigation mechanism through its policy conditions or through encouraging parallel measures that are deemed to reduce the occurrence, severity, or impact of a natural hazard. Insurance is expected to reduce property owners' vulnerability and exposure to flood hazards as a result of changes in behaviors and attitudes, thus contributing to speedy recovery and resilience while redistributing the costs of damage across the population and through time.

Insurers cannot control any hazard directly, although they rely, to a greater or less degree, on predictions and sea defense activities. In this context, when integrating insurance schemes into the Sources, Pathways, Receptor, Consequences (SPRC) model (see Chapter 2 and Zanuttigh (2011)), the most obvious link is the impact of the insurance market on consequences. In particular, by increasing risk awareness—by linking the cost (premium) to the risk level—and by creating incentives for mitigation actions, there is a direct impact on the expected damages and the number of people flooded as the result of a flood event.

The main intervention mechanism of insurance in the context of flooding is in its redistribution of the costs of damage across the population and through time. Insurance against flooding as a financial mechanism (1) may serve as a means of communicating risk and (2) may serve to persuade people to integrate it in their decision making by making its purchase compulsory in flood-prone zones (Dawson et al., 2011).

Insurance arrangements for flood risk may require households to undertake measures that mitigate damage or that stimulate households to undertake precautionary measures voluntarily (e.g., requiring valuable items to be kept above flood level or constructing flood-proof buildings). These measures may limit damage during floods and thus may be complementary to traditional flood protection (Kunreuther & Pauly, 2006). Moreover, insurance schemes can influence consequences by refusing coverage in a flood-prone area, making it more likely that risk-averse households and firms will chose not to establish in flood-prone areas.

In this way, insurance schemes reduce the consequences through their policy conditions and through encouraging self-help measures, which are deemed to reduce the occurrence, severity, or impact of a natural peril, and also by influencing the structure and development of economic activity on coastal areas prone to flooding (Botzen, Aerts, & van den Bergh, 2009). Therefore, it is important for insurance to provide the right incentives/disincentives to the residents of the coastal areas to modify their behavior in relation to the local levels of risk (Treby, Clark, & Priest, 2006).

The effectiveness of insurance as a mitigation measure depends on a range of factors, some of which are outlined in the following list. Decision makers should consider these factors when evaluating and implementing insurance schemes:

- The existence of actuarially effective premiums;
- How much the insurers are regulated by government;
- Existing policy arrangements for compensating flood damage;
- Issues that affect human behavior and determine an individual's risk perception and awareness;
- Whether insurance provides the right incentives/disincentives to the residents of the coastal areas to change their behavior in relation to risk;
- Overconfidence in "hard" measures such as dikes, resulting in the phenomenon of underinsurance in flood-prone areas;
- Early warning systems and evacuation plans that reduce flood damages and, consequently, flood insurance claims;
- The degree of exposure of decision makers to the victims of the hazards and, hence, to the emotional involvement with flood victims;
- Participation of communities in measures of flood preparedness (e.g., establishing a system to detect floods and warn people about them) and risk communication within a Community Rating System that could enable inhabitants to acquire credits and achieve flood insurance premium rate reductions.

Market incentives are possibly the most effective way of changing social behavior. Thus, flood insurers could play a critical role in risk reduction and avoidance (mitigation) via the use of financial (dis)incentives (Treby et al., 2006): lower deductibles and/or lower premiums for properties that take action to reduce their exposure to flood risk, such as flood proofing; bonuses for nonclaims; premium pricing related to risk, placing the onus on clients to assess their reaction to the known risk as highlighted by premiums that are high or indeed lacking; resilient reinstatement (i.e., reconstruction undertaken as a result of insurance payouts that aims to reduce the risk of future losses); compensation and rehousing in an area of lower flood risk. However, for this (and the previous) strategy to be adopted requires a degree of government intervention.

Nevertheless, a government's intervention is of crucial importance because policies interact with the formation of risk behavior and are capable of creating incentives to both households and investors (Botzen et al., 2009). Hence, apart from exploring further a community's risk perception conditional on the factors mentioned previously, it is also important to acknowledge the interlinkage in coastal policy design. Last, although the role of insurance as a vulnerability reduction mechanism has been emphasized, it should also be acknowledged that its effectiveness is conditional on a number of parameters, such as socioeconomic and physical changes, and governance arrangements.

5.2.2 Enhancing Resilience through Insurance

In this section, implementation guidelines associated with resilience enhancement of coastal systems with respect to flood insurance are provided. As presented in Table 5.1, Wardekker et al. (2010) offer an interesting approach to climate change adaptation under uncertainty. They suggest that local actors could apply principles of resilience to make the system less prone to disturbances and to enable rapid and flexible responses. This approach is better capable of dealing with surprises than traditional predictive approaches because it is more flexible, and adaptation can be more suitable and better tailored to local situations than rigid top–down regulations.

Insurance may play a central role in terms of resources and means (the omnivory principle in Table 5.1). This means that multiple different approaches can be used alongside each other. In terms of insurance, this could mean hedging outside the insurance market. It is also possible to allocate part of the insurance fees to finance protection measures in the most affected areas. In case of a huge payout, diversification can be achieved by ensuring access to alternative funds/financial resources. Another way of achieving diversification is by helping people to implement different measures to protect their properties (special precautions) and helping local authorities to implement protection measures (using, for example, spatial planning and evacuation plans). Other suggested parallel measures are the promotion of the development and effectiveness of social networks in terms of cohesion to reduce risk and the size of payouts, and the promotion of public awareness on natural disasters and on the link between community vulnerabilities and policy options. Last, it is also important to promote an alliance and to establish a common emergency plan with the public or private sector, or organizations.

Furthermore, insurance schemes should contribute to feedback loops (homeostasis principle, see Table 5.1) to counteract disturbances and stabilize the system. Insurance companies should be able to update their contracts and operations given previous experiences (their own and those garnered from around the world) so they are tailored to local conditions. More specifically, insurance schemes should be able to adopt measures that minimize the risk of the most vulnerable. Insurance companies should be able to (re)integrate local participation at different stages of the design of an insurance scheme after an event and to update knowledge regarding scientific data, research developments, and scenarios related to climate change, flooding, and so forth. They should also be able to cooperate with local authorities to update their protection measures given previous experiences (and so decrease the probability of risk and manage the size of payouts). Hence, cooperative or community (instead of individual) insurance schemes may reinforce social cohesion. Monitoring is a key issue to provide multiple feedback loops to counteract disturbances. Thus, there should be a monitoring capacity of insurance schemes on the impact they have on people with different levels of vulnerability, and a capacity to monitor the premium rate, the deductible level, and the insurance cap level to maintain demand for their policies.

Insurance implementation must allow for a fast rate of movement of resources through the system (high-flux resilience principle, see Table 5.1). This ensures fast mobilization of resources to cope with perturbations. In terms of insurance implementation, this means collaborating with different insurance companies to work together and make the claims process faster. Therefore, the aim should be to achieve a high standard in the quality of information and clarity given in the conditions under which a payout is received so it is done more quickly. A competitive insurance market can provide a wide availability of insurance coverage and contribute to a fast rate of movement of resources by ensuring a spreading of the risk and higher coverage of the population. Other related measures include keeping only a certain percentage of high-risk contracts among clients, allocating payouts and giving priority to the most affected or according to social criteria (e.g., income), and encouraging collaboration with local authorities (having already established a contingency plan) and with nongovernmental organizations, and so forth, to increase reliability and timeliness in case of flooding. Last, the facilitation of payouts is important in terms of a quick response that avoids bureaucracy and time-consuming paperwork.

To avoid a top-heavy insurance system (flatness resilience principle, see Table 5.1), there should be direct and easy personal contact of each beneficiary with the insurance institution. This can be promoted by insurance schemes tailored according to local conditions and the needs and characteristics of local stakeholders and households. Therefore, a participative process in identifying uninsurable high-risk zones is always to be desired, and local participation in the design of insurance schemes to increase flexibility and response should be encouraged.

Buffering (as a resilience principle, see Table 5.1) capabilities can be achieved by linking the insurance premium to measures that promote risk reduction (e.g., flood-resistant functions on ground level) through economic incentives (reducing premiums for those who invest in mitigation or loss-reduction measures). Also, insurance companies and/or the government could keep a reserve fund to enable them to cope financially with natural disasters without collapsing. Ensuring that the money is available to pay for the losses can be achieved through indexed-based or parameterized catastrophe bonds. Insurance firms could also purchase an indemnity contract against claim payments more than a certain amount. Other measures include the possibility of sharing the risk of a huge payout by cooperating with other insurance companies, and the ability of insurance institutions to offer payouts that give priority to the most affected or according to social criteria (which means that at least the insured part of population is offered relief, and resources are allocated directly to the more disadvantaged). A cap on the maximum insured value could be set or an increase in deductible levels stipulated so that the payouts from any disaster would decrease. Nevertheless, all these measures require certain standards to be met (e.g., building codes) before issuing an insurance policy (the mitigation measure).

Last, having several alternative options is better in terms of resilience (redundancy resilience principle, see Table 5.1). It is always useful to have a backup plan in case something goes wrong, because there are always unseen circumstances that may disrupt activities and services. These overlapping functions are evident in the case of insurance programs. For example, if insurance companies fail, the government or the European Union intervenes with regard to residents and businesses (and this can be both in contracting and in reimbursing). Insurance companies can also offer the possibility of both individual and group/cooperative contracts (the risk-pooling option) for businesses, households, or communities. In a more global setting, international organizations and nongovernmental organizations can develop alarm systems that can help if the government fails. On the other hand, the insurance market may offer the possibility to participate in both multicountry and national "risk-pool" policies so if one fails, the other could counterbalance the effects. Furthermore, insurance is usually coupled with regulations and standards policies.

5.3 Spatial (Land Use) Planning

5.3.1 Land Use Planning: Structuring Space to Reduce Vulnerability

Spatial planning has the potential to reduce future impacts through controlling the type and extent of property built in flood or erosion risk areas and therefore to restrict the impact that hazards bring at the coast. This might, for example, be effected by reducing inappropriate future development and by changing the standards or future use of current development (Deboudt, 2010; Pottier, Penning-Rowsell, Tunstall, & Hubert, 2005). Such measures could reduce the need for, or the design standards of, the kinds of major coastal defense structures that are now common, and that may need to become more widespread with future climate change-induced sea-level rises.

The use of the SPRC model (see Chapter 2) highlights the potential that spatial planning has as a mitigation option rather than providing evidence of how it currently operates. There is considerable effort still needed to incorporate Flood and Coastal Erosion Risk Management (FCERM) effectively into spatial plans (Richards, White, & Carter, 2008). Spatial planning is concerned primarily with the idea of exploring and making changes with the nature of the pathway, receptor, and, therefore, consequences rather than with the source (except perhaps marginally in the creation of urban "heat islands" that might exacerbate storminess and hence flooding). There is some potential to influence the pathway through the creation of "green areas" that would allow the flow of floodwater, or by the development of structures that inhibit the flow, but spatial planning and urbanization are most likely to have an impact through changes to the nature of the receptors within the flood plain. Spatial planning offers the potential to reduce or prevent development in areas most at risk of flooding and so reduce potential exposure. It may also be used to change the nature of buildings in a risk area so that they are more able to withstand and recover quickly from floods (see the case of HafenCity in the Elbe estuary, Section 7.5). This may be through physical measures or through changing use (e.g., example moving key functions to higher floors). There may also be an effect at a communal level; spatial planning could perhaps be used to influence where to locate essential services such as water, gas, electricity, transportation hubs, hospitals, and other emergency services so they are less likely to be interrupted by flooding or similar events. Planning could also be used to ensure the provisions and maintenance of suitable evacuation routes. It may also be possible to ensure that large communal buildings, which might be used as rescue centers, are suitably sited.

All of these measures have the potential to change the nature of the consequences. There is also the potential for changes that lead to the following:

- fewer people exposed to the risk;
- fewer properties exposed to the risk;
- more people evacuated safely in the event of an incident;
- more essential services maintained during an incident;
- faster restoration of essential services where they have been disrupted;
- shorter recovery times in restoring buildings and infrastructures to their usual use.

In an ideal world, spatial planning would take into account flood and coastal erosion risk, and decisions would be made accordingly so that future risk is reduced. In reality, the situation is much more complex. The integration of FCERM into spatial planning is incomplete and inconsistent. Spatial planning has to balance many competing demands at a range of timescales, and residents may resist changes for a variety of reasons. Section 5.3.2 offers guidelines for addressing some of these issues.

5.3.2 Land Use Planning and Resilience

The mitigation option proposed is to ensure that an assessment of FCERM risk is included in spatial planning at a strategic and a local scale to reduce current risk and to restrict the buildup of potential damage from flood and coastal erosion in the future. Despite some moves toward integrating FCERM into spatial planning across Europe, this is far from being achieved effectively. Awareness of recent legislation is limited, and integration not fully realized. The suggestions in the following paragraphs provide ways in which FCERM may be integrated more effectively into the spatial planning process to achieve greater flood resilience. They are applicable at a number of different levels; some need tackling at a policy level whereas others could be adopted by individual planning teams.

First, planning teams should be comprised of more than professional spatial planners (in a move toward the omnivory resilience principle, see Table 5.1), and should include experts in conveying complex and uncertain information (e.g., future flood risk), experts in participatory community engagement, and experts in resilience.

Second, spatial planners should incorporate a more complex understanding and use of the concept of resilience into their working practice. The concept of resilience is being used at the theoretical level in planning, and in relevant areas such as flood/disaster-related research. The issue, however, is how to ensure it is being used by planning practitioners. In the United Kingdom, for example, individual and community resilience are being incorporated into "emergency management," as is seen clearly in the Cabinet Office's approach (https://www.gov.uk/government/uploads/system/uploads/attachment_data/file/60678/Cabinet-Office-continuous-improvement-strategy.pdf) and in the management plans of the Environment Agency (http://a0768b4a8a31e106d8b0-50dc802554eb38a24458b98ff72d550b.r19.cf3.rackcdn.com/geho0711btze-e-e.pdf).

Third, spatial planners should learn from and make use of the existing knowledge and expertise to ensure that genuine, participatory engagement is taking place. This will also lead to increase in homeostasis (allowing for more feedbacks to stabilize the system) and flatness (making the system less top heavy through subsidiarity). Community engagement is a key strand of the current understanding of subsidiarity in policymaking, not only for planning and FCERM, but also much more widely. Therefore, there exists a wide range of expertise in this area that could be harnessed usefully, which would avoid costly mistakes, would build on existing skills and knowledge, and would be, potentially, a relatively inexpensive way to meet a key risk reduction requirement. In particular, efforts must be taken to ensure that engagement is a genuinely participatory process. Token forms of engagement that do not move beyond simple consultation may lead to cynicism, loss of confidence in the planning system, avoidance of planning regulations, and conflict.

Fourth, spatial planners should adopt a more nuanced understanding of place identity (in other words, meaning and significance of places for their inhabitants and users), collective memory (shared accumulated experience and history), and how they relate to resistance to change. This should be linked to community engagement processes and visioning. Collective memory can be an important constraint to building resilience. For example, some residents may be attached to a particular view (place identity) of their locality and thus resistant to some of the changes suggested to improve resilience. This is true of any locality, but coastal communities often have a special resonance for people, many of whom may be fiercely protective against unwanted changes. Conversely, collective memory of flooding or erosion may also enhance resilience through a shared knowledge of how to cope, and its loss can be a problem.

Fifth, a better understanding of the processes involved could enable the spatial planning process to build on and enhance existing resilience. There is a need to move beyond the current view that people simply do not like change. A more nuanced approach to understanding attachment to particular places might reveal which aspects are most valued and reveal those where change can be made. Understanding how such identities come to be constructed would enable spatial planners to intervene at points that are likely to have the most impact. The spatial planning process could contribute significantly to resilience building by creating new ways of exploring memory and linking this with envisioning processes—making links between "backward" and "forward" thinking.

Sixth, more focused and comprehensive training and capacity building needs to be developed for planners to improve their understanding about the nature of hazards and the role of land use (spatial) planning in reducing hazard risks, and to ensure that hazards are "mainstreamed;" into their daily work (Glavovic, Saunders, & Becker, 2010).

Seventh, "climate proofing" spatial development (Birkmann, 2007) would strengthen links between FCERM and spatial planning. It is problematic that spatial planning processes rarely take into account the extreme event.

Eighth, flexibility and variation could be incorporated further within the planning system. Instances of rigid inflexibility of FCERM policy is seen as one of the main reasons for the tensions created and why people may be prepared to build illegally in well identified high-risk areas. Furthermore, if community engagement is to be taken seriously, and people are to be genuinely involved in the decision-making process, then some variation in outcomes is to be expected, and innovative solutions would suggest going outside normal procedures.

The TE2100 Flood Risk Management Plan is an interesting example of a plan that is designed to be flexible enough to cope with various climate change scenarios (EA, 2012). Led by the UK Environment Agency, the Thames Estuary 2100 project (TE2100) was established in 2002 with the aim of developing a long-term tidal flood risk management plan for London and the Thames estuary. Public consultation was used and the comments received were fed into the final draft. The plan sets out recommendations for managing flood risk across the estuary through to the end of the century, and it contains recommendations on what actions the Agency and others need to take in the short term (2020s), medium term (2050s), and long term (2080s). The plan is based on current guidance on climate change, but is adaptable to changes in predictions of sea-level rise and climate change throughout the century. This illustrates that it is possible to plan flexibly at the coast or in estuaries for a variety of scenarios in a way that takes local views into account.

Last, special coastal zone designations must be used to create momentum for resilience building (Chouinard, Baztan, & Vanderlinden, 2011). The fact that Integrated Coastal Zone Management does not have effective links to spatial planning is a key issue. Creating special spatial planning designations for at-risk areas at the coast might be one practical means of facilitating more focus on coastal issues and mutual learning (Bastien-Daigle, Vanderlinden, & Chouinard, 2008). There is evidence that such a designation can create momentum to work with local communities for new, innovative ideas.

5.4 Business Continuity Planning

5.4.1 Reducing Businesses Vulnerability through Recovery Planning

Coastal flood risk zones are often the sites of business activities. Waterfront sites or locations close to the coast are attractive for cargo storage and handling, marine transport, marine businesses and businesses linked to them, tourism businesses, and coastal urban settlement business and service functions. Coastal flooding leads to these businesses being vulnerable to physical damage caused by floods as well as to the disruptive effects of consequential losses (e.g., business interruption, loss of value added and loss of business to flood-free competitors) (Penning-Rowsell et al., 2013).

Business disruption and response planning (BDRP) has the potential to reduce business disruption in a variety of ways, including through (1) enhancing "resilience thinking," (2) reducing or eliminating ill-considered location decisions, (3) developing contingency plans linked to early warnings designed to avoid or minimize direct physical and consequential disruptive impacts, and (4) spreading and sharing risks and consequences (Barnes, 2001; Hiles, 2011; Snedaker, 2007; Wallace & Webber, 2010).

A first element to consider here is supply chain vulnerability. Almost all businesses are backward and forward linked—backward to suppliers and forward to customers—and businesses often depend on a supply chain in which those who supply them with raw materials/components are dependent on other businesses, and in which they in turn may be the supplier of another business (Parker, 2007). A business located in a flood risk zone but that is not flooded may have its business operations disrupted when a supplier in a different flood risk zone (possibly in another country) is disrupted by flooding. BDRP has the potential to reduce supply chain vulnerability by using critical impact analysis tools in which (1) dependencies that exist both internally and externally to achieve critical objectives of the business are identified and (2) steps are taken to diversify supply chains with flood risks in mind. Businesses should be strongly encouraged to adopt these approaches (Syed & Syed, 2003).

Second, utility and transportation disruption vulnerability must be taken into account. Electricity supply outages caused by flooding (or accompanying high winds), and damage to electricity distribution stations and networks, are frequent causes of business interruption and economic loss in coastal zones. Some businesses may also be dependent on the continuous functioning of other utilities (e.g., public water supplies) that may be equally vulnerable to flood damage or contamination. Similarly, businesses depend on a transport infrastructure (often the first to be flooded) for their employees to get to and from work, and for supplies and products to be transported. BDRP should be much more widely adopted because it has the potential to avoid or minimize these vulnerabilities by identifying the critical dependencies and, for example, by developing backup power supplies and alternative transportation routes and/or modes.

BDRP knowledge and skills concerning vulnerability are critical for small and medium enterprises (SMEs). Business continuity planning (BCP) is most used by large businesses with professionally trained managers who have been educated in BCP. SMEs have a much greater proportion of owners/managers without the benefit of this education and who do not fully comprehend the potential value of BCP to reducing flood (or any hazard) disruption. Increasing the proportion of SME owners and managers with professional skills in BCP can therefore help reduce the disruptive impacts of floods (Business Continuity Institute, 2005).

Furthermore, businesses are particularly vulnerable if their business data (e.g., data on customers, debtors, supplies, stock, accounts, and so on) are damaged, lost, or otherwise unavailable during a flood (Snedaker, 2007). Data vulnerability needs very careful attention.

Last, the lives of customers are at risk when flooding threatens. For example, retail stores may have hundreds of customers and their vehicles onsite at any one time. Similarly, the lives and welfare of staff may be at risk. The risk is high in rapid inundation zones (e.g., low-lying areas behind seawalls that may be overtopped or breached).

Considering all the issues identified here, the following recommendations on implementing measures to reduce key vulnerabilities are made:

- When balancing decisions about (1) growth and development, and (2) risks, spatial planning processes should take account of the potential for business damage and disruption consequences caused by flooding in coastal risk zones, including the consequences for individual businesses and the wider consequences of business losses on local coastal economies;
- As they evolve, businesses should consider relocating the whole or part of their operations (critical business elements in particular) to more elevated locations, including the following options: beyond the coastal flood risk zone to lower risk locations in the coastal flood risk zone or to higher elevations in situ;
- Businesses located in coastal flood risk zones should be strongly encouraged to prepare generic business disruption and response (BDR) plans that take into account the risks from all relevant sources of flooding (i.e., sea flooding, surface water flooding, tidal fluvial flooding). These BDR plans should be tailored to the type of business and risks involved, and should include risk assessment, business impact and dependency analysis, disruption mitigation, and recovery, and they should be proportionate to the degree of risk, potential consequences, and the resources available to the firm;
- Corporate (i.e., group) businesses that have branches located in coastal flood risk zones should develop generic and flood-specific contingency plans for the temporary relocation of a branch threatened with flooding or a branch that has been flooded to maximize business continuity. These plans should consider relocating employees temporarily and should include plans to accommodate transferred workers (Richardson, Gordon, & Moore, 2008, pp. 253–278, chap. 14);
- Individual businesses located in coastal flood risk zones should, when feasible, identify temporary or mobile buildings to which parts of their operations could be transferred temporarily to maintain business continuity. This is particularly important for businesses that provide essential services to the community, including vulnerable groups. Examples are doctor's surgeries and dispensing pharmacies;
- Where feasible, businesses should organize building space so that nonessential and noncritical business functions are on the ground floor, and essential and critical functions are on higher floors. When practicable, businesses should also identify alternative site access and transport routes in and out of the area among employees' homes, suppliers, and customers to reduce the vulnerability of their business sites to flooding disruption;
- Where available, businesses should subscribe to a flood warning service to create sufficient lead time for BDR plans to be implemented effectively (Kreibich et al., 2011). Lead time may be enhanced by monitoring severe weather warnings for coastal zones that may be used to bring key staff or emergency procedures to a

greater state of readiness than would otherwise be the case should a flood warning be disseminated subsequently. This will save valuable time in the event of a flood and should result in reduced disruption;

- Businesses should develop flat internal organizational and responsibility structures—with as few intermediaries as possible and with appropriate discretion to make damage and disruption-saving responses located at the "local" level—to allow early warnings and updated warnings to be communicated rapidly and to allow timely emergency decisions to be made. Normally, incoming flood warnings should be received by one responsible individual who is able to disseminate the warning rapidly and internally to those who need it. These arrangements must be available on a 24/7 basis;
- In flood risk zones, wherever feasible, businesses should acquire equipment that is mobile and should ensure that critical facilities, such crisis management centers, are portable;
- Businesses located in coastal flood risk zones should have flood insurance with coverage for both direct physical damage and consequential loss (i.e., business interruption loss). The following alternatives should be considered: a business consortium indemnity scheme, private insurance coverage, and corporate (business group) internal "self-insurance";
- Businesses should protect themselves from supply chain disruption by developing parallel or contingency arrangements with alternative suppliers that are not at risk from flooding and/or should require businesses that have supply contracts with them to be able to demonstrate their flood resilience plans. Supply chains need to be "switchable";
- Businesses should consider ways of protecting themselves and their brands from "leakage" or loss of customers in the event of flooding. Ways of achieving this include regularly storing products at an alternative location outside the flood risk zone and establishing a means of communicating professionally with customers to reassure and update them;
- Businesses should establish backup business data storage facilities in a location outside the flood risk zone, and data should be backed up regularly. Adequate backup electricity generation and distribution arrangements should be made by businesses to maximize business continuity during flood events. When feasible, similar arrangements should be made for other business-critical utility supplies. Businesses should acquire knowledge and skills in BCP so that appropriate and effective BDR plans can be created (Wallace & Webber, 2010);
- Businesses that have customers on their premises (e.g., large retail outlets) should make contingency plans to protect the lives of their customers as well as their staff at times when there is a flood threat. Special attention should be given to flows of people and vehicles, bottlenecks, basements, car parks, exits, and underpasses. They must be able to communicate with customers and staff rapidly and

simultaneously using a range of signage and communication media designed to reach everyone, and staff should be rehearsed in evacuation procedures suitable for flooding (these procedures may be different for other hazards, such as high wind). Failure to do so could pose serious legal liability and business/brand consequences as well as adverse human and social consequences. These plans need to be based on a risk-to-life assessment.

5.4.2 Enhancing Resilience through Business Recovery Planning

5.4.2.1 Resilience through a Collective, Nondeterministic System Approach to Coastal Safety

Businesses have the potential to collaborate and to combine to "buy" or "acquire" collective security in the same way that (1) businesses and coastal towns/local/regional economies and (2) businesses and local communities do. However, with some notable exceptions, related mainly to their supply chains, all too often businesses exhibit tunnel vision when it comes to an ecosystem perspective (discussed next) and the potential for resilience that it presents.

Business companies do not exist in isolation. They are part of nested ecosystems: (1) the business ecosystem (of coastal settlements and business environments extending beyond the flood risk zone), (2) the local and regional economic ecosystem, and (3) the coastal settlement, local, regional population, and community ecosystem. These ecosystems exhibit multiple interdependencies among businesses, among businesses and their employees and customers, and among businesses and the economies of which they are part. The consequences of flooding of businesses will ripple through other parts of these ecosystems, creating financial and other implications on a scale from minor to major (Parker, 2007), in the same way that flooding of employees' and customers' homes will impact local businesses (in possibly both negative and positive ways).

Business forums (e.g., business breakfast clubs, business workshops) led by local chambers of commerce, local and regional trade associations, and government departments should be used to communicate flood risk using a cognitive pathway sequence (THESEUS OD4.8, 2013; Vorst, 2010, pp. 15–21). A cognitive pathway is a succession of steps that allows for the sharing of information, with the goal of a progressive increase of knowledge sharing that, in this case, commences with consequences for businesses and that generates a shared understanding of the probabilistic nature of flooding and a progressive change of perception.

These forums should also be used to encourage engagement and to develop an understanding of ecosystem interdependencies, the consequences of flooding on them, practical and innovative ways in which resilience may be enhanced, and practical forms of collaboration and constructive partnership to build collective resilience. The relevance of such an approach to shareholder and auditor perspectives should be explored by drawing them into discussions, because they usually have financial security and ethical interests. Leadership may also be required from local community leaders, the flood risk management agency, and/or local councils, who should be encouraged to draw business representatives into building collective local flood resilience using graphic examples of interdependencies.

A particular window of opportunity exists to stimulate system/ecosystem perspectives, as described in the previous paragraph, in the immediate aftermath of a flood either when businesses have experienced adverse consequences, when such consequences are apparent in neighboring coastal settlements, or when affected businesses are willing to share their experiences.

5.4.2.2 Integrating Coastal Flood Risk Resilience with Spatial and Economic Planning

Although Sections 5.3.1 and 5.3.2 dealt with spatial planning as a central governance activity, in this section the influence of the outcomes of spatial planning on business recovery are discussed. Coastal flood risk should be considered alongside and integrated with other spatial and economic planning issues such as transport, housing, economic and employment stimulation and growth, natural resources, economic and social regeneration, biodiversity, the historical environment, and the management of other hazards. Policies should recognize the positive contribution that flood risk avoidance and resilience management can make to the development of sustainable communities, including improved local amenities and better overall quality of life.

The adverse, localized, and wider disruptive consequences of floods on businesses and, in turn, on local economies and communities—which are often poorly understood and gauged in advance of flooding—should be a variable that is factored explicitly into coastal planning decision making concerning business development (Penning-Rowsell et al., 2013). This aspect should be drawn into professional education for planners so they have a more nuanced understanding of changing flood risks (including the more extreme risks) and flood risk-reduction opportunities.

Practical steps toward this objective include recognizing (1) the tension that exists among spatial and economic planning, their growth and development focus, and flood and coastal risk management, and (2) the opportunity that a specific focus on resilience for businesses presents for harmonizing these goals and solving planning problems.

It is important to ensure that spatial planning supports coastal flood risk management and that illegal business development does not take place. An effective working relationship is required between local development planning and control agencies and coastal flood risk management agencies, including some staff sharing and making the best use of expertise and information that is available so that new business development proposals may be assessed in terms of both planning and flood risk management criteria with a view to adopting a scale of potentially multiple harmonizing interventions designed to alleviate yes/no development decision dilemmas. This scale should include (1) movement of proposed developments to areas with the lowest flood risk or beyond the flood risk zone, (2) the substitution of less vulnerable development types for those incompatible with the degree of flood risk, (3) implementation of measures to make buildings resistant to floods (using property elevation, and/or automatic and/or manual resistance measures) and to apply flood resilience treatments to building interiors, and (4) measures designed to mitigate loss of access and disruption to properties and local economies and communities.

5.4.2.3 Designing (and Redesigning) Businesses to be Resilient to Floods (as Well as to Other Hazards)

Architects currently design buildings with particular objectives in mind—for example, to minimize their carbon footprint, to minimize their visual intrusion, to minimize sound transmission, to maximize floor space, and so on. The same notion can be applied not only to designing buildings that are resistant or resilient to physical flood damage, but also to the design of businesses that are resilient to physical damage and to consequential or disruptive loss (Bowker, Escarameia, & Tagg, 2007). Indeed, the same design principle may be applied to business parks and to communities comprising businesses. However, flood-resilient design principles are currently rarely applied in this manner in part because architects, developers, and planners know little about the different disruptive impacts of floods and the disruptive commonalities that may be discerned.

Therefore, business leaders, architects, developers, planners, builders, and flood risk managers need to be drawn together to identify (1) what is already known through their pooled experience about how business buildings and operations may be designed or redesigned to enhance their resilience to floods and other common hazards, (2) the research and development that may be required to establish firmer proposals, (3) the regulatory codes of practice (e.g., for architects, developers, builders, business managers) that are required, and (4) new training that may be drawn up to feed new flood-resilient designs into practice. Existing building regulations and codes of practice should be examined critically to identify ways in which they may be strengthened to increase business resilience to flooding.

5.4.2.4 Insurance

In Section 5.2.2, it was noted that the resilience of the coastal system as a whole could be increased by the design and practice of insurance schemes. Insurance has a role to play at the business level, and a key entry point is that BCP and insurers should be encouraged, or required, to adopt a systems perspective that stimulates business resilience.

A virtuous circle needs to be created by using insurance constructively for business damage and loss in which (1) insurance premiums reflect the degree of (changing) flood risk and (2) insurance is made available only to businesses that demonstrate regularly that they have taken steps to minimize their potential losses by creating and adopting a BCP that conforms to prescribed standards (but that is also proportionate to the degree of risk, consequences of flooding, and size/resources available to the business). Alternatively, a higher premium could be charged to companies that do not demonstrate that they have BCPs in place. This may be easier to achieve when flood insurance is provided by a business consortium or local insurance consortium or by the state than by private insurers, although they, too, could be coerced by the state into adopting these principles.

It may prove difficult to achieve but, when possible, flood risk consequence assessment and related insurance coverage (if it exists) should not be "individualized" completely and it should factor in the wider, secondary risks, consequences, and costs of business disruption on coastal towns and their communities.

5.4.2.5 Integration into Flood Crisis Response and Evacuation Plans

Postflood recovery and evacuation (described as activities in their own right, respectively, in Sections 5.5.5 and 5.5.7) have a direct impact on the efficiency of BCPs. Businesses located in and adjacent to coastal flood risk zones should be as fully integrated as possible into flood crisis response and evacuation plans. It is important that all available, necessary financial, human, and material resources are available in a timely manner when crisis response and evacuation becomes necessary. This means the resources of the local business, government, and community system need to be drawn together and used collectively to achieve the desired outcome. There are two elements. First, businesses should be fully prepared to enact their flood crisis plans and to evacuate their own premises that, in some cases, may involve many staff and/or customers. Second, businesses and others within the community (including the military) should be prepared to offer each other mutual aid to maximize the positive effects of crisis response and evacuation.

Practical steps to ensure a timely, systems-oriented, resilient approach to flood crisis response and evacuation should be based on principles of (1) appropriate flows and volumes of timely information and (2) fast, timely mobilization of resources. The principles are as follows.

Flood crisis and evacuation plans should be premised on flat organizational systems that integrate businesses and that make decision-making processes as short, as uncomplicated, and as fast as possible; and that transmit and accept flood warnings rapidly, provide accurate and reliable flood information updates, including quick "in-event" feedback; and identify the threats to businesses (including their staff and customers) as well as to municipal and utility facilities. The plans should also be based on (1) threats to transportation facilities and to the residential population (including tourists), including the linkages among them (e.g., traffic flows to and from business premises that may conflict with population evacuation flows); (2) the collective resources of the entire community, which should include business organizations that have technical, equipment, machinery, and transportation resources that may be useful in a flood crisis and during evacuation; and (3) agreement over the roles assigned to each player, including businesses (some businesses adjacent to flood risk zones may be able to act as a refuge or evacuation center). Business organizations should also be prepared to participate in community flood crisis and evacuation exercises as well as undertake their own "internal" exercises. All resources should be focused on moving resources and people quickly, and on plans that identify, in advance, all potential bottlenecks and areas of friction that may slow down crisis response and evacuation.

5.5 Postflood Recovery, Vulnerability, and Resilience

Postcrisis management includes the processes, policies, and procedures related to preparing for recovery and ensuring the continuity of an infrastructure that is vital to an organization after a natural or human-induced disaster. Postcrisis management has a strong impact on how effectively a flood is managed.

What seems crucial for vulnerability reduction in terms of effectiveness and efficiency is how the flood risk is understood in the logical framework of intervention, and how the different instruments for immediate and long-term risk reduction are chosen. Thus, the central element for addressing and reducing vulnerability in postcrisis management is the implementation of a method for interventions. A participatory approach is required. Involvement of local stakeholders and organizations is recommended to increase trust and communication effectiveness. Risk perception and safety culture among coastal managers and citizens needs to be investigated and, when necessary, addressed. The level of flood preparedness needs to be investigated. This involves the degree to which people know what to do in case of flooding and know where they can acquire relevant information, and whether drills and education programs have been implemented properly. It also involves checking the adequacy of the skills, resources, and facilities of first responders and rescue personnel.

How a community perceives nonstructural flood mitigation measures is highly relevant to the appropriate integration of such measures with traditional structural measures such as flood barriers. For example, in some communities a warning system can be perceived as needed only when "push" and "pull" technologies (THESEUS OD4.8, 2013) are used together. Similarly, mobile barriers are unlikely to be completely effective unless warning systems are available and if these systems are associated with high levels of trust.

Postcrisis management is characterized by the presence of qualitative and quantitative changes in the SPRC structure (Samuels, 2006; Zanuttigh, 2011) (see Chapter 2). Four main mitigation solutions may be identified in the causal chain, and all of them modify the receptors by changing the expected consequences of the event. They are as follows.

1. Information, education, and training: The implementation of safe practices, training, and increased effectiveness in dissemination systems influence citizens' capacity to act and react during emergencies. Furthermore, these factors condition the effectiveness of nonstructural measures such as warning systems and evacuation plans. The effectiveness of these plans is also related to people's cultural perception of risk and to the skills of disaster managers. Education and dissemination plans, therefore, need to be developed that take into account the characteristics of the local population. Demographic data, data on mean levels of education, and data on the diffusion of the Internet may be used to guide education and dissemination plans, and to target those groups that require the most attention. Clearly, in a village community with a concentration of elderly people with low levels of education but well-developed social capital (e.g., good social networks), the education and warning dissemination strategy should rely more on face-to-face information and informal social networks than on other means. A flood risk and flood warning dissemination process focused on Web technologies alone will probably be a failure in such a community, and traditional modes of alerting people using members of local associations or groups is likely to be more effective.
2. Warning systems: Warning systems provide flood alerts and should be designed to allow people to react before water reaches urban areas. They can affect the amount of physical losses (death or injuries), psychological consequences (stress levels), and economic losses (damages). As said earlier, to be effective, warning systems have to be implemented together with information/education interventions that explain what warning messages mean and what should be done when a warning is delivered. The analysis of channels of flood risk and warning information should be undertaken in "quiet" periods between flood events to support and build preparedness levels. Ideas concerning the creation of shared meanings in warning and alert messages need to be tested first by using volunteers. Questions about how alert messages are best conveyed should be developed using focus groups that involve both volunteers and coastal managers. If passing information by word of mouth is considered to be a primary channel of information, informal dissemination channels and stakeholder participation can be used to control this channel.
3. Critical facilities: According to Wu, Yarnal, and Fisher (2002), the survival of a community depends on the essential services provided by critical facilities such

as schools, utilities, hospitals, shelters, fire and rescue departments, communication hubs, and transportation depots. These facilities are not only crucial to the everyday functioning of a community, but also they provide extremely important and necessary services during emergencies. Indeed, the total number of these facilities subject to flood risk is considered to be a useful indicator of the overall community's vulnerability to flood hazards (Kleinosky, Yarnal, & Fisher, 2007; Wu et al., 2002). The notion of critical facilities can be adopted at different levels. On the one hand, facilities are an instrument for social planning; they introduce the idea that damages to particular buildings could condition the responses of the community. On the other hand, facilities are a useful instrument for emergency and emergency planning as used in the United States (Kleinosky et al., 2007).

4. Evacuation plans: Evacuation plans affect potential physical and human losses, and the recovery of the community. Lessons learned from the response to Hurricane Katrina as well from three flood basins of the Tisza river in Hungary (Vari, Linnerooth-Bayer, & Ferencz, 2003) documented the importance of evacuation plans. Thevenaz and Resodihardjo (2010) highlight the importance of contingency plans (which include evacuation plans) as a condition that enables effective community emergency response during disasters.

Effective community involvement is required to reduce vulnerability. Furthermore, the roles and responsibilities of municipal stakeholders need to be defined in a contingency plan. Special attention should be given to their different social values before, during, and after the flood to increase accountability levels and to establish a clear partnership between citizens and institutions. Different instruments and strategies have to be developed to enable effective feedback mechanisms regarding interventions while providing complete information capable of activating self-help and self-protection strategies. The complementary use of traditional dissemination tools and new technologies is recommended to reach as many people as possible.

It should be emphasized that the involvement of local stakeholders in planning ad hoc meetings in public spaces, especially during critical periods of the year (e.g., the early winter season) is important. Such events need to be used to inform and educate people about safety measures to adopt in households (e.g., emergency packs) and enterprises (e.g., recovery business plans). Recommended actions should be explained in clear, simple language, whereas the importance of cooperative behavior during the different phases of floods should be highlighted to reinforce community self-help strategies. Similarly, discussion groups should be used so that questions may be raised and explanations provided, and these discussions should be supported with open-access learning material.

The implementation of warnings should integrate effective technology designed to gain additional flood lead time and to provide instructions on how to respond safely and effectively to the warning. First, formal and informal channels of warning dissemination need to be used together, and their combined use needs to be codified in terms of good practice. This is especially the case in areas with a high presence of tourists, because tourists typically do not have experience of local flood conditions and are likely to be unaware of the flood risk and flood warning codes. Indeed, the delivery of alert messages must rely on employees and owners of hotels or campsite hosts who have direct contact with the public. This requirement should be supported properly by the long-term establishment of a legal framework that recognizes community responsibility and the integration of formal and informal networks in warning dissemination processes.

Thus, community resilience can be improved with the development of multihazard early warning systems (UNISDR, 2005). A multihazard approach can address simultaneously flood as well as other coastal hazards. For example, flood alerts could be used during tourist season for heat waves or fires, as long there is no danger of them being confused. The implementation of warning systems requires an additional focus on local features, especially in the gray area between agency procedures and citizens. Specific "pre-alert" codes have to be created for high-risk areas to allow the creation of "cushion" zones for first evacuation or faster emergency intervention supported by the population. Simple instruments such as checklists for face-to-face operators can be useful in maximizing reaction and persuasion capacities, according to the priorities defined by end users for warning systems.

Bottom–up strategies could be used to gain information in relief and rehabilitation phases, because the social media provide a means of relevant information acquisition and data refreshing. Similarly, instruments to provide feedback on the ground, such as smart phones and related applications, may be used to provide real-time updates on actions implemented during the postdisaster period.

Increased coordination among levels of intervention (Ferraris, 2007, Nivolianitou & Synodinou, 2011) is recommended for the management of risk in critical facilities. Personnel working in buildings should be involved in regular drills with emergency operators to verify communication procedures and the different "learn-by-doing" practices related to the evacuation of facilities and to the use of emergency barriers (Simpson, 2002). Moreover, multifunctional spaces and buildings can provide specific water-retention areas that reduce overall damages in core zones, whereas disturbance-proof, low-elevation spaces and ground floors with nonessential functions may be created. The diffusion and use of materials that are resistant to water, heat, and ice drift in critical buildings may be encouraged by financial incentives for such measures. Last, some particular suggestions for the adoption of evacuation plans may be made from psychosocial perspectives. Strategies should be adopted for community participation (Newport & Jawahar, 2003) in all evacuation planning phases.

As much as feasible, all local associations and agents should be involved in raising risk awareness and trust levels. Community-based approaches (Simpson,

2001) could be used for determining places where information on safety measures should be positioned, as well as determining suitable symbols to be used in evacuation signals. Community-based training programs should be developed that are designed to teach residents basic emergency response skills using drills and simulations (Simpson, 2001).

5.6 Risk Communications at the Crossroads of Vulnerability Reduction and Resilience Enhancement

Risk communication, as a transverse activity of risk governance, may be framed as part of several, quite distinct, causal chains, some of which have already been touched in the preceding sections (THESEUS OD4.8, 2013). First, overexposure to risk has often been framed as a poor understanding of the causes and consequences of floods. Under this hypothesis, knowledge gaps are viewed as the key factor explaining poorly designed policies and overexposure. Yet this framing, relying on what is known to be the "knowledge gap hypothesis," does not withstand empirical analysis (Kahan et al., 2012). Although robust knowledge is a necessary condition, the sharing of this knowledge is not a sufficient condition. Recent analyses (Pennings & Grossman, 2008; THESEUS OD4.8, 2013) demonstrate that beyond knowledge, attitudes toward risk in general, and toward coastal risk in particular, can also be explained by issues of diverging material and moral values. In the light of these results, it is now widely accepted, if not yet often implemented, that risk communication entails the creation of safe, deliberative space where knowledge may be shared in a way that is respectful and attuned to risk stakeholders' understanding and paradigmatic representations, material constraints and value, and moral values.

In other words, risk communication entails addressing the three following challenges (Kane, Vanderlinden, Baztan, Touili, & Claus, 2014):

1. Heuristic diversity: high diversity of experience with the coastal system and associated risk, leading to a variety of understandings of the way the coastal system functions (i.e., addressing potential knowledge discrepancies)
2. Diversity in material terms: differing priorities regarding what should be protected and what deserves attention (i.e., making sure that the risk issue corresponds whether you are a risk scientist, a risk manager, or a member of the general population at risk)
3. Diversity in normative terms: differing moral statements regarding how values may be threatened by the risk or by the risk governance/mitigation options envisioned (i.e., envisioning risk management as a potentially value-laden activity leading to the assessment of mitigation not only in scientific or material terms, but also in moral terms)

In coastal settings, these challenges may be specified. First, stakeholders associate risk with the modified state of the receptor or the consequences of flooding and erosion. Very seldom do stakeholders mention the probabilistic nature or flooding and erosion risk. Therefore, it seems safe to consider that, for most coastal stakeholders, risk "equals" consequences (Touili et al., 2014). Furthermore, work with stakeholders indicates that normative claims are central to risk governance for stakeholders. The following line of thought is followed most often when envisioning coastal risks. First, risk management is seen as a political process; as such, it entails the weighing of many factors, some of them external to the risk governance process (such as the odds of a politician of being reelected). Therefore, coastal risk stakeholders consider that authorities will act only if their proposal is accepted by the affected population or by the affected economic agents, regardless, sometimes, of the efficiency of the choice being made. Acceptability by affected stakeholders and economic agents is seen as contingent on the redistributive nature of the decision to be made (THESEUS OD4.8, 2013). Therefore, coastal flooding and erosion boils down to the normed acceptability of the options envisioned. Stakeholders understand that a specific risk may very well not be managed in a way that makes sense in terms of increased safety, but that does make sense in terms of social acceptance.

Although the normative statement made here is essentially associated with risk management options, it seems critical from the onset to allow as much space as possible for deliberations about the nature of the risk, and on its redistributive nature (as opposed to the redistributive nature of its mitigation). This allows for a clearer understanding of the normative challenges associated with risk management and associated options. This is especially true when normative statements regarding the consequences of a specific risk enter in conflict with a normative statement regarding the management options that are envisioned.

Another challenge in terms of risk communication lies in the sharing of the probabilistic nature of the source. This challenge has two dimensions. First, the probabilistic nature of events with relatively low occurrence is often not well understood by stakeholders. Second, society tends to be forgetful of events with low occurrence. Within a cognitive pathway (see Section 5.4.2.1), the sharing of the probabilistic nature of an event is, therefore, a critical hinge point.

Risk communication should therefore be envisioned as a cognitive pathway that addresses the following challenges:

- the cognitive chain is C-R-P-S for the stakeholders whereas the causal chain underlying the conceptual integrated risk assessment model is S-P-R-C;
- The probabilistic nature of the averse event is a subsidiary to other considerations for stakeholders whereas it is critical for risk managers;
- The causal evidence embedded in the SPRC model is completely secondary for the stakeholders (their priority lies in the normative claims associated with the

consequences under scrutiny); yet, for risk managers, the causal evidence lies at the core of their interventions.

To address these challenges, the deliberative space should be structured around a CRPS–probabilistic dimension–SPRC loop. First, consequences are envisioned and associated normative, pertinence, and evidence declarations are identified and discussed (THESEUS OD4.8, 2013). The same is then done for the receptor, the pathway, and the source. After this first sequence, a shared understanding of the probabilistic dimensions must be achieved. Then, the SPRC sequence is followed in this order to envision all associated mitigation possibilities, documenting, again, normative, pertinence, and evidence dimensions.

After the cognitive loop just described is completed, the options for risk mitigation will be generated. The generation of these options will be contextualized by the various pertinence, evidence, and normative claims that have been collected. This allows for an initial assessment of the choices that are possible in terms of societal acceptance. A critical dimension of the framework lies in the identification of a risk indicator portfolio that makes sense for all involved. This is another element that calls of a clear expression of the stakeholders' pertinence, evidence, and normative claims.

This risk communication scheme, which is deeply structured around an SPRC–vulnerability-centered approach contributes to the resilience principles outlined in Table 5.1.

By providing a series of feedback loops during the course of the cognitive pathway that allows for information flow between stakeholders, risk managers, and stakeholders, the proposed risk communication scheme will increase the homeostatic features of the coastal system. By giving access to and mobilizing several knowledge bases (scientific, public, and expert), the proposed risk communication scheme will enhance the system's omnivory. The proposed communication scheme will also, by giving a space for the expression of all stakeholders' values, contribute to the system's flatness—flatness that, in turn, will be more efficient through the appropriation of robust knowledge by stakeholders.

Implementation-wise, the following elements need to be taken into account. Stakeholders that will be affected by the risk management envisioned should be involved in the process as early as possible. Involvement includes the proactive management of iterative exchange and communications. All people involved follow a cognitive pathway where new knowledge is acquired. This knowledge acquisition process is the proposed risk communication scheme. This cognitive pathway is designed explicitly to take stock of the current knowledge of risk perception. Ideally, this cognitive pathway should be discussed explicitly with all stakeholders to integrate it into their individual and collective heuristics.

At each step of the cognitive pathway, and for all information that is mobilized, the answer to the following questions should be gathered and associated with the information under scrutiny (this should be documented unfiltered, and exchanged among stakeholders):

- What is the phenomenon that is under scrutiny and why does one find it important or not?
- What are the causal linkages expressed?
- On what basis are these causal linkages expressed?
- Are there ways of associating judgment with this information?
- Is this information threatening to values held dear by a group of stakeholders?

Moving up the cognitive pathway sequence, the most critical dimension lies in the sharing of the meaning associated with the probabilistic nature of the source. The critical recommendation here is *not to proceed* until all stakeholders develop a common understanding of the probabilistic nature of the source. An understanding of how future probability density functions are obtained should be part of this sharing process. One of the key challenges here lies in the nature of climate change. First, attribution issues may very well be questioned. These questions must be faced. Second, accepting climate change may be a challenge for some. When envisioning mitigation options, they should be assessed not only in terms of risk reduction, but also in terms of compatibility with stakeholders' values and beliefs. These stakeholders' values and beliefs should be documented as part of the indicator system. If the mitigation option conflicts with these values and beliefs it is critical not to proceed before deliberation occurs to assess the mitigation option acceptability. After the cognitive pathway sequence has been iterated once, all stakeholders should convene and assess collectively whether their perceptions have changed in terms of causality, relevance, and norms. This is the point where the deterministic and nondeterministic characteristics of the coastal system must be discussed. This is also the point where the blending of resilience enhancement and vulnerability reduction measures should be discussed collectively.

5.7 Evacuation Plans

5.7.1 Evacuation Plans and Vulnerability Reduction

Evacuation planning is a risk mitigation option designed to protect people from the effects of a flood (Jonkman, Vrijling, & Vrouwenvelder, 2008). It helps to reduce the vulnerability of coastal areas by providing some directions about how to prevent people from drowning when all other structural measures fail.

Evacuation planning is part of an emergency management strategy. In this regard, it supplements other classic mitigation options "in real time"—that is, when the event actually strikes. By reducing the number of people in the exposed area at the start of the disaster, evacuation planning reduces the vulnerability of the area and the consequences of the flood.

The methodology created for planning a preventive mass evacuation can be effective when the disaster can be predicted with sufficient forewarning (usually at least 24 h). This is normally the case for coastal storms, when meteorological and hydrological elements responsible for a flood can be forecast by numerical models combined with real-time observations (Morel, Hissel, & Bouchrit, 2011). Fully accurate predictions with precise water levels and velocities are not necessary as long as the models allow identification of the part of the coastal region that is likely to be flooded and the extent to which it will be flooded.

Preventive evacuation cannot be improvised. To be effective, the authorities should have a clear idea of which areas are expected to flood, how many people must be evacuated, how many of them will not be able to do so by themselves, and where people should go to be safe (Keys & Opper, 2002). Many problems are likely to arise during an evacuation because of equipment or staff not being available to fulfill the needs of evacuees, or because of traffic jams on routes used by evacuees fleeing from the exposed area. Disseminating the warning and advice for an evacuation—and for them to be heard and understood by as many people as possible in a short amount of time—is a major issue. If these problems are not addressed fully before a flood disaster, the warning period (or lead time) may be insufficient to allow all residents to leave the exposed area. Sound evacuation planning must address all these issues so that everything is ready when a disaster is anticipated.

Ensuring evacuation is rapid and effective usually leads to the number of human casualties being reduced (Jonkman et al., 2008; Lumbroso, Gaume, Logtmeijer, Mens, & van der Vat, 2009). Goods and animals (e.g., pets) may also be included in evacuation plans if it is felt necessary, although priority should be given to humans. The indirect effects of a flood will decrease as a consequence in much the same way as direct effects. Firms may still need to cease activity during a flood, but if their employees are safe and unharmed, they may be able to return to work more rapidly after an event, limiting business disruption time. Also, if there are no human casualties, the attractiveness of the exposed region for residents and tourists is less likely to suffer. The psychological impact of the event will also be reduced.

However, evacuation is a complicated and risky process that brings with it a number of negative issues that must be taken into account when deciding to use such a loss mitigation option (Litman, 2006; Pine, Marx, Levitan, & Wilkins, 2003). The first complication is that people who are forced to leave their homes by car or by using a public transportation option are placed temporarily at a greater level of risk before they leave the exposed area. For example, in addition to the risk of motor vehicle accidents, there is the potential for them to be trapped in traffic jams when flooding commences. The probability of survival is then lower than if the evacuees had stayed at home. Also, if the evacuation is not prepared carefully, shelters may lack essential items such as food, potable water, medicine, and so forth. Such events can cause panic, which increases quite rapidly when evacuees encounter difficulties. Last, making people leave their homes is not always easy and has a psychological cost no matter how it is done.

A key step in preparing an evacuation plan is collecting data. To make an evacuation plan as realistic as possible and to take advantage of all the facilities in the coastal region, the plan has to provide a clear understanding of the composition of the coastal region. The number of inhabitants and their spatial distribution is a key variable; other important parameters include the location of public facilities that may be used as shelters, the characteristics of transportation routes, and the media available to communicate to the people. Such data may be collected either from national databases or by surveys of residents, including face-to-face interviews (Hissel et al., 2014).

The evacuation plan should be established during "quiet" times, when no storm is likely to occur. The plan should present a sorted and prioritized list of actions to be taken prior to the flood and, as soon as evacuation is decided, together clear allocation of responsibilities for actions (Beaulieu & Marchand, 2001). It is recommended that different lists of actions be formulated for different kind of risks or combination of risks that may be anticipated (Morel et al., 2011). It is infeasible to address all possible situations, because, even if the meteorological and hydrological processes can be reasonably well predicted, dealing with the behaviors of human beings often generates the unexpected. Much may occur during an emergency that is impossible to predict and that can change the way the evacuation evolves, or even prevents it from evolving. One cannot predict such uncertain events, but at least the evacuation planner should assess the possible events that can interfere with an evacuation. Scenarios may then be prepared to reflect a wide scope of possible occurrences. Each scenario should contain a set of assumptions that describe how the evacuation evolves. The meteorological conditions responsible for the disaster are obviously included among these values, but in addition it is possible to include factors such as the likely number of people choosing to evacuate by vehicle, the season (which affects the number of tourists in the coastal region), and the time of the day, because this is likely to affect the extent of the reception of an evacuation order.

An evacuation plan must identify clearly the structure of the emergency response and the organizations involved in the process, along with the responsibilities of each key stakeholder (Beaulieu & Marchand, 2001; EMA, 2005). This is not always established clearly in national regulations, but there is no time to discuss or argue about this during a disaster. Therefore, everyone involved in emergency management should agree to their roles during a crisis before it actually happens. Even if public authorities are the first to be concerned with emergency management, private organizations may also be involved (Beaulieu & Marchand, 2001). To guarantee their involvement during disasters, contracts with private organizations, which describe their role during the crisis, should be sought and signed on before a crisis occurs.

With all the data collected, the list of actions to be carried out before the crisis may be prepared. Each action is associated with one or more of the organizations involved in emergency management. The evacuation plan is the collection of the documents associated with the steps that described earlier. This plan should be communicated carefully to all those who will contribute to emergency management. Part of the plan should be disseminated to citizens so they are prepared to receive an evacuation order and understand what it means. The plan must be tested during exercises and must also be updated on a regular basis.

5.7.2 Evacuation Plans Associated with Resilience Enhancement

Evacuation planning contributes to the coastal system homeostasis (Table 5.1). First, the data collection stage includes interviews with stakeholders and citizens, and surveys of the population. Not only are such actions necessary to generate realistic action plans, but also they contribute to risk and evacuation awareness of the population at risk. Inhabitants who are better informed about the risk are more likely to accept the instructions of local authorities when a flood is approaching, and they may also be more inclined to take individual measures to protect their homes against the flood to supplement the measures taken by local authorities. Evacuation plans usually insist on the need to educate people about appropriate response behaviors when confronted by a flood threat. Well-prepared plans should consider citizens as actors in their own safety (Litman, 2006; Townsend, 2006). Inhabitants are part of the protection system; emergency management units often have insufficient time and resources to take fully into account all individual situations, but they are able to provide advice and recommendations, and provide assistance to those residents who need it most. Apart from this, they have to trust citizens and hope they will behave in a responsible manner. In this sense, an evacuation plan based on the methodology described in (Hissel et al. 2014) helps to create a feedback loop between authorities and organizations involved in crisis management, as well as the citizens who are included in the emergency management process.

Effective evacuation planning relies on a considerable amount of data about flood risk and those exposed to it, but this information is usually insufficient to predict reliably whether a plan is likely to produce the expected results, which is to bring all people to a safe place before the flood actually strikes. To this aim, lessons learned from previous floods and from realistic emergency exercises are invaluable (EMA, 2005; Morel et al., 2011). Exercises will pinpoint issues with actions plans and will help planners understand which kind of unforeseen events might hinder their application. Action plans can also benefit from homeostasis of the system. For example, a feedback loop between emergency management and urban planning, incorporating risk-aware spatial planning of new districts, may facilitate evacuation by providing well-equipped shelters close to every district that are capable of accommodating the number of people expected. Acquiring information about the way the evacuation is taking place also helps authorities to adapt time their action plans in real time to the current situation.

Evacuation plans include considerations regarding the system's omnivory and, in this way, contribute to this resilience principle. The very basic process that a flooded city is unable to deliver is providing its residents with the primary goods they need to survive: food, potable water, medicine and medical equipment, warm places to sleep, and safety through law enforcement, hygiene, and power. A sound evacuation provides a new way of fulfilling these needs when people are relocated in shelters that offer decent accommodation. The functioning of the system depends on the capacity of people to move from place to place, for example, to see their relatives and to keep working. This may be seriously compromised in the case of a major disaster in which a city is mostly under water. However, if only a part of a city's territory is flooded, the evacuation plan should include measures to supplement normal transportation options that are disrupted because of a flood. Evacuation planning may also benefit from different strategies regarding the ways by which people are brought to safety. In fact, a mass evacuation is seldom a solution, and it is likely to put more people in danger than a more targeted and selective evacuation. One should consider different kinds of relocation based on the actual characteristics of the flooded districts (Kolen, Kok, Helsloot, & Maaskant, 2013). In some places where the flood duration is not too long, a vertical evacuation to higher levels may be rapid and effective; in other districts where water levels and velocities are lower, in-place sheltering may be an appropriate option. These alternative strategies are likely to have less psychological impacts than others, and a lower concentration of people on evacuation routes also involves a lower probability of critical accidents occurring. Preventive horizontal evacuation should be used only when there is no alternative.

The high-flux principle for resilient systems is associated with the fact that a resilient system should provide a rapid response to threats and changes. This is the main purpose of evacuation planning as a mitigation option. Indeed, when faced with an extreme event involving a large flood extent, regardless of whether they are well prepared, the authorities will always tend to decide on evacuation. However, collection of data before a crisis and the planning of actions beforehand when there is time to consider consequences and risks, can expedite decision-making processes during an event. Here, planning is key to speeding up the flux of information in real time. An evacuation plan should also be based on establishing a warning system based on detection and forecasting of a flood. Modern information and communication technologies should be harnessed alongside some low-tech ones to keep informed those citizens and organizations responsible for emergency management (Litman, 2006). By following these recommendations, flows of information are likely to be more rapid. Evacuation benefits from high flux in the system, when this flux allows for a rapid dissemination of risk information and evacuation orders across the population.

Flatness, as a resilience principle, is the characteristic of an organization that is not overly hierarchical and heavy. The mitigation option related to evacuation planning facilitates the introduction of this characteristic in the system by clearly identifying the persons and organizations responsible for the different stages of the emergency response and by promoting citizen involvement in a coordinated response. If all those responsible for the different actions comprising evacuation plans are clearly identified, this reduces the need for hierarchical pyramids to be involved in making decisions, thus producing a faster reaction. The involvement of citizens is crucial because they are usually willing to help the authorities and their relatives, but they do not always know how to be useful, as has been shown during recent events (Litman, 2006). Thus, a well-informed public, the actions of which are channeled toward priority actions through preventive information and planning, can prove to be highly useful in supplementing the actions of regular emergency organizations (Keys & Opper, 2002).

Evacuation planning may contribute to an improving buffering quality by identifying places that are most likely to get flooded during an event and focusing on evacuating them very rapidly when the event is forecast. Those districts are then less vulnerable to the flood, and the evacuated areas can be used as a buffer to reduce the level of flooding in other districts. Thus, an evacuation in multiple stages can be planned, during which the most exposed but less vulnerable places are evacuated first and the most vulnerable places are protected for a longer time. However, this strategy only works if one can estimate the dynamics of the flood and if the arrival of the flood is not so fast that it becomes impossible to distinguish between evacuation sectors. In evacuation plans, it is advisable to elevate vital functions as much as possible (EMA, 2005). The shelters and transportation networks, which play a central role in the evacuation process, should be protected against rising water levels and elevated above the level of the adjacent ground. Nonessential functions or flood-resistant infrastructures can remain at ground level. To this end, evacuation planning should be combined with risk-aware spatial planning.

Evacuation planning contributes to improving the redundancy of the flood-prone area when it is combined with other nonstructural measures, with the same purpose of limiting the number of human casualties of a flood event, such as early warning, educating people about the consequences of the risk, and so on. Evacuation planning also benefits from redundancy in all the components of the system. For instance, the provision of different transportation options for people who wish to evacuate can reduce the number of individual vehicles on roads and may help avoid traffic congestion. The addition of a number of shelters spread across a territory also contributes to reduce the risk of a major problem should the authorities be unable to accommodate all evacuees. Multiple communication media help in addressing more people in disseminating evacuation orders and information about which shelter to move to and which route to take. Using several sources of energy helps to avoid a total power blackout in shelters. Therefore, redundancy is a crucial element of an effective evacuation plan.

5.8 Conclusions

The mitigation options presented here share the potential to increase coastal safety in the face of climate change. From general principles—reducing vulnerability and increasing resilience—key criteria need to be met to maximize risk reduction.

The first element central to all options lies in the need to adopt approaches that incorporate stakeholders in the implementation process. Involving stakeholders reduces vulnerability by increasing insurance scheme efficiency, accepting spatial planning processes, reinforcing the implementation of evacuation plans and BCPs, and allowing for better organized postflood recovery planning. It is central to develop communication schemes allowing for the development of a shared understanding of coastal risks. Involving stakeholders also increases the resilience of coastal systems by (1) contributing to the coastal system homeostasis by fostering multidirectional information flow, (2) promoting high flux through the multiplication of information transfer channels, and (3) increasing flatness by creating bypasses in hierarchical governance systems.

A second element that is central to all nonstructural mitigation options is the fact that they increase safety through a reduction in consequences. As such, they cater to the priority of individuals at risk. This has a double benefit. First, it allows the correct prioritization of investments in cognitive and material resources. Second, by focusing on the central concern of those at risk, it contributes to their involvement in the mitigation choices.

A third element that nonstructural mitigation options share is the fact that they interact strongly, showing the potential to transcend the sum of their individual contributions. Insurance schemes that are designed according the discussion in Section 5.2.2 must internalize the premiums' existing land use, BCP, and other individual-level risk-reduction behavior. Land use planning must take into account business continuity planning, postflood recovery and evacuation planning, and vice versa. Risk communication applies to all mitigation options.

These three characteristics lead to three overarching recommendations in the implementation of nonstructural mitigation options:

1. Participatory approaches to mitigation choice and implementation should be favored as much as possible.

2. When envisioning coastal risk management, the entry point should be the reduction of consequences.
3. When envisioning mitigations options, they should always be thought as a part of a system comprising several mitigation possibilities.

When facing the choice of mitigation options, it is critical to move stepwise by envisioning key questions. Considering specific mitigation options, these questions are focused on site specificities. A clear understanding of the site specificities in terms of insurance, spatial planning, business continuity planning, postflood recovery, risk communication, and evacuation plan should be conducted as part of the risk assessment phase (see Chapter 2).

On the insurance front, effectiveness is conditional on the current insurance framework within which a specific community or region is embedded. If formal insurance does not exist, historical compensation schemes (mostly public) should be explored to determine how they may be improved by following the approaches described in Section 5.5.2, including how a more formal system may be put into place. If formal insurance exists, they should be assessed in the light of the discussion in Section 5.5.2. For land use planning, planners should be involved in a formal exercise that evaluates the inclusion of FCERM within their practice. In terms of BCP, SMEs should be approached to assess their capacity for implementing the BCP, and the existence of BCPs for large businesses should be confirmed. The central elements emphasized in Section 5.5.4 should not be overlooked. The recommendations associated with postflood recovery are an integral part of any risk assessment procedure. In terms of risk communication, there is an emphasis on the importance of not focusing solely on knowledge gaps as a key driver for risk communication. A much richer approach is needed for which not only evidence is discussed, but also norms and material constraints are considered. Last, when envisioning evacuation plans, a data needs and availability assessment should be conducted. After these steps have been taken, an explicit assessment should be undertaken of the most adaptable options and their synergies.

The options presented in this chapter may give the impression that guidance is lacking on how to respond to a flood threat at the site level (i.e., that guidance is lacking decisions regarding a given measure, or combination of measures, and their implementation). This is the result of a choice made by us, through our collective experience as social scientists. We recognize that, in terms of nonstructural options, such recommendations cannot be made. The diversity of institutional coastal settings means that establishing dichotomous choices is misleading at best and, potentially, counterproductive. Nonstructural mitigation options should therefore be regarded as highly adaptable approaches to the fostering of safer coasts. All nonstructural options have a potential role, provided that stakeholder participation is nurtured, that the central entry point is the reduction of consequences, that cross-influences among mitigation options are always analyzed, and that vulnerability and resilience are envisioned jointly.

References

Adger, W. N. (2006). Vulnerability. *Global Environmental Change – Human and Policy Dimensions, 16*(3), 268–281. http://dx.doi.org/10.1016/j.gloenvcha.2006.02.006.

Barnes, J. C. (2001). *A guide to business continuity planning*. London, UK: J Wiley and Sons.

Bastien-Daigle, S., Vanderlinden, J.-P., & Chouinard, O. (2008). Learning the ropes: lessons in integrated management of coastal resources in Canada's maritime provinces. *Ocean & Coastal Management, 51*(2), 96–125. http://dx.doi.org/10.1016/j.ocecoaman.2007.04.006.

Beaulieu, P., & Marchand, J.-F. (2001). *Guide de planification stratégique des opérations d'évacuation en milieu hautement urbanisé*. Publication préparée pour Bureau de la protection des infrastructures essentielles et de la protection civile, Gouvernement du Canada. Groupe de Recherche Interdisciplinaire en Gestion de l'Environnement (GRIEGE), Université de Québec. http://publications.gc.ca/collections/Collection/D82-42-2003E.pdf. Accessed 05.11.13 (in English).

Berkes, F., Colding, J., & Folke, C. (Eds.). (2003). *Navigating social ecological systems, building resilience for complexity and change*. Cambridge, UK: Cambridge University Press.

Birkmann, J. (2007). *Climate change and vulnerability: Challenges for spatial planning and civil protection*. Forum DKKV/CEDIM: Disaster Reduction in Climate Change. Karlsruhe, Germany: Karlsruhe University http://www.unisdr.org/files/8020_03Birkmann1.pdf. Accessed 28.10.13.

Botzen, W. J. W., Aerts, J. C. J. H., & van den Bergh, J. C. J. M. (2009). Willingness of homeowners to mitigate climate risk through insurance. *Ecological Economics, 68*(8–9), 2265–2277. http://dx.doi.org/10.1016/j.ecolecon.2009.02.019.

Bowker, B., Escarameia, M., & Tagg, A. (2007). *Improving the flood performance of new buildings. Flood resilient construction*. London, UK: Department for Communities and Local Government, Department of Environment, Food and Rural Affairs, Environment Agency and CIRIA.

Business Continuity Institute, B. C. R. (2005) (www.thebci.org).

Chouinard, O., Baztan, J., & Vanderlinden, J.-P. (2011). *Zones côtières et changements climatiques (in French)*. Québec, Canada: Presses de l'Université du Québec.

Dawson, R. J., Ball, T., Werritty, J., Werritty, A., Hall, J. W., & Roche, N. (2011). Assessing the effectiveness of non-structural flood management measures in the Thames Estuary under conditions of socio-economic and environmental change. *Global Environmental Change-Human and Policy Dimensions, 21*(2), 628–646. http://dx.doi.org/10.1016/j.gloenvcha.2011.01.013.

Deboudt, P. (2010). Towards coastal risk management in France. *Ocean & Coastal Management, 53*(7), 366–378. http://dx.doi.org/10.1016/j.ocecoaman.2010.04.013.

EA. (2012). *Thames Estuary 2100: Managing flood risk through London and the Thames estuary*. London, UK: Environment Agency. http://www.environment-agency.gov.uk/homeandleisure/floods/125045.aspx. Accessed 19.11.13.

EMA. (2005). *Evacuation planning*. Australian emergency manual series, No. 11. Dickson, ACT, Australia: Emergency Management Australia, Attorney-General's Office, Australian Government http://www.em.gov.au/Documents/Manual11-EvacuationPlanning.pdf. Accessed 05.11.05.

Ferraris, D. P. (2007). The role of coordination and the ability of interaction among civil protection operators into international experiences, in case of tunnel accident and flood risk. In *Pre-emergencies, preparedness, management, communication and lesson learnt in emergencies* (pp. 11–56).

Glavovic, B. C., Saunders, W., & Becker, J. (2010). Realising the potential of land-use planning to reduce hazard risks in New Zealand. *Australasian Journal of Disaster and Trauma Studies*. Online journal http://trauma.massey.ac.nz/issues/2010-1/glavovic.htm. Accessed 28.10.13.

Hiles, A. (2011). *The definitive handbook of business continuity management*. London, UK: J Wiley and Sons.

Hissel, F., Morel, G., Pescaroli, G., Graaff, H., Felts, D., & Pietrantoni, L. (2014). Early warning and mass evacuation in coastal cities. *Coastal Engineering, 87*, 193–204.

Jonkman, S. N., Vrijling, J. K., & Vrouwenvelder, A. C. W. M. (2008). Methods for the estimation of loss of life due to floods: a literature review and a proposal for a new method. *Natural Hazards, 46*(3), 353–389. http://dx.doi.org/10.1007/s11069-008-9227-5.

Kahan, D. M., Peters, E., Wittlin, M., Slovic, P., Ouellette, L. L., Braman, D., et al. (2012). The polarizing impact of science literacy and numeracy on perceived climate change risks. *Nature Climate Change, 2*(10), 732–735. http://www.nature.com/nclimate/journal/v2/n10/abs/nclimate1547.html#supplementary-information.

Kane, I. O., Vanderlinden, J.-P., Baztan, J., Touili, N., & Claus, S. (2014). Communicating risk through a DSS: A coastal risk centred empirical analysis. *Coastal Engineering, 87*, 240–248.

Keys, C., & Opper, S. (2002). On the proper conceptualisation of the warning, evacuation and community education tasks in the context of planning for dam failure. In *Presented at the Australian national committee on large dams (ANCOLD) 2002 Glenelg, Australia*. http://www.ses.nsw.gov.au/resources/research-papers/dams/on-the-proper-conceptualisation-of-the-warning.

Kleinosky, L., Yarnal, B., & Fisher, A. (2007). Vulnerability of Hampton roads, Virginia to storm-surge flooding and sea-level rise. *Natural Hazards, 40*(1), 43–70. http://dx.doi.org/10.1007/s11069-006-0004-z.

Kolen, B., Kok, M., Helsloot, I., & Maaskant, B. (2013). EvacuAid: a probabilistic model to determine the expected loss of life for different mass evacuation strategies during flood threats. *Risk Analysis, 33*(7), 1312–1333. http://dx.doi.org/10.1111/j.1539-6924.2012.01932.x.

Kreibich, H., Seifert, I., Thieken, A., Lindquist, E., Wagner, K., & Merz, B. (2011). Recent changes in flood preparedness of private households and businesses in Germany. *Regional Environmental Change, 11*(1), 59–71. http://dx.doi.org/10.1007/s10113-010-0119-3.

Kunreuther, H., & Pauly, M. (2006). Rules rather than discretion: lessons from Hurricane Katrina. *Journal of Risk and Uncertainty, 33*(1–2), 101–116. http://dx.doi.org/10.1007/s11166-006-0173-x.

Litman, T. (2006). *Lessons from Katrina and Rita, Victoria transport policy institute*. Victoria, BC, Canada: Victoria Transport Policy Institute. http://www.vtpi.org/katrina.pdf. Accessed 31.10.13.

Lumbroso, D., Gaume, E., Logtmeijer, C., Mens, M., & van der Vat, M. (2009). *Evacuation and traffic management*. Floodsite Report T17-07-02. Delft, The Netherlands: Delft Hydraulics http://www.floodsite.net/html/partner_area/project_docs/T17_07_02_Evacuation_and_traffic_management_D17_1_V4_4_P01.pdf. Accessed 05.11.13.

Morel, G., Hissel, F., Juan Jia, X., & Bouchrit, A. (2011). Aide à la planification des évacuations de masse en cas d'inondation ou de submersion majeure. In *Presented at the 7th Géorisques: Plan communal de sauvegarde, retour d'expériences et bilan, outils de gestion de crise 26–27 January*. Montpellier 3: Université Paul-Valéry.

Newport, J. K., & Jawahar, G. G. P. (2003). Community participation and public awareness in disaster mitigation. *Disaster Prevention and Management, 12*(1), 33–36. http://dx.doi.org/10.1108/09653560310463838.

Nivolianitou, Z., & Synodinou, B. (2011). Towards emergency management of natural disasters and critical accidents: the Greek experience. *Journal of Environmental Management, 92*(10), 2657–2665. http://dx.doi.org/10.1016/j.jenvman.2011.06.003.

Parker, D. J. (November 2007). *Evaluating business interruption losses caused by flooding*. London, UK: Presentation to the Lighthill Risk Network.

Penning-Rowsell, E. C., Priest, S., Parker, D. J., Morris, J., Tunstall, S. M., Viavattene, C., et al. (2013). *Flood and coastal erosion risk management: A manual for economic appraisal*. London, UK and New York, USA: Routledge.

Pennings, J. M., & Grossman, D. B. (2008). Responding to crises and disasters: the role of risk attitudes and risk perceptions. *Disasters, 32*, 434–448.

Pine, J., Marx, B., Levitan, M., & Wilkins, D. (2003). Comprehensive assessment of Hurricane shelters: lessons from Hurricane Georges. *Natural Hazards Review, 4*(4), 197–205. http://dx.doi.org/10.1061/(ASCE)1527-6988(2003)4:4(197).

Pottier, N., Penning-Rowsell, E., Tunstall, S., & Hubert, G. (2005). Land use and flood protection: contrasting approaches and outcomes in France and in England and Wales. *Applied Geography, 25*(1), 1–27. http://dx.doi.org/10.1016/j.apgeog.2004.11.003.

Renn, O. (2008). *Risk governance: Coping with uncertainty in a complex world*. London, UK: Earthscan.

Richards, J., White, J., & Carter, J. (2008). Local planning practice and flood risk management in England: is there a collective implementation gap? *Environnement Urbain/Urban Environment, 2*, 11–20.

Richardson, H. W., Gordon, P., & Moore, J. E. (Eds.). (2008). *Natural disaster analysis after Hurricane Katrina. Risk assessment, economic impacts and social implications,* Chapter 14 (pp. 253–278). Cheltenham, UK: Edward Elgar.

Samuels, P. G. (2006). A European perspective on current challenges in the analysis of inland flood risks. In J. Schanze, E. Zeman, & J. Marsalek (Eds.), *Flood risk management: Hazards, vulnerability and mitigation measures*. Dordrecht, The Netherlands: Springer Netherlands.

Simpson, D. (2001). Community emergency response training (CERTs): a recent history and review. *Natural Hazards Review, 2*(2), 54–63. http://dx.doi.org/10.1061/(ASCE)1527-6988(2001)2:2(54).

Simpson, D. M. (2002). Earthquake drills and simulations in community-based training and preparedness programmes. *Disasters, 26*(1), 55–69. http://dx.doi.org/10.1111/1467-7717.00191.

Snedaker, S. (2007). *Business continuity and disaster recovery planning for IT professionals*. Burlington, MA, USA: Syngress and Elsevier Publishing.

Syed, A., & Syed, A. (2003). *Business continuity planning methodology*. Washington DC, USA: Sentry.

THESEUS OD4.8. (2013). *Structured portfolio of tested operational innovative tools and protocols for policy and management purposes of coastal flooding risks*. http://www.theseusproject.eu/resources/documents/Deliverables/Official-Deliverables-(OD)/OD4.8-(M36)-Structured-portfolio-of-tested-operational-innovative-tools-and-protocols-for-policy-and-management-purposes-of-coastal-flooding-risks./.

Thevenaz, C., & Resodihardjo, S. L. (2010). All the best laid plans … conditions impeding proper emergency response. *International Journal of Production Economics, 126*(1), 7–21. http://dx.doi.org/10.1016/j.ijpe.2009.09.009.

Touili, N., Baztan, J., Vanderlinden, J.-P., Kane, I. O., Diaz-Simal, P., & Pietrantoni, L. (2014). Public perception of engineering-based coastal flooding and erosion risk mitigation options: lessons from three European coastal settings. *Coastal Engineering, 87*, 205–209.

Townsend, F. F. (2006). *The federal response to Hurricane Katrina: Lessons learned*. Washington DC, USA: The White House. http://library.stmarytx.edu/acadlib/edocs/katrinawh.pdf. Accessed 05.11.13.

Treby, E. J., Clark, M. J., & Priest, S. J. (2006). Confronting flood risk: Implications for insurance and risk transfer. *Journal of Environmental Management, 81*(4), 351–359. http://dx.doi.org/10.1016/j.jenvman.2005.11.010.

UNISDR. (2005). *The Hyogo Framework for Action 2005–2015: Building the resilience of nations and communities to disasters*. Geneva, Switzerland: United Nations Office for Disaster Risk Reduction. http://www.unisdr.org/we/inform/publications/1037. Accessed 05.11.13.

Vari, A., Linnerooth-Bayer, J., & Ferencz, Z. (2003). Stakeholder views on flood risk management in Hungary's Upper Tisza Basin. *Risk Analysis, 23*(3), 585–600. http://dx.doi.org/10.1111/1539-6924.00339.

Vorst, H. C. M. (2010). *Evacuation models and disaster psychology*. In *Procedia engineering* (Vol. 3). http://ac.els-cdn.com/S187770581000473X/1-s2.0-S187770581000473X-main.pdf?_tid=c4d78278-72e8-11e3-9ea7-00000aacb35d&acdnat=1388583161_9b031dd08a472dabd69e4da0165bef2e.

Wallace, M., & Webber, L. (2010). *The disaster recovery handbook: A step-by-step plan to ensure business continuity and protect vital operations, facilities and assets*. New York, USA: AMACOM.

Wardekker, J. A., de Jong, A., Knoop, J. M., & van der Sluijs, J. P. (2010). Operationalising a resilience approach o adapting an urban delta to uncertain climate changes. *Technological Forecasting and Social Change*, 987–998.

Wu, S. Y., Yarnal, B., & Fisher, A. (2002). Vulnerability of coastal communities to sea-level rise: a case study of Cape May County, New Jersey, USA. *Climate Research, 22*(3), 255–270. http://dx.doi.org/10.3354/cr022255.

Zanuttigh, B. (2011). Coastal flood protection: what perspective in a changing climate? The THESEUS approach. *Environmental Science & Policy, 14*(7), 845–863. http://dx.doi.org/10.1016/j.envsci.2011.03.015.

Chapter 6

Toward Sustainable Decision Making

Phoebe Koundouri[1], Barbara Zanuttigh[2], Osiel González Dávila[1], Jean Paul Vanderlinden[3], Susan Hanson[4], Robert Nicholls[4], Simon Hoggart[5], Richard C. Thompson[5], Wout de Vries[6], Edmund Penning-Rowsell[7]

[1] *Department of International and European Economic Studies, Athens University of Economics and Business (AUEB), Athens, Athens, Guyancourt, Guyancourt,Greece*

[2] *Department of Civil, Environmental and Materials Engineering (DICAM), University of Bologna (UniBo), Bologna, Italy*

[3] *Cultures, Environments, Arctic, Representations, Climate Research Center, Universite ́de Versailles Saint-Quentin-en-Yvelines (UVSQ), Guyancourt, France*

[4] *Faculty of Engineering and the Environment, University of Southampton (Soton), Southampton UK*

[5] *Marine Biology and Ecology Research Centre, School of Marine Science and Engineering, University of Plymouth (UoP), Plymouth, UK*

[6] *INFRAM International BV (INFRAM), Marknesse, The Netherlands*

[7] *Flood Hazard Research Centre, Middlesex University (MU), Hendon, UK*

6.1 Introduction

In this chapter, we analyze the issue of sustainable decision making for successful coastal flood management. In this framework, sustainable development is a key concept. This is defined as a pattern of resource use that aims to meet human needs while preserving the environment so that these needs can be met not only today, but also for future generations (WCED, 1987). The concept implies the consideration of spatial and temporal dimensions, system components (i.e., economic, social and environmental aspects), and the interactions between these components. On the one hand, development is a dynamic process and a function of its dimensions and

Coastal Risk Management in a Changing Climate. http://dx.doi.org/10.1016/B978-0-12-397310-8.00006-3

components. On the other hand, sustainability is complex and adaptive and can be achieved via innumerable paths, yielding contrasting results for system components. Decision making is also a dynamic process and is carried out at different levels of society, considering cultural, social, economic, institutional, political, and environmental differences. This process requires the design of a strategy, the definition of policies, and the implementation of actions. As a result, different types of information are required at different stages of the decision-making cycle (Winograd & Farrow, 2013). Thus, key to successful coastal flood management is the use of mitigation techniques that are appropriate for the local circumstances. This is best achieved if all alternatives are reviewed to identify the most efficient individual or suite of options for consideration by stakeholders and decision makers.

Different mitigation options change the consequences of flooding in different ways; engineering generally changes the amount or extent of flooding, whereas planning can change the nature of the flooded area and therefore the consequences.

Considering the long-term and large-scale aspects of coastal risks, the development of defense strategies should be based on a hierarchical planning approach, ranging from a top-down national (master) planning level to the analysis of individual flood-prone areas and the consideration of specific strategies and measures within these areas (Holman, Rounsevell, Berry, & Nicholls, et al., 2008, pp.1–187). The master plan should facilitate the necessary communication between coastal defense managers, contingency planners, and crisis managers and warrant the continuity of coastal risk management policies, the protection measures, and the related operational procedures (i.e., in terms of financing, institutional arrangements, legal rights/obligations, and operational responsibilities).

In general, the development of coastal protection strategies involves the following steps.

6.1.1 Specification of Detailed Regional Scenarios for the Time Period of the Decision (Often 50–100 years)

Regional scenarios should provide more specific and detailed information with respect to:

- the regional translation of the various aspects of climate change (to be included in the regional specification of hydraulic loads);
- spatial and infrastructural developments based on specific regional development potential, existing plans, and specified development priorities; and
- already planned developments in flood/erosion protection systems.

The regional scenarios should reflect the main scope of possible developments and capture the major uncertainties in developments driving future risks and the possible effects of measures and strategies to reduce these risks.

6.1.2 Problem Assessment and Identification of Promising Mitigation Measures

Problem assessment refers to establishing the extent of future coastal risks and carrying out a flood risk assessment for the relevant regional scenarios, as specified in step 1.

The feasibility of mitigation measures to be considered depends on many factors related to the natural and socioeconomic characteristics of the coastal area, the existing flood protection system, and the coastal management context (as affected by political, institutional, and cultural conditions). Within the directives provided by the master plan, a screening exercise should be performed to identify possible measures that are more plausible or promising than others, given the specific characteristics and conditions pertaining to the coastal area considered.

6.1.3 Analysis and Evaluation of Alternative Strategies

Alternative strategies regarding the protection and management of the coastal area should be based on various combinations of promising mitigation measures, each strategy representing a logical and coherent mix of measures. Assessment of the relevant impacts of the alternative protection and management strategies in terms of societal costs and benefits is required. To deal with the various (large) uncertainties, the impacts of possible strategies need to be considered for different regional scenarios. The evaluation should not so much aim at selecting the "best" strategy within a specific scenario, but rather to identify the most "robust" strategy, showing an acceptable performance (in terms of meeting required objectives or achieving anticipated benefits) across all relevant scenarios. Hence, the primary aim of the evaluation would be to minimize the risk of selecting a wrong strategy.

The development of decision-support methods is an important part of selecting and assessing mitigation options. Generally, they are unable to determine the best option or provide detailed option applicability or placement. They can, however, identify, examine and explore mitigation options by evaluating their relative efficiency, equity, and sustainability in determining risk levels and potential consequences. This is particularly important when selecting mitigation strategies under uncertain future conditions.

6.2 Efficiency, Equity, and Sustainability of Mitigation Options

6.2.1 Hydraulic Efficiency

The efficiency of engineering solutions consists of the degree of coastal protection they offer, and it can be evaluated specifically in terms of reduction of incident wave height, reduction (or stabilization) of sediment transport, reduction of wave run-up on the beach, and wave overtopping at the sea bank and of frequency and magnitude of inland flooding.

Results and methodologies to estimate both wave transmission and coastal erosion are reported already in Chapter 3 for some kind of defenses and interventions. The use of numerical and physical models may help to predict in extreme and ordinary meteo-marine climate conditions the hydromorphological consequences of each engineering mitigation option and their suitability to accomplish the overall design objectives (Burcharth, Hawkins, Zanuttigh, & Lamberti et al., 2007; Zanuttigh et al., 2005).

Estimated waves and currents allow, for instance, evaluation of the following:

- the near-shore wave transmission;
- the wave run-up on the beach and the overtopping at the sea bank;
- the water residence time inside the protected cells to assess water recirculation and quality;
- the current patterns and intensities, in particular at gaps and roundheads, to verify far field erosion and bathing safety; and
- the wave loads at structure, to verify structure stability and to assess possible detachment of colonizing organisms.

Estimated sediment transport allows, for instance, evaluation of the following:

- beach reshaping because of intense storms, and therefore the error that may be accounted for when modeling the flood process without modeling the beach retreat during the storm;
- the "global" sand volume balance for the protected cells the required frequency and quantity of periodic nourishments;
- the formation of local scour that may produce structure instability to redesign a proper toe protection or structure extension; and
- the erosive/depositional patterns and their rate to identify the level of disturbance to the assemblages.

Even in case detailed modeling results are available, there is the need to synthesize hydraulic maps, historical data, and modeling outputs through qualitative and/or quantitative indicators. This offers the chance to combine hydraulic effects with social, economic and ecological effects to be presented in Sections 6.2.2–6.2.4.

Different methods are available in the literature to qualitatively assess the hydraulic vulnerability of coastal stretches to erosion and flooding, including also the presence of structures and defenses. The approaches by EUROSION (2004, 44 pp.) and by Gornitz, Daniels, White, and Birdwell (1994, pp. 327–338) are selected because of their simplicity and adaptability to diverse coastal environments.

The method proposed by EUROSION (2004, 44 pp.) is based on a simplified Driving forces—Pressure—State—Impact—Responses model (DPSIR) and aims at representing the current and future pressure factors as well as the potential impact of erosion and flooding to assets located in the coastal areas. To provide a proxy of the areas that may be affected, it introduces the concept of the Radius of Influence of Coastal Erosion (RICE). By accounting for sea-level rise, subsidence and other parameters such as tide, extreme sea storms, bottom, and shoreline morphology, the RICE is defined as all areas within 500 m from the shoreline extended to areas lying below −5 m. Once the RICE is defined, the approach considers 13 indicators, nine being related to pressure and four related to impact of coastal flooding and erosion in present and future conditions (Table 6.1). Indicators may have to be properly tuned considering the specific site under analysis (Martinelli, Zanuttigh, & Corbau, 2010).

The method by Gornitz et al. (1994, pp. 327–338) defines objective and univocal criteria for the classification of littoral vulnerability, providing also useful indications for integrated management plans (Simeoni, Tessari, Gabbianelli, & Schiavi, 2003)

A given coastal stretch is first described through a set of variables belonging to the following main groups:

- marine-weather conditions, that have a significant importance for characterizing the hazard to the extreme event;
- physical conditions (shoreface width, width and height of the emerged beach, mean grain size, and pressure of the use) that characterize the beach system in terms of accommodation and mitigation capacity against the inundation;
- coastline evolution (recent and historical shoreline trends, shoreface evolution) that indicates the beach system behavior within the short and long term;
- subsidence of the coastal territory that affects coastal submersion; and
- typologies of the defense structures along the coast and inland that identify the beach passive mitigation answer to flood.

The degree of submersion tendency or vulnerability is then synthesized by the use of indices, specifically:

- The potential vulnerability V_p. It expresses the degree of geomorphological, sedimentary, and anthropogenic tendency to submersion without accounting for the existing protection. The evaluation of V_p is done through a multiple regression such as:

$$V_p = v_1 k_1 + v_2 k_2 + v_3 k_3 + \ldots + v_n k_n \tag{6.1}$$

  where v is the codified value of the variable following a proper division into classes and k is the variable weight according to its contribution to the system vulnerability;
- The total efficiency of the structures IED is defined as:

$$\mathrm{IED} = \mathrm{IES} + \mathrm{D}_i \tag{6.2}$$

  where IES is the Efficiency and Stability Index for natural dunes and D_i is the efficiency of each defense structure.

  The efficiency of each defense structure D_i is represented by:

$$\mathrm{D}_i = d^* V_{\mathrm{pmax}} / e_{\mathrm{max}} \tag{6.3}$$

TABLE 6.1 Indicators Used by the EUROSION (2004) Approach

Methodology for Rating European Regions in Terms of Coastal Erosion and Flooding			
Indicator	**0 points**	**1 point**	**2 points**
Pressure Scoring			
1. Relative sea level rise (best estimate for the next 100 years)	*<0 cm (per region)*	*Between 0 and 40 cm (per region)*	*>40 cm (per region)*
2. Shoreline evolution trend status	*Less than 20% of the shoreline is in erosion or in accretion (per region)*	*Between 20% and 60% of the shoreline is in erosion or in accretion (per region)*	*More than 60% of the shoreline is in erosion or in accretion (per region)*
3. Shoreline changes from stability to erosion or accretion between the 2 versions (CORINE Coastal Erosion (CCEr) and European Coastal Erosion Layer (CEL))	*Less than 10% of the shoreline changes between the 2 versions (CCEr and CEL)*	*Between 10% and 30% of the shoreline have changed between the 2 versions (CCEr and CEL)*	*More than 30% of the shoreline have changed between the 2 versions (CCEr and CEL)*
4. Highest water level	*Less than 1.5 m*	*Between 1.5 and 3 m*	*More than 3 m*
5. Coastal urbanization (in the 10-km land strip)	*Urban areas (in km^2) have increased of less than 5% between 1975 and present*	*Urban areas (in km^2) have increased of 5–10% between 1975 and present*	*Urban areas (in km^2) have increased of more than 10% between 1975 and present*
6. Reduction of river sediment supply (ratio)	*Ratio between effective volume of river sediment discharged and theoretical volume (i.e., without dams) is superior to 80%*	*Ratio between 50 and 80%*	*Ratio is less than 50%*
7. Geological coastal type	*>70% of "likely nonerodable" segments*[1]	*"likely nonerodable segments" between 40 and 70%*	*<40% of "likely nonerodable segments"*
8. Elevation	*<5% of the region area lies below 5 m*	*Between 5 and 10% of the region area lies below 5 m*	*>10% of the region area lies below 5 m*

TABLE 6.1 Indicators Used by the EUROSION (2004) Approach—cont'd

Methodology for Rating European Regions in Terms of Coastal Erosion and Flooding			
Indicator	**0 points**	**1 point**	**2 points**
9. Engineered frontage (including protection structure)	*<5% of engineered frontage along the regional coastline*	*Between 5 and 35% of engineered frontage along the regional coastline*	*>35% of engineered frontage along the regional coastline*
Impact Scoring			
10. Population living within the RICE	*<5000 inhabitants per region*	*Between 5000 and 20,000 inhabitants per region*	*>20,000 inhabitants per region*
11. Coastal urbanization (in the 10-km land strip)	*Urban areas (in km^2) have increased of less than 5% between 1975 and present*	*Urban areas (in km^2) have increased of 5–10% between 1975 and present*	*Urban areas (in km^2) have increased of more than 10% between 1975 and present*
12. Urban and industrial living within the RICE	*<10% of the land cover within the RICE is occupied by urban and industrial areas (per region)*	*Between 10 and 40% of the land cover within the RICE is occupied by urban and industrial areas (per region)*	*>40% of the land cover within the RICE is occupied by urban and industrial areas (per region)*
13. Areas of high ecological value within the RICE	*<5% of areas of high ecological value within the RICE per region*	*Between 5 and 30% of areas of high ecological value within the RICE per region*	*>30% of areas of high ecological value within the RICE per region*

where d = the original value of the class of the structure, V_{pmax} = theoretical maximal V_p, and e_{max} = maximal class of efficiency relative to the structure. The values of e_{max} are assigned based on structure position (in the sea, on the beach, etc.), typology (groins, breakwaters, barriers, etc.), and construction material (rocks, concrete, etc.). The value d is attributed based on the presence of the defense in the given coastal stretch;

- The IES for natural dunes is defined as

$$\text{IES} = \sum V_i/n \tag{6.4}$$

where V_i represent the dune-related variables: dune crest height, dune seaward slope (=height to seaward width ratio), dune system long-shore continuity (=dune long-shore length to reach long-shore length ratio), conservation state (presence of breaches, blow outs), and vegetal cover (=covered surface to whole surface ratio). The value of n is given by the sum of the maximum values attributed to the variables and is used to normalize IES in the range 0–1.

Finally, the actual vulnerability V_r of each coastal stretch is obtained by considering that the potential vulnerability is mitigated by natural and artificial defenses.

$$V_r = V_p - \text{IED} \tag{6.5}$$

The degree of risk (R) depends both on actual vulnerability V_r and on socio-economic value E of the same area:

$$R = V_r \cdot E \tag{6.6}$$

where E represents the evaluation (through defined figures) of social, economic, and natural values of the exposed zones or the cost in monetary terms of direct and/or indirect impacts. According to the results and to the global characteristics of the coastal area under examination, the values of the risk may be finally divided into categories.

A quantitative analysis requires the definition of quantitative indicators, such as the flooded area to the examined area ratio, the beach retreat to the beach width ratio, the wave transmission coefficient, etc. These parameters can be obtained from simplified and detailed numerical models. Threshold values should be defined that of course are site-specific. An application of this type of is presented in Section 7.6, where hydraulic, social, economic, and ecological indicators are derived from simulations of different scenarios and defense strategies performed with THESEUS Decision Support System (see Section 6.6).

Besides the assessment of the effects induced by a coastal structure, the reliability of the structure itself should be considered and a scenario analysis of the probability of defense failures and related consequences should be considered (Naulin, Kortenhaus, & Oumeraci, 2011).

The event probability of failure, conditional on the applied load (the so-called fragility concept), is usually combined with flood system risk analysis models (Gouldby, Sayers, Mulet-Marti, Hassan, & Benwell, 2008) as the Source-Pathway-Receptor-Consequence (S-P-R-C) model outlined in Chapter 2. This type of failure can be determined by analyzing historical failure data and by probabilistic calculation of the limit states. In a real case, the loads in the limit state functions are nearly always functions of multiple variables and the problem has to be solved by means of Riemann integration or Monte Carlo simulations.

The reliability tools currently available (Kortenhaus & Kaiser, 2009; Morris et al., 2008, pp. 581–591) represent flood defense reliability as a snapshot in time. However, time-dependent processes in the hydraulic climate (e.g., water levels and wave conditions) as well as the behavior of flood defense properties (e.g., crest levels, vegetation, erosion) can lead to time-dependent defense reliability. The incorporation of such processes within a reliability analysis allows the explicit consideration of processes that may reduce (e.g., deterioration from history of loading) or increase (e.g., growth of vegetation) the structural stability of flood defenses in time (Gouldby et al., 2008). This can be extremely important when considering future flood defense reliability and may allow emergent failure processes to be revealed (e.g., the deterioration of a structure may trigger new failure mechanisms which need to be assessed).

6.2.2 Environmental Perspective

Environmental considerations relate to the ecological aspects of flooding. The selection of any mitigation option will affect local habitats, species, and ecological interactions. This can result in the loss of natural habitat, the modification of existing habitats, and/or the creation of new habitats. Habitats such as saltmarshes are naturally fairly insensitive to frequent saline intrusion, whereas others such as grazing marsh and pasture are highly sensitive prolonged seawater inundation, especially if such flooding has not previously been experienced.

To assess the vulnerability of ecosystems to changes in stresses and to disturbances, an index was adopted within the THESEUS project (THESEUS OD3.3, 2012). This provides a rapid and standardized method for characterizing vulnerability across coastal systems and identifies issues that may need to be addressed in order to reduce vulnerability. By looking at combinations of factors, ecosystem vulnerability can be assessed. Such factors are the inherent ecosystem characteristics, the natural drivers that act upon the ecosystems, human use of the ecosystem, and the effects of climate change.

Vulnerability of habitats is dependent on:

1. which part of a particular habitat area will be a subject to the unfavorable impact and which species will be affected and
2. the degree of sensitivity of habitats/key species to unfavorable impact/hazard.

Table 6.2 illustrates an Environment Vulnerability Index (EVI) based on determining to what degree the habitat is affected by flooding using a categorical method; short-term and seasonal processes are represented by categories 1 and 2; and for long-term processes it is assumed that habitats will have permanent physiological consequences (e.g., species composition or extent). To establish the thresholds, best scientific judgment is used based on published information relating to the habitats/

TABLE 6.2 Habitat Vulnerability Categories and Descriptions

	Effect of Driver on Habitat			
	Negligible	Transient Effect	Moderate Effect or Semipermanent Change	Permanent Effect/Change
Environmental vulnerability index	0	1	2	3
Habitat/key species	Negligible impact to habitats/ species	Changes within the range of receptor's natural seasonal variation and full recovery is likely within a season	Changes are beyond receptor's natural seasonal variation. Partial recovery is possible within several seasons, but full recovery is likely to require human intervention	Changes are so drastic that natural recovery of receptor is very unlikely without human intervention

species affected together with experiments, where appropriate. The advantage of using an index is that allows key step changes or tipping points (Scheffer et al., 2009; see also Rietkerk, Dekker, de Ruiter, & van de Koppel, 2004) in a habitat properties to be captured because it is these dynamic changes of an ecosystem that is of interest rather than the transition from one ecosystem to another.

The primary consideration should be the socioeconomic need to defend a coastline cross-referenced with the options that might effectively be used as part of an appropriate defense strategy. In other words, it is unlikely that the natural environment per-se will be a driver for coastal defense. Rather, it will be affected by defenses and management and the type and extent of such effects can be influenced by proper consideration of alternative options from an environmental viewpoint. In the broadest sense these options include do nothing (Hoggart et al., in press), the use of soft defense options such as saltmarshes (Bouma et al., 2013) or sand dunes (Hanley et al., in press), and incorporation of habitat into hard defenses (Firth et al., 2014).

Before reaching decisions, it is important from an environmental perspective to quantify on a case-by-case basis the ecological consequences of the various defense options. This needs to include an assessment of existing habitats and species in terms of local, national, or international rarity; potential ecosystem goods and services contributed (including any flood defense services); and the value of the natural habitat for tourism or for commercially important species. There needs to be an accompanying assessment of habitat and species under the modified scenarios resulting for each of the different defense options. Weighting the various options is to some extent a political decision which will inform the tradeoffs between the relative importance of the habitats and species lost versus those gained. From a purely ecological perspective, one might say that it is not possible to improve on nature so any change to the natural environment will be deleterious. However, the majority of our coastlines and inshore waters, and most of those in areas of interest for coastal protection, are already heavily modified by urbanization, fishing, tourism, and a range of other challenges, so due consideration of the defense options may allow opportunities to help mitigate for other problems. For example, to increase the spatial extent of a relatively rare habitat such as a saltmarsh while at the same time offering coastal protection or to provide habitat on a hard engineering structure for species that are locally rare or exploited.

It is essential to recognize there is no universal best option and a case-by-case ecological assessment needs to be made. Alongside this, it is also important to be aware of the limitations to any predicted outcomes. The natural environment is inherently very variable, influenced by natural variations in environmental factors including seasonality and long-term weather patterns, and often leading to stochastic consequences for biota such as interannual variability in recruitment of juveniles. In addition, there are a wide range of other anthropogenic factors, such as climate change, eutrophication, and species invasions, acting in isolation and in combination to influence the natural environment. Hence, although our ability to anticipate generic outcomes is good, it can be very difficult to make precise predictions on species composition and abundance or the timescales associated with some of the anticipated ecological changes.

6.2.3 Economic Perspective

Coastal areas provide a diverse array of goods and services that directly or indirectly translate to economic services and values to the human population. Any interventions in terms of flood risk reduction measures will interfere with the delivery of these goods and services. Some may be enhanced and others adversely affected. In economic terms, what we seek are efficient solutions to these risk problems at the coast, namely those measures where the outputs in terms of risk reduction are maximized in relation to the costs of those intervention measures (i.e., the "inputs" necessary for that intervention).

Taking a broad view of impacts, these include both tangible flood damages, indirect flood losses (i.e., the disruption of communication and other links in the economy), and so-called "intangible" effects in terms of the trauma and health impacts of flooding and the disruption of people's lives during the recovery period. However, many of the values associated with beneficial functions are difficult to assign a monetary value and are therefore consequently often ignored in planning and decision making. This can often lead to decisions that turn out to be unwise. But research has developed approaches to circumvent this situation, and as a result stated preference approaches (i.e., the construction of the market for the goods through the use of questionnaires) can be used to establish some appropriate valuations. These approaches include the Contingent Valuation Method (CVM) and Choice Experiments (CE) that provide economic valuations of preferences so that the costs and benefits of different policy designs can be compared.

In terms of cost, we need to think about the capital costs of the intervention measures and the maintenance costs for continuing to see the measures maintain their effectiveness. Each of these costs has a different time profile, with capital costs typically front-ended to pay for an intervention measure, and maintenance costs following behind. But that is a traditional engineering approach to flood risk management at the coast, whereas the costs of nonstructural flood risk reduction measures such as warning systems or spatial planning have a more even profile during their lifetime. Notwithstanding this point, some of the costs of nonstructural flood risk management measures are difficult to disentangle from their attribution to other parts of society. For example, spatial planning at the coast is not just about flood risk reduction, but also about environmental protection, landscape conservation, heritage protection, and economic regeneration efforts. The whole cost of the spatial planning system cannot be attributed to flood risk management, but rather should be apportioned between the different objectives of that spatial planning system.

Similarly, many of the benefits of flood risk reduction measures are not just the flood risk reduction achieved. The building of promenades that incorporate sea walls to defend against flooding at coastal locations brings recreation and leisure opportunities as benefits that need to be counted. Coastal defense structures may also enhance navigation at the coast, or in estuarine situations, and bring improved port facilities. The artificial nourishment of beaches, to provide protection to the sea walls behind them and therefore protection to urban areas behind those walls, brings opportunities and benefits to those seeking seaside recreation. Indeed, one of the principal benefits or coastal protection is the construction of such recreational resources. Quantification may be difficult, but that does not deny the importance of the benefit obtained.

What these examples show, of course, is that flood risk reduction measures at the coast should not be considered in isolation from other economic and social functions active at the relevant locations (ICE and RIBA, 2010). In simple terms, making coasts safer brings other advantages as well as safety. A comprehensive economic evaluation of the sustainability of flood risk mitigation options need to take all these considerations seriously.

But the choice of intervention measures is not just about economics. Economists, in this respect, tend not to consider the distributional consequences of investment decisions. Rather they concentrate on the efficiency of those intervention measures. But the relevant stakeholders will be interested in distributional effects, in terms of who pays for intervention measures and who benefits from them. This raises the question of equity and "fairness." In this regard, there are "procedural fairness" considerations, as to whether stakeholders' concerns are all considered within the process of making decisions about intervention measures, and "outcome fairness" considerations, for example in terms of whether an equitable degree of risk reduction is provided to all the stakeholders at risk. Such social justice considerations are relatively new in flood risk management (Johnson, Penning-Rowsell, & Parker, 2007), but reflect a growing realization of the substantial cross-subsidies between those who pay for these measures and those who benefit from them.

Flooding is a natural event that occurs in variable social, economic, and governance settings and it is the setting that determines planning response strategies. This includes situations where, despite the potential consequences, a flood is either not perceived as a significant risk or is a risk worth taking (Aven & Renn, 2010). This has been largely ignored in technical risk estimates (Burns, 2007; Renn, 2008), yet it determines what are appropriate, acceptable, and realistic management decisions over both the short and long term. It is therefore essential to involve the full range of stakeholders and the public when making decisions on flood management (de Boer, Wardekker, & van der Sluijs, 2010; Dilling & Lemos, 2011).

This has the advantage of (1) understanding what is important to those "on the ground," (2) building an appreciation and/or understanding of the flood system and the steps required to manage it—those affected will participate most effectively and accept decisions if they understand the purpose of the process, and (3) encouraging the scientific community to communicate in an effective manner, particularly as public perception of risk is often not driven by facts, or by what is understood as facts by risk analysts and scientists (Renn, 2008).

6.2.4 Society's Perspective

Social vulnerability is a complex phenomenon and no single measure comprehensively covers the whole spectrum of how vulnerability is manifested (Adger, Brooks, Bentham, & Agnew, 2004; Adger, Hughes, Folke, Carpenter, & Rockström, 2005).

The Social Vulnerability Index provides a comparative spatial assessment of human-induced vulnerability to environmental hazards (Cutter, Boruff, & Shirley, 2003; Wisner, Blaikie, Cannon, & Davis, 2004). The Social Vulnerability Index is based on a large set of measurable variables that can be grouped into major common factors such as: population structure, gender, income, socioeconomic status, and renters (www.csc.noaa.gov/slr). Analysis and mapping of social vulnerability should also consider critical facilities or resources (such as schools, hospitals, and transportation) to help prioritize potential hazard mitigation.

More recently, social vulnerability is modeled in Decision Support Systems (see Section 6.6) considering two main aspects: (1) the damages to critical facilities (CFs) and (2) the expected number of fatalities. Flood damages to society also include psychological consequences that are mainly qualitative in nature and are hard to be translated in linear functions with quantitative outputs for practical and ethical reasons (Tapsell, 2011). Continued development of social vulnerability methods is to be expected.

CFs are defined as "the primary physical structures, technical facilities and systems which are socially, economically or operationally essential to the functioning of a society or community, both in routine circumstances and in the extreme circumstances of an emergency" (UNISDR, 2009). On the one hand, the notion has been adopted recently in disaster management, and is related to the creation of geographic information systems (GIS) maps on community vulnerability (e.g., DEFRA, 2005; FEMA, 2007); on the other hand, CFs have been applied in the development of priority lists for the effective reactivation of buildings after disasters and applied emergency management (e.g., Hillsborough County–Florida, 2009).

The impact of the flooding process on CFs is estimated following three steps.

1. Ranking of critical facilities
   In the THESEUS project (Zanuttigh et al., 2014), a rank was derived based on the function of buildings in relation to social vulnerability (Hillsborough County–Florida, 2009). Considerations were made both in terms of use in emergency management, function in ordinary activities and community aggregation, and symbolic function. The corresponding approximated social value (ASV) was derived and is reported in Table 6.3, with values from 1 (low) to 5 (high). The final output is an overall view of possible intangible damages in the range 0–100. Even if it maintains high levels of uncertainty, it is one of the first attempts to provide to end users the possible effects of floods on the community and individuals.
   The ASV also provides a reactivation list in reverse order because the highest values are supposed to receive priority in emergency interventions for reducing social damages. From the perspective of land use planning, the adoption of such an approach should lead to the identification of possible relocation of high scoring buildings to safer areas or encourage measures to increase building's resilience. Similarly, higher scores indicate where efforts for higher education and training of personnel should be concentrated and where emergency measures such as mobile barriers could be deployed with maximum effectiveness.
2. Estimation of physical damage for structures
   The damage scale is estimated based on flood depth and duration.
   Following the method by Schwarz and Maiwald (2008), the damage grade is related to the flood depth (De) through a nonlinear function.
   Intuitively, the effects on society and structures are inversely proportional to flood duration (D), if one excludes flash flood phenomena. Long-duration floods, even

TABLE 6.3 Ranking Values and Factors Required to Estimate the CSD

ASV	Associated Social Vulnerability Factors: Definition
5	Critical structures that if involved could compromise the emergency action, the coordination chain, public safety and public health in the long term. For example, hospital and emergency facilities. Depending on local features, main military facilities, power plants, and institutions can be included in this category
4	Facilities that provide significant public services and should be activated within 24 h. For example, nurseries, major water and sewer facilities, fire and police stations, schools, and park facilities used to support critical purposes can be included
3	Facilities that provide important public services but should be sequent to critical facilities are ranked 4 and 5 points. Main centers of aggregation, education, or prayer that are important for symbolic belonging to the community. Some particular place that links those features to economics can be included, too
2	Facilities that provide public services but that are less critical for the community. Common storage areas and sport centers can be included depending on the context. Literature on social capital can be used also as reference
1	Places which value are mainly symbolical, but can influence anyway the overall amount of social damages. For example, particular community areas of meditation and prayer
	Depth-Induced Damage
Factor De	Depth range from Schwarz and Maiwald (2008): has to be adapted to the site
1	0.1–0.5 m
2	0.6–1.5 m
3	1.6–2.5 m
4	2.6–5 m
5	>5 m
	Duration-Induced Damage
Factor D	Flood duration
1	Hours
2	Days
3	Weeks
	Seasonality
Factor S	Definition
1	Low seasonality
2	High seasonality

(Continued)

Table 6.3 Ranking Values and Factors Required to Estimate the CSD—cont'd

	Collateral Social Damage Scale
Score	Definition
0	No collateral social damage
1–10	Possible malfunctions in citizen's ordinary life are possible but can be prevented. The damage is limited and could be managed with experimented procedures and stakeholders activation. The situation could require more details about which critical facilities involved and planning of alternative solutions
11–20	Malfunctions in citizens' life are expected. The damage is still limited but diffused (or high and very concentrated), and requires higher mobilization for the rehabilitation process
21–30	Social damages are concrete and visible. A major involvement of local relief and reprise resources is expected. The presence of external help is suitable and should be activated in advance to avoid higher losses
31–50	Massive social damages in ordinary period or medium involvement of critical infrastructure in high touristic period. Massive damages could be managed with timing alert and planning, but the presence of external help is absolutely needed. Long times for reactivation of services and community reprise should be prevented
51–100	Exceptional damages, calamity. The situation could have terrible social damages and should be mediated with external help and cooperation at the highest level possible. Very long times for reactivation of services and community reprise should be prevented

if relatively limited in space, produce greater impacts on social functions: a bridge blocked might be a nuisance for an hour, whereas it could compromise trade routes or tourism activity if blocked for a week. Therefore the following scenarios (corresponding to different scores, see Table 6.3) should be considered: (1) short D (hours), (2) medium D (days), and (3) long D (weeks).

3. Definition of touristic impact

The geographic features that determine the social vulnerability are related both to the physical structures and to the situation where the action is settled (Cutter, 1996). In many coastal areas, one of the most relevant variables affecting the ordinary social pattern should be considered the presence of tourism. It can be presumed that not all the tourists have previous experiences in flooding, and that if a flood happens with a large number of tourists in place, critical infrastructures may suffer higher pressure and warning messages may face more problems in their dissemination. The tourist presence can be represented through a value reflecting seasonality S; this factor will act as a final scale multiplier, where low season ($S = 1$) denotes ordinary conditions and high season ($S = 2$) implies that the effects will be exacerbated. Timing is also key and, for example, the seasonality of tourism and flooding should be considered, which may not coincide.

The collateral social damages (CSD) are finally estimated as:

$$\text{CSD} = \Sigma_I \text{ASV}_i \cdot \text{De} \cdot \text{D} \cdot \text{S} \lim_{x \to \infty} \tag{6.7}$$

The value of CSD is related to a common scale to allow exportability to other case studies and comparison of the results. The scale is also reported in Table 6.3.

For tangible social damages, we derived a function of life losses and injuries (NI) from Penning-Rowsell, Floyd, Ramsbottom, and Surendran (2005)

$$\text{NI} = (H^*\text{AV})/(\text{Pa} + \text{ID}) \tag{6.8}$$

where H is the hazard rate, AV is the area vulnerability, Pa is the sensitive population (age <14 years and >65 years), and ID is the number infirm/disabled/long-term sick people.

The value of H is computed in each cell of the domain as

$$H = \text{NI} \cdot y \cdot v \cdot \text{DF} \tag{6.9}$$

where N is the number of people involved in the flood, y is the flood depth, v is the flood velocity, and DF is the debris factor equal to 1 for the Mediterranean and 2 for the ocean.

The area vulnerability (AV) is derived as:

$$\text{AV} = \text{W} + \text{Fo} + \text{Na} \tag{6.10}$$

where W denotes the warning, Fo is the speed of onset of flooding and Na is the nature of the flooded area; see Table 6.4.

The value of Na can be derived from statistical demographic data or can be schematized based on Penning-Rowsell et al. (2005). If statistical data are available, their main use should be identified and impact levels from 1 (low) to 3 (high) are attributed as shown in Table 6.5. Because social patterns determine the impact levels of special attributes, three main scenarios were identified: day, night, and touristic periods. Higher impact was attributed to residential areas when people are generally at home sleeping (night), whereas zones identified for schools and education are vulnerable when children are in classes (day). Finally, tourist resorts are most susceptible during holidays (touristic period).

The percentage of the population aged (Pa) can be derived from demographic data (ISTAT, 2009 for the case of Cesenatico, presented in Section 7.6) or referred to national middle average. The final value of Pa should be conformed to a common value of 50 as: NPa: $\times 50 = \text{Pa}:50$, $\times 100 = n\text{Pa} * (100/\text{Pa})$.

The percentage of ID can be set based on perception or on the national average.

TABLE 6.4 Ranking Values and Factors Required to Estimate Life Losses and Injuries

W	Not Present	Present But Not Implemented	Present and Working Well
	3	2	1
So	Slow flooding (many hours)	Gradual flooding (an hour or so)	Rapid flooding
	1	2	3
ID	Low presence	Medium presence	High presence
	10%	25%	50%
Na	Touristic season	Day	Night
Residential area	2	1	3
Tourist area	3	2	1
Manufacturing	2	3	2
Common or religious area	2	3	1
Education area	1	3	1
City center	3	3	3
Parking and green	1	1	1

TABLE 6.5 An Eight-Step Process to Inform the Compliance of New Modification with the Water Framework Directive

Step	Action
1	Collect up-to-date water body baseline data
2	Collect proposed scheme baseline data
3	Preliminary assessment
4	Design and options appraisal
5	Detailed impact assessment
6	Apply Article 4.7 tests
7	Reporting
8	Follow-up post-project appraisal work

(Source: EA, 2011.)

The values for the ID factors are synthesized in Table 6.4. In general, this function provides an overall count of people that could be subject to death or injuries. These two aspects were not distinguished because too many external variables such as local lifestyle, wealth, or public health services influence the final output of life losses, and the uncertainties are high.

For economic vulnerability analysis, major sectors of economy and the primary centers of activity in those sectors need to be identified. These economic centers are areas where flooding can have major impacts on the local economy.

An EVI was proposed (Guillamont, 2009), based on the composition of the following seven indicators: (1) population size; (2) remoteness; (3) merchandise export concentration; (4) share of agriculture, forestry, and fisheries in gross domestic product; (5) homelessness resulting from natural disasters; (6) instability of agricultural production; and (7) instability of exports of goods and services.

However, within a multicriteria analysis (MCA; see Section 6.3.2), where social and economic impacts must be distinguished and separately weighted, this index turned out to be inadequate because it combines social and economic indicators. Instead, where detailed data on economic activities in gross domestic product terms are available, a consistent approach based on incomes for each economic land use can be adopted (e.g., hotels are evaluated in terms of annual gross domestic product, houses are evaluated in terms of annual rents, and beaches are evaluated in terms of annual willingness to pay to preserve them).

The overall economic consequences (EC) of flood in terms of flood depth and flood duration can be estimated by applying the following expression (Zanuttigh et al., 2014):

$$\mathrm{EC} = \mathrm{vij} \cdot \mathrm{bj} \cdot \mathrm{Fd} + \mathrm{vij} \cdot \mathrm{aj}\sqrt{\mathrm{Fy}} \tag{6.11}$$

where vij are the values of land uses in euro/m^2/year from census statistical data; Fd is flood duration and Fy is flood depth; aj are proportionality constants as functions of Fy that are normalized for each land use j at the maximum value of Fy in a given year for a storm with a fixed return period Tr, assuming different reference percentage of damage depending on the use; and bj are proportionality constants as functions of Fd that express the expected period to restore economic activities as a factor of duration, depending on the land use and are normalized to annual incomes with the days/year. Note that flood velocity is not considered and hence is assumed to be irrelevant.

Beach losses can be derived by combining the beach value function with the beach loss because of erosion. The value function can be derived from specific surveys, see for instance the surveys carried out within the DELOS (Zanuttigh et al., 2005) and THESEUS projects (Diaz et al., in press, Zanuttigh et al., 2014).

Alternatively, a consistent approach based on market values of infrastructures may be used. Note that it is theoretically possible to move from an income approach to an infrastructure approach under a standard set of assumptions about market competition.

6.3 Evaluation of Mitigation Options

To evaluate costs and benefits from a pool of different mitigation options, valuation methods are required (both marketed and nonmarketed). There is a range of methods and techniques that can help decision-making concerning investment appraisal, and cost-benefit analysis (CBA) is one of them (DEFRA, 2009). Similarly, there is a range of techniques that can be used to quantify the environmental impacts of a number of policy options. This is an important component of the data that the decision maker requires.

An important theoretical approach for capturing and describing costs and benefits is the total economic value (TEV) framework. This tool considers the full range of impacts a mitigation option has on human welfare. The way to derive TEV is from preferences of individuals. Such preferences are elicited using stated preference methods and revealed preference methods (see Figure 6.1). Revealed preference methods use data regarding individuals' preferences for a marketable good based on market-based and surrogate market-based methods. Surrogate market-related methods include travel cost methods and hedonic pricing. Stated preference methods use structured questionnaires to elicit individuals' preferences for a given change in a natural resource or environmental attribute. The CVM and CE are included in this category. The CVM is based on the development of a hypothetical market or scenario in which the respondents to a survey are given the opportunity to state their willingness to pay (WTP) or willingness to accept. Because WTP and willingness to accept values are contingent on the hypothetical market, this method is called CVM (Koundouri, Dá;vila, Stithou, & Stuitver, 2013).

In a CE framework, the good in question is broken down into its component attributes. Then, a set of combinations of such attributes and levels is presented to respondents and they must state their preferred combination. (Bennett & Adamowicz, 2001; Birol & Koundouri, 2008) Flooding, land loss, and their impact on water resources are important sources of concern.

An example of how to use this approach is found in Díaz-Simal, Koundouri, Rulleau, and Remoundou (2013). They used a choice experiment to elicit the WTP for avoiding climate change challenges (i.e., environmental and health risks in marine environments) via the payment for mitigation measures. The experiment was implemented in Santander, Spain (Section 7.9), a coastal region with vulnerability to marine dynamics and the effects on its beaches (and their role as crucial locations for social and touristic activities), loss of marine biodiversity, and a surge in exposure to medusas and other dangerous species present on the beaches that have motivated restrictions of bathing activities due to health risks. They followed a split-sample approach to elicit the value people place on improvements in biodiversity and recreational opportunities and reductions in the health risks associated with the presence of jellyfish species in the short, medium, and long run.

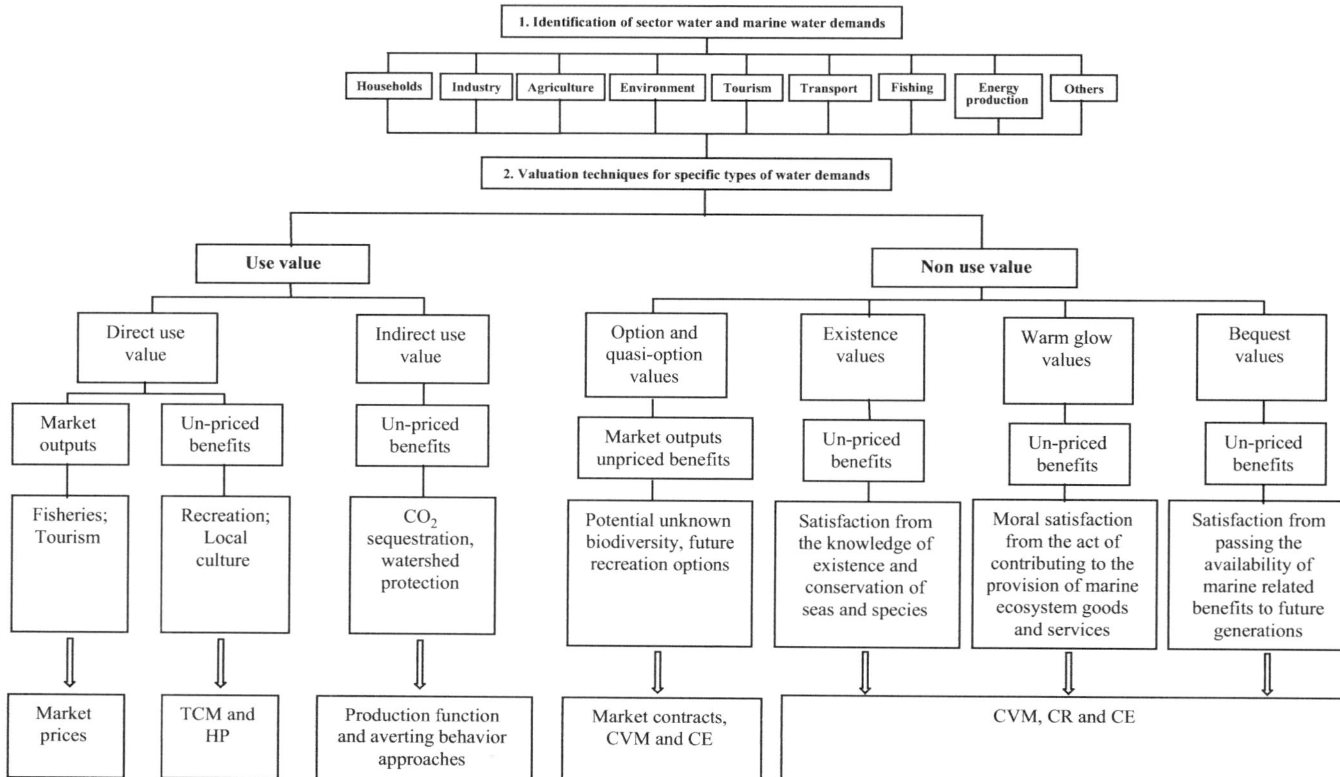

Figure 6.1 Economic valuation methods.
Source: Adapted from Pearce and Moran (1994) and Remoundou et al. (2009).

In a CE, the good under valuation is described in terms of its characteristics, attributes, and the levels these attributes take (Bennett & Blamey, 2001, p. 269). Price is usually included as an attribute because this allows the evaluation in monetary terms of the marginal value of the other attributes. Then, the respondents are requested to choose their most preferred option among different combinations of levels of attributes that are shown to them. The results show that people place a positive value on increased biodiversity and recreation opportunities in all the considered time frames and imply that the present value of future biodiversity and recreation related benefits increases with the time frame.

Another important finding is that people are willing to accept health risks from the presence of jellyfish. The monetary estimations under their exercise could inform the assessment of a long-run CBA to investigate whether different planned mitigation measures are economically efficient.

CBA assesses the monetary social costs and benefits of an investment project over a time period in comparison to a well-defined baseline alternative. In this way, the costs and benefits of a project are evaluated and compared and the long-run economic efficiency of implementing such project is assessed. In this framework, the estimated economic values are aggregated over their relevant populations and added to capture the TEV generated by the investment project. If the total benefits exceed the total costs of the project, then it is considered to be profitable (Koundouri et al., 2013). CBA is part of a more general procedure named environmental assessment. This is because CBA is concerned with a particular "product" (the cost-benefit ratio or net present value as the measure of the economic return from that investment). Environmental assessment is first and foremost more concerned with a process of incorporating information on all environmental attributes, values, and changes into the decision-making sequences.

On the other hand, MCA attempts to quantify in some way all aspects of environmental and economic significance related to a particular decision, and weight them so that a simple range of indices can be developed that capture all adverse and beneficial results—or potential adverse and beneficial results—from an investment decision (DEFRA, 2009). Therefore:

- CBA can provide a sophisticated means of comparing very different investments and outcomes by reducing them all to a common monetary form. It is limited to consideration of those impacts to which a monetary value can be attached but it leads to a simple set of parameters on which choices can be made (e.g., the benefit-cost ratio);
- MCA can be used to broaden the scope of analysis, but brings significant difficulties in terms of determining the appropriate weights to use or the different criteria involved;

- Environmental assessment is broader still, but lacks the precision (or apparent precision) of CBA, and can avoid the kind of discipline involved in quantification that both CBA and MCA bring to the decision-making process;
- Decision makers should not reject any vehicle which clarifies the decision that they are to make, but excessive simplification brings dangers as well as advantages.

6.3.1 Cost-Benefit Analysis

As mentioned in the previous introduction to Section 6.3, CBA is a systematic process for calculating and comparing benefits and costs of a project, decision, or government policy (hereafter, "project"). CBA has two purposes:

1. To determine if the project is a sound investment/decision (justification/feasibility).
2. To provide a basis for comparing projects. It involves comparing the total expected cost of each option against the total expected benefits, to see whether the benefits outweigh the costs, and by how much.

CBA is related to, but distinct from cost-effectiveness analysis. In CBA, benefits and costs are expressed in monetary terms, and are adjusted for the time value of money, so that all flows of benefits and flows of project costs over time (which tend to occur at different points in time) are expressed on a common basis in terms of their net present value (NPV). If the project is expected to have long-run impacts on the local economy and ecology, its sustainability is to be tested using a long-run cost CBA, and the NPV of the project is estimated using different discount rate schemes (Birol, Koundouri, & Kountouris, 2010). The NPV results reveal whether the net benefit generated by the investment project is positive and significant well into the future. CBA helps to predict whether the benefits of a policy or a set of measures outweigh their costs, and by how much relative to other alternatives (i.e., one can rank alternate policies in terms of the cost-benefit ratio). Generally, accurate CBA identifies choices that increase welfare from a utilitarian perspective. Assuming an accurate CBA, changing the status quo by implementing the alternative with the lowest cost-benefit ratio can improve Pareto efficiency. An analyst using CBA should recognize that perfect evaluation of all present and future costs and benefits is difficult, and although CBA can offer a well-educated estimate of the best alternative, perfection in terms of economic efficiency and social welfare are not guaranteed.

The following is a list of steps that comprise a generic CBA.

1. List the alternative risk management measures at the coastal site.
2. List stakeholders that need to be involved.
3. Select value measurement(s) and measure all cost/benefit elements.
4. Predict outcome of cost and benefits over relevant time period.

5. Convert all costs and benefits into a common currency.
6. Apply a discount rate.
7. Calculate net present value of project options.
8. Perform sensitivity analysis.
9. Adopt the recommended choice of measures.

Uncertainty in CBA parameters (as opposed to risk of project failure, etc.) can be evaluated using a sensitivity analysis, which shows how the results—and hence the recommended choice of measures—respond to parameter changes.

The two economic criteria to be used in comparing different options are:

1. Cost-benefit ratio: the ratio of the present value of all of the streams of benefits over the present value of all of the streams of costs; and,
2. NPV: the difference between the present value of all of the streams of benefits and the present value of all of the streams of costs.

Brent (1990) provides the following rules for the use of the cost-benefit ratio and NPVs:

1. When the options are mutually exclusive: e.g., in a flood risk management project with options with, say, different standards, select the project with the highest NPV; and
2. When there is a budget constraint: e.g., when the issue is which projects to include in a national program, select the projects that maximize the benefit-cost ratio of the program as a whole.

6.3.2 Multicriteria Analysis

Multiple-criteria decision-making or MCA explicitly considers multiple criteria in decision-making environments rather than the single criterion of NPV. The reason for this approach is that there are typically multiple conflicting criteria that need to be evaluated in making decisions. Cost or price is usually one of the main criteria; some measure of quality is typically another criterion that is often in conflict with cost.

A key aspect of this approach is the need for greater consideration of social and environmental impacts within appraisal. MCA is used to factor in societal and the environmental considerations in flood risk management appraisal, in addition to the traditional focus on protecting against property damage from flooding. As with CBA, the approach also seeks to place value on the environmental and social impacts and benefits of flood risk management options. It remains an economic approach, however, to the extent that these wider impacts are evaluated in terms of the "worth to society" as expressed often in monetary terms. Factoring in social and environmental outcomes can change the view of which options have the biggest benefits compared with the more traditional CBA. The difficulty of the problem originates therefore from the presence of more than one criterion. There is no longer a unique optimal solution to an MCA problem that can be obtained without incorporating preference information and normally one has to "tradeoff" certain criteria for others.

There are many different approaches to MCA, and no one is necessarily superior to another. Some approaches seek to convert environmental values into monetary values, whereas others avoid this approach and simply score and weight particular aspects of the resource problem being investigated. Some have a mixture of methods, with some criteria being evaluated in monetary terms, whereas the others are left unquantified. This diversity of methodologies should not be seen as a weakness, but as a strength of this approach to decision making.

A scoring and weighting methodology can be used to estimate monetary values for those impacts that are difficult to measure in monetary values (EA, 2010). It has been developed for use in UK flood and coastal erosion risk appraisals. The method is based on three main steps:

- Step 1: scoring of project impacts;
- Step 2: weighting of these impacts; and
- Step 3: calculating implied monetary values for the intangible impacts.

This includes a step 4 of verifying the implied values.

After identifying the decision alternatives and the relevant criteria to be assessed a full MCA includes, scoring, weighting, and finally the combination of these factors into an overall value for each alternative (Communities and Local Government, 2009).

The methodology forms part of an approach to project appraisal, relying on options that have been identified elsewhere and impacts (benefits and damages) that have been described using an Appraisal Summary Table, see EA (2010). The approach requires scores to be assigned to each option and weights to be assigned to each category of benefit. Once all the scores and weights have been assigned, implied monetary values can be calculated as a method for estimating the value of the intangible benefits (where intangible benefits are defined as those that are difficult to value in monetary terms, such as environmental and social benefits) relative to the tangible benefits (where tangible benefits are defined as those that can be readily valued in monetary terms, such as property damages avoided).

Scoring and weighting can, therefore, provide a less costly alternative to stated preference and other willingness-to-pay techniques as a method for directly eliciting monetary values for those project impacts that are otherwise difficult to estimate in monetary terms. It is therefore most useful when the intangible impacts are likely to be significant and hence influence the choice of preferred option.

Implied values could also be used alongside benefits transfer values as a method of validation/verification that the estimated benefits are reasonable. Overall, the use of implied values could help to keep the costs of appraisal down while allowing all of the benefits to be valued. The outputs from step 3, described previously, can be designed to be compatible with the current method for identifying the preferred option since they enable the NPV, cost-benefit ratio and incremental cost-benefit ratio to be calculated.

6.4 Legislative Constraints

In addition, there are a number of legislative constraints to option choice. Every country in the world has a different approach to this, designed to fit local circumstances and/or reflecting the current national legislative position regarding environmental protection. Particularly prescriptive is the situation in the European countries that are members of the European Union. Here a decade of legislative process has accorded particular status to many environmental sites, and attached to this designation fairly stringent conditions about their modification. Preeminent in this situation is the Water Framework Directive (WFD) and other European legislation, translated into national legislation by each national government, which virtually prohibits interference with particular resources such as water bodies, habitats, and related designated areas.

These examples of legislative constraints usually override consideration of economics, so that conventional economic appraisal of investment decisions becomes redundant. This is not to say that risk assessments and economics are unimportant, but the traditional approach is one that seeks a cost-effective solution, minimizing costs, which complies with the relevant legislative imperatives.

These legislative constraints are discussed before examining the economic and the deliberative methods. The involvement of stakeholders is essential within this process as opinions will vary according to the site and personal experiences. This emphasizes the point that we are making value judgments, rather than this being an exact science. However, it is possible to provide general methods to assist this judgmental process.

6.4.1 An Example of a Legislative Override—the European Directives

6.4.1.1 Water Framework Directive

The WFD is a European Directive (European Commission, 2000) that introduces a new strategic planning process designed to manage, protect, and improve the water environment (EA, 2011). The Directive means that interventions in the water environment are strictly controlled, including at the coast, and this will affect all selection criteria.

The purpose of the WFD is to establish a framework for the protection of water bodies (including terrestrial ecosystems and wetlands directly dependent on them) which aims to:

- prevent further deterioration;
- enhance their status;
- promote sustainable water use;
- reduce pollution; and
- mitigate the effects of floods and droughts.

In this context river basin management plans are statutory plans for protecting and improving the water environment. They describe the main issues for the water environment within each river basin district. They tell us, at a local level, which measures the competent authority (the Environment Agency in England and Wales) and others need to implement to achieve the objectives of the WFD.

The WFD requires organizations such as the Environment Agency in England to aim to achieve good status or potential in all water bodies. For surface waters, this means:

- good ecological status in water bodies; or
- good ecological potential (GEP) in water bodies designated as artificial or heavily modified water bodies (AWB/HMWB); and
- good chemical status.

GES is the WFD default objective for all water bodies and is defined as a slight variation from undisturbed natural conditions. This term includes both the hydrological and geomorphological characteristics that can support a healthy functioning aquatic ecosystem. GEP is the WFD objective for AWB/HMWBs and are designated for a specific uses, such as recreation, flood risk management, or urbanization. Water bodies are designated as AWB/HMWBs when the level of modification in these water bodies means the biological status is not able to achieve good ecological status or the uses for which the water body has been modified are still needed and cannot be achieved through "other means."

The AWB/HMWB designation accepts that the biological status of the water body has been impacted by its modification and so the alternative objective of GEP is set: GEP is the best ecological status an AWB/HMWB can achieve without compromising the use for which it was designated. No WFD action can be taken on these water bodies that will have a significant adverse impact on its use. So a water body that has been designated as having a flood risk management use should maintain that use. Only when all the relevant mitigation measures have been put in place can an AWB/HMWB be said to have reached GEP.

The WFD includes an obligation to prevent deterioration in the overall status of water bodies, referred to as "no deterioration." New activities such as flood alleviation schemes could lead to deterioration. This may lead to a water body failing to meet its ecological objectives.

For new Defra and Environment Agency flood and coastal erosion risk management schemes any hydromorphological impacts need to be fully assessed to establish if they will cause deterioration or prevent the achievement of ecological objectives. To do this, a WFD assessment needs to be made, for which an eight-step process has been developed by the Environment Agency in England to help assess the compliance of new modifications with the WFD (Table 6.5). Exceptionally, there may be situations in which it is not possible for a scheme to be designed to prevent deterioration in ecological status/potential. Under these circumstances, the project needs to satisfy the exemptions criteria set out in Article 4.7 of the Directive. These criteria are summarized below:

- All practicable steps or measures are taken to minimize the impact;
- The reasons for the modification are explained in the river basin management plans;
- The reasons for the modification are of overriding public interest and/or the benefits to human health, safety, or sustainable development outweigh the benefits of achieving WFD objectives;
- The benefits of the modifications cannot be achieved by another means (i.e., they are not technically feasible or are disproportionately costly).

6.4.1.2 The European Habitats Directive and the Birds Directive

The Habitats Directive, together with the Birds Directive, forms the cornerstone of Europe's nature conservation policy. It is built around two pillars: the Natura 2000 network of protected sites and the strict system of species protection. The Directive protects more than 1000 animal and plant species and 220 so-called "habitat types" (e.g., special types of forests, meadows, wetlands), which are of European importance. The protection of these sites is a legislative requirement, overriding consideration of the costs and benefits of any coastal protection measures.

The Habitats Directive (more formally known as Council Directive 92/43/EEC on the conservation of natural habitats and of wild fauna and flora) is a European Union directive adopted in 1992 as an EU response to the Berne Convention (EC 1992). The Directive led to the setting up of a network of Special Areas of Conservation, which together with the existing Special Protection Areas form the Natura 2000 network across the European Union. Article 17 of the directive requires EU Member States to report on the state of their protected areas every six years. The first complete set of country data was reported in 2007.

The Birds Directive bans activities that directly threaten birds, such as the deliberate killing or capture of birds, the destruction of their nests and taking of their eggs, and associated activities such as trading in live or dead birds, with a few exceptions (listed in Annex III; III/1 allows taking in all Member States; III/2 allows taking in Member States in agreement with European Commission).

The directive also recognizes that habitat loss and degradation are the most serious threats to the conservation of wild birds. Therefore, it places great emphasis on the protection of habitats for endangered as well as migratory species (listed in Annex I), especially through the establishment of a coherent network of special protection areas comprising all the most suitable territories for these species. Since 1994, all special protection areas form an integral part of the NATURA 2000 ecological network and the protection of these sites overrides any consideration of the costs and benefits of any relevant coastal protection measures.

6.4.2 Further Examples of Legislative Override: the USA and Japan

In the United States, there are several federal statutes passed by Congress and signed into law by the President that are central to the Office of Water's mission. In addition, Presidential Executive Orders play a central role in a number of Office of Water activities. These executive orders are legally binding orders that direct the Environmental Protection Agency and other federal agencies in their execution of established laws and policies. Thus, for example, the Clean Water Act is the cornerstone of surface water quality protection in the United States, designed to achieve the broad goal of restoring and maintaining the chemical, physical, and biological integrity of the nation's waters so that they can support the protection of fish, shellfish, wildlife, and recreation.

Specific to the US coast, the Beaches Environmental Assessment and Coastal Health Act of 2000 authorizes the Environmental Protection Agency to award grants to eligible states, territories, and tribes to develop and implement beach water quality monitoring and notification programs for coastal and Great Lakes recreational beach waters. The grants also help these governments to inform the public about the risk of exposure to disease-causing microorganisms in the water at the nation's beaches. The Coastal Zone Act Reauthorization Amendments Section 6217 addresses nonpoint pollution problems in coastal waters, requiring states and territories with approved Coastal Zone Management Programs to develop Coastal Nonpoint Pollution Control Programs. The Endangered Species Act provides a program for the conservation of threatened and endangered plants and animals and the habitats in which they are found. The lead federal agencies for implementing the Endangered Species Act are the U.S. Fish and Wildlife Service and the U.S. National Oceanic and Atmospheric Administration (NOAA) Fisheries Service. Federal agencies are directed to carry out programs for the conservation of threatened and endangered species. Thus, federal agencies must consult with

NOAA Fisheries Service and/or Fish and Wildlife Service on activities that may affect a listed species. The Marine Protection, Research, and Sanctuaries Act (also known as the Ocean Dumping Act) prohibits the dumping of material into the ocean that would unreasonably degrade or endanger human health or the marine environment.

In Japan, the Basic Environment Law sets out basic principles and directions for formulating environmental policies, enacted in 1993. In the same year, the National Action Plan for Agenda 21 was submitted to the United Nations. In December 1994, an action plan called the Basic Environment Plan was adopted as the most important measure introduced under the Basic Environment Law. The plan systematically clarifies the measures to be taken by the national and local governments as well as actions to be carried out by citizens, businesses, and private bodies. It also defines the roles of parties involved and the ways and means for effectively pursuing environmental policies. The Japanese Nature Conservation Law, which provides the basic frames for nature conservation, gives powers to protect and manage natural resources and natural ecosystems in cooperation with other related laws. In accordance with this law, locations are designated and established as nature conservation areas including areas that preserve and maintains their valuable natural ecosystems in rivers, lakes, marshes, sea coasts, and marine areas with valuable wildlife.

6.5 Decision Making

6.5.1 Deliberative' Decision-Making

Consensus or "deliberative" decision making is a group decision-making process that seeks the consent of all participants. Consensus may be defined professionally as an acceptable resolution, one that can be supported, even if not the "favorite" of each individual.

Consensus is defined as (1) general agreement and (2) group solidarity of belief or sentiment. It has its origin in the Latin word cōnsēnsus (agreement), which is from cōnsentiō, meaning literally feel together. It is used to describe both the decision and the process of reaching a decision. Consensus decision making is thus concerned with the process of deliberating and finalizing a decision, and the social and political effects of using this process.

As a decision-making process, consensus decision making aims to be:

- agreement seeking: A consensus decision-making process attempts to help everyone get what they need;
- collaborative: Participants contribute to a shared proposal and shape it into a decision that meets the concerns of all group members as much as possible;

- cooperative: Participants in an effective consensus process should strive to reach the best possible decision for the group and all of its members, rather than competing for personal preferences;
- egalitarian: All members of a consensus decision-making body should be afforded, as much as possible, equal input into the process. All members have the opportunity to present, and amend proposals;
- inclusive: As many stakeholders as possible should be involved in the consensus decision-making process;
- participatory: The consensus process should actively solicit the input and participation of all decision makers.

Consensus decision making is an alternative to commonly practiced adversarial decision making processes or one based mainly on economic considerations. Proponents claim that outcomes of the consensus process include:

- better decisions: Through including the input of all stakeholders, the resulting proposals may better address all potential concerns;
- better implementation: A process that includes and respects all parties and generates as much agreement as possible sets the stage for greater cooperation in implementing the resulting decisions;
- better group relationships: A cooperative, collaborative group atmosphere can foster greater group cohesion and interpersonal connection.

6.5.2 Decision Making and Local Perception of Risk

Risk is both real and socially constructed. Residents who choose to live in area exposed to coastal erosion and floods are often assumed to take this risk knowingly and willingly. However, the views and opinions of those potentially affected by any flood event should be considered throughout the risk assessment process in addition to those involved in flood risk management to answer relevant issues such as: how can consequences be dealt with and how are consequences ranked? For such a participatory approach, selection of the stakeholders for interview should be as representative as possible within the confines of time and availability.

Stakeholders have a wide concept of risk which is often "grounded" within their knowledge and experience of flooding and erosion events. By systematically collecting this information through group and individual interviews, the insight derived is more to resemble "reality" providing an improved understanding of flood risk in the local community, and is therefore a meaningful guide for decision makers (Strauss & Corbin, 1998). Issues and values raised in these discussions can also be addressed throughout the science-based appraisal process, allowing important cross-fertilization to occur.

Key to this process is the analysis of statements made by individuals. Coding transcripts of recorded interviews provides the opportunity to compare responses to

TABLE 6.6 Common Coding for Risk Perception Studies

	Category	Description
Perception of flooding	Relevance claim	Quotes where the interviewee states what is important
	Evidence claim	Quotes where the interviewee establishes a cause–effect link related to coastal flooding
	Normative claim	Quotes where the interviewee states what is good, acceptable, and tolerable regarding coastal flooding risk management options
Emerging themes	Uncertainty	Quotes where the interviewee states the role of uncertainty
	Future	Quotes where the interviewee states his or her beliefs about future states of the coastal flooding risk related issues
	Options	Quotes where the interviewee states his or her beliefs about coastal flooding risk mitigation options

identify emergent themes as well as how risk perception is articulated. Statements or claims frequently belong to three categories: (1) relevance, (2) evidence, and (3) normative (see Table 6.6) (Renn, 2008). Relevance claims are expressing what matters to society, what are the important phenomena that should receive our attention. Evidence claims express causal linkages and are influenced by knowledge and are potentially associated with a need for science-based information. Finally, normative claims express what is good, tolerable, and/or acceptable. These three claim categories can be intertwined to produce specific attitudes toward risk.

The undertaking of this process helps to:

1. clearly define national and regional coastal risk management goals in a broad and long-term perspective;
2. reduce and better manage uncertainty identify methods to explicitly include uncertainty in all decisions relating to coastal management;
3. further develop the integrated planning approach to manage coastal risks, including scenario development, coastal risk assessment methodology, linking short- and long-term time horizons, and different geographical scales; and
4. continue the cooperation and learning process, similarities in coastal problems and possible solutions, and the commonality in methodological approaches.

6.6 Decision Support Systems

A spatial decision support system (DSS) is a computer-based software tool that can assist decision makers in their decision process. Such a DSS is an exploratory tool that allows assessing the conditions of a system under a variety of scenarios and the consequences of different adaptation and mitigation measures. A DSS will generally integrate the relevant environmental models, database and assessment tools, coupled within a graphic user interface. Spatial problems such as flood and erosion risk require a GIS approach. GIS is a set of computer tools that can capture, manipulate, process, and display spatial or georeferenced data facilitating spatial data integration, analysis, and visualization (Burrough & McDonnell, 1998). These functionalities make GIS tools useful for efficient development and effective implementation of DSS within the management process. For this purpose, GIS tools are used either as data managers (i.e., as a spatial geo-database tool) or as an end in itself (i.e., media to communicate information to decision makers). The use of GIS for coastal zone management has expanded rapidly during the past decade. (Bartlett & Smith, 2004, 300 pp.; Sheppard, 2012, 514 pp.; Wright & Bartlett, 2000; Wright, Dwyer, & Cummins, 2011, 350 pp.)

Based on a review of a range of existing DSSs that deal with coastal areas (Table 6.7), the main objectives of these tools are the analysis of vulnerability, impacts and risks, and the identification and evaluation of related management options to support robust decisions for sustainable management. Specifically, the objectives of the examined DSS tools are concerned with three major issues (with examples in brackets from Table 6.7):

- the assessment of vulnerability to natural hazards and climate change (DIVA, RegIS, CVAT, DESYCO, KRIM, Coastal Simulator);
- the evaluation of present and potential climate change impacts and risks on coastal zones and linked ecosystems to predict how coastal regions will respond to climate change (RegIS, CVAT, Coastal Simulator); and
- the evaluation or analysis of management options for the optimal use of coastal resources and ecosystems through the identification of feasible measures and adequate coordination of all relevant users/stakeholders (COSMO, WADBOS, SIMCLIM, RAMCO).

The THESEUS project (www.theseusproject.eu) built on this experience by developing a comprehensive GIS-based intended as a vehicle for communication, training, forecasting, and experimentation (Zanuttigh et al., 2014).

The tool is based on the SPRC model described in Chapter 2 and supports an assessment of the change in risk from a range of scenarios and selection of the most appropriate intervention measures from an available portfolio of engineering, ecological and social measures.

It filled in the gap among the existing tools, based on the following pillars.

- It provides seamless integration across disciplines: physics, engineering, ecology, social sciences, and economy;
- It considers intermediate spatial scales (10–100 km) and short-, medium-, and long-term time spans (1–10–100 years);

TABLE 6.7 Review of Existing Exploratory Tools That can be Used for Supporting Decisions Applied to Coastal Areas. These GIS-Based Tools Perform Scenario Construction and Analysis.

Name	Year	Reference	Processes	Functionalities
COSMO	1992	Feenstra, Programme, and Milieuvraagst (1998)	Sea-level rise	Problem characterization (e.g., water quality, coastal erosion) Impact evaluation of different development and protection plans Multicriteria decision analysis Ecosystem-based
Coastal simulator	2000–	Mokrech et al. (2009) Dawson et al. (2009)	Storm surge Flooding Coastal erosion Sea-level rise Socioeconomic scenarios	Environmental status evaluation Risk analysis Management strategies identification and evaluation Uncertainty analysis Integrated risk assessment
CVAT	1999–	Flax, Jackson, and Stein (2002)	Multihazard Extreme events Storm surge	Hazard analysis Social, economic, and environmental vulnerability indicators Mitigation options analysis Risk analysis at regional scale
DESYCO	2005–2010	Torresan et al. (2010)	Sea-level rise Storm surge Flooding Coastal erosion Water quality	Impacts and vulnerability analysis Adaptation options definition Multicriteria decision analysis Regional risk assessment
DIVA	1999–	Vafeidis et al. (2008) Hinkel and Klein (2009)	Sea-level rise Coastal erosion Storm surge Flooding Wetland loss and change Salinization	Environmental status evaluation Impact analysis Adaptation options evaluation Cost-benefit analysis
KRIM	2001–2004	Schirmer, Schuchardt, Hahn, Bakkenist, and Kraft (2003, pp. 269–273)	Sea-level rise Extreme events Coastal erosion	Environmental status evaluation Adaptation measures evaluation Information for nontechnical users Risk analysis

RegIS	2003–2010	Holman et al. (2008), pp. 1–187	Coastal and river flooding Wetland loss and change Sea-level rise Emission scenarios Socioeconomic scenarios	Implementation of DPSIR conceptual model Management measures evaluation Impact analysis Integrated risk assessment Information for nontechnical users
RAMCO	1996–1999	De Kok, Engelen, White, and Wind (2001) http://www.riks.nl/resources/papers/RamCo2.pdf	Socioeconomic scenarios Coastal and river flooding Policy options Impact of human activities Integrated management	Environmental status evaluation Management measures evaluation
SimCLIM	2005–	Warrick et al. (2009)	Sea-level rise Coastal flooding Coastal erosion	Environmental status evaluation Impact and vulnerability evaluation Adaptation strategies evaluation Cost/benefit analysis
WADBOS	1996–2002	van Buuren, Engelen, and van de Ven (2002)	Socioeconomic scenarios Policy options Impact of human activities Integrated management	Socioeconomic, hydrological, environmental, ecological data Socioeconomic, ecological, landscape models Management measures identification and evaluation
CLIMSAVE	2010–2013	Harrison et al. (2013)	Emission scenarios Agriculture Forests Water resources Coastal and river flooding Urban development	Implementation of DPSIR conceptual model Impact analysis Adaptation strategies
THESEUS	2010–2014	Zanuttigh et al. (2014)	Sea-level rise Coastal flooding Coastal erosion Socioeconomic scenarios	Hydraulic, social, economic, ecological vulnerability Combination of engineering, social, economic, and ecologically based mitigation options Multicriteria analysis High-resolution risk assessment

From Zanuttigh et al. (2014).

- It allows diverse portfolios of mitigation options such as engineering defenses (i.e., barriers, wave farms, etc.), ecologically based solutions (i.e., biogenic reefs, sea-grasses, etc.) and socioeconomic mitigations (i.e., insurance, change of land use, etc.);
- It supports decision making based on a balance between deterministic models and expert, discussion-based assumptions;
- It uses an open source approach—based on a specific request from the European Commission—to maximize the availability and uptake of the tool. The technological framework selected for the development is Microsoft. NET 4.0, a solid software platform which includes a wide class library for common tasks such as data access, user interface, or network communications. DotSpatial, a free, open source set of libraries for .NET, is the GIS component of the DSS to easily incorporate spatial data, analysis and mapping into an application. A relational database management system has been implemented with SQLite engine to store the site-specific data and information.

THESEUS DSS is developed on top of an integrated simulation model suitable for performing "What if" analyses based on scenarios. By means of this kind of analysis, the user tries to find out how management strategies and scenario sensitive variables and parameters influence risk at the selected coastal site. The policy analysis mainly focuses on the consequences of changing coastal management options.

Figure 6.2 gives an overview of the structure of the integrated model at the most synthetic level. The integrated model is the actual calculation kernel of THESEUS DSS. It contains relations in the form of mathematical equations, formal rules, or transfer functions representing the real-world processes.

The primary end-users are intermediate-level coastal managers who need to make sound evidence-based decisions regarding spatial planning and coastal protection.

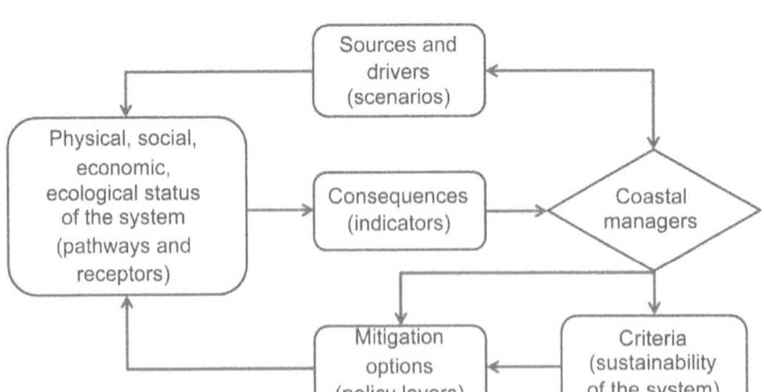

Figure 6.2 System diagram view of THESEUS DSS. *From Zanuttigh et al. (2014).*

The main foundation of this DSS is that it has to be "open and parametric," not only in terms of source code and technology but also in terms of usability. This software is designed to be easily modified and distributed across many sites with many diverse characteristics: this requires adequate flexibility in terms of configuration parameters and input materials.

The DSS should also be "interactive" so that users can explore a combination of scenarios, whereas being trained in interdisciplinary risk assessment, including the best (i.e., sustainable) solution or combination of solutions for risk mitigation. Here sustainable means protecting the coast while preserving its socioeconomic development and the integrity of the ecosystem services.

The inclusion and participation of relevant stakeholders (coastal managers) is essential to test the outcomes of the modeling, to identify the most relevant parameters and related scenarios to be included in the analysis and to evaluate adaptation options (Dessai & Hulme, 2004). To maximize the utility of THESEUS DSS, the stakeholders gave their input on:

- definition of the site boundaries;
- identification of critical pathways of the existing management that may lead to failure and are worthy of further investigation;
- usefulness of output indicators for each of the meta-models;
- appropriateness of the mitigation measures to be included in future coastal management strategies for a given site;
- site-specific relevance of the social, economic, and environmental components of risk; and
- functionality and user-friendliness of the interface.

Following Holman et al. (2008, pp. 1–187), the setup of the tool considered two key points.

1. Intuitive and interactive design of the GUI and possibility as follows.
   a. The physical layout of the tool should closely mirror the conceptual model (i.e., the SPRC components).
   b. The user should be able to vary the input parameters through sliders to analyze the potential changes induced by different scenarios or mitigation strategies.
   c. "Realistic" and plausible ranges of values for a given parameter should be used to give guidance on the uncertainty associated with a scenario.
   d. The users should be allowed to save and compare the graphical outputs from more than one model or scenario.
2. Balance of simplified modeling assumptions and speed to promote the use of the tool for testing different combinations of mitigation options by:
   a. avoiding extensive or prolonged model setup has been avoided and
   b. providing rapid outputs.

Figure 6.3 The viewer at the startup for the Cesenatico site, Italy; see details in Section 7.6.

THESEUS DSS operates at high resolution to provide geographic specific outputs. Although users should be encouraged to study the detailed maps, this output is not suitable for direct application, nor should it be confused with the policies that would accomplish those outcomes and judged based on the avoided monetary damage only. Therefore although the intermediate maps of specific results (e.g., flood depth, land value loss, life losses) are shown with their own scale, the results of (hydraulic, social, economic, and ecological) vulnerability and the overall risk assessment map are given as normalized quantitative indicators.

Based on these guiding concepts, on the experience gained from other tool development (and specifically RAMCO and RegIS) and on the feedback from stakeholders, the interface for each site consists of a viewer at start up (Figure 6.3), where the user can visualize the input data (bottom elevation, habitats map, land use map, etc.) and evolves to the following four screens, each with a different purpose.

1. Definition screen: this allows the user to define the name of the test and write a short description; he or she can also load the settings of a previously performed analysis.
2. Scenarios screen (Figure 6.4): this allows the user to select among climate, social, environmental, and economic scenarios. The user can adopt preset scenarios defined by scientists; in this case, the default set of input parameter values for each preset scenario allows a rapid model setup. The user can also create their own scenarios by directly changing the input parameter values used in the models. This enables the user to become familiar with the most significant parameters related to the site-specific scenarios and to explore the effects of uncertainty in any scenario, which cannot be defined by a single set of unique values.

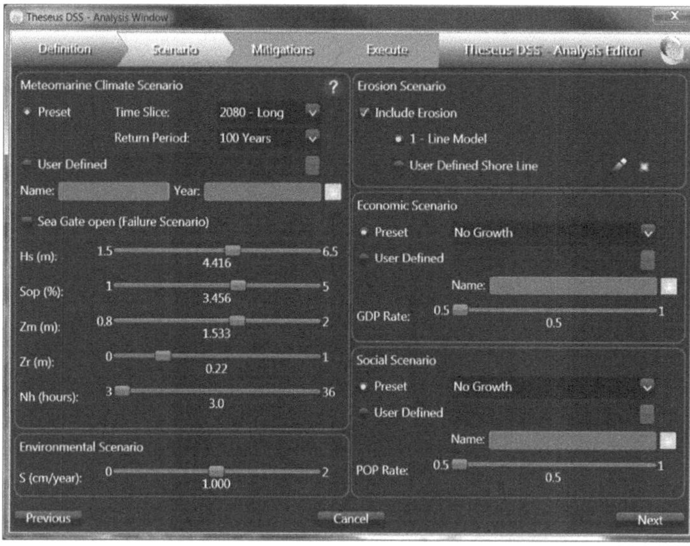

Figure 6.4 Scenarios screen. *From Zanuttigh et al. (2014).*

3. Mitigations screen (Figure 6.5): this allows the user to include:
   a. engineering mitigations, such as wave farms, barriers, floating breakwaters, sea walls, and nourishments;
   b. ecologically based mitigations, such as management or construction of dunes, reinforcement of saltmarshes, and creation of biogenic reefs; and
   c. economic and social mitigations such as evacuation plans, land use change and zoning (for instance, managed realignment), insurance scheme.

   When selecting a mitigation option for which size and location has to be defined (for instance, a biogenic reef, a breakwater, a managed realignment), the user can: (1) include the shapefile prepared by the scientists with the suggested configuration of the mitigation (position, extension, design parameters); (2) upload a shapefile and enter the design parameters; or (3) draw the mitigation directly from the graphic user interface. For other mitigations, such as insurance schemes or evacuation plans, the user can interact by modifying the insurance premium value, the percentage of evacuated people or the destination of a given area.
4. Execution screen (Figure 6.6): this guides the user through the analyses to be performed based on the selections made in the previous windows; these analyses include the following steps: (1) modeling of the physical processes (erosion, flooding); (2) modeling the impacts on the environment, the society and the economy; (3) assessing the global hydraulic, social, and environmental vulnerability; and finally (4) assessing the risk. It also imposes constraints on the analysis.

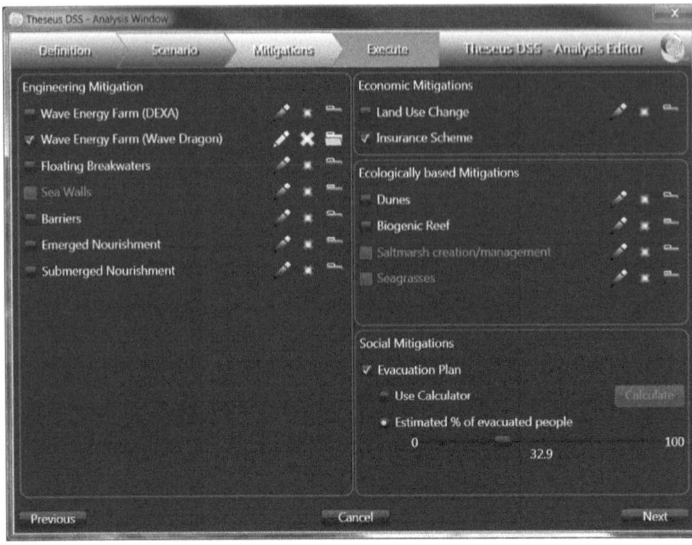

Figure 6.5 Mitigation screen. *From Zanuttigh et al. (2014).*

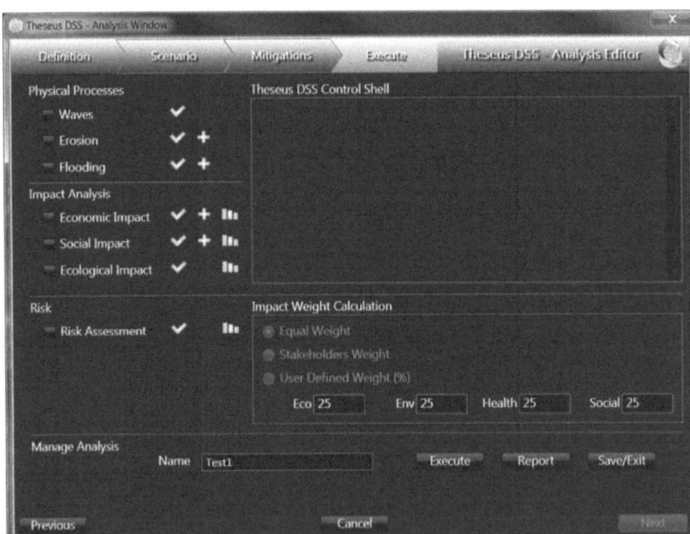

Figure 6.6 Analysis screen. *From Zanuttigh et al. (2014).*

For instance, if the user does not include the erosion process in the Scenarios screen, he or she cannot flag the corresponding analysis to be run in the Execution screen. Let us suppose that the user changes the settings of the analysis just performed by including for instance a new mitigation in the Mitigation screen. When he or she is back at the Execution screen, he or she will be forced to rerun the flooding model if the mitigation is such that it affects the physical processes (for instance, a seawall or a dune), whereas the flooding model will be hidden if the mitigation does not interfere with the physical processes (for instance, an evacuation plan or a change of land use).

Hydraulic, social, and economic vulnerability maps are generated, being vulnerability assessed as:

$$\text{vulnerability} = \text{exposure} = -\text{resilience}$$

where exposure is the value at risk (De Vries, 2011) and resilience is the damage that will not alter the main functions of human and physical systems in equilibrium in discrete times and at local scale (De Bruijn, 2004).

Appropriate impact functions were developed to link economic (see Section 6.2.3), social (see Section 6.2.4) and ecological (see Section 6.2.2) data to hydraulic parameters, such as: beach retreat, flood depth, flood duration, and flood velocity. These functions allow obtaining the maps of social, economic, and ecological consequences, each one expressed with the typical unit. The economic losses are divided into the losses in the urban area, expressed as euro/m^2, and into the beach losses, expressed as euro/m. The social losses are derived in terms of life losses, expressed as a percentage of the number of expected deaths of the local population in the area, and of CF losses, expressed as a percentage of the functionality loss of each CF (see the example in Figure 6.6).

A normalization procedure of each map of consequences is then carried out to obtain a 1–4 scale, in which 1 = low, 2 = medium, 3 = high, and 4 = very high impact. The normalization is performed by dividing the local values of the consequences by the corresponding site-specific thresholds that are obtained by comparing the consequences of different scenarios with the historical experience and/or data available in the sites (i.e., through a process that involves both stakeholders and experts). Site-specific threshold values for low, medium, high, and very high impact are defined for each relevant parameter: flood depth, velocity and duration; beach retreatment; beach and land use value losses; and life and CF losses. The normalized ecological vulnerability map is directly derived from the calculated values of the EVI, by associating the EVI 0–3 scale to the 1–4 vulnerability scale.

The hydraulic vulnerability map is derived from a weighted average with equal weights of the normalized maps of flood depths, velocities, and durations. The economic vulnerability map is obtained by a spatial combination of the normalized beach losses and of the normalized inland value lost, with the two areas being complementary. The social vulnerability map is derived as an equally weighted combination of the normalized maps of life losses and CF losses.

Social, economic, and ecological vulnerability maps are then combined through a weighted procedure to obtain the overall risk map (Figures 6.7 and 6.8). Within this additive combination, the hydraulic vulnerability is not explicitly considered to avoid duplication, because it is already indirectly included through the social,

Figure 6.7 Example of impact on critical facilities (%). Long-term (2080) scenario with return period (combined wave and storm surge statistics) Tr = 100 years. *From Zanuttigh et al. (2014).*

Figure 6.8 Example of integrated risk map, scale from 1 to 4 (from low to very high impact). Long-term (2080) scenario with return period (combined wave and storm surge statistics) Tr = 100 years. *From Zanuttigh et al. (2014).*

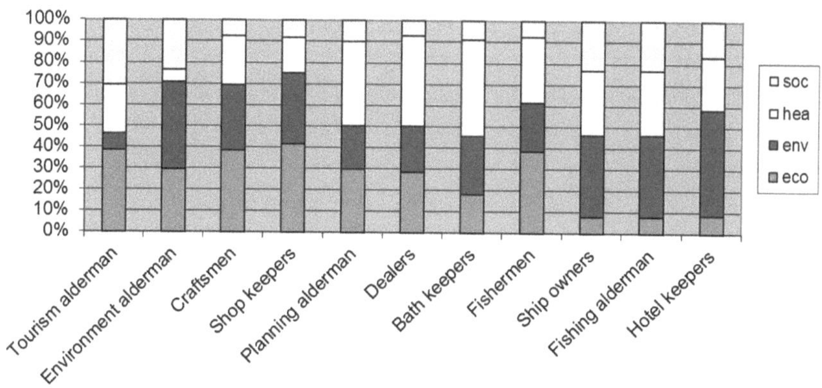

Figure 6.9 Weights assigned by stakeholders in Cesenatico, Italy (see Section 7.6) to impact on human health (hea), infrastructures and activities (soc), environment (env), and economics (eco). *From Zanuttigh et al. (2014).*

economic, and ecological vulnerabilities that are all estimated on the basis of selected hydraulic parameters (flood depth, velocity, and duration).

In the generation of the risk map, the users have the chance to select equal weights, their own weights, or to use the results of the surveys carried out in THESEUS study sites (see Figure 6.9 for the synthesis of the results in the Italian case study, Section 7.6). Within these surveys, the stakeholders were asked to rank three cards where the three titles referred to the represented main issues (economic, environmental, and social). Some items were clarified with some examples, for instance: the "economic" card shows "houses, tourism, fishery, ..."; the environmental card presents "pine forest, biodiversity, animal species, habitats, ..."; and the social card shows "social cohesion, meeting facilities, sports, psychological distress, fatalities, injuries, ...". Stakeholders were then asked to insert one or more blank cards between the ordered cards to stress relative differences in importance attached to each issue or group of issues.

The normalization procedure suggested by Kodikara et al. (2010) led to obtain the relative weights for each stakeholder and consequently the weights of each criterion were estimated as the average values.

An example application of the tool for decision making purposes is given in Section 7.6 for the study site of Cesenatico, Northern Adriatic Sea, Italy.

6.7 Conclusions

The key to successful coastal flood management is the use of mitigation techniques that are appropriate for the local context. Thus, the development of coastal protection strategies should involve the specification of detailed regional scenarios, an appropriate problem assessment, and identification of promising measures and the analysis and evaluation of alternative strategies.

During the decision-making process, the range of possible flood mitigation alternatives should be reviewed to identify the most efficient individual or suite of options.

The efficiency of engineering solutions consists of the degree of coastal protection they offer, and it can be evaluated specifically in terms of reduction of incident wave height, reduction (or stabilization) of sediment transport, reduction of wave run-up on the beach and wave overtopping at the sea bank, and of frequency and magnitude of inland flooding.

From an environmental perspective, it should be noted that the selection of any mitigation option will affect local habitats, species, and ecological interactions. To assess the vulnerability of ecosystems to changes in stresses and to disturbances, an index was adopted within the THESEUS project (see THESEUS OD3.3, 2012). This provides a rapid and standardized method for characterizing vulnerability across coastal systems and identifies issues that may need to be addressed to reduce vulnerability.

From an economic perspective, marketed and nonmarketed valuation methods are required to value costs and benefits from a pool of different mitigation options.

Nonmarket valuation methods are useful to assign monetary values for environmental goods and services that are not traded in actual markets. Such market price data are missing and purchasing behavior is observed within the context of a hypothetical market using the methods described in Figure 6.1. On the other hand, there is a range of methods and techniques that can help decision-making concerning investment appraisal. For example the CBA (it assesses the monetary social costs and benefits of an investment project over a period in comparison to a well-defined baseline alternative) and the MCA (it considers multiple criteria in decision-making environments rather than the single criterion of net present value).

From society's perspective, social vulnerability can be modeled in DSSs considering damages to CFs and the expected number of fatalities (see Section 6.6). It should be noted that flood damages to society also include psychological consequences that are mainly qualitative in nature and are hard to be translated into linear functions with quantitative outputs.

In any case, the decision maker should recall that there are several legislative constraints on option choice, and every country has a different approach to this, designed to fit local circumstances and/or reflecting the current legislative position regarding environmental protection (examples in the European Union are the WFD and the European Habitats and Birds Directive).

Finally, an important part of selecting and assessing mitigation options is the development of decision-support methods because they can examine different mitigation options by evaluating their relative efficiency, equity, and sustainability in determining risk levels and potential consequences. This is important when selecting mitigation strategies under uncertain future conditions.

Spatial DSSs can assist decision makers in their decision process. Among these, the THESEUS project (www.theseusproject.eu) developed a comprehensive GIS-based DSS intended as a vehicle for communication, training, forecasting, and experimentation. It supports an assessment of the change in risk because of a range of scenarios and selection of the most appropriate intervention measures from an available portfolio of engineering, ecological and social measures.

References

Adger, W. N., Brooks, N., Bentham, G., & Agnew, M. (2004). *New indicators of vulnerability and adaptive capacity, technical report 7, Tyndall centre for climate change research.*

Adger, W., Hughes, T., Folke, C., Carpenter, S., & Rockström, J. (2005). Social-ecological resilience to coastal disasters. *Science, 309*, 1036–1039.

Aven, T., & Renn, O. (2010). *Risk appraisal. Risk management and governance.* Springer Berlin Heidelberg.

Bartlett, D., & Smith, J. L. (Eds.). (2004). *GIS for coastal zone management.* London: Routledge.

Bennett, J., & Adamowicz, V. (2001). Some fundamentals of environmental choice modelling. In J. Bennett, & R. Blamey (Eds.), *The choice modelling approach to environmental valuation.* Cheltenham: Edward Elgar Publishing Limited.

Bennett, J., & Blamey, R. (2001). *The choice modelling approach to environmental valuation.* Cheltenham, UK: Edward Elgar.

Birol, E., & Koundouri, P. (2008). *Choice experiments in Europe: Economic theory and applications.* Edward-Elgar Publishing, Wally Oates and Henk Folmer's 'New Horizons in Environmental Economics' Series, ISBN 9781845427252.

Birol, E., Koundouri, P., & Kountouris, Y. (2010). Assessing the economic viability of alternative water resources in water-scarce regions: combining economic valuation, cost-benefit analysis and discounting. *Ecological Economics, Elsevier, 69*(4), 839–847.

Bouma, T. J., van Belzen, J., Balke, T., Zhu, Z., Airoldi, L., Blight, A. J., et al. (2013). Identifying knowledge gaps hampering application of intertidal habitats in coastal protection: opportunities and steps to take. *Coastal Engineering, Elsevier.* http://dx.doi.org/10.1016/j.coastaleng.2013.11.014.

Brent, R. J. (1990). *Project appraisal for developing countries.* New York, USA: Harvester Wheatsheaf.

Burrough, P. A., & McDonnell, R. A. (1998). *Principles of geographical information systems.* Oxford University Press.

Burns, W. J. (2007). *Risk perception: A review.* Create report—Office of Naval Research Grant No. N00014-05-1-0630. Los Angeles, CA, USA. Center for Risk and Economic Analysis of Terrorism Events, University of Southern California http://create.usc.edu/research/54570.pdf. Last accessed February 2012.

Burcharth, H. F., Hawkins, S. J., Zanuttigh, B., & Lamberti, A. (2007). *Environmental design guidelines for low crested coastal structures.* Elsevier, ISBN 978-0-08-044951-7.

Communities and Local Government. (2009). *Multi-criteria analysis: A manual.* London: Department for Communities and Local Government.

Cutter, S. L. (1996). Vulnerability to environmental Hazards. *Progress in Human Geography, 20*(4), 529–539.

Cutter, S. L., Boruff, B. J., & Shirley, W. L. (2003). Social vulnerability to environmental hazards. *Social Science Quarterly, 84*(2), 242–261.

Dawson, R. J., Dickson, M., Nicholls, R. J., Hall, J., Walkden, M. J. A., Stansby, P. K., et al. (2009). Integrated analysis of risks of coastal flooding and cliff erosion under scenarios of long term change. *Climatic Change, 95*(1–2), 249–288. http://dx.doi.org/10.1007/s10584-008-9532-8.

de Boer, J., Wardekker, J. A., & van der Sluijs, J. P. (2010). Frame-based guide to situated decision-making on climate change. *Global Environmental Change-Human and Policy Dimensions, 20*(3), 502–510. http://dx.doi.org/10.1016/j.gloenvcha.2010.03.003.

De Bruijn, K. M. (2004). Resilience and flood risk management. *Water Policy, 6*, 53–65.

De Kok, J. L., Engelen, G., White, R., & Wind, H. G. (2001). Modelling land-use change in a decision-support system for coastal-zone management. *Environmental Modeling & Assessment, 6*(2), 123–132.

De Vries, D. H. (2011). Temporal vulnerability in hazardscapes: flood memory-networks and referentiality along the North Carolina Neuse River (USA). *Global Environmental Change, 21*, 154–164.

DEFRA, Environment Agency. (2005). *Flood warning for vulnerable Groups: Measuring and mapping vulnerability.* R&D Technical Report W5C-018/4. London: DEFRA.

DEFRA. (2009). *Appraisal of flood and coastal erosion risk management. Defra policy statement.* London, UK: Department for Environment, Food and Rural Affairs (DEFRA). https://www.gov.uk/government/uploads/system/uploads/attachment_data/file/69419/pb13278-erosion-manage-090619.pdf.

Dessai, S., & Hulme, M. (2004). Does climate adaptation policy need probabilities? *Climate Policy, 4*, 107–128.

Díaz-Simal, P., Koundouri, P., Rulleau, B., & Remoundou, K. (2013). *Valuing climate change mitigation in coastal environments exposed to extreme natural hazards: A choice experiment simulated for different time horizons.* DEOS Working Papers, Athens University of Economics and Business.

Dilling, L., & Lemos, M. C. (2011). Creating usable science: opportunities and constraints for climate knowledge use and their implications for science policy. *Global Environmental Change-Human and Policy Dimensions, 21*(2), 680–689. http://dx.doi.org/10.1016/j.gloenvcha.2010.11.006.

EA. (2010). *Flood and coastal erosion risk management appraisal guidance. FCERM-AG.* Bristol, UK: Environment Agency. http://a0768b4a8a31e106d8b0-50dc802554eb38a24458b98ff72d550b.r19.cf3.rackcdn.com/geho0310bsdb-e-e.pdf.

EA. (2011). *Flood and coastal risk management–An introduction to delivering the water framework directive.* Operational Instruction 871_11. Bristol, UK. London, UK: Environment Agency Erosion Sector, Defra.

European Commission. (1992). *Habitats directive council directive 92/43/EEC on the conservation of natural habitats and of wild fauna and flora.* Brussels, Belgium: European Commission. http://ec.europa.eu/environment/nature/legislation/habitatsdirective/index_en.htm.

European Commission. (2000). *Directive 2000/60/EC establishing a framework for community action in the field of water policy.* http://eur-lex.europa.eu/legal-content/EN/ALL/?uri=CELEX:32000L0060.

EUROSION. (2004). *Living with coastal erosion in Europe: Sediment and space for—Part III—Methodology for assessing regional indicators.* www.eurosion.org.

Feenstra, J. F., Programme, U. N. E., & Milieuvraagst, V. (1998). *Handbook on methods for climate change impact assessment and adaptation strategies.* United Nations Environment Programme.

FEMA. (2007). *Risk management series-design guide for improving critical facility safety from flooding and high winds.* Washington DC: FEMA.

Firth, L. B., Thompson, R. C., Abbiati, M., Airoldi, L., Bohn, K., Bouma, T. J., et al. (2014). Between a rock and a hard place: environmental and engineering considerations when designing flood defence structures. *Coastal Engineering.* http://dx.doi.org/10.1016/j.coastaleng.2013.10.015.

Flax, L. K., Jackson, R. W., & Stein, D. N. (2002). Community vulnerability assessment tool methodology. *Natural Hazards Review, 3*, 163–176.

Gornitz, V. M., Daniels, R. C., White, T. W., & Birdwell, K. R. (1994). The development of a coastal risk assessment database: vulnerability to sea-level rise in the US Southeast. In C. W. Finkl (Ed.), *Journal of coastal research: Vol. 12. Coastal hazards: Perception, susceptibility and mitigation*. Florida.

Gouldby, B., Sayers, P., Mulet-Marti, J., Hassan, M., & Benwell, D. (2008). A methodology for regional-scale flood risk assessment. *Proceedings of the Institution of Civil Engineers-Water Management, 161*(3), 169–182.

Guillaumont, P. (2009). An economic vulnerability index: Its design and use for international development policy. *Oxford Development Studies, 1469–9966, 37*(3), 193–228.

Hanley, M. E., Hoggart, S. P. G., Simmonds, D. J., Bichot, A., Colangelo, M. A., Bozzeda, F., & Thompson, R. C. (2014). Shifting sands? Coastal protection by sand banks, beaches and dunes. *Coastal Engineering, 87*, 136–146.

Harrison, P. A., Holman, I. P., Cojocaru, G., Kok, K., Kontogianni, A., Metzger, M., et al. (2013). Combining qualitative and quantitative understanding for exploring cross-sectoral climate change impacts, adaptation and vulnerability in Europe. *Regional Environmental Change, 13*, 761–780.

Hillsborough County—Florida. (2009). *Hillsborough county local mitigation strategy appendix K-1 critical facilities*. Hillsborough County. http://www.hillsboroughcounty.org/DocumentCenter/Home/View/1126.

Hinkel, J., & Klein, R. J. T. (2009). Integrating knowledge to assess coastal vulnerability to sea-level rise: the development of the DIVA tool. *Global Environmental Change, 19*(3), 384–395.

Holman, I. P., Rounsevell, M. D. A., Berry, P. M., & Nicholls, R. J. (Eds.). (2008), *Climatic change: 90(1/2). Development and application of participatory integrated assessment software to support local/regional impacts and adaptation assessment*.

Hoggart, S. P. G., Hanley, M. E., Parker, D. J., Simmonds, D. J., Bilton, D. T., Filipova-Marinova, M., & Thompson, R. C. (2014). The consequences of doing nothing: the effects of seawater flooding on coastal zones. *Coastal Engineering, 87*, 169–182.

ICE and RIBA. (2010). *Facing up to rising seas: Retreat, defend or attack?* Institution of Civil Engineers and the Royal Institution of British Architects. Downloadable at www.buildingfutures.org.uk/.../Facing_Up_To_Rising_Sea_Levels.pdf.

ISTAT, 2009. http://www3.istat.it/dati/catalogo/20100526_00/Avvio2009.pdf, in Italian.

Johnson, C., Penning-Rowsell, E., & Parker, D. (2007). Natural and imposed injustices: the challenges in implementing 'fair' flood risk management policy in England. *Geographical Journal, 173*, 374–390. http://dx.doi.org/10.1111/j.1475-4959.2007.00256.x.

Kodikara, P. N., Perera, B. J. C., & Kularathna, M. D. U. P. (2010). Stakeholder preference elicitation and modelling in multi-criteria decision analysis–a case study on urban water supply. *European Journal of Operational Research, 206*(1), 209–220.

Kortenhaus, A., & Kaiser, G. (2009). Lessons learned from flood risk analyses at the North Sea coast. *Journal of Coastal Research, SI 56, Proceedings ICS*, 822–826. ISSN 0749–0258.

Koundouri, P., Dávila, O. G., Stithou, M., & Stuitver, M. (2013). *Methodology for integrated socio-economic assessment of offshore platforms: Towards facilitation of the implementation of the marine strategy framework directive*. Available as working paper at: http://wpa.deos.aueb.gr/wpa_show_paper.php?handle=1304.

Martinelli, L., Zanuttigh, B., & Corbau, C. (2010). Assessment of coastal flooding risk along the Emilia Romagna littoral, Italy. *Coastal Engineering, 57*(11–12), 1042–1058 (Elsevier).

Mokrech, M., Hanson, S., Nicholls, R. J., Wolf, J., Walkden, M., Fontaine, C., et al. (2009). The Tyndall coastal simulator. *Journal of Coastal Conservation: Planning and Management*. http://dx.doi.org/10.1007/s11852-009-0083-6.

Morris, M. W., Hassan, M. A. A. M., Kortenhaus, A., Geisenhainer, P., Visser, P. J., & Zhu, Y. (2008). Modelling breach initiation and growth. In P. Samuels, S. Huntington, W. Allsop, & J. Harrop (Eds.), *Flood risk management: Research and practice*. Taylor & Francis Group.

Naulin, M., Kortenhaus, A., & Oumeraci, H. (2011). Reliability analysis and breach modelling of coastal and estuarine flood defences. ISGSR 2011. In N. Vogt, B. Schuppener, D. Straub, & G. Bräu (Eds.), *Bundesanstalt für Wasserbau*, ISBN 978-3-939230-01-4 http://vzb.baw.de/e-medien/geotechnical-safety-and-risk-2011/PDF/5%20Flood%20Defence/5_03.pdf.

Penning-Rowsell, E. C., Floyd, P., Ramsbottom, D., & Surendran, S. (2005). Estimating injury and loss of life in floods: a deterministic framework. *Natural Hazards, 36*, 43–64.

Renn, O. (2008). *Risk governance: Coping with uncertainty in a complex world*. London, UK: Earthscan.

Rietkerk, M., Dekker, S. C., de Ruiter, P. C., & van de Koppel, J. (2004). Self-organized patchiness and catastrophic shifts in ecosystems. *Science, 305*, 1926–1929.

Scheffer, M., Bascompte, J., Brock, W. A., Brovkin, V., Carpenter, S., Dakos, V., et al. (2009). Early-warning signals for critical transitions. *Nature, 461*, 53–59.

Schirmer, M., Schuchardt, B., Hahn, B., Bakkenist, S., & Kraft, D. (2003). KRIM: climate change risk construct and coastal defence. In *DEKLM German climate research programme-proceedings*.

Schwarz, J., & Maiwald, H. (2008). Damage and loss prediction model based on the vulnerability of building types. In *4th international symposium on flood defence: Managing flood risk, reliability and vulnerability. Toronto, Ontario, Canada, May 6–8, 2008.*

Sheppard, S. R. J. (2012). *Visualizing climate change: A guide to visual communication of climate change and developing local solutions*. London: Routledge.

Simeoni, U., Tessari, U., Gabbianelli, G., & Schiavi, C. (2003). *Sea storm risk assessment in the Ravenna littoral (Adriatic Sea, Northern Italy)*. In E. Ozhan (Ed.), *Proceedings of the sixth international conference on the Mediterranean coastal environment, October 7–11, 2003, Ravenna, Italy, 1–3*; (pp. 2223–2234).

Strauss, A. L., & Corbin, J. M. (1998). *Basics of qualitative research: Techniques and procedures for developing grounded theory*. Sage Publications.

Tapsell, S. (2011). Socio-psychological dimensions of flood risk management. In G. Pender, & H. Faulkner (Eds.), *Contained in flood risk science and management*. Oxford, UK: Wiley-Blackwell.

THESEUS OD3.3. (2012). *Natural habitats for coastal protection and relevant multi-stressor coastal risks*. Report and European Scale overview http://www.theseusproject.eu/resources/documents/Deliverables/Official-Deliverables-(ODhttp://www.theseusproject.eu/resources/documents/Deliverables/Official-Deliverables-(OD)/OD3.3-(M18)-Natural-habitats-for-coastal-protection-and-relevant-multi-stressor-coastal-risks.-Report-and-European-Scale-Overview./.

Torresan, S., Zabeo, A., Rizzi, J., Critto, A., Pizzol, L., Giove, S., et al. (2010). Risks assessment and decision support tools for the integrated evaluation of climate change impacts on coastal zones. In *International congress on environmental modelling and software modelling for environmental sake, fifth biennial meeting, Ottawa, Canada.*

UNISDR. (2009). *UNISDR terminology on disaster risk reduction*. United Nations New York.

Vafeidis, A. T., Nicholls, R. J., McFadden, L., Tol, R. S. J., Hinkel, J., Spencer, T., et al. (2008). A new global coastal database for impact and vulnerability analysis to sea-level rise. *Journal of Coastal Research, 24*(4), 917–924. http://dx.doi.org/10.2112/06-0725.1.

van Buuren, J. T., Engelen, G., & van de Ven, K. (2002). *The DSS WadBOS and EU policies implementation*.

Warrick, R. A. (2009). Using SimCLIM for modelling the impacts of climate extremes in a changing climate: a preliminary case study of household water harvesting in Southeast Queensland. In *18th world IMACS/MODSIM congress, Cairns, Australia, July 13–17, 2009.*

Winograd, M., & Farrow, A. (2013). Sustainable development indicators for decision making: concepts, methods, definition and use. In *Encyclopedia of life support systems (EOLSS), developed under the auspices of the UNESCO*. Oxford, UK: Eolss Publishers. http://www.eolss.net.

Wisner, B., Blaikie, P., Cannon, T., & Davis, I. (2004). *At risk: Natural hazards, people's vulnerability and disasters*. London: Routledge.

World Commission on Environment and Development (WCDE). (1987). *Our common future: The brundtland report*. Oxford: Oxford University Press. First definition and implications on sustainable development concept.

Wright, D., & Bartlett, D. (2000). *Marine and coastal geographical information systems*. London: Taylor and Francis.

Wright, D. J., Dwyer, E., & Cummins, V. (Eds.). (2011). *Coastal informatics: Web atlas design and implementation*. Hershey, PA: IGI-Global. http://dx.doi.org/10.4018/978-1-61520-815-9. ISBN13: 9781615208159.

Zanuttigh, B., Martinelli, L., Lamberti, A., Moschella, P., Hawkins, S., Marzetti, S., et al. (2005). Environmental design of coastal defence in Lido di Dante, Italy. *Coastal Engineering, 52*(10–11), 1089–1125 (Elsevier).

Zanuttigh, B., Simcic, D., Bagli, S., Bozzeda, F., Pietrantoni, L., Zagonari, F., et al. (2014). THESEUS decision support system for coastal risk management. *Coastal Engineering*. http://dx.doi.org/10.1016/j.coastaleng.2013.11.013.

Chapter 7

Case Studies Worldwide

Coastal Risk Management in a Changing Climate. http://dx.doi.org/10.1016/B978-0-12-397310-8.00007-5

7.0

Introduction

Susan Hanson [1], Robert Nicholls [1], Barbara Zanuttigh [2]

[1] *Faculty of Engineering and the Environment, University of Southampton (Soton), Southampton, BJ, UK*

[2] *Department of Civil, Chemical, Environmental and Materials Engineering, University of Bologna (UniBo), Bologna, Italy*

In this chapter, a number of case study sites show how, by developing a holistic view of the existing coastal flood system, assessment of the impacts of flooding and erosion can be carried out, further research needs identified and appropriate mitigation options selected. The sites (location in Figure 7.1) were selected to represent a range of different social, economic, environmental, and hydraulic vulnerability conditions: estuaries/deltas/open coasts and urban/industrial/tourist areas (see Table 7.1).

Following the same approach for describing the coastal system, i.e., the source-pathway-receptor-consequence (SPRC) approach, see Chapter 2, each case study site includes the following: estimate of hazard scenarios related to climate variability and change, analysis of the impact of flooding and erosion on infrastructures, environment and human activities, and preliminary selection and design of mitigation options based on local policy constraints and conditions. The analysis basically followed in each site the same generic procedure as described in Figure 7.2.

Table 7.2 summarizes the research undertaken in each of the case study sites. From the assessments, it is clear that there are some issues that are common across the sites:

1. balancing the functioning of the coastal system with socioeconomic use is a key aim of future mitigation,
2. establishing the expectations and risk perceptions of the local population should be fundamental to the coastal management decision-making process,
3. extreme water levels and flood extents are expected to increase over this century, mainly as a consequence of sea-level rise (SLR),
4. historical coastal management decisions can restrict the number and type of future mitigation options, and
5. considering a full range of mitigation options, individually and in combination, ensures the most effective way to preserve coastal areas and enhance their resilience.

Judicious sediment management is essential for many mitigation options. For the open coast sites of Varna (Section 7.2), Cesenatico (Section 7.6), Cancun (Section 7.7), and the Hel Peninsula (Section 7.8), where tourism is important to the local economy, mitigation options aim to optimize beach extents to both maintain the attractiveness of the area but also to provide a barrier to flood waters. For the Hel Peninsula, Cancun, and the Gironde (Section 7.4), there is also the need to manage the possibility of major change in the coastal system because of a breach in the local dune systems. For Cesenatico, there is the peculiar challenge of facing subsidence in low-lying areas, a challenge that is particularly exacerbated by climate change. In the Scheldt (Section 7.1), Yangtze (Section 7.3), and Elbe (Section 7.5) estuaries, understanding the water and sediment regimes within the estuary provides information that can be used in the development of appropriate mitigation option.

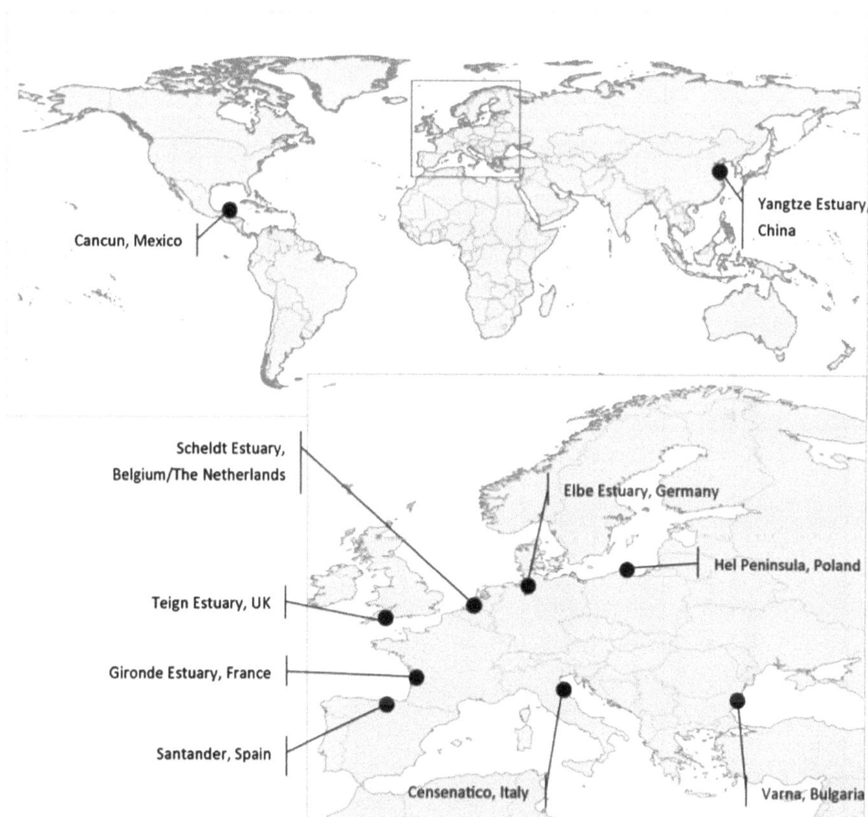

Figure 7.1 Location of the case study sites.

TABLE 7.1 Synthesis of the Most Relevant Characteristics of the Case Studies

Site Name and Section Number	Location	Geomorphological Setting	Relevant Urban Center	Habitats	Important Activities
7.1 Scheldt estuary	North Sea, Belgium/ Netherlands	Large estuary	Antwerp	Intertidal flats and marshes	Harbor, fisheries, power plants, industry
7.2 Varna	Black Sea, Bulgaria	Open coast with small inlets	Varna	Beach, dunes	Tourism, harbor, industry, power plants
7.3 Yangtze estuary	East China Sea, China	Large estuary	Shanghai	Intertidal flats and marshes	Harbor, industry
7.4 Gironde estuary	Atlantic Ocean, France	Large estuary	Bordeaux	Beach, dunes, intertidal flats	Harbor, marina, fisheries, refinery, nuclear power station
7.5 Elbe estuary	North Sea, Germany	Large estuary	Hamburg	Intertidal flats	Harbor, fisheries, nuclear power plant, industry
7.6 Cesenatico	Mediterranean Sea, Italy	Delta plain	Cesenatico	Beach, intertidal flats	Tourism, marina
7.7 Cancun	Caribbean, Mexico	Open coast, lagoon	Cancun	Beaches, lagoon	Tourism
7.8 Hel Peninsula	Baltic Sea, Poland	Spit	Władysławowo, Jastarnia, Hel	Beach dunes	Harbor, fishery, industry, tourism
7.9 Santander Bay	Gulf of Biscay, Atlantic Ocean, Spain	Open coast and spit enclosed bay/estuary	Santander	Beach, dunes, intertidal flats	Harbor, fisheries, airport, tourism
7.10 Teign estuary	English Channel, United Kingdom	Open coast and small estuary	Teignmouth, Newton Abbot	Beach,intertidal-flats, hard substrates	Harbor, fisheries, railway, port

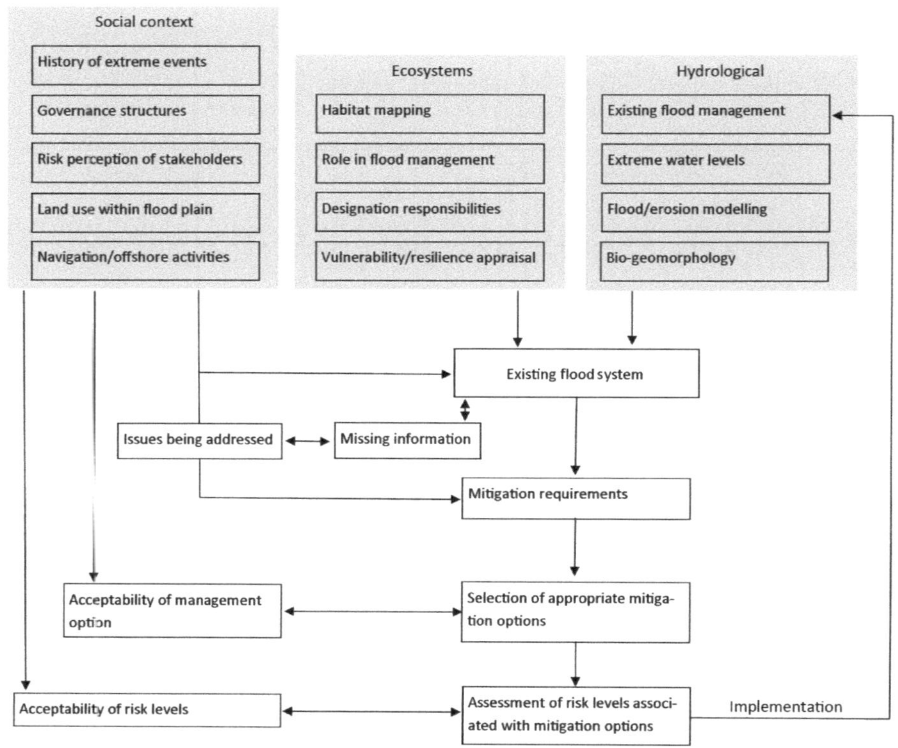

Figure 7.2 Flow diagram showing the steps each case study has undergone.

TABLE 7.2 Summary of Coastal System Knowledge Gained (in Addition to Climate-Induced Influence on Extreme Water Levels) and Mitigation Options Across the Case Study Sites

			Coastal System		
	Site	Main Issue	Additional Knowledge	Mitigation Options	Location in the SPRC Model
7.1	Scheldt estuary	Balancing safety with navigation and ecosystems	Combined coastal and river extreme water levels		Source
			Habitat resilience		Pathway
			Channel morphology		Pathway
7.2	Varna	Maintenance of beach areas	Sediment dynamics	Offshore submerged and floating breakwaters	Source
				Beach nourishment	Pathway
			Habitat vulnerability		Pathway
			Governance structure		Receptors/consequences
7.3	Yangtze estuary	Habitat resilience	Habitat vulnerability	Sediment management	Pathway
7.4	Gironde estuary	Flood warnings and evacuation	Dune breach scenarios	Real-time evacuation planning	Pathway
			Real-time extreme water levels		Source
					Consequences

7.5	Elbe estuary	Flooding of unprotected areas	Channel morphology	Artificial sand banks	Pathway
7.6	Censenatico	Maintenance of the beach and protection of the urban area from flooding	Subsidence	Beach and shore nourishment	Pathway
			Habitat vulnerability	Wave energy converter	Source
			Governance structure	Breakwaters (emerged ad submerged)	Source
				Evacuation planning	Consequences
				Insurance	Consequences
7.7	Cancun	Interaction of mitigation with natural processes	Sediment dynamics	Dune nourishment and restoration	Pathway
				Planning (spatial and building)	Receptors/consequences
				Hard engineering	Pathway
7.8	Hel Peninsula	Interaction mitigation with natural processes	Sediment dynamics	Offshore submerged breakwater	Source
		Stakeholder engagement		Beach nourishment	Pathway
				Dune stabilization	Pathway
			Stakeholder engagement	Planning (spatial and building)	Receptors/consequences
7.9	Santander	Functionality of infrastructure, beach maintenance	Probability density analysis	Beach nourishment	Pathway
				Submerged breakwater	Pathway
				Raising elevation of structures	Pathway
				Seagrass farming	Source
7.10	Teign estuary	Stakeholder acceptance of mitigation options	Stakeholder engagement	"Hand-stitched" management solutions	Pathway/consequences

7.1

Extreme Estuarine Water Levels and Channel Morphology, Effect on Flood Extents and Habitat Resilience: The Scheldt Estuary, Belgium and the Netherlands

Jaak Monbaliu [1], Qilong Bi [1], Tjeerd J. Bouma [2], Erik A. Toorman [1], Jim van Belzen [2], Patrick Willems [1]

[1] *Hydraulics Section, Civil Engineering Department, KU Leuven (KUL), Leuven, Belgium*

[2] *Royal Netherlands Institute for Sea Research, Spatial Ecology Department (NIOZ), Yerseke, The Netherlands*

7.1.1 Description of the Site

The Scheldt estuary discharges into the North Sea having passed through both the Netherlands and Belgium. It is relatively heterogeneous, with rural, urban, and industrial areas clustered along its banks and has been reshaped over the centuries to satisfy human needs (Meire et al., 2005). The scale of economic activity in the Scheldt basin has increased profoundly in the past century because of increasing human population and wealth. This has resulted in a significant increase in anthropogenic pressure on the estuary and a need for flood protection (Meire et al., 2005). There are records of heavy floods since medieval times, but the more recent floods of 1953 in Zeeland (the Netherlands) and 1976 in the Belgian part of the Scheldt basin resulted in the current flood defense measures.

Consequently, the estuary has a long history of flood management in both Belgium and the Netherlands that has resulted in the development of the Delta plan (The Netherlands, http://www.deltawerken.com/The-Deltaplan-/92.html) and the Sigma plan (Belgium, http://www.sigmaplan.be/en/sigma-plan). These generated extensive systems of dikes designed to protect against extreme events and accommodate climate change in the form of SLR.

The ports of Vlissingen, Terneuzen, Ghent, and Antwerp and the industrial activities and services connected to them are of crucial importance for the welfare of a wide region, with dredging an essential part of maintaining access. Increasing ship size and draught made deepening of the fairways necessary. Deepening programs were carried out in the 1970s, 1990s, and very recently between 2005 and 2010. The effect of this on both flooding and the sediment regime of the estuary is a concern, given the new management aims in the estuary. There is a long tradition of negotiation between the Netherlands and Flanders/Belgium and a well-established governance framework for decision-making. More recently, a bilateral agreement on a common policy and management has been undertaken that moves away from the focus on protection to an approach that gives equal weighting to (1) free access to the different ports, (2) safety, and (3) nature development (http://www.vnsc.eu/english.html).

7.1.2 The Scheldt Estuary Flood System

Running through three countries, the Scheldt catchment area is 22,116 km^2. The length of the river, from its origin in France, passing through Belgium to the mouth in the southwest of the Netherlands, is 355 km (Baeyens, van Eck, Lambert, Wollast, & Goeyens, 1997). The trajectory from the mouth (0 km at Vlissingen, Flushing, the Netherlands) to 160 km at Gent (Ghent, Belgium) where tidal propagation is artificially stopped by a weir is referred to as the Scheldt estuary (Figure 7.3). Tidal forces are strong along this part of the river, and fresh river water and salty seawater are responsible for a salt gradient. Because of this gradient and the river's extent, the Scheldt is often considered the last real estuary in Western Europe (Baeyens et al., 1997; Meire et al., 2005). The Scheldt estuary is both situated in the Netherlands and in Belgium and can be subdivided in three main sections: the lower estuary, or Western Scheldt, with a unique multichannel system; the upper estuary (with the harbor of Antwerp); and the fresh tidal river systems (Figure 7.3). The literature on the Scheldt estuary is abundant. Baeyens et al. (1997), Meire et al. (2005), and results from THESEUS research (THESEUS OD1.15, 2012; Monbaliu et al., 2014) provide a good overview.

7.1.2.1 Existing Policy and Management Strategies

Despite a continuous dialogue between the Netherlands and Flanders about the management of the Scheldt estuary, the approach to risk and the plans and actions for protection against flooding are country-dependent. In the Netherlands, flood protection for the Western Scheldt is taken care of by the Delta Plan; in Flanders, this is taken care of by the Sigma Plan. Recently the Sigma Plan has been revised and a social cost-benefit analysis study has been carried considering different mitigation options including the construction of a storm surge barrier, the use of the Eastern Scheldt as a controlled flood area, heightening and strengthening of dikes and quay walls, and the creation of additional space for the river by constructing flood plains (both controlled flood areas without nature development and areas with reduced tide with nature development); see De Nocker et al. (2006), RA-IMDC and Vito (2004) and Broekx, Smets, Liekens, Bulckaen, and De Nocker (2011). In the Netherlands, the Delta Commission presented new advice on coastal safety to the Dutch Cabinet in 2008 by drafting a Delta Program to implement its recommendations for a climate-proof Netherlands (Deltacommissie, 2008).

ProSes (2005) stated in its treatise on the development of the Scheldt that Flanders and the Netherlands are committed to developing a safe, accessible, and natural Scheldt Estuary. The Flemish–Dutch Scheldt Commission is the instrument for the realization of these common objectives. Plans have always been revised and adapted according to new insights and needs. The Long Term Vision 2030 (LTV 2030, 2001) for the Scheldt estuary has been developed based on the concepts of accessibility, ecology, and safety. The LTV 2030 outlines a development for the Scheldt estuary with a continuous dialogue between navigability, safety, and nature values. A methodology has been developed with key indicators for communicating if LTV objectives are met. Examples of indicators are population pressure, socioeconomic importance of ports, safety against flooding, loads of pollutant substances, threats to biodiversity, etc.: for a full list, see VNSC (2011), pp. 165. Underneath these indicators, test parameters, calculation parameters, explanatory parameters, and evaluation criteria are defined. Positive evaluation requires no further action. Negative evaluation is followed by actions if system knowledge is considered sufficient or by an additional research action if knowledge is considered insufficient. Note that the methodology for evaluation is also subject to evaluation of its applicability and has for the first time been tested for the 2009 reference situation (Ministerie van Infrastructuur en Milieu et al., 2013).

7.1.2.2 Meteomarine Climate: Present Conditions and Long-term Trends

SLR has been 1.69 mm/year since 1927 based on data from 1925 to 2000. An increased rate of SLR has been reported in according to Ozer, van den Eynde, and Ponsar (2008) so that the most common estimate used in impact studies is an increase of 0.6 m by 2100 compared with the 1990 level. Many studies include an extreme scenario, not that these studies consider such a scenario very likely but as a "what if" scenario. Here, an extreme scenario of 2 m for SLR has been adopted.

Tidal range and high waters have also been increasing quite dramatically. It is an adverse effect of SLR and anthropogenic pressure (deepening for navigation and/or sediment management which includes dredging, dumping and extraction strategies). The rate of increase also increases toward upstream sections. This is clearly seen from the graph in Figure 7.4.

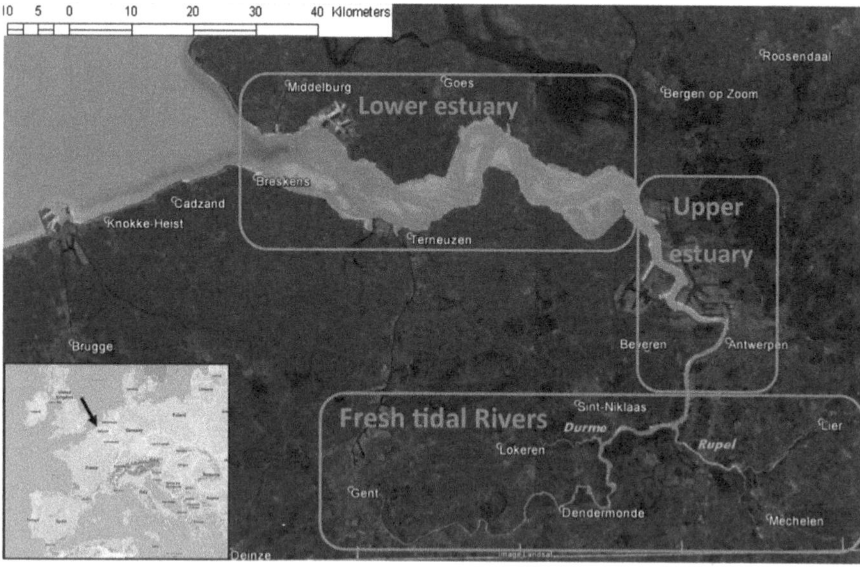

Figure 7.3 Scheldt estuary from the North Sea to inland Belgium.

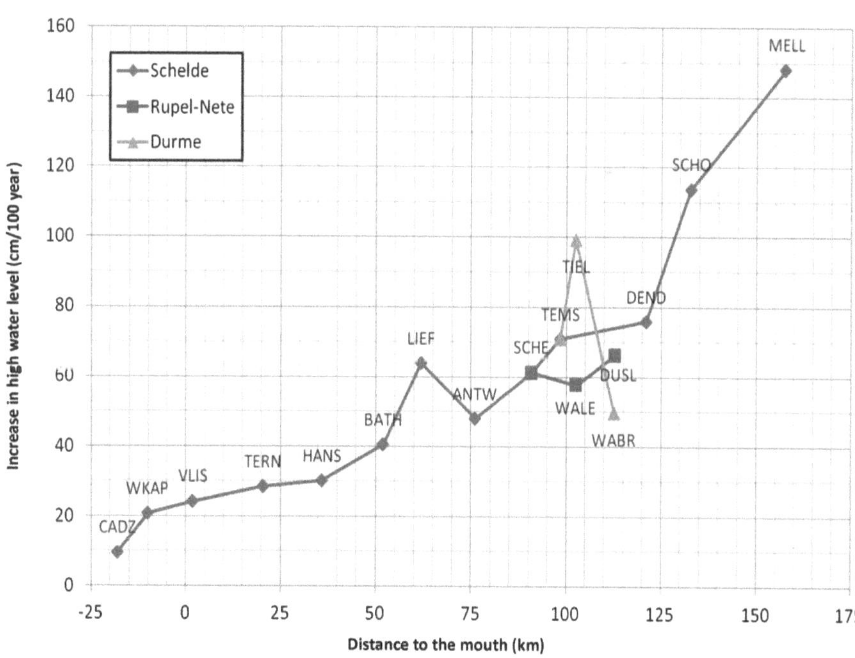

Figure 7.4 Trend of the increase in average high-water levels (in cm/100 year) in function of the distance to the mouth (average taken between 1960 and 1975 and ending in 2009). *Image from T2009 Report Scheldt Estuary, Ministerie van Infrastructuur en Milieu et al. (2013).*

High storm surges and extreme rainfall are at least partially correlated. Future surge and rainfall scenarios were implemented by applying quantile perturbations to the available historical time series (Ntegeka, Baguis, Boukhris, Willems & Roulin, 2008; Ntegeka, Roulin, Baguis, & Willems, 2014; Willems, 2013; Willems & Vrac, 2011). To take into account the correlation between surges and rainfall, the methodology for perturbing rainfall and evapotranspiration series has been extended toward perturbing surges correlated with rainfall (Ntegeka, Decloedt, Willems, & Monbaliu, 2012).

Waves are important for erosion of intertidal areas and are an important player in overtopping of dikes, but have not been investigated for the more upper parts of the estuary because there is no wave penetration from the North Sea and local wave generation is limited from fetch restriction. For the lower reaches of the Scheldt estuary, this aspect deserves more attention—also in view of possible changes in future wind direction and the influence of ship waves that can play an important role in the erosion process of tidal flats (Schroevers, Huisman, Van der Wal, & Terwindt, 2011).

Relatively little is known about suspended sediments. In the T2009 assessment report on the Scheldt estuary (Ministerie van Infrastructuur en Milieu et al., 2013), an increase of suspended sediments has been reported in the zone of maximum tidal water levels. Higher concentrations were particularly noticed in more recent years. Although there is no indication that this is more than a response to the recent deepening and to changes in water quality from the purification of the waste water of Brussels, there is some concern that this might be an early signal of a regime shift similar to the one seen in the Ems river (Winterwerp, 2011). Such a regime shift toward an up-river transport of fine sediments would have significant consequences for nature values in the estuary. Better knowledge through monitoring and better modeling capabilities of the fate of fine sediments is therefore essential for the management of the estuary.

7.1.2.3 Coastal Habitats

Downstream of the Dutch-Belgian border to the mouth is the estuary called Western Scheldt (Westerschelde). The Western Scheldt is more diverse in terms of its habitats compared with the river section, as the estuary consists of multiple deep and shallow gullies alternate with saltmarshes and mud- and sandflats. All wetlands in this estuary are protected under the Ramsar convention and EU legislation, Natura-2000 (Baeyens et al., 1997; Meire et al., 2005).

The increasing use of the estuary has led to pollution, modification, and stabilization of the navigation channels, and changes in land use have altered the initial natural landscape elements and biodiversity. These have either been severely degraded or lost entirely. In particular, large parts of the saltmarsh, mudflat, and other wetland areas have been claimed and transferred into agricultural land. For instance, a 6040-ha estuary has been reclaimed since 1900, a decline of 56% between 1900 and 1990. With these changes the estuary lost much of its dynamic nature (Meire et al., 2005).

Compared with their original extent, tidal saltmarshes are mainly present as marginal fringes along the Scheldt estuary (apart from the "Verdronken land van Saeftinghe"), but some saltmarsh vegetation is establishing on the mudflats in the middle of the estuary (Figure 7.5). Further degradation of the wetlands in the estuary, saltmarshes as well as mud- and sandflats, is undesirable. Additional deepening of the navigation channel, shipping traffic and relative SLR might cause further degradation of these habitats in the estuary (Temmerman, Govers, Wartel, & Meire, 2004; Kirwan & Temmerman, 2009). Although protected under Ramsar and Natura-2000 legislation, there is substantial economic pressure to further deepen the navigation channels to allow larges ships to enter the harbor of Antwerp. Because of the influence of the river on the estuary, there is a salinity gradient, which influences the vegetation composition of the saltmarshes along the Western Scheldt. Near the mouth, mainly saltwater species can be found.

The nonvegetated areas in the intertidal are mainly mud- and sandflats. The mudflats contain diversity in benthic species and are valuable foraging and resting grounds for many bird species, such as waders. Furthermore, seals use these mudflats for resting. Shellfish and lugworms are the most important species in the mudflat habitats, but in some parts diatoms are also important, stabilizing the mudflat surface preventing fine sediments from eroding (Weerman et al., 2010).

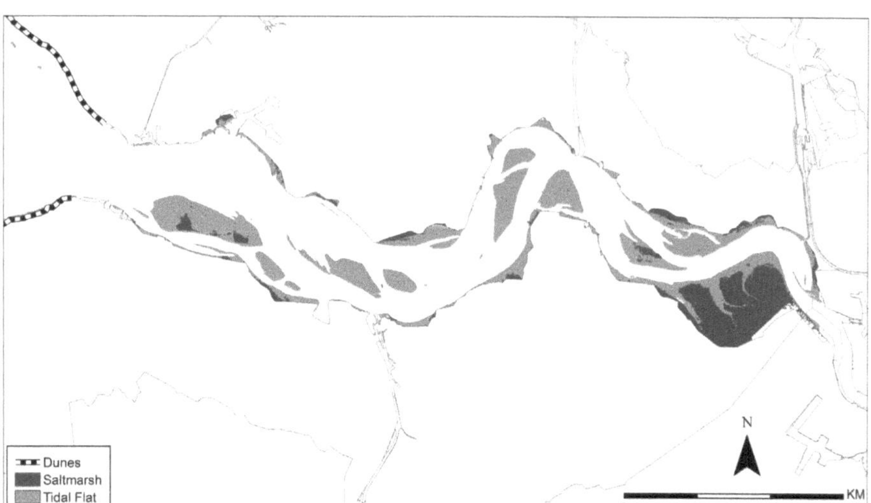

Figure 7.5 Coastal ecosystems along the Western Scheldt.

There is active maintenance dredging of the navigation channels, which affects the mudflats. Moreover, there are ongoing experiments with deposition of dredged material close to and at mudflat edges. These experiments are carried out under responsibility of the water and infrastructure administration.

Dunes, and adjacent beaches, are mainly found in the mouth of the estuary west of Vlissingen and Breskens. Here, these coastal ecosystems are the most important defense against extreme events, fulfilling the same role as the dikes that can be found further upstream. Large sections of these dunes have been reconstructed and reinforced during the past five years, because they were assessed to be weak links in the coastal defense during the implementation of the Delta law. These reinforcement works are almost complete. Along with the reinforcements, some additional freshwater wetlands have been developed behind these reinforced dunes.

Species that structure biogenic reef habitats, such as mussels, oysters, or *Sabalerria*, seem to play no role of importance in the Western Scheldt anymore. This is in contrast to the nearby Eastern Scheldt (Oosterschelde) where, for example, oysters occupy a substantial area of about 6.82% of the intertidal mudflat, an area that is still increasing (Smaal, Kater, & Wijsman, 2009). Nevertheless, after range expansions in other parts of the Netherlands (Eastern Scheldt, Wadden Sea (Smaal, Stralen, & Craeymeersch, 2005, Smaal, Kater, & Wijsman, 2009)), on some mudflats there are commercially exploited mussel beds as the Pacific Oyster (*Crassostrea gigas*) has colonized mudflats in the Western Scheldt. Along the whole Western Scheldt, Pacific Oysters can be found attached to hard structures (e.g., dikes, breakwaters, groins) and new oyster reefs are starting to form on mudflats west of Hoofdplaat (personal observation, van Belzen, 2011).

7.1.3 Application of the SPRC Model

A detailed discussion of the application and benefits of an SPRC approach for the Scheldt estuary can be found in THESEUS OD1.15 (2012) and Monbaliu et al. (2014). Here the main aspects are pointed out. For the Scheldt estuary, risk modeling was carried out using a tow-level SPRC approach. At the first level, a simplified linkage diagram captures the entire Scheldt, including both the Belgian and the Dutch parts. This indicated the reliance of flood protection on the presence of a well-developed dike system and the need for an investigation of the consequences of dike breaches. This was based on the existing lizard flooding model (http://flooding.lizard.net), and the impact is illustrated in Section 7.1.5.2. At a second level, a refined diagram for the Dendermonde area was constructed. This area was selected because of the combination of flood sources: downstream surges and risk from upstream rainfall runoff. The climate models predict wetter winters and dryer summers, whereas downstream water levels are rising. The correlation between extreme surges and extreme rainfall, which can lead to increased probability of flooding in the future, should be therefore explicitly incorporated in the scenarios used for flood modeling in this area (see Section 7.1.4).

The SPRC analysis showed that safety has been historically addressed by the existing dike system, evacuation plans, etc., which are well developed. However, the new, more balanced, management aims (safety, navigability, and nature development) recognize that there are strong interrelationships and that any flood mitigation or management option has implications to the other. Within THESEUS, three particular aspects of flooding within the estuary were identified that would benefit from further investigation. The first issue was, considering the extensive dike system, the potential for a dike breach in the Western Scheldt. The consequences are described in Section 7.1.5.2.

The second aspect focused on Dendermonde, Belgium, where the SPRC highlighted the presence of two flood sources extremes that could coincide producing a more extensive flood event. A methodology to assess safety levels and consequent risk levels by taking into consideration the correlation between surge and freshwater runoff in the future climate scenarios would therefore be beneficial. Also, the role of the anthropogenic pressure represented by a deepening of the estuary is considered. Flood sources and drivers and their influence on future risk conditions are described in Section 7.1.4. Flood maps for the Dendermonde area are shown in Section 7.1.5.1.

The third aspect focused on knowledge gaps in the description of hydrodynamics and sediment mechanics. This is of particular importance to predict the impact of the efforts that are made to maintain navigability. It has potential consequences toward flooding and toward the resilience of intertidal habitats particularly in the Western Scheldt.

Considering that the Scheldt is already heavily managed, the focus for this study site was not on defense options as such but on

1. how to improve the methodology to assess safety levels and consequent risk levels by taking into consideration the correlation between surge and freshwater runoff under future climate and channel deepening scenarios and their effects toward flooding;
2. how physically based conceptual changes to the friction model used in the hydrodynamics and sediment transport modeling can alter sediment fluxes substantially; and
3. the implications of the above for the understanding of ecosystem resilience.

7.1.4 Flood Sources and Drivers for the Dendermonde Case

Hydrographs (discharge measurements or numerical estimates at inland locations) and limnographs at the coast (water levels at the mouth of the estuary) are required as input to the hydrodynamic model boundary conditions. To map out the flood zones for given protection levels, synthetic hydrographs and limnographs were constructed and implemented following the method and algorithm developed by Willems, Vaes, Popa, Timbe, and Berlamont (2002). The algorithm comprised an extreme value analysis for a range of durations after which Discharge–Duration–Frequency (QDF) or Water level–Duration–Frequency (HDF) curves are developed. The QDF durations are then used to construct hydrographs; the HDF curves to construct limnographs, both at particular security levels. The composite hydrographs have the interesting feature that all flows for different durations have an identical return period. Given that the response time of the river to upstream rainfall runoff varies along the length of the river, composite hydrographs have the advantage of simulating river states (e.g., water levels, discharges)–for a given protection level (return period)–at any location along the river with one single simulation (because they compose all response times in one single runoff event). The same applies for the downstream water levels and composite limnographs. For the Scheldt application, the composite hydrographs were computed for each upstream main tributary river and combined with composite limnographs for the downstream water level boundary at the mouth of the estuary (at Vlissingen).

Whether the occurrence of high changes for surges coincide with the high inland changes for runoff flows was tested, taking into account the correlation between both under current and future conditions. This was done by comparing the surge at the coast for the same set of climate models with inland peak flow changes. Because surge extremes tend to occur mostly during October–March, surges occurring in this season acted as a guide for understanding the future coastal–inland covariations. Some climate models showed highest impacts at Dendermonde, resulting from a combination of high surges with high upstream runoff flows. Other models indicate that the mean surge changes in autumn would be combined with high, mean, or low peak flow changes. This led to different scenarios of combinations of climatic changes to the up- and downstream hydro- and limnographs; more details in THESEUS OD1.17, 2011).

This approach resulted in three main types of climate scenarios for further use: inland scenarios, coastal scenarios, and correlated inland-coastal scenarios. For the inland type, three climate change scenarios for rainfall and evapotranspiration were generated: a low, a mean, and a high scenario. These were converted in changes to the rainfall-runoff flows for the different upstream river catchments by means of catchment runoff models. The different climate scenarios use perturbed time series of rainfall and evapotranspiration as input to these conceptual runoff models. These scenarios are based on earlier work by Ntegeka, Baguis, Boukhris, Willems and Roulin (2008) and Ntegeka, Roulin, Baguis and Willems (2014), and use a large ensemble set of climate model simulations available for Belgium after statistical downscaling (Willems & Vrac, 2011; Willems et al., 2012; Willems, 2013). Because the study focused on flooding, two scenarios were used: the mean and the high (see Table 7.3). The coastal scenarios relate to SLR and storm surge. Three scenarios are considered: a mean scenario with an SLR of 60 cm and a 6% increase in surge levels, a high scenario with an SLR of 60 cm, and an increase of 21% in surge levels and an extreme scenario with an SLR of 200 cm and an increase in surge levels of 21% (see Table 7.3 and Ntegeka et al., 2012).

Next to the climate change scenarios, some deepening scenarios along the estuary were simulated. Deepening is the consequence of anthropogenic pressure (i.e., dredging in order to improve the accessibility to the port of Antwerp) and can be seen as a mitigation option against a possible decline in economic activities (harbor and aggregate extraction). However, within an SPRC analysis, the linkages show that it may have consequences for safety and nature values and should therefore be incorporated into any modeling setup.

The deepening scenarios are based on information regarding the historical deepening of the Scheldt since 1950 and simple extrapolation following the same trend until 2050. The deepening is carried out more than 110 km from Hansweert (50 km from the mouth) to Ghent (160 km). This was implemented in the hydrodynamic model by changing the topography along this 110 km of the river, which includes about 700 cross-section profiles in the MIKE 11 model of the Scheldt River (Truong, 2013). Based on the bathymetry of the hydrodynamic model, which in the original was based on the reference year 2000, the cross-sections were adapted to be roughly representative for these changes of mean depth below +0 m TAW (Belgian reference level). Historical records show a lowering of the bottom level nearly all along the estuary and a consequent increase of the water depth below a reference water level (taken as 0 m TAW) with an order of magnitude of 25–30% between 1950 and 2000 (see Figure 4.77 in Ministerie van Infrastructuur en Milieu et al., 2013). For the deepening scenarios, it is assumed that this trend would more or less continue and therefore an increase in volume below the reference water level 25% by 2050 and 45% by 2100 were taken as an estimate to look at the impact, being well aware that extrapolation of the deepening of the cross-section is rather simplistic and, for example, does not take into account possible differences in dredging, dumping, and aggregate mining strategies in the two periods. Anyway, all cross-sections in the MIKE 11 Scheldt model were adjusted to accommodate for these future changes. To verify if the model reproduces historical trends in water level changes along the estuary, the cross-section volumes below the reference water level were reduced by 25%.

TABLE 7.3 Climate Change Scenarios for 2071–2100

	Upstream Flow	Sea-Level Rise (SLR)	Surge
S1: Extreme	High: 30%↑	2 m↑	21%↑
S2: High	High: 30%↑	0.6 m↑	21%↑
S3: Mean	Mean: 16%↑	0.6 m↑	6%↑

As a result of the change in water depth along the river Dender, changes in flood areas were studied for floodplains along the lower reaches of the river Dender. Figure 7.6 shows the risk factor, calculated as the ratio in flood volume between the historical climate conditions and the climate scenario. The maximum value for the risk factor is 8 for the high scenario. This means that the 50-year flood volume at the Dendermonde area for the future climate is approximately eight times the volume for the current climate conditions. This is for the pessimistic scenario when high inland flow changes and high tidal water level changes occur; other scenarios produce lower impacts.

This is because a given increase in inland flows and tidal water levels results in a larger relative change for lower return periods. For the 1000-year return period, the flood volume at the Dendermonde area for the future climate is approximately 4 times the volume for the current climate conditions. This increase is mainly caused by the rise in tidal water levels. The contribution of the inland flows to the change in the volume is small; the risk factors only slightly increase (slightly higher than one) for the inland flow changes only.

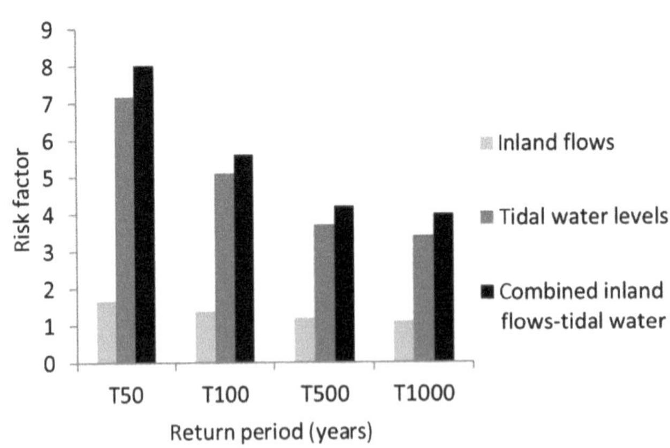

Figure 7.6 Risk factor for different return periods for the high scenario for changes in inland flows only (light gray); changes in tidal water levels only (dark gray); the combined-correlated changes in inland flows and tidal water levels, for the Dendermonde area (river chainages at 47 km to 49 km along the Dender).

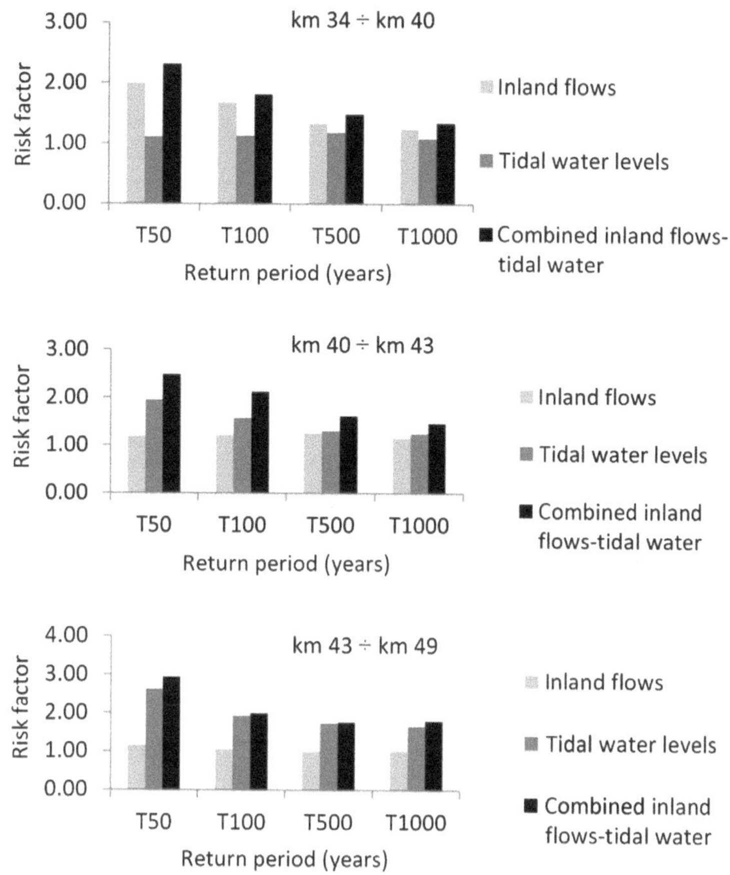

Figure 7.7 Risk factor for different return periods for the mean scenario for changes in inland flows only (light gray); changes in tidal water levels only (dark gray); the combined-correlated changes in inland flows and tidal water levels, for different reaches along the downstream Dender (different km chainages).

Figure 7.7 shows the same risk factor, but for the mean climate scenario and for different reaches along the downstream Dender. Along the Dender, reach with chainages km 34 until km 40, this risk factor is about 2 for the mean scenario, and the change in inland flows again contributes most to it. The changes in downstream water levels only contribute little (risk factor 1.11). The risk factor for the combined inland-coastal change is 2.31. For the next 3 km downstream reaches of the river, the risk factor for the combined changes increases to 2.48, with significant contributions from both the coastal changes (risk factor 1.95) and the inland changes (risk factor 1.19). Along the most downstream reaches of the Dender, the inland changes do not contribute and the risk factor is completely determined by the coastal changes.

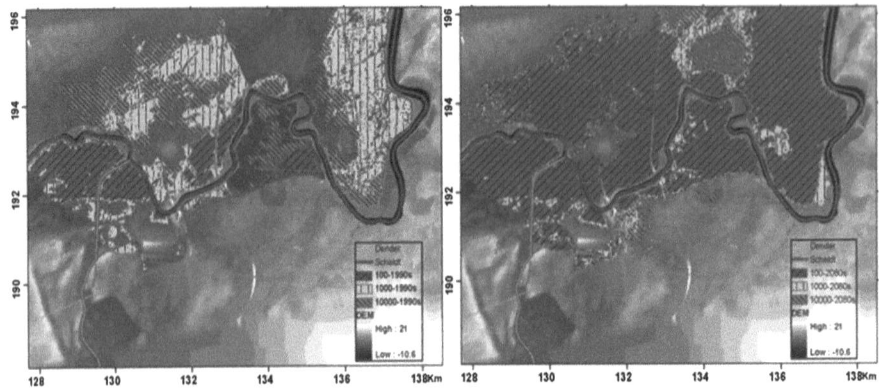

Figure 7.8 Flood hazard maps (1990s left figure, 2080s right figure) for the Dendermonde area for return periods of 100, 1000, and 10,000 years after simulation of the correlated upstream composite hydrographs and downstream limnographs.

A slight increase is in flood risk at Dendermonde is found in the deepening scenarios. The risk factor is 1.28 for 2050s deepening scenario, and 1.87 for the 2100 deepening scenario. This is due to the tidal wave propagating more inland, leading to higher water levels at Dendermonde. When the deepening and high climate scenarios are combined, the risk factor for the 2100s goes up to 2.12 for the mean climate change scenario and 6.82 for the extreme case and a return period of 100 years, which is about 2–7 times higher than the risk for current climate conditions (2000s). Figure 7.8 shows that the spatial extent of flooding substantially increases from the extreme climate scenario changes. The effect of the changes in river bathymetry (deepening scenarios) is less noticeable (for the assumed climate scenario). Note again that climate scenarios have high uncertainty and that these findings have to be interpreted with care.

Figure 7.9 shows the changes in water levels along the downstream reaches of the Dender for the pre- and postdeepening (dredging) conditions. The figure shows that there a slight increase in the water levels along the 10 km most-downstream reaches of the Dender for the postdredge scenario, but the increase is strong when combined with the high climate scenario.

7.1.5 Flood Mapping and Impacts

As indicated in Section 7.1.3, for flood mapping and impacts the focus is (1) on the Dendermonde area where the two flood sources (extreme rainfall and extreme tide and surge including SLR) are combined and (2) the lower estuary (Western Scheldt) where breach scenarios are investigated.

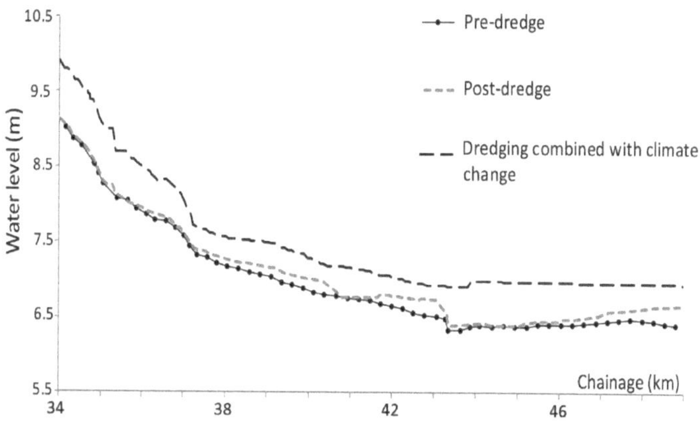

Figure 7.9 Water level changes along most downstream reaches of the Dender River for the separate and combined deepening (dredging) and extreme climate scenarios.

7.1.5.1 Dendermonde Area

For the hydrodynamic modeming, a MIKE 11 implementation was used. A quasi-two-dimensional (2D) approach was taken. The main river and the downstream part of the tributary rivers were modeled one dimensionally. The floodplains were schematized as a network of spills and one-dimensional (1D) flood branches. Particular focus was given to the floodplains in the Dendermonde region and the surrounding areas upstream along the Dender between Aalst and Dendermonde. To efficiently study the flood hazard along the Scheldt, upstream river flows (rainfall-runoff from the inland catchments) and the downstream water levels (North Sea at Vlissingen) were implemented in the form of composite hydrographs and limnographs representative for a given security level (see Section 7.1.4).

Figure 7.8 shows results of the hydrodynamic simulations for the Dendermonde area, in the form of flood hazard maps for return periods of 100, 1000, and 10,000 years. The figure also makes a comparison between the flood hazard maps for the current climatic conditions (1990s) in the left figure, and the same maps for future climatic conditions (2080s) in the right figure. The climate change scenario simulated account for the high scenario changes in SLR, surges, and upstream flows (S2 in Table 7.3) and the correlation between these changes.

Further investigation was made of the changes along the downstream part of the Dender River, where the impacts of the correlated coastal-inland changes were found to be highest (higher than along the Scheldt). Also the separate contributions of the inland and coastal changes were studied. Figure 7.10 shows the longitudinal profile of the change in water depth along the 15 km most downstream reaches of the Dender for the separate inland and coastal changes versus the combined changes for given return periods.

As seen in both Figures 7.8 and 7.10, the coincidence of high inland flows and high tidal water levels has a significant impact on the flood hazard along the Dender. For the simulation results for a return period of 500 years and the high climate scenario, water depths increase by around 120 cm at Dendermonde for the combined inland-coastal change scenario, compared with current climate conditions. Figure 7.10 shows how this transfers to water depth changes at other places more upstream along the Dender. The figure also shows the separate contributions of the upstream inland changes and the downstream coastal changes in these total changes.

It is also seen in Figure 7.20 that from chainage at 34 km to 40 km of the Dender, the inland flows are the major driver controlling the increase in water depth. The contributions of the tidal water levels are insignificant for that reach. Both the coastal changes and inland flow changes contribute to the rise in water depth along the next 4 km downstream of the Dender. The coastal changes drive the change in the water depth along the most downstream part of the Dender (the 5-km most downstream part). The combination of the tidal water levels and inland flows causes the highest water depths, tidal waters being the dominant factor.

For the deepening scenarios, a decrease of about 25% in mean bed level with reference to the 0 m TAW (Belgian reference level) is assumed. This is a rough extrapolation for the next 50 years (2000–2050) of the trend determined from the past 50 years (1950–2000) (Figure 7.11), and gives changes in the flood hazard as shown in Figure 7.12. An example of the combined impact of climate scenarios and dredging scenarios is shown in Figure 7.13.

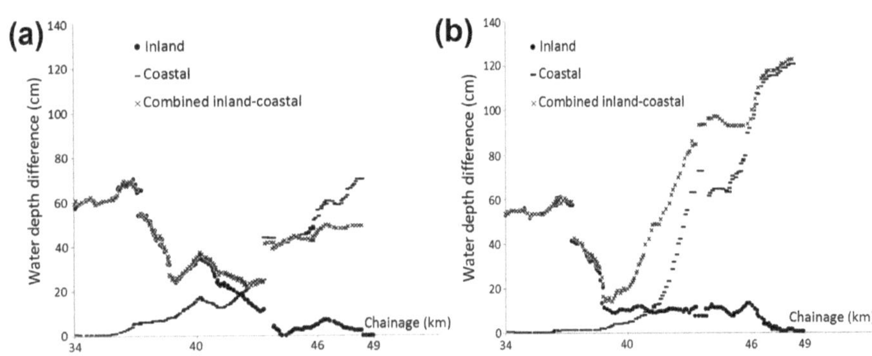

Figure 7.10 Difference in water depth between current conditions and the high climate scenario along the 15 km most downstream reaches of the Dender for the separate inland and coastal changes versus the combined changes, for return periods of 100 years (left) and 500 years (right).

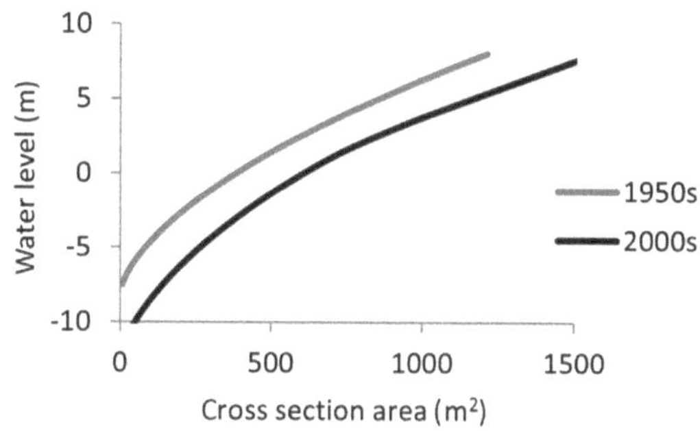

Figure 7.11 Water level versus cross-section area at Zeeschelde chainage 71.307 in predredging (1950s) and postdredging (2000s).

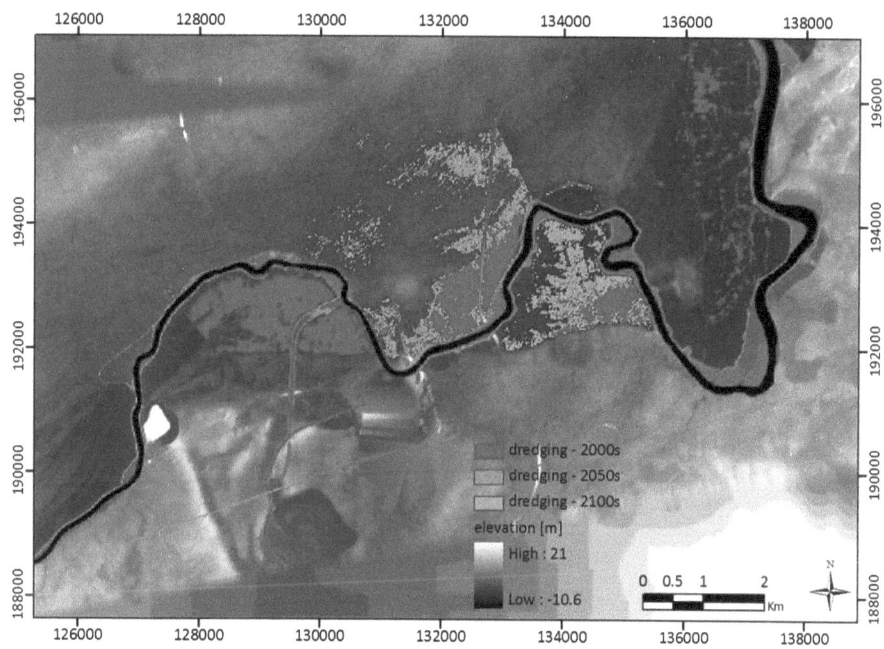

Figure 7.12 Flood hazard maps for a return period of 100 years at the Dendermonde area after simulating deepening (dredging) scenarios (2050s and 2100s).

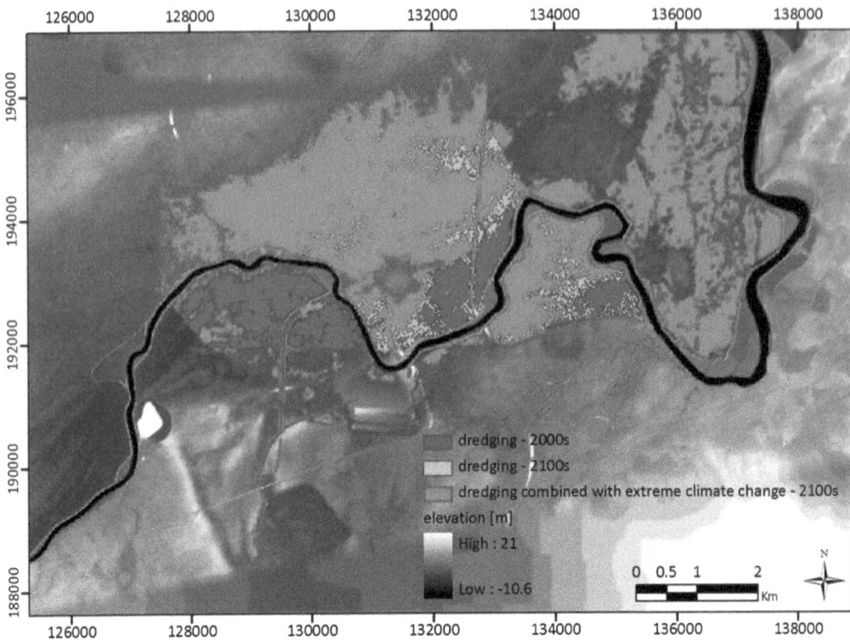

Figure 7.13 Flood hazard maps for a return period of 100 years at the Dendermonde area after simulating the combined impacts of deepening (dredging) and extreme climate scenarios (2050s and 2100s).

7.1.4.2 Lower Estuary

For the lower estuary (see Figure 7.3), the consequences of flooding are mapped based on Lizard flooding data (http://flooding.lizard.net). Along the Western Scheldt, the impact of a dike breach will be limited to a dike ring. If a dike breaches, a whole polder area within the dike ring will be completely flooded (see Figure 7.14 with flooding depths), reducing the probability of a second breach. Thus, it is not likely that all dikes will breach during one storm.

The Western Scheldt is a relatively heterogeneous in terms of land use, with some rural, urban, and industrial areas clustered along its banks. This means that impact of flooding can be very different, depending on the economic, social, or ecological activities that are present in a certain dike ring, as can be seen in the analysis of the damage costs (Figure 7.15). Damage can range from less than a million to more than a one thousand million euros. The biggest losses are found in urban and industrial areas. This analysis is based on lizard flooding data for current conditions (no future climate scenarios). The number of people that are exposed when a dike fails is also confined to individual dike rings. Each dike ring has a different level of occupation; this results in a heterogeneous outcome in the estimated number of victims (Figure 7.16). The highest impact (1404 people) will occur if the dike ring protecting the city of Vlissingen en Middelburg (near the mouth of the estuary) breaches.

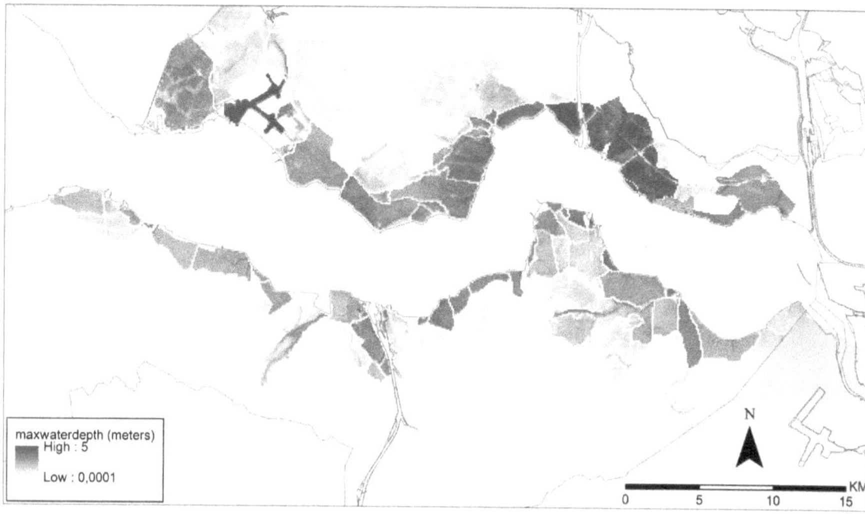

Figure 7.14 Maximum water depth if dike breaches under present conditions. *Data source: Lizard flooding.*

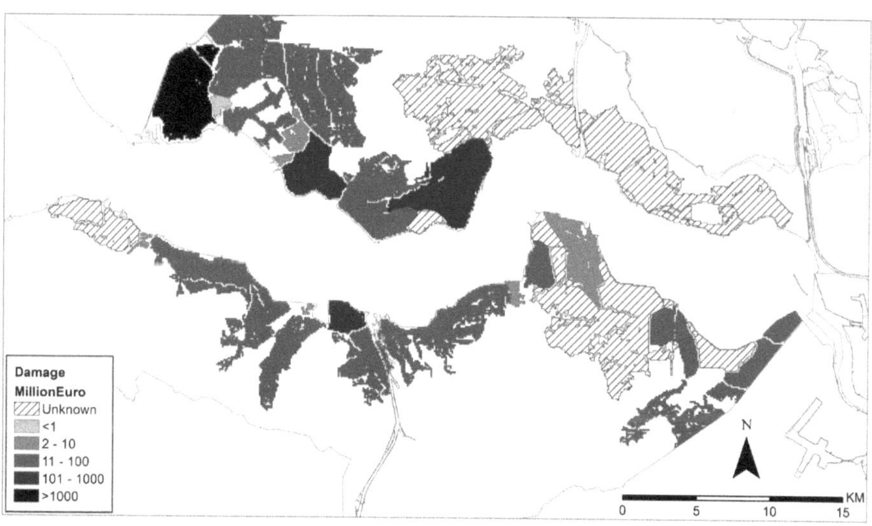

Figure 7.15 Economic damages in million euros when a dike is breached. *Data source: Lizard flooding.*

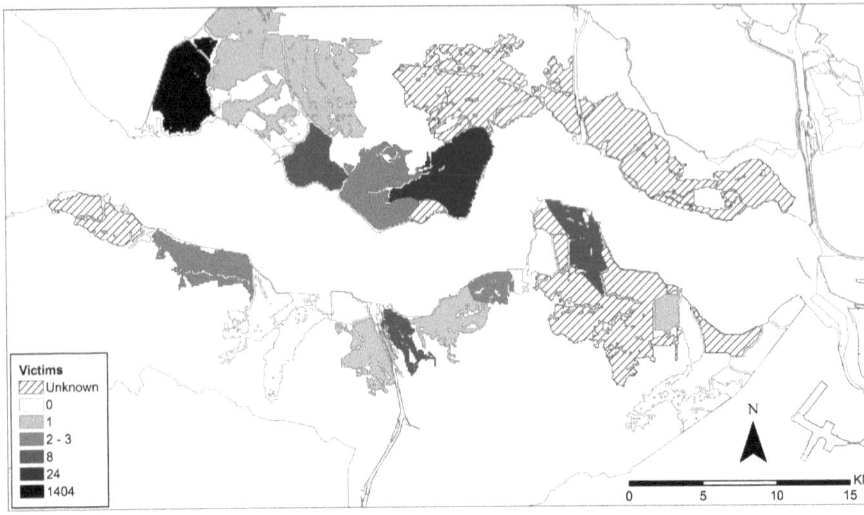

Figure 7.16 Number of people exposed when a dike fails along the lower estuary. *Data source: Lizard flooding.*

A breach would also have consequences for the terrestrial ecosystems located within the dike ring because flooding is not part of the natural dynamical processes in these habitats. The ingress of seawater for sustained periods could have severe impacts because most fresh water habitats and species cannot cope with salt water. As the lower estuary contains a number of internationally (e.g., Natura-2000) important areas and red list species, a dike breach in the lower estuary may have severe consequences for some rare species (see Figure 7.17).

7.1.6 How Sediment Dynamics May Alter Conclusions on Flood and Erosion Risks

The vulnerability of the Western Scheldt with regard to flood and erosion risks has been highlighted during the accompanying studies for the most recent deepening (LTV 2030, 2001). In the preparation phase, a solution has been sought to reconcile the economic interests with the preservation of the unique ecological values, mainly determined by the area of intertidal flats and flood plains, and the safeguarding against flooding. Deepening (i.e., removal of sediments by dredging) reduces the resistance for tidal penetration and disturbs the previous morphodynamic quasiequilibrium. The medium- and long-term effects are typically studied using numerical models for hydrodynamics and sediment transport.

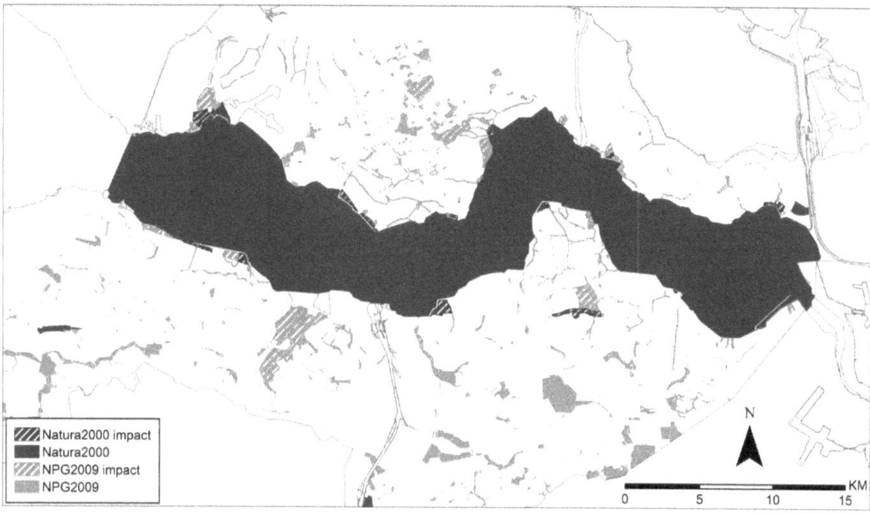

Figure 7.17 Natural areas affected with a dike is breach. *Data source: Lizard flooding.*

However, too many modeling studies do not account for the two-way coupling between water movement and sediment dynamics. The new approach, described in Section 3.5.3.3, has been implemented in a new fully coupled sediment transport model for the Scheldt estuary to demonstrate that the predicted natural morphodynamic response of the system is highly dependent on the treatment of the subgrid scale physical processes near the river bottom, whereas the predicted hydrodynamic variables (velocities and water levels) are found to be much less sensitive. However, in the long run, for time scales of decades, the high sensitivity on sediment dynamics may generate very different morphological conditions and bathymetries, which eventually may affect predictions for flood and erosion risks.

For studying sediment transport processes in the Western Scheldt, a 2D depth-integrated numerical model has been created using the TELEMAC system TELEMAC-MASCARET (2013) (Figure 7.18). The nonuniform model grid, consisting of 67,689 linear triangular elements and 37,527 nodes (Figure 7.19), has been optimized such that iso-depth lines are followed and grid refinement is applied where details, such as creeks, require this and where steep bathymetric slopes are encountered.

To deal with the complexity within the research domain and improve the prediction accuracy, the model has been set up as realistically as possible, i.e., including two-way hydrodynamic-sediment transport coupling, mixed sand-mud sediment transport (bedload transport as well as suspended load in the water column) and a dynamic nonuniform bed composition (Figure 7.20).

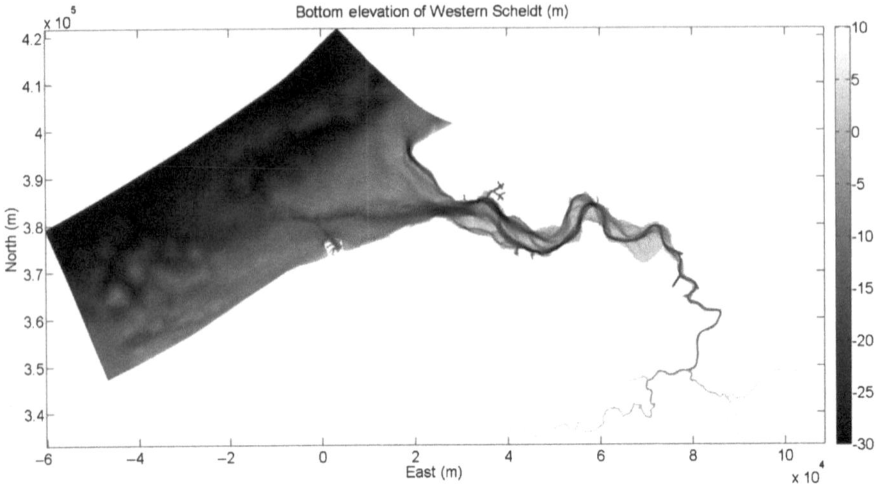

Figure 7.18 The model domain and bathymetry of the Scheldt estuary.

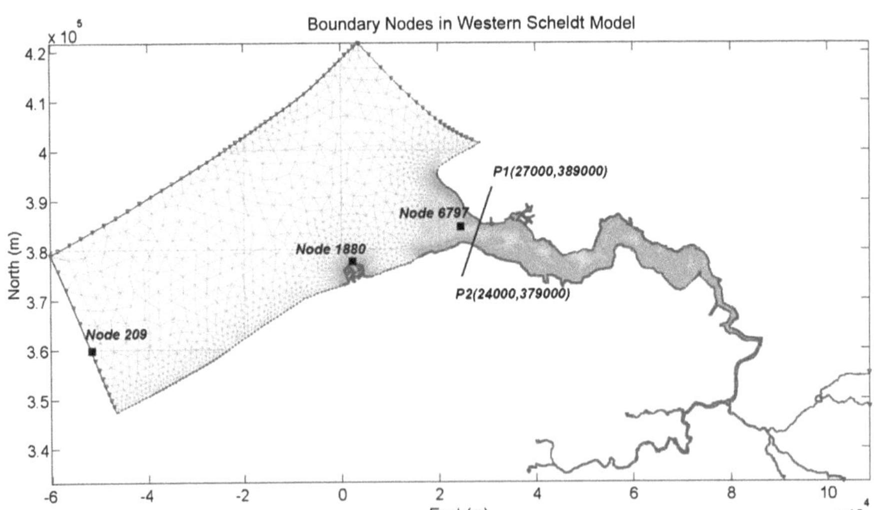

Figure 7.19 Mesh and North Sea open boundary (indicated with black triangles) of the Scheldt model. Model output nodes and the cross-section where model fluxes are calculated are indicated.

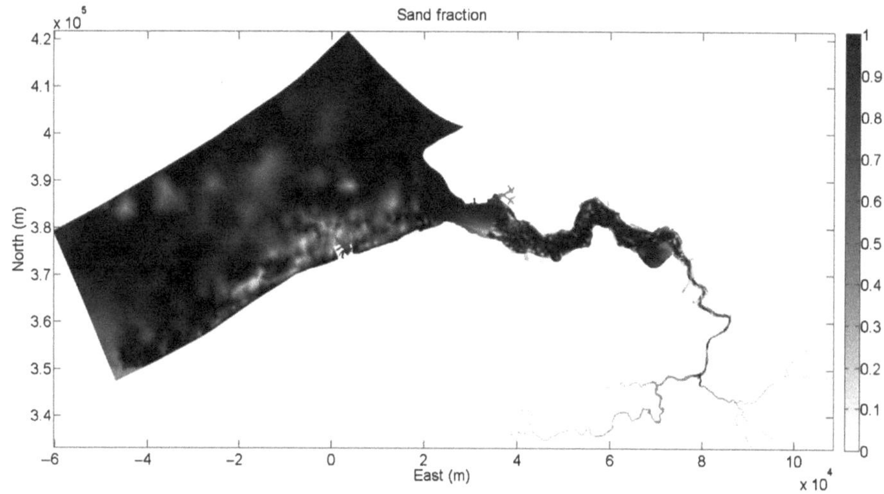

Figure 7.20 A sand fraction distribution map used as initial bed composition (combined data from Rijkswaterstaat, the Netherlands, and RBINS-MUMM, Belgium).

Wave effects are important in the coastal area around the mouth because they mobilize fine sediments found in the high turbidity area around Zeebrugge (node 1880 in Figure 7.19), which may be imported by the tide into the estuary. Therefore, wave effects have also been incorporated in the latest version of the model. With waves included, the model is able to represent the wave–current–sediment interactions, also called three-way coupling as well as the mechanisms mentioned previously.

Numerical simulations have been carried out with the KU Leuven-Télémac Western Scheldt model to compare and analyze the model results obtained for different friction laws. Three bottom-friction laws have been implemented and applied to the same test case (identical model setup and initial conditions), partially calibrated and validated with field data from 2009. The first model applies a constant roughness coefficient, which at present is the most current practice in engineering studies. The second model employed a depth-dependent formulation, obtained by integration of the standard logarithmic velocity profile (Prandtl-von Karman). The third model is an extension of the second, including viscous stresses, effective bed roughness (dissipation by separation vortex generation), and additional dissipation mechanisms that become important in high-concentrations as found above the bed (i.e., interparticle friction and collisions, and particle wake turbulence) resulting in a bed shear stress predictor (Section 3.5.3.3). In each case, the computed friction has been converted into an equivalent Chézy roughness coefficient.

Because the new bed shear stress closure Toorman & Bi, 2012, 2013a and 2013b is constructed as the superposition of the turbulent and the laminar part, it allows the simulation of both transient conditions and the transition during drying or wetting of tidal flats Toorman, 2011. Nikuradse's empirical roughness model is implemented allowing to account for both hydraulic smooth and hydraulic rough conditions. The latter is important because the tidal induced oscillatory flow may fulfill either condition, depending on the phase of the tidal cycle.

An additional advantage of the new friction law is that its value automatically tends to infinity when the water depth goes to zero. Therefore, the drying and flooding of intertidal flats can now be modeled without specifying an inundation threshold. This proved to work very well for the many intertidal areas in the Western Scheldt. The benefit is that intertidal morphodynamics can now be modeled at great detail for the first time. Erosion and deposition in these areas can now be estimated with much higher accuracy, as well as their contribution to the overall net fluxes.

Another innovative feature is the new dynamic deposition criterion. The critical shear stress for deposition is no longer taken constant, but related to the available energy for suspending particles. Its instantaneous value is obtained from a local suspension capacity condition. It is no longer a pure material parameter but a function of sediment concentration, settling velocity, water depth, and bed shear stress (Toorman, 2002).

The sensitivity of the model to the bed friction model has been evaluated by intercomparison of the computed sediment budget at the mouth of the estuary. This has always remained a point of discussion, because previous model results did not match with estimates based on field data, even opposite directions have been claimed. Near-bottom sediment transport, which is hard to model and which escapes from most measuring techniques, most likely plays an important role in explaining the differences. The comparison of the three friction models (Figure 7.21) indeed results in very different evolutions of the net sediment budget calculated at the mouth over a cross-section from roughly Vlissingen to Cadzand (P1-P2 in Figure.7.19).

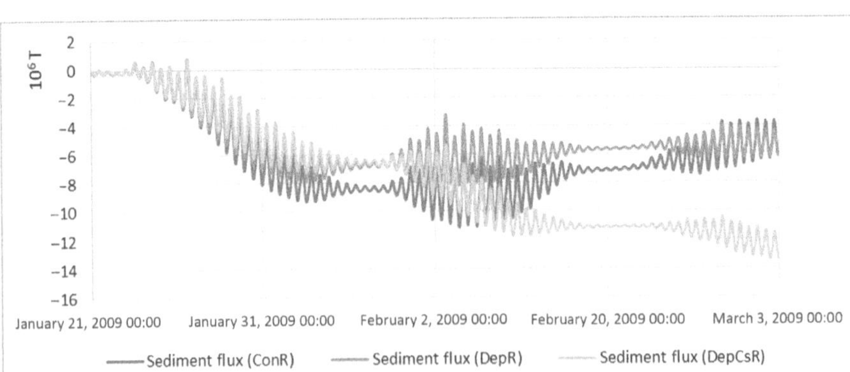

Figure 7.21 Comparison of the computed accumulated sediment flux over the period January 21, 2009 to March 2, 2009 (simulation started at January 1, 2009, day 21 is taken as reference) for three bottom-friction models (ConR = constant Chézy, DepR = depth-dependent, DepCsR = depth and sediment load-dependent). The effect of sediments on the bottom friction (model DepCsR, light gray) clearly is significant.

These differences are also reflected in significant differences in local erosion–sedimentation patterns. However, despite the many monitoring campaigns, mainly focussing on hydrodynamics, hardly any time-series data on SPM before 2012 are available. Therefore, the validation of the model for the simulated period (of 3 months) in 2009 is incomplete. It will require longer term simulation runs (>1 year) and a careful evaluation of parameter sensitivity before justified conclusions can be drawn on how the bed friction law affects the long-term morphodynamic response of the estuary. Furthermore, the modeling approach for sand-mud mixture transport in the Télémac sediment module Sisyphe needs further validation. Therefore, at present it is still not possible to estimate the long-term evolution of the estuary and its implications with regard to flood and erosion risks with sufficient confidence. This was also pointed in Ministerie van Infrastructuur en Milieu et al. (2013).

7.1.7 Conclusions and Recommendations for Risk Assessment Studies

It is recognized that the Scheldt estuary is already heavily managed. In 2001, a more holistic view of the estuary was adopted and a long-term vision was published (LTV 2030, 2001) where the overall management aim for the Scheldt estuary was described in terms of (1) free access to the different ports, (2) safety, and (3) nature development. The SPRC model as worked out in THESEUS project provided a framework that encourages this type of approach and allowed to identify some particular issues of concern: (1) the combination of flood sources and the expected role of climate change and anthropogenic pressure with as example the impact on the Dendermonde area where both coastal surge and upstream rainfall-runoff are of concern; (2) the need for improved process knowledge when dealing with sediment concentrations and sediment transport; and (3) the possible impact on the ecological functioning and the value of the ecosystem to mitigate inverse effects.

It was demonstrated that including the aspect of correlation between inland rainfall and coastal surges in climate change impact studies is necessary to make a good risk assessment at least for those parts of the estuary where both sources have a similar impact on flood risk. The methodology to include future climate effects in both oceanic (SLR, tide, and surge) and land component (rainfall-runoff) of estuarine flow can easily be incorporated in an SPRC approach and will result in more realistic impact studies.

It was also recognized that anthropogenic pressure can and should be included in impact studies. Here it is exemplified by a rather crudely implemented deepening of the estuary in line with past observation. This has been done to some extent already for the design and implementation of the Delta and Sigmaplan. Although there is a large uncertainty toward the topographic evolution of the estuary, extrapolation of observed trends in bathymetric evolution can give a first estimate. Although such an extrapolation is speculative and uncertain because it did not take into account possible changes in deposition strategy of dredged material nor in strategy for aggregate extraction, it gives an indication of the relevance compared with, for example, those expected from climate change effects. In that sense, it puts an order of magnitude on the uncertainty involved.

Process knowledge on sediment mechanics (erosion/deposition/material in suspension) remains an issue of concern. Sediment dynamics have a great influence on the ecosystem functioning in general (e.g., water quality, light penetration, stability of multichannel systems), and, more specifically, the sediment bed variability will have a large impact on the response and resilience of coastal habitats and saltmarshes in particular. The new physically based approach to model bottom roughness, as described in Section 7.1.5, results in very different sediment fluxes compared with a more traditional approach. This has a direct impact on the ability to perform quantitative modeling of the morphodynamic response of coastal and estuarine areas to specific events as well as to long-term trends. Improvements in modeling will result in improvement in interpretation of measurements (still very scarce), improvement in measuring strategy and instrumentation, and in better recommendations toward monitoring.

The response and resilience of the coastal habitats in the Scheldt estuary are to a large degree linked to changes in the geomorphology and sediment dynamics of the estuary. Forcing of the habitats by tidal currents and waves will change if the morphology is altered and can have a large impact on the coastal habitat. For example, research in the estuary (see Chapter 4) has shown that the level of short-term variability (from within a spring-neap tidal cycle to within season) in bed level is an important determinant of reestablishment of saltmarsh vegetation. If the variability is too high, vegetation gets uprooted and will not establish or can trigger the large-scale erosion of saltmarsh by forming a scarp (Bouma et al., 2014). As a result of changes in the morphology of tidal flats in front of saltmarshes, wave conditions change and can increase the bed level variability, thereby hampering marsh establishment and/or enhancing erosion. Similarly, increased exposure to shipping waves, from the entrance of more and bigger ships, can contribute to the increased forcing on the bed.

As all relevant coastal habitats in this estuary are soft-bottom habitats (dunes, salt-marshes, tidal flats and shellfish reefs) they will to a large degree behave in a similar way as saltmarsh vegetation and changes in the sediment variability (although with different thresholds) will determine their response and resilience. Thus, the ability to improve the prediction of sedimentation and erosion patterns plays a key role in understanding and predicting the response and resilience of the coastal habitats in the estuary. The accurate estimation of sedimentation and erosion patterns is further complicated, because coastal habitats consist of foundation species that are able to alter environmental conditions. For example, vegetation can dissipate wave energy, divert flow, or increase the critical threshold at which sediments erode because of binding of roots and rhizomes (Christianen et al., 2013). In a similar way, oysters and mussels can dampen wave energy and divert flow. This alters the sediment balance locally and therewith alters the response and resilience of these coastal habitats to some degree. Although not included at the moment, the model should be extended in the future to account for such biotic effects, such as energy dissipation by intertidal vegetation to improve our ability to foresee how the functioning and resilience of the coastal habitats will be affected by management decisions and/or by changes in sea levels and weather patterns.

Acknowledgments

The authors acknowledge the Hercules Foundation and the Flemish Government department EWI for providing access to the Flemish Supercomputer Center. The hydrodynamic simulations for the rivers Scheldt and Dender were conducted by KU Leuven in cooperation with the Flanders Hydraulics Research division. Also the NEVLA model data have been made available by Flanders Hydraulics Laboratory (Flemish Ministry of Mobility and Public Works). We also gratefully acknowledge Lizard Flooding for providing the data to map the consequences of flooding in the Western Scheldt area, Kirsten Beirinckx from Maritieme Toegang (Flemish Ministry of Public Works) for providing Figure 7.4 and Figure 7.6, and Hana Ortega Yamamoto for careful screening and for producing many of the figures.

7.2

Mitigating Flood and Erosion Risk using Sediment Management for a Tourist City: Varna, Bulgaria

Ekaterina Trifonova [1], Nikolay Valchev [1], Stoyan Keremedchiev [1], Iliyan Kotsev [1], Petya Eftimova [1], Valentina Todorova [1], Tsenka Konsulova [1], Valentina Doncheva [1], Mariana Flipova-Marinova [2], Stoyan Vergiev [2], Jordan Petkov [3], Rosen Nikolaev [3], Wout de Vries [4], Rodolfo Silva [5], Nataliya Andreeva [1], Panagiota Galiatsatou [6], Dimitrichka Kirilova [1], Yannis Krestenitis [10], Alexander Polonsky [7], Ioannis Androulidakis [10], Katerina Kombiadou [10], Ralf Weisse [8], Edgar Mendoza [5], Gerardo Durán [5], Theofanis Karambas [6], Theoharris Koftis [6], Panayotis Prinos [6], Sergey Kuznetsov [9], Yana Saprykina [9]

[1] *Institute of Oceanology, Bulgarian Academy of Sciences (IO-BAS), Varna, Bulgaria*

[2] *Department of Natural History, Varna Regional Museum of History (IO-BAS), Varna, Bulgaria*

[3] *Department of Mathematical Sciences, University of Economics, Varna, Bulgaria*

[4] *INFRAM International BV (INFRAM), Marknesse, The Netherlands*

[5] *Instituto de Ingeniería, Universidad Nacional Autónoma de México (UNAM), Coyoacán, DF. México*

[6] *Laboratory of Hydraulics and Hydraulic Works, Aristotle University of Thessaloniki (AUTh), Thessaloniki, Greece*

[7] *Marine Hydrophysical Institute - Ukrainian National Academy of Sciences (MHI), Sevastopol, Ukraine*

[8] *Helmholtz-Zentrum Geesthacht Centre for Materials and Coastal Research (HZG), Geesthacht, Germany*

[9] *Coast & Shelf Laboratory, P.P. Shirshov Institute of Oceanology of Russian Academy of Sciences (IORAS), Moscow, Russia*

[10] *Laboratory of Maritime Engineering and Maritime Works, Department of Civil Engineering, Aristotle University of Thessaloniki (AUTh), Greece*

7.2.1 Description of the Site

Varna is the third largest city in Bulgaria and covers an area of 205 km^2 and, according to the monthly tables of population for September 2013 (www.grao.bg), has a population of 372,279; effectively more than 70% of the population of the

Varna province concentrated on less than 6% of the land area. In addition, the population within the city can increase up to 1 million during the tourist season (Municipality of Varna, 2011). Located on the Western Black Sea coast, tourism is an important aspect of the economy, with more than 57,000 bed-places available along the coast (NSI, 2013), with many people attracted by the sandy beaches (see Figure 7.22), the wider natural environment, and a rich historical and cultural heritage.

The service industry, including tourism, dominates the gross added value of the district of Varna (62%), followed by industry (35%) and agriculture (3%) (Varna Municipality–www.varna.bg). Supporting the tourist industry, the Varna region has an international airport, railway, and motorway networks and well developed water-, power-, heat-, and gas-supply systems. Varna also has one of the largest harbors in the country, hosting the Bulgarian Navy forces' headquarters and a large naval base on the artificial island in Varna Bay. The economic activity in the Varna region generated BGN 3,820,067 thousand in 2007, which represents around 6.75% of the gross national product and 60.7% of the domestic product generated in the Northeastern region of Bulgaria.

Figure 7.22 Beaches in Varna are heavily used during the summer months. *de Vries et al., 2011.*

Figure 7.23 Erosion caused during a storm in February 2012 at Varna Beach. *Eftimova, 2012.*

However, exposed to the wave climate of the Black Sea, extreme events can cause significant effects. During a storm in February 1979, the significant wave height of 4.58 m was reached, holding at above 4 m for approximately 24 hours, as wind speed had risen up to 24 m/s and maintained high values between 15 and 20 m/s more than 3 days (Trifonova, Valchev, Andreeva, & Eftimova, 2010). The long-lasting winds from a constant direction gave rise not only to very rough seas, but also to extreme storm surge in the coastal zone; a maximum rise of 1.4 m was registered at tide gauge station in Irakli. As a consequence, whole beaches were flooded; at Golden Sands, the promenade was destroyed (Stakev, 1980); the low-lying spit was breached at Pomorie and around 400 houses and several storehouses were inundated, and a number of electric transmission facilities and telephone channels were damaged (Grozdev, 2006); and at Primorsko Beach, erosion (thickness of eroded layer more than 1 m) was so substantial that the foundation of the pier, located near the international youth center, became exposed (Stakev, 1980). Similarly, a storm in March 2010 had significant impacts; broken and missing boats e.g., in, for example, Shabla and Rusalka, damage to seasonal beach structures (Rusalka), and beach front properties (e.g., Kranevo). Dikes were also breached between Kranevo and Ekrene and in Pomorie. In Shkorpilovtsi, as a result of the wave action, the upper part of the profile between the shoreline and 2 m depth was intensively eroded; the average thickness of the eroded layer here reached 0.42 m and the maximum was 0.92 m (Trifonova et al., 2010). More recently, a storm in February 2012 caused damage on Varna Beach (see Figure 7.23).

These events have highlighted the need for the authorities to develop sustainable coastal management plans that are effective in reducing the impacts of coastal flooding and erosion while still preserving the natural assets for which the area is known. The most undesirable consequences of storm flooding would be inundation of the promenade with implications for the cliff erosion at Varna Maritime Park and the initiation of landslides there. This infrastructure (the promenade) would not only be inundated, but also partially or completely destroyed, which would interrupt the road link with the affected sites and consequently hamper the recovery of the region after the flood. Because beaches in Varna are popular tourist destinations, reducing the recreational potential of the area because of the flooding of beaches and the destruction of roads and tourist infrastructure would negatively affect the revenues of the tourist industry. In such a case, the local economy will suffer from a reduction in income caused by business interruption during both the extreme event and the recovery period. Furthermore, a significant investment will be needed for the infrastructure recovery/rebuilding.

Varna is located in the Western Black Sea sector and spreads along the coasts of the homonymous bay. The Black Sea is almost tideless and historical records of the extreme hydroclimatic events show that for the Western Black Sea, shelf storm waves and sea-level fluctuations are the most dangerous sources of flood hazards (Andreeva et al., 2011; Trifonova et al., 2010; Trifonova, Valchev, Andreeva, & Eftimova, 2012; Valchev, Trifonova, Andreeva, & Eftimova, 2009).

Interestingly, in maps from the nineteenth century, the current beaches of Varna (between the Port and Pochivka areas) were shown as a cliff coast (Dachev, 2003). The sand strips were either completely absent or of insignificant width and those recorded were found exclusively at ravines' mouths (the ravine system used to cross landslide areas, providing a sediment supply to the Varna beaches during rainstorms). Fast industrialization in the Varna region at the end of the nineteenth century necessitated the construction of a port, which was completed in 1906. The interruption of the northeast to southwest long-shore sediment flux by the main port area caused rapid beach growth in front of the existing cliffs (Rozhdestvenski, 1971). These artificially created beaches have remained intact until the expansion of Varna and construction of a coastal protection complex occurred during the late 1970s and early 1980s. This reduced the natural nourishing of the beach and activated the landslide/landslip processes of Varna coastal cliffs. The coastal protection complex consists of four cross-shore constructions, built consecutively between 1981 and 1985, which extended seaward to a depth of between 2 and 3 m, and a 2900-m-long dike constructed directly on the existing sand strips. The building of the dike, Y-shaped constructions, and modifications to the local ravine system finally interrupted the natural sediment supply to the beach. After the construction of the coastal protection complex and the resultant sediment deficit, the northern part of Varna beach gradually retreated and the adjacent sea floor lowered.

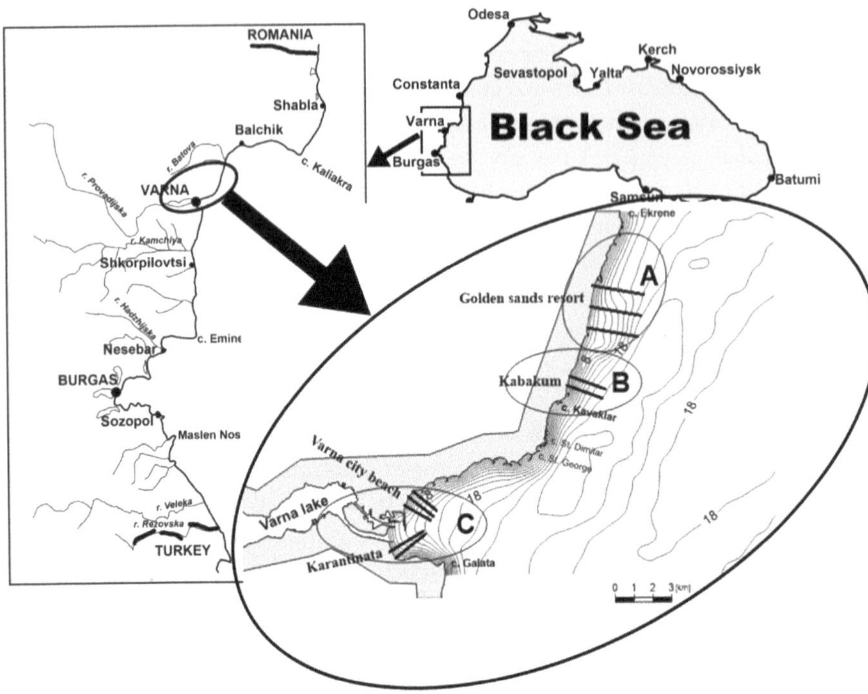

Figure 7.24 Location of beaches susceptible to flooding in Varna.

In addition, changes in the positive sediment balance of Varna Bay are related to rapid inland urbanization processes. During the beginning of the past century, ravines passed through sparsely populated areas covered with vineyards, orchards, and natural land and were not modified by human intervention. This allowed the delivery of eroded sediments to the coastal system during extreme events. Urbanization of the areas in the vicinity of the ravines caused the regulation of water outfalls and revetment of the ravine's lower stream. As a consequence, the amount of sediment was drastically decreased, affecting the natural nourishment of Varna beaches and initiating erosion and flooding susceptibility (THESEUS OD1.10, 2012). Four beaches are known to be subject to flooding: Golden Sands (A), Kabakum (B), Varna Central, and Karantinata (C); see Figure 7.24.

7.2.1.1 Existing policy and management strategies

Four Ministries of the Republic of Bulgaria currently have responsibilities for coastal flood management, including the application of existing policies and the development of new strategies and policies. At the national level, the *Ministry of Environment and Water* is responsible for the development of the national coastal zone policy under the Water Act amendment and the transposition of the EC Floods Directive in Bulgaria (Republic of Bulgaria National Assembly, 2010). It is also responsible for the management and protection of all areas of conservation importance according to the Protected Areas Act and the Biodiversity Act (Republic of Bulgaria National Assembly, 1998, 2002). In Varna, this Ministry is represented by the Basin Directorate for water management in the Black Sea region and the Regional Inspectorate on Environment and Water. In addition, the *Ministry of Transport* (in Varna, represented by the Executive Agency of the Maritime Administration and Port of Varna) is responsible for development of other national policies, such as those on ports and (navigational) waterways; the *Ministry of Interior* (in Varna represented by the Department of Civil Protection [public safety]) is responsible for the disaster management; and the *Ministry of Regional Development and Public Works* (in Varna represented by the Geoprotection Agency) is responsible for development of the Bulgarian coastal spatial planning Master Plan.

From the coastal flood management point of view, there are appropriate sets of institutional, societal, and legislative entities responsible for flood protection in place (de Vries et al., 2011), but the coordination between them sometimes lacks efficiency. Currently, there is no party responsible for integration across the ministries, their institutions, and those affected (THESEUS OD1.15, 2012), although exchange of knowledge on flood defense types and free access to information (e.g., weather forecasts, flood warnings, historic data) would be beneficial.

Historically, flood management within Bulgaria has concentrated on fluvial flooding. The current, incomplete, Master Plan that was developed for the entire Bulgarian Black Sea Coast in the early 1980s, unfortunately did not include management strategies with regard to coastal waters or adjacent coastal territory. This has meant that the current law on waters does not function in the way it was originally intended (THESEUS OD1.15, 2012). Similarly, the Spatial Development Plan of Varna dates back to 1984 and had a scope of 30 years. Amendments to that plan were added in 1999. During the process of the development of the new Varna Master Plan, experts with wide ranges of expertise (including specialists in coastal flooding) were invited for consultation, in spite of the fact that current legislation does not obligate planners to do so. Within this report, the basis of spatial planning of the coastal zone and adjacent aquatic areas relies on hydro-climatic conditions (wind, wave, and sea-level regime at extreme storm conditions), level of geodynamic activities, potential technical interventions and functional efficiency of hydraulic structures, environmental objectives, and cost-benefit analysis. As a result of this analysis, the area is divided into so-called red and blue zones defined by the Bulgarian Ministry of Regional Development and Public Works. In these zones, new infrastructure needs to be accompanied with special preventive measures including water storage, soil management, and structural land use.

An approximate figure for the annual coastal maintenance cost for Bulgaria is up to €1 million maximum per year (Policy Research Corporation, 2009). Although the State owns the beach and the coastline, financing of coastal defenses is currently the responsibility of any investor who wishes to build in the coastal zone. When an investor wants to construct a building (hotel, etc.), the requirements are that it should not be located on the beach and that, when needed, the initiator takes care of the required flood defense investments. From a financial point of view, this prevents some investments. It has therefore been proposed that a financing framework similar to the French system be used where funding is provided (1) 30% by the state, (2) 30% by the municipality, and (3) 40% by the investor.

7.2.1.2 Meteomarine Climate: Present Conditions and Long-term Trends

Past storm climate of the Western Black Sea showed a well-pronounced trend for worsening until the early 1980s. Afterwards, a period of storm intensity decrease followed, which is mainly attributed to the overall storm duration cut, involving a shift of main stormy winds' direction from ENE-NE and NNE-N. The storminess was found to exhibit quasidecadal variability. The first decade of the twenty-first century was marked by relatively mild storm conditions (Valchev, Trifonova, & Andreeva, 2012). As for joint occurrence probability of storm surge and wave height (period), the most likely pair of their 100-year return period annual maxima, obtained through bivariate Extreme Value Analysis (EVA), is 0.62 and 5.82 m 8.13 s, respectively (Table 7.4).

TABLE 7.4 Annual Maximum Wave Heights, Wave Periods, and Annual Maximum Storm Surge for Past Storm Climate

Return Period	Eastern Direction			Northeastern Direction			Southeastern Direction		
(Years)	H_s (m)	T (s)	SS (m)	H_s (m)	T (s)	SS (m)	H_s (m)	T (s)	SS (m)
5	5.22	6.86	0.28	3.79	5.31	0.26	3.92	6.05	0.29
20	5.62	7.47	0.4	4.48	5.8	0.32	4.9	6.8	0.39
50	5.74	7.84	0.53	4.83	6.01	0.37	5.38	7.21	0.5
100	5.82	8.13	0.62	5.01	6.12	0.43	5.79	7.53	0.58

Mean sea level trend was assessed on the basis of sea-level time series spanning a 130-year period (1880–2009). It was found to be positive because of both global SLR and neo-tectonic land sinking. The observed hazardous land subsidence trend is relevant only for the most inward part of Varna Bay, whereas for the rest of the study site it is not determinant. The present SLR trend (excluding subsidence) was estimated to be 1.2 mm/year.

Projections of wind, wave, and surge climate toward the end of twenty-first century were estimated using the European domain wind and sea level pressure output of the CLM (data stream 2) regional climate model under the global International Panel of Climate Change (IPCC) emission scenarios A1B and B1. Information on various climate projections of air temperature and precipitation for the region was extracted from several climate models output, namely GFDL-CM2.1, MIROC, MRI RCM, and ECHAM5. The ensemble mean projection for the selected model(s) and emission scenarios were calculated as well. Regional wave and surge gridded fields were obtained be means of dynamical downscaling. With respect to waves, it was carried out using a coupled system of third-generation spectral WAve Models (WAM) and Simulation WAves Nearshore (SWAN) at 1/30° resolution, whereas for storm surge a high-resolution (1/20°) 2D storm surge model was employed (Krestenitis, Androulidakis, Kontos, & Georgakopoulos, 2011). Mean SLR was obtained through a statistical downscaling procedure using long-term tide gauge records at Varna tide-gauge station.

The climate change in the Western Black Sea region was not found to take a catastrophic course in the next century. Most of the models showed regional warming and drying, which is particularly pronounced in summer. At the same time, the majority of precipitation trends are not significant, especially in winter. The typical atmospheric pattern, characterized by presence of anticyclonic area over the northern part of the Black Sea and cyclonic conditions over its southern part, resulting in severe storms, as known from the past climate, will continue to cause intense and persistent winds from the eastern half. Mean sea level is expected to rise with 7, 18, and 30 cm in short-, mid-, and long-term horizons, respectively (Weisse et al., 2014). The climate projections show that the frequency and duration of storm events will increase but not drastically. Relatively sharp storm intensification is predicted only in short-term horizon that afterwards will undergo a decrease to grow up again with a long-term horizon peak. There are well-defined areas of mean wind velocity (northern part of shelf) and mean significant wave height (southern) increase: averaged over the entire Western Black Sea shelf area, they are 0.34 m/s and 8 cm, respectively (Valchev et al., 2012).

Storm surge projections revealed some differences between the two considered scenarios. Under A1B, the storm surges are not particularly high, resulting in a decreasing trend of the peak sea level height of the order of 20 cm per century. Contrastingly, scenario B1 reveals higher surges with increasing centenary trend by the order of 10 cm, in particular over the northern part of the western shelf. The highest surges (>70 cm) are predicted along the central coastal area, which is distinguished also for the most frequent occurrence of high surges (>30 cm) as well. Therefore, the coastal flooding risk appears to be high for the mild-slope areas (Table 7.5), Valchev et al. (2012).

7.2.1.3 Coastal Habitats

Coastal sand strips form unique habitats, with no equivalent elsewhere in Bulgaria. The vicinity of the sea and the specifics of the sandy substrates create conditions for the development of vegetation cover that includes a number of rare and protected species of great value to science. Being a zone of intense activity, these fragile natural habitats are extremely vulnerable and sensitive to the impact of waves and storms.

Panorama and Karantinata beaches are located within Varna coastal area in its northern and southern portions, respectively. Despite being adjacent to the highly urbanized "Golden Sands" resort and Asparuhovo district of Varna, they serve as representative examples of a natural sheltered (Panorama) and a sand spit (Karantinata) beach types that used to be widely distributed at Varna coastal region in the past.

On these beaches, two terrestrial habitat types were identified in accordance with the nomenclature set in Annex 1 of Council Directive 92/43/EEC (Habitats Directive), namely 1210 (annual vegetation of drift lines) and 2110 (embryonic shifting dunes). Furthermore, 14 dominant vascular plants and 21 associations were mapped during the field campaigns (Vergiev, Filipova-Marinova, Trifonova, Kotsev, & Pavlov, 2013). Three, *Eryngium maritimum, Lactuca tatarica,* and *Centaurea arenaria,* are listed in the Red book of Bulgaria as endangered. These three psammophytic plants as well as the two habitat types identified require protection in accordance with the Biodiversity Act of Republic of Bulgaria. *Leymus racemosus* and *Carex ligerica* play important roles as habitat engineers because of their abilities to fix loose sandy substrates and withstand storm-related disturbances.

TABLE 7.5 Annual Maximum Wave Heights, Wave Periods and Annual Maximum Storm Surge for Future Storm Climate

	Return Period	Eastern Direction			Northeastern Direction			Southeastern Direction		
Scenarios	(Years)	H_s (m)	T (s)	SS (m)	H_s (m)	T (s)	SS (m)	H_s (m)	T (s)	SS (m)
A1B	5	3.96	6.4	0.7	2.48	4.45	0.7	2.95	5.3	0.7
	20	4.88	7.28	0.88	2.97	4.89	0.88	3.75	6.21	0.87
	50	5.47	7.84	0.99	3.28	5.16	0.99	4.26	6.7	0.98
	100	5.9	8.25	1.08	3.51	5.37	1.08	4.64	7.07	1.05
B1	5	4.12	6.52	0.71	2.53	4.45	0.7	3.12	5.44	0.71
	20	5.09	7.48	0.87	3.04	4.96	0.87	3.92	6.21	0.87
	50	5.71	8.09	0.98	3.37	5.25	0.98	4.43	6.7	0.98
	100	6.18	8.54	1.05	3.61	5.47	1.05	4.81	7.07	1.05

The most vulnerable associations are those formed by *Crambe maritima*, *Medicago marina*, and *E. maritimum* as leaf decay is initiated after just 24 h of inundation. Nevertheless, associations of *C. maritima*, *M. marina*, and *E. maritimum* demonstrate strong abilities to recuperate during the next vegetative season; therefore, over the longer term no permanent damages have been observed. Sand plant strip associations at Panorama and Karantinata beaches are in very good ecological condition despite being adjacent to highly urbanized areas. The good ecological condition and high conservation value make the Panorama and Karantinata beaches suitable to be included in the existing "Golden Sands" and "Galata" Natura-2000 protected sites (Habitats Directive).

The role of fouling assemblages (*Mytilus rocky zoocenosis*) on a T-shaped groin in Varna bay for biodiversity increase in comparison with the fouling complex of the natural rocky reef (Aladzha bank in front of Golden Sands resort) under recent environmental conditions along the Bulgarian coastal zone was assessed. Shore protection structures play a positive role for the epifauna development, providing a hard substrate and contributing for the biodiversity increase.

The impact of coastal structures on mediollitoral sand habitat of *Donacilla cornea* (Poli, 1791) was studied to test the hypothesis that groins have a negative impact on *D. cornea* populations and associated macrobenthic communities in the mediolittoral. The negative impact of existent "hard" coastal structures, which led to changes in sand grain size and beach morphodynamics, was demonstrated in increased abundance of tolerant and opportunistic species; deteriorated ecological status according to biotic indices; decreased average abundance, biomass, and length of the indicator species *D. cornea*.

7.2.2 Risk Assessment

7.2.2.1 Application of the Source-Pathway-Receptor-Consequence Model

Focusing on those beaches known to be liable to flooding (see Figure 7.24), an SPR analysis was carried out for Varna central beach and Karantinata beach. The latter is a constituent part of "Galata" Natura-2000 protected site, designated in accordance with the Birds Directive (www.natura2000bg.org). Conceptual models show the extent of the potential flood system, mutual dependencies, and interactions among its various elements of the flood system.

For the Varna area, which is tideless and has no large rivers, the main source of flood waters are the extreme meteorological events (unusual storm surge levels combined with reinforced wave impact upon the shore) occurring between mid-fall and mid-spring (see Section 7.2.1.2). Flood sources for present and future climate are presented by synthetic time series (storm prototypes), elaborated on the basis of about 200 real events reconstructed by means of hindcasting (Valchev, Davidan, Belberov, & Palazov, 2007; Valchev, Davidan, Belberov, Palazov, & Valcheva, 2008) and storm surge reconstruction calculated using the model developed by (Posada, Simmonds, Silva, & Pedrozo, 2008) forced with synthetic wind fields. Thirty-six scenarios include significant wave heights, peak wave period, and storm surge for three storm directions (NE, E, and SE) and four return periods (5, 20, 50, and 100 years). Storm scenarios for present climate conditions are calculated using extremes presented in Table 7.4 and for future climate in Table 7.5. Duration of storm events is presented in Table 7.6.

Both beaches and their adjacent area which are rather flood-prone because of their low topography are not protected against flood by engineering structures. The susceptibility of the beaches to erosion can affect flood extents as it reduces the wave attenuation capacity of the beach. There are two constructions in close proximity to beaches: main port construction at the southern tip of Varna beach and a jetty at the Northern tip of the Karantinata beach. The main port construction protects the inner harbor area because the jetty was constructed to defend the neighboring navigable canal, so neither construction protects the beaches from flooding. The area behind Varna beach is partly protected by "The Sea Baths" construction. Local topography and especially steep slope between the Maritime Park and promenade do not allow flooding of the area behind the promenade in any storm conditions. The area behind Karantinata Beach is partly backed by a pre-Bulgarian earth rampart dating from late VII-early VIII c. AD (Popov & Mishev, 1974, pp. 228).

Potentially susceptible to inundation are the Sea Baths and promenade adjacent to the beach (Figure 7.25). In addition, there are several beach bars, restaurants, and surf clubs situated along the promenade, which are also at risk. With few exceptions, almost the entire rear line of Varna Central Beach is separated from the Promenade by a bar and restaurant entertainment complex. This consists of separate premises, built either as temporary constructions on the beach or within converted parts of the former Sea Baths. Storm-related flood events with a return period of between 5 and 10 years do not reach the promenade, but are able to damage the bar complex. In addition to the permanent structures shown on the map, parasols are erected on Varna Central beach during the summer season. These would be flooded under the majority of storm-induced flood events. The SPRC schematic and linkage diagram (Figure 7.25, right panel) was elaborated on the basis of a detailed Digital Terrain Model (DTM) and satellite images. The Maritime Park of Varna, shown in Figure 7.25 as the Green Park area, may be flooded only through the promenade. The location of the bar and restaurant entertainment complex at the rear part of the beach and the relatively high elevations above mean sea level are the main factors that protect the Maritime Park from floods caused by severe storms.

TABLE 7.6 Duration of Storm Events for Present and Future Climate Scenarios A1B and B1 in Hours

Return Period (Years)	Eastern Direction			Northeastern Direction			Southeastern Direction		
	Present	A1B	B1	Present	A1B	B1	Present	A1B	B1
5	83	92	94	71	73	73	72	79	81
20	86	105	108	77	79	80	80	90	92
50	87	113	116	80	83	84	84	96	99
100	88	118	122	81	86	88	88	101	104

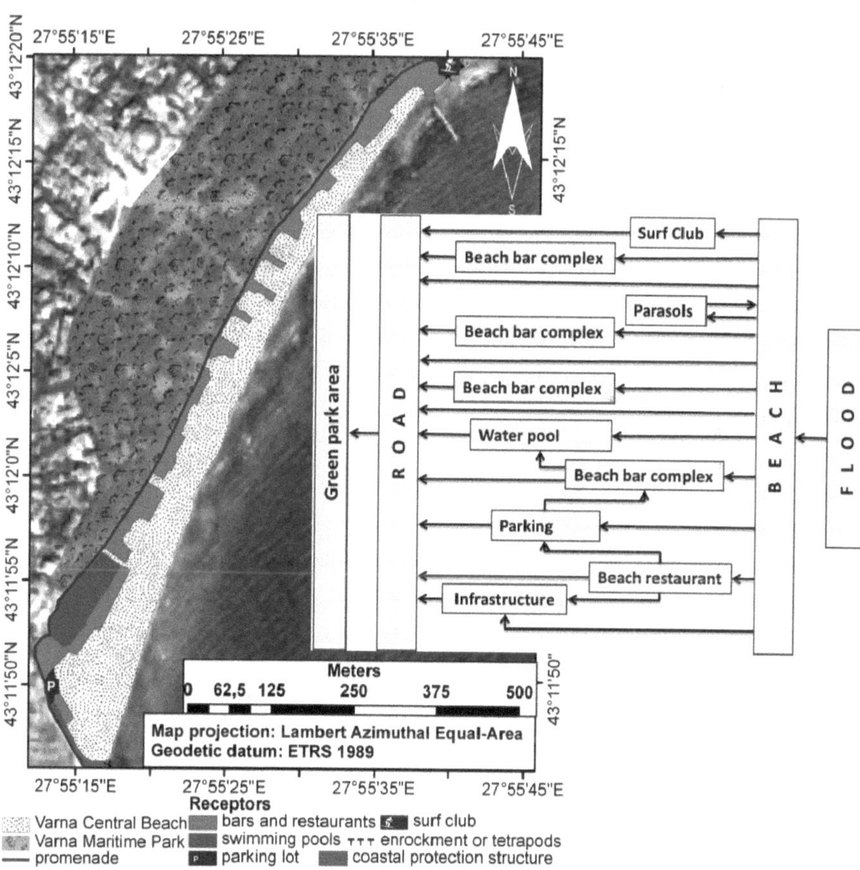

Figure 7.25 Sea-Level Rise (SPRC) map and system diagram of Varna Central beach.

7.2.2.2 Flood Mapping

Flood and erosion, caused by storms, are usually manifested at the same time, although their impact depends on different factors. Damage caused by flooding tends to increase with exposure time and depth increases, whereas damage caused by erosion increases proportionally to the volume of the sand body washed away from the beach.

Flooding consequences for Varna region are presented following two different approaches using cross-shore profiles. In the first, the flood maps show the flood extent occurred at maximum storm development associated with a storm with given return period and expressed as the percentage of total beach width. In the second approach, the period during which specific sections of the beach (corresponding to a given elevation) would be flooded in the course of a storm event is estimated. In this case, results are presented by set of curves and corresponding analytical functions. Using the concept of the *maximum flood extent* is appropriate for estimation of the expected damage to moveable or nonmovable property at the beach, infrastructures, etc., the resilience of species to flood being strongly dependent on the flood duration. Combination of flood model results and maps of plant associations allowed determination of the time-span, during which certain location at the subaerial beach profile would remain inundated during a storm of a given return period. Results are presented in terms of intensity-duration-frequency (IDF) curves relating flooding extent (elevation) and duration for four return periods: 5, 20, 50, and 100 years.

Maximum flood end erosion during storm event was calculated for each prototype storm using Morphodynamical Model of Institute of Oceanology of Bulgarian Academy of Sciences (Trifonova, 2007) along each cross-shore profile shown in Figure 7.24. The flood extent is considered as superposition of storm surge, wave setup, and wave run-up, and maximum erosion is considered as maximum retreat of beach scarp. Changes in sea level from climate change in the Varna coastal region were accounted for the estimation of flood scenarios for short- (2020), mid- (2050), and long-term (2080) horizons because the time series for each prototype were adjusted with respect to the prediction for mean SLR.

In Figure 7.26, an example of *IDF* curves for Karantinata Beach is presented. During an Eastern storm with a return period of 100 years all objects, the area located below 2 m elevation will be flooded for at least 48 h. This means that *C. maritima* located between 1.2 and 1.8 m will be in dangerous conditions for 24 h). All flood curves are approximated with polynomial function of fifth degree.

Differences in flood intensity between A1B and B1 climate scenarios are insignificant. As a whole, flood scenarios related to B1 are more severe than those to A1B, which is a consequence of rougher wave forcing for models.

Flood and erosion maps in terms of maximum flood extent and maximum beach scarp erosion were constructed for present and future climate conditions. Results for Varna central beach calculated for present climate conditions are presented in Figure 7.27, and all results are summarized in Table 7.7.

Results of manipulative experiment – Decomposition						
Karantinata beach southern part	leaves		stems		roots	
	beginning	complete	beginning	complete	beginning	complete
Crambe maritima	2	5	8	14	10	14

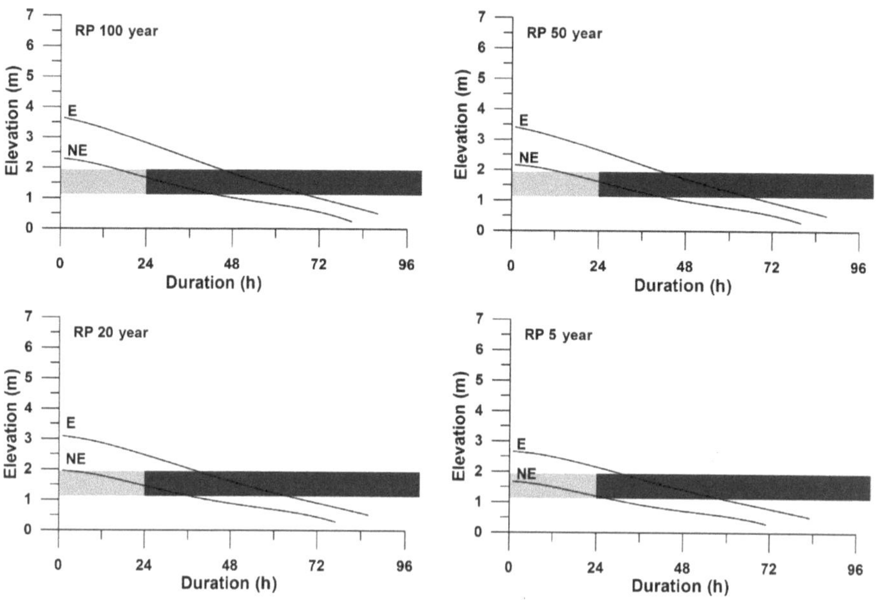

Figure 7.26 Intensity-Duration-Frequency (IDF) curves for Eastern storm and for Karantinata Beach (profile T268).

Based on the analysis of flooding maps and IDF diagrams it can be concluded that flood extent because of Eastern storms considerably exceeds the one caused by NE prototype storms for all four subareas for both present and future conditions.

7.2.2.3 Flood Impacts

The impact of flood events on coastal system in Varna is manifested in mischief for ecosystem, temporary flooding of beaches and adjacent areas, beach erosion, damage for infrastructure, and connected to the later business interruption.

Sandy beaches in Varna study site form habitats with intrazonal psammophytic and halophytic vegetation, including a considerable number of protected species. These are extremely vulnerable to storm impacts, which may result in several negative consequences (e.g., direct wave-related uprooting and burial, initiation of leaf, stem, and root decomposition processes from long-term inundation). To assess the degree of vulnerability, data about beach plants distribution, and storm-related flood extents with different return periods was collected, including a full inventory of the vascular flora, identification of protected species, and their conservation status (THESEUS OD1.15, 2012).

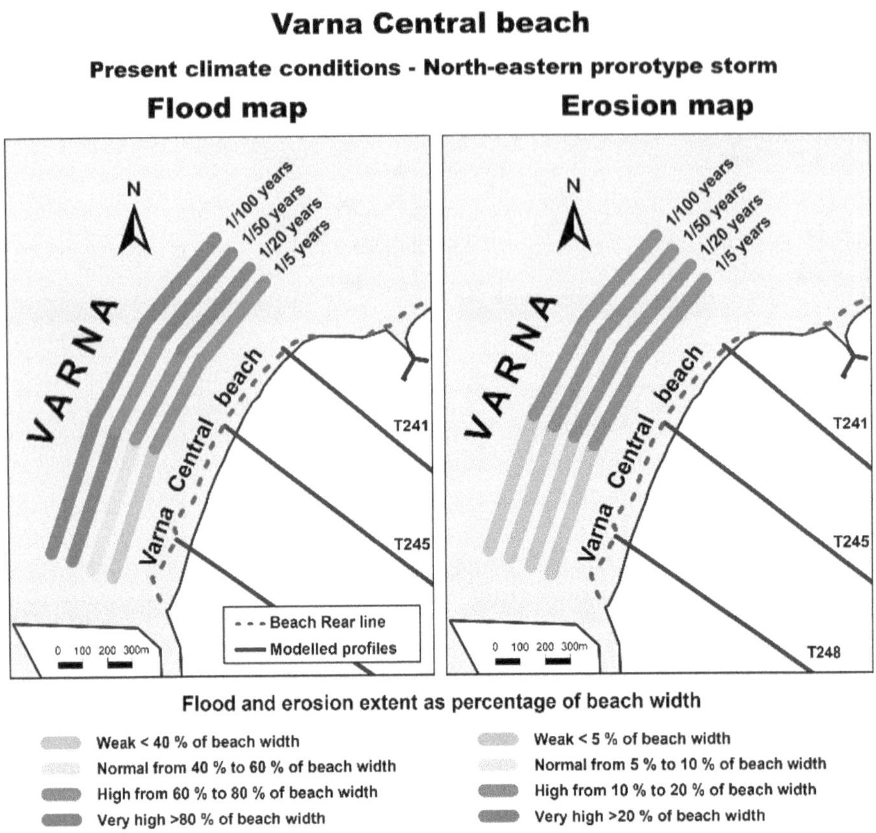

Figure 7.27 Varna central beach flood and erosion maps.

Using this information, an environmental vulnerability index (EVI) (see Chapter 6.2) with values from 0 to 3 was created which represents the step change categories of vulnerability of coastal habitats to inundation impact (see Table 7.8).

The EVI was assessed for all studied species separately based on the thresholds described in Table 7.4. Critical flood duration for initiation of decay processes is compared to the thresholds and the corresponding EVI are assigned. The EVI for the whole habitat corresponds to the EVI of "the weakest element" in the assemblage. The threshold between EVI 0 and 1 is set according to leaves decay of *E. maritimum, M. marina, C. maritima, and Artemisia vulgaris*; between 1 and 2 is the root decay of *L. tatarica* and *A. vulgaris*.

The permanent sea-level change will squeeze the sand strip habitat area thereby resulting in loss of terrestrial habitat and gain of new aquatic habitat. The largest resistance to flooding was registered for *L. racemosus* ssp. *sabulosus, Phragmytes australis,* and *C. ligerica*. A characteristic feature of these species is the growth of the stem and root sprouting from the seventh day after flooding

TABLE 7.7 Width of Flooded Beach in %

			Wave Approach											
			Eastern				Northeastern				Southeastern			
			Return Periods											
Beaches		Profiles	5	20	50	100	5	20	50	100	5	20	50	100
Present climate conditions	Golden Sands	*41*	48	72	91	97	31	38	46	56	37	58	79	94
		52	73	93	100	100	56	68	76	85	63	85	100	100
		63	68	76	84	88	56	68	68	72	64	72	80	84
	Kabakum	*94*	41	67	84	88	24	29	39	45	27	47	78	84
		99	48	73	91	91	19	30	34	39	22	41	64	75
	Varna Central	*241*	92	100	100	100	71	83	88	96	79	96	100	100
		245	79	96	100	100	63	75	79	88	71	88	100	100
		248	96	100	100	100	19	50	85	100	40	100	100	100
	Karantinata	*264*	34	52	72	82	12	32	42	48	25	47	61	76
		268	89	100	100	100	65	81	92	100	76	100	100	100
Future climate conditions scenario A1B	Golden Sands	*41*	54	84	98	99	35	48	62	71	39	65	81	93
		52	85	100	100	100	71	88	100	100	76	100	100	100
		63	76	84	92	96	68	72	80	80	68	80	84	88
	Kabakum	*94*	45	82	88	94	27	41	49	67	31	51	82	86
		99	53	100	100	100	22	41	56	69	36	64	77	100
	Varna Central	*241*	96	100	100	100	79	92	100	100	83	100	100	100
		245	88	100	100	100	71	88	96	100	79	96	100	100
		248	100	100	100	100	23	71	100	100	42	100	100	100
	Karantinata	*264*	50	73	83	100	41	55	69	78	42	67	78	97
		268	100	100	100	100	81	100	100	100	97	97	97	97

(Continued)

TABLE 7.7 Width of Flooded Beach in %—Cont'd

			Wave Approach											
			Eastern				Northeastern				Southeastern			
			Return Periods											
Beaches		Profiles	5	20	50	100	5	20	50	100	5	20	50	100
Future climate conditions scenario B1	Golden Sands	*41*	58	88	98	99	35	49	62	71	41	67	82	94
		52	88	100	100	100	71	88	100	100	78	100	100	100
		63	76	84	92	96	68	72	80	80	72	80	84	88
	Kabakum	*94*	47	84	90	96	27	43	49	65	35	55	82	86
		99	58	100	100	100	22	42	56	69	38	66	100	100
	Varna Central	*241*	100	100	100	100	79	92	100	100	88	100	100	100
		245	92	100	100	100	71	88	96	100	79	96	100	100
		248	100	100	100	100	25	73	100	100	54	100	100	100
	Karantinata	*264*	52	74	84	100	42	55	67	76	44	68	79	97
		268	100	100	100	100	84	100	100	100	97	97	97	97

TABLE 7.8 Environmental Vulnerability Index (EVI) Assessment for Sand Strips

EVI Index	Effect on Habitat	Consequences	Thresholds	Frequency	Duration
0	Negligible	–	A single/partial immersion of less than 2 days. No decay processes are observed.	Single event: return period 1 year	<2 days
1	Transient	Changes within the range of Receptor's natural seasonal variation and full recovery is likely within a season. Leaves decay only	Total immersion of plants for 2–7 days. Initiation of leaves decay process at some species. Recovery is possible during the next season due to vitality of stems and roots.	Single event: return period 16 year	From 2 to 7 days
2	Moderate	Changes are beyond Receptor's natural seasonal variation. Partial recovery is possible within several seasons, but full recovery is likely to require human intervention. Decay of leaves, stems, and roots	Total immersion of plants for more than 7 d. Initiation of stem decay process at some of the plants. Recovery is possible during next season from seed banks.	Single event: return period more than 16 year	More than 7 days
3	Permanent	Changes are so drastic that natural recovery of receptor is very unlikely without human intervention. Conversion of terrestrial into marine habitats.	Permanent change in inundation regime. Total immersion of plants for more than 17 days. Habitat conversion		

onwards. Preservation of *L. racemosus* ssp. *sabulosus* and *P. australis* contributes to improved beach stability during coastal storms.

The recent heavy storm of February 2012 lasted 3 d. In the period of its maximum development at a depth of 21 m (9 km from the coast), the wave height reached 4.8 m and the sea level rose to 0.6 m. Only 4500 m^2 of 51,400 m^2 of the beach were flooded, which is about 9% of the total beach area. The poststorm field survey showed significant erosion at Varna Beach: the maximum thickness of the eroded layer was 0.80 m, whereas at Karantinata Beach it reached 0.60 m.

During a storm in February 2012, the beach–bar complex (see Figure 7.25) stopped working for at least 1 week, and the recovery period during which they worked with reduced capacity lasted 2–4 weeks. Some of the temporary facilities near the beaches were destroyed to such an extent that they are not recoverable.

7.2.2.4 Key Source Drivers of Future Risk Conditions

The present meteomarine climate and available historical records indicate that the joint action of storm surges and waves constitutes the most significant source of flood risk for Varna coastal area. Furthermore, the analysis of future climate projections substantiates the importance of these drivers even if they would not worsen drastically. A slight increase of storminess is expected with well-defined peaks in short- and long-term horizons. It is known that the magnitude and damage caused by eastern storm events exceeds those hitting from northeast and southeast. Moreover, the flood extent caused by eastern storms surpasses the one caused by southeastern and northeastern storms (Trifonova et al., 2014).

To evaluate the relative contribution to flood of storm waves, storm surges, and SLR at Varna beaches, a comparative analysis of drivers and consequences (flooding) was carried out. It revealed that in present conditions, contribution of surge to flood varies from 9 to 41% with respect to the 100-year eastern storm. The contribution of SLR for present conditions is set to zero and, with the expected increase in long-term horizon of 0.30 m, it would represent 10% of the maximum flood level in the 2080s. Consequently, over time, the contribution of wave run-up and surges diminishes at the expense of SLR. The ranking of drivers for present and future climate conditions in Varna is shown in Table 7.9.

TABLE 7.9 Contribution of Flood Components for Present and Future Climate Conditions

		Contribution to Total Flood Level in %			
Rank	Driver	Present	2020	2050	2080
1	Wave run-up	59	59	56	55
2	Surge	41	38	37	35
3	Sea-level rise	0	3	6	10

7.2.3 Mitigation Measures

7.2.3.1 Selection of Mitigation Options

Preliminary selection of the mitigation options is driven by the economic activity in the Varna area (the need to maintain tourist environment—the receptors). This means that the construction of dikes is not appropriate. The focus is therefore on ways to mitigate the expected rise in mean sea levels and future storm intensity in ways that can change both the Source and Pathway of extreme events while promoting the existence of the beach. Submerged and floating breakwaters both have the potential to affect the wave climate (source) before it reaches land and artificial beach nourishment, by maintaining beach volume, is able to change the Pathway. In Varna, submerged and/or floating breakwaters are suggested in front of the central and northern sections of the beach, where the flood extent in February 2012 reached 100% of the beach width (Table 7.22). Restoration of the recreational potential of the beaches is possible by artificial nourishment. This involves considerable investment but it is a viable mitigation option within the pathway to decrease or avoid flooding. In Varna, the nourishment is suggested for the northern section of the beach, where the beach width at present tends toward zero.

Mitigation options suggested for Varna Beach, a set of reasons for their introduction, and main challenges to Bulgarian coastal engineering community for their implementation are listed in Table 7.10.

TABLE 7.10 Lists of Mitigation Options Together with the Rationale and the Main Challenges

Mitigation Option	Rationale	Challenges
Submerged detached breakwater	Decrease the wave impact on the beach Allow water circulation between the barrier and the beach Provide habitat for fouling communities	Lack of wide practice in construction among Bulgarian coastal engineers
Floating breakwaters	Decrease the wave impact on the beach Easy decommissioning and seasonality in application (it is possible to be removed during the high tourist season) Reduction of incident wave height	Lack of wide practice in construction among Bulgarian coastal engineers
Emerged beach nourishment	Restoration of beach area Increasing the regional tourist potential Protect property from flooding and erosion Highly appreciated by Varna society	Limited practice in realization among Bulgarian coastal engineers Lack of experience in maintaining of nourishment projects

7.2.3.2 Preliminary Design of Selected Mitigation Options

The preliminary design of engineering mitigation options depends on their location. The description of submerged and floating breakwaters includes recommendation for proper depth for their installation.

Two variants (depth 4 and 5 m) for the submerged breakwater and three variants (depth 6, 7, and 8 m) for the floating breakwater were investigated.

The suggested location for nourishment is the Northern portion of Varna Central beach, as the different variants of nourishment in terms of new beach width have different spatial extent.

The bathymetry in Varna bay is shown in Figure 7.28, where the beach of Varna, the area proposed for nourishment are highlighted (gray, red, and blue lines), and potential positions for variants of submerged (dashed lines) and floating (dot-and-dashed lines) breakwaters are shown.

Emerged beach nourishment (see Section 3.5) is suggested to reduce beach flooding and protects the beach from further erosion. It is proposed the replenishment activities to be implemented at the Northern part of the beach (Profiles 19–27)—the total amount of additional sand volume is calculated to be 92,000 m^3. An example of initial and nourished profiles is shown in Figure 7.29. Parameters of the emerged nourishment are summarized in Table 7.11.

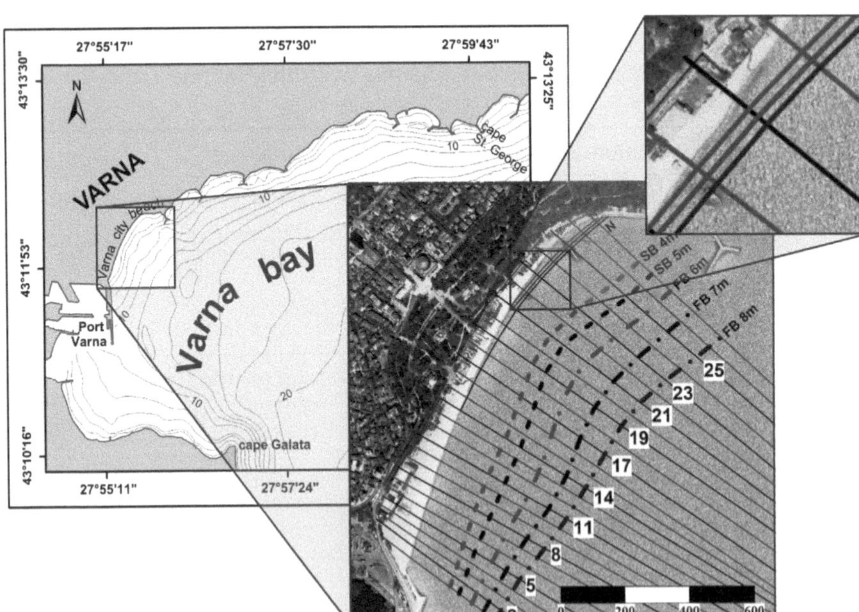

Figure 7.28 Bathymetry of Varna bay and location of proposed mitigation options: N—nourishment, SB—submerged breakwaters and FB—floating breakwaters.

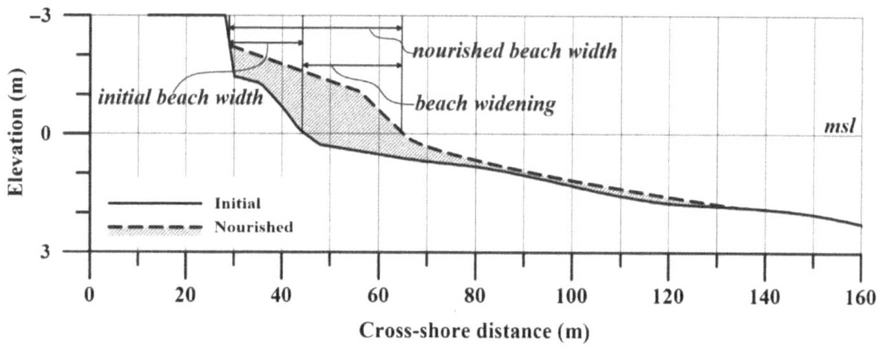

Figure 7.29 Cross-shore beach profile 23 (see Figure 7.28) with and without nourishment.

TABLE 7.11 Parameters for Design of Selected Mitigation Options

Emerged Beach Nourishment	
Depth of closure	7 m
Sediment size	0.3–0.4 mm
New beach width	40–60 m
Sand nourishment volume	130–630 m^3/m
Permeable Pontoon	
Depth	6–8 m
Off-shore distance	300–400 m
Formulae for wave transition coefficient	3.1.9, 3.1.10 (this book)
Crest width	5 m
Rouble Mound Detached Submerged Segmented Breakwaters	
Depth	4–5 m
Off-shore distance	180–200 m
Formulas for wave transition coefficient	van der Meer (2005)
Crest width	5 m
Breakwater length	200 m
Gap width	60 m
Free-board	1 m

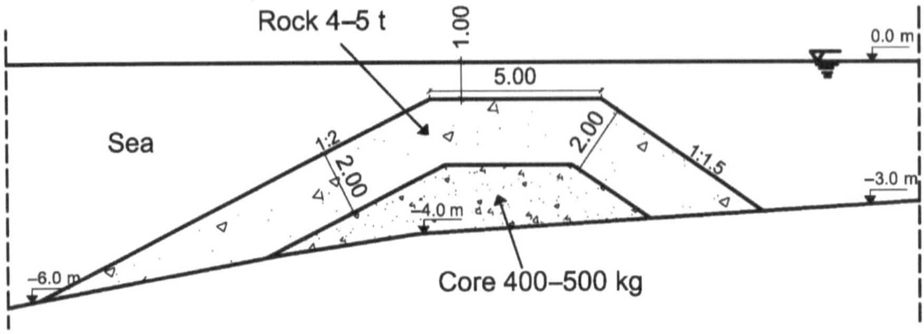

Figure 7.30 Typical cross-section of the proposed submerged breakwater.

Minimal value of the new beach width is determined with respect to the needs of recreational beach capacity. The range of grain size is determined with respect to the available borrow sites. Sand volume for nourishment (per 1 running meter of beach length) depends on sediment size and initial bathymetry because the average volume is 380 m^3/m. The nourished profiles are generated using initial profile features and Dean's formulae (Dean, 1977).

The construction of a submerged breakwater (see Section 3.2) is suggested to reduce incident wave heights by 40–50%. Using well-known rubble mount structural design formulas, the following characteristics are proposed: weight of the cover layer stone: 4–5 t, weight of the core stone: 400–500 kg, and width of the cover layer: 2 m. The cross-section of the structure is given in Figure 7.30. Parameters of the rouble mound detached submerged segmented breakwaters are summarized in Table 7.11.

The permeable pontoon (see Chapter 3.1 and Figure 3.1.3) is suggested to reduce incident wave heights by 20–40%. Permeable pontoon is presented in Figures 3.1.1 and 3.1.3. Parameters of the floating breakwater suggested for Varna are summarized in Table 7.11.

7.2.3.3 Ecological Considerations for Engineered Mitigation

Because the Bulgarian Black Sea coast needs urgent measures for protection, the structures should be designed in a manner that will minimize the negative environmental impact of the formation of stagnant zones. The wedgeclam *D. cornea* is considered as an indicator species for the Black Sea midlittoral sands habitat—a subtype of the natural habitat type of Community importance 1140 Mudflats and sandflats not covered by seawater at low tide (Micu, Zaharia, Todorova, & Niţă, 2007; Todorova, 2012). *D. cornea* is also sensitive to decreased oxygen and increased organic load; therefore, changes in the hydrological conditions that lead to decreased water exchange are detrimental to the species growth and development. Suggested engineering mitigation options allows the water exchange and therefore will not affect *D. cornea* population.

The most significant changes over time in relation to environmental factors are due to the introduction and mass development of the alien predator *Rapana venosa* in the past 20 years. By diminishing the population of the main habitat forming species *Medicago galloprovincialis*, it contributes to negative changes in epifauna, mainly in the area with shore-protection structures and stagnant zones formed at the T-shaped groin, which is reflected in the reduction of pollution-sensitive species and biodiversity (Konsulova & Doncheva, 2014). The comparative study between T-shaped groins in Varna bay and unprotected natural rocky Aladzha bank shows that the negative consequences on the fouling assemblages on T-shaped groin is expressed in reduction of pollution-sensitive species mainly at the landward stations. The seaward stations are characterized by better development of habitat forming species and epifauna on all parameters: the percentage cover of the main species *M. galloprovincialis*, which determines the higher numbers, biomass, and epifauna biodiversity. Suggested engineering mitigation options do not favor the stagnant zones formation, which create favorable conditions for development of the predatory snail *R. venosa*, a major threat to habitat forming black mussel *M. galloprovincialis* at present. To achieve maximum positive environmental effect, implementation of submerged breakwaters (Saito, Watanuki, Nishigori, & Hanzawa, 2002), similar to the natural rocky Aladzha Bank, exposed to the prevailing in the region E-wind waves and NE winds could be recommended.

7.2.4 Planning the Best Combination of Defense Strategies

Because the mitigation options are suggested to minimize the negative consequences of flooding and erosion at Varna, the initial criteria for their comparison are linked to the flood patterns. When planning the best combination of variants mitigation options the flood extent is used.

Combinations of the following mitigation options were compared: (1) for beach nourishment, six variants of nourishment; (2) for the floating breakwater, three variants of the breakwater locations; and (3) for the submerged breakwater, two variants of breakwater locations. A "do nothing" scenario was included as a benchmark case. For the latter, flood and erosion were estimated using the original bottom profiles while sources were defined using present and future climate scenarios (Tables 7.4 and 7.5).

The flood extent calculated for the "do nothing" option versus combinations of nourishment with either floating breakwater or submerged breakwater shows that for present and future climate conditions the best solution is the nourishment up to 60 m beach width with sand sized at $D_{50} = 0.4$ mm combined with the submerged breakwater at 5 m depth.

7.2.5 Overall Conclusions and Recommendations

For the Varna region, which is tideless with an absence of large rivers, the main sources of flood hazard are the extreme meteorological events (unusual storm surge levels combined with reinforced wave impact upon the shore). Two storms in March 2010 and February 2012 showed that the current protection measures are insufficient to protect the coast from flood and erosion and the trends in the development of the Varna coastal system discussed here, demonstrate that further mitigation options need to be considered.

By researching existing management structures and governance, present and future climate, flood and erosion conditions, coastal habitat spreading and mapping, resistance of coastal plants to flood conditions, and the interaction between coastal seabed habitats and coastal structures an holistic appreciation of the flood and erosion system has been established. This allowed a series of flood and erosion maps to be generated in response to expected trends in present and future climates and potential changes in the beach areas calculated to identify the most vulnerable coastal section (the northernmost part of Varna beach). Eastern storms present the most considerable risk for the entire coastal zone of Varna region. The most undesirable consequences of flooding would be inundation of the promenade which has implications for the initiation of landslides and cliff erosion at Varna Maritime Park. This infrastructure (the promenade) could not only be inundated, but also partially or completely destroyed, which would interrupt the road link with the affected sites and consequently hamper the recovery of the region after the flood.

Because the beaches in Varna are a popular tourist destination, any loss of area from either erosion or flooding of roads and tourist infrastructure would negatively affect the revenues of the tourist industry. In the Varna region, issues related to the coastal protection and beach restoration should therefore be formulated with two complementary goals. Mitigation options should (1) protect movable and nonmovable property against flood and erosion while also (2) maintaining the northern and central parts of the beach to meet the needs of the tourist-valued region. To achieve this, improving the coordination between the range of institutional, societal, and legislative entities involved in coastal management would be beneficial.

Taking into consideration the future climate, the submerged breakwater should be combined with the beach nourishment at the northernmost part of the Varna Beach.

However, the need for wider beaches in the central and northern areas for tourism may also require that beach nourishment should also be carried out in these areas. From the ecological point of view, engineering interventions should also not cause negative environmental impacts.

Acknowledgments

The research leading to these results has received funding from the European Community's Seventh Framework Programme under grant agreement n° 244104 (THSEUS Project). We would like to thank Prof. Fernando Mendes from University of Cantabria for supporting our work in flood mapping, Dr. Siddharth Narayan from the University of Southampton for support in SPRC diagram elaboration, and Dr. Barbara Zanuttigh, University of Bologna, and Prof. Robert Nicholls, University of Southampton, for strong support received by Varna team during the whole project. Finally, we thank Susan Hanson for her useful suggestions to improve the manuscript.

7.3

Coastal Wetlands in the Yangtze Estuary, China

Liquan Zhang, Zhongyuan Chen, Lifang Cui

State Key Laboratory of Estuarine and Coastal Research, East China Normal University (SKLEC), Shanghai, China

7.3.1 Description of the Site

As the largest (1,800,000 km^2) and longest (6300 km) river in east Asia, the Yangtze River ranks fifth globally in terms of water discharge (900 km^3/year) and, until recently, fourth in terms of sediment load (470 Mt/year) (Milliman & Farnsworth, 2011) (Figure 7.31). With an increasing watershed population (>460 million) and more than 50,000 dams, ranging in size from dams on farmers' fields to dams more than 100 m high, built throughout the Yangtze's watershed since 1950, the Yangtze River is also one of most heavily impacted rivers in the world (Yang, Milliman, Li, & Xu, 2011).

The Yangtze estuary is located on the Eastern coast of China, which is neighbored by the Hangzhou Bay to south, Jiangsu province to North, and opens to the East China sea (30°50′–31°50′N; 121°0′–122°05′E). The estuary coast is 542 km long, including 284 km coast of the islands in the estuary (Figure 7.31). It has a semidiurnal tide and the tidal amplitude of the estuary system ranges from 2.4 to 4.6 m and averages 2.6 m (Gao & Zhang, 2006). The Yangtze estuary is a typical mesotidal estuary with multiorder bifurcations, shoals, and sand bars. Because of its large sediment supply, the Yangtze Delta has prograded more than 200 km into its incised river valley since the mid-Holocene. The Holocene sediment thickness in the river mouth area reaches about 60 m (Hori, Saito, Zhao, & Wang, 2002; Li et al., 2002) forming extensive shoals and tidal flats (Gao & Zhang, 2006). The intensive interactions between runoff and tidal currents and hydrodynamic processes of the estuary have created a unique sedimentlogical pattern among the estuaries of the world (Yang, Milliman, Li, & Xu, 2011).

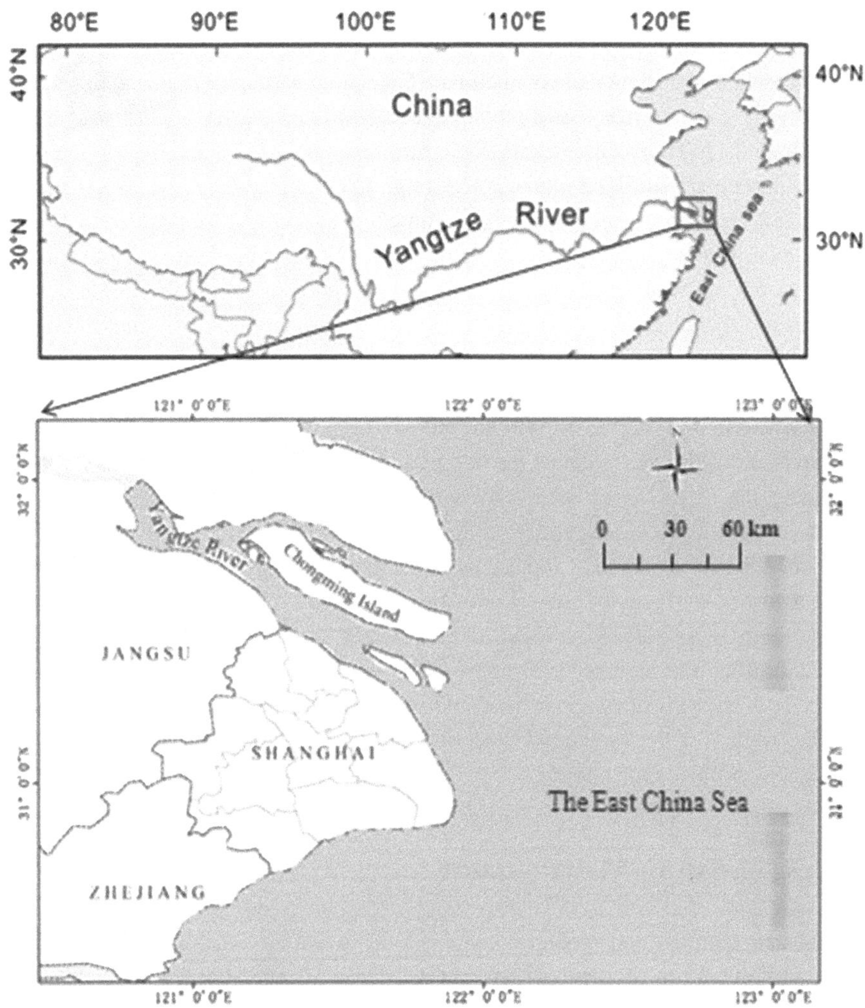

Figure 7.31 Location of the Yangtze river estuary, China.

In China, coastal areas are vital economic hubs in terms of settlement, industry, agriculture, trade, and tourism as well as being of great environmental significance. Shanghai is a notable example of this as the Shanghai metropolitan region has a land area of 6340 km^2 and is surrounded on three sides by the East China Sea, the Yangtze estuary, and Hangzhou bay (Figure 7.31). The region is an important industrial, scientific, technological, cultural, and economic center, and China's largest harbor city. With 1.2% of the total population of China, Shanghai generates 4.9% of the nation's industrial output and 4.2% of the national gross domestic product (GDP). The city is also the fastest growing region among all major Chinese cities with the population increasing from 5.2 million in 1949 to more than 20 million in 2010 (Wang, Gao, Xu, & Yu, 2012). However, the elevation of the city generally lies between 3 and 4 m (using the local

Wusong bathymetric benchmark) with significant areas below 3 m and existing sea walls are often constructed in excess of 6 m to prevent flooding (Wang et al., 2012).

In recent decades, the prospect of climate change, in particular SLR and its impacts on low-lying coastal areas, has focused worldwide attention on coastal ecosystems (Nicholls et al., 2007; Solomon et al., 2007; Stralberg et al., 2011) and the services they provide, particularly in flood management (e.g., Ervin et al., 2012; Fu, Wang, Xu, & Yan, 2013). The enormous quantities of sediment produced by the Yangtze river have created extensive areas of coastal wetlands in the estuarine region and the Yangtze estuary is one of the largest estuarine alluvial wetlands in the world. These coastal wetlands are dynamic and provide valuable ecosystem services including habitat of biodiversity, production, water purification, and carbon sequestration as well as coastal protection (Gao & Zhang, 2006; see also Chapter 4). For the latter, the extensive wetlands help to mitigate the impacts caused by SLR and storm surges, either by acting as a physical barrier that reduces the height and speed of storm surges and waves or stabilizing the shoreline to reduce rates of erosion (see Chapter 4). However, the Yangtze estuary is more susceptible to SLR than other coastal areas because of land subsidence, reduced sediment input, low elevations and gentle slope, and it is facing potential loss of wetlands and shoreline erosion as a result of storm surge, flooding, and saltwater intrusion (State Oceanic Administration, 2012; Wang et al., 2012). The rate of urbanization has also imposed substantial pressures on the coastal wetlands adjacent to the city both in terms of space demand and other environmental damage. To consider the effectiveness of the wetlands as part of a flood management plan, it is essential to assess their resilience to change.

7.3.2 Existing Management

Coastal management and protection in the Yangtze estuary is the responsibility of three administration bodies within the Shanghai municipal government; the Shanghai Water Authority and Oceanic Bureau are responsible coastal defense and coastal wetlands management, the Shanghai Forestry Bureau is responsible for the management of coastal wetland reserves, and the Shanghai Environment Protection Bureau is responsible for the coastal environment management.

The major stresses related to human activities on the coastal wetlands in the Yangtze Estuary are urbanization, reclamation, invasive species, overfishing, and environmental pollution (Yuan, Zhang, Xiao, & Huang, 2011; Zhou & Xie, 2012). As a fast-growing mega-city, the accelerating urbanization of Shanghai needs more land for socioeconomic development. According to the reports of the Shanghai Water Authority (2010), 1040.9 km^2 of coastal wetlands were reclaimed between 1949 and 2010 (Table 7.12). Large-scale reclamation poses, therefore, a significant pressure on the preservation of coastal wetlands It is a great challenge to continue sustainable socioeconomic development while maintaining and preserving healthy coastal wetlands in the Yangtze Estuary.

TABLE 7.12 Reclamation of Coastal Wetlands (km^2) in the Yangtze Estuary, 1950–2010

Period	1953–1959	1960–1969	1970–1979	1980–1989	1990–1999	2000–2005	2005–2010	Total
Area	136.58	251.92	152.67	79.69	194.15	172.6	53.3	1040.9

(Data from Shanghai Water Authority, 2010).

Recent public education and dissemination activities on the considerable environmental and ecological benefits associated with the coastal wetlands in the Yangtze Estuary have focused concern and attention on coastal wetland conservation. As a result, the importance of the coastal wetlands in the Yangtze Estuary has been recognized widely by both the local government and citizens with the establishment of wetland nature reserves (since the 1990s) playing a vital role for the conservation of biodiversity and various ecosystem services of coastal wetlands in the estuary. Two national nature reserves, Chongming Dongtan National Birds Nature Reserve (242 km^2) and Jiuduansha National Wetland Nature Reserve (402 km^2), were listed in *Chinese Protected Wetlands* in 1992 and were designated as internationally important under the Ramsar Wetlands Convention in 2001 and as a national nature reserve in 2005. The area covered by nature reserves amounts to 25% of the total area of the current coastal wetlands in the Yangtze estuary. By designating the coastal wetland as nature reserves, a legal safeguard for the conservation of coastal wetlands has been established, preventing further large-scale reclamation within the estuary. In addition, several restoration projects for degraded coastal wetlands have recently been carried out (e.g., Yuan et al., 2011).

7.3.3 Coastal Wetlands in the Yangtze Estuary

According to the Ramsar International Wetland Conservation Treaty in 1971, wetlands are defined as ".... areas of marsh, fen, peatland or water, whether natural or artificial, permanent or temporary, with water that is static or flowing, fresh, brackish or salt, including areas of marine water the depth of which at low tide does not exceed six metres" and "wetlands may incorporate riparian and coastal zones adjacent to the wetlands, and islands or bodies of marine water deeper than six meters at low tide lying within the wetlands" (Mitsch & Gosselink, 2007). There are two different types of coastal wetlands in the Yangtze Estuary, i.e., mainland and island wetlands (Figure 7.32). According to the reports of Shanghai Water Authority (2010), the total area of coastal wetlands, stretching from the inland dike and moving seaward toward the −5 m isobaths, in Shanghai between 1990 and 2008 amounted to 2318.5–2892.5 km^2 (Table 7.13).

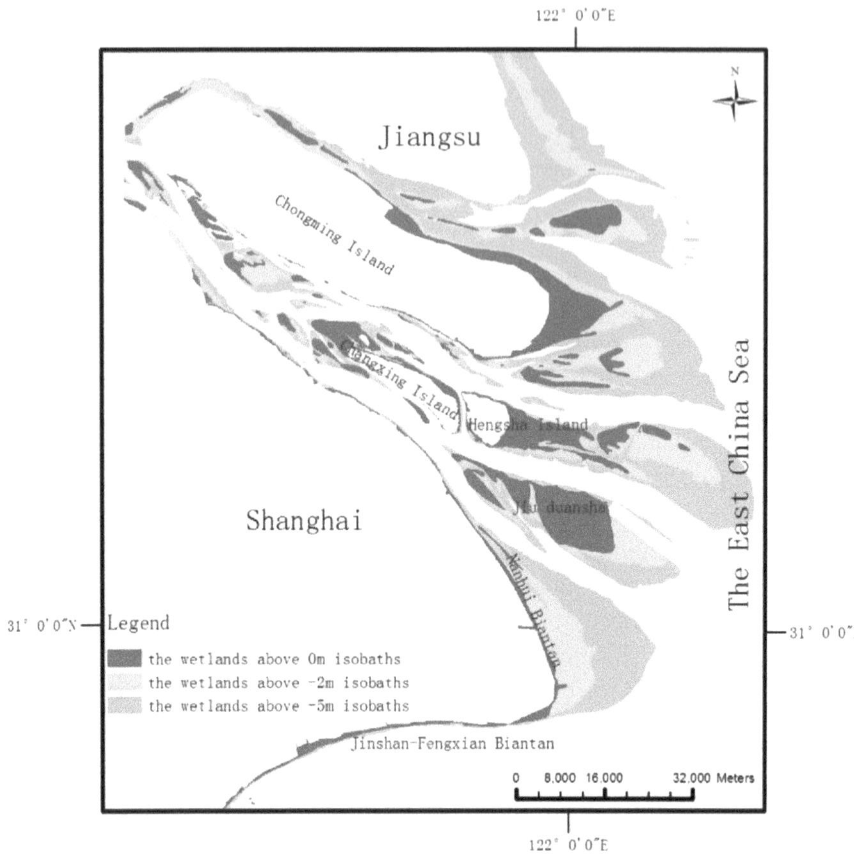

Figure 7.32 Spatial distribution of coastal wetlands in the Yangtze estuary.

TABLE 7.13 Total Area of Coastal Wetlands in the Yangtze Estuary, 1990–2008 (km²)

	Elevation Zone			
Year	>0 m	−2 ~ 0 m	−5 ~ −2 m	Total
1990	828.6	713.2	1350.7	2892.5
2000	841.4	723.3	1221.8	2786.4
2004	753.0	691.9	1152.8	2597.8
2008	559.0	666.5	1093.0	2318.5

(Data from Shanghai Water Authority, 2010).

The coastal wetlands in the Yangtze Estuary are dynamic and develop very fast owing to sedimentation of the huge amount of silt brought by the Yangtze river. By their very nature, the coastal wetlands in the Yangtze Estuary provide high ecosystem services such as improving water quality, mitigating climate change, and conserving biodiversity. They are also an important destination and stopover point for migratory birds along the pathway in the Asia Pacific. Each year, millions of birds arrive for staging and wintering; for example, the number of geese coming for wintering reaches more than 30,000, four species are listed in the National first-class categories for state protection: white stork (*Ciconia boyciana*), hooded crane (*Crus monacha*), black stork (*Ciconia nigra*), and white-tailed eagle (*Haliaeetus albicilla*). More than 10 species water birds exceed 1% level of the population of East Asia and Australia. The coastal wetlands in the Yangtze Estuary are also the channel for migratory fish, which provide the spawning and feeding grounds for many species of fish, especially for the Chinese sturgeon (Zhou & Xie, 2012).

Zonation of coastal wetland is widely recognized in the Yangtze Estuary and distinct zones of habitat, related to elevation, can be identified within the tidal flats in the region (Figure 7.33) (Zhang & Yong, 1992). The tidal flats closest to the low water mark, at an elevation below 2.2 m are characterized by mudflats that are devoid of any higher plants (angiosperms). The tidal flats located in between 2.2 and 3.0 m elevation are dominated by *Scirpus mariqueter* communities. Above 2.9 m, plant communities are dominated by *Phragmites australis*, with some small areas of *Imperata cylindrica, Suaeda glauca, Juncus setchuensis*, and *Carex scabrifolia*, the latter tending to border the various creeks crossing the saltmarshes. An additional species in this zone is an exotic species *Spartina alterniflora* (smooth cordgrass), which is a perennial saltmarsh grass and native to the Atlantic and Gulf coasts of North America. This species has been widely introduced to many coastal and estuarine regions of the world as a species for ecological engineering due to its great capacity to reduce tidal wave energy, mitigate erosion, trap sediments, and protect the seacoast (Yuan et al., 2011). It was introduced to the coastal wetlands in the Yangtze Estuary during the 1990s. Over the past 20 years, this species has gradually invaded large areas formerly covered by *Phragmites australis* and also started to invade the upper parts of the *S. mariqueter* zone (Li & Zhang, 2008).

The saltmarsh vegetation is one of the most important components of coastal wetland in the Yangtze estuary and provides high biodiversity and various ecosystem services. The spatial and temporal dynamics of saltmarsh vegetation in the Yangtze Estuary were studied for the period between 1990 and 2008 and the results are summarized in Table 7.14 (Huang, 2009).

The total area of saltmarsh vegetation by 1990 in the Shanghai region was 17,881.9 ha, and *P. australis* community was dominant, accounting for 14,100.4 ha (80%) of the total area of saltmarshes. However, almost half of *P. australis* saltmarsh had been lost during the 1990s and 2000s as a result of the 1990s reclamation. At the same time, an exotic species *S. alterniflora* was introduced in 1990s and began to invade to the wetlands. By 2003, the total area of saltmarsh vegetation had expanded to 21,953 ha with the area of *S. alterniflora* accounting for 4276 ha. After 2003, the total area of saltmarshes decreased mainly from intense reclamation. By 2008, both *S. mariqueter* and *P. australis* communities declined almost by half, whereas *S. alterniflora* community continued to increase reaching 5698 ha, one quarter of the total area of saltmarsh vegetation (Figure 7.34 and Table 7.14).

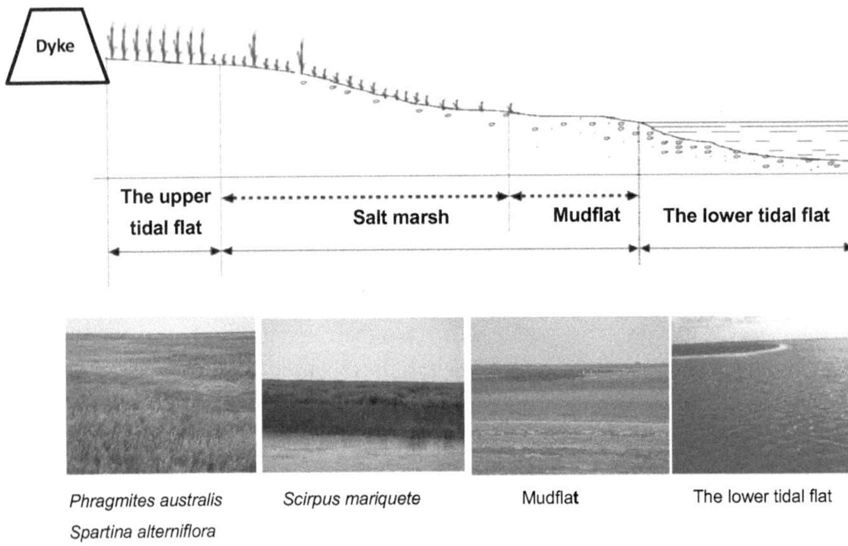

Figure 7.33 Zonation of coastal wetland along the elevation gradient in the Yangtze estuary.

TABLE 7.14 Changing Area of Saltmarsh Vegetation (ha) in the Yangtze Estuary (1990–2008)

Year	1990	2000	2003	2008	Typical Elevation Range
Scirpus mariqueter	3781.5	4833.9	7602.2	4234.7	2.2–3.0 m
Phragmites australis	14,100.4	7085.6	10,075.4	5617.5	2.9–4.0 m
Spartina alterniflora	0	1902.8	4275.8	5697.9	2.7–4.0 m
Others	0	0	0	2764.7	2.9–4.0 m
Total	17,881.9	13,822.2	21,953.5	18,314.8	

Based on Huang (2009).

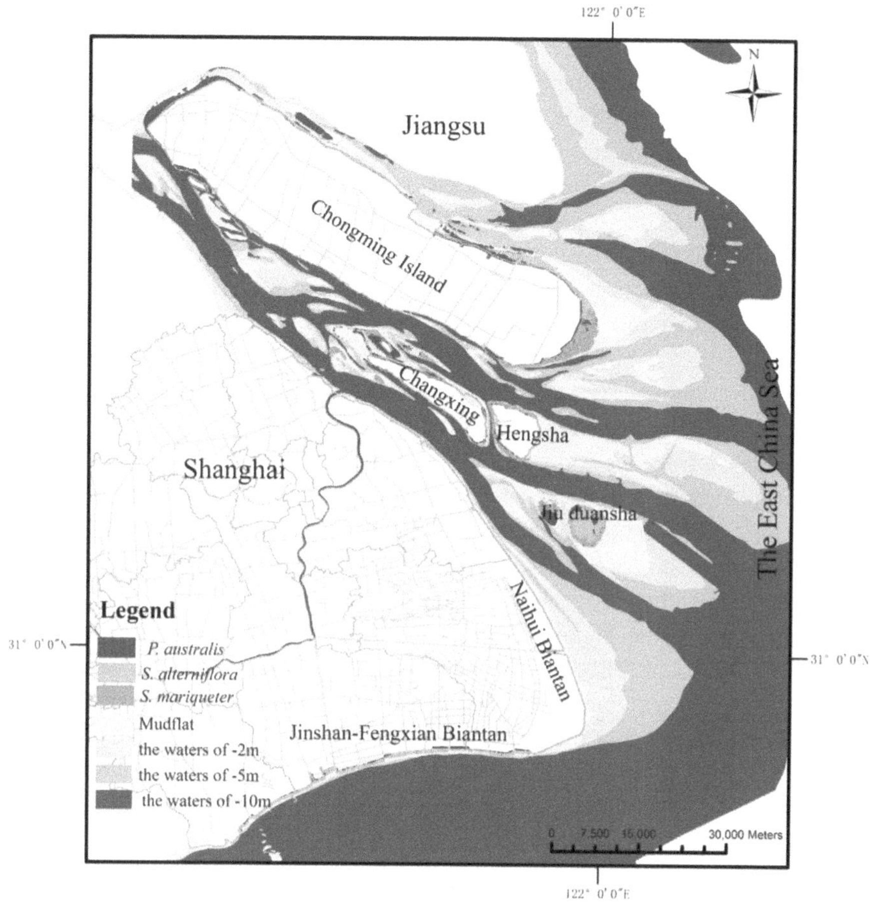

Figure 7.34 Distribution of saltmarsh vegetation in the Yangtze estuary (year 2008).

7.3.4 Present Climate and Environmental Conditions and Future Trends

The region of the Yangtze Estuary has a Northern subtropical monsoon climate, with an average annual temperature 16 °C, summer temperatures average 28 °C, whereas winters are cold with an average temperature of 4 °C. Average annual precipitation is approximately 1200 mm, with 60% of rainfall occurring during May to September. Typhoons occur during summer and autumn. Typhoon storm surges are one of the most devastating natural disasters affecting the Yangtze estuary, as because of the shallow continental shelf, the surge is amplified considerably as it approaches land, leading to severe flooding along the coast. The elevation of the city generally lies between 3 and 4 m with significant areas below 3 m and existing sea walls are often constructed in excess of 6 m to prevent flooding (Wang et al., 2012). Table 7.15 shows the major storm events (Grade $\geq$10, maximum average surface wind speed $\geq$24.5–32.6 m/s) and the associated losses in the Yangtze estuary since 1949. The impacts of such events can be expected to increase with the effect of climate change on extreme water levels.

Except for several exceptionally wet or dry years, notably 1954 (wet) and 1978 (dry), the Yangtze River water discharge (1950–2010) has remained relatively constant since the 1950s (Figure 7.35), averaging about 900 km^3/year measured at Datong station, 600 km from the river mouth, just upstream of the tidal influence. The sediment discharge past Datong station, by contrast, fell from ~490 Mt/y in the 1950 and 1960s to ~150 Mt/y after the closure in 2003 of the Three Gorges Dam (TGD) (Figure 7.35). The TGD is currently the world's largest dam, 185 m in elevation above the mean sea, and has generated a huge impact on the eco-hydro-morphological setting of the river and estuary (Yang et al., 2011); with abundant sediment supply from the Yangtze river, the coastal wetlands of the estuary had rapidly prograded in the recent centuries; However, an obvious decrease in the progradation rate occurs after the construction of TGD (Yang et al., 2011).

Data from the Chinese State Oceanic Administration indicate that the coastal sea level in the Yangtze Estuary is clearly rising. The monitoring results indicate that the coastal sea level in the Yangtze estuary during 2001–2010 was higher than that during 1991–2000 and 1981–1990, at about 14 and 47 mm, respectively. The coastal sea level in the Yangtze Estuary in 2012 was the highest in the past 30 years, about 112 mm higher than that of the mean value of 1975–1993. The mean rate of SLR in the past 30 years (1980–2010) reaches 2.6 mm/year and it is predicted that the coastal sea level in the Yangtze Estuary may rise by 98–148 mm over the next 30 years (State Oceanic Administration, 2011).

Within the estuary, land subsidence could exacerbate the negative impacts of SLR. Natural conditions and anthropogenic activities are the two primary causes of this process. The former includes neotectonic movement and natural settlement of poorly consolidated soils. The latter includes excessive pumping of groundwater, construction, and other activities (Wang et al., 2012). The crust of the Yangtze river Delta and adjacent areas shows a phase of gradual subsidence since the Pliocene and the average neotectonic subsidence rate is 1.5 mm/year (Wang et al., 2012). However, there is a long history of anthropogenically induced subsidence in the urban area of Shanghai, where excessive groundwater extraction started in the 1920s and continued through the 1960s. Between 1921 and 1965, the average cumulative subsidence of the city center was 1.76 m, with a maximum of 2.63 m. Since between 1966 and 1985, anthropogenic subsidence has been mitigated through artificial recharge and was stabilized at an average of 0.9 mm/year. However, as a result of rapid urban development between 1986 and 1997, subsidence of the downtown area increased rapidly, at an average rate of 10.2 mm/year. Between 2001 and 2007, subsidence rates again began to decrease (Wang et al., 2012).

TABLE 7.15 List of Major Typhoon Storm Surges in Yangtze Estuary since 1949

Tropical Cyclone Number	Year	Beaufort Scale	Category	Maximum Storm Set Up (m)	Direct Economic Loss (Million Yuan, RMB)	Fatalities (Number)	Sea Walls and Banks Destroyed (km)	Farmland Flooded (km^2)	Houses Destroyed (Number)
4906	1949	10	Severe tropical storm	1.5	N/A	18,000	N/A	1389	63,200
5612	1956	12	Typhoon	2.5	1.784	0	0	0	0
6207	1962	12	Typhoon	2.03	50	49	2.75	93	10,000
7413	1973	11	Severe tropical storm	1.6	2.854	1	64.7	0	76
7910	1979	10	Severe tropical storm	1.42	0.782	5	24	20	NA*
8114	1981	12	Typhoon	1.7	13.298	6	28	47	7000
9711	1996	11	Severe tropical storm	2.35	63.49	7	99.1	500	4000
0012	2000	11	Severe tropical storm	1.65	12.2	1	30	179	200
0205	2001	11	Severe tropical storm	1.73	70	6	NA	NA	500
0509	2005	11	Severe tropical storm	NA	135.8	7	0	56	15,600

*NA, no data available.
Based on data from Wang et al. (2012).

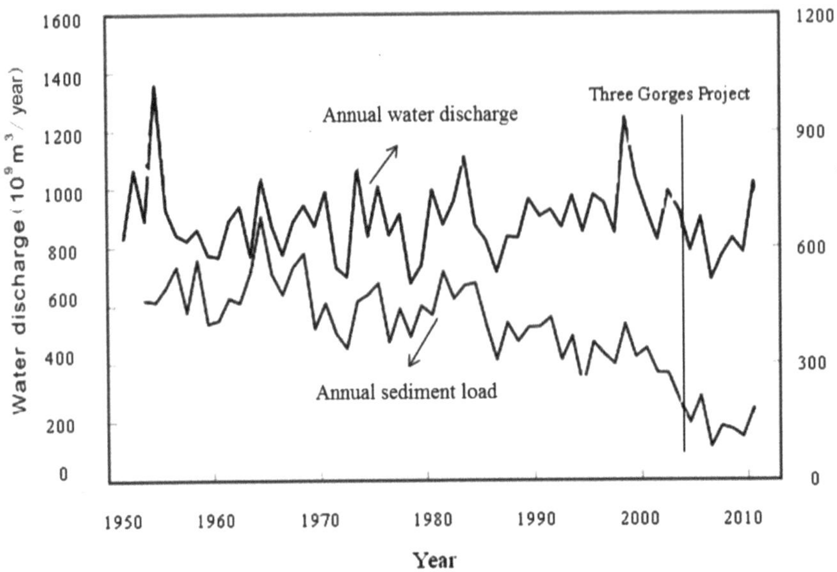

Figure 7.35 Annual water and sediment discharge, 1953–2010, for the Yangtze river measured at Datong station (the farthest downstream gauging station). *(Based on Yang et al., 2011).*

7.3.5 Vulnerability Assessment of the Coastal Wetlands under SLR

Coastal wetlands, as important parts of coastal ecosystem, are particularly sensitive to SLR. Studying the responses of coastal wetlands to climate change to assess the impacts of climate change on coastal wetlands and formulate feasible and practical mitigation strategies are the important prerequisites for securing the coastal zone ecosystems. For the Yangtze estuary, the possible impacts of SLR induced by climate change on coastal wetlands were initially analyzed using the SPRC model (see Chapter 2).

7.3.5.1 Habitats Within the Source-Pathway-Receptor-Consequence Model

Within the larger flood system analysis, the coastal habitats of the Yangtze were classified as components both as a Pathway for flooding to the urban areas of Shanghai and as Receptors as they will exhibit change as a result of the external influences in the flood system. They also have an internal SPRC specific to the habitats and species within the ecosystem. As the aim is to look at potential change in habitat areas, an SPRC model (see Chapter 2) can be developed specifically with these as the Receptor. The conceptual model was worked out and is shown in Figure 7.36. In this model, Source is the factors of impact on the coastal wetlands, namely SLR caused by climate change; Pathways are the route(s) between the Source and the Receptors, including subsidence and sedimentation; Receptors are where impacts may be found, which are the coastal wetland ecosystem and habitats in the Yangtze estuary; and consequences are the environmental damage/improvement that may result from SLR.

Source analysis: The Primary Sources of a flood event are weather-related phenomena that generate water that could cause flooding. The Source of SLR is climate change–related phenomenon that can change the water level and sedimentation conditions of the coastal zone, which directly affects the survival and distribution of coastal wetlands. Located in the intertidal zone, the coastal wetlands are particularly sensitive to SLR. The monitoring results indicate that the coastal sea level in the Yangtze Estuary has risen quickly in the past 30 years and it is predicted that the coastal sea level in the Yangtze Estuary may continuously rise in the future, which will have an important impact on the coastal wetlands in the Yangtze estuary.

Pathway analysis: The Pathway for SLR is the link between the Source and the Receptor and provides the route for rising water level. Subsidence can increase the rate of SLR, which will further exacerbate the impacts of SLR on coastal wetlands. The rate of subsidence is affected by the earth's crust vertical movement, deltaic compaction, and human activity such as ground water abstraction. High sedimentation rate carried by rivers and open sea to coastal wetlands may mitigate or even offset the impacts of relative SLR, whereas negative sedimentation rate would further exacerbate the impacts and result in coastal erosion and loss of coastal habitats.

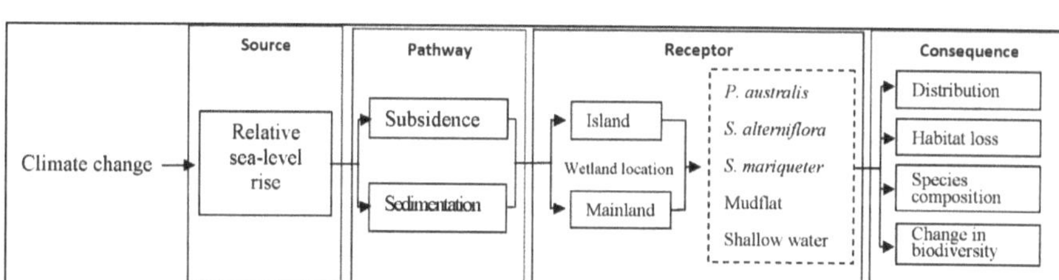

Figure 7.36 Framework of source-pathway-receptor-consequence (SPRC) model for vulnerability assessment of coastal wetlands in the Yangtze estuary under sea-level rise (SLR).

Receptors analysis: Receptors are the coastal wetland that would be impacted by SLR event within the case study boundaries. The coastal wetlands in the Yangtze estuary could be divided into mainland coast and island/shoal coast wetlands by geographic location. According to the key characteristics of the habitats/ecosystems, these locations could be further divided into the individual species receptors of *P. australis, S. alterniflora,* and *S. mariqueter* community, mudflat, and shallow water. For performing a spatial analysis for the wetland vulnerability, a broad-scale classification of receptors for the coastal wetlands in the Yangtze Estuary is presented in Figure 7.37.

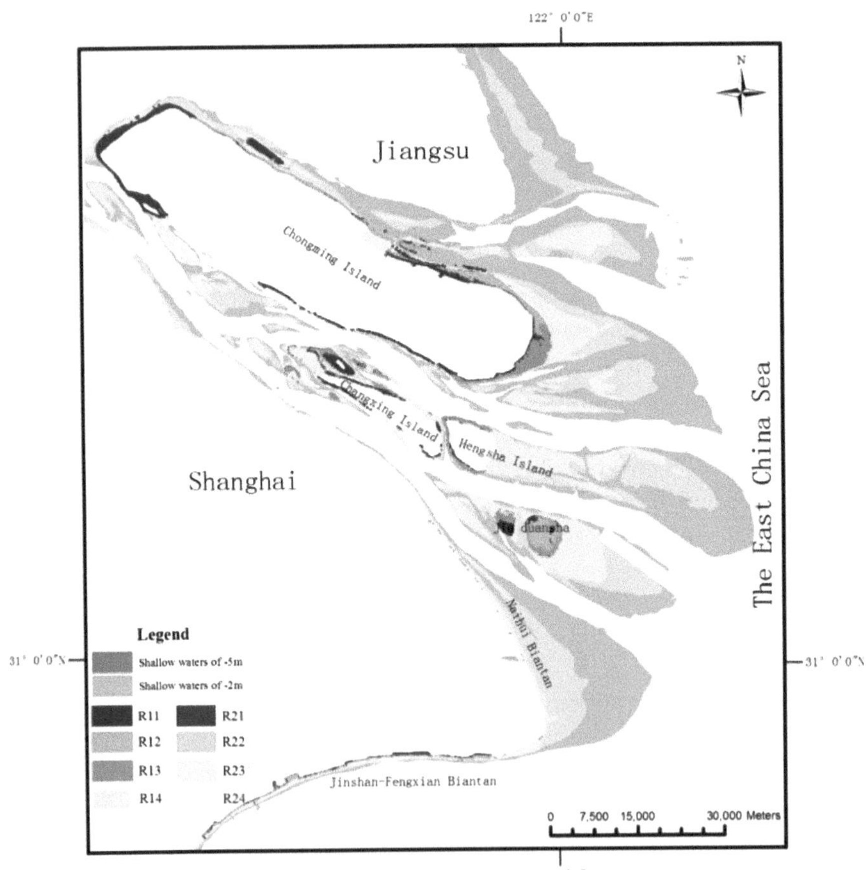

Figure 7.37 Spatial distribution of the receptors of coastal wetland in the Yangtze estuary.

Consequences analysis: Consequences are the habitat damage/improvement on the coastal wetlands that may result from SLR. Consequences are the result of the location and magnitude of exposure (i.e., severity of SLR), the susceptibility of the receptor (e.g., effectiveness of any natural response within the system), and the adaption to the stresses (i.e., ability to "absorb" the event). Consequences can be expressed quantitatively by category or descriptively. Distinct zonation of habitat, related mainly to elevation (see Figure 7.33), can be identified within the coastal wetland in the Yangtze Estuary and each type of habitats has their own tolerance range of environment conditions. If the relative SLR exceeds the tolerance range of the habitat/ecosystem, it will change the ecosystem structure and function, and ultimately result in habitat lost.

7.3.5.2 Development of an Environmental Vulnerability Index

The coastal wetlands in the Yangtze Estuary are extremely vulnerable to the relative SLR, which may result in negative consequences or even habitat lost. To build the EVI (see Chapter 6) for the coastal wetlands in the Yangtze Estuary under SLR, different scenarios of SLR rate and subsidence rate were selected as the key parameters of exposure to SLR. Because they are the most sensitive factors for habitat resilience, habitat elevation and inundation threshold (the mean daily inundation duration of habitat) were selected as the key parameters of sensitivity to SLR. The sedimentation rate was selected as the parameter of habitat adaptation to SLR (Table 7.16).

Rate of SLR: Three scenarios of SLR were adopted for vulnerability assessment on the coastal wetlands in the Yangtze Estuary; the present SLR trend (1980–2010) of 2.6 mm/year (State Oceanic Administration, 2011) and the IPCC (SRES) A1FI scenarios of 2.6–5.9 mm/year (Meehl et al., 2007). These parameters were used for the three sets of projections over the short-term (2030), mid-term (2050), and long-term (2100) (Table 7.17).

TABLE 7.16 Parameters for Coastal Wetland Vulnerability Assessment in the Yangtze Estuary

System	Aspect Layer	Parameter Layer	Unit
Coastal wetland	Exposure	SLR rate	cm/year
		Subsidence rate	cm/year
	Sensitivity	Habitat elevation	cm
		Mean daily inundation duration of habitat	h/day
	Adaptation	Sedimentation rate	cm/year

TABLE 7.17 Maximum Sea-Level Rise in the Yangtze Estuary Under Different Climate Change Scenarios

Scenarios	Period	Sea-Level Rise (m)
Present trend	2010–2030	0.052
	2010–2050	0.104
	2010–2100	0.234
IPCC A1FI	2010–2030	0.118
	2010–2050	0.236
	2010–2100	0.531

Habitat elevation: Each wetland ecosystem has its distinct distribution range and specific requirement of habitat (Huang, 2009). According to the map and elevation data from the resent survey of coastal wetlands in the Yangtze Estuary (Zhou & Xie, 2012), a digital elevation model was build (Figure 7.38(a)).

Subsidence rate: The Yangtze river Delta and adjacent areas shows a phase of gradual subsidence and the average neotectonic subsidence rate is 1.5 mm/year. Based on the annual subsidence map of the coast (Wang et al., 2012), a raster-based map of annual subsidence rate of the coastal wetlands in the Yangtze Estuary was created (Figure 7.38(b)), using Ordinary Kriging spatial interpolation function of ArcGIS software.

Mean daily inundation duration of habitat: Each coastal wetland ecosystem distributed along the gradient of elevation has its distinct daily inundation duration (i.e., the inundation threshold). According to the tidal table published by Chinese State Oceanic Administration in (2012) and the distribution of coastal wetland ecosystems in the Yangtze Estuary (Zhang & Yong, 1992, Huang, 2009), a mean daily inundation duration for *P. australis* community, *S. alterniflora* community, *S. mariqueter* community, and mudflat can be calculated (Table 7.18), which were 7, 10, 11, and 15 h/day, respectively.

Sedimentation rate: Long-term monitoring data (1997–2010) were available from the Yangtze Estuary Bureau of Hydrology and Water Resource Survey. This was used to create a map of annual sedimentation/erosion rate for the coastal wetlands in the Yangtze Estuary (Figure 7.38(c)).

The EVI for the coastal wetlands is calculated for the three sets of projections for the SLR (Table 7.16) using the following equation:

$$\text{EVI} \in \text{Rang}(T_{\text{sl}} - T) \quad (1)$$

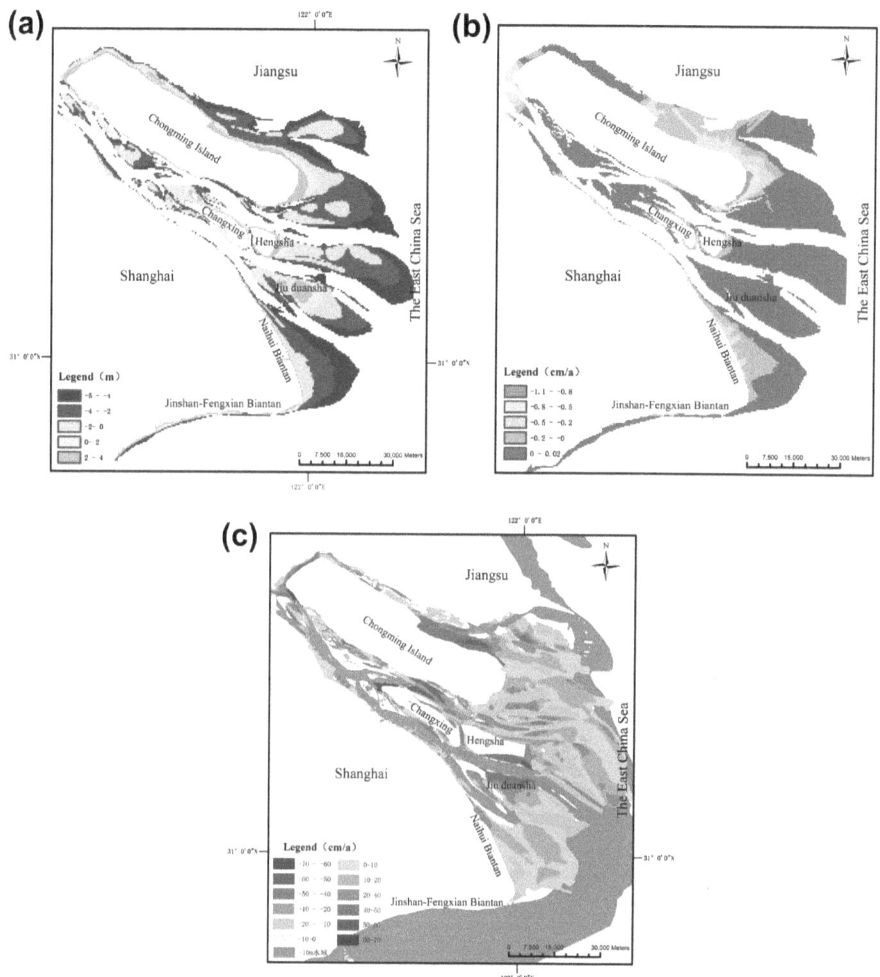

Figure 7.38 A digital elevation model. (a), distribution of subsidence rate, (b) sedimentation rate (c) for the coastal wetlands in the Yangtze Estuary.

where T is the daily inundation duration of the coastal wetland habitats (i.e., *P. australis, S. alterniflora, S. mariqueter* communities, and mudflat) obtained from Table 3.3.7; T_{sl} is the daily inundation duration of the corresponding habitats under different scenarios of SLR and projection sets, which can be calculated as follows:

$$T_{sl} = m(f(tide), f(e)) \quad (2)$$

TABLE 7.18 Inundation Threshold (Mean Daily Habitat Inundation Duration) for the Coastal Wetland Habitats under the Scenarios of Sea-Level Rise (SLR)

		Inundation Threshold Mean Daily Habitat Inundation Duration (h/day)			
SLR scenario	Projection	Mudflat	*S. mariqueter* habitat	*S. alterniflora* habitat	*P. australis* habitat
Present sea level		15	11	10	7
	2030	15	12	11	7
Present trend	2050	16	12	11	9
	2100	16	12	11	10
	2030	16	12	11	9
A1F1	2050	16	12	11	10
	2100	20	15	14	11

where m is the intercept of the daily tidal range under different scenarios of SLR and projection sets f(tide) on the coastal wetland habitat elevation $f(e)$; the f(tide) is a function of relative SLR rate(R_{rsl}) changed with the time (t), i.e., the number of years for the projection:

$$f(\text{tide}) = \frac{dR_{\mathrm{rsl}}}{dt} \tag{3}$$

The relative SLR rate R_{rsl} can be calculated by the equation as follows:

$$R_{rsl} = R_{\mathrm{sl}} - R_{\mathrm{sub}} + R_{\mathrm{sed}} \tag{4}$$

where R_{sl} is the absolute SLR rate; R_{sub} is the subsidence rate; and R_{sed} is the sedimentation rate.

The calculated EVI values were then graded in to a system of EVI scores from 0 to 3 (Table 7.19). The threshold between EVI scores 0 to 1 is according to Table 7.18 (i.e., the daily inundation duration of mudflat was 4, 5, and 8 h longer than *S. mariqueter*, *S. alterniflora*, and *P. australis* habitats, respectively). When EVI$\in$ [$\leq$0], the EVI score is 0, i.e., the inundation duration of coastal wetlands is within their tolerance range and the effect on habitat is negligible; when EVI$\in$ [0, 4], the EVI score is 1, i.e., the effect on habitat is moderate; when EVI$\in$ [4, 5], the EVI score is 2, i.e., the effect on habitat is moderate; when EVI$\in$ [> 5], the EVI score is 3, i.e., the effect on habitat is permanent.

TABLE 7.19 EVI Assessment for the Coastal Wetlands in the Yangtze Estuary

EVI Score	0 None	1 Low	2 Moderate	3 High
Inundation Threshold (h/day)	≤0	[0, 4]	[4, 5]	> 5
Impact on habitat/key species	Negligible impact to habitats/species	Transient effect (no long-term change anticipated)	Moderate effect/semipermanent change	Permanent effect/change

7.3.5.3 Spatial Assessment of the EVI in the Yangtze

Based on the establishment of a geo-referenced database on a geographic information systems (GIS) platform, vulnerability assessment of the coastal wetlands in the Yangtze Estuary under SLR could be performed on a unified geographic coordinate framework. In the GIS platform, the receptors of coastal wetlands (Figure 7.37) were assigned as the basic assessment units. Before any EVI index could be calculated, the categorical data such as the receptors of coastal wetlands had to be translated into numerical data and a raster-based map to be created using Ordinary Kriging spatial interpolation function of ArcGIS. The quantitative parameters of habitat elevation, habitat daily inundation duration, subsidence rate, and sedimentation rate (Figure 7.38 and Table 7.18) were then assigned for each receptor unit. According to the EVI calculation process (Eqn (1–4)), the EVI values for each receptor unit of coastal wetlands in the Yangtze Estuary under different scenarios of SLR and projection sets were calculated and then graded in to a system of EVI scores (Table 7.19). These results were then projected onto a unified geographic coordinate frame to present the spatial vulnerability assessment maps.

7.3.5.4 Short-term Vulnerability: 2010–2030

The results of vulnerability assessment of the projection 2010–2030 for the coastal wetlands in the Yangtze Estuary under the scenarios of present trend and A1F1 are presented in Figures 7.39(a) and (b), respectively. In general, the coastal wetlands in the Yangtze Estuary in the short term (2010–2030) have an EVI score of 0, i.e., the impact of SLR on habitats/species of coastal wetlands is negligible (Table 7.20). The coastal wetlands with an EVI score of 1 (low impact) amount to 7.4% and 8.7% under the scenarios of present trend and A1F1, respectively. The percentage of coastal wetlands in the EVI score of 2 (moderate impact) is 0.3% only (Table 7.20). These coastal wetlands with an EVI score of 1 and 2 occur mainly at the southern, southeastern, and western coasts of Chongming Island as well as the coast of Jinshan Biantan (Figures 7.39(a) and (b)). The areas of low vulnerability mainly occur in the southeastern coast of Chongming island as the subsidence rate is relatively higher and the sedimentation rate is low or even negative. Even though classified as low vulnerability, these wetlands still show evidence of erosion (see Figures 7.38(b) and (c)).

7.3.5.5 Medium-Term Vulnerability: 2010–2050

The results of vulnerability assessment of the projection 2010–2050 for the coastal wetlands in the Yangtze Estuary under the scenarios of present trend and A1F1 are showed in Table 7.20. In general, the coastal wetlands in the Yangtze estuary in the mid-term (2010–2050) are mainly categorized with an EVI score of 0 (i.e., the impact of SLR on habitats/species of coastal wetlands is negligible). Compared with the results in the short-term, however, the percentage of coastal wetlands with an EVI score of 1 (low impact) in the mid-term increases to 10.7% and 10.5% under the scenarios of present trend and A1F1, respectively. The percentage of coastal wetlands with an EVI score of 2 (moderate impact) also increases to 0.5% and 2.4%, respectively. A total of 0.2% of the coastal wetlands has an EVI score of 3 (high impact) under the A1F1scenario (Table 7.20). The spatial distribution of the coastal wetlands with an EVI score of 1 is similar to that in the short term, which also occurs mainly at the southern, southeastern, and western coasts of Chongming Island as well as the coast of Jinshan Biantan. With the further rise of sea level, some coastal wetlands with an EVI score of 2 or 3 occur in the areas where the subsidence rate is relatively higher and the sedimentation rate is low or even negative (Figures 7.38(b) and (c)).

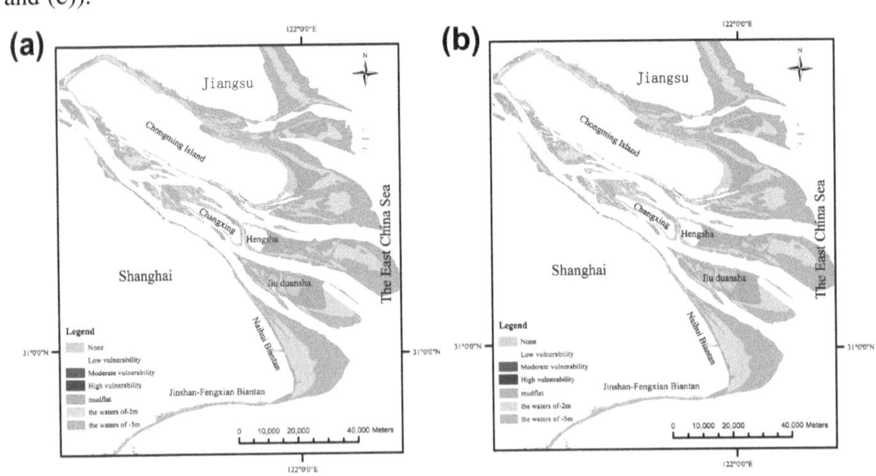

Figure 7.39 Spatial vulnerability for coastal wetlands in the Yangtze Estuary over the short term (2010–2030) for sea levels of present trend (a) and A1F1 (b).

7.3.5.6 Long-term Vulnerability: 2010–2100

The results of vulnerability assessment of the projection 2010–2010 for the coastal wetlands in the Yangtze estuary under the scenarios of present trend and A1F1 are presented in Figures 7.40(a) and (b), respectively. In the long-term projection (2010–2100), the percentages of coastal wetlands with an EVI score of 0, i.e., the impact of SLR on habitats/species of coastal wetlands is negligible, are 85.7% and 82.4% under the scenarios of present trend and A1F1, respectively. Coastal wetlands with an EVI scores of 1 (low impact) and 2 (moderate impact) amount to 7.3% and 2.0%, 7.1% and 2.4% under the scenarios of present trend and A1F1, respectively, whereas the percentage of coastal wetlands with the highest EVI scores of 3 (high impact) has increased considerably to 5% and 8.1%, respectively (Table 7.20). These coastal wetlands with an EVI score of 3 occur mainly at the southern, southeastern, and western coasts of Chongming Island and the western coast of Hengsha Island as well as the coast of Jinshan Biantan (Figures 7.40(a) and (b)), where the combined effect of SLR, low sedimentation rate, and land subsidence give rise to the high risk of habitat loss.

7.3.6 Mitigation Measures

The combined effects of SLR, land subsidence, and reducing sediment discharge may considerably increase the risk of habitat extent or loss of the coastal wetlands in the Yangtze estuary (particularly in 2050 and 2100).

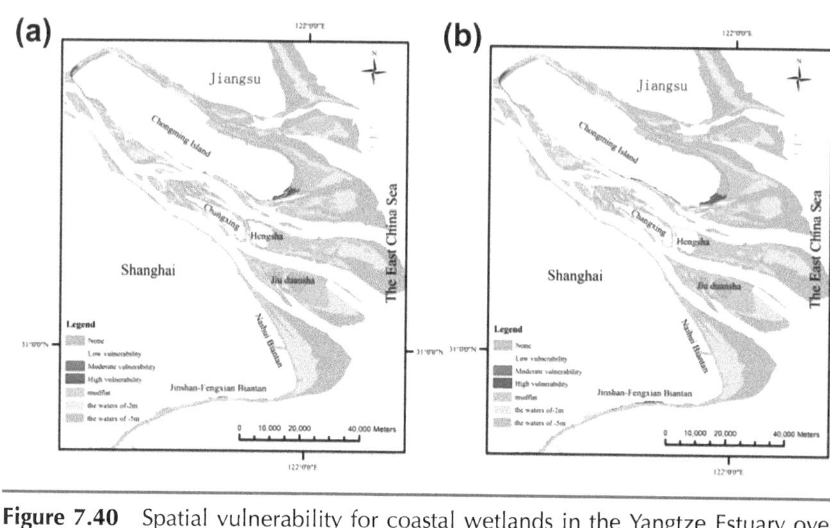

Figure 7.40 Spatial vulnerability for coastal wetlands in the Yangtze Estuary over the long term (2010–2100) under sea levels of present trend (a) and A1F1 (b).

TABLE 7.20 Percentage of Wetland Area in Each EVI Category Over the Short-Term (2010–2030) Medium-Term (2010–2050), and Long-Term (2010–2100) Under the Sea-Level Rise Scenarios

Projection	Sea-Level Rise Scenarios	Percentage of Vulnerable Wetlands (%)			
		EVI = 0	EVI = 1	EVI = 2	EVI = 3
Short term	Present trend	92.3	7.4	0.3	0
(2010–2030)	A1F1	91.0	8.7	0.3	0
Medium term	Present trend	88.8	10.7	0.5	0
(2010–2050)	A1F1	86.9	10.5	2.4	0.2
Long term	Present trend	85.7	7.3	2.0	5.0
(2010–2100)	A1F1	82.4	7.1	2.4	8.1

The management of the suspended sediment concentration in the coast zone is a crucial mitigation option. If the suspended sediment concentration in tidal water is high enough, the reduction in flow and wave energy by saltmarsh vegetation could enhance sedimentation and allow the coastal wetlands keep pace with SLR (Morris, Sundareshwar, Nietch, Kjerfve, & Cahoon, 2002; Kirwan & Temmerman, 2009). As the sediment discharge from the Yangtze River fell from ~490 Mt/year in the 1950 and 1960s to ~150 Mt/year after the closure of the TGD (Yang et al., 2011), it is essential to consider the implementation of appropriate management techniques such as building groins, recreating and extending coastal wetlands, reducing land subsidence, and controlling reclamation and invasive species.

7.3.6.1 Management of Sedimentation by Groins

As the elevation of the mega-city Shanghai generally lies between 3 and 4 m, sea walls were built along the whole Yangtze estuary to protect the lower-lying areas. Coastal erosion occurs mainly in the coasts of southern Chongming island and Jinshan Biantan in the Yangtze Estuary, which is caused by the decline of suspended sediment concentration and hydrodynamic conditions. Hundreds of groins were built to conserve coastal wetlands along the shores of Chongming island, Changxing Island, and Hengsha island. Groins used in the Yangtze Estuary are constructed perpendicular to the shore to trap and reduce the alongshore transport of silts. The height of groins usually is dependent on the annual mean water level during the dry season, which is about 0.4–0.6 times the mean width of local wave tape. In general, the siltation of groins is decided by their direction, height, length, and distance. The height of the groins varies depending on the desired amount of silt bypassing. High-profile groins will effectively block the transport of silts, whereas low-profile groins will allow the storm surge and waves to transport silts over the structure. Groins spaced too far apart lead to excessive erosion between structures and groins spaced too close together generate strong currents that limit the amount of silt deposited in their lee. Although groins have been widely used, a large number of scientific questions still remain to be answered, and the feasibility and effectiveness of these management measures need to be studied and verified (see Chapter 3).

7.3.6.2 Recreating and Extending Coastal Wetlands

Sediment dynamics are a key factor in lateral coastal wetland development. Both the onset of lateral saltmarsh erosion and the (re)establishment of pioneering saltmarsh vegetation enabling lateral extension are largely dependent on the sediment bed level variability of the exposed mudflat. The short-term variability of the bed level is especially crucial in determining the survival and growth of saltmarsh vegetation (Zhu, Zhang, Wang, Schwarz, & Ysebaert, 2012). Saltmarsh vegetation can promote sedimentation by slowing flow and reducing wave energy (Shi, Yang, Wang, Bouma, & Zhu, 2012). Planting saltmarsh vegetation (*P. australis* and *S. alterniflora*) combined with building groins after large-scale reclamation have been used as a mitigation measure to recreate and extend coastal wetlands in the Yangtze estuary from the late 1970s. After such a project in Naihui Biantan in the 1990s, the annual vertical accretion of sedimentation reached 0.3 m and the lateral extension rate seaward increased from 23 to 100 m/year (Li et al., 2007). It is therefore an effective management option to mitigate the impacts of SLR on coastal wetlands in the Yangtze Estuary.

The Yangtze estuary is also the important navigation channel for cities along the estuary and further upstream in the Yangtze River. Each year, 70 Mt/y of sediments are dredged to maintain the deep-water navigation channel. A first-phase project of using dredging to direct mudflat or saltmarsh development was carried out in the eastern coast of Hengsha Island during 2003–2006. The result of this project indicated that the lateral extension of mudflat seaward reached 36 km^2 and the vertical accretion of coastal wetlands reached 0.7 m (Li et al., 2007). It is therefore also an effective soft management option) to consider for further mitigation of the impacts of SLR on coastal wetlands in the Yangtze estuary.

7.3.6.3 Reducing Land Subsidence

The Yangtze estuary and adjacent areas show a phase of gradual subsidence, which will further exacerbate the impacts of SLR on coastal wetlands (see Section 7.3.5). The rate of this land subsidence can be reduced by limiting groundwater extraction and increasing groundwater recharge (Wang et al., 2012). Water management policies to reduce the rate of land subsidence are therefore an important mitigation option to reduce the local impacts of SLR.

7.3.6.4 Controlling Reclamation and Invasive Species

In a fast growing mega-city such as Shanghai, the accelerating urbanization needs more land for socioeconomic development. According to the reports of Shanghai Water Authority, 1040.9 km^2 of coastal wetlands in the Yangtze Estuary has been reclaimed between 1949 and 2010 (see Table 7.12). This large-scale reclamation caused significant loss of coastal wetlands and has exacerbated the vulnerability of the remaining coastal wetlands from relative SLR by restricting landward movement of habitats. It is a great challenge to balance conservation of coastal wetlands with sustainable social and economic development.

The coastal wetlands in the Yangtze Estuary also face a potential threat of invasive species. *S. alterniflora,* a non-native plant species introduced in the 1990s has expanded rapidly, at faster rate than the native species *S. mariqueter* and *P. australis*. As a consequence, the area of monodominant *S. alterniflora* community now accounts for almost one third of the total intertidal saltmarsh in the Yangtze estuary. Outcompeting native plants in this way can threaten the native ecosystems and cause a decline in the wider native species richness. A demonstration project, established at a coastal wetland reserve, showed a technique of cutting plus managed waterlogging can be an effective measure for controlling the large scale invasion of *S. alterniflora* on the coastal wetlands in the Yangtze estuary (Yuan et al. 2011) and will be a vital management option for conservation of the coastal wetlands in the future.

7.3.7 Conclusion

The Yangtze Delta in China is vital economic hub in terms of settlement, industry, agriculture, trade, and tourism as well as of great environmental significance. In recent decades, the prospect of climate change, in particular SLR and its effects on low-lying coastal areas have generated worldwide attention to coastal ecosystems. Coastal wetlands, as an important part of coastal ecosystems, are particularly sensitive to SLR. To study the responses of coastal wetlands, assess the impacts of climate change on coastal wetlands, and formulate feasible and practical mitigation strategies are the important prerequisites for securing the coastal zone ecosystems. In this study, taking the coastal wetlands in the Yangtze Estuary as a case study, the possible impact of SLR on coastal wetlands was analyzed by adopting the SPRC model. A methodology of EVI spatial assessment was developed for three time-scale scenarios of SLR (the present trend and IPCC A_1F_1). The results show that the combined effects of SLR, land subsidence, and reducing sediment discharge give rise to the high risk of a considerable decrease or even habitat loss for the coastal wetlands in the Yangtze estuary (particularly in 2050 and 2100). Without proper mitigation options, the potential decreases and loss in habitats and ecosystem services of the coastal wetlands in the Yangtze estuary is inevitable. There are mitigation options that have been shown to be effective in the Yangtze estuary and should be continued into the future, but these need reinforcing by appropriate policies to manage sedimentation, rehabilitating and recreating wetland habitat, and controlling reclamation and other human activities. The SPRC model adopted, the methodology developed, and the results presented in this study can quantitatively and spatially assess the vulnerability of coastal wetlands in the Yangtze estuary under the impact of SLR. The EVI assessment approach shows promise as a methodology to support the successful management of coastal and estuarine systems, and assist with the objective and quantitative assessment of the vulnerability of coastal wetlands undergoing the impacts of SLR elsewhere.

7.4

Managing Risk in a Large Flood System, the Gironde Estuary, France

François Hissel [1], Juan Baztan [2], Amandine Bichot [3], Olivier Brivois [4], Didier Felts [5], Hugues Heurtefeux [6], Idrissa Oumar Kane [7], Vanessya Laborie [1], Goneri Le Cozannet [4], Carlos Oliveros [4], Guirec Prévot [1], Bénédicte Rulleau [7], Philippe Sergent [1], Hassan Smaoui [1], Blanche Thierry [1], Nabil Toulli [7], Jean Paul Vanderlinden [7]

[1] *Centre d'études Techniques Maritimes et Fluviales (CETMEF), Margny Lès Compiègne, France*

[2] *Marine Science for Society, Cultures, Environments, Arctic, Representations, Climate research Center, Université de Versailles Saint-Quentin-en-Yvelines (UVSQ), Guyancourt, France*

[3] *EID Méditerranée (EID), Montpellier, France*

[4] *Bureau de Recherches Géologiques et minières (BRGM), Orléans Cedex, France*

[5] *Coordonnées des Centres d'études techniques de l'Equipement (CETMEF) Bordeaux Cedex, France*

[6] *Entente Interdépartementale pour la démoustication du littoral méditerranéen (EID), Montpellier Cedex, France*

[7] *Cultures, Environments, Arctic, Representations, Climate Research Center, Université de Versailles Saint-Quentin-en-Yvelines (UVSQ), Guyancourt, France*

7.4.1 Description of the Site

The Gironde is a highly dynamic macrotidal estuary located on the Bay of Biscay in the southwestern part of France. It is the largest estuary in Europe, covering an area of 635 km^2. It is commonly regarded as extending approximately 75 km from the coast to the confluence of the Dordogne and Garonne rivers at Ambès although tidal influence is still observed near La Réole, 170 km from the coast.

Being exposed to the storm systems generated over the Atlantic, the risk of flooding has always been a major concern for the authorities along the Gironde estuary. Champion (1862) reported flooding of the Garonne and Dordogne in 1212, 1310, 1425, 1523, 1536, and 1542. The most damaging flood occurred in April 1770, when about 24,000 km^2 were flooded causing about 4147 million livres of damage for the city of Bordeaux alone. In response, the king (Louis XV) allocated special funding to help repair the city. From this point, measures were taken to limit the consequences of flooding. However, they did not prevent significant flooding occurring in 1835, 1855, 1856, 1875 (when there were 500 fatalities), and in 1930 when floods caused the destruction of 1000 houses and more than 300 deaths (Champion, 1862). More recently, significant floods have occurred in December 1981 (mainly from strong river discharges with high tidal amplitude), Lothar and Martin storms in 1999 (when the nuclear power station was affected), and Xynthia in 2010 (flood damage and fatalities). These later storms have had a great influence on the flood risk management in France, providing the impetus for a number of new policies and tools.

In comparison with the majority of European estuaries (e.g., the Elbe and Scheldt Estuaries; Sections 7.5 and 7.1, respectively) and despite its history of flood management, the Gironde remains a relatively natural system. It has substantial areas of mudflats, sandbanks, dunes, sand beaches, and lagoons. The surrounding arable land provide for the cultivation of wheat, rye, maize, oats, tobacco, and, importantly, grapes. At the same time, the estuary is an important navigational route for shipping and cruise liners, and a range of marinas, container ports, and docks can be found along the length of the river (e.g., Verdon, Pauillac, Blaye, Ambès, Bassens, and Bordeaux). Energy infrastructure is also found in the estuary, notably the Pétrole du Bec refinery and Blayais nuclear power station. Population is spread throughout the estuary but the highest densities are reached around the city of Bordeaux.

With this variety of uses, the key challenge for the management of the estuary against flooding is, in line with national policy, to equitably reduce the impact of flooding while maintaining the natural functioning of the estuary. Some solutions have already been implemented to increase the protection level of local assets, but the effects of climate change may lead to an important increase of the costs of the risk management policy. Therefore new solutions, including nonstructural options, need to be considered to guarantee a sufficiently low level of vulnerability for the estuarial regions. The most important challenge is to identify policies which improve the protection levels where appropriate while minimizing impacts for less inhabited areas.

7.4.2 Existing Coastal Policy and Management Strategies

Repetitive historical floods have led to an early adoption of preventive policies and protection measures and there are 433 km of engineered defenses, the Gironde department alone having 130 km of dikes or levees (Coastal Wiki, www.coastalwiki.org/wiki/Gironde_estuary). Dike management is very fragmented; for instance, there are more than 400 owners for a stretch of 20 km, each providing different levels of protection of between 10 and 100 years. The latter standard (best defense) is used especially in densely populated areas. Different standards of maintenance also occur; after Xynthia, an inspection of the state of all dikes along the estuary showed that about 50% of dikes were considered to be in good condition, 30% in a moderately good state, and 20% in bad condition (THESEUS OD1.15; see appendices of Theseus Consortium, 2012).

The French legal framework for risk prevention is established at the national level. The base legislation dates back from 1935 when the first plans with floodable surfaces were drawn. These were mainly used to avoid building that would worsen the threat in those zones. The national flood policy is largely driven, in contrast, by the fact that flood losses are insured collectively rather than by the private individual, as explained later. Coastal protection is primarily addressed at the local level. Regions can undertake actions in the framework of state-regional planning contracts and the state encourages the creation of risk prevention plans. However, previous experience (see Penning-Rowsell et al., 2014) shows that those policies still lack coordination at the scale of the estuary.

7.4.2.1 PAPI: A Concerted Tool for the Management of the Estuary

In the Gironde estuary, recent events and especially Xynthia raised awareness of the need for a joint approach to risk mitigation. SMIDDEST, a syndicate of municipalities and local authorities, was established in 2001 and is building a consensual strategy for risk mitigation shared by all stakeholders on the estuary. One of the first actions of SMIDDEST with the national government was the development of a large flood database on Gironde Référentiel Inondations de la Gironde (RIG), including a risk assessment and a numerical model of the estuary. This tool served as a basis for the preparation of the action plan for the prevention of flooding, introduced in 2003.

The French Flood Prevention Action Programme introduced PAPI (Programme d'Action pour la Prévention des Inondations), the first coordinated policy for the reduction of the risk, including structural mitigation measures and nonstructural options to limit the vulnerability of the exposed areas (SMIDDEST, 2012). Within this, one aim is to finance investments to manage or build new flood-defense infrastructure transferring flooding flows from high vulnerability areas (usually urban areas) to lower vulnerability areas (natural land and farmland) in order to reduce overall risks at the scale of the river basin (Erdlenbruch, Thoyer, Grelot, Kast, & Enjolras, 2009).

As a result, the 32% of dikes along the estuary are now managed by SMIDDEST and other syndicates of municipalities in a more homogeneous way (THESEUS OD1.15, 2012). The management strategy of the estuary devised in the PAPI process also deals with numerous aspects of risk mitigation: structural protection and managed realignment for the most part, but nonstructural measures such as the raising of risk awareness, monitoring and early warning, emergency management, spatial planning, individual vulnerability reduction, dynamic slowing down of flows, and management of hydraulics works are considered as well.

7.4.2.2 Spatial Planning for Risk Prevention

The Barnier law in 1995 created the currently applied regulatory framework of the PPR (Risk Prevention Plans). These plans are developed at a basin level by the Directions Régionales de l'Environnement, de l'Aménagement et du Logement, which has responsibility for environmental risk planning. A risk prevention plan contains a summary of the studies carried out to assess flood risk as well as a map with different colors to delineate zones where building is possible, or subject to particular restrictions. Some documents may describe additional constraints for buildings in each zone. When the PPR is signed by the representative of the state, its provisions are added to those of other documents (e.g., economic development, territorial consistency) and applied to the local spatial planning policies. Although this often places restrictions on development, it can ultimately prevent any building works.

Although the framework for PPRs is consistent and comprehensive, its practical application has proved to be slower than expected. In 2011, more than 10 years after the establishment of the system, only the 37% of the municipalities at risk approved a PPR to regulate spatial planning and building (D^2SOU project, 2011).

7.4.2.3 Monitoring and Emergency Management

In addition to the prevention of flooding, a high priority is given to emergency planning in response to a developing event. This requires monitoring of riverine and marine conditions, which is carried out by flood forecast services, part of the Ministry of Ecology.

The Ministry is responsible for developing and using models to predict floods based on water levels and rain observations. When a flood is likely to occur, they warn the representative of the state at the department level (prefect) who, together with mayors in municipalities, is responsible for emergency management and population safety. To speed up emergency management operations, plans are prepared in advance by the municipalities (PCS, Plans Communaux de Sauvegarde [municipal plans for safeguarding]) to evaluate the risks and their impacts and list the actions that should be carried out to protect population and goods.

7.4.2.4 Mandatory Insurance

In France, government policy on flood insurance is based on a public–private partnership. Insurance coverage against flooding and other natural hazards is mandatorily included in building and home contents insurance at a level set by law (currently 12%). In return, the government provides a guarantee in case flood damage exceeds the insurance companies' ability to cover the claims (CCR, 2011). When a flood or any other kind of natural event occurs, the state and its local representative (prefect) is responsible for evaluating the damages and declaring the event as a natural disaster at the municipality level. From there on, the insurance pays for damage. This may be a reason for the relatively low safety levels adopted for coastal protection in France, because the population is always reimbursed when a disaster occurs.

7.4.3 Flood System in the Gironde Estuary

7.4.3.1 Water Levels in the Gironde Estuary

The Gironde is subject to very strong tidal currents. Because the funnel shape of the estuary, the tidal amplitude increases when it propagates inland. For average tides, the amplitude is about 3.1 m at the mouth and goes up to 4.2 m in Bordeaux before falling again. The tide is strongly asymmetric, all the more so upstream, with the ebb tide lasting for about 2/3 of the semidiurnal period.

The Garonne River mainly flows over modern alluvial fields limited by outcrops of early Miocene (Mignot, 1969). Between Ambès and Blaye, it is framed by Eocene hills on its right bank and by modern alluvial fields whose width reaches 9 km. From Mortagne to the ocean, Cretaceous sediments form a natural barrier to the estuary. On the left side, modern alluvions and small Eocene hills can alternately be found. The Pointe de Grave (spit) near the mouth creates a bottleneck that decreases the width of the estuary significantly and has a strong influence on the incoming tide. The bed is covered by relatively consolidated sand and mud but, below these, it is usual to find much more solid elements made of rock, gravels, and compact mud. The estuary has several islands downstream of the river confluence at Bordeaux. These islands and sandbanks were always present in the estuary but undergo major changes over time, which also involves changes in the navigation channel. During the last survey in 2009, the largest island had an area of 2 ha.

The cumulated river discharge from both rivers (Dordogne and Garonne) ranges between 50 and 2000 m^3/s that, during flood events, can reach greater than 5000 m^3/s (Figure 7.41).

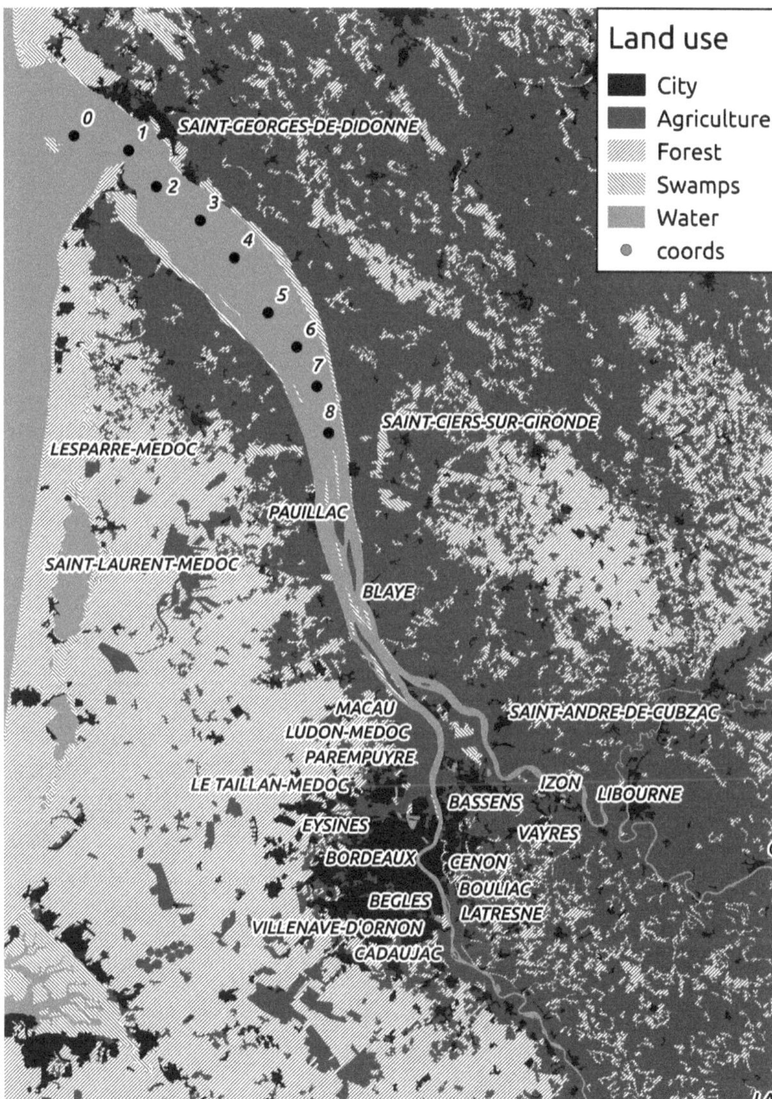

Figure 7.41 The Gironde estuary with its land use.

TABLE 7.21 Climate Scenarios for Gironde Pilot Site

Input Data of Models		Reference Period 1960–2000	A1B Scenario	B1 Scenario
Primary data (high sea)	Sea level	Observations	ONERC	
	Winds and pressures	Observations	CLM/SGA	
	Riverine discharges	Observations	Reproduction of reference period	
Secondary data (obtained by simulation for the mouth of the estuary)	Surge	Observations	Simulated based on simplified law	
	Astronomic tide	FES 2004		
Results	Waves	Tomawac simulation		
	Sea levels	Telemac simulation		

Water levels in the Gironde depend both on the propagation of the tidal wave from downstream—therefore on the mean sea level, tidal amplitude and storm surge—and on the discharges of the two rivers that meet upstream, Dordogne and Garonne. Wind and atmospheric pressure also have an influence as they can cause surfaces to move from the "horizontal." Extreme flood waters are caused by a combination of several high values for those parameters. Waves contribute to flooding downstream, very close to the mouth of the estuary.

The synthesis of the climate scenarios here examined is reported in Table 7.21.

Sea Level

National regulations have been established to provide a common basis for technical studies investigating the potential impacts of climate change (ONERC, 2010). According to these regulations, three scenarios should be considered for SLR between 2010 and 2100. The optimistic scenario foresees a 40-cm rise by 2100, the pessimistic one—used for the simulations discussed here—60 cm SLR, whereas the extreme one goes for 100 cm rise (Figure 7.42). Between 2010 and 2100, SLR is calculated using a parabolic function where the peak occurs in 1870 at 18 cm below the 2000 level (see Figure 7.42). These scenarios are applicable for all the coasts of France. The regional variability of future SLR (Slangen, Katsman, van de Wal, Vermeersen, & Riva, 2012) and the potential impacts of ground subsidence or uplift to relative SLR (Woppelmann, Miguez, Bouin, & Altamimi, 2007) are presently not included in these scenarios. Still, the approach of ONERC is one of the first initiatives to integrate consistent scenarios of future SLR within existing risk prevention policies.

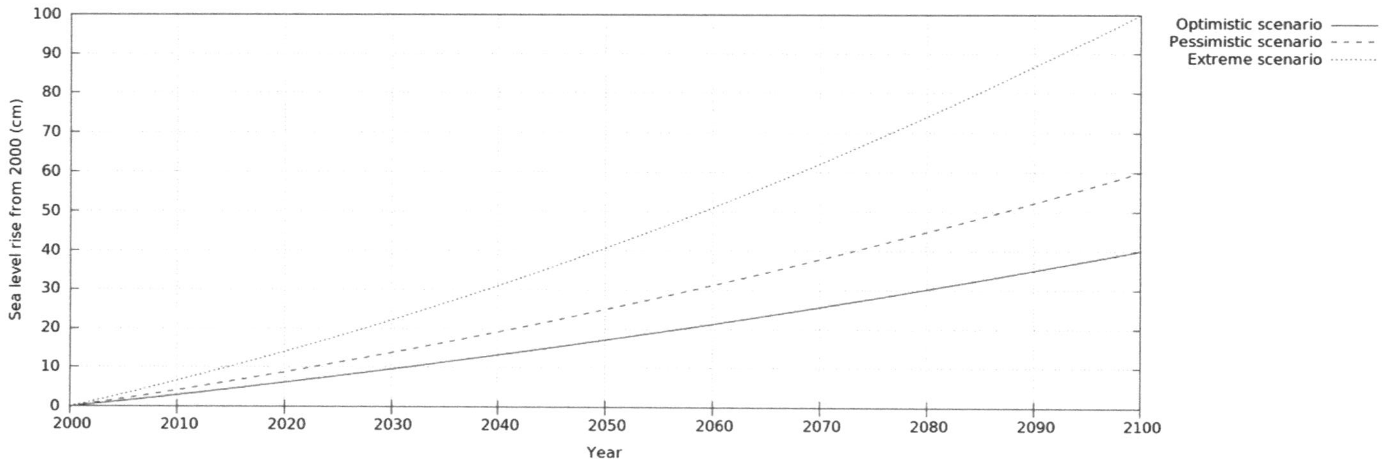

Figure 7.42 Sea-level rise used for climate change scenarios in the Gironde estuary. Pessimistic scenario (center) is used for flood mapping.

Storm Surges

Offshore storm surges were estimated for the twenty-first century using the CLM/SGA database (CLM/SGA, 2010) of winds and pressures over the sea through a simplified model (Laborie, Hissel, & Sergent, 2012a). This simplified model assumes the instantaneous storm surge in Le Verdon (mouth of the estuary) is related to the latter variables averaged on a domain around the mouth of the estuary through the following relation where α, β, and γ are numerical parameters, v is the spatially averaged speed of wind over the mouth of the estuary (from CLM/SGA database), and p is the spatially averaged pressure over the mouth of the estuary (also from CLM/SGA database):

$$\delta = \alpha v^2 + \beta p + \gamma \tag{5}$$

Coefficients α, β, and γ were estimated according to a best-fit approximation applied on a set of events for which both winds, pressures, and water levels were available. For this regression, 10 events were used from 1981 to 2006, each lasting for about two weeks. To be consistent with storm surges observations, observed pressures and winds from Météo-France were also used for those 10 events (Figure 7.43).

Waves

Two models were used to study evolution of waves in Gascogne Gulf and Gironde estuary with a background of climate change (Morellato & Benoît, 2010). Both are based on a finite element mesh grid and use the Tomawac software (http://www.opentelemac.org) Telemac-Mascaret 2013. The external grid, "ocean model," has a resolution between 1 and 0.25° along the French coasts. It provides limit boundaries to the second model which has a finer resolution, of between 0.25 and 0.03° in the Gironde estuary.

Winds used to force both models come from two sources: the Global climate model ECHAM5 (Max-Planck-Institut für Meteorologie, 2010) (1.8° every 6 h) and the European climate model CLM/SGA forced with ECHAM5 model (0.165° every hour). The models were run both for current conditions (hindcast from 1960 to 2000) and for future conditions of the twenty-first century (from 2001 to 2100) based on two IPCC scenarios for climate change—A1B and B1 (Meehl et al., 2007).

By comparing the present period with future ones, it appears that the changes in wave amplitudes for any future scenarios are not that important (Morellato & Benoît, 2010). By 2100, Figure 7.44 shows that the mean wave height will fall by 5% from 1.5 m today at the mouth of the estuary. Conversely, as shown in Figure 7.45, the extreme wave heights (100-year return period) may increase slightly by 3% over the same period. Other simulations of future waves are in agreement with this conclusion, but also suggest a northerly shift of the direction of waves (Charles, Idier, Delecluse, Deque, & Le Cozannet, 2012).

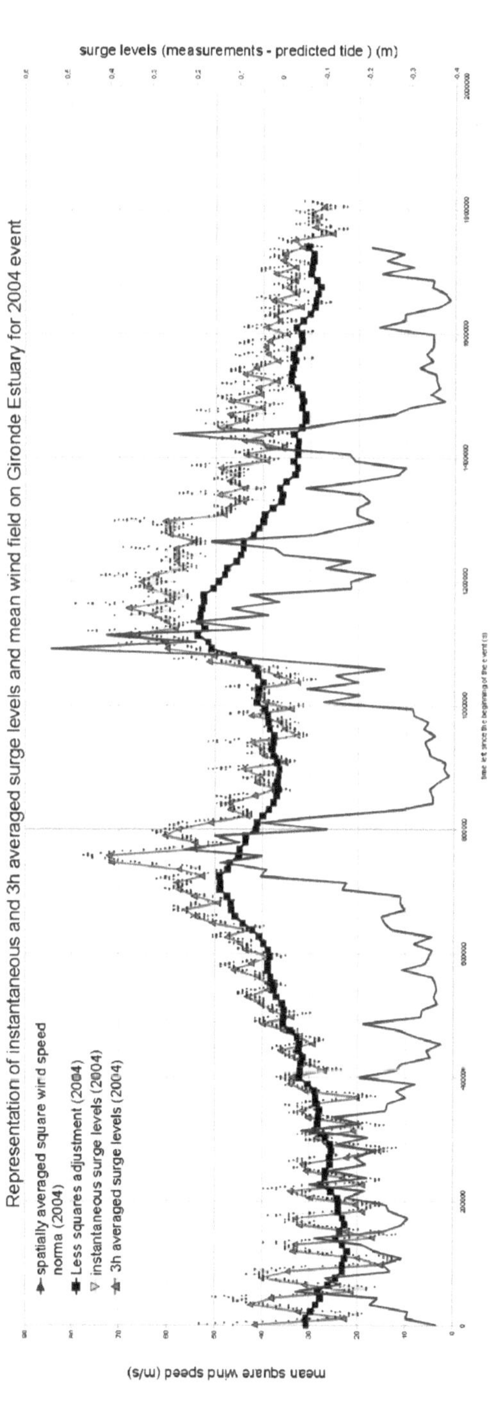

Figure 7.43 Example of comparison of observed water levels in Le Verdon and water levels computed with the simplified model for one event in 2004.

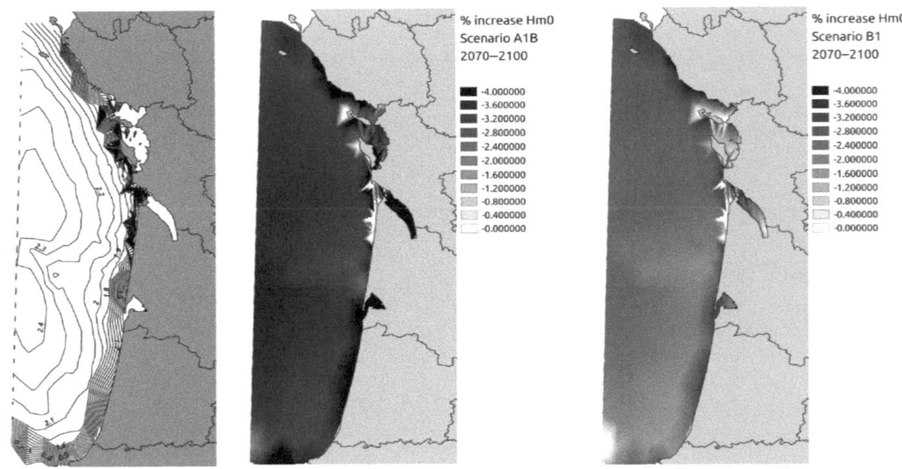

Figure 7.44 Change in mean wave heights between 1960–2000 and 2070–2100 for two scenarios of climate change (left is current situation 1960–2000).

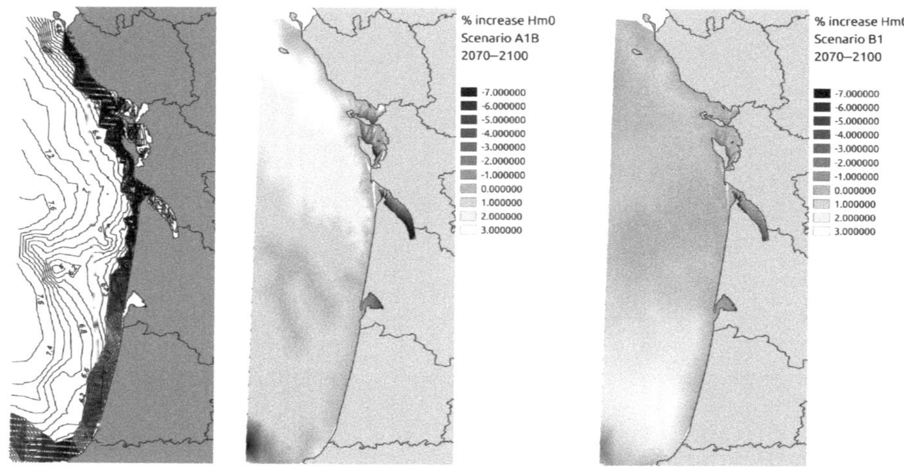

Figure 7.45 Change in 100-year extreme wave heights between 1960–2000 and 2070–2100 for two scenarios of climate change (left is current situation 1960–2000).

Considering each scenario separately, the extreme values of wave heights for A1B scenario increase by 3% at the north of the estuary and decrease at the south by 1% whereas for B1 scenario the extreme values of wave height increase by 1% only. The results also show that the number of storm events decreases and there is no significant change in wave direction. In summary, climate change is likely to bring fewer storms in the Gironde estuary, but those storms will be slightly higher than today.

Coastal Water Propagation

Because the Gironde estuary is convergent, the tidal wave (including its storm surge component) amplifies when it propagates from the mouth of the estuary toward upstream. This phenomenon may be hindered by two riverine discharges from the Dordogne and Garonne, which meet 90 km from the mouth near Ambès. This complex hydraulic process needs to be understood before accurately evaluating water levels along the estuary. In achieve a better insight of the propagation phenomenon, a 2D shallow water model based Sogreah, 2010 on the Telemac software Telemac-Mascaret, 2013 was used and adapted to simulate the flood dynamics in the flood plain accounting for dike overflow. It is assumed however that existing dikes do not breach during a flood event. The model was calibrated over real observations between 1960 and 2000, and run between 2010 and 2100 using as input the hydraulic loads established for sea level, storm surge, and riverine discharges.

The results of the model (Figure 7.46 and Table 7.22) show that extreme water levels along the estuary tend to increase with time. Looking in detail at local results, the 100-year return period level of water levels in the estuary grows by 35 cm near Ambès (where the two rivers meet) from 4.57 to 4.93 m and by 20 cm in the Garonne near Bordeaux from 5.13 to 5.33 m.

Riverine Discharges

Two rivers meet in the Gironde estuary near Ambès: Garonne from the south and Dordogne from the east. A minor river, the Isle, flows in the Dordogne at Libourne close to the confluence, but its discharge is low compared with the other rivers.

The average combined river discharges of Garonne and Dordogne is 30 billion m^3 per year at the confluence in Ambès. At this point, the river discharge is in the same range of values as the oscillating tidal volume. The discharge of Garonne may, exceptionally, reach a value of more than 5000 m^3/s, but usually does not exceed 4000 m^3/s with an average discharge of 620 m^3/s. In summer, low flows may lead to discharges of less than 200 m^3/s. The Dordogne's discharges are lower and seldom exceed 2000 m^3/s, with a yearly average value of 270 m^3/s at Bergerac.

A statistical analysis of the likely extreme discharges of those rivers has been carried out using a peak-over-threshold method based on daily discharge observations from January 1, 1875, to December 31, 2009. The extreme values of the distributions (with a threshold of 3000 m^3/s for Garonne and 1000 m^3/s for Dordogne) were fitted on a general Pareto distribution with an unbiased probability weighting moment estimator. This allowed the estimation of parameters for the statistic distribution of discharges for both rivers. The resulting curves give the expected discharge with respect to the return period (Figure 7.47). Using this method, the 100-year return period discharge is 5534 m^3/s for Garonne in La Réole and 2287 m^3/s for Dordogne in Pessac.

The large uncertainties around atmospheric conditions as well as land use and management changes in the estuary mean that it is difficult to estimate how river discharge could change in the future. It is also expected that the relative impact of riverine discharge will remain low compared with the seawards tide propagation (at least downstream of Ambès). Thus, when applied to the future period (2010–2099), the flood model for the estuary was simply fed upstream by the same discharges series as observed during the 1960–2000 period (Table 7.23).

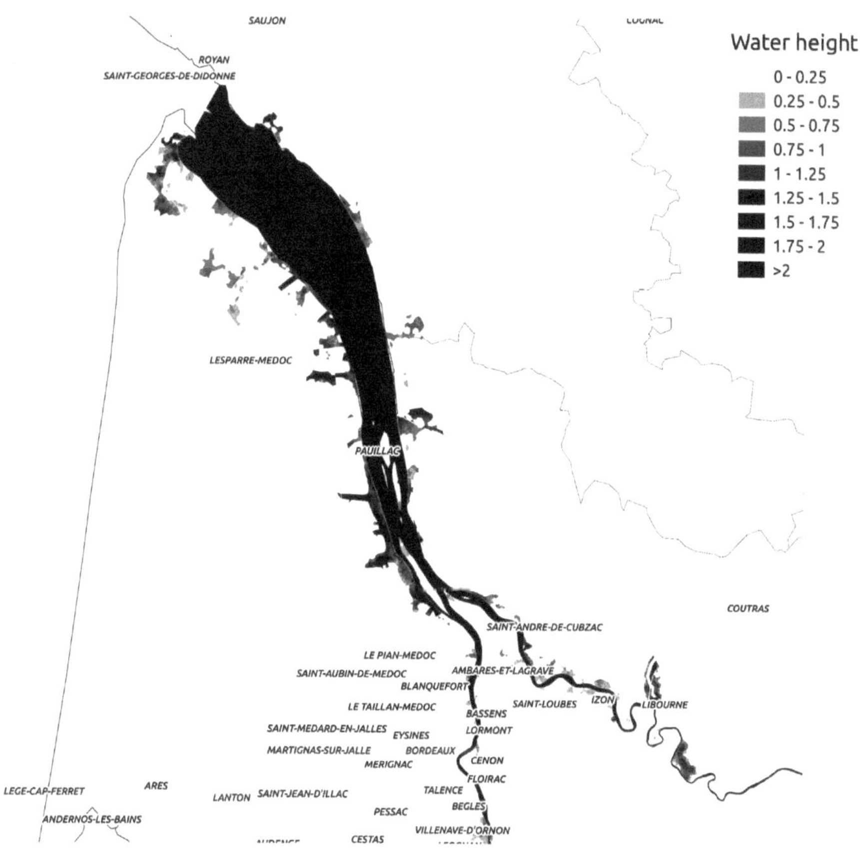

Figure 7.46 Extreme 100-year return period storm surges levels as calculated by the Telemac model for the 2070–2100 period.

TABLE 7.22 Percentage of Flooded Areas with Respect to the Total Area Subject to the Flood Risk (867 km^2), for a 100-Year Return Period Flood in the A1B Scenario, and for Different Time Periods

Time Period	Percentage of Flooded Area	Percentage of Area Flooded Over 0.3 m Depth
1960–2009	24.2	20.7
2010–2040	22.2	16.8
2040–2070	26.0	24.0
2070–2100	27.9	25.7

7.4.3.2 Social Vulnerability

The Local Economy and Population

The estuary is mostly a rural region, surrounded by agricultural fields. Within the flood-prone area (which covers about 800 km^2), the majority is agriculture and meadows (644 km^2) including 80 km^2 of vineyards and 54 km^2 of forest. The 8% of the highly valued total vineyard of Bordeaux's region is located in the flood-prone area. Urban areas (cities, infrastructure, etc.) cover 65 km^2. The area is quite heterogeneous: among the 220,000 people resident in the maximal flood-prone area, 164,000 live in Bordeaux or its environs. In addition, the city of Bordeaux welcomes 2.5 million tourists per year, especially during summer and early autumn.

Demographics in Gironde are generally in line with national rates. The growth rate of the population is +1% per year, with a birth rate of 1.08% per year and a death rate of 1% per year. The migration rate is +0.9% per year. About 16.7% of the population is age 14 or younger; 17.1% are between 15 and 29; 19.7% are between 30 and 44, 20.8% between 45 and 59, and 15.2% between 60 and 74.

The economy is rather mixed. Ten percent of the labor market work in agriculture and fisheries, 7.5% in building and public works, 12.5% in other industries—with some major firms specializing in aeronautics, metallurgy, and aerospatial—and 75% are in the tertiary (service) sector. Tourism is well developed and represents a significant part of this latter sector. The unemployment rate is close to 10% in the whole region, while the GDP per capita is €25,000 per year on average, with strong variations in urban areas. The area of Bordeaux for example reaches a GDP per capita of €44,850 per year.

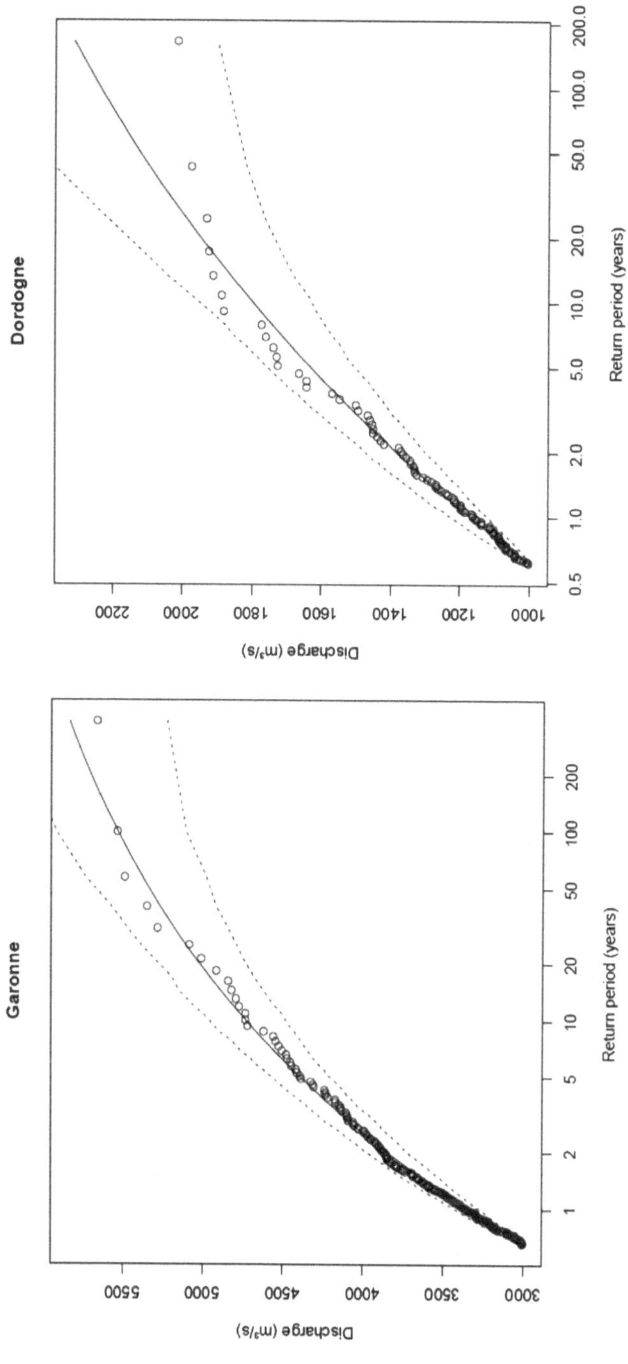

Figure 7.47 Extreme riverine discharges estimated (curves) and observed events (circles) for Garonne in La Réole (left) and Dordogne in Pessac (right).

TABLE 7.23 Return Periods of Extreme Events for Dordogne and Garonne, Upstream of Gironde Estuary

Return Period (year)	Dordogne Discharge (m^3)	Garonne Discharge (m^3)
10	1791	4731
20	1955	5020
25	2005	5104
50	2152	5337
100	2287	5534

Local Perception of Risk

The local population is aware of the flood risk because of a number of events that took place over the past few decades. In February 2010, 27 municipalities were flooded during the Xynthia storm event and two major cities were partly flooded in May 2008. In December 1999, the whole region was severely hit by the Lothar and Martin storms, including Bordeaux and Ambès. However, most of the damages at that time were the consequences of the strong winds, rather than of the flood itself. In December 1981, a major disaster occurred and 142 municipalities were declared to be affected. Other events, e.g., May 1982, March 1988, February 1995, and April 2005, had fewer impacts.

Interviews conducted on site with stakeholders and local managers (THESEUS OD1.15, 2012) showed that risk is mostly associated with its consequences, seldom with the probabilistic nature of its drivers. Risk management is perceived as relevant firstly because it is already being addressed and secondly because the capacity to do so exists with current technology and knowledge. The main reason for a disaster is not considered to be the physical phenomenon but a consequence of bad policies and human activities. Except for a few cases, it is generally acknowledged that the preferred approach to risk management is to spread the costs of any risk prevention policy across the general public, even if the schemes are only locally beneficial.

7.4.4 Flooding in the Estuary

In order to assess flood impacts within the estuary and neighboring territories, the flood system was analyzed and modelled using current and future scenarios of water levels at the mouth of the estuary, water discharges, and atmospheric conditions defined in Section 7.4.2.1.

The Gironde estuary, as well as the other two estuaries presented in this book: Section 7.1, the Scheldt, and Section 7.5, the Elbe, is a large varied estuary. Therefore, a two-nested-scale SPRC analysis (see Chapter 2) was adopted: one at the estuary level focusing on flooding and one at a local level near the presqu'île of Verdon looking at both flooding and erosion (see Monbaliu et al., 2014; Narayan et al., 2014).

In the estuary, the main sources are the sea and the two rivers Dordogne and Garonne. The sea can cause flooding either directly (west areas of the estuary) or through the propagation of tides and storm surges into the estuary. This is always combined with the two other sources, the river discharges from the Garonne and Dordogne. The influence of tides is predominant downstream while the influence of river discharges is more important upstream. Waves are only important right at the mouth of the estuary. For all other parts upstream, only water levels are involved in the flooding processes.

Most of the estuary is covered with dikes used to protect inhabited areas against flooding. An inventory of those dikes together with their state and their characteristics (crest level, type) was carried out in 2009 by SOGREAH, 2010. This inventory helped to identify some possible paths that the water can take to flood vulnerable areas. However, because of the complexity of the system with a large number of possible breaching involved in a flooding process and some interactions between them, only the overtopping and overflowing of dikes were considered as flooding modes in this analysis, and no possible dike breaching was simulated.

At the same time, the most vulnerable areas in the flood prone zone were identified based on the Corine land-use map.

7.4.4.1 Flood Modeling

The flood mapping carried out in the estuary is based on the application of a 2D shallow-water model for surge and tide propagation within the estuary (see Section 7.4.3). It also takes into account riverine discharges and atmospheric conditions. Waves are quickly damped when entering in the estuary and therefore only have a minor influence on floods. The mapping shown hereafter therefore does not take them into account. A nearshore wave model was also developed to get more insight on specific regions near the mouth of the estuary where wave run-up may increase the flood risk.

The Telemac 2D shallow water model of the estuary was customized to consider floodplains. Those were linked to the meshing of the estuary using weir transfer functions to estimate at each time step the volumes of water exchanged between the sea and the floodplain (Laborie et al., 2012a). These functions evaluate the water levels in the receptor and consequently update the levels in the river bed. Dikes are represented by weir functions that compute the total volume of water that flows over the dike crest. Classic empirical equations are used for this step. For an emerged weir, the discharge Q is:

$$Q = \mu\sqrt{2g}(z_a - z_w)^{3/2} \tag{6}$$

where z_a is the level of water in front of the dike and z_w is the crest level of the dike. For a drawn weir, the total discharge is estimated by:

$$Q = \frac{3\sqrt{3}}{2}\mu\sqrt{2g}(z_b - z_w)\sqrt{z_a - z_b} \tag{7}$$

with the same notations as previously and z_b being the level of water behind the dike. Other protection structures such as drains are represented by source and pit terms in the model. This choice of representing structures allows for an easy change of their characteristics in the model. It is thus possible to change the level of a dike to test a new protection scheme, or to fully remove a dike from the model to analyze the impacts of a breach.

The model was used with both current data (1960–2009) and meteomarine climate scenarios (see three time periods in Table 7.21) and was able to provide current and future flood maps. The model was initially run over a period of 140 years between 1960 and 2100 Laborie et al. (2012b). Following the meteomarine climate scenarios, flood maps for each period (present, short-term future, mid-term, long-term future) were created using a statistical analysis based on the peak-over-threshold method and a Gumbel distribution fitted to the raw results. Finally, the maps were compared to see how the flood hazard would change in the future.

The results of the simulations (Figure 7.48) show that the intensity of flood hazard will grow in the future, and so probably will the risk from a growth of the assets in the region. Most of the increase will occur before 2040. The wind and pressure conditions at the mouth of the estuary as taken from CLM/SGA database (see Table 7.21) do not change significantly between 2040 and 2100 and so the flood risk will remain relatively constant. Some places in the estuary will see the increase of the 100-year flood level of the 100-year flood increase by more than 1.5 m. Fortunately, more inhabited areas such as the Bordeaux region will not experience such a rise in risk levels. Places that are likely to be flooded more often are mainly located in the Western part of the estuary, downstream of Ambès.

7.4.4.2 Analysis of Impacts

Social Impacts

A flood has a range of social consequences. In the short term, the population may be shocked by the disaster but the long-term consequences are still more important, with a decrease of the attractiveness of the region and less trust in the authorities (see Chapter 6). At the same time, a flood usually helps to raise the awareness of the threat among the population (see Section 5.5). It is difficult to estimate this type of consequence within a quantitative analysis and therefore the social impacts considered here are limited to the flooding of critical infrastructures and a very basic estimation of life losses (see the methodology described in section 6.2.4 and also the application in Cesenatico, Section 7.6).

A database of critical infrastructures (monuments, administrative buildings, schools) was built for the estuary. When cross-referenced with the flood maps, it helps to provide a list of all critical infrastructures within the flood zone.

To estimate life losses, it is not sufficient to cross-reference a database of population with the flood maps. If the flood is anticipated, the authorities usually launch emergency plans to protect the population. With an advance warning of at least 12 h, it may be possible to evacuate the population if necessary. By taking such procedures into account (see Section 7.4.6), for the Bordeaux region and Ambès using the 100-year return period flood map from the long-term future period (2070–2100), see Figure 7.49, results show that the fatalities are expected not to be high in those two cities as there would be enough time enough to evacuate everyone before the flood arrived. However this requires a careful preparation of an evacuation plan before the event and a good communication to the population to prepare them to evacuate in case it is needed. Such a communication might include regular evacuation exercises.

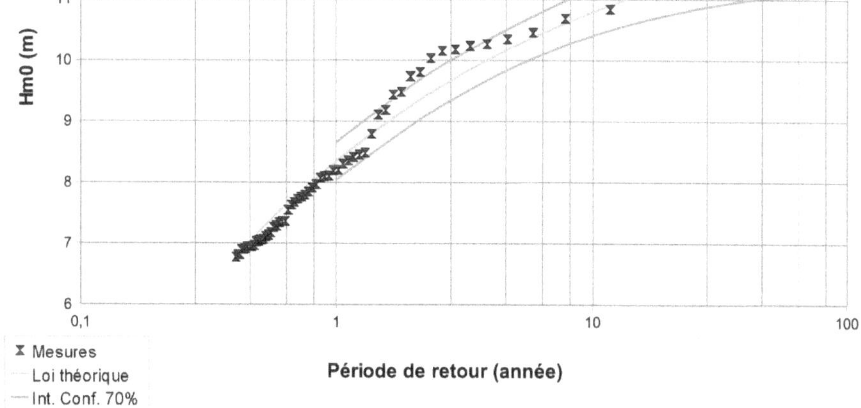

Figure 7.48 Extreme wave height at Lat: 45°49.140′ N – Long: 1° 30.540′ W – Prof : 29.10 m.

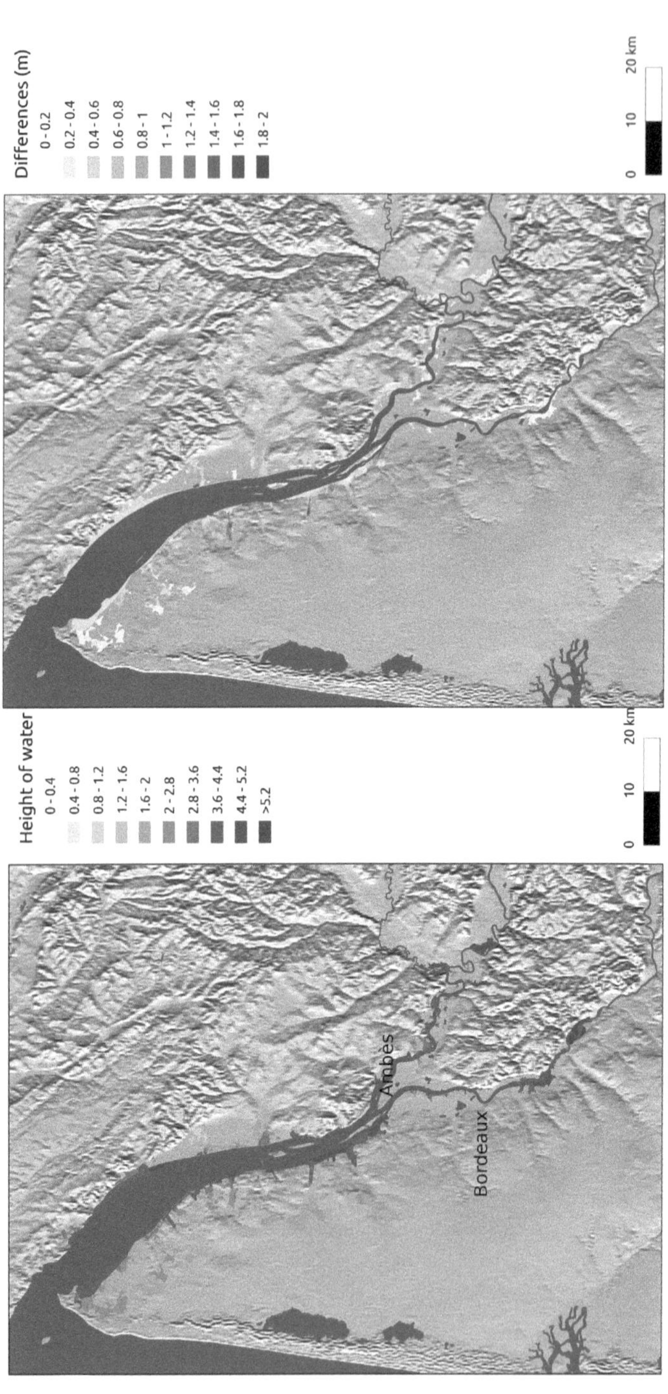

Figure 7.49 100-year return period flood maps in the Gironde for the long-term future period (2070–2100): height of water levels to the left and difference with flooding at present conditions current levels to the right.

Economic Impacts

Because of the large uncertainties about the response of the economic system to a disruption, only marketable damages were considered in the economic impacts of a flood. A systematic approach was used to estimate the material damages of a flood in each receptor area, as defined in the SPRC framework. The model takes into account both the base value of the flooded land with its settlements, and the damages it bears during the flood through a damage rate. The latter is a function of the flood characteristics, namely flood height, velocity, and duration (see Sections 6.2.3 and 6.2.4). The same damage functions were used as the ones from the Cesenatico study site.

Due to its size, the area presents an important economic activity in different sectors (industrial, commercial, agricultural, tourist). The base value of the land therefore depends on the location and the economic background of the region. Those values were estimated from real estate values in inhabited area, official national values for agricultural lands, and the method (EDATER, 2001) elsewhere.

More local impacts were identified at a smaller scale near the mouth of the estuary (see Figure 7.50): a road near the shore in Talmont, the harbor of Royan with direct damages to the boats and a cutoff of the ferry line to Le Verdon, submersion of some districts of the town of Royan, a holiday village in La Palmyre, the port facilities in Le Verdon (with damages on the container terminal, and on structures—berths and cranes), and erosion of the beaches near Soulac.

7.4.4.3 Key Source Drivers of Future Risk Conditions

The SPRC analysis of the Gironde estuary showed that the flood risk in the estuary can be caused by three different elements: storm surges, riverine discharges, and the strength of the wind over the estuary itself. All past events resulted from the combination of two of these drivers. Changes in wave climate will not be a key source driver of future risk conditions because the analysis carried out showed that their height would not increase considerably during this century, and that their direction will not change significantly. Although this is true for offshore waves, an increase of the mean sea level may still cause an additional increase to the hydraulic loads on the protection structures because of the higher nearshore depth and consequently lower damping of waves. However, their impacts will still remain confined on the mouth of the estuary. Riverine discharges may change in the future but in a limited manner. However, flood events are expected to become more frequent and more intense in the future mainly because of an increased mean sea level. Sea level rise and storm events will therefore be the key source drivers of future risk conditions.

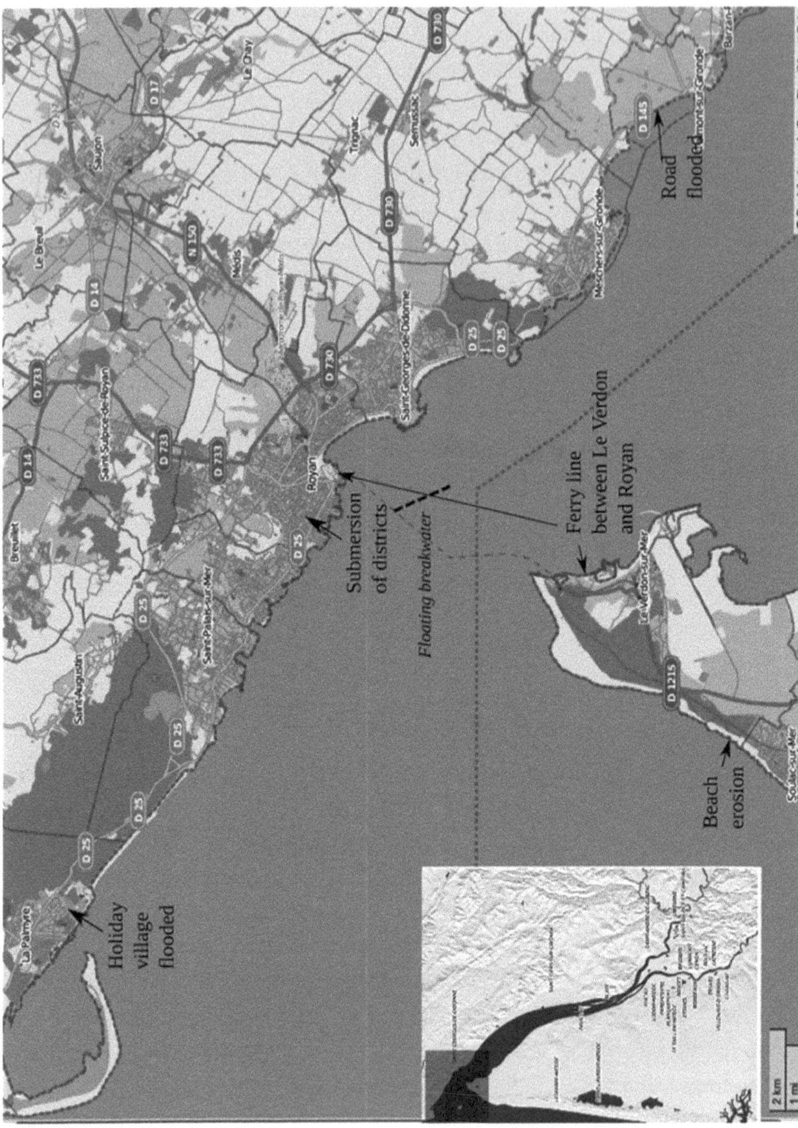

Figure 7.50 Local impacts at the mouth of the estuary.

7.4.5 Mitigation Measures

Despite its long history of flooding management, the Gironde estuary is not limited in the number of mitigation options that can be considered. Areas that are the most vulnerable to flooding are urban areas, with the most important one being the region of Bordeaux, and other smaller cities spread along the estuary such as Ambès, Pauillac, Blaye, and Royan. Industrial areas, some of which are located north of Bordeaux or near Ambès, also need protection. Other large areas of the estuary are still sparsely inhabited and provide opportunities for techniques such as managed realignment that alter the geomorphology and hydraulics of the estuary to reduce water levels. The use of such techniques is consistent with existing legislation through the Conservatoire du Littoral (http://www.conservatoire-du-littoral.fr/). After the 1999 floods for instance, the Conservatoire du Littoral bought some fields near Mortagne-sur-Gironde on the northern bank of the estuary, about 50 km from the mouth. These fields were used for growing cereals and the Conservatoire du Littoral decided, after a flood event, not to restore their protection, leading to the first managed realignment mitigation scheme in the estuary. Social and economic measures are also considered of high interest for risk management policy in the estuary. Based on the lack of preparedness in previous flood events, the improvement of emergency forecasting and planning for evacuation is essential.

In this section, the following mitigation options are discussed further:

- at the estuary scale: managed realignment can help to dampen the tidal wave when it enters the estuary, and therefore reduce the level of high waters upstream;
- locally: engineered structures. Reinforcement of dikes near Royan, La Palmyre, Le Verdon, and Talmont using a range of techniques and the building of a floating breakwater at Talmont to protect an exposed road; and
- locally: evacuation, early warning, and emergency response.

7.4.5.1 Managed Realignment

Managed realignment is designed to improve the damping of the tidal wave when it enters the estuary. The Gironde estuary is classified as a strongly dissipative and strongly convergent estuary, and Savenije (1998) proposed a simplified expression for the tidal damping in such an estuary:

$$\frac{\mathrm{d}H}{\mathrm{d}x} = \frac{H}{h}\left(\frac{\frac{h}{b} - f\frac{g}{C^2}F\sin\in}{1 + \frac{H}{h}\frac{1}{2F\sin\in}}\right) \tag{8}$$

where H is the amplitude of the tidal wave, h is the depth, x is the curve distance to the mouth of the estuary, b is a shape parameter describing the convergence length of the estuary, f is a friction factor, g is the gravity field, C the Chezy coefficient, ε is the phase difference between high water slack and high water, and F is the Froude number. It is here considered that the width of the estuary is an exponential function of the distance to the mouth:

$$B = B_0 \exp\left(\frac{-x}{b}\right) \tag{9}$$

If the tidal amplitude is important compared with the depth of the estuary (*H/h* close to *1*) and there is no difference of friction between low water and high water (second term in the numerator of the fraction is negligible) and the phase difference of high water slack is low (ε close to 0), then the equation becomes:

$$\frac{\mathrm{d}H}{\mathrm{d}x} = \frac{h}{b} \tag{10}$$

This Eqn (9) means that there is an amplification of the tidal wave because of the shape of the estuary. This was also verified by simulating the propagation of a large range of tidal waves in the estuary with the Telemac model (Laborie, Hissel, & Sergent, 2013). Results show that the damping does not depend on the amplitude, because all curves (which differ by the amplitude at the mouth and mean level) are "parallel," at least to a certain distance to the mouth (about 70 km, not far from the confluence of Dordogne and Garonne). After this point, riverine discharges start to play a role and the tide propagation model is not applicable.

What the equation also reveals is that the amplification of the wave can be limited by increasing the value of b, the convergence length that means widening the estuary upstream relative to its mouth.

To extend the flood plain upstream, a new managed realignment option is proposed by lowering the design levels of protection structures at four sparsely inhabited places in the middle of the estuary (see Figure 7.51). The new design level is just above the level of water reached along the dikes during the 1999 storm (i.e., no upgrade in protection for those zones as a result of climate change is considered). Any population or buildings located in the propose realignment areas will need to be considered for relocation or other adaptive measures (e.g., flood proofing).

7.4.5.2 Reinforcement of Dikes

Local assets around the cities near the mouth (Le Verdon, Royan, La Palmyre, Talmont) are already protected by dikes that may not provide a sufficient level of protection especially when considering the impacts of climate change (i.e., a combination of higher sea levels and higher waves). Most current technical recommendations explain how to develop a new protection system, but they do not address the upgrading of existing dikes; indications are given in Chapter 3 and in Burcharth,

Andersen, and Lara (2014). Here, a reinforcement of existing protection structures is proposed to maintain protection levels under the effects of climate change.

The reinforcement options consist mainly of the addition of a crown wall on top of existing dikes, the addition of a new rock layer to increase the stability of current structures, or the setup of a berm or a detached breakwater to attenuate the waves before they reach the armor stones (see Table 7.24).

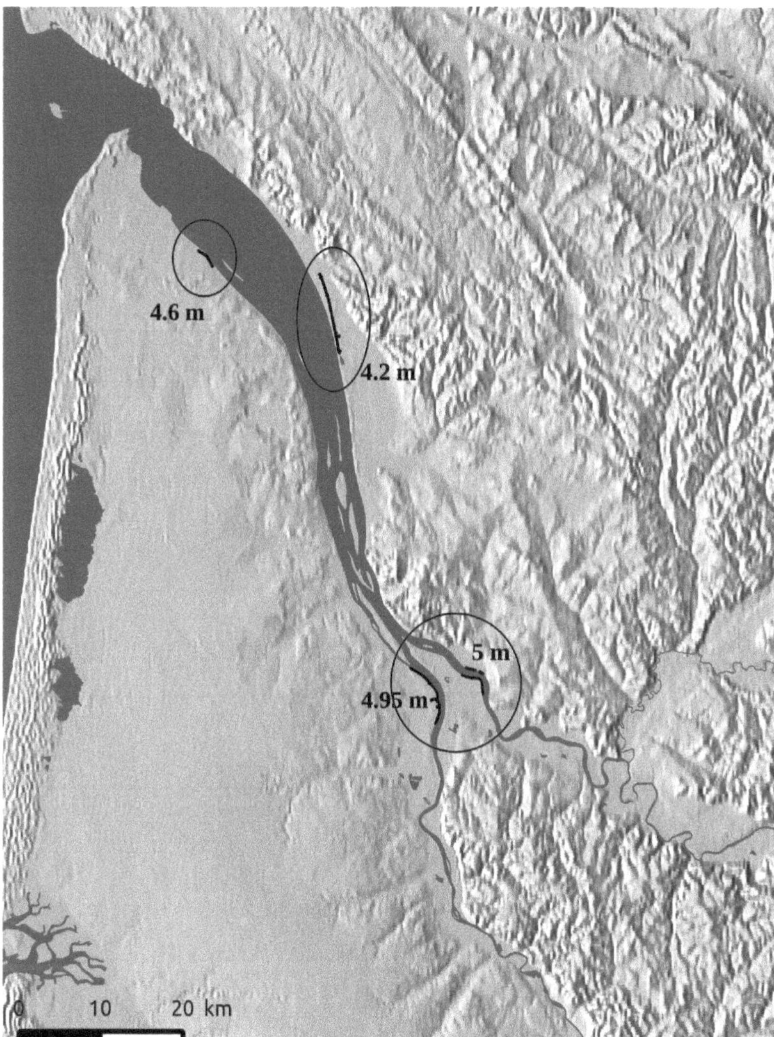

Figure 7.51 Location of changes of design levels of dikes for the managed realignment strategy.

TABLE 7.24 Location of Improved Coastal Structures and the Consequences Being Addressed

Site	Expected Consequences of Flooding	Type of Reinforcement
Talmont 	Cut off of the road along the shore	Crown wall New rock layer
Royan Harbor 	Disruption of ferry service to Le Verdon Damages to yachts Damages to structures	Crown wall Berm New rock layer Detached breakwater
Royan town 	Submersion of districts Erosion of beaches	Crown wall on the top of beaches

(Continued)

TABLE 7.24 Location of Improved Coastal Structures and the Consequences Being Addressed —Cont'd

Site	Expected Consequences of Flooding	Type of Reinforcement
La Palmyre 	Submersion of holiday village	Crown wall Berm New rock layer
Le Verdon Harbor 	Disruption of ferry service to Royan Boat sinking	Crown wall Berm New rock layer
Soulac-sur-Mer 	Submersion of districts Erosion of beaches	Wave dragon

7.4.5.3 Floating Breakwater

This mitigation option addresses a road located west of Talmont, very close to the river, protected only by a narrow pocket beach. The road may suffer from flooding as a consequence of erosion of the beach, induced by the waves propagating upriver from the mouth. The aim of this mitigation is to reduce the average wave height incidence at the beach, reduce the erosion, and ultimately the possible flooding of the road. The tentative length is 1 km (50 modules 20 m long). The floating breakwater (FB) system is composed of very large pontoons, with π-type cross-section and is moored with piles. The FB is installed at a water depth of 4 m (Figure 7.48).

The offshore average wave climate incident the mouth of the estuary is shown in Figure 7.44 and extremes in Figure 7.45. Average waves are essentially characterized by significant wave height Hs in range 0–5 m, peak wave period Tp in range 4–12 s, and average direction in range 255–300°. This information is generally and easily available for a preliminary design. In this specific case, it was derived from AME-MOC (2010).

The FB system (see sketch of system in Section 3.1.2) is composed of 50 modules, each being 20 m long and moored with two piles. The modules are connected to each other with special connectors (cables and rubber fenders). The resulting breakwater is 1000 m long, aligned and parallel to the shoreline.

The proposed dimensions are: water depth $h = 4.0$ m, width $w = 8.0$ m, total height $hs = 2.7$ m, freeboard $Fr = 0.7$ m, draft $d = 2.0$ m (d_1 : 1.4 m, d_2 : 0.6 m), length $L = 70.0$ m. For the FB system described previously, the calculated transmission coefficient varies, $kt = 0.66$. The actual values vary depending on the incident wave period. The mooring system is formed by two piles of a 400-mm diameter, $L = 10$ m, installed laterally (see Section 3.1.2 for more information about floating breakwaters).

7.4.5.4 Emergency Response: Evacuation Plans and Early Warning

One of the aims of risk management policies is to reduce the vulnerability of assets and urban centers along the estuary without losing the natural functioning of the physical system. However, as these are rather sparsely distributed, simply increasing the standard of protection structures does not appear to be a cost-effective solution for the estuary, especially in the long term when levels of water are expected to grow significantly for extreme events. To be effective, improving design levels would need to be applied on substantial lengths of coast. The legal status of the current protection structures and various ownerships also do not facilitate this operation. Therefore other solutions must be devised to ensure an adequate protection levels.

As far as social measures are concerned, early warning and emergency procedures in France can still benefit from various improvements. Emergency management policy has changed since the beginning of the twenty-first century with the introduction of mandatory emergency plans at the level of the municipalities and a more rational organization of emergency services at the national level. However, as shown by recent events, evacuation is still seldom used as an option to ensure people's safety. Carefully prepared evacuation plans could help local authorities include this option in their emergency management plans.

Early Warning and Emergency Response

One feature of the 1999 and 2010 storms was the unpreparedness of the authorities for the extent and impact of the storms. The development of more warnings, using actual real-time data, with comprehensive actions plan describing all actions that should be carried out before the flood occurs, would be beneficial to emergency evacuation and response planning. In the Gironde estuary, a tool based on the Osiris software developed by the river basin authorities of Loire and Meuse in France and CETMEF after the eponymous FP5 project has been developed.

Osiris-Inondation (www.osiris-inondation.fr) is mainly designed for local decision makers (at the level of the municipality or the group of municipalities) and intends to provide them with operational solutions to better use of forecast data, generally based on water levels or discharges at official stations. The software then converts this information to useful data about the potential impacts and how the available resources can be used to limit the damages. The software has two modes of operation: one is made to help the authorities prepare their emergency management plans (PCS), and the other is intended to be used during a flood event to support the implementation of the plan, actions being triggered when the predicted level of water exceeds some predefined threshold.

To configure the software, all the main land users (main industries, public buildings, schools) had to be identified and characterized, using as much as possible data from national or local databases. Actions will be launched when those elements at stake are flooded. For each category, danger thresholds were defined in terms of level of water that would initiate predefined actions. The preparation of the action plans then uses the realistic scenarios of floods described in Section 7.4.4.1. The output of the software is a PCS that complies with the format prescribed by the French government. The action plan may be used either as a roadmap for staff in the emergency response unit.

For the responsive mode, the software can automatically import predicted water levels so that it selects the predefined scenario which is the best suited to the actual situation (the user can select a more extreme scenario if required). A sorted list of recommended actions is then generated and the software follows the real-time implementation of the action plan. The updated plans can be used by people in the field or stakeholders during an event to check on responsibilities and whether all the protection actions have been carried out according to the plan.

Evacuation Plans

Evacuation of an area is likely to be essential in the event of a significant flood, particularly when no other option, such as local emergency shelters or moving to higher floors, is available. It is therefore better to plan for, rather than simply respond to, such an occasion.

The Gironde estuary is well suited for applying the methodology for evacuation planning. First, models already exist at an operational level and are able to predict future flooding with a rather good accuracy and confidence, and with sufficient delay to anticipate the emergency procedures (24 h). Then, a lot of transportation options are available to bring people to safe places, which makes evacuation a relevant and not too expensive a solution. This may also be better accepted by the population as it does not require a decrease in the safety standards of some parts of the estuary i to improve the protection of the city (as for managed realignment).

In the Gironde estuary, two main areas are both densely populated and with few outbound routes and so they are therefore subject to evacuation planning: the city of Bordeaux, where many people can be flooded, and the presqu'île of Ambès, which only has one access route. Figure 7.52 shows that, in an extreme event, particularly if SLR is considered, a large part of Bordeaux (La Bastide and Le lac districts) and the majority of presqu'île of Ambès may be flooded. Appropriate evacuation planning should therefore focus on those areas.

Hissel et al. (2014) present an evacuation plan of Bordeaux that includes the consideration of possible in-place sheltering which was omitted from the official documents (Plan communal de sauvegarde, ville de Bordeaux). The methodology recommended the division of the flooded area into several sectors.

For each sector, the buildings that need to be evacuated and those where in-place sheltering can be applied were identified. In-place sheltering is an option that should always be considered because it strongly reduces the psychological impact of a real evacuation. This is however only viable in specific conditions and for a short duration. Possible shelters outside the flooded zone were also located and their capacity checked to ensure enough room to accommodate all evacuees. Evacuation routes were considered to be the most reasonable road that may be taken by vehicles between homes and shelters (not necessarily the shortest, but one that uses well-known and well-used roads).

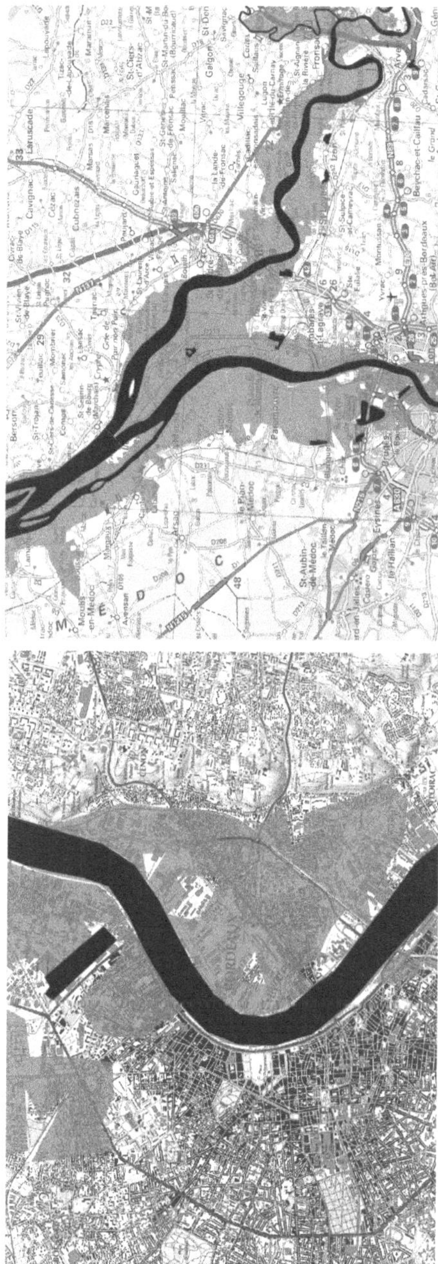

Figure 7.52 Expected flooded areas in Bordeaux (left) and Ambès (right) under a scenario of 1-m SLR (extreme scenario in ONERC, 2010).

The number of people who may need to be evacuated because of flooding is 2941 in Ambès, 2052 in Saint-Louis de Montferrand, and 1080 in Saint-Vincent de Paul. These people can be relocated in Ambarès, the nearest municipality outside the exposed area, then Artigues-près-Bordeaux and public buildings in neighboring villages.

To see if the evacuation plan would be successful, an evacuation model was applied. Data required included population statistics (number, ages, social group), the transportation network, and, crucially, the behavior of inhabitants in case of an evacuation order. Data on the latter were gathered using an online survey but reasonable assumptions still needed to be made (for example, time needed for the individuals to prepare to evacuate, proportion of the population who does not want to comply with the evacuation order). These were based on published experiences in different countries. In both regions, it was found that all the people threatened by the flood could be safely evacuated if well prepared beforehand. This shows that in the Gironde estuary, evacuation is mostly a matter of raising awareness among the population of the need to accept the potential need to leave their homes. The geographical configuration of the region makes it relatively easy to implement the evacuation once it has been decided and accepted by the population.

7.4.6 Strategic Management (in the Short, Mid, and Long Term)

In the short term, the priority, also stated by local authorities, is to improve the protection of the main assets of the estuary without decreasing the level of protection of secondary assets. The construction of new defense structures should not worsen the risk hazard exposure of in other parts of the estuary. To this purpose, the current action plans for the prevention of flooding (PAPI) proposes 78 actions to reinforce dikes near the main assets. These actions consist mainly in the upgrading of the design levels of dikes, or the creation of fuse dikes (i.e., dikes that are designed to open when the level of water goes beyond a certain threshold). This will lead to a quick reduction of the level of risk in major places. Such actions have not been tested in this work but other studies have been performed by the national authorities to assess their efficiency (Référentiel Inondations de la Gironde). The upgrading or design of dikes should, for the future, be carried out to recognize the possibility of an extreme events, which would cause higher levels of flood than previously experienced, caused by the coincidence of high winds, high riverine discharges, higher mean sea levels, and increased storminess from climate change.

Over the longer term, new hydraulic simulations discussed in Section 7.4.3 show that storms will be more frequent or intense along the estuary. An increase in of the significant wave potential height of the 100-year event by more than 1.5 m can be expected in some places.

Upgrading dikes to maintain the existing standard of protection may be a costly and nondesirable option. Because in the Gironde estuary, compared with the other sites discussed here, land use is not so intensive, there are opportunities in spatial planning for proposing new mitigation options, including managed realignment of the estuary. A combination of risk-aware spatial planning and an improvement of emergency management could significantly contribute to a mitigation of the risks. This may however break the constraint of not worsening the hazard anywhere in the estuary. For example, if some fields are allocated for managed realignment schemes for flood mitigation purposes, consultation with the stakeholders affected is essential.

Finally, places located close to the mouth of the estuary (Le Verdon, La Palmyre, Royan, Talmont) will need special attention to ensure a minimal level of protection as a rise in sea level will have direct consequences in addition to increasing wave heights. To this purpose, some special options were designed to improve the protection of the elements at stake.

7.4.7 Policy Implications and Overall Conclusions and Recommendations

A well-designed regulatory framework already exists in France to cope with natural risks. It is especially the case with the regulatory framework (PPR system to avoid new buildings in flood-prone areas), the insurance system that ensures a good spread of the costs of risk across the population, and local procedures that encourage stakeholders to participate in the preparation of action plans for the prevention of flooding (PAPI system). However, most of the risk management strategy in the estuary is based on current knowledge of the hazard and its consequences, and is mainly focused on the protection of current assets. Climate change is seldom taken into account, or it is taken account of through a "safety coefficient." A long-term management of risk should go beyond those practices and include state-of-the-art knowledge about the expected impacts climate change in the next century.

The work described here has shown that climate change will have serious consequences on the level of floods in the estuary. Even if there is no clear trend in the number and frequency of storms in the future, the expected SLR combined to the amplification potential of the estuary alone will lead to stronger events in already floodable areas, with 100-year water levels rising about 20–40 cm in inhabited areas near Bordeaux and Ambès, whereas the 100-year flood extent will grow about 50% toward the end of the century.

Knowledge about the effects of climate change is available, however. It is still subject to a lot of uncertainties, due to the different future scenarios for the rise of carbon dioxide emissions, future temperature change and to the uncertain development of the economy of the region, but some evidence is available about what can be expected in the future. The current tools—PAPI for instance —are well adapted to the planning of a long-term defense strategy, but they need to integrate this long-term vision into today's plans.

In order to maintain the current level of safety while keeping the natural functioning of the estuary in a period of climate change, fully relying on traditional system is not enough any longer. New innovative systems for flood protection and prevention should be designed to tackle all the stages of a risk prevention policy, from the building of new environmentally friendly protection structures such as wave-energy converters, dikes upgradable for a fair price, and dikes more resistant to overtopping to socioeconomic adaptation measures aimed at decreasing our vulnerability to the flood involving risk-aware spatial planning and a better emergency management. The development of these approaches will take time and effort so, to be in time for the future risk, preparations should start now, initially with the integration of climate change in the development of the local PAPI.

7.5

Flood Hazard Mitigation in a Heavily Modified Estuary, the Elbe Estuary, Germany

Jens Kappenberg [1], Janina Sothmann [1], Nino Ohle [2], Dagmar Schuster [2], Ge Jianzhong [3]

[1] *Helmholtz-Zentrum Geesthacht Centre for Materials and Coastal Research (HZG), Geesthacht, Germany*

[2] *Hamburg Port Authority (HPA), Hamburg, Germany*

[3] *State Key Laboratory for Estuarine and Coastal Research (SKLEC), East China Normal University, Shanghai, China*

7.5.1 Description of the Site

Occupation and use of the Elbe estuary over many centuries has resulted in the estuary being greatly modified from its natural state; land claim for agriculture has gradually restricted the Elbe river and, as the largest port in Germany, the maintenance of shipping access to the port has required constant adjustment of the river's morphology. Both of these have resulted in significant changes to the flood system. In addition, Hamburg and its environs has always adapted to tides and extreme high waters with the expectation that flood risk levels will always be minimized, and has therefore created an extensive protective network of dikes.

The majority of development in the Elbe estuary is situated behind these dikes reducing flood risk but the HafenCity in Hamburg is a new development of 157 ha of the old harbor area of Hamburg, Germany. This development is situated on the river side of the existing dikes and remains exposed to the possibility of flooding. Management of this exposure therefore requires consideration of the design of the site itself (flood protection and resilience) and the functioning of the wider estuary (flood propagation).

Because further dikes were not considered suitable, the HafenCity itself has been designed to be able to continue functioning even during a flood event using a dwelling mound flood protection concept (e.g., Hoeksema, 2006) where ground levels for residential and business areas are raised to above the expected flood level, in this case +7.5 m above German datum (NN). This is supported by other methods of flood management such as flood gates in the lower levels of the buildings, which are the responsibility of the land owners. However, some areas remain exposed to flooding (see Figure 7.53) and any further reduction in risk is related to the mitigation of future flood levels that entails estuary-wide investigations.

This assessment of flood risk looks at the extent and probability of flooding for the HafenCity district of Hamburg, how this may change over time as a result of changes in sea levels and how this risk could be mitigated by further modification of the estuary.

The Elbe estuary is a multipurpose area that has historically supported a range of activities. Fishery is a traditional use of the estuary, with prawn fishing typical in the mouth of the estuary. Eel (*Anguilla anguilla*) and the spring-caught stint fish (*Osmerus eperlanus*) are economically the most important. The Elbe irrigates and drains the surrounding agricultural areas, maintaining the very fertile soils in the area that generate produce with international reputation (particularly the orchards in Lower Saxony close to the city of Hamburg). The tidal river Elbe is also relevant for recreation and tourism. About 120 marinas are situated along the estuary and also large cruise ships are becoming ever more common. In addition to natural recreation, water sports, angling, and beach life, especially at the mouth, play an important role.

Figure 7.53 Inundation of a part of the HafenCity/Speicherstadt during a storm surge in December 2013. *(Photo: Christian Hans, Germany).*

However, the most economically important aspect of the estuary is its function as a shipping channel to the Port of Hamburg that directly and indirectly employs about 166,000 people in the Hamburg metropolitan region and 275,000 in Germany. Annually, more than 10,000 ships, half of them container ships, arrive at the port from worldwide destinations to distribute the goods to the hinterland via railway, truck, and inland navigation vessels. Other ports are located along the Elbe estuary (Glückstadt, Stade, Cuxhaven, and Brunsbüttel) as well as heavy industry such as chemicals and oil and power generation.

7.5.2 Physical Characteristics

The Elbe estuary as a typical coastal plain estuary is naturally importing sediment both from the catchment and the sea. The entire catchment basin of the Elbe covers 148,268 km^2. The long-term average yearly freshwater run-off is approximately 700 m^3/s in a range from 150 m^3/s up to 4060 m^3/s. The most frequently freshwater run-off is in a range of 300–400 m^3/s. Because it is also quite sheltered from wave action, the Elbe study site is mainly affected by siltation and erosion is negligible. The Elbe estuary is defined in the west by the islands of Scharhörn and Trischen at the border to the North Sea (Elbe-km 740) and in the east by the city of Hamburg (Elbe-km 610). As long as no storm-tide conditions prevail, the tidal influence of the Elbe estuary is limited by the Geesthacht weir at Geesthacht (Elbe-km 588). The Elbe km is measured from the German Czech border (Elbe-km 0). Figure 7.54 shows an overview of tidal river part of the Elbe.

Tides are semidiurnal with a marked diurnal asymmetry. The mean tidal range decreases from 3 m at the inlet to 2.7 m 50 km up-estuary, rises again toward the port of Hamburg, where it attains its maximum of 3.6 m and then falls to 2 m at the Geesthacht weir. Spring high tides exceed neap high tides by about 0.7 m from the inlet to Hamburg, whereas the difference between spring and neap low tides can only be felt up to 50 km from the mouth. The estuary can be classified as a mesotidal and partially mixed coastal plain estuary. The influence of wind during a storm event with winds from a northwesterly direction leads to considerable increases in the tidal high water level along the Elbe, sometimes producing significant storm surges (see Figure 7.55). Although there is a slight increase in storm surge height during the past 50 years, this is well below the existing dike heights.

Although the water level and the salinity in an estuary have a uniform large-scale distribution, current velocity is a parameter that varies strongly with space and time. This factor leads to the "tidal pumping" of sediments from the lower reaches of the Elbe to areas further upstream, resulting in sedimentation and insufficient water depths in the navigation channel and the harbor basins.

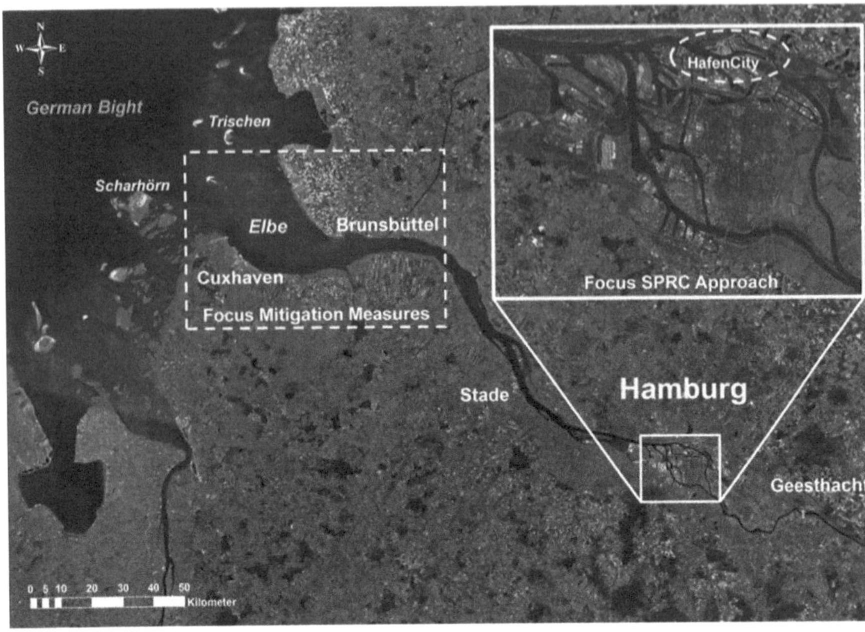

Figure 7.54 Tidal river of the Elbe from Geesthacht to the North Sea. *(Source: Brockmann Consult GmbH ©2003).*

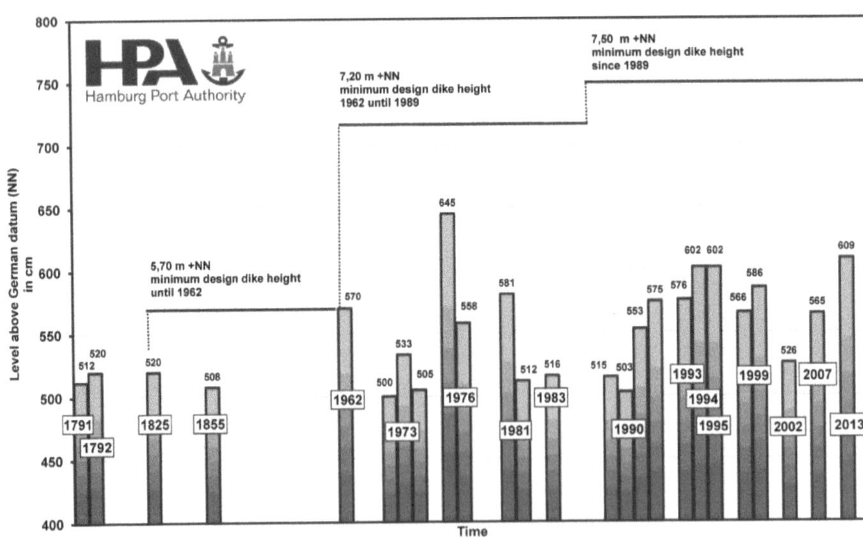

Figure 7.55 Storm surge events above +5.00 mNN (German datum) and dike heights since 1791.

7.5.3 Historical Development of the Estuary

7.5.3.1 Morphological Development

From the eleventh century onwards, the natural landscape began to change because of agricultural and pasture farming, increase in settlement, diking, and hydraulic engineering. This resulted in turning the estuary into a cultivated manmade landscape, restricting the river and reducing the river's flood plain. By dewatering the flood plain areas and by uncoupling the hinterland from the sedimentation processes, surface levels were unable to maintain pace with the continually rising water levels, making further dewatering more and more difficult. In total, 209 km^2 foreshore areas along the estuary have been diked since 1955 (see Figure 7.56) and, over a similar period, 180 km^2 of foreshore areas and flood plains of the tidal Elbe river have been lost. With the construction of river barriers, the foreshore areas of the tributaries were also no longer available as flood plains leading to a further loss of ecologically valuable intertidal areas.

In addition, civil engineering measures such as the development of port facilities, provision of navigation access to the port of Hamburg, and flood protection have also restricted the natural development of the estuary, including the adjustment to SLR. Without the construction of dikes and barriers, the marshlands would have increased across the whole cross-section of the river Elbe (Figure 7.57).

As a result of these modifications, the sediment regime in the estuary has also changed. The tidal river Elbe flows through a glacial valley mostly consisting of clay covered by fine silt. Deep exploration boreholes show the existence of peat, sand, gravel, and glacial drift. Usually the composition is coarse-grained at greater depths. Because of the marine influence, coarse and fine sands can be found up to a large extent at the estuary mouth. Sandy material dominates the river bed and is characterized by large ripples up to 5.8 m height. Once the ripples extend into the shipping channel, dredging has to be carried out to safeguard the navigation. Alluvial material is only found in areas with low-flow velocities, as it is the case in harbor basins.

The combined effect of the adaption of the shipping channel for large container vessels, the realignment of the flood defense system, and the lack of any mitigation within the estuary led to an increase of the tidal energy further upstream. Consequently, the sediment load within the water column is higher and the extent of intertidal areas has decreased over the past 50 years. This caused changes in the nature of the soil and the biota within the flood plains. With all the erosion and sedimentation taking place in this very dynamic system, the estuary mouth has broadened during the past two decades. The actual reasons are not clear yet, but this expansion of the river mouth has allowed more tidal energy to enter the tidal Elbe system. These described morphological changes are strongly related to a series of hydrodynamic changes, which in turn feed back into the morphodynamics.

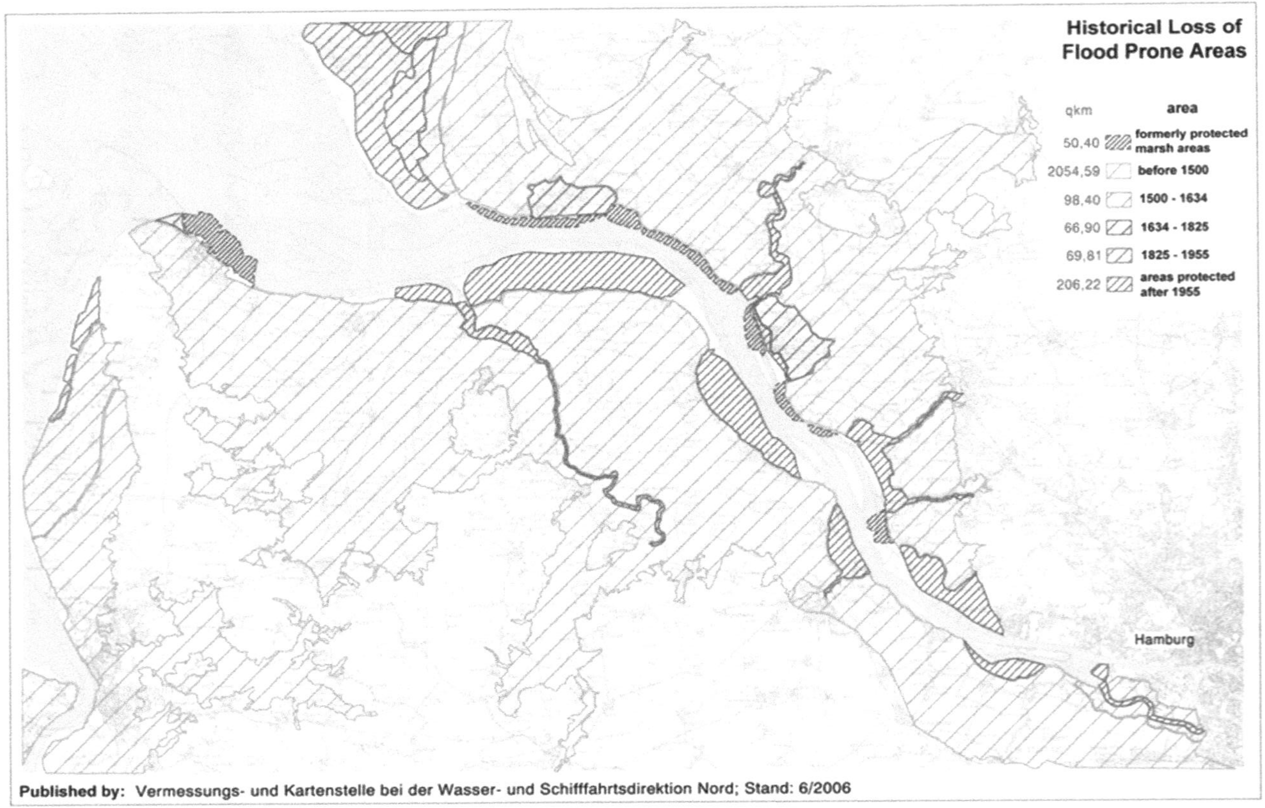

Figure 7.56 Loss of flood prone area over time, based on Glindemann, Thode, Dücker, and Witte (2006).

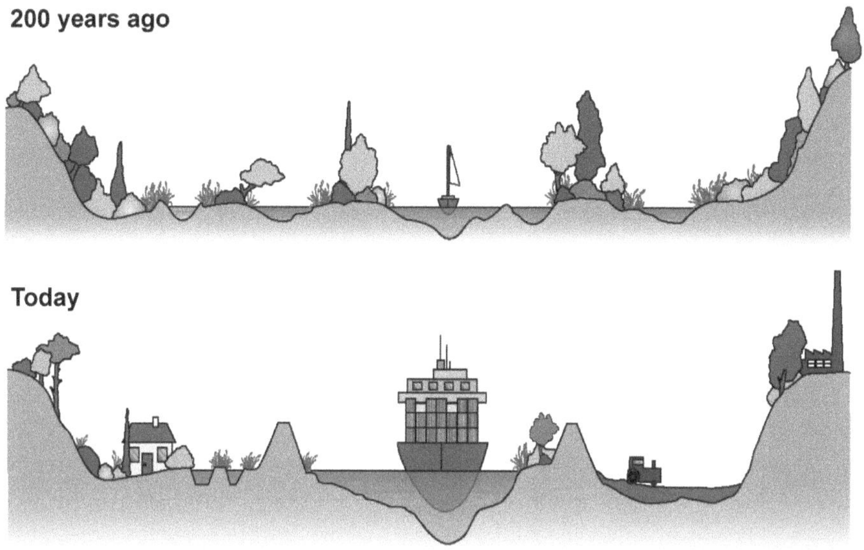

Figure 7.57 Manmade change in the Elbe estuary based on Hochfeld and Meine (2011).

7.5.3.2 Hydrodynamic Development

The hydrodynamic development of the tidal parameters within the estuary is characterized by an increase in the high water level and a decline of the low water level. This development is more significant further upstream. The maximum tidal amplitude is recorded at the tide gauge at St. Pauli in Hamburg where the current average is 3.6 m 150 years ago the tidal range was about 2.0 m in St. Pauli (Figure 7.58).

Until 1960 there was a slight decline of the tidal amplitude visible from the mouth up to Hamburg. With the construction of the weir at Geesthacht in 1956–1960, the tidal border was fixed and the tidal wave was reflected, which resulted in greater tidal amplitudes at Hamburg. The majority (66%) of the increase in tidal range is related to the movement of the low water level. The most significant change of the low water level was after the deepening of the channel in 1967 (12 m) and 1979 (13.5 m). A deepening in 1999 (14.5 m) did not have such a clear effect, which leads to the suggestion that the estuarine system could be considered stable at the moment. There are many more factors that influence the hydrology. For example, the siltation in anabranches, the backfilling of port basins, the construction of barriers, the decreasing freshwater flow, and the anticipated climate changes, especially SLR.

Obviously the flow velocities correlate directly with the water level gradient. So far any increase in flow velocities could not be determined from available data; there are very few comparable data to evaluate because of the constantly varying tidal characteristics and the lack of data from within the deep channel. Logically though, the steeper gradient between Cuxhaven at high tide and Hamburg at low tide has led to increasing flow velocities during the past few decades. These assumptions were validated by the results of a three-dimensional (3D) model developed by the Federal Waterways Engineering and Research Institute. Plüß and Heyer (2007) showed that increasing flow velocities mainly concentrate in the deep channel, whereas flow velocities in the anabranches continually decrease, leading to sedimentation in these areas. The results of the 3D sediment transport model also show that the flood tide stream has increased far more in comparison to the ebb tide stream, which explains the observed tidal pumping effect. This effect is even greater if the freshwater flow from upstream falls below 700 m^3/s.

The brackish water zone (where the salt water from the North Sea mixes with fresh river water) has moved further upstream during the past few decades. This is due to the widening cross-section at the mouth, letting more tidal energy enter the system as well as the constantly decreasing freshwater flow. The turbidity zone extends up to 20–40 km between the Elbe-km 660 and 700. The location can vary up to 25 km depending on the tidal circumstances and the flushing force from upstream. Bergemann (1995) concluded that the border of the turbidity zone has moved upstream between 5 and 20 km from 1953 until 1994.

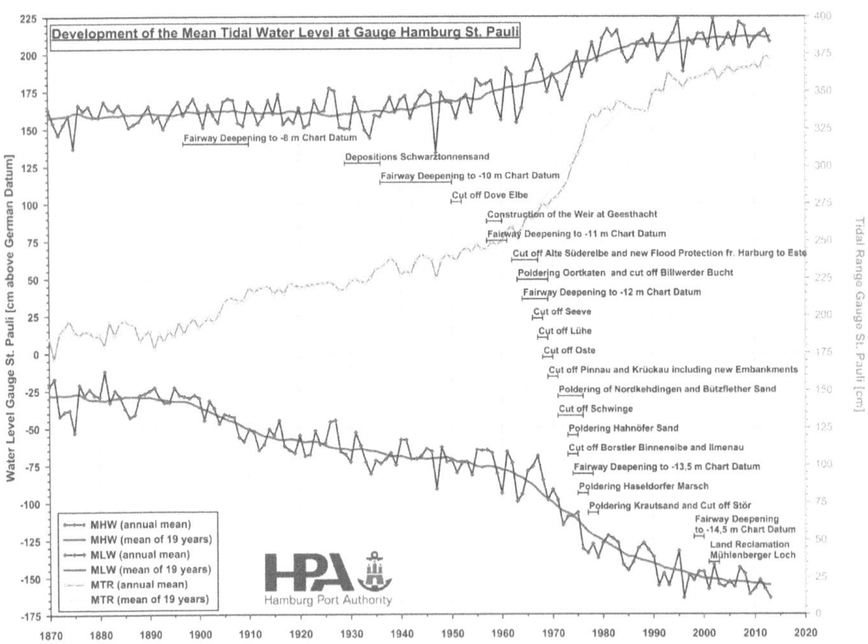

Figure 7.58 Development of the mean high water and mean low water at the tide gauge at St. Pauli, Hamburg, as annual values and 19-year-average values, with timing of estuary management interventions.

7.5.3.3 Coastal Habitats

The postmodification coastal habitats in the estuary can be divided into subtidal, intertidal, and supratidal habitats according to their height in relation to mean low and mean high water. In Herrling and Niemeyer (2006), deep and shallow water zones are categories of the subtidal habitats and siltation zones and marshes build supratidal areas.

Wildlife in these intertidal areas has adapted to the local flood duration. Subtidal habitats give shelter to animals and plants that can cope with being permanently flooded or are able to swim and can cope with different current velocities. Organisms that are unable to bear high currents, however, need permanent flooding, and are found in the adjacent shallow water zones. Close to the shore are the marshes. Between tidal habitats and marshes, the siltation zone is described as a transition zone.

7.5.3.4 Existing Flood Policy and Management Strategies

In Germany the coastal protection is organized in a framework management plans that are based on laws of the federal states. The general policy is a no failure strategy based on a deterministic analysis of historic extreme events and scenarios of the future SLR. In principle the flood protection policy is distinguished between the federals states and other governmental institutions, even in one federal state. The key players are:

- Hamburg: Hamburg Agency of Roads, Bridges and Waters is authorized to operate for the Authority for City Development and Environment for the public flood defense and the Hamburg Port Authority for the private flood defense;
- Schleswig Holstein: The Ministry of Agriculture, environment and rural;
- Lower Saxony: Ministry of Environment.

These institutions are responsible in their federal state for the flood defense works (keeping the area safe by building and maintaining flood defense works). The waterway maintenance (such as dredging, navigation provisions) is delegated in Hamburg from the federal government to the city of Hamburg (and then to Hamburg Port Authority). Downstream, the waterway maintenance is again the responsibility of the federal government.

Within each federal state, several dike boards take care of maintenance and have a role in disaster management (dike watch). The interior ministries are responsible for disaster management in their respective states. Within the federal states, the institutions have the same policy, but between the federal states the policy differs. Representatives of the federal states meet regularly to evaluate the differences between the policies of each state and identify commonalities. Moreover the federal government of Germany and EU directives play an important role and give the framework for legislation and financing.

7.5.4 The Flood System within the Elbe Estuary

The hinterland in the Elbe estuary is protected to a very high standard by embankments and levees. These coastal protections are designed for a predicted storm surge in the year 2085, including and allowance for climate change.

7.5.4.1 SPRC Approach

A detailed estuary-wide SPRC analysis is impracticable for the Elbe because of the overall extent of the Elbe estuary (length of more than120 km) and the division of the responsibilities along the estuary among the federals states and their subordinated authorities, making the sourcing of the required data difficult.

Nevertheless, a first attempt to describe the flooding risk in the framework of the SPRC approach has been undertaken in the German XtremRisk project (Burzel et al., 2010; http://www.xtremrisk.de/). A simplified version was therefore applied to highlight any areas remaining exposed to flooding. All the receptor areas R_i include the infrastructures and monetary values behind the levees/dikes. These receptor areas are protected by coastal defense systems and are endangered by flooding from storm surge, if those coastal defense structures (levees/dikes) are in a bad condition or breach during surge events. The latter is, however, considered very unlikely.

This analysis indicated that the consequences for any flood even would be concentrated in the HafenCity (see Figure 7.54), with its different flood management strategy. A more detailed SPRC analysis was therefore carried out for this limited area. HafenCity is located between the Norderelbe and the public flood protection line (main dike line) and is an old harbor area, which is now a development area for a new city district. The elevation of the area ranges from +4.4 m to +7.2 m above sea level, and is thus within the flooding area of the Elbe.

The SPRC model of HafenCity highlights the sensitive receptors, which in some cases were not identified in the flood maps generated by numerical models. This reflects that it is virtually impossible to include all the linkages and small-scale structures of the SPRC model within the layout of numerical models. Therefore the SPRC approach can enable a better assessment of possible consequences of floods. The sensitivity analysis of the receptor area can also be useful for the optimization of evacuation routes and plans. Moreover the results of the SPRC analysis can be used in the next construction stages of HafenCity. Further details of the simplified SPRC approach of the HafenCity can be found in Monbaliu et al. (2014).

7.5.4.2 Meteomarine Climate: Present Conditions and Long-term Trends

As a basis for the estimation of potential hazards in the study site (e.g., future SLR, storminess), the knowledge of the prevailing meteomarine climate under recent and climate change conditions is essential.

Extreme sea levels at the mouth of the Elbe estuary primarily occur from a combination of high astronomical tides, storm surges, changes in mean sea level, and severe sea states. Changes in any of these factors may thus significantly modify the risk associated with extreme sea levels. Apart from decadal variability, severe sea states and storm surges do not show systematic changes over the past decades up to a century. On the other hand, mean sea level has increased over the past 100 years at a rate comparable to that for global mean sea level; that is, about 1.8 mm/year. In total, extreme sea levels thus increased over the past about 100 years with the increase basically driven by corresponding changes in mean sea level (see Figure 7.59).

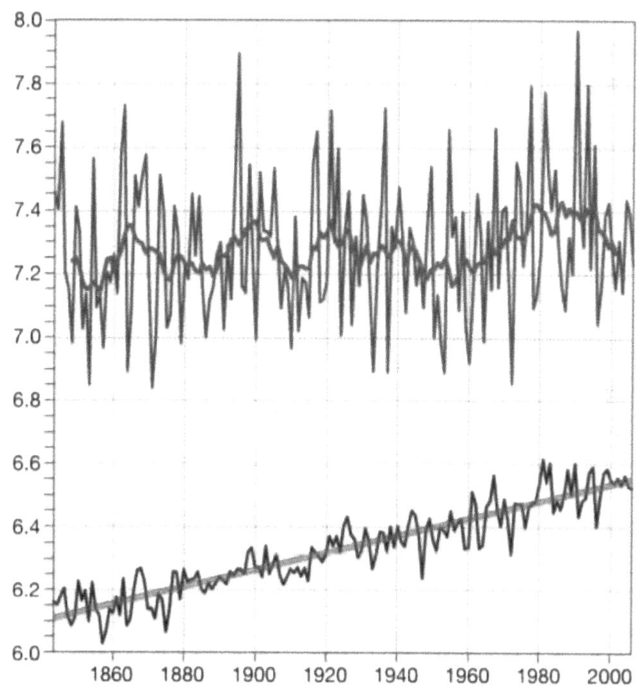

Figure 7.59 Annual mean high water levels (black) and linear trend (gray) in m at the mouth of the Elbe in Cuxhaven, Germany (bottom), and the corresponding difference in meters between the 99th percentile and the annual mean high water levels (top, black) together with an 11-year running mean (top, gray). The latter reflects the changes in storm surge levels and corresponding storm activity. Although for annual mean high waters a clear upward trend can be inferred, the corresponding changes in storm-induced water levels show pronounced interannual and decadal fluctuations but no clear long-term trend.

TABLE 7.25 Overview of Investigated Mean Sea-level Rise

Time Horizon	2025	2055	2085
Mean sea-level rise	30 cm	50 cm	90 cm

TABLE 7.26 Overview of Investigated Climate Change Scenarios

Climate Change IPCC Emission Scenario	Investigated Time Horizon
A1_B1	2010–2039, 2040–2069, 2070–2100
A1_B2	2010–2039, 2040–2069, 2070–2100
B1_1	2010–2039, 2040–2069, 2070–2100
B1_2	2010–2039, 2040–2069, 2070–2100

Based on these findings future mean sea-level conditions at the seaward boundary of the Elbe study site were used as indicated in Table 7.25.

Changes in meteorologically driven sea levels and wave heights derived from four different climate change simulations were analyzed (see Table 7.26). It was found that changes were generally small when compared with natural variability and consistency in climate change signals among the different simulations was small. It should be noted, however, that the size of the ensemble considered is relatively small and that results obtained from studies using forcing fields from different climate change simulations find either somewhat larger changes (e.g., Woth, 2005) or note large uncertainty ranges comprising increases as well as decreases in future storm surge risk (e.g., Sterl, van den Brink, de Vries, Haarsma, & van Meijgaard, 2009).

7.5.4.3 Key Source Drivers of Future Risk Conditions

The main source key source driver of the future risk is the mean SLR as shown in Table 7.25. There will be no significant changes (exceeding the natural variability) in waves and surges. The river flow will not significantly be altered, but its seasonality will change (higher flow in winter, lower in summer).Because the current standard of protection, the flood risk for the poldered areas will not be influenced. For the receptor area HafenCity only SLR will increase flood risk in the future. Table 7.27 shows the ranking of the importance of the different sources (SLR, storm surge, wind characteristics, wave height, and river flow) on the flood risk along the Elbe estuary.

TABLE 7.27 Ranking of the Source Components for Three Future Climate Scenarios

Water Level Component	Time Horizon 2025 (+30 cm)	2055 (+50 cm)	2085 (+90 cm)
Sea-level rise	Low	Low	Moderate
Surge	Low	Low	Low
Wave	Low	Low	Low
River flow	Low	Low	Low

7.5.5 Flood Mapping for HafenCity

To investigate the consequences of present and future hazards, the inundated areas in the receptor area HafenCity have been calculated with a numerical model (FVCOM) using the surge and sea-level scenarios described in the previous section as boundary conditions. All model simulations were performed by the State Key Lab of Estuarine and Coastal Research of the East China Normal University in Shanghai, China. For a comprehensive overview and a description of FVCOM, see Ge et al. (2013).

The bathymetry of the year 2006 in Hamburg, HafenCity, and adjacent regions is shown in Figure 7.60. It indicates that the Elbe river is deeper near the western boundary and shallower upstream in the eastern part. Some places with significant low-lying elevation are also shown in Figure 7.60, especially along the main river channel.

The local bathymetry and the land elevation of the HafenCity are based on actual data of the year 2010, which show that the HafenCity is a region still under construction. Therefore, some areas surveyed based on today's (ground) level may change in elevation in the future. As a result, some of the flooded areas indicated here may be different in 2025, when the development of the HafenCity is completed.

The HafenCity is subject to the astronomical tide of Elbe estuary and the freshwater discharge from upstream. High-frequency water level data at six gauge stations (Figure 7.61), including Seemannshöft, St. Pauli, Harburg, Bunthaus, and Zollenspieker, are used to both provide boundary conditions for the hydrodynamic model and for validation purposes.

These water level time series internally contain the astronomical tidal oscillation, the influences of the river discharge and sea surge. The data at the downstream station Seemannshöft are selected as the water elevation boundary condition. The upstream boundary is located at Geesthacht, where the freshwater discharge as measured 50 km upstream at Neu Darchau forces the model. Boundary conditions for the mid- and long-term scenarios are derived from the investigations of the climate change conditions in (see Section 7.5.2.4). For the calculation of the future flooding scenarios the expected SLR (see Table 7.25) is superimposed to the boundary conditions for the 1999 storm surge.

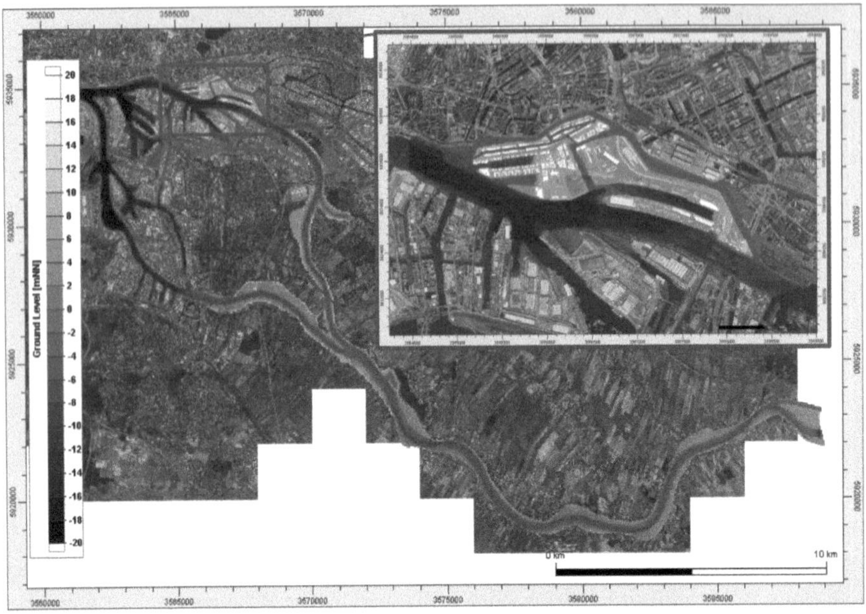

Figure 7.60 Bathymetry and surface elevation in and upstream of Hamburg and detailed view of the bathymetry and land elevation around the HafenCity.

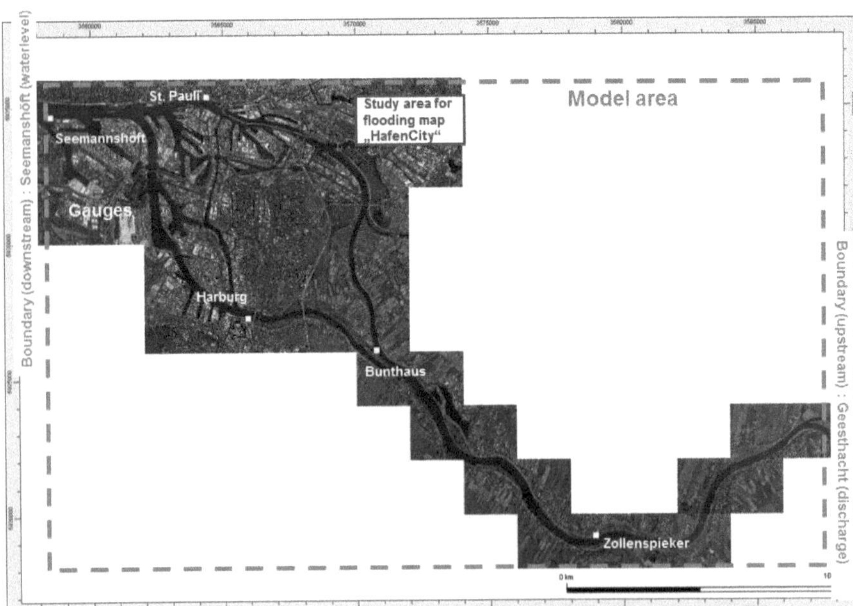

Figure 7.61 Distribution of Gauge Stations along the Elbe River in the Vicinity of Hamburg.

For the actual 1999 storm surge (without additional SLR), the flood map at peak water level is shown in Figure 7.62. The main submerged areas are located near the low-lying areas of Baakenhafen (1), Oberhafen/Oberhafenkanal (2), and Grasbrookhafen (3). Under this storm impact, the flooding process also threatened the city shorelines of east HafenCity and single parts of the old warehouse district (4). In the western part of HafenCity (5), floodings occur in an area, which is not elevated to the design levels jet.

For the simulation of inundated areas in the HafenCity under scenario conditions for mid- and long-term time scales, a single SLR of 30, 50, and 90 cm is taken into account. The flooding map of the 30-cm scenario (2025) is shown in Figure 7.63. The map illustrates that most of the flooding occurs in areas that are already flooded under present conditions.

These areas will be always flooded even with the dwelling mound flood management approach. The areas represent streets and areas adjacent to the low-lying canals and embankments or historical building, which cannot be moved on to the artificial dwelling hills. The additional impact of mid- and long-term scenarios result in higher water depth in the flooded areas. The most impacted sections are the region between the Baakenhafen (1) and Oberhafen/Oberhafenkanal (2).

In the long-term scenario (2085), the flooding near the Magdeburger Hafen and Norderelbe (3) increase, additionally. Because these areas are presently still under construction the underlying elevation data will not be realistic for the mid- and long-term scenarios. Nevertheless the planning for this area takes into account a future SLR of 80 cm.

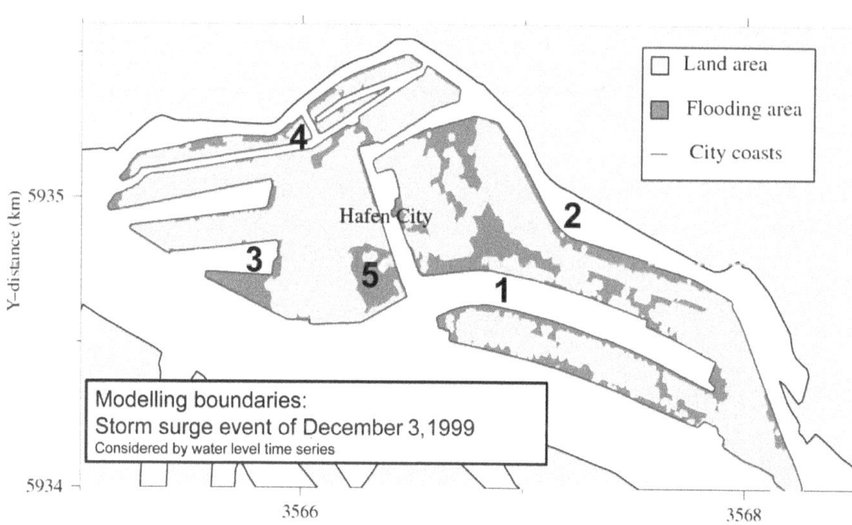

Figure 7.62 Flood map around the HafenCity at surge peak of storm in 1999.

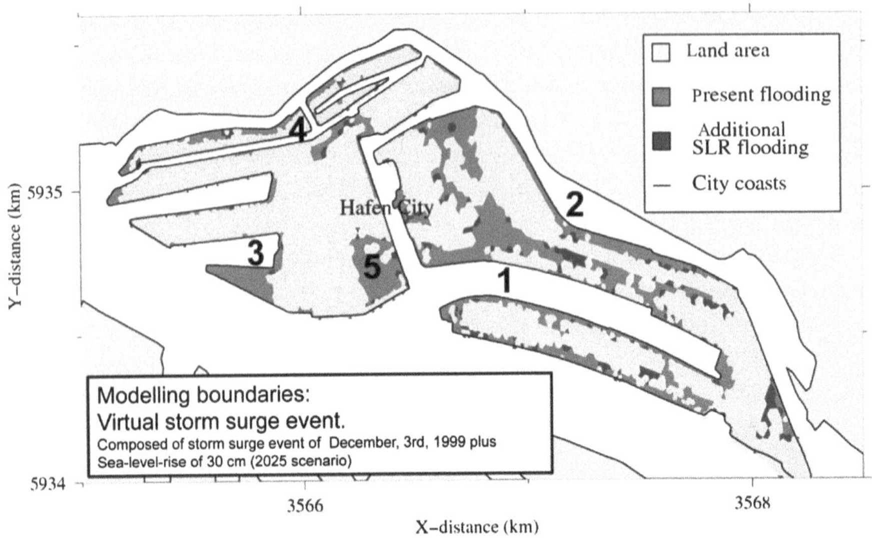

Figure 7.63 Predicted flooding map under the 30 cm sea-level-rise scenario in 2025 during storm peak impact.

TABLE 7.28 Maximum Flooding Height and Area at the Three Sea-Level Rise Scenarios

	2025 (+30 cm)	2055 (+50 cm)	2085 (+90 cm)
Flooding height (m)	6.209	6.409	6.795
Flooding area (m^2)	314,015	355,774	409,804

In summary, the peak flood levels will rise according to the respective SLR, whereas the flooding area will increase by 18% (2025), 34% (2055), and 54% (2085). The maximum flooding height and area for the three SLR scenarios is shown in Table 7.28.

7.5.6 Mitigation Measures

7.5.6.1 Preliminary Selection of Mitigation Options

The traditional flood mitigation option used in the Elbe estuary has been dike rising and dike upgrade.

From a conceptual point of view, the following set of innovative mitigation measures to reduce flood risks are feasible in the Elbe estuary:

- Depoldering (managed realignment);
- Wind and ship wave attenuation devices;
- Artificial sandbanks in the mouth of the Elbe.

From these three options, only the construction of artificial sandbanks has been investigated in some depth. In addition an exemplary attempt to investigate the effectiveness of wave attenuation devices in the inner and sheltered estuary just downstream of Hamburg was carried out.

7.5.6.2 Wave Attenuation Devices

As a part of this investigation, a number of wave attenuation devices were constructed, improved, and tested. Most of these were applicable on the open shore. One device—suitable for estuarine conditions—is the Blow-Jet (see Figure 3.1.7). To assess how this device may work for management and protection of a beach at Wittenbergen (Figure 7.64), numerical simulations were obtained. The hazard at Wittenbergen is a combination of high tides, ship-generated waves, and a solid concrete structure placed at the upper part of the beach. The result is an addition of the incident and reflected waves that generates a considerable "splash."

The numerical model used to estimate the propagated wave fields is the WAve Propagation On the coast model developed at the Universidad Nacional Autónoma de México (Silva, Borthwick, & Taylor, 2005, 2006b). In total eight wave conditions were run to understand the study area and be able to suggest shore protection alternatives. For the shore protection, a chain of Blow-Jet devices were investigated. They should be constructed outside the navigation channel. The dimensions of the whole investigated wave attenuation device chain are 40 m long, 8 m wide, and of variable height. For best hydraulic performance (shore protection and energy conversion), the orientation of the structure was chosen to be perpendicular to the wave fronts generated by ships traveling along the navigation channel.

Figure 7.64 Satellite view and image of Wittenbergen beach. *Source A, FHH, Landesbetrieb Geoinformation und Vermessung, Spring 2013; B Nino Ohle, Germany.*

The installation of the 18 blow-jet wave attenuation devices results only in a local (100 m) reduction of the wave height in the hazard area.

7.5.6.3 Design of Artificial Sandbanks in the Mouth of the Elbe Estuary as a Mitigation Option

The basic idea of the this mitigation approach is to reduce water levels along the estuary and in the city of Hamburg, including HafenCity, by introducing artificial sandbanks in the mouth of the Elbe estuary. The function of these sandbanks would be to reduce the amplitude (energy) of the tidal wave by increased partial reflection and dissipation. Two kinds of sandbank layouts were investigated:

- Emerged sandbanks (above tidal high water);
- Submerged sandbanks (below tidal water).

The criteria for the selection of the several sandbank scenarios were:

- Nautical implications;
- Geomorphology aspects;
- Availability of sandy material for construction; and
- Stability and sustainability of the layout.

In the following the different sandbank layouts are described in terms of location and size. Figure 7.65 shows the locations of the investigated emerged and submerged sandbank scenarios (see also Table 7.29). The stability and sustainability of the layout is mainly influenced by the wave and current attack in the construction area.

Because the wave propagation within coastal area of the Elbe estuary is strongly influenced by the coastal morphology with its islands, bars, shoals, and channels the wave attack was simulated with numerical model SWAN (www.swan.tudelft.nl, TU Delft). For the numerical simulations with SWAN alongside the bathymetry of the study area boundary conditions for wind and water level conditions were needed to calculate the wave conditions within the model. These boundary conditions were laid down for the sensitivity analysis as shown in Table 7.30.

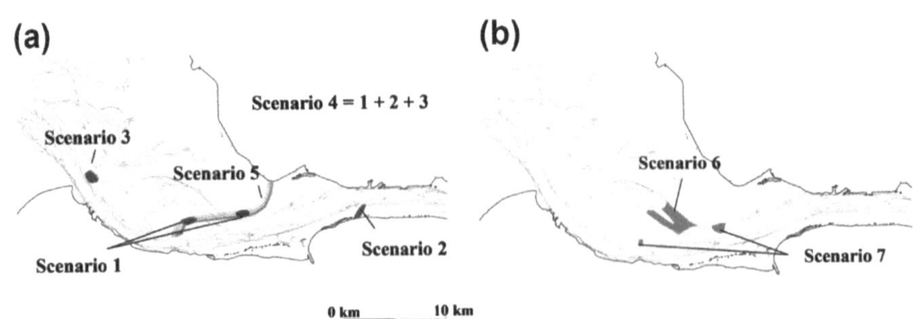

Figure 7.65 Location of emerged (a) and submerged sandbanks (b).

TABLE 7.29 Volume of Sand Needed for the Construction of the Investigated Scenarios in the Mouth of the Estuary

Scenario	Scenario Name	Volume of Sandbanks ($\times 10^6$ m^3)	Volume Above Mean Low Water ($\times 10^6$ m^3)
1	Medemsand and Neufelder Sand	3.6	3.6
2	Brunsbüttel	2.0	2.0
3	Cuxhaven	3.1	3.0
4	All sandbanks	9.0	8.3
5	Dam	33.9	24.3
6	Medemsand and Neufelder Sand + underwater deposition	15.8	3.6
7	Submerged sandbanks	6.1	0.1

TABLE 7.30 Boundary Conditions of the Sensitivity Analysis

Water level	0 mNN, 1 mNN, 2 mNN, 3 mNN, 4 mNN, 5 mNN, 6 mNN
Wind velocity	8 m/s, 12 m/s, 16 m/s, 20 m/s, 24 m/s, 28 m/s, 32 m/s
Wind direction	210–330° in 10° steps
Significant wave height	Varying between 2 and 10 m

In the first sensitivity test, the wave heights at the boundary of the model were changed between 2 and 10 m. The wave periods were chosen to fit from 6 to 14 s depending on the wave height. The investigations were performed for three wind directions 270, 300, and 330°. This resulted at the sand on the island Medemgrund (Scenario 1) for a water level of 3 mNN to only a slight difference between the varying input wave heights (see Figure 7.66). From a water level of 4 mNN larger deviations are found up to 11 cm. There is the greatest difference not continuous between the smallest input shaft height and the greatest input shaft height.

The sand islands in the Wadden of Neufeld show similar variations, as the sand of the Medemgrund. Again the wave heights at the island show almost no variation for different input wave heights at a water level of 3 mNN. The same results were determined with a water level of 4 mNN. Discrepancies between 2 and 5 cm in the wave height show mainly the dependence on wind direction (see Figure 7.67).

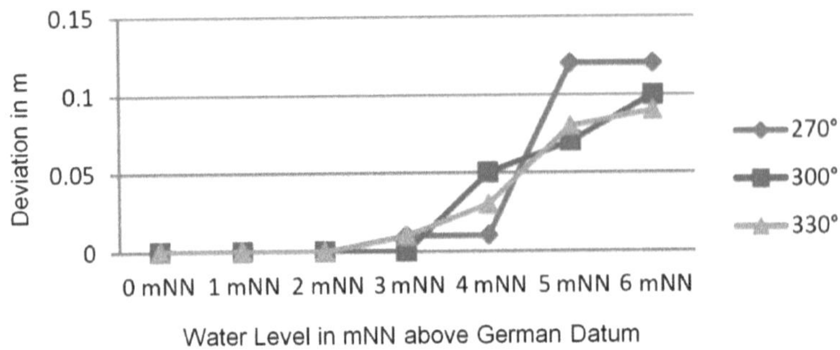

Figure 7.66 Maximum deviation at different input wave heights (sand island Medemgrund). *(Based on Dröse, 2012).*

The wave simulations were also conducted with different water levels. It was shown—in contrast to the different levels of input wave heights—that they have a greater influence on the wave height at the respective area of the islands. It can be shown, that the influence of water level at the two islands of scenario 1 is particularly large (see Figures 7.68 and 7.69), as the area around the islands is very flat, so that at low water levels the influence of ground friction is high, which leads to lower wave heights.

The influence of the wind velocity on the wave height was examined. Therefore the boundary conditions were not changed and were fixed to an input wave height of 4 m and a wave period of 8 s. The water levels were assumed to be 4 m above sea level and the wind direction was controlled at 300°. With these assumptions a dependency of the wave heights and the wind velocity can be shown. The wave height is in accordance with increasing wind velocities higher within all areas of the model and can be investigated with an upper limit between 1.6 and 1.8 m near the islands of the scenario 1 (see Figure 7.70).

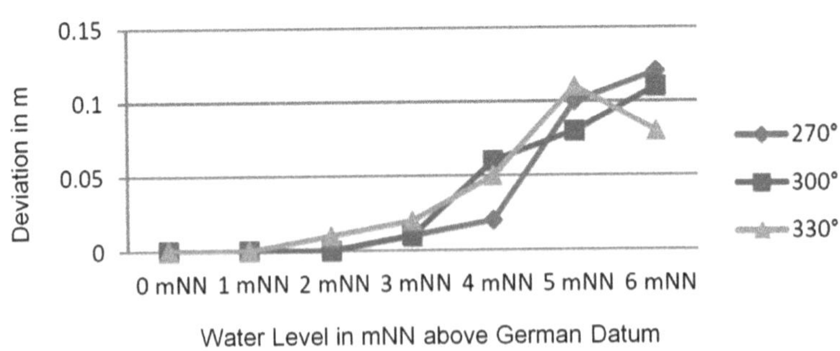

Figure 7.67 Maximum deviation at different input wave heights (sand island Neufeld). *(Based on Dröse, 2012).*

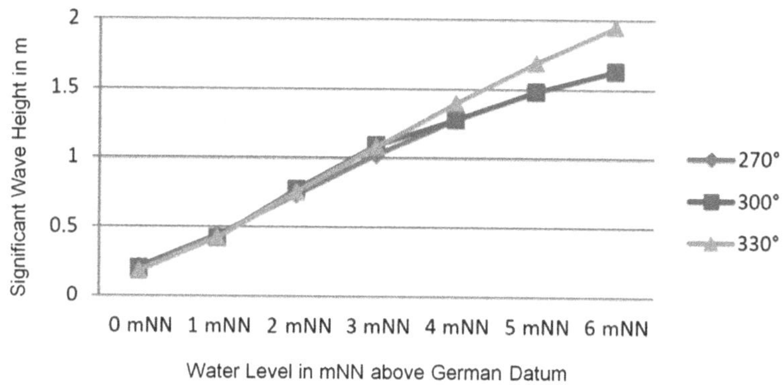

Figure 7.68 Influence of the water level on the wave height (sand island Medemgrund) *(Based on Dröse, 2012).*

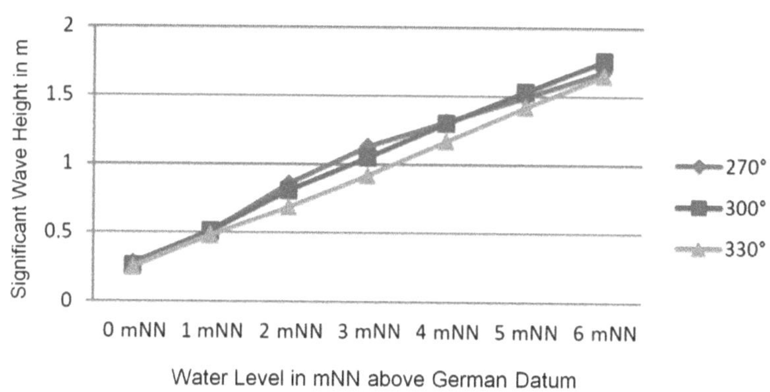

Figure 7.69 Influence of the water level on the wave height (sand island Neufeld) *(Based on Dröse, 2012).*

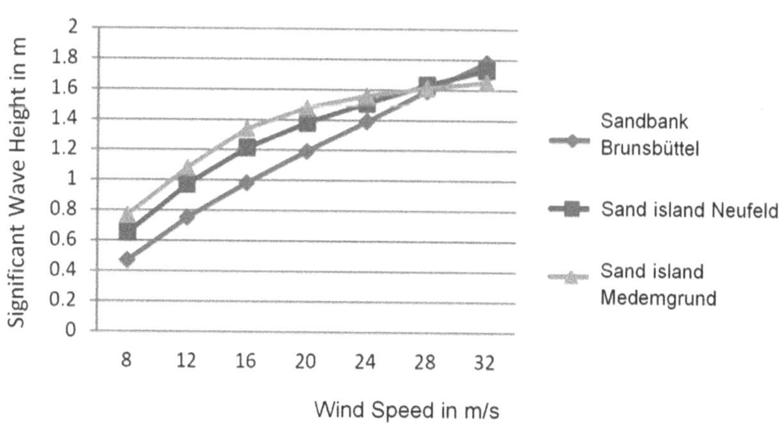

Figure 7.70 Influence of wind speed on the wave height. *(Based on Dröse, 2012).*

For the individual study areas, the wave heights were determined as a function of wind direction. With a fixed water level of 4 mNN, a wind velocity of 20 m/s, a significant wave height of 6 m as input wave height at the boundary and a corresponding wave period of 10 s the highest significant wave height of 1.4 m can be found at the island Medemgrund with a wind directions achieved by 310° (see Figure 7.71).

For the sand on the sand island Neufeld with the same previously described boundary conditions, the maximum significant wave height of 1.3 m be analyzed at wind direction of 280° (see Figure 7.72).

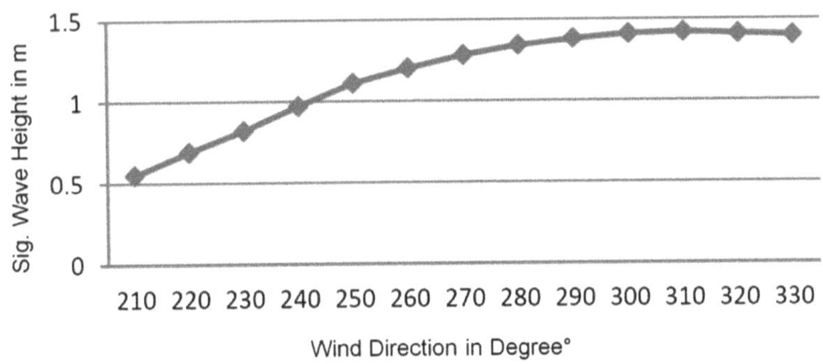

Figure 7.71 Significant wave heights as a function of wind direction (sand island Medemgrund). *(Based on Dröse, 2012).*

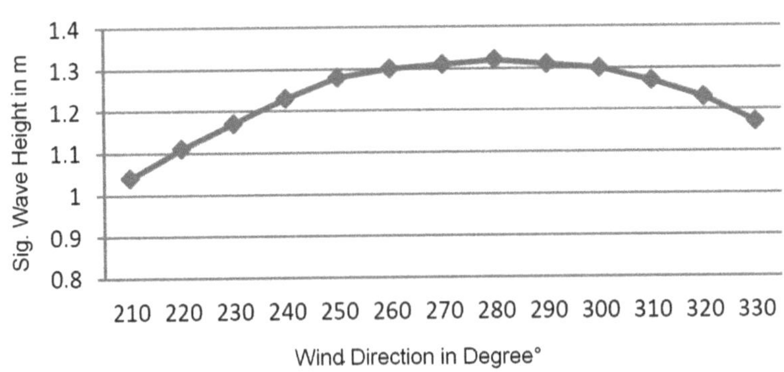

Figure 7.72 Significant wave heights as a function of wind direction (sand island Neufeld). *(Based on Dröse, 2012).*

Investigation on the Stability and Sustainability of the Layout Due to Currents

The maximum tidal currents in the construction area were computed by means of the high-resolution 2DH hydrodynamic model TRIM (Cheng, Casulli, & Gartner, 1993). The model for the Elbe estuary is a rectangular grid with a spatial resolution of 50 m, which allows the investigation of the effects of the coastal structures in the mouth of the Elbe on the currents and water levels in the inner estuary (e.g., the port of Hamburg). All numerical simulations were carried out for the different sandbank (see Figure 7.65 and Table 7.29) and boundary conditions, including storm surge conditions and climate change options.

The model results shows that the tidal flood currents are generally higher than the ebb currents. The maximum flood currents were ~2 m/s in the investigated scenario 1 for both sandbanks. The calculated ebb currents amount up to ~1 m/s and vary between at the different locations. While the flood current field is almost homogenous in space, the maximum ebb currents occur at the southern tips of the sandbanks.

Compared with the wave attack the influence of the tidal currents is of second-order importance for the stability of the sandbanks. The analysis also shows that the orbital velocities are negligible and therefore these were not further considered in the analysis.

Summary on Stability and Sustainability

Overall, the wave heights depend significantly on the individual test points on the water depth at the location and the wind speed respectively the wind direction within the model area. The input wave height turns out as little or nonsignificant, for the wave height within the study area. Nevertheless for the overall sustainability of the sandbanks also the active large-scale morphodynamics in the mouth of the Elbe estuary have to be taken into account.

7.5.6.4 Efficiency of Artificial Sandbanks in the Mouth of the Elbe Estuary as Mitigation Option

Introduction

The efficiency of the several artificial sandbank scenarios was investigated by comparing time series of high water levels (HWL) and maximum current velocities at four monitoring stations along the Elbe estuary (see Figure 7.73). For the purpose of mitigating the HWLs in Hamburg the most important of these stations is Hamburg-Teufelsbrück, which represents the situation in the receptor area.

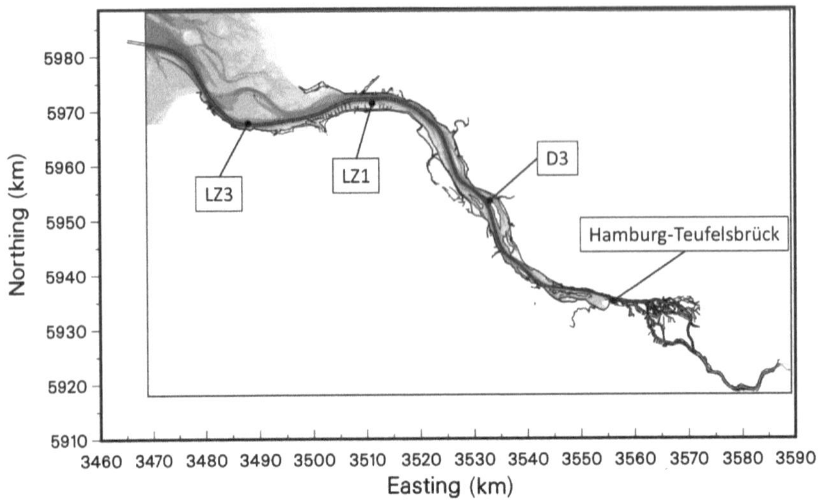

Figure 7.73 Locations of the four monitoring stations.

Effect of Various Sandbank Scenarios in the Receptor Area Hamburg

The effects of the seven different sandbank scenarios (year 2006 condition) and scenario 4 under climate change (SLR) conditions in the receptor area Hamburg are summarized in Table 7.31. The changes with respect to the reference situation (no sandbanks) are given as decreases; a negative value indicates an increase.

Effect of Scenario 4 (All Sandbanks) Various Sandbank Scenarios in the Receptor Area Hamburg

Figure 7.74 shows exemplary the effects of the combination of scenario 4 (all sandbanks) on HWL (a) and maximum current velocity (b). This scenario combines the scenarios 1 to 3 and was analyzed for today's condition as well for the climate change scenarios.

The results for the present condition (year 2006) are displayed in Figure 7.74 for all four monitoring stations along a transect of the Elbe estuary from the mouth to Hamburg. The curves show the time series of differences between the sandbank scenario and the reference situation, which have been smoothed by a 14-day moving average. In this scenario the fluctuation in HWL is most pronounced at the outer stations LZ3 and LZ1. There is a general decrease in HWL of some 2–3 cm at all stations. This is attended by an increase in maximum current velocities at the outer stations (1–3 cm/s) and a decrease of one–2 cm/s at the inner stations D3 and Hamburg-Teufelsbrück.

TABLE 7.31 Summary of the Effects of the Various Sandbank Scenarios on Tidal High Water Level and Maximum Tidal Current Velocity at Station Hamburg-Teufelsbrück under Current Conditions (Reference Year 2006) as well as Scenario 4: "All Sandbanks" under Future Climate Change Conditions

	Effect of Mitigation on Flood Waters in Hamburg			
Scenario No.: Year	Decrease in HW Level (Annual Mean)	Decrease in HW Level (SD)	Decrease in Maximum Current Velocity (Annual Mean)	Decrease in Maximum Current Velocity (SD)
1: 2006	0 cm	2 cm	0 cm/s	1 cm/s
2: 2006	0 cm	2 cm	0 cm/s	1 cm/s
3: 2006	1 cm	3 cm	0 cm/s	1 cm/s
4: 2006	3 cm	3 cm	1 cm/s	1 cm/s
5: 2006	13 cm	4 cm	<1 cm/s	1 cm/s
6: 2006	1 cm	2 cm	0 cm/s	1 cm/s
7: 2006	2 cm	3 cm	1 cm/s	1 cm/s
4: 2025 (+30 cm)	2 cm	2 cm	−11 cm/s	1 cm/s
4: 2055 (+50 cm)	2 cm	2 cm	−10 cm/s	1 cm/s
4: 2085 (+90 cm)	2 cm	1 cm	−9 cm/s	1 cm/s

7.5.6.5 Conclusions on the Examined Mitigation Measures

The two mitigation measures investigated in this study (wave attenuation devices and artificial sandbanks) have a different application area. Wave attenuation devices may be useful for reducing the ship wave attack on exposed beaches of limited length in the inner estuary. They have to be placed in such a way that they will fulfill their function without impeding the navigation. In this area, they are not effective in mitigating storm surges.

As for the artificial sandbanks, the results of the investigations suggest that such structures in the mouth of the Elbe estuary may dampen the tidal energy to some extent. Their efficiency in this respect does not so much depend on their geographical position, but on their elevation. Nor does the effectiveness in general increase with the introduced amount of sand (e.g., the volume of the sandbanks). For the investigated scenarios the reduction in HWL in Hamburg amounts to less than 3 cm compared with a tidal range of 3.6 m. For the SLR scenarios in the next century, the reduction of HWL is even smaller, but accompanied by significantly increased current velocities (10%). In general dissipation or redirection of the incoming tidal energy at the river mouth is not achieved to an extent which would be relevant for flood protection in Hamburg. There are local modifications of the current fields and the water elevations in the vicinity of the sandbanks, but the tidal wave in the estuary as a whole is very little affected. Only the highly hypothetical dam scenario produces the desired reflection and damping of the tidal wave and a 10% decrease of HWL in Hamburg. Keeping in mind the nautical and environmental constraints, it must be stated that there are narrow margins for constructions in the mouth of the Elbe to influence the high water levels in and near Hamburg. Other possible beneficial effects of the sandbanks like for example their influence on sediment transport or as intertidal habitats and nesting places for birds have not been investigated, but may have a positive impact.

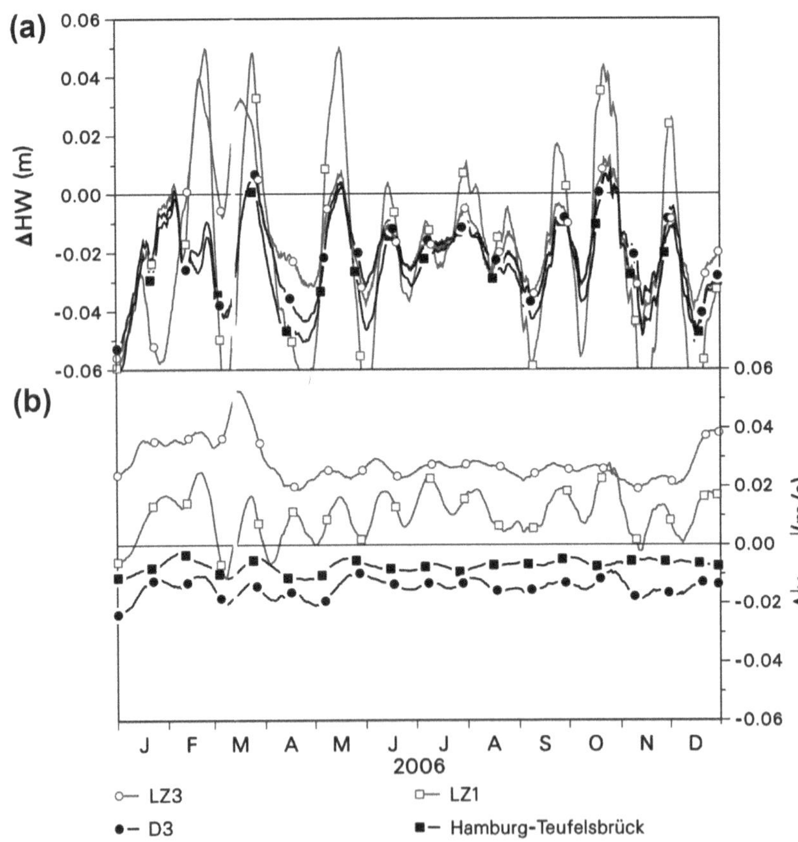

Figure 7.74 Effects of scenario 4: "all sandbanks" on tidal high water level (a) and maximum tidal current velocity (b) at stations LZ3, LZ1, D3, and Hamburg-Teufelsbrück. Curves represent differences between the sandbank scenario and the reference situation in 2006.

The HafenCity is an example of an area without traditional flood protection, intentionally left exposed to flooding hazards. The "mitigation measures" are in the design and construction, namely placing the new buildings on dwelling mounds or providing individual flood defense for every structure.

7.5.7 Strategic Management

The Elbe estuary will need to have a long-term strategic policy particularly because of its underlying premise that the population always has the highest safety priority and its legacy of flood management by dikes. The current governance structure ensures that coordination between authorities occurs regarding the implementation of the EU Flood Directive in Germany that has been incorporated in the National Water Act. On the other hand, the specific design and management of flood protection differs between the federal states. Through the Bund/Länder-Arbeitsgemeinschaft Wasser, where all government employees of the federal states who are involved in flood safety regularly meet) the work is coordinated. In addition, the federal states have water laws that match the national and international laws. Also dike laws are present and transferred into the appropriate master plans.

7.5.7.1 Existing Proactive Planning in the Elbe Estuary

The effect of climate change on flood events in the Elbe estuary is mainly related to changes in mean sea level. Drier summers, wetter winters, and change in storminess may also be expected but these do not appear to be a major influence, neither does subsidence, sedimentation, or erosion, except possibly further deepening of the navigation channel. Existing flood defense works along the Elbe have been designed such that they can easily be reinforced when needed. All defense works are designed with a time horizon of 100 years and this proactive approach has ensured that potential SLR has already been considered to 2085. In addition, dike design can be relatively easily adapted as sea levels rise as space has been reserved for future dike reinforcement and foundations allow for a future dike heightening (an extra 0.8 m in the Port of Hamburg area and 1.0 m in the other two federal states).

Because of the availability of significant, high-quality data sets, the design approach of coastal defense systems on the tidal Elbe is based on a full deterministic approach, and is based on the concept of design loads (at the Elbe wave heights and water levels). Precise, historic, single values are used as design parameters and there is therefore no probabilistic design on return periods. As these are only calculations for the design water level and not for entire flood protection design, a return period for the flood protection itself cannot be given. Moreover the design levels are under revision and, for example, are being upgraded in the city of Hamburg at the moment.

Flood defenses are financed by several stakeholders. In Hamburg the National State finances 70% of the budget and the city of Hamburg the remaining 30% (around €30 million a year). In Lower Saxony, the budget is divided in the same ratio: Lower Saxony finances €65–70 million a year. In Schleswig–Holstein, 40% of the flood defense projects is paid through the EU, 40% by the national state, and 20% by the federal state (around €60 million a year).

7.5.7.2 Other Risk Management Options

Other flood risk management techniques are also used in the Elbe estuary and Hamburg in particular. For example, in Hamburg construction of any buildings on flood protection structures is not allowed and, in Schleswig–Holstein, a distance of 20 m has to be left open along the dikes to allow for future reinforcement and raising. In Lower Saxony, this distance amounts to 50 m. An accurate flood warning system is also in operation, so that the respective floodgates are closed at certain water levels. In addition, evacuation plans are available and regularly tested. In the HafenCity, Hamburg, a different flood management approach is being followed. Flood safety is realized through adaptive building with flood defenses incorporated in buildings and clear evacuation routes as well as building at higher levels, keeping economically valuable materials safe. Infrastructure is also created at a high level, allowing life to proceed more or less as normal during a flood. The design allows for higher water levels and is therefore robust for the coming decades.

Because of the historical development of the estuary, the role of habitats in the mitigation of flood levels is limited although, with every new development, the demands of the nature and environmental are assessed through Environmental Impact Assessments and protection requirements need to be met. A key future legislative change for flood protection in the Elbe estuary is that the Port of Hamburg area will no longer be exempt from the requirement to compensate for habitat loss from dike construction. This will be likely to have implications for the type of mitigation option selected in the future. Additionally managed realignment options (or depoldering) are discussed and could be considered when traditional mitigation measures will not suffice.

7.5.8 Overall Conclusions and Recommendations

7.5.8.1 Long-Term developments

In Germany the flood protection incorporates possible effects of climate change, mainly SLR and higher runoff. In case the urban pressure would increase further (Hamburg), the spatial plans and design of the flood defense works should guarantee flood safety in the coming century. The adaptive building of the HafenCity could prove to be a promising alternative to the traditional flood defense approach in the urban area. In the more agricultural areas of Schleswig Holstein, it is expected that the urban pressure will decrease further. Environmental impacts from construction measures along the Elbe include the loss of ecologically highly valuable areas such as mudflats, naturally flooded areas, and shallow water zones and the increase of oxygen depleting processes in the estuary leading to fish die-off during summers. The latter effect will be amplified in the projected scenario of strong climate change due to higher temperatures.

7.5.8.2 Transfer to Other Estuaries

The results of the chosen mitigation by artificial sandbanks show that significantly reduced flood levels in Hamburg can only be achieved by the obstruction of a substantial part of the cross-section at the mouth of the estuary. In general this will conflict with the objectives of navigability and nature protection. In other comparable estuaries conditions for the constructions of sandbanks may be more favorable (e.g., smaller cross-sections at the mouth, less ecological value). In Europe the estuaries of the Humber, Thames, Weser, and Ems may provide further test cases for such investigations.

Acknowledgments

The following contributions are gratefully acknowledged: HafenCity SPRC-approach, Siddharth Narayan, Faculty of Engineering and the Environment, University of Southampton, UK; HafenCity flood modeling, Zhongyuan Chen and Jianzhong Ge, State Key Laboratory for Estuarine and Coastal research State Key Lab of Estuarine and Coastal Research), East China Normal University, Shanghai, China; Wittenbergen investigation of the wave attenuation device, Rubí Esmeralda Martínez Martínez, Edgar Gerardo Mendoza Baldwin and Casarín Rodolfo Silva, Coordinación de Hidráulica, Instituto de Ingeniería, Universidad Nacional Autónoma de México, México D.F.; artificial sandbank sustainability investigations (wave modeling), Christian Dröse, Fachgebiet Küstenwasserbau, Universität Rostock, Rostock.

7.6

Risk Assessment and Mitigation in a Low-lying Coastal Area: Cesenatico, Northern Italy

Barbara Zanuttigh [1], Luca Martinelli [2], Fabio Bozzeda [3], Marina Antonia Colangelo [3], Luca Pietrantoni [4], Fabio Zagonari [5], Gabriele Prati [6]

[1] *Department of Civil, Chemical, Environmental and Materials Engineering, University of Bologna (UniBo), Bologna, Italy*

[2] *Padova University, ICEA Deptartment. Part of the Consortium for coordination of research activities concerning the Venice lagoon system (CORILA), Padova, Italy*

[3] *Dip. Scienze Biologiche, Geologiche Ambientali, University of Bologna – Ravenna Campus (UniBo), Ravenna, Italy*

[4] *University of Bologna (UniBo), Filippo, Bologna, Italy*

[5] *Department of Economics, University of Bologna (UniBo), Rimini, Italy*

[6] *University of Bologna (UniBo), Filippo, Bologna, Italy*

7.6.1 Characterization of the Site

The Emilia Romagna littoral is located in the northeast of Italy (Figure 7.75) and comprises 130 km of low and sandy coast, most of which are strongly urbanized.

The Emilia Romagna beaches face the Northern Adriatic Sea, a relatively shallow epi-continental shelf with low tidal amplitude. A general erosive tendency is mainly caused by the reduced sediment transport rates of the rivers and by the increased anthropogenic subsidence. Subsidence, eustatism, and erosion of dunes pose a serious threat for coastal flooding.

Cesenatico municipality is included in the province of Forlì-Cesena. The site is famous for its marina and is a well-known tourist resort with a sandy beach rich in bathing facilities.

7.6.1.1 Existing Policy and Management Strategies

The extension of the Port of Rimini strongly interfered with the littoral dynamics, inducing an erosive pattern in the northern beaches, aggravated by the reduced sediment discharge of the Marecchia River (Figure 7.75). The erosion gradually moved toward Cesenatico, 20 km northward, as a consequence of the reduction of long-shore sediment transport due to the continuous series of groins built between Rimini and Cesenatico in the period 1947–1997. In addition, the extraction of water, initially associated to land reclamation and, later, for agriculture and industrial reasons, induced severe subsidence, with peaks of 50 mm/y in the 1980s. A special law in 1980 regulating water extraction, and the building of the Ridracoli dike (Figure 7.75) and water supply network, proved to be effective in reducing the subsidence rate.

Cesenatico coastline is approximately 7 km long and is divided by the harbor jetties into a northern and a southern area, and by a groin to the north of the jetties into a northern and a central area. To face the beach erosive tendency and the flooding events, the following management is present (Figure 7.76):

- Northern area: unprotected;
- Central area: protected by a sandbag–submerged barrier 0.8 km long, 12 m wide, and 250 m distant from the shoreline; it includes the canal harbor that for water levels exceeding 0.9 m a.s.l. is closed by sea gates, the so-called "Porte Vinciane," 2.0 m high a.s.l.; the sea gates are combined with a pumping system to ensure the seaward urban drainage;
- Southern area: emerged barriers, crest level 1.5 m a.s.l.; soil dike, integrated into the urban use of the back beach, 20 m wide, 1 m high, 1.4 km long, starting from the southern jetty (extending southward);
- The capacity of the drainage system has been recently increased and a bypass system and a series of expansion basins were also built.

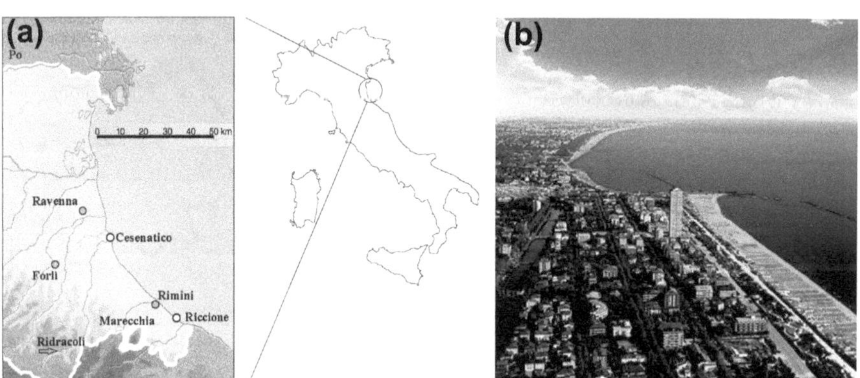

Figure 7.75 (a) Location of Cesenatico along the Emilia Romagna coast, Italy. (b) Aerial side view of Cesenatico littoral from the south.

Figure 7.76 Aerial view of Cesenatico (c), subdivided into the areas recalled in the text, and pictures of the existing management (a) semisubmerged barrier; (b) sea gates at the Canal Harbor entrance; (d) the "Gardens"; (e) and emerged barriers.

A decennial coastal plan was recently published addressing the problem of integrated coastal zone management in the Emilia Romagna Region (Preti et al., 2009). However, this plan does not account in a systematic way for climate variability (i.e., extremes) and climate change effects. The approach is essentially driven by the consequences of major events: based on the damages to beaches and bathing facilities, on historical trends, surveys, and climate studies, the regional government has recently proposed, in agreement with municipalities, medium-term plans for beach nourishment along the whole coastline (Penning-Rowsell et al., 2014). The regional government needs of course to receive funding from the Ministry of the Environment to cover the expenses of the intervention.

Nourishment has been selected by the coastal authorities and local managers as the preferred technique to face both erosion and flooding mainly for the following reasons: maintenance of wide beaches for tourism and recreation; low environmental impact; low aesthetic impact. Although the impact of emerged barriers and groins has been recognized, all management plans keep in place the existing structures due to high costs for their removal and the kilometers of barriers that were constructed in the past.

The impact of the creation of "winter dunes" by moving the sand in front of the bathing facilities and therefore reshaping the beaches is usually not tackled. These operations lead to undesired impact on the benthic communities that are very important indicators of water quality and therefore also bathing allowance.

Finally, so far other measures such as insurance, evacuation plans (implemented so far in Cesenatico only), landscape planning, or proactive citizen defense (for instance by means of temporary flood defenses) have not been included yet in a systematic management protocol but have been used as exceptional measures in local rehabilitation designs.

7.6.1.2 Meteo-Marine Climate: Present Conditions and Long-term Trends

Wind fields were derived from an ensemble runs of regionally downscaled IPCC A1B (IPCC, 2007) emission scenarios (2010–2039, 2040–2069, 2070–2100), and a control period (1960–1990), available at the World Data Centre for Climate, Hamburg, Germany.

In these simulations, the regional climate model COSMO-CLM (see http://www.clmcommunity.eu/) was used in a dynamic mode at a spatial resolution of approximately 18 km (Keuler et al., 2009) to downscale six hourly outputs from the global climate model ECHAM5/MPIOM (Max Planck Institute Ocean Model). These wind fields were used as atmospheric forcing of the Shallow Water Hydrodynamic Finite Element Model that comprises a hydrodynamic core module coupled with a wind wave, a sediment transport, a bed load, and a dispersion module (Umgiesser, Canu, Cucco, & Solidoro, 2004; Bellafiore & Umgiesser, 2010). The Shallow Water Hydrodynamic Finite Element Model was implemented in the northwestern end of the Adriatic Sea and provided the time series of significant wave height (H_s), wave direction, peak wave period (T_p), and mean surface elevation including astronomical tide and surge (Z). Such data were extracted in for a point located offshore Cesenatico (Weisse et al., 2014), and have been statistically characterized (THESEUS OD1.15).

Yearly maxima events (considered as average of 1 h) were identified based on the maximum H_s or maximum Z. Each event was then also characterized by the corresponding peak off-shore wave steepness s_p and mean wave direction θ.

By observing the correlation matrix, it was decided to separate the data into two sets based on wave directions: one set is characterized by θ in the range $-60°$N to 90°N (the southern sector) and the other is complementary (θ from 90°N to 300 °N).

The probability of an extreme wave condition from the southern sector is approximately 20%.

Climate variables were successfully fitted to Weibull (H_s) and Gaussian (s_p, θ, Z) distributions.

Wave steepness and wave height are only weakly correlated and were considered to be independent. The effect of wave direction was already accounted for by separating the data into two sets and was not further investigated. The strong positive correlation between H_s and Z was evaluated by a conditioned probability rule. The synthesis of the results of these analyses, relative to the fitting and related scenarios, is reported in Table 7.32 and Table 7.33, respectively.

TABLE 7.32 Results of the Fitting of Yearly Maximum Surge Level Z (Surge Including Tide), and Conditioned Values of Significant Wave Height $H_s|_Z$, Wave Steepness s_p, and Mean Wave Direction θ. Extraction Point: 44.259701°N, 12.527600 E

$h = 14.7$ m	Z (m)	$H_s\|_Z$ (m)	s_p (–)	θ (°)
Distribution type	Normal	Normal	Normal	Normal
Mean	1.14	Equation 1	0.028	100
Standard deviation	0.1	0.045	0.009	6

TABLE 7.33 Climate Scenarios Considering Surge Z as the First Variable in the Statistics. Conditioned Values of Significant Wave Height ($H_s|_Z$) and Peak Wave Period (T_p); Wave Steepness s_p is Assumed to be Equal to 3.96%

$h = 14.7$ m; $s_{op} = 4\%$	Z (m)	$H_s\|_Z$ (m)	T_{op} (s)
Tr = 2 years	1.14	2.12	5.85
Tr = 5 years	1.23	2.35	6.17
Tr = 10 years	1.28	2.47	6.33
Tr = 20 years	1.32	2.57	6.45
Tr = 25 years	1.33	2.60	6.49
Tr = 30 years	1.34	2.63	6.52
Tr = 50 years	1.36	2.69	6.59
Tr = 100 years	1.39	2.76	6.69

The typical storm duration was found to be on the of order 12 h. The typical rise and fall time of the storms were also analyzed, leading to the results that in at least 1 case of 30, the rise time from $H_s = 0.5$ m to, $H_s = 3.5$ m and, the fall time from $H_s = 3.5$ m to $H_s = 0.5$ m have similar duration of order 1 h.

To analyze the storminess trend, the H_s with 100-year return period is considered: H_s increased from 4 m in 1960–1990 to 4.6 m in the 2010–2039 scenario and decreases in the subsequent scenarios. Therefore the projected increase during 2010–2039 reflects climate variability rather than a systematic trend due to climate change.

7.6.1.3 Coastal Habitats

Only few natural habitats are present in Cesenatico, the most important being represented by sandy beaches. Other habitats are some scattered vegetated patches of limited naturalistic value. The building of tourism facilities substituted the dunes and altered the beach equilibrium.

The effects of breakwaters and the additional impact from bulldozing and scraping during the winter season in the intertidal zone were investigated (Hanley et al., 2014) and the main ecological consequences can be summarized as follows.

- All physical and abiotic characteristics (organic matter, grain size, beach length) vary depending on the degree of beach protection, and specifically they increase in case of emerged structures with respect to submerged structures;
- The total number of individuals and number of taxa differs between the central and the southern areas. In particular, very low abundance and number of species were recorded at high and medium tidal level in the area protected by emerged structures;
- Changes in the composition of benthic assemblages can be considered also as consequence of grooming and bulldozing.

Overall, it is worth noting that the response of the benthic community depends on many physical and chemical factors and that the relationships among all the components are indeed very complex and not completely understood.

7.6.1.4 Society and Economy

Cesenatico is a popular tourist resort. It is home to the Marine Museum (Museo della Marineria), where historic fishing boats are displayed in the canal. The town also features a 118-m-high skyscraper, which for a few years was among the 30 highest buildings in Europe. The total inhabitants in Cesenatico were 25,375 in 2009. The average components of families (household) are about 2.31 people. The population of Cesenatico has been growing in the past 150 years, from 5593 in 1861 to 15,878 in 1961 and 25,412 in 2011.

The impact of this coastal stretch for the Italian economy can be summarized with a few figures (for year 2006) relative to tourism activities: 41 M person/day in the period May–September, 3384 hotels, 154,000 employees, and a gross income per year of €9.8 billion (Preti et al., 2009).

Table 7.34 synthesizes the most relevant economic activities and related income and Table 7.35 is the classification of land use types and related land use values in the three areas of Cesenatico (see Figure 7.76).

Based on survey data (THESEUS OD1.15, 2012), the citizens of Cesenatico feel they are moderately exposed to coastal erosion and flooding risk (Figure 7.77). In other words, they feel that this risk is significant but they do not feel they are adequately informed or prepared to cope with such risks. About 30% of the citizens are not aware of the evacuation routes in case of a flood and only a minority of citizens (18%) is aware of the alert warning system by mobile phone (through SMS), despite it is generally rated to be a useful application.

According to the surveys, the main risks associated to coastal erosion and flood are the damages to tourism and beach recreation (29%), the damage to the ecosystem (28%), the flooding of the city center (23%), and saltwater ingression (15%).

The coastal defense that is perceived to be most effective is sand nourishment (Table 7.36), followed by transversal barriers and submerged breakwaters. The sea gates are considered to be an inefficient defense. Among innovative mitigation options, insurance and restrictive land use planning are the preferred ones, whereas converting bathing establishment and retreating from the coastline are not considered to be acceptable measures.

TABLE 7.34 Economic Activities and Related Income in Cesenatico

Activity	M€/year	% of GDP
Tourism	123	25% of municipality GDP

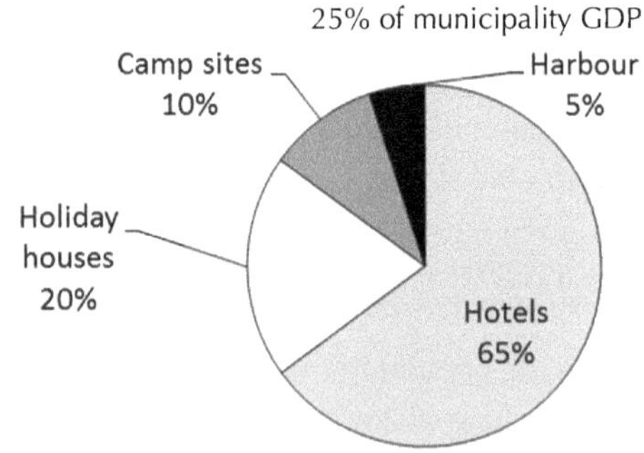

Activity	M€/year	% of GDP
Fishing harbor and infrastructures	52	2% of municipality GDP
Private services	355	47% of municipality GDP
There are no industrial firms in the area examined	87	18% of municipality GDP

TABLE 7.35 Classification of Land Use Types and Related Land Use Values in the Three Areas of Cesenatico

Land Use Type	Land Use Value (€/m²)		
	Northern Area	Central Area	Eastern Area
Residential homes	130	165	140
Holiday homes	191	217	199
Historical buildings	180	180	180
Hotels	152	152	152
Camp sites	61	0	0
Tourism harbor and infrastructures	–	97	–
Fishing harbor and infrastructures	–	179	–
Private services	3554		

(see Figure 7.76).

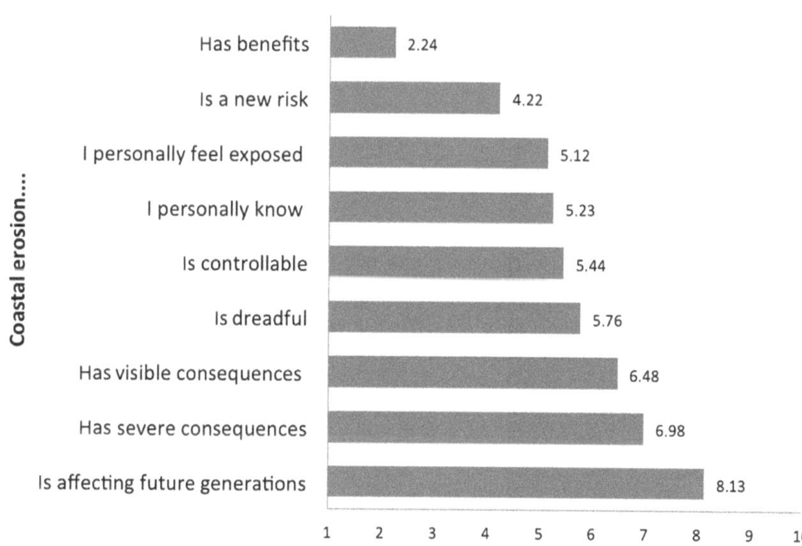

Figure 7.77 Perception of coastal erosion risk in Cesenatico residents (mean value from 1 = "no risk" to 10 = "extremely high").

TABLE 7.36 Perception of Mitigation Options in Cesenatico Residents (Mean Value from 1 = "Not Effective" to 10 = " "Extremely Effective").

	Perception of Effectiveness
Existing mitigation options	
Sand nourishments	4.9
Transversal barriers	4.3
Submerged breakwaters	4.2
Vincian gates	3.4
Innovative mitigation options	
Insurance coverage	8
Restrictive land use planning	7.4
Reutilizing buildings on the coastline	6.4
Relocating buildings further inland	6.2
Converting bathing establishments	5.2
Retreat from the coastline	5

7.6.2 Risk Modeling

7.6.2.1 Application of the SPRC Model

Because defense planning measures are usually taken at municipality level, the site extension is defined specifically by the municipality border. The landward boundary follows the railway track. Therefore in this case the area under analysis with the SPRC model does not correspond to a physiographic unit, as suggested in Chapter 2. This implies that the typical sediment long-shore transport budget in the area is assumed as a pathway and is considered to be constant in the analysis (i.e., defense strategies at the site boundaries or low levels of sediment transport from rivers are not expected to change with time).

Key receptors include the beach for tourism activities and ecological value (i.e., benthic communities), the urban area for social and economic activities, the marina that is crucial for the economy in the area, the railway that is the essential transportation route, and finally pinewoods and agricultural areas.

The identified pathways consist of the existing engineered management (already described in Section 7.6.1.1), and of the geomorphological response that in this case is the sedimentary budget from/to adjacent areas (as motivated in the previous paragraph).

Drivers are climate change, natural and anthropogenic subsidence that may significantly change over the time. Sources in order of importance are storm surge, runoff discharge, and extreme wave events.

The methodology selected for assessing the relative impact of climate change on the Sources of the flood system is the development of risk multipliers (THESEUS OD1.1.0, 2012). The estimated change in inundation probability due to changes in the sources is taken to be equivalent to the change in flood risk. Based on considerable previous analysis, the major effects of higher water levels is increased inundation frequency of the existing flood plain with flood plain expansion being a secondary effect. Hence, to a first order, change in inundation probability translates into a change in risk, all other factors being equal.

The evaluation of the risk multiplier is based on the comparison among the extension of the flooded area associated with a given risk probability. In this case, the selected risk probability is 0.01% (i.e., storm surge event with return period of 100 years at the current status and in the short-, medium-, and long-term scenarios).

Values are reported in Table 7.37, which includes also two combinations of the Sources to emphasize the highly nonlinearity of the site response to risk multipliers.

The area indeed is flat and the maximum beach elevation (i.e., the natural beach limit in front of the urban area) coincides with the elevation of the street and urban areas immediately behind the beach. This elevation is of about 1.5 m; therefore, the contribution to the risk and to its change produced by waves (through wave run-up) is relevant only when the beach is fully submerged, i.e., when water level (given by a combination of SLR, subsidence, storm surge level) is higher than 1.5 m. Because subsidence rate is much greater (so far it is 1 cm/year) than SLR, subsidence and time play the most relevant roles in the change of exposure to the flood hazard.

Ranking of the drivers is reported in Table 7.38.

7.6.2.2 Flood Mapping

Wave structure interaction and wave run-up on beaches and/or dunes require complex models capable to deal with steep and emerged slopes such as 3D RANS-VOF models (see Chapter 3). To limit the computational effort while modeling a number of scenarios, the approach adopted here combines an original procedure (Martinelli, Zanuttigh, & Corbau, 2010) developed for propagating waves from off-shore to the sea bank (i.e., beach/land boundary, whatever it is) with the MIKE 21 Flexible Mesh modeling suite developed by DHI for propagating the flood starting from the shoreline.

TABLE 7.37 Risk Multipliers for the Po Delta Site. The Risk Multipliers Indicate the Multiplication Factor of the Inundated Area Forecasted at Present (2010), with Tr = 100 years. Mean Sea Level Considers Both Sea Level Rise and Land Subsidence. Climate Scenarios as in Table 7.33

Driver Type	Name	2020	2050	2080
Source	Waves	1.0	0.9	1.0
Source	Surge	0.8	1.0	1.3
Source	Mean sea level	1.1	1.4	2.1
Source	River flow	Not quantified (expected low)		
Combined sources	Waves + surge	1.1	1.2	2.0
Combined sources	Waves surge + mean sea level	2.0	2.5	3.0

TABLE 7.38 Ranking Table for the Secondary Sources for the Po Delta Site

Importance of Secondary Sources in Flood Risk Over Time		
2020s	**2050s**	**2080s**
Mean sea level	Mean sea level	Mean sea level
Waves	Surge	Surge
Surge	Waves	Waves
River flow	River flow	River flow

Waves are transferred for a given tidal range from offshore to the shore, following the method by Goda (2000) and including wave reduction due to structures where applicable (van der Meer et al., 2005). The 2% wave run-up on the beach is then estimated from the formula by Stockdon, Holman, Howd, and Sallenger (2006), where the original formulation is modified to include the transmitted wave height inshore the mitigations and the local wavelength (see Chapter 3, -Section 3.5.1.2 for details regarding this procedure). A random-phase Gaussian process is finally generated having 2% characteristic value consistent with the value of the 2% run-up derived from this formula. This process is combined with the constant tidal and storm surge level to produce the varying water level in time that is provided as input to the flooding model.

The flooding model Mike 21 therefore adopts as "shoreline" boundary condition this water level varying in time and a zero velocity condition. The flooding process is represented as a dam break and wave celerity is indirectly represented by the added wave run-up component (i.e., the potential wave energy at the shoreline). The other lateral and inland boundaries are set to be closed boundaries and the computational domain is enlarged to avoid spurious reflection effects.

As regards the presence of the river and the channels, the average annual discharges are given along the river/channel sections at the inland boundary. The water levels at the river/channel outlets—that are in principle unknown—can be assessed through a preliminary MIKE 21 coupled wave-current FM simulation on a coarse grid extending from off-shore until the same inland boundary of the flooding model. These simulations show that sea conditions and specifically storm surges are prevailing, and therefore basically the water level at the river/channel outlets very rapidly increases up to the value given by the sum of the storm surge level and the wave set-up level. A detailed description of these simulations can be found in the Appendix to THESEUS OD1.15 (2012).

In Cesenatico—as in most places along the Emilia Romagna region—the dominant flooding parameter is the storm surge level. Flooding is rapid but is characterized by relatively small flooding depth so that evacuation plans may be very efficient. The maintenance of the urban drainage system (sea gates opening/closure, bypass, and channel banks) and of the beach width is essential for the safety of the urban areas.

An example of the flooding depth and extension for a medium-term scenario is reported in Figure 7.78.

While looking at these results, it is worth remembering that the beaches are not reshaping during the storm and therefore the severity of the flood consequences may be underestimated.

The more probable failure scenarios are associated to missing closure of the sea gates and/or insufficiency of the drainage system.

7.6.2.3 Habitat Vulnerability

As for sandy beaches, the intertidal beach zone was considered to be the major ecological receptor of flooding events. Within this habitat, the benthic component was specifically analyzed because it is generally considered as the most significant ecological indicator of the marine habitat status.

The response of the habitat to selected parameters, such as wave agitation, beach slope, sediment grain size, turbidity, etc., was assessed by means of a newly developed method Fuzzy Bayesian Method (FBM) which is a combination of fuzzy logic and a naïve Bayes (Bozzeda, 2013).

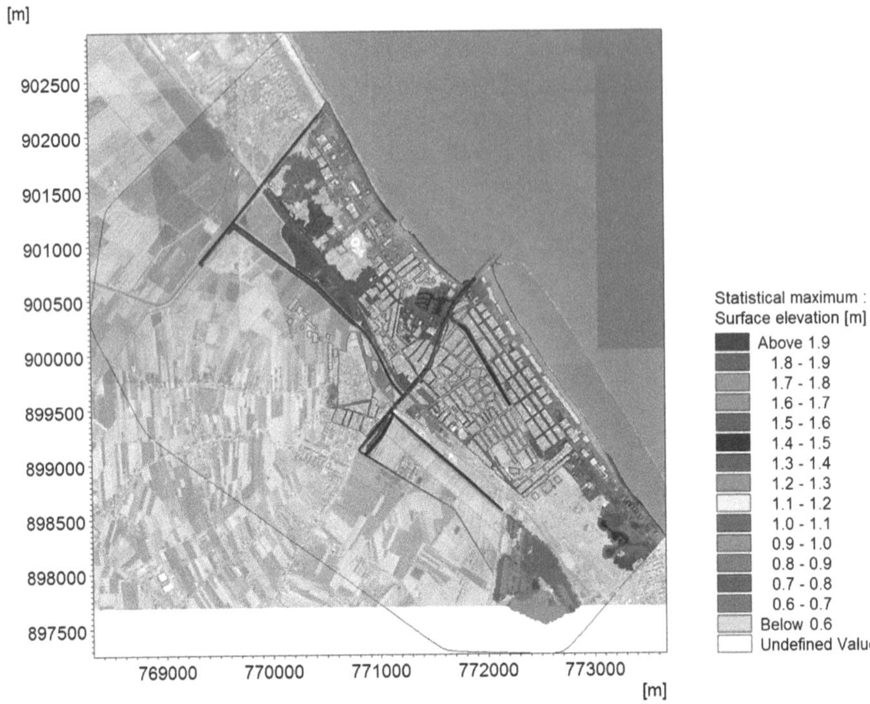

Figure 7.78 Maximum surface elevations for Tr = 10 years. Medium term: 2050. Scenario with sea gates closed, operating pumping system, and no subsidence. Results obtained with Mike 21 FM. Details in THESEUS OD 1.15 (2012).

The following measured data were used to train the FBM

- environmental variables (i.e., the input data): median particle diameter, sorting coefficient, organic matter percentage content (TOM%), intertidal zone length, tidal level, beach slope, wave height, riptide velocity, and beach deposit index (McLachlan & Dorvlo, 2005);
- biotic/target variables of macrofauna communities (i.e., the expected output): macrobenthic abundance.

All these data were considered as the set of "benchmark" values for establishing the EVI thresholds.

The impacts of flooding events on the macrobenthic abundance was then simulated for the whole set of climate scenarios reported in Table 7.33, by changing in the FBM the corresponding input values of H_s, T_p and Z.

Based on the application of the FBM and on surveys in the area, the following empirical curves are obtained for beaches EVI_B and for pinewoods EVI_P, respectively

$$\text{EVI\_B} = \text{B1y}^2 + \text{B2y} + \text{B3} \tag{11}$$

$$\text{EVI\_P} = \text{P1h}^3 + \text{P2h}^2 + \text{P3h} + \text{P4} \tag{12}$$

where y = water depth (m) and h = duration of the flood (h). The specific coefficients are reported in Table 7.39 for both benthos and pinewood.

As for the pinewoods, they resulted as highly tolerant to salt water and marine spray, whereas the roots are very sensitive to salt water intrusion and therefore even a short-term exposure would produce a high impact (i.e., presumably nonrecoverable).

To better evaluate the relevance of climate change on the assessment of the ecological impacts, modular simulations were carried out with and without SLR. The results are shown in Figure 7.79 in terms of the number of macrobenthic individuals (i.e., the organisms whose size is greater than 0.5 mm) as a function of the storm return period Tr. When SLR Z_r is disregarded from the scenarios in Table 7.33, the benthic abundance appears particularly vulnerable (Figure 7.79(a)), i.e., it rapidly decreases with increasing wave height Hs with Tr. When Z_r is included (Figure 7.79(b)), the abundance tends to be constant and even to increase with increasing Tr. This peculiar result is due to the shift from inter-tidal to sub-tidal communities, the latter being characterized by greater abundance and by species that have to face different wave impact.

In summary, from an ecological point of view, macrobenthos as a whole is relatively robust to flooding events in terms of total abundance, even though it is affected by wave action (Figure 7.79).

TABLE 7.39 Values of the Parameters to Evaluated the EVI Coefficients for Benthos (B) and Pinewoods (P), see Eqns (11) and (12) for each Scenario and Sea Level Rise

Scenario	2010	2020	2050	2080
SLR	0	0.7	0.13	0.22
B1	0	0	0	1.333
B2	0.097	0.038	0.223	0.019
B3	−0.082	−0.014	−0.228	2.241
P1	−0.00006	−0.000006	−0.000100	−6E-07
P2	0.001	0.001	0.001	0.000
P3	0.06	0.06	0.08	0.096
P4	−0.4	−0.4	0.5	1.1

See climate conditions in Table 7.33.

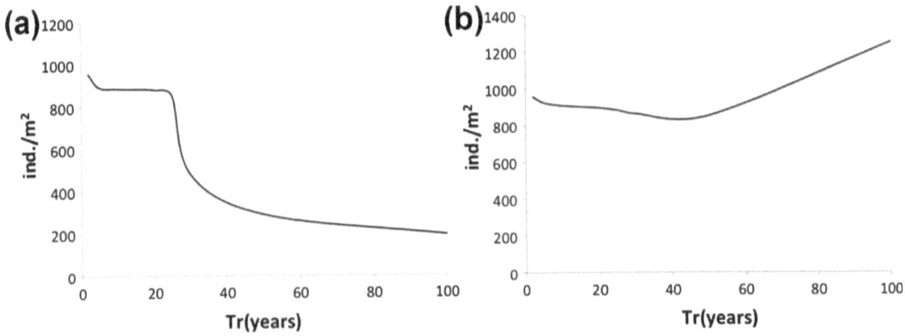

Figure 7.79 Response of the macrobenthos abundance to wave height and flooding events for different return times Tr (see Table 7.33) (a) in absence of sea level rise and (b) in presence of sea level rise.

7.6.2.3 Analysis of Social Impacts

The analysis of social impact is mainly focused on two issues: (1) the damages to critical facilities (CFs) and (2) the expected number of fatalities. It is worth remarking that flood damages to society include also psychological consequences that are mainly qualitative in nature and are hard to be translated in linear functions with quantitative outputs for practical and ethical reasons (Stanke, Murray, Amlot, Nurse, & Williams, 2012; Zanuttigh et al., 2014).

CFs are defined as the primary physical structures, technical facilities, and systems that are socially, economically, or operationally essential to the functioning of a society or community, both in routine circumstances and in the extreme circumstances of an emergency. On the one hand, the notion has been adopted recently in disaster management and has been applied in the development of priority lists for the effective reactivation of buildings after disasters and applied emergency management.

The social impact on CFs has been estimated following these steps.

1. *Ranking of CFs.* A rank was derived based on the function of buildings in terms of use in emergency management, function in ordinary activities and community aggregation, and symbolic function. The corresponding Approximated Social Value (ASV) was evaluated, with values reported in Table 7.40 from 1 (low) to 5 (high). In Cesenatico, it was possible to identify up to 80 CF: 4 with ASV = 5, 10 with ASV = 4, 45 with ASV = 3, 15 with ASV = 2, and 6 with ASV = 1. The ASV also provides a reactivation list in reverse order, as the highest values are supposed to receive priority in emergency interventions for reducing social damages. In the perspective of land use planning, the adoption of such an approach should lead to the relocation of high-scoring buildings to safer areas or encourage measures to increase buildings resilience. Similarly,

higher scores indicate where efforts for higher education and training of personnel should be concentrated and where emergency measures such as mobile barriers could be deployed with maximum effectiveness.

2. *Estimation of physical damage for structures.* The damage scale was estimated based on flood depth and duration. Following the method by Schwarz and Maiwald (2008), the damage grade is related to the flood depth (De) through a nonlinear function. Intuitively, the effects on society and structures are inversely proportional to flood duration (D), if one excludes flash flood phenomena. Long duration floods, even if relatively limited in space, produce greater impacts on social functions: a bridge blocked might be a nuisance for an hour, whereas it could compromise trade routes or tourism activity for a week. Therefore the following scenarios have been considered: (1) short D (hours), (2) medium D (day/days), and (3) long D (week/weeks).
3. *Definition of tourist impact.* In Cesenatico, one of the most relevant variables affecting the ordinary social pattern is tourism. It can be presumed that not all the tourist had previous experiences of floods, and that if a flood happens with a large number of tourists in place CFs may suffer higher pressure and warning messages may face more dissemination problems. The tourist presence has been represented through a value reflecting seasonality S; this factor will act as a final scale multiplier, where low season ($S=1$) denotes ordinary conditions and high season ($S=2$) implies that the effects will be exacerbated.
4. *Estimation of collateral social damages.* A final estimation of the impact has been computed following this function

$$\mathrm{CSD} = \sum_{1} \mathrm{ASV_i} \cdot \mathrm{De} \cdot \mathrm{D} \cdot \mathrm{S} \tag{13}$$

TABLE 7.40 Ranking Values of Critical Facilities

Approximate Social Value	Definition
5	Critical structures that if involved could compromise the emergency action, the coordination chain, public safety, and public health in the long term. For example, hospital and emergency facilities.
4	Facilities that provide significant public services and should be activated within 24 h. For example, nurseries, major water and sewer facilities, fire and police stations, schools, and park facilities used to support critical purposes.
3	Facilities that provide important public services. Main centers of aggregation and education that are important to the community.
2	Facilities that provide public services but that are less critical for the community. For example, common storages, sport centers.
1	Places that value is mainly symbolic.

The social impact on the expected number of fatalities has been estimated following these steps. The following function of life losses and injuries (NI) was derived from Penning-Rowsell, Floyd, Ramsbottom, and Surendran (2005)

$$NI = (H^{*}AV)/(Pa + ID) \tag{14}$$

where H is the hazard rate, AV is the area vulnerability, Pa is the sensitive population (age <14 years and >65 years), and ID is the number of sick and disabled people. Table 7.41 reports the factors used to estimate life losses and injuries.

The function (14) provides an overall number of people subject to death or injuries. These two aspects were not distinguished as too many external variables such as local lifestyle, wealth, or public health services influence the final output of life losses, and the uncertainties are high.

7.6.2.4 Analysis of Economic Impacts

A consistent approach based on incomes for each economic land use was adopted (Zanuttigh et al., 2014) for estimating the economic consequences (EC) e.g., hotels are evaluated in terms of annual GDP, houses are evaluated in terms of annual rents, beaches are evaluated in terms of annual willingness to pay to preserve them.

The EC in the inland area ECI are supposed to be dependent on flood depth and duration following the formula:

$$ECI = vij \cdot bj \cdot Fd + vij \cdot aj\sqrt{Fy} \tag{15}$$

where vij are the values of land uses in euro/m^2/year from Census statistic data; Fd is flood duration and Fy is flood depth; aj are proportionality constants that are normalized for each land use j based on the maximum value of Fy in 2050 for a storm return period of $Tr = 100$ years, assuming different reference percentage of damage depending on the use (for instance, 50% damage for buildings/homes/hotels, 25% damage for harbors); bj are proportionality constants that express the expected period to restore economic activities depending on Fd and on land use (for instance, a value of 30 is set for hotels and of 20 for private services) and are normalized to annual incomes with the days/year. Note that flood velocity is assumed to be irrelevant based on results of detailed mathematical modeling (Section 7.6.2.2).

The land use value losses are then combined with the EC along the beach, ECB, due to erosion. The value of beach loss was derived from a choice experiment exercise carried out in Santander, ES (details about the site in Section 7.9). By adapting the results of the survey Diaz et al. (2012) to the current population and to the GDP per capita prevailing in Cesenatico, the intangible value attached to beach is ECB = 1.47 €/m^2/year.

TABLE 7.41 Factors Used to Estimate Life Losses and Injuries

Value	Description
H	Hazard rate: $H = NI \cdot y \cdot v \cdot DF$ Where N is the number of people involved in the flood, *y* is the flood depth, *v* is the flood velocity, and DF is the debris factor equal to 1 for the Mediterranean and 2 for the ocean.
AV	Area vulnerability: $AV = W + Fo + Na$ Where W denotes the warning, Fo is the speed of onset of flooding, and Na is the nature of the flooded area.
Pa	Percentage of the sensitive population (age <14 and >65 years) Pa, derived from demographic data of Cesenatico (ISTAT, 2009).
ID	Percentage of infirm/disabled/long-term sick population ID, based on data from the municipality of Cesenatico.

7.6.3 Identification of Mitigation Options

The mitigation options most suited for Cesenatico were identified by considering jointly

- the existing management, and therefore the costs induced by any change of what is already in place;
- the historical management of the coastal area in the Emilia Romagna region;
- the feedback from stakeholders during surveys and focus groups (see recommendations for risk assessment outlined in Chapter 2); and
- the perceived effectiveness of existing or potential mitigation options from public at large, see Table 7.36.

Table 7.42 includes the selected mitigation options, the rationale to propose them, and the main challenges posed by these solutions. More details about each kind of these engineering, ecologically based social and economic solutions are provided in Section 7.6.4.

7.6.4 Preliminary Design of Mitigation Options

7.6.4.1 Engineering Solutions

Past experience, environmental constraints, and end-users feedback suggest the traditional nourishment with blond sand (carried out on the emerged beach) as one of the most effective and suited mitigation option against erosion. All the engineering mitigation alternatives here proposed consider some sort of periodic sand nourishment, either on the emerged or on the submerged beach.

TABLE 7.42 Lists of Mitigation Options Together with the Rationale and the Main Challenges

Mitigation Option	Rationale	Challenges
Wave farm	Low aesthetic impact Reduction of incident wave height Production of wave energy Low sensitivity to water level variation Increase of offshore biodiversity	Low effectiveness for coastal protection purposes Low structure reliability under large storms Immature technology
Floating breakwaters	Low aesthetic impact Easy decommissioning Reduction of incident wave height Low sensitivity to water level variation Increase of near-shore biodiversity	Low effectiveness during storms Low structure reliability under large storms
Low-crested detached barrier	Reduction of incident wave height Low visual impact Increase of near-shore biodiversity	Increase of wave setup Sensitivity to sea level rise
Nourishment on the submerged beach	Availability of resource at low cost (offshore dredged sands) Low environmental impact compared with hard defenses Low aesthetic impact Low impact on benthos compared with emerged nourishment Reduction of incident wave height	Limited benefit to shoreline advancement
Nourishment on the emerged beach	Temporary increase or maintenance of beach width Recreational value Low environmental impact compared with hard defenses Low aesthetic impact Reduction of wave run-up	Limited availability of compatible resource High sensitivity to negative impacts in case of unsuited nourishment material
Dune reconstruction	Significant reduction of wave overtopping Compatible with landscape	Feasible only where sufficient space is available Not easy to reinforce or maintain (breaching experienced)
Heightening of Vinciane sea gates	Significant reduction of the expected flooding of historical city center	Combined bypass systems required Combined periodic dredging plans of the canal harbor entrance required Temporary reduction of harbor accessibility Negative social perception of the existing gates
Evacuation plan	Reduction of casualties Promotion of risk awareness	Ineffective communication Unclear responsibilities during crisis management
Insurance scheme	Promotion of rapid business recovery Market incentive mechanism (to reduce premium) Promotion of urbanization of low-risk (= low premium) areas Redistribution of damage	Unpopular introduction of adequate laws and regulations Careful estimate required of the additional cost paid by society to cover insurance profit

When performed on the emerged beach, the designed sand diameter is similar to the existing one, characterized by a $d_{50} = 0.2$ mm, and the target beach advancement is 30 m based on historical data (Preti et al., 2009)).

Because the availability of inland sandy stocks of large diameter sand is rather limited, whereas there are large quantities (more than 50 Mm^3) of fine sand ($d_{50} = 0.1$ mm) available from offshore pits, the option of nourishment on the submerged beach is also investigated. The fine sand may only be placed at a depth that where its stability is assured, and used to reduce the incident wave energy (see Chapter 3). Specifically, the option considers that the designed submerged nourishment is quite significant (i.e., 300 m^3/m, and the fine material is placed at isobaths −3 m b.s.l.).

The vulnerability of the site from both flooding and erosion, the widespread use of detached breakwaters to defend the surrounding coastal stretches, the social perception of this kind of defense and the high costs for their removal suggest that the engineering management should consider to maintain and maybe modify the existing breakwaters.

Therefore, most of the selected portfolios consider to maintain and reshape the existing breakwaters, to obtain a more homogeneous cross-section and a better protection of the northern area. The rationale of protecting the whole area with barriers is to minimize the losses of the nourished sand and to reduce the run-up on the beach, thus decreasing the flooding risk. The expected wave run-up is further reduced thanks to the combination of the breakwaters with nourishment interventions.

The innovative option of installing an array of wave energy converters is also considered, with the twofold function of harvesting energy and protecting the coastline. The available energy in Cesenatico is indeed quite limited, on the order of 2 kW/m (Martinelli, Pezzutto, & Ruol, 2013). The DEXA device is selected considering the relatively high efficiency also in case of mild climate (see Chapter 3). Based on the typical wave length in the area, the device cross-shore length is fixed to be equal to 30 m (to achieve a device length to peak wave length ratio close to 1), i.e., approximately in scale 1:2 compared with the hypothetical prototype developed for the Danish North Sea. The minimum water depth allowing for the installation of such a device, considering the draft and the moorings, is not less than 10 m. The minimum feasible size for the array consists of seven lines of staggered devices, displaced in a 200 × 200 m grid. Each device is formed by a combination of reinforced concrete, ballast, and steel, whose weight is 110 t. The rated power/unit is 30 kW and the expected annual power production/unit (assuming a cost of electricity of 0.2 €/kWh) is 17’500 €/y/unit.

The main design parameters for the nourishment interventions, the submerged and emerged breakwaters and the array of DEXA devices are summarized in Table 7.43. Layouts combining engineering solutions are provided in Figures 7.81–7.83.

TABLE 7.43 Design Parameters of Selected Engineering and Ecologically Based Mitigation Measures. Symbols Refer to the Layouts of the Portfolios of Mitigation Options Reported in Figures 7.81–7.83

Nourishment on the Emerged Beach		Nourishment on the Submerged Beach	
d_{50}	0.20 mm	d_{50}	0.10 mm
Nourishment volume	210 m^3/m	Cross-shore location	from −3.0 m to −4.0 m
Beach advancement Bc	30 m	Nourishment per m	300 m^3/m
Length Lb	5100 m	Nourishment extension Lb	5100 m
Estimated duration	5.0 y	Width of nourishment Bc	75 m
Submerged barrier		**Emerged barrier**	
Crest freeboard Rc	−0.5 m s.w.l.	Crest freeboard Rc	1.0 m s.w.l.
Barrier length Lb	2100 m	Barrier length Lb	3000 m
Depth at the structure toe h	3.0 m	Depth at the structure toe h	3 m
Crest width of barrier Bc	15.0 m	Crest width of barrier Bc	3.0 m
Gap width to barrier length ratio	30%	Gap width to barrier length ratio	0.3
Barrier offshore slope	1:2	Barrier offshore slope	1:2
Array of wave energy converters			
Weight of single wave energy converters (DEXA)	110 tons	Footprint of each device	200 × 200 m
Unit width and length	15 × 30 m	Height	1.5 m
Water depth	10 m	Distance from shore *Dp*	8000 m
N. Of lines	7, staggered	Overall K_T for T = 5 s	0.6
Array width Bc	1400 m	Array length	2000 m
Dune			
Inshore distance from the shoreline	40 m		
Crest width Bc	2.5 m	Height	2.0 m

7.6.4.2 Dunes

Coastal dunes protection and restoration is made from ecological engineering techniques, using natural processes to help coastal ecosystem to regenerate and to improve its efficiency to fight against erosion.

The results of dune restoration depend on the following different factors:

- the sediment characteristics;
- the techniques used for sand trapping;
- the rate of sediment exchange between the beach and the dune after nourishment; and
- the space available seaward of human structures.

The recommendations from other similar interventions carried out in the Northern Adriatic can provide useful inputs (Bezzi, Fontolan, Nordstrom, Carrer, & Jackson, 2009; Calabrese, Lorito, & Perini, 2010). A dune system with medium vulnerability is characterized by a foredune beach 40–70 m wide and a crest 1.5–2.5 m high.

The purposes of dune reconstruction in Cesenatico are mainly (1) to build a natural reinforcement of the beach to defend the inland area from waves overtopping the beach bank; (2) to set up a sand "reservoir" along the beach to mitigate the effects of erosion induced by severe storms; and (3) to recreate a valuable ecological habitat.

In Cesenatico, the northern area appears the most suitable site because of beach width and absence of bathing facilities. The selected dune design parameters are reported in Table 7.43.

A common practice to restore dunes is to import sand by nourishment and bulldozing and then transferring the sand from a low to a high level. This method however may reduce the intertidal beach width and impact the benthic communities. This impact can be decreased if the bulldozing is carried out in winter when benthic populations are not in the reproductive phase and massive settlement by organisms does not occur.

Vegetal cover development is the best guarantee of securing and protecting sand dunes. In Cesenatico, it is recommended to follow the protocol developed for dune reconstruction and restoration at the Bevano river outlet (Calabrese et al., 2010). The species to be used for planting have to be autochthonous and transplanted from sites where there are wild populations available within the same bioclimatic region.

In the Emilia Romagna region, the most suitable species for dune stabilization are the two perennial plants *Agropyron Junceum* and *Ammophila littoralis*. The plants have to be in good condition and with a well-developed root system and the has to be performed in late autumn, when the temperatures still allow the vegetative activities and the water from precipitation or condensation is enough to guarantee the absence of great water stress.

Furthermore, the vegetation distribution on the dunes has to be established according to the natural zonation of each species. Historical information, photos, or similar local beach environments may help to choose the best zonation. To ensure that the vegetation can grow and develop, the planted area has to be protected with a fence until the dune stabilization and the access to the beach has to follow established paths that minimize the interruption of the continuity of the dune system. A continuity of at least 300 m length of the dune system is one of the key parameters required to maintain a medium vulnerability level (Calabrese et al., 2010).

7.6.4.3 Evacuation Plans

The setup of an evacuation plan is a nonstructural mitigation option aimed at getting a maximum number of people from risk prone areas to safe places (shelters in place or outside risk areas, upper floors) in a limited time.

Hissel et al. (2014) proposed a methodology meant to clarify all the stages of an evacuation and the thinking process that can lead to a robust evacuation plan.

In Cesenatico, the software "Serapia" has been used to calculate evacuation times. This tool estimates the time needed for evacuating an area under different circumstances and can therefore assist in developing an effective evacuation strategy. The total evacuation time is divided into three time intervals: (1) the time it takes between the flood alert and the moment that inhabitants leave their homes; (2) the time it costs evacuees to travel from their home to a congested exit point; and (3) the time it takes to drive through this congested point. For this estimation, the model needs as an input the following four sources of data.

1. *The number of people to evacuate.* In Cesenatico a significant percentage of their inhabitants will be able to leave the evacuation area with their own cars whereas others (elderly, residents of hospitals and nursing homes) will have to be evacuated with buses or special vehicles. In our model (Hissel et al. 2014), 9645 people must be evacuated in case of a strong event to 18 evacuation sectors of the city. According to available data, in 2012 around 15,000 cars were registered in Cesenatico, with average occupancy of 2.3 people per car.
2. *The time between the flood alert and the departure from home.* The duration of this time interval depends on how fast all inhabitants are informed, how well they are prepared, the extent they are aware of the urgency of the evacuation, and their willingness to leave home. A minor percentage of inhabitants will refuse to evacuate.
3. *The travel time from the departure location and the location of the exits of the evacuation area.* This depends on travel distance and average velocity. During evacuation, traffic behavior is expected to differ from normal circumstances.
4. *The chances of congestion.* Congestion is likely to occur when many cars have to leave the area in which there are a small number of exits. Cesenatico is a relatively small town with many road exits. Five exit points have been identified to lead people toward shelters. Two safe routes with larger road capacity can be identified in the directions to the north and to south, respectively: Viale Saffi and SS Adriatica (Figure 7.80). The capacities of evacuation lanes are going to be reduced by 50%.

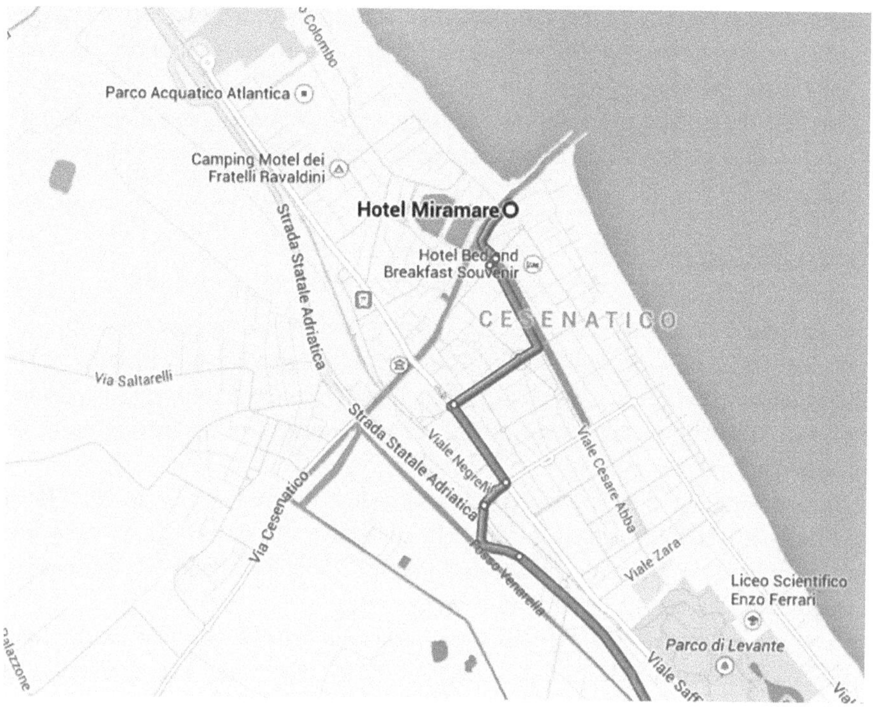

Figure 7.80 Exit route from the historical urban area and the harbor area toward the south of Cesenatico.

According to Hissel et al. (2014), the following methods can be adopted to shorten the needed evacuation time:

- decrease the total number of vehicles leaving the area by raising the number of inhabitants per vehicle. This can be achieved through information campaigns;
- shorten the departure time by an effective warning system. In a previous survey, inhabitants in Cesenatico reported that they felt moderately informed about flooding risk in Cesenatico (the mean value being M = 4.25, in a range from 1 to 10, where 1 means lowest and 10 highest). The municipality's alert system by mobile phone (SMS) was generally considered a useful instrument (M = 7.34, range from 1 to 10) but it was known only by 20.1% of citizens and a small minority was effectively registered to the service. A more effective warning system can be developed, by improving the existing measure (acoustic signal from the harbor area);
- shorten the traveling and waiting times by means of a better distribution of vehicles toward the different exits. In the Plan for Civil protection in Cesenatico, 18 safe areas have been identified and they are efficiently distributed in the area.

However, according to the survey (Hissel et al., 2014), the majority of participants did not know evacuation routes (61.9%) and safe places. Information campaigns are therefore needed; and

- increase the road capacity, for instance by using some roads in one direction only. Such measures require effective organization and they do not seem advisable in this case.

7.6.4.4 Insurance

The estimated economic damages are spatially dispersed because they are very high close to the railway, at the inland boundary, and high at the beach, at the water/land boundary. However they are also functionally concentrated (i.e., banks and markets). Therefore, the development of a dedicated insurance scheme is suggested.

Indeed, although the private sector is totally absent nowadays, in the 1996 flood the Emilia-Romagna Regional government covered 100% firms with an estimated damage larger than 5 million Italian liras (around 2500 euros) and 5% households, respectively, with the 42% and 34% coverage rates.

It is assumed here that a compulsory insurance scheme, without exemptions, where the premium represents a proportion of the average yearly exposure, either of property values (direct impacts) or of the business activities (indirect values) or of both.

The average premium (AP) is evaluated as

$$AP = c(AC + AI) \qquad (16)$$

where AC are the administrative costs, AI is the average indemnity, and c is the commercial viability that usually ranges from 0.5 to 1.0.

Specifically, here $c = 1$ to maximize insurance coverage; AC is given by the product of the average cost (€10 per family), the family size and the total population (see the figures reported in Section 7.6.1.4); $AI = ECI/Tr$, where ECI is derived from Eqn (15) and Tr is the return period associated to the selected storm scenario (Table 7.33).

Note that all the losses are estimated in terms of land use values (GDP).

In other words, a heterogeneous agent (as opposed to representative) perspective and a financial (as opposed to economic) approach are adopted. In particular, the standpoint of the individual household or organization involved, at local rather than at national level, allowed us to measure the financial direct and indirect losses to individuals and organizations that are affected rather than the national economic losses caused, and to consider, for example, the income loss in one particular retail shop, even if the trade this represents is likely to be deferred in time or transferred to another retail outlet.

7.6.5 Selection of the Portfolio of Mitigation Measures

Based on the rationale outlined in Section 7.6.3, Table 7.42, and on the site-specific preliminary design of the mitigation options carried out in Section 7.6.4, a limited number of portfolios, i.e., combination of mitigation options, was selected. Table 7.44 includes five portfolios:

- Portfolio 1 (Figure 7.81) is an engineering solution only (Section 7.6.4.1), based on the traditional management of the area, and therefore includes detached breakwaters and nourishment on the emerged beach; the breakwaters are reshaped with respect to the existing ones (with prolongation in the northern unprotected area) but the difference of the breakwater submergence is maintained;
- Portfolio 2 (Figure 7.82) is again an engineering solution only (Section 7.6.4.1), but it is considered to be innovative with respect to Portfolio 1, thanks to the introduction of the wave energy farm; in this case the reshaped breakwaters are lowered to reduce their visual and environmental impact, leading to better water recirculation in the protected cell; again, the nourishment is made on the emerged beach as well as in Portfolio 1;
- Portfolio 3 consists of social and economic measures only (i.e., the evacuation plan; Section 7.6.4.3) and the insurance scheme (Section 7.6.4.4);
- Portfolio 4 (Figure 7.83) is meant to be a purely ecologically friendly solution; the protection of the northern area is performed by means of a dune system (Section 7.6.4.2) and the nourishment is carried out on the submerged beach (Section 7.6.4.1), to limit the environmental impact and the costs; and
- Portfolio 5 is somewhat a mix of the previous portfolios, combining all the types of mitigations with a very rough estimation in mind of potential costs and benefits. From an engineering perspective, the existing structures are maintained and only reshaped and extended without changing the submergence to avoid the costs of removal, lowering, and at the same time the reduction of protection from coastal erosion. The nourishment is carried out on the emerged beach to allow the immediate perception of the nourishment temporary effects and promote recreational activities. The wave energy farm is not included considering the high installation costs and low energy production, and also the insufficient development level of energy storage technology leading to unfeasibility of the design in the area. The ecologically based solution of creating dunes in the northern area is incorporated to increase the resistance of the beach to wave run-up and flooding. Both social and economically based options are also included (i.e., the evacuation plan and the insurance scheme, respectively) as cost-effective and climate-proof solutions.

When the evacuation plan is included, its efficacy is limited to the 50% of the population at the selected time slice as a cautious assumption.

TABLE 7.44 Selected Portfolios of Mitigation Options

	Barriers		Wave	Nourishment				Evacuation
Portfolio#	Emerged	Submerged	Farm	Emerged	Submerged	Dunes	Insurance	Plan
1	South	North, Central		x				
2		South, North	x	x				
3							x	x
4					x	x		
5	South	North, Central		x		x	x	x

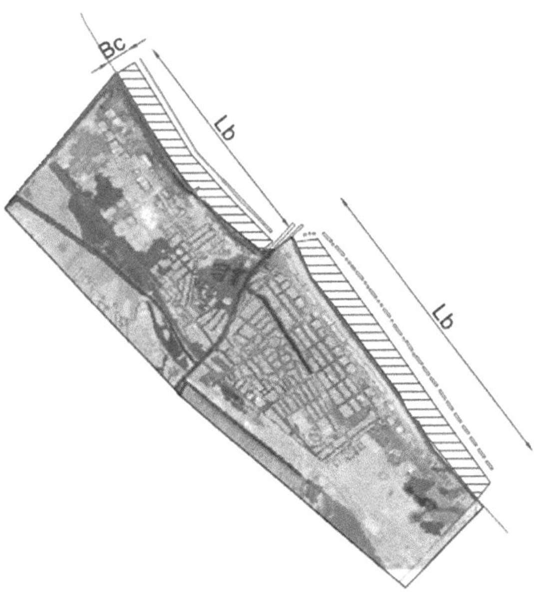

Figure 7.81 Layout of portfolio 1. Nourishment on the emerged beach, reshaping of the existing rubble mound emerged barriers in the southern area, construction of similar rubble mound–submerged barriers in the northern and central areas, with removal of the existing sandbag barrier.

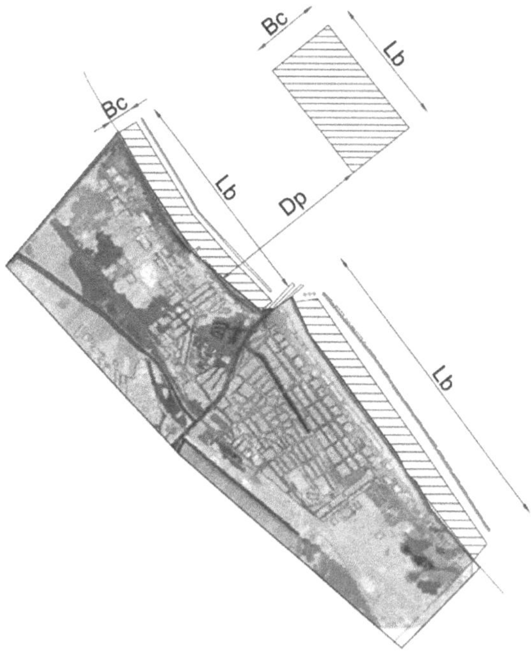

Figure 7.82 Layout of portfolio 2. Nourishment on the emerged beach, reshaping and lowering of the existing rubble mound–emerged barriers to submerged barriers in the southern area, construction of similar rubble mound submerged barriers in the northern and central areas, with removal of the existing sandbag barrier, installation of nearshore arrays of wave energy converters in the northern area.

7.6.6 Impacts of the Portfolio of Mitigation Options

The hydraulic, social, economic, and ecological impacts of the portfolio of mitigation options in Table 7.44 were assessed by means of the decision support system delivered by the THESEUS project (see Chapter 6, Section 6.6; Zanuttigh et al., 2014).

The following indicators were selected to quantitatively synthesize the hydraulic, social, economic, and ecological vulnerability maps derived from the decision support system.

1. Hydraulic vulnerability:
   a. 1a: percentage of the flooded area with respect to the total area under investigation;
   b. 1b: percentage of the flooded area characterized by flood depth greater than 0.5 m with respect to the total area;
   c. 1c: percentage of the flooded area characterized by flood velocity greater than or equal to 0.5 m/s with respect to the total area;

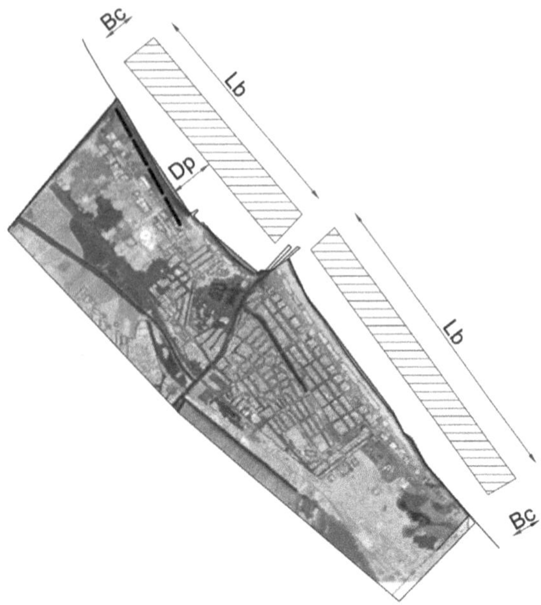

Figure 7.83 Layout of portfolio 3. Nourishment on the submerged beach, removal of the existing barriers in the southern and central areas, and building of a dune in the northern area.

   d. 1d: percentage of the flooded area characterized by flood duration greater than or equal to 6 h with respect to the total area;
   e. 1e: percentage of beach retreat.

2. Ecological vulnerability:
   a. 2a: percentage of the area where the EVI for benthos is greater than or equal to 2, with respect to the total area covered by benthos;
   b. 2b: percentage of the area where the EVI for benthos is greater than or equal to 2, with respect to the total area covered by pinewood.

3. Social vulnerability:
   a. 3a: percentage of CFs interested by a loss greater than or equal to 20%, with respect to the total number of CFs,
   b. 3b: percentage of life losses greater than or equal to 1/1000, with respect to the total number of people.

4. Economic vulnerability:
   a. 4a: percentage of the flooded area characterized by land value losses greater than or equal to 30% of the total value loss,
   b. 4b: percentage of beach loss; actually this indicator is exactly the same as the indicator 1e by definition. However in this way it was disregarded the

identification of a threshold minimum value of beach extension that implies a relevant beach loss. This is indeed in agreement with the fact that for simplicity the beach value is considered to be homogeneous over all the beach width.

5. Risk indicator: percentage of the area characterized by the overall risk level greater than or equal to 2.

Of course the threshold values here selected for each indicator, for instance: the value of 2 for number 5, the risk indicator, or the value of the 30% for number 4a, the land value losses, are site dependent. These values were identified based on the results obtained for the nondefended cases and for the different climate scenarios (Table 7.33).

The results for the selected portfolios of mitigation options (Table 7.44) and a few climate scenarios are compared in Table 7.45. To consider a strategic planning, only the long-term climate scenario (2080s) was simulated. This scenario was combined with two storms characterized by very different return periods (i.e., frequent, Tr = 10 years and exceptional, Tr = 100 years) to verify both the effects of SLR and increase of storm intensity. The impacts of the portfolios are compared with the case of no mitigations (i.e., with the erosion and flooding that would occur in Cesenatico in absence of any kind of management).

The impacts of the mitigation options are discussed here in terms of the criteria already identified in Chapter 6, Section 6.2.

- Hydraulic efficiency.
  Portfolios 1 and 2 are essentially equivalent in case of the frequent storm, whereas Portfolio two appears to be more effective in reducing the extension of the flooded area and the flood depth (and therefore duration) in case of the exceptional storm.
  Portfolio 3 of course does not mitigate flooding or erosion and therefore the results of the simulations essentially correspond to the benchmark cases without structures or other kind of protection schemes.
  Portfolio 4 offers a modest degree of protection that is more appreciable in case of the more intense storm.
  Even if the wave farm is not included, portfolio 5 provides slightly better performance than portfolios 1 and 2, thanks to the presence of the dunes in the northern area.
- Ecological sustainability.
  Benthos and pinewoods show different response dynamics that are coherent with the knowledge about these habitats (see Section 7.6.1.3). The benthos sensitivity to changes of the wave height is appreciable (for storms with return periods of 100 years instead of 10 years), whereas it is modest in presence of SLR (by changing the time slice from 2010 to 2080).

TABLE 7.45 Synthesis of the Consequences of the Portfolios of Mitigation Options in Terms of Indicators, Portfolio of Mitigation Options (see Table 7.42), and Climate Scenarios (See Table 7.33). Where the Number of the Portfolio is not Indicated, the Simulated Condition Corresponds to the Absence of Any Kind of Defense

Scenario	2010	2080	2080	2080	2080	2080	2080	2080	2080	2080	2080	2080	2080
Tr	100	10	100	10	10	10	10	10	100	100	100	100	100
Portfolio#				1	2	3	4	5	1	2	3	4	5
Indicators													
1a	13.56	13.65	13.58	3.88	3.6	13.65	13.51	3.44	6.06	3.51	13.58	12.42	2.42
1b	5.29	6.56	7.35	0.44	0.39	6.56	0.46	0.10	3.51	0.46	7.35	4.82	1.82
1c	2.47	2.61	2.48	0.29	0.27	2.61	0.31	0.00	0.49	0.31	2.48	1.58	0.49
1d	0.52	0.38	0.54	0.05	0.05	0.38	0.06	0.00	0.09	0.06	0.54	0.17	0.08
1e	30.81	39.00	42.76	34.09	20.39	39.00	27.48	21.95	36.84	38.47	42.76	39.42	32.94
2a	1.84	2.74	12.92	0.00	0.00	2.74	0.00	0.00	5.90	5.70	12.92	2.30	1.77
2b	1.61	0.23	1.99	0.00	0.00	0.23	0.00	0.00	1.35	0.86	1.99	1.70	1.35
3a	41.91	37.36	43.79	25.53	30.53	9.29	35.87	7.41	31.85	31.86	11.91	40.01	9.01
3b	24.29	21.02	31.04	12.51	14.03	9.49	15.33	6.33	13.92	15.92	10.41	20.32	8.21
4a	23.17	20.28	31.45	15.18	15.18	5.10	15.38	4.85	21.86	17.02	7.43	22.43	7.08
4b	30.81	39.00	42.76	34.09	20.39	39.00	27.48	21.95	36.84	38.47	42.76	39.42	32.94
5a	3.33	3.96	4.03	3.33	2.19	3.10	2.89	1.72	3.01	2.29	2.98	3.12	2.01

The impact on benthos is reduced by a factor of 2 in case of hard defense measures, i.e., Portfolios 1 and 2, if compared with the unprotected cases (Portfolio 3). The lowest impact on benthos is obtained with Portfolios 4 and 5.

The low values of the EVIs obtained for pinewoods prove that this habitat is resilient to the flooding and erosion conditions in Cesenatico. One determinant factor for the EVI is indicator 1d, the greater the flood duration the greater the EVI (of course the relationship is nonlinear).

Portfolio 2 seems to offer the best protection to the pinewood, whereas the effects increase by about a factor of 2 in case of Portfolios 1 and 5, being however still a factor of 2 less than in the unprotected case.

- Social equity.
  Portfolios 1 and 2 include hard defenses that reduce the impact on society (in terms of damage to CFs) of about 1/2 than in the unprotected corresponding cases. Portfolio 3 that includes insurance and an evacuation plan only reduces the impact on CFs and population of about 1/4 and—as expected—the impact decreases again for Portfolio 5, when hard engineering solutions are combined with the socioeconomic ones.
- Economic efficiency.
  Portfolios 1 and 2 provide an almost equal reduction of the economic impacts, with a higher protection offered by the latter one for more intense storms. Portfolio 3 almost eliminates the economic impact in the area, and this effect is of course reinforced in presence of other mitigations (see Portfolio 5). Indeed, this 100% efficiency of the insurance scheme is due to the fact that each land use is assumed to be totally covered for its possible losses, by paying an insurance premium and by making people outside Cesenatico bearing the whole cost. Therefore a yearly insurance premium is much smaller than the related possible losses, although the former depends on the latter, together with the administrative costs.
  Based on the overall risk assessment indicator, Portfolio 5 offers the greatest reduction of risk in the examined area. This is quite obvious, because a solution based on an integrated multidisciplinary approach is the most suited to excel in a multicriteria comparison. This indication of the optimal portfolio of mitigation strategies can be used as a basis for a future detailed design process (Zanuttigh et al., 2005).
  From the detailed analysis of the impacts, it is evident that hard structures placed in front of Cesenatico are essential for the safety of this area. It is also imperative to boost the implementation of insurance plans and increase risk communication for efficient evacuation plans.

7.6.7 Strategic Management and Policy Implications

Tourism is the market force driving the Emilia Romagna coastal management and it should therefore be preserved if not enhanced. It is recognized that erosion is the most critical threat to the economy, jeopardizing the site recreational value. With this perspective, the management essentially consists of the optimization of beach nourishment (i.e., the only measure that bathing owners, hotel owners, coastal managers, and many citizens consider as the most useful measure) for maintenance or widening of the beach width, keeping high-quality sand, attracting tourists, and improving recreational value. This poses social constraints to designs involving hard structures and innovative measures. In practice, the defenses at the sea should be characterized by low aesthetic (submerged structures only) and environmental impact, so that, for example, new emerged barriers cannot be accepted. The use of other kind of mitigations (e.g., insurance, evacuation), mainly focusing on reduction of coastal flooding risk, or the synergy of such mitigations with existing management has not been considered yet since the main target design objective is widening/maintaining the beach.

The market should drive the attention on the role of business recovery actions, insurance, land use planning, etc. and the combination of mitigation options that so far is poorly known and not practiced.

A decennial coastal plan for this area (Emilia Romagna Littoral) has been recently published addressing the problem of integrated coastal zone management (Preti et al., 2009)). However, climate change and variability are usually not tackled and despite the effort in performing an integrated coastal zone management, legal issues and risk governance show many shortcomings at present.

With respect to the distinction made by Renn (2008), governance is a combination "adversarial" approach and "fiduciary" approach under the following main aspects: (1) stakeholder involvement is considered as essential at least from a theoretical point of view and in the preliminary debate regarding the policy framework changes and/or new design plans; (2) different institutions/universities/public bodies try to emerge in the policy arena by disseminating their role and/or research outcomes; and (3) decision-making at the operative level is confined to a group of few people who usually base the decision on historical data and on their expert judgment, while trying to account for the suggestions of the groups 1 and 2.

The following legal deficiencies for risk management can be identified

- lack of national resources/budget;
- lack of a national clear chain of responsibilities, and consequent need to stress the role of United Nations platforms tackling risk at national scale;
- lack of a national plan to prioritize the area of intervention considering impact-based indicators (following World Bank and similar approaches to identify hot spots);
- lack of coordination between projects, which make them more expensive than would be the case if parties would join forces (for instance, combined dredging and nourishment interventions);

- lack of harmonization of monitoring plans, so that costs of surveys are too high and results are not suited to a large scale integrated planning; and
- lack of prioritization in testing innovative structures, not coordinated at regional scale.

Innovation in risk management is boosted by an increased sensitivity to potential environmental impact of traditional coastal engineering schemes designed to reduce risk from erosion or flooding (Penning-Rowsell et al., 2014). This, coupled with the increased possible threat from the impacts of climate change in increasing sea levels, requires rethinking of measures to reduce risk, and indeed a further emphasis on a philosophy of flood risk management rather than flood defense. Technical issues concerned with risk assessment and risk reduction choices are not central to the process of innovation with regard to risk management, but that institutional culture, traditions, and capabilities are of greater significance. The real capacity for radical changes of institutional management approaches and the low level of public risk perception prevent innovation more than the lack of risk information or a poor understanding of the performance of innovative measures.

7.7

Balancing Flood and Erosion Risk with Landscape Sustainability: Cancun, Mexico

Monique Marie Villatoro, Mireille del Carmen Escudero, Edgar Mendoza, Rodolfo Silva

Instituto de Ingeniería, Universidad Nacional Autónoma de México (UNAM), Coyoacán, DF. México

7.7.1 Description of the Site

Mexico has more than 23,000 km of coastline that, excluding insular nations, makes it one of the countries with the greatest coast-to-area ratios (Ortiz Pérez & de la Lanza Espino, 2006). The Mexican coast, although underdeveloped compared with those of regions such as Europe, plays an important role in the local GDP: 36% of the national GDP is produced by coastal states (CIMARES, 2009) through tourism, fishing, transport, and energy industries. However, 76% of the coast is composed of sandy and shingle beaches, which are among the most dynamic types of coasts and thus extremely vulnerable to both natural processes and human interventions. The increasing economic importance of this coastline and its subsequent development, make it vital to understand how it will respond to changes induced by future anthropogenic and natural pressures. Already, the combination of lack of long-term planning and regulation (through ignorance or disregard for them) coupled with a limited understanding of the nature and magnitude of the changes induced by coastal structures and/or modification of the sediment supply and/or reducing resilience has resulted in severe erosion problems at various points along the Mexican coast.

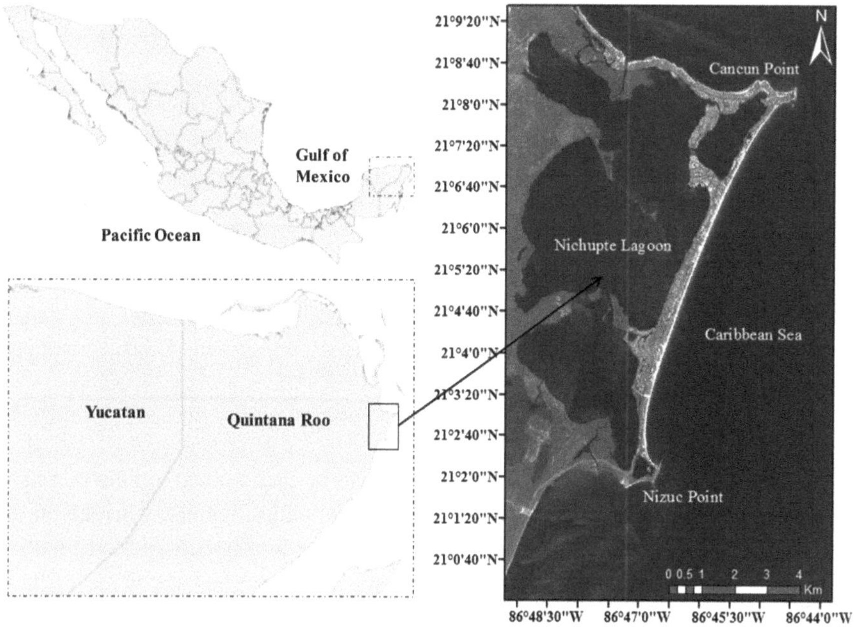

Figure 7.84 Location of the study site, Cancun, Mexico.

Of particular interest for this study is the case of Cancun, a beach resort located on the Mexican Caribbean, on the northeastern coast of the Yucatan Peninsula, in the state of Quintana Roo (see Figure 7.84). Cancun was built on a lagoon-barrier system formed by Nichupte Lagoon, Punta Cancun, Punta Nizuc, and a dynamic barrier island that separates the lagoon from the Caribbean Sea. Its ongoing development started in the 1970s on the northern part of the lagoon-barrier system (Cancun Point), and has now covered the entire length of the barrier island.

Cancun is the largest tourist development in Mexico, with 181 hotels, more than 35,000 hotel rooms, and 380 restaurants (INEGI, 2010; Mendoza, Silva, Enriquez-Ortiz, Mariño-Tapia, & Felix, in press). It attracts 45% of all international visitors to the country, with an average spending of between 2 and 3 billion USD per year (SEDETUR, 2008), providing 35% of Mexico's tourist income. Around 660,000 people live in Cancun, all of whom depend directly or indirectly on tourism. A summary of the most relevant socioeconomic parameters for the adequate estimation of the vulnerability of this site is presented in Table 7.46.

The predominant land uses within the barrier island are urban, characterized by buildings, marinas, and sports facilities mainly serving the tourism industry. The density of development in this area is extremely high (up to 95% of the land is covered) and exposed directly to the action of extreme winds, waves, and storm surges, with slopes varying between 1:0.05 and 1:0.12, and a maximum elevation of 4 m.

TABLE 7.46 Socioeconomic Profile of Site (Information Valid for 2010 (INEGI, 2010))

Information	Potential proxy for socioeconomic vulnerability assessment
Population	661,176 permanent inhabitants (3.1% annual increase, with 54% of the population being from other parts of the country) and 4 million tourists per year
Resources	GDP per capita is 7500 USD (9.6 billion USD total for the state of Quintana Roo) with a 5% unemployment rate
Tourism is the main activity, representing	35% Of total national tourism income 100% of local income (directly or indirectly) 80% of the population is employed in tourism
Demographics	Average age is 25, with 30.1% of the population between 15 and 29 years, 28% younger than 15 years and 4% older than 60 years

The economic exploitation of Cancun seems to be paradoxical, on the one hand its economic value and tourist appeal are due to its natural assets, whereas on the other it is that very natural environment that makes it highly vulnerable to natural and/or artificial perturbations. It is therefore not surprising that the value of these beaches as income generators has decreased with time as well as the average expenditure and number of visitors. For example, Félix Delgado, Mendoza, Silva, Rivera Arriaga, and Palacio (2008) indicated that from 1999 to 2004 a decrease in the economic performance of the tourist development of Cancun was registered and as a consequence a reduction of around 20% in its economic contribution to the state. In addition, from 2008 to 2009 the mean hotel occupation rate fell from 71% to 60%, whereas the room prices at three-star hotels also fell, revealing an increasing demand for cheaper facilities (Balderas, 2009).

7.7.2 Characterization of the Flood System

7.7.2.1 State of the System and Existing Management Strategies

Erosion problems in the region were first noticed after Hurricane Gilbert in 1988, when, for the first time, the unregulated character of the resort development was discussed. Until then, any attempt to manage the coast in Cancun had relied solely on the existing regulatory framework, which is complex and difficult to apply because of the many laws and government levels it entails. It consists mainly of (1) the distribution of rights and responsibilities among the state, municipalities, and citizens, regarding the environment and the territory, as stated in the Constitution; (2) a series of federal laws and regulations dealing with the environment, national resources, national waters, and the 20-m strip of land that constitutes the federal littoral zone, among others; (3) international treaties; and (4) municipal land use plans.

Figure 7.85 Two sections of the barrier island development, looking north (left) and a central section (right), *From CFE (2008).*

Unfortunately, the ambiguous and unintegrated nature of these regulations resulted in indiscriminate development of the barrier island and the lagoon; the majority of the hotels were built over the dunes and seasonal inlets (Figure 7.85); destruction of the dune system severely limited the ability of the beach to recover after storm events, whereas permanent closure of seasonal inlets affected the flushing cycles of the lagoon, particularly during storm events (Table 7.47). For more details, see Félix Delgado et al. (2008) on the evolution of the altered system, and Silva et al. (2006a), for the effect of extreme meteorological events. There is evidence that erosion and overtopping of the dunes and the barrier has occurred periodically since its formation. Moreover, the reduced resilience of the beach-dune system and its inability to recover after storms has caused chronic erosion (160,000 m^3/y) (Dibajnia et al., 2004; Silva et al., 2006a, Diez, Esteban, & Paz, 2009) and coastline recession of up to 35 m after single events, as reported by Martell, Mendoza, Mariño, Silva, and Escalante (2012) and Silva et al. (2012), which has increased the risk for people and infrastructure.

To deal with the erosion, local authorities have implemented a series of measures, such as two emergency beach nourishments in 2006 and 2009 (Guido Aldana, Ramírez Camperos, Godínez Orta, & Cruz León, 2009) and the construction of a terminal groin at Punta Cancun (CFE, 2001, pp120), see Figure 7.86. Nevertheless, the long-term effects of such a structure on the morphology of Cancun beach are not yet known. Although important advances in the knowledge of the sediment transport patterns have been made (Martell et al., 2012; Gonzalez-Leija et al., 2013; Ruiz, Silva, & Posada, 2013), the littoral processes occurring during calm and extreme conditions, and how they balance in the long term, need to be better understood to propose an efficient management strategy. The risk of erosion is considered to be closely linked to the risk of flooding, given that a healthy, wide beach-dune system can be the best protection against flooding. Therefore, although consideration of both is essential, in the case of flooding this will often be done indirectly, through the estimation of coastline recession. It is expected that the findings presented here will provide the government, as well as the local engineering community and decision makers, with valuable information for coastal management.

7.7.2.2 Geomorphological Setting

Cancun is located on a 12-km long, 100–400 m wide barrier island that separates a shallow coastal lagoon (4–7 m deep) from the Caribbean Sea, one of many similar lagoon-barrier systems found along the coast of the Yucatan Peninsula. The lagoon is home to mangroves, several fish species, and migratory birds. The barrier island itself is formed by an active beach-dune system deposited over a series of relict dunes and is backed by mangroves and a series of creeks on the lagoon side.

The geological characteristics of the Yucatan Peninsula play a dominant role in local coastal processes. This region is a karst platform, composed of carbonate and soluble rocks, such as limestone and dolomite. Because of its extreme karst nature, there are no rivers in the central and northern parts of the peninsula, resulting in a lack of supply of terrigenous sediment to the coast. This means that most of the sand found on this coast is of marine origin (fragments of shells and corals) (Ruíz Martínez, 2009; Alcerreca, Silva, & Mendoza, 2013), which, by nature, is limited. The sources of sediment for this barrier-lagoon system have been identified as coral reef sands from the south and oolitic sand banks located north of Cancun. The analysis of bathymetric data shows that the continental shelf in front of the lagoon-barrier system is very narrow, leading to a very steep beach profile (Figure 7.87). This, in turn, causes the approaching waves to break closer to the coast line, and with more energy, than in dissipative coasts. On the other hand, according to an analysis of Cancun paleodunes (Loucks & Ward, 2001), after breaching and before the beach-dune equilibrium profile could be reestablished, the area in front of the eroded dune was filled in with new beach and eolian sands. During this time, vegetation grew on the intradune marine storm layer and a thin caliche layer developed. This shows that, even before urban development took place, the beach-dune system required a relatively long period to rebuild its fair weather profile and cover the eroded surface. Breaching episodes, along with the flushing provided by natural inlets, were crucial to the hydrodynamics of the lagoon and the preservation of the fragile ecosystems it sustained. Unfortunately, land reclamation, construction of marinas, closure of seasonal inlets, and dredging of navigation channels has severely altered the equilibrium of the system.

TABLE 7.47 Chronological Description of Development and Coastal Protection Activities

1970	Stable lagoon-barrier system
1976	Beginning of large-scale construction over the dune system
1988	Largest loss of sand ever recorded (Hurricane Gilbert)
1999	First records of unregulated coastal protection structures
1999–2004	Natural recovery of beach volume
2004–2005	Three category 4 and 5 hurricanes hit the area (Ivan, Emily, and Wilma)
2006	First government-led beach renourishment (Ministry of Tourism and the National Electricity Company)
2007	Placement of geotextile shore-parallel breakwaters; Hurricane Dean rearranges renourishment material into a natural beach profile
2010	Second government-led beach renourishment (Ministry of Tourism, Natural resources and the environment Ministry and the National Electricity Company); construction of terminal groin at Punta Cancun

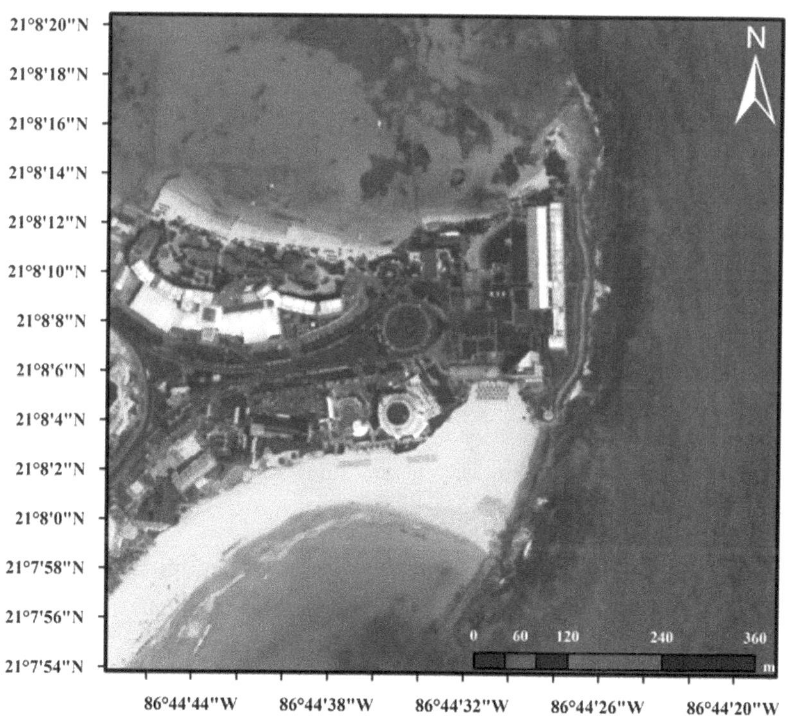

Figure 7.86 Terminal groin constructed in 2009 on the northern end of the barrier island (Cancun Point).

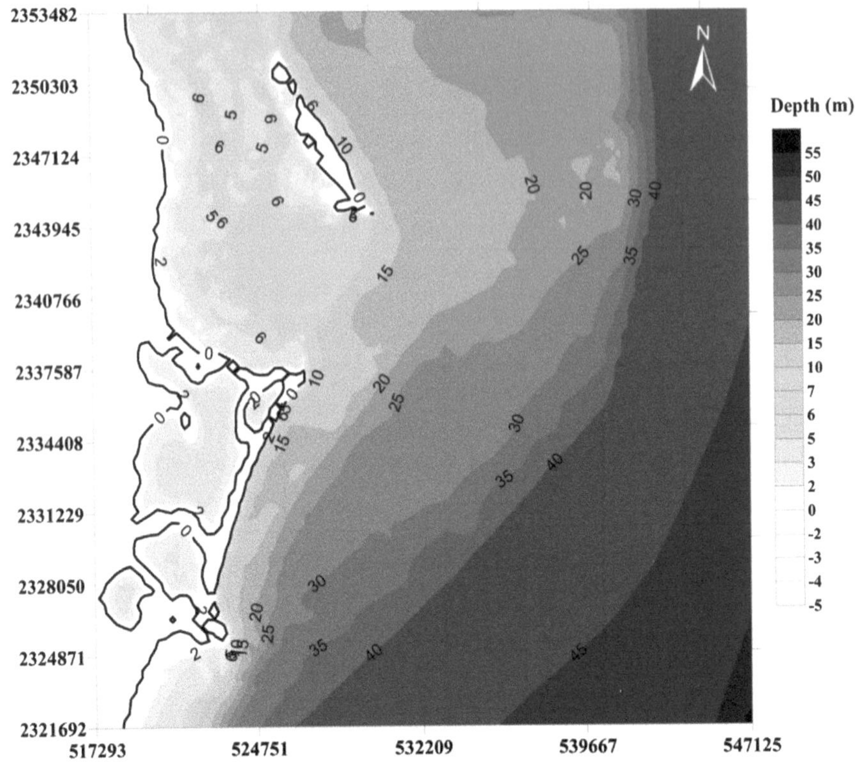

Figure 7.87 Bathymetry of the continental shelf in front of the study site (produced with data from the Ministry of Transport and Infrastructure of Quintana Roo).

7.7.2.3 Meteomarine Climate: Present Conditions and Long-term Trends

Cancun is subject to year-round wind attack from the E and ESE, which is the direction of the longest available fetch in the area. It also receives strong winds from the N and NE (Nortes), during winter; and hurricanes, during summer. The wave climate in the study site is closely related to these wind patterns, as expected. A summary of wind, wave, tide, and current conditions for both prevailing and extreme scenarios is presented in Table 7.48.

Probability density functions (pdf) that best describe the characteristics of the wave climate were obtained from analysis of hindcast data (Silva et al., 2008) to identify future wave climate trends (increase or decrease in significant wave height, storm occurrence and magnitude, etc.). The trend analysis was performed in three phases: one for the totality of the data under the storm threshold condition ($H_t = 2$ m), a second for extreme conditions, and a third corresponding to individual storm parameters identified through the peak over threshold (POT) method (Schlichting & Gertsen, 1999). No statistically significant trends were observed for the first and second part of the analysis (all data and extreme conditions), which is in agreement with the results obtained in other studies (Kunkel et al., 2012; Weinkle, Maue, & Pielke, 2012). A perceived increase during recent years is attributed the occurrence of several hurricanes between 1988 and 2007 and is not an indication of a real long-term trend resulting from climate change. The last part of the analysis, consisting of linear regression of the specific storm parameters, such as number of storms per year (n_{storm}), storm duration (dur_{storm}), maximum wave height (H_{smax}), peak period (T_p) and approach direction (θ), and maximum storm surge (η) shows that nor is there a trend in the duration or magnitude of the storms for the whole period.

Given the lack of a distinguishable trend in wave height, storm duration, or magnitude, the only adjustment required for the generation of future scenarios is the addition of the predicted sea level. Future sea-level values were estimated on the basis of the IPCC (2007) recommendations (Meehl et al., 2007), which state that a global SLR between 0.5 and 0.7 m is expected to occur over the next century. The use of the IPCC recommended global trends is justified as the region is not subjected to any tectonic activity that might alter the expected values and there is a lack of long-term sea-level measurements in the area from which to derive local estimates. The simulated sea-level values for the year 2050 and 2099 were corrected by +0.25 m and +0.5 m, respectively, to account for potential SLR.

TABLE 7.48 Prevailing and Extreme Meteomarine Conditions

		Extreme	
	Calm (Prevailing Conditions)	Nortes	Hurricanes
Season	Year around	Winter	Summer
Wind direction	E and SE (90–150°)	N and NE	E and SE
Waves	Directions E and ESE (54% of waves) H_s = 1–1.5 m (90% of waves) T_p = 4–9 s (63% of waves)	Directions N and NE H_s up to 5 m T_p up to 10 s	Directions E and SE H_s up to 12 m T_p up to 12 s
Tides	Microtidal (0.2 m neap and 0.6 m spring)	Microtidal (0.2 m neap and 0.6 m spring)	Storm surge of up to 2 m
Littoral currents	NNE 0.4–0.7 m/s (CFE, 2001, pp120)	SSW	NNE
Tidal currents	Semidiurnal and parallel to coast 0.15 m/s (Kjerfve, 1994) at 3 m deep		

7.7.3 Application of the SPRC Model

7.7.3.1 Sources of Risk

The sources of risk at Cancun are related to wind-driven waves, currents, and water levels (prevailing and extreme). For the purpose of this analysis, the sources are divided into those occurring during prevailing or "calm" conditions and those of extreme events, and they are incorporated into the SPRC model as scenarios.

The scenarios constructed were of two types:

1. Monte Carlo–simulated yearly conditions with the added predicted SLR for 2020, 2050, and 2099. These scenarios consist of sequences of hourly sea states (waves and sea level) simulated from the probability distribution functions (PDF) that best describe the behavior of the parameters involved. In this case, the PDFs were estimated separately for the hindcast values of H_s, T_p, and θ, representative of calm conditions, and for the maximum values corresponding to each storm identified, such as dur_{storm}, H_{smax}, T_p, θ, and η. The simulated conditions for both calm and storms were then joined to construct a year's worth of data. The observed trends in the frequency and magnitude of storms (Figure 7.88) were incorporated alongside the predicted SLR into the resulting scenarios. A summary of the main parameters for each scenario is presented in Table 7.49. The statistics in the top part of the table correspond to the totality of the hourly conditions, whereas the bottom part corresponds only to the storms.

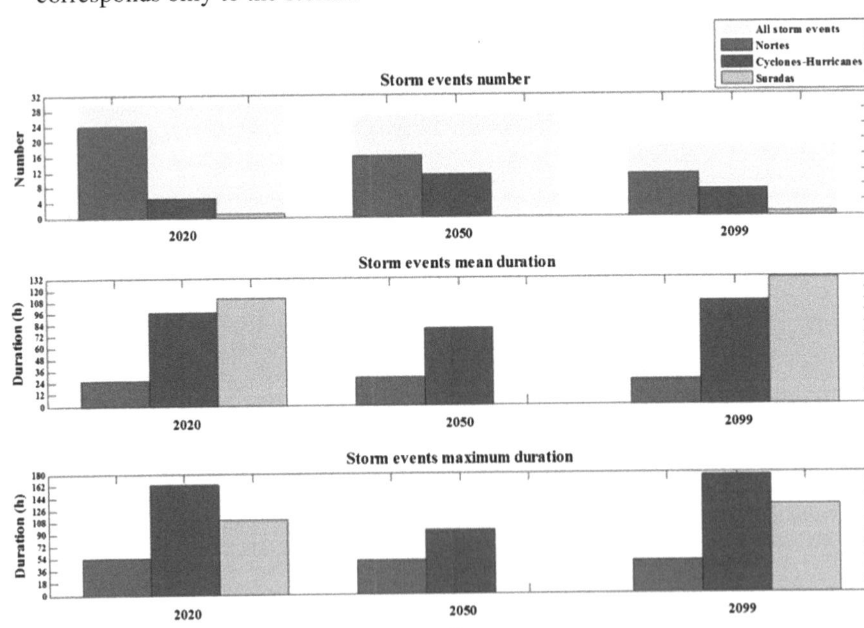

Figure 7.88 Storm trends for scenarios.

TABLE 7.49 Summary of the Main Characteristics of the Scenarios Constructed

	2020		2050		2099	
	Mean	Max	Mean	Max	Mean	Max
H_s	1.3 m	6 m	1.3 m	4.2 m	1.4 m	6.1 m
T_p	6.7 s	12.7 s	6.6 s	12.8 s	6.7 s	12.2 s
η	0.03 m	0.55 m	0.27 m	0.48 m	0.54 m	1.12 m
No. storms	30 Nortes	Hurricanes	28 Nortes	Hurricanes	19 Nortes	Hurricanes
dur_{storm} (max)	55 h	163 h	50 h	95 h	48 h	174 h
H_{smax} (max)	2.7 m	7.3 m	2.6 m	4 m	2.6 m	7.9 m
T_p (max)	8.4 s	9.9 s	8.3 s	9.1 s	8.3 s	10.1 s
α (mean)	55°N	118.8° N	58.8°N	105.9°N	53.8°N	143.3°N

2. The maximum expected storm for return periods of 1/20, 1/50, and 1/99, also considering the expected SLR. In this case, the following conditions were used: H_s of 10.05 m and a duration of 2 h for the 2020 storm; 12.28 m and 7 h for 2050; and 14.08 m and 7 h for 2099. Wave height values were extrapolated from the best fit PDF of the hindcast wave data, whereas storm duration was estimated from the statistical relationship between wave height and storm duration.

7.7.3.2 Pathways of Risk

In this case, the identified pathway of risk is the beach-dune system, which stands between the sources (waves, storm surge, etc.) and the receptors of risk (infrastructure, population, ecosystems). The evaluation of the potential vulnerability of the pathway was based on a conceptual model of sediment transport and overall morphodynamics proposed and validated later in this document through numerical modeling. The conceptual model allows the qualitative identification of the areas more prone to erosion, and thus more vulnerable, from the characteristics of the littoral currents and the presence of features such as vortexes, which induce seaward transport. It was produced for both calm and extreme conditions (see Figures 7.89 and 7.90) because, to evaluate the vulnerability effectively, it is important to understand how these might balance out in the long term.

In calm conditions (characterized by waves between 1 and 1.5 m high, with periods of 4–9 s and an approach direction of approximately 120°), the wave-induced littoral drift induces a net transport of sand in the NNE direction, which supplies the barrier with sands from the coral reefs located south of the system (see Figure 7.89). Some of this sediment is deposited in the beach-dune system and some is lost offshore and to sand banks located north of the system. This behavior has been reported by several studies (CFE, 2001, pp120; Diez et al., 2009). However, recently, Gonzalez-Leija et al. (2013) highlighted the presence of a marked divergence in sediment transport near the southern end of the sand barrier. This divergence induces a constant southerly longshore movement near Punta Nizuc, while also generating strong cross-shore transport at the point of this divergence. According to Gonzalez-Leija et al. (2013), during "Nortes," when there are particularly strong waves (>1 m) from the ENE, this divergence is pushed northward, to the middle section of barrier island. When the divergence occurs along the middle section of the barrier island, it generates northward transport along the northern half of the barrier and southern transport along the southern half. Another smaller and weaker divergence occurs during these conditions, near Punta Cancun, at the northern end of the barrier. It is possible that during Nortes, sediment is supplied from the sand banks north of the barrier island and transported southward toward the beach at some sections and offshore. Even though the magnitude of these events is greater than the prevailing conditions, their short duration is not enough to generate a yearly net southward transport, although it is thought that this alternating pattern might have an effect on the recovery of the beach-dune system.

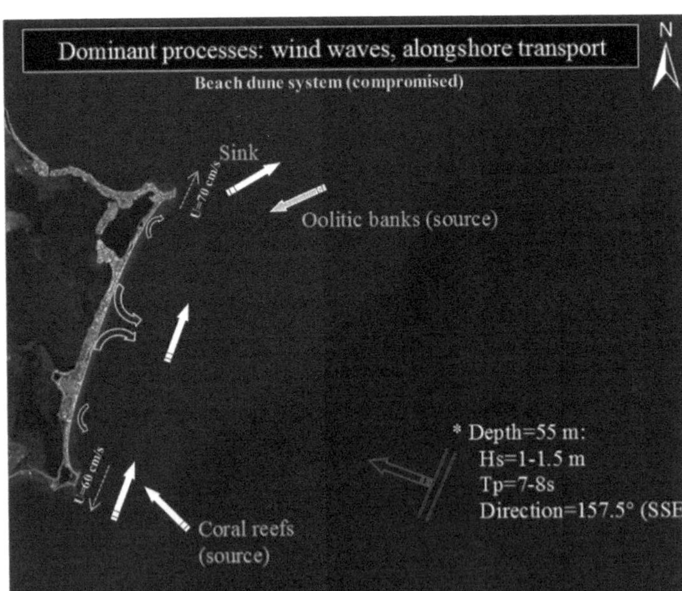

Figure 7.89 Conceptual model of the sediment budget for calm conditions.

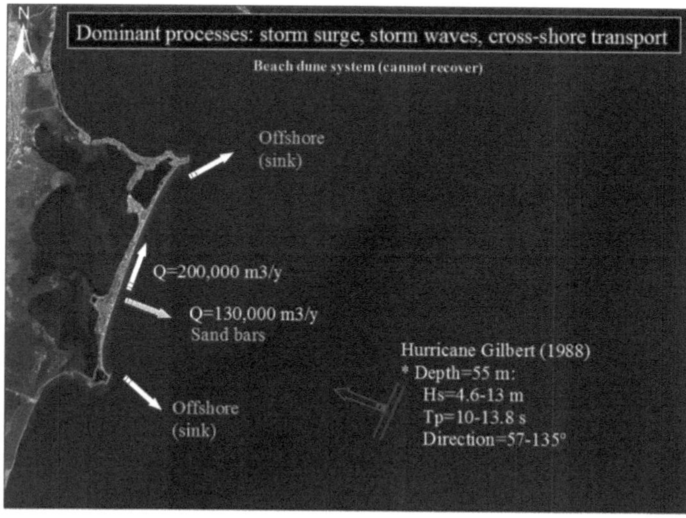

Figure 7.90 Conceptual model of the sediment budget during extreme events.

There is no agreement in results from previous studies on the net transport rates and resulting morphological changes during calm conditions. Most of the studies have used empirical long-shore transport formulations or volumetric methods for their estimations, with the reported values varying widely. A study carried out by the Engineering Institute from the National University of Mexico between 1989 and 1990 (Instituto de Ingenieria, 1989, pp74; Instituto de Ingenieria, 1990, pp56) estimated a net long-shore transport rate of 7000 m^3/y northward and 610,000 m^3/y southward, and cross-shore transport of one–1.7 million m^3/y (Diez et al., 2009). Later on, between 2000 and 2001 (no extreme events occurred during that period), the National Electricity Company (CFE, 2001, pp120) estimated an average annual sediment loss from the beach/dune system of between 110,000 and 160,000 m^3/y, with a net northward transport. No estimates exist on the magnitude of the potential supply of sand to the barrier system, and how much of this is lost during calm conditions.

During extreme events and because of the higher sea levels caused by storm surge, higher waves break even closer to the coast line, setting a large amount of sediment in suspension. This is then carried offshore by the very strong wave-induced currents. The configuration of the Caribbean Sea and the effect of the Coriolis force mean that the majority of hurricanes that cross the coast of Quintana Roo approach from the ESE, inducing waves and resultant transport in the northward direction (Figure 7.90). Few estimates on sand losses from extreme events have been published; Hurricane Gilbert (1988) is reported to have induced transport rates of 200,000 m^3 northward and 130,000 m^3 seaward (Diez et al., 2009), while a record loss of 7 million m^3 is quoted by Mendoza et al. (in press) for Hurricane Wilma (2005) and 1.8 mm^3 is reported by Martell et al. (2012) for Hurricane Dean.

7.7.3.3 Receptors and the Consequences of Risk

The receptors of risk are the residential and tourist infrastructure, the population, and the ecosystems existing on the island. The vulnerability or degree of exposure of the identified receptors of risk was estimated on the basis of the physical, socioeconomic, ecological, and cultural characteristics of the study site, through the following equation:

$$Vulnerability = \frac{(\mathrm{aS} + \mathrm{bE} + \mathrm{cC})}{beach\_response\_capacity} \qquad (17)$$

where *S*, *E*, and *C* refer to the socioeconomic, ecological, and cultural components of the vulnerability, respectively; *beach_response_capacity* refers to the resilience of the system; and *a*, *b*, *c* are weighting parameters. The values of the weighting parameters ($a = 3$; $b = 2.5$; $c = 2$) were established according to the relative importance of each component of the vulnerability, considering the interests of the local population and visitors, and the contribution of each component to the relevance of Cancun as the foremost resort in the country. The resulting vulnerability index, according to Eqn 17, can vary from 0 to 7.5.

The variables used to estimate the different components of the vulnerability (socioeconomic, ecological, and cultural) and the *beach response capacity* were organized from less vulnerable to more vulnerable (in a scale of 0–1), and were determined on the basis of the characteristics of the area (the variables that best described the area were chosen) and the information available (Tables 7.50 and 7.51).

The socioeconomic component of the vulnerability was evaluated as

$$S = 0.35^{*}land\ values + 0.1^{*}economic\ units + 0.2^{*}employees + 0.35^{*}occupied\ rooms,$$

where the weights for each variable were obtained by normalizing the values to 1, e.g., the number of employees in a polygon was divided by the highest number of employees in any polygon, thus obtaining a range of values for each variable from 0 to 1. In the case of the ecological component, the values assigned to each variable were 1 for beach, mangrove, and native vegetation, respectively, and 0.3 for golf courses and introduced vegetation. Beach, mangrove, and native vegetation were assigned the same weight because they all fulfill an equally important role in coastal protection. Finally, the weights of the variables corresponding to the cultural component were assigned according to the values assigned by the National Institute of Anthropology (INAH, 2013) for their cultural importance; 1 for archaeological zones, 0.8 for Mayan ruins, and 0.2 for an archaeological museum.

TABLE 7.50 Variables Used to Define the Socioeconomic, Ecological, and Cultural Components of the Vulnerability

Socioeconomic Component	Ecological Component	Cultural Component
1. Land And property values (Burad, A. personal communication, July 2013)	1. Golf courses	1. Archaeological Museum
2. Number of economic units per polygon (DENUE, 2010)	2. Introduced vegetation	2. Mayan ruins
3. Number of employees per polygon (DENUE, 2010)	3. Native vegetation	3. Archaeological zone
4. Importance of tourism (number of occupied hotel rooms per polygon)	4. Mangroves	
	5. Beach (no vegetation)	

TABLE 7.51 Variables Used to Define the Beach Response Capacity and their Categories

State of the Dune System	Building Density	Building Orientation
1. Unaffected by infrastructure	1. High	1. Favorable
2. Infrastructure behind dune	2. Medium	2. Unfavorable
3. Development in progress over dune	3. Low	3. Not applicable
4. Infrastructure over dune	4. Not applicable	

The resilience, or beach response capacity, was estimated quantitatively as a function of three variables that are directly related to the degree of anthropogenic influence: state of the dune system, building density, and building orientation (Table 7.51). The value resulting from *State of dune system* + *Building density* + *Building orientation* ranges from 1 to 4. *State of the dune system* describes the degree of degradation of the dunes as function of the presence and position of infrastructure in reference to the dune system, so that a system unaffected by infrastructure is assigned a value of 1, infrastructure behind the dunes is given a value of 0.6, 0.4 corresponds to a development being built over the dunes, and a value of 0 to infrastructure built directly over the dunes. *Building density* considers the level of pressure exerted by coastal infrastructure. It is divided into high (0), medium (0.3), low (0.6), and not applicable, when the area evaluated is not on the beach front. Last, *building orientation* refers to the position and design of buildings along the beach front, where a favorable orientation (0.6) would be one that makes the building less vulnerable to the hazards, as opposed to an unfavorable orientation (0), and not applicable (1) that is used for areas that are not along the beach front.

The resultant annual vulnerability index (ranging from 0 to 7.5, Eqn (17)) is shown in Figure 7.91. As expected, the most vulnerable areas are those immediately next to the water front, together with the most developed parts of the barrier island. The exposure was found to vary seasonally, depending on the number of tourists and the resulting economic activity. The most vulnerable periods, which are winter and summer (high tourist seasons), coincide with the seasons of Nortes and hurricanes, respectively. This increases the risk of damage due to erosion and/or flooding for both population and infrastructure.

7.7.3.4 Pathway Modeling

To validate the proposed conceptual morphodynamic model for the evolution of the beach/dune system (Pathway), numerical modeling was carried out. The model selected is a 1D model, adapted from the method proposed by Miller and Dean (2004), in which the shoreline at a given time is estimated from representative values of significant wave height and sea level. The equilibrium shoreline change, Δyeq, is based on the equilibrium beach profile theory and a Bruun-type conservation of volume argument (Miller & Dean, 2004).

$$\Delta yeq = -w(t)*\left(\frac{0.106.H_s(t)+\eta}{Bh+2.H_s(t)}\right); w*(t) = \left(\frac{db}{ADean}\right)^{1.5} \tag{18}$$

where H_s is the significant wave height at the breaking point, η is the sea level, Bh is the berm height, db is the breaking depth, and $ADean$ is Dean's parameter (Dean, 1987).

The calibration of the accretion/erosion rate parameters k_{ac}/k_{er}=[$t-1$] was performed through the comparison of results from the XBeach model and measured beach profile data (17 profiles). After the parameters for XBeach were calibrated, this model was used to calibrate Miller and Dean's equation. To do this, a series of stationary sea states modelled with XBeach were used to:

- estimate hourly values of *yeq*, for different levels of breaking wave height and sea level (storm surge);
- adjust the *k* coefficient (accretion or erosion), so that the normalized mean squared error between the shoreline position estimated with the equation and the shoreline position modeled with XBeach (once calibrated) is equal to 0;
- obtain the function $k = f(yeq)$;
- estimate the coefficients for a set of wave and sea-level conditions.

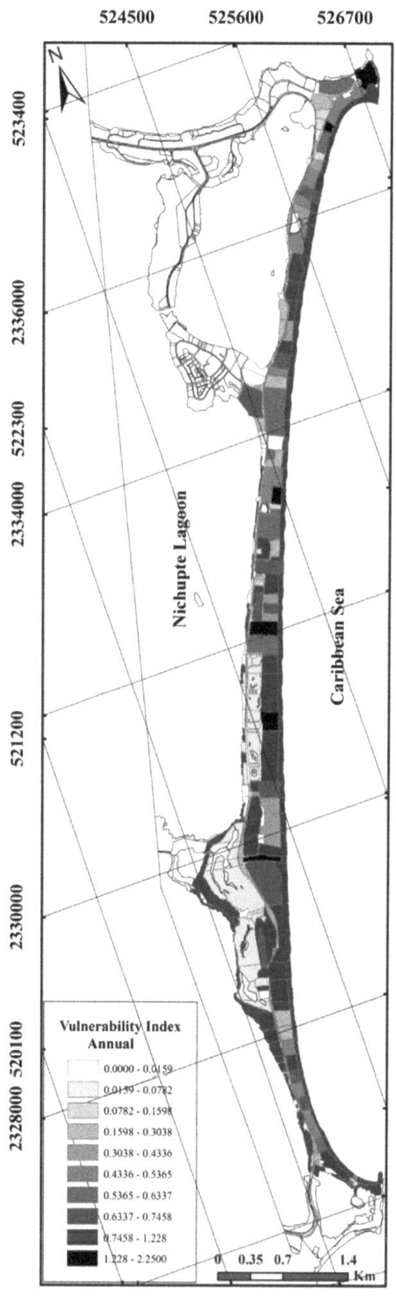

Figure 7.91 Annual vulnerability index estimated from the physical and socioeconomic characteristics of the study site (ranges from 0 to 7.5).

The inputs for the calibration process of Miller and Dean's equation are:

- time series of sea level and significant wave height at breaking;
- time points of the modelled (XBeach) shoreline position associated with a specific significant wave height and sea level;
- the equilibrium/initial shoreline position, *yeq/y0* (m);
- the mean grain size, D_{50} (m);
- the berm height, *Bh* (m);
- the time at which the shoreline position is estimated, *t* (h).

7.7.3.5 Implementation of the Model

Future beach evolution was evaluated for three scenarios, corresponding to 2020, 2050, and 2099 (see Sources of Risk), along 17 profiles distributed down the entire frontage (Figure 7.92). In addition, the maximum expected storm for each year was also modeled.

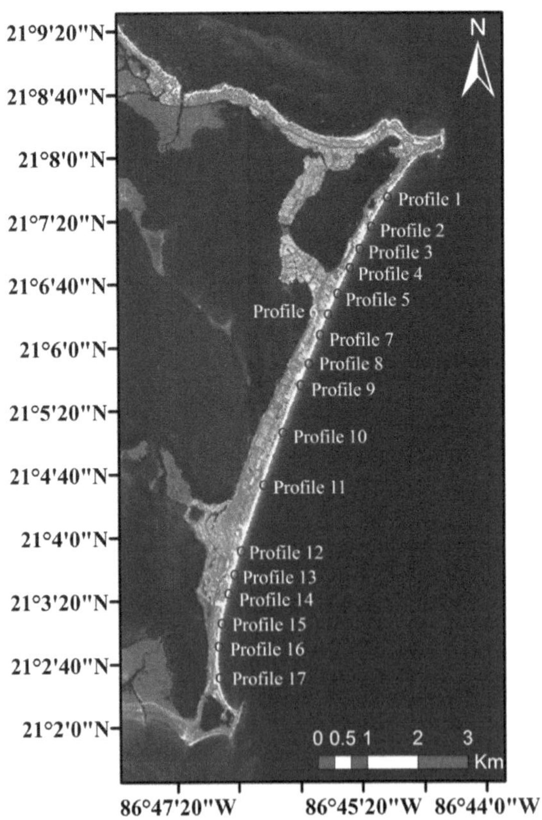

Figure 7.92 Location of the profiles where Miller and Dean's equation was applied.

TABLE 7.52 Total Shoreline Retreat (m) Per Profile for Each Year Modeled

	Cross-Shore Profile Number																
	1	2	3	4	5	6	7	8	9	10	11	12	13	14	15	16	17
2020 $\Delta\eta = 0$ m	1.9	2.0	2.0	2.7	2.8	3.2	3.7	4.3	9.8	5.2	5.3	3.0	3.4	2.9	2.7	3.1	3.2
2050 $\Delta\eta = 0$ m	1.8	1.8	2.3	3.6	4.2	4.8	4.6	5.4	6.7	7.3	7.7	3.8	3.9	3.3	3.3	4.2	4.8
2050 $\Delta\eta = 0.25$ m	3.0	6.3	4.7	7.7	6.9	8.6	5.9	8.5	8.8	7.6	11.6	6.7	6.9	5.6	5.3	7.9	7.5
2099 $\Delta\eta = 0$ m	1.9	1.8	1.9	2.0	2.1	2.2	4.1	4.4	14.1	5.0	4.5	2.8	4.5	2.9	2.0	2.1	2.7
2099 $\Delta\eta = 0.5$ m	12.9	10.0	10.8	8.8	8.2	9.1	8.3	9.5	9.8	8.9	10.6	8.9	9.4	10.1	7.2	9.1	8.6

Only one Monte Carlo simulation (series of hourly data simulated for 1 year) was performed for each yearly scenario, and therefore no probability of occurrence can be associated with these conditions. The objective of modeling these yearly scenarios is to provide insight into the effect of the predicted SLR as well as the seasonal variations throughout the year. To achieve the latter, the analysis for each year was performed in different steps to better understand the behavior of the beach in response to the identified forcing, such as northerly storms, calm conditions, or hurricanes. The procedure consisted of running each wave condition, including sea level (from the beginning of December), and using the morphological output of each run as the initial position of the coast for the following condition. This allows the identification of not only erosive and accretive patterns, but also how they balance on a yearly basis.

The resultant shoreline retreat modeled for the three identified years, for each profile (Figure 7.92), is shown in Table 7.52. Each year considered the 2010 coastline as the initial position of the coast. Given that there were no specific storm trends, the main difference between the 3 years evaluated is the added sea level (from IPPC estimates), which was +0, +0.25, and +0.50 m, for 2020, 2050, and 2099, respectively. All three scenarios caused important erosion along the frontage by the end of the period considered, with identifiable areas of stability and of acute erosion (Figure 7.93). The results are in agreement with recent studies (e.g., Gonzalez-Leija et al., 2013), which have also identified that the most severe erosion occurs in the central part of the barrier island.

The seasonal analysis for each year has provided a better understanding of the changing hydrodynamic patterns along the frontage that, in turn, result in morphological conditions that change throughout the year. It was found that northerly events (December–March) induce a stable trend in the northern section of the beach, erosion in the central section, and deposition in the south. After the winter, there is a period characterized by calm conditions between April and

June. This often allows the system to recover or, at least, to remain stable. The model was not able to reproduce beach recovery at every point for every year. This could be a limitation of the model, but even with some periods of recovery, the beach-dune system is so highly damaged that forecasting a long-term erosion process seems appropriate. Finally, the period from July to October is often marked by a series of tropical storms and hurricane events, which can induce severe erosion. The trends reported herein for all seasons were similar for each scenario, but with a greater tendency to erode with an increased sea level.

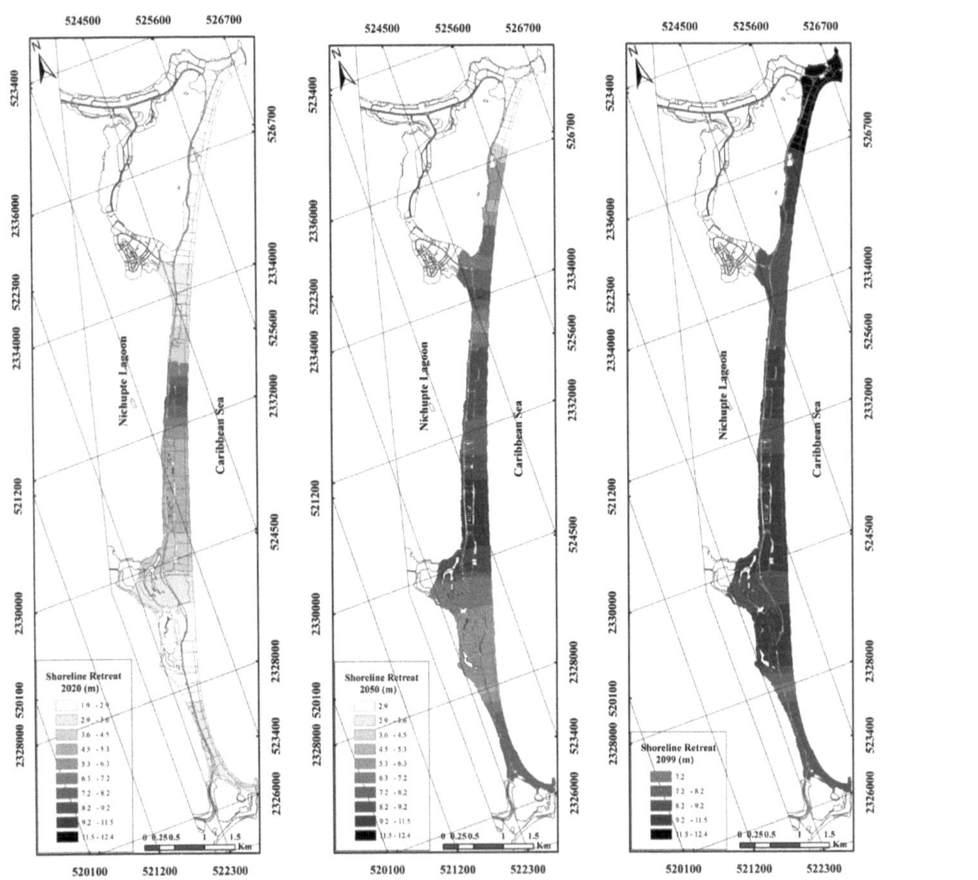

Figure 7.93 Shoreline retreat (m) for 2020 (left), 2050 (center) and 2099 (right).

For the simulation of the probability of erosion due to the maximum expected storm for return periods of 1/20, 1/50, and 1/99, the following conditions were used: H_S of 10.05 m and a duration of 2 h for the 2020 storm; 12.28 m and 7 h, for 2050; and 14.08 m and 7 h, for 2099. The results show that although severe erosion was simulated along most of the frontage for all three yearly scenarios (Figure 7.93), it is not the number nor the magnitude of storms that determines the extent of the potential damage, but other factors, such as the duration of the storms and even more so, the sea level. According to the results of the maximum expected storm simulations, the shoreline retreat once the ≈ 100 years storm has occurred is similar to that observed after the 2005 hurricane season that had storms with that same return period (erosion of up to 30 m was simulated in some areas for the 2099 storm). This means that the effect of the estimated SLR is not as relevant as the intensity and duration of the extreme events. Nevertheless, it has to be said that an increase in the sea level may reduce the return period of the storms, allowing higher waves to reach further inland.

7.7.3.6 Analysis of Impacts

High risk of erosion and flooding is not a condition that any important tourist resort, such as Cancun, can afford to ignore. Erosion in itself is a serious problem, but when the impacts of losing the dry beach area are analyzed, the implications of the situation become even more apparent.

One of the consequences of the ongoing beach erosion problem in Cancun is the increased risk of flooding during storms, which results in more severe and more frequent flooding of the urban areas. Flooding of urban areas causes great social problems, such as a halt to normal activities, from commerce and tourism, to schools and hospitals (as observed during Wilma, 2005). These, although short term, can cause pressure on the local residents who cannot always recover afterward, permanently affecting the social fabric of the city. This process is already being observed in Cancun, making it clear that what one might have considered to have mostly an economic impact, can also have a huge impact on society, outweighing by far the damage that local infrastructure or ecosystems may have suffered. In addition, over the longer term the reduction in the quality of the beaches leads to a reduction in the appeal of the resort and the income this generates; fewer tourists, lower room rates, a reduction in the amount and expenditure of tourists, and so on. Ecologically, the effects of future flooding and erosion are also noticeable, particularly for future scenarios, and it can be expected that rising sea levels will put further pressure on the already delicate balance of the lagoon ecosystem. However, the self-adaptation potential of these areas is limited because of the unregulated development of the area, which has destroyed the majority of the dunes and mangroves as well as some land reclamation within the lagoon. If the system is not allowed to adapt after storm conditions (by opening new inlets or migration of the sand barrier, for example) because of the restrictions of the existing infrastructure, it is likely to disappear, further reducing the tourist value of the area.

7.7.3.7 Key Source Drivers of Future Risk Conditions

To make recommendations for risk management and mitigation, it is useful to consider the relative importance of the different drivers on risk levels.

In the case of Cancun, the hazard probability for a 20, 50, and 99 year storm was multiplied by the annual vulnerability. The latter was assumed to be constant in time, because the layout and density of the development is not expected to change in the future, given that 95% of the land has already been developed. Furthermore, the identified seasonal changes in vulnerability are negligible. As expected, given the higher hazard probability, the risk associated with erosion and flooding is greater for a 20 year storm scenario, than for a 99 year storm (Figure 7.94).

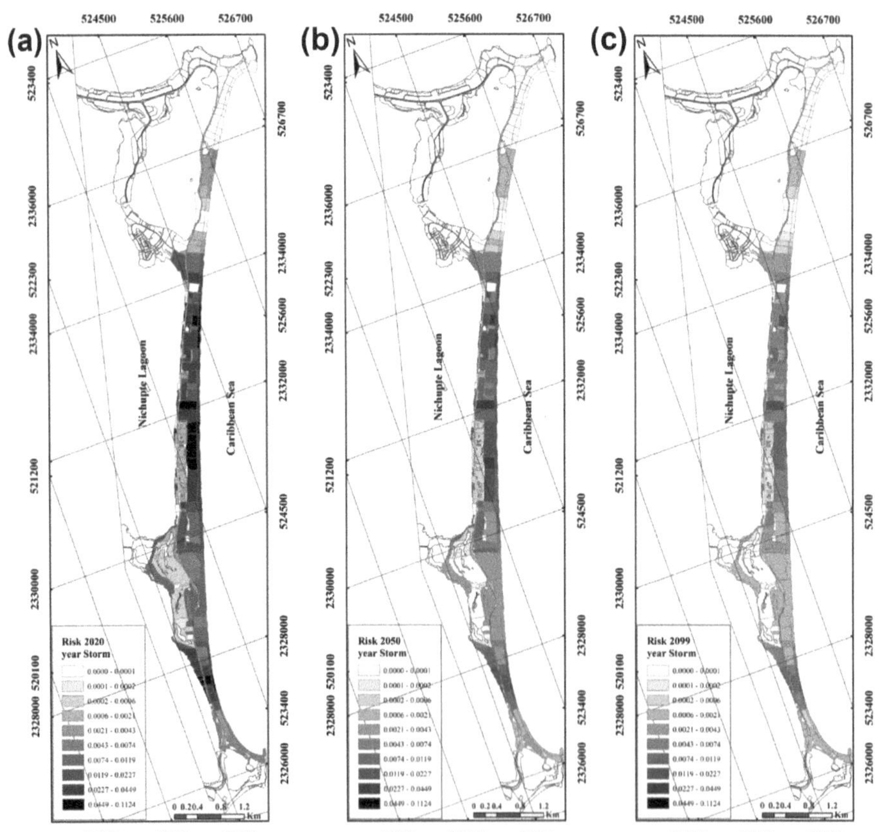

Figure 7.94 Modeled risk (as function of hazard probability and vulnerability) for (a) a 20 year storm, (b) a 50 year storm, and (c) a 99 year storm.

From the overall analysis of risk (as a function of both hazard probability and vulnerability), it can be concluded that the key source drivers of future risk conditions in Cancun are a rise in sea level from climate change, an increase in human activities that further compromise the resilience of the beach/dune ecosystem, and an increase in the density and exposure of the population. Although no indication of an increase in the magnitude and frequency of storms could be determined for future scenarios from the analysis of hindcast data, this possibility should not be excluded entirely from any risk management strategy proposed. A key driver to be considered in future scenarios though is the duration of storms, which was found to be a more important determinant to the amount of damage caused by a storm, than its magnitude.

7.7.4 Risk Mitigation Measures

The integrated risk analysis performed offers not only a better understanding of the present state of the system and the main processes driving coastal erosion and long-term morphological change, but also allows appropriate mitigation options to be identified. Because the main drivers of flood risk in Cancun are factors beyond the control of local managers, emphasis should be put not in the prediction of future trends in storminess, but in how to improve the resilience of the coastal communities. These measures have to be based on governance, in particular on land use planning, construction codes (focused on resilience to flooding), and ecological guidelines, complemented with appropriate response measures in terms of emergency services and civil protection strategies. These considerations are the foundation of the mitigation options proposed herein.

To define thresholds of acceptable risk for future scenarios, one must consider social variables that may be affected with each level of risk, such as jobs and housing, alongside the physical variables. Essentially, for Cancun, the quality of the beach is the key to its economic viability. The level of acceptable risk for the beach and infrastructure must therefore include other ecological and social variables on which the preservation of the infrastructure hinges. That is, the analysis of the level of acceptable risk should include the consideration of the consequences of allowing that level of risk. From this perspective, and taking into account previous disasters in Cancun, the level of risk permitted is very low. This means that the "no action" option is not viable and cannot be considered.

Possible mitigation options are:

1. *Retreat or removal of facilities where possible.* The majority of the infrastructure built on the barrier island was placed directly over the dunes. This has weakened the system and limited its ability to adapt to extreme events,

resulting in accelerated beach erosion. Removing or retreating the infrastructure would allow for the regeneration of the dune system and eliminate the reflective effect of the structures. This would help to stabilize the beach-dune system; however, it would also have some important negative economic consequences.

2. *Construction of flood-resistant buildings (modification where possible).* This strategy focuses on the adaptation and resilience of the development. It accepts that frequent flooding will occur and deals with this instead of trying to stop it. This is achieved by constructing buildings that can cope with flooding for extended periods, with the least damage possible, while still preserving some level of operation and assuring the well-being of both employees and tourists.
3. *Periodical beach renourishments.* This alternative would allow the beach to retain its width and would reduce the consequences of erosion considerably. Success would depend on the availability of nearby offshore sand banks, with enough material of similar characteristics to the sand in Cancun. Furthermore, given the frequency of extreme events, renourishment activities must be flexible enough to allow for emergency renourishments. The downside to this alternative is the very high economic cost and well as the visual impact during the activities (that could be offset by performing the nourishment during off-peak tourist periods).
4. *Coastal dune restoration/reconstruction/construction of a new dune.* Even though there is not enough space left to undertake dune regeneration, ideally, the system would benefit greatly from having dunes to dissipate wave energy and serve as a reservoir of sand for the beach after storms. The environmental services provided by the dunes are a very efficient risk mitigation strategy. This option should be complemented with regulations regarding the use of the dunes (access to the beach, recreational activities, etc.) to secure their preservation.
5. *Improving the water quality of the lagoon system.* Improving water quality and circulation in the lagoon would help reduce the pressure posed by recreational activities on the beach while allowing for better flushing between the lagoon and the sea (breaching).
6. *Hard engineered structures.* The purpose of the implementation of hard engineered structures, either submerged or placed strategically at erosion hot spots, is to hold the line and maintain the beach width. This option must be evaluated carefully because poorly designed structures can further aggravate the erosion problem. However, when properly planned and implemented, this can be one of the most efficient alternatives to reduce the vulnerability of the system.

7.7.5 Strategic Management

7.7.5.1 Planning the Best Combination of Defense Strategies

The individual actions listed in Section 7.7.4 (A to F), when combined, can be used to generate a risk-management strategy that can then be evaluated on its applicability to the particular site in question (see Table 7.53 for an analysis of these strategies).

The analysis of the potential effects of combining the proposed mitigation options allows the identification of not only which alternatives would not be effective on their own, but also which are best implemented in conjunction with others to improve their efficiency. It should be noted that this analysis is derived from the authors' judgment, and from it, it can be established that the most appropriate strategy to mitigate the risk of erosion and flooding in Cancun is A + D + F (priority), followed by B and E as complementary measures. That is, to remove or retreat any infrastructure that is currently too close to the water front, to reconstruct the dune system along most of the frontage, and to complement these measures with a submerged breakwater, which would dissipate wave energy and allow the system to regain its natural balance more efficiently. This strategy would significantly reduce the risk of erosion and flooding, by protecting both the beach-dune system and the infrastructure behind it. It would also be less costly than carrying out periodic beach renourishments. This strategy could be implemented if the regulations set in the new municipal land use and ecological guidelines were retroactive. That is, if the legislation regarding construction over dunes (now forbidden) could be applied to developments predating these guidelines. Unfortunately, Mexican law rarely allows this and the application of this strategy would be contingent on a very efficient communication strategy that positively involves stake holders. As complementary measures, and to assure the quality of the resort in the long term, other ecological issues should be addressed, such as the water quality of the lagoon ecosystem and the resilience of the community as a whole.

7.7.5.2 Policy Implications

Coastal management strategies are intrinsically linked to laws and regulations, from land use planning to ecological directives, which define and delimit the scope of any erosion management action. Thus, knowledge of existing regulations is the key to successful planning and implementation of a coastal management strategy in any country or region. On the other hand, the lack of clear rules, as is often the case in developing countries such as Mexico, can make the work of coastal planners and those in charge of dealing with coastal erosion problems much more difficult. Often the problem is not a lack of laws, but that these have been created independently from each other, without considering their joint application, sometimes even resulting in incompatibility or contradictions.

TABLE 7.53 Analysis of Potential Risk Mitigation Strategies. The Actions Labeled in the First Column Refer to the List Given in Section 7.7.4

Actions	Strategy
A + A	*Retreat or removal of facilities along the entire frontage:* One key issue found to have aggravated the natural erosive processes along the barrier island is the construction of infrastructure directly over the frontal dunes. Therefore, to allow the beach-dune system to regain its natural balance it is only logical to propose to set back the first line of buildings. This strategy, although one of the best in terms of addressing the causes of the problem, would not be feasible in terms of its economic costs. Besides, it is possible that given the state of the beach-dune ecosystem, it would not be able to recover through such set back alone.
A + B	*Retreat or removal of facilities where possible and construction/modification of flood resistant buildings:* This strategy is a variation on the previous one, in that it allows the flexibility to establish a comparative cost-benefit analysis between completely retreating and adapting the infrastructure to cope with frequent flooding, for individual cases. It still addresses, in part, the main cause of the problem, which is a combination of exposure and limiting the natural sedimentary balance between the beach and the dunes.
A + C	*Retreat or removal of facilities where possible and periodic beach renourishments:* The benefits of this strategy are twofold: on the one hand the system is allowed to recover its resilience while protecting the infrastructure; on the other, the process of regaining an equilibrium state can be aided by the availability of sand, provided by the renourishment. The problem, in this case, is the limited availability of sand banks from which to obtain the sand for renourishment. As established previously, the only source of sand in the area is biogenic material from broken shells and detritus, and the chemical precipitation of calcareous sand (from the lack of rivers), therefore, not only are offshore sand banks extremely limited, but extracting sand from them carries innumerable ecological consequences, in addition to altering the hydrodynamics in the area. Also, it has been demonstrated that offshore sand banks can themselves provide shelter from wave action on coastal areas.
A + D	*Retreat or removal of facilities where possible and dune restoration:* Dune restoration can only be undertaken if there is enough space between the infrastructure and the beach, and retreating or removing infrastructure that is currently over the dunes would allow this. These two options are thus complementary and would constitute a very efficient risk mitigation strategy due to the environmental services provided by the dune, as explained previously.
A + E	*Retreat or removal of facilities where possible and improving the water quality of the lagoon system:* These two options combined would result in reduced pressure on the beach, because more activities would be undertaken on the lagoon side allowing for the beach system to recover naturally after the infrastructure has been removed. The downside is the high economic cost.
A + F	*Retreat or removal of facilities where possible and construction of hard engineering structures:* As with strategy A + D, these two options should be undertaken in combination. Construction of a submerged structure would dissipate wave energy and allow the system to recover more efficiently after the infrastructure has been removed from the top of the beach.

TABLE 7.53 Analysis of Potential Risk Mitigation Strategies. The Actions Labeled in the First Column Refer to the List Given in Section 7.7.4 —Cont'd

Actions	Strategy
B + B	*Construction of flood-resistant buildings:* Although it is important to improve the resilience of existent infrastructure, this option on its own would not constitute an efficient risk mitigation strategy. It should instead be used in combination with another of the options proposed.
B + C	*Construction of flood-resistant buildings and periodical beach renourishments:* Maintaining a healthy, wide beach, although very costly, would reduce the risk of flooding, so that buildings would only become flooded during the most extreme events. It is in those extreme events that their ability to cope with flooding would become a valuable asset.
B + D	*Construction of flood-resistant buildings and dune restoration:* This strategy achieves the same objectives as the previous one, but at a reduced cost, because dune restoration would be significantly less costly than periodic beach renourishments. The limitation would be the space in which to carry out the reconstruction of dunes, which is not existent in the present resort layout.
B + E	*Construction of flood-resistant buildings and improving the water quality of the lagoon system:* A combination of these two options would not significantly mitigate the risk of erosion and flooding.
B + F	*Construction of flood-resistant buildings and hard engineering structures:* This strategy achieves the required objective of significantly reducing the risk of erosion and flooding at the study site, while also addressing the resilience of the development. Although, its efficiency would be improved by combining it with one further option, such as periodic beach renourishments or dune restoration, to assure the beach regains its stability.
C + C	*Periodical beach renourishments:* Implementation of periodical beach renourishments on their own is the management strategy of choice at many tourist resorts in developed countries, because there is no visual impact of a hard structure nor is there discontent of property owners when they are required to make modifications to infrastructure. These measures can be very effective in maintaining the beach width and reducing the risk of erosion and flooding, but also have great ecological impacts in the long term, as well as being a very expensive scheme to run.
C + D	*Periodical beach renourishments and dune restoration:* These two options are complementary because dune restoration requires additional space and artificially nourishing the beach would provide this space, and ensure enough material for the beach-dune system to regain its stability. In addition, the ecological services provided by the dunes would mitigate the risk of erosion and flooding very efficiently. The disadvantages of this option are the cost of performing periodical beach renourishments along with the environmental consequences.
C + E	*Periodical beach renourishments and improving water quality of the lagoon ecosystem:* This strategy would efficiently reduce the risk of flooding and erosion while offsetting the environmental impact of offshore sand extraction by improving the lagoon ecosystem. Its main negative aspect: high cost.

(Continued)

TABLE 7.53 Analysis of Potential Risk Mitigation Strategies. The Actions Labeled in the First Column Refer to the List Given in Section 7.7.4 —Cont'd

Actions	Strategy
C+F	*Periodical beach renourishments and hard engineering structures:* Construction of hard protection structures, such as terminal groins and/or submerged breakwaters would assure that the sand provided by beach renourishments stays within the littoral cell for longer periods, thus reducing the overall costs of the strategy. It is also a very efficient strategy to mitigate risk. Its downside: environmental impact.
D+D	*Dune restoration:* Unfortunately, this mitigation option cannot be undertaken with the current layout of the resort. There is not enough space to reconstruct the dunes.
D+E	*Dune restoration and improving water quality of the lagoon ecosystem:* As with the previous strategy, there is not enough space to reconstruct the dunes with the current layout.
D+F	Dune restoration and hard engineering structures: Same as previous.
E+E	*Improving water quality of the lagoon ecosystem:* This option, although necessary, would not contribute enough to risk mitigation if implemented on its own. Improving the water quality of the lagoon ecosystem as a complementary environmental action alongside any integral risk management strategy is recommended.
E+F	*Improving water quality of the lagoon ecosystem and hard engineering structures:* This strategy would fulfill the objective of mitigating the risk of erosion and flooding while improving the overall health of the system by reducing pressure on the beach.

In the case of Cancun, a planning instrument has recently been approved concerning municipal land use and ecological issues (it sets guidelines on construction near the dune and mangrove ecosystem, density of development, and so on). This will operate in combination with existing federal laws, such as the "Ecological equilibrium law"and the "National Law of Natural Resources," which only provide general guidelines as to how to plan and conduct a residential/tourist development in a coastal area.

The strategies proposed here raise key issues such as the distribution of responsibilities between governments and developers when coastal erosion problems arise, retroactive application of the law as well as the importance of establishing efficient communication strategies to disseminate regulations to stakeholders.

7.7.5.3 Overall Conclusions and Recommendations

The role of hurricanes in flooding and erosion of the coast of Cancun has been studied through various approaches; nevertheless, the role of the prevailing hydrodynamic conditions, which can also induce important littoral currents, had not been properly quantified. The dramatic, observed losses after extreme events have caused erosion to be attributed solely to the effect of the extreme event; however, the role of the normal littoral drift and the balance between calm conditions and extreme events was still not fully understood. The modeling of the effects of both prevailing and extreme conditions reported here was first carried out separately, and then simultaneously to understand how they balance in the long term and to provide robust data to propose appropriate risk management and mitigation strategies. This approach could be applicable also at other sites where significant littoral drift occurs during calm conditions as well as during storms, such as the Adriatic west coast, for example.

Knowing that the actual state of the beach system at Cancun is highly altered, with very low or null recovery capacity, it is easy to conclude that the risk level for almost any receptor is high. The stability that the renourished beach has shown from 2010 to 2013 may lead one to think that the risk level has been reduced, but the results obtained from the numerical modeling and the experience acquired from the 2006 beach fill show that the beach, which now has no coastal dune, could be rapidly removed by an intense storm. In this sense, the risk of erosion, despite the action undertaken in an attempt to mitigate it, has not been reduced. Thus, all the expected damage and losses remain highly probable.

Actions need to be taken for the physical, ecological, and social conditions to be kept sound and improved. Even though the analysis performed indicates that the frequency and magnitude of storms is not expected to increase, the system was found to be highly vulnerable to extreme events; if a storm with a return period close to 100 years occurs, the devastation will be similar or greater than that seen in 2005 (more than 7000 million USD in damage and eight people dead in the Yucatan Peninsula). On the other hand, it is important to point out that, although the growth of the tourist development is limited because it has covered most of the sand barrier, urban development around the lagoon is not controlled, Also, there is no way of knowing whether the newly proposed municipal land use and ecological guidelines will actually be enforced. Therefore, there is no information as to whether in 2020, 2050, and 2099, there will be more land reclamation conducted within the lagoon or if the urban settlement will increase or decrease in size. In any case, if present indiscriminate development trends continue, the system will only become more rigid and, thus, the risk of erosion and flooding for the scenarios simulated can only be expected to increase.

Several mitigation strategies can be implemented; none of them represents a definite solution and some of them may reduce the risk of erosion while inducing undesirable social consequences. In any case, the combination or concatenation of strategies seems to be the best alternative for achieving a sound physical system. Enforcing the appropriate regulations and good management is also necessary to keep the resilience of the system in the face of unexpected hazards.

A particular point of conclusion is that in an age of climate change and rising sea levels, coastal communities should prepare themselves for the possibility of more frequent and more severe erosion and flooding. This means focusing on resilience and adaptation, rather than prediction alone. Some examples of mitigation strategies proposed in this document suggest building residential and tourist infrastructure in such a way that it can withstand high water, or, where possible, to try to rebuild the natural defenses that dunes represent. It is recommended that such building methods are written explicitly into building regulations and development planning codes for coastal areas if the mistakes made in Cancun are to be avoided at other resorts in Mexico and elsewhere in the future.

Acknowledgments

The authors would like to express their gratitude to the organizations that allowed the use of their data for this project, especially the Ministry of Transport and Infrastructure of the State of Quintana Roo for providing bathymetric information.

7.8

Flood and Erosion Management on a Dynamic Spit: the Hel Peninsula, Poland

Beata Kowalska [1], Jacek Lendzion, Mirosław Mętus [2], Rafał Ostrowski [3], Ida Stanisławczyk [1], Piotr Szmytkiewicz [3], Marzenna Sztobryn [1], Elżbieta Zawadzka-Kahlau [1]

[1] *Institute of Meteorology and Water Management O/M Gdynia – National Research Institute (IMGW), Gdynia, Poland*

[2] *University of Gdańsk, Institute of Geography (IMGW), Gdańsk, Poland*

[3] *Institute of Hydro-Engineering of the Polish Academy of Sciences (IBW PAN), Gdańsk, Poland*

7.8.1 Description of the Site

The Hel Peninsula, actually a 36-km-long sandy spit, is located on the Southern Baltic Sea at the western edge of the Gulf of Gdansk (Figure 7.95). The Baltic Sea is a semiclosed tideless sea, temporarily covered by ice. The Peninsula is characterized by low-lying sandy beaches that are of particular value for tourism. It has also some role in regulating the wave climate in the Gulf, providing shelter to the harbors of Gdynia and Gdansk. The whole ecosystem of the Peninsula is also nationally and internationally recognized as an important area and designated as part of the European NATURA 2000 network.

The Peninsula is mainly fed by long-shore drift from the west. Active management of the spit is a locally important issue as, due to its specific shape, there is the risk of breaching or other damage that can lead to significant impacts on the local population and economy. This is particularly the case at the root of the Peninsula whose width is only 200 m.

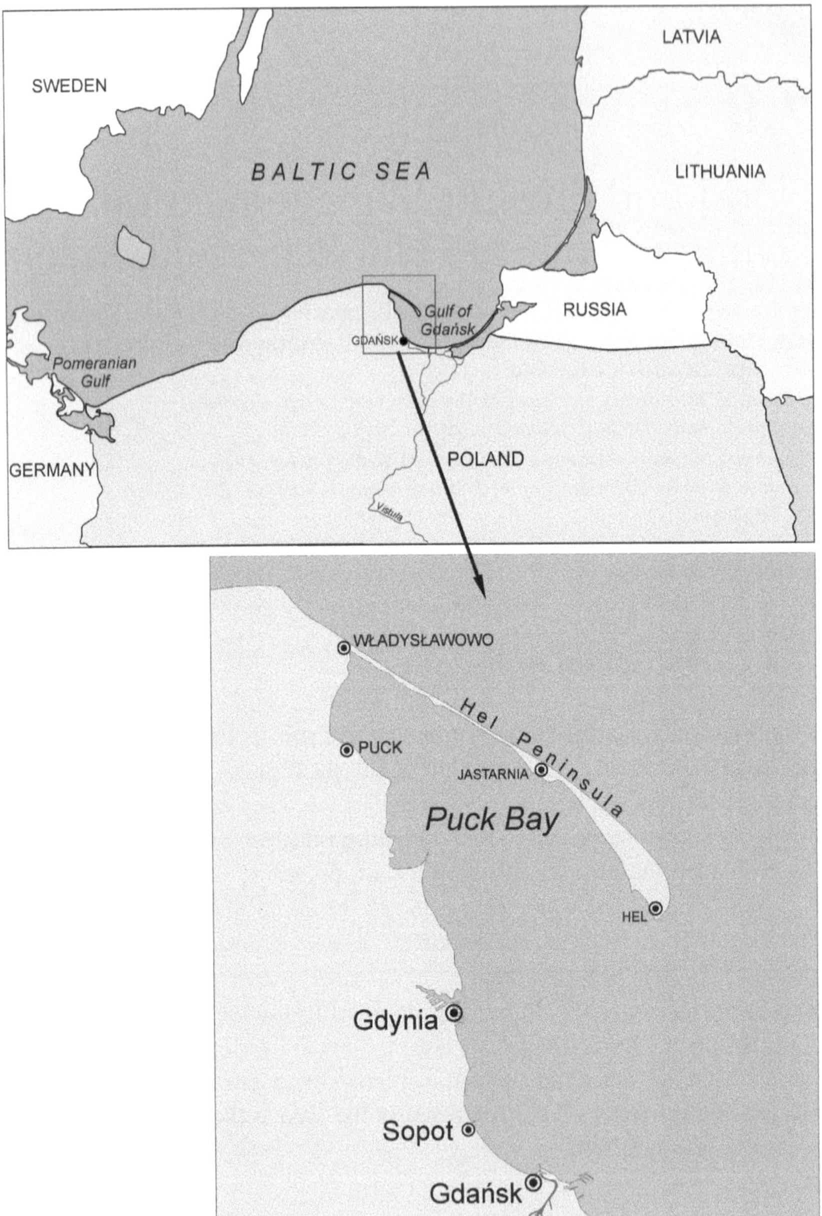

Figure 7.95 Location of the Hel Peninsula, Poland.

Figure 7.96 Characteristics of the Hel Peninsula coastline: (left) open coast and (right) Puck Bay side (photos by B. Kowalska).

Storm wave inflows and short-term overflows occurred in the past. (Tomczak & Domachowska, 1999, p. 99). At present a major breach of the peninsula was a concern in the past few decades after flood events in 1970s and 1980s and recent reports on the consequences of expected SLR. The main concerns for the Hel Peninsula are therefore twofold: the erosion of the dune system that protects low hinterland against flooding and, on the Bay side, moving ice banks that are dangerous for buildings, trees, and roads (Basiński, 1995).

The coast is currently protected against open sea hazards mainly by natural and artificial dunes as well as artificial beach nourishment supplemented by traditional groins, sea walls, and revetments. The coast on the more sheltered Puck Bay side is either natural (e.g., reed bed habitats) or managed with occasionally implemented measures such as stony revetments, land reclamation, and sections of dikes (protecting the built up areas). Figure 7.96 shows the different characters of the Hel Peninsula coastline: the open sea coast with wide beaches after nourishment and along Puck Bay where reed beds have been removed by camp managers to allow easy access for windsurfers.

With the expected increase in extreme water levels, the future management of the spit against flooding and erosion will involve the consideration of additional mitigation measures while maintaining the tourist attraction of the area.

7.8.2 The Current Flood System

As described in Chapter 2, the assessment and mitigation of risks require understanding the flood system as it responds to a range of planned and unplanned interventions. In addition to hydrological aspects (including climate change impact on future flood events) and the role of ecosystems, the understanding of the social context of flooding (the existing governance and perceptions of the local community) enables the evaluation of the type, scale, and importance of flooding in the area as well as identifying acceptable levels of risk and mitigation options.

7.8.2.1 Social and Economic Background of the Hel Peninsula

Local administration for the Hel Peninsula is shared among three municipalities: Hel, Jastarnia, and a part of Wladyslawowo. The Wladyslawowo municipality is a popular seaside holiday resort and a small urban conurbation located toward the root of the spit, whereas the other two municipalities, Hel and Jastarnia, are located on the spit itself. In social terms, the area of Hel Peninsula belongs to a region of Poland with a strong cultural identity. The Kashubians have experienced many hazards (natural, seaborne, or political) and have preserved their own cultural traditions, remaining very close to the natural environment. Fishery, with major fishing ports in Wladyslawowo and Hel and two smaller ones situated on the Puck Bay coast, remains crucial for the local economy and social life, but since the beginning of the twentieth century, it has been affected also by tourism development.

Geographically remote, the area covered by the municipalities constitutes 0.02% of the total area of Poland. On the peninsula itself the resident population is approximately 8000 inhabitants, showing a decreasing trend over recent decades (see Table 7.54). However, during each summer, there are up to 50,000 tourists and daily visitors (particularly at weekends), representing 2.3% of the national overnight stays. The majority of these visitors are domestic, with overseas visitors contributing only 13,300 of almost 1 million overnight stays registered in 2007. Because of this, tourism dominates the employment in the three municipalities, with approximately 12,000 people employed in hotels, restaurants, and tourism-related businesses (including retail).

TABLE 7.54 Basic Socioeconomic Data of the Hel Peninsula Municipalities

Socio-Economic and Related Factors	Poland	Hel Peninsula Municipalities			Detailed Analysis Area*
		Jastarnia + Hel	Wladyslawowo	Total	
Land area (km^2)	312,000	30.0	38.1	68.1	2.18
Total population (1000)	38,512	7.9	15.3	23.2	0.4
Population density (people per km^2)	123.4	254.4	396.6	340.4	172.5
Unemployment (% of active age population)	13,2	8.0	7.1	7.5	n/a

*See Section 7.8.2.3.

7.8.2.2 Existing Coastal Policy and Management Options

Coastal protection and management in Poland is the responsibility of the National Administration, represented in regions by three Maritime Offices in Gdynia, Slupsk, and Szczecin. These offices cooperate in issuing decisions for any investment within the coastal zone with respective local and regional administration. The Maritime Offices have also some competence in terrestrial and maritime spatial planning and in the protection of the maritime environment.

The coastal zone in Poland is legally defined in the Act on Marine Areas of the Polish Republic and Maritime Administration (1991). This includes two important subzones, termed the "technical belt" and "protective belt" (Cieślak, 2005). The technical belt is the area that directly separates the sea and the land from each other; it is an area intended for keeping the coast in a condition "consistent with the needs of safety and environmental protection." The protective belt is defined as "an area in which human activity has a direct influence on the status of the technical belt area." The local authorities and maritime administration responsibilities are mutual in the coastal zone, but the balance in the roles is different in specific subzones (see Figure 7.97). In the technical belt, the Maritime Office has decisive responsibility for spatial management and land use, whereas for the protective belt areas the maritime administration issues (binding) opinions for proposed decisions of local authorities.

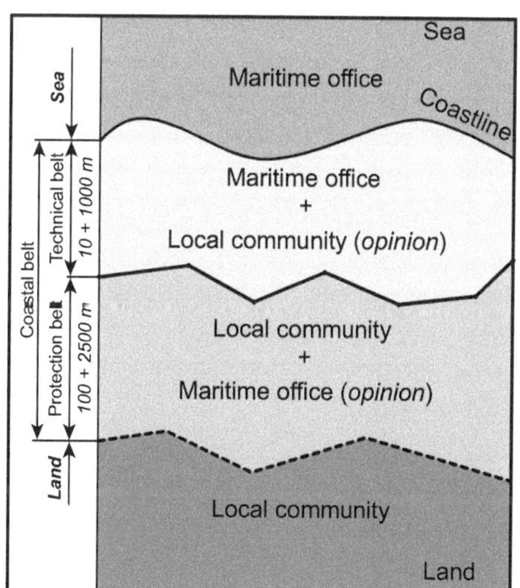

Figure 7.97 Cross-shore distribution of responsibility for governing development and use of land within the coastal zone (based on Cieślak, 2005).

Coastal protection policy is defined by an Act of the Polish Parliament (2003) that established a long-term "Coastal Protection Program." The Program is supervised by the Minister in charge of maritime economy and implemented by respective Maritime Offices. The Program established the principles and methods of active coastal protection for particular sections of the Polish coast, specifying the preferred (but not exclusive) use of soft management techniques. The basic assumption of the coastal protection strategy was to maintain the shoreline position for the 30% of the coast as it was in 2000. For the remainder, managed retreat should be applied. Another underlying assumption is to ensure flood and erosion safety of the hinterland and prevention of beach loss according to local plans.

To achieve these aims, the Act secured an overall budget for 2005–2023 to cover technical measures as well as the necessary monitoring and research. The Program is financed from the government's central budget, with a minimum contribution of PLN 25.5 million per year (approximately 6 million euros). Supplemental funds are also sourced; in recent years Poland, as full member of the EU, has applied for structural funds devoted to the protection of the sea coast and a number of protective investments (on the coast of Gulf of Gdańsk in particular) were financed. The actual amount allocated in each year has to be determined and approved on an annual basis; about 5% of the total funding was foreseen for hard measures, about 65% for mixed measures (e.g., combination of artificial beach fills and erection or modernization of coastal hard structures), and about 30% for artificial beach nourishment only. As a rule, the essential and indispensable measures undertaken in every section of the coast should be based on the monitoring of coastal development and research activities concerned with the issue of coastal protection.

Locally, the coastal defense system of Hel Peninsula has evolved over time. Between 1946 and 1970, along 12.3 km of the Peninsula's seaward coastline, 161 groins were built and at the root of the peninsula a 420-m concrete seawall was constructed to protect the power plant site. After a catastrophic storm in 1983, the middle part of Hel Peninsula was protected by 1500-m-long soil dike. In more recent years, the decreasing effectiveness of existing coastal defenses has been observed, resulting in the erosion along nearly 80% of the first 23.5 km of the coastline (Zawadzka, 1996; Boniecka & Zawadzka, 2006).

After a pilot stage between 1984 and 1988, which tested different technologies and sources of sediment, beach nourishment has been practiced regularly since 1989 and the volume of nourished sand along the Hel Peninsula reached 15.4 million m^3 by 2010 (more than 50% of total volume used along the whole Polish coast). The most intensively nourished area was the root section of the Peninsula (along the first 4 km), where a volume of 510 m^3/m was placed in 1989–1993, and another volume of 250–300 m^3/m was nourished in 2003–2008. These nourishments allowed the restoration of the shore profile, and safety standards were achieved by raising the height of the dune foot to 1.8 m above mean sea level. Differences in the sand sources affected the design of the artificial beach (Zawadzka, 1996). Grain size parameters for material initially borrowed from

Puck Bay indicated that it could be moved rapidly during average conditions of wave and current. Since the mid-1990s, therefore, coarse sand from open sea (dredged at a 20-m water depth) and some sediment from the bypass system of the Port of Wladyslawowo has been used to counteract the loss of sediment on the western side of the port breakwaters. Although the area still shows erosive tendency, the extensive nourishment has helped to improve the beach quality and therefore to maintain the significant recreational value of the coast.

In addition to coastal engineering, the development of spatial planning strategies are defining rules to be followed by developers in the coastal zone with the final planning decision in individual cases lying with the Maritime Administration. Emergency response plans are also being prepared at the county and regional level, to provide a framework for local response.

7.8.2.3 Social Vulnerability and Attitude to Risk

For coastal risk management, it is important to understand the local population's attitude to risk, including the way they understand the science and existing management structures. For the Hel Peninsula, a group of 26 stakeholders (including regional and local coastal managers, spatial planners, environmental managers, business representatives, and nongovernmental organizations) was interviewed to establish current perceptions of the coastal flood risk (THESEUS OD 1.15, 2012) and the planning and management instruments used so far (Table 7.55).

From the interview material, local understanding of risk was classified according to three types of claims: the evidence, normative and relevance claims (Vanderlinden, 2013, Chapter 2 of this volume). Almost half of the claims referred to the interviewees knowledge regarding the cause and condition of risk generation (the evidence claims), there was a lot of vagueness and bias shown in the way coastal risks were placed in the value scale of the interviewees (normative claims), whereas claims on the relevance (i.e., predominantly not high importance of the flood/erosion risks) were more clear but less frequent.

TABLE 7.55 The Structure of Stakeholders Interviewed at Hel Peninsula

Coastal administration	8%
Other public administration	19%
Planning and research	23%
Sustained development/ environmental nongovernmental organizations	23%
Public utilities and local business	27%

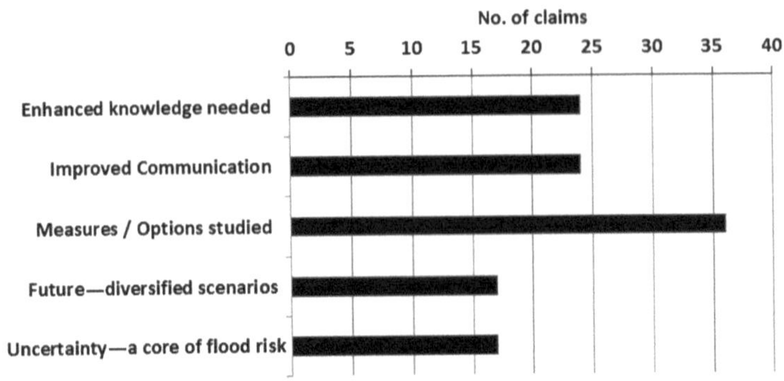

Figure 7.98 Concerns for risk management strategy highlighted by stakeholders.

Further analysis identified specific factors contributing to risk perception (Figure 7.98). Most often interviewees relate their answers to presently used risk management (mitigation options). The consequences of flooding and erosion are often subject of controversies, which may explain why a substantial number of statements (approximately 23%) highlighted the need for improved communication between the involved agents (managers) and the affected groups. The same number of statements also point at the need for deeper knowledge of natural or social processes that contribute to the holistic perception of risk level and effective strategies to minimize their consequences. Relatively little concern was expressed for the future (expectations, plans, etc.) or for uncertainty, especially in terms of probability of a not fully controlled natural process.

The majority of interviewees perceived either natural processes alone or a combination of natural process and human actions as the cause of most negative consequences of coastal flood risks.

The analysis showed that the perception of risk varies significantly with the background education and actual role in the coastal management. The officers involved in the coastal protection stressed the risk of coastal flood and change as important task of crisis management (particularly those responsible for maritime administration) but claimed the situation is under control because of the measures applied. The representatives of business sector assessed the coastal risks and their consequences as significant, however not top priority issue. Those representing local community perceived, coastal flood/erosion risks and their consequences as meaningful but, in general, not the most important concern locally.

The environmental nongovernmental organizations and scientists working on flood risks have seen it as part of natural processes; they were aware of the interrelation between the coastal processes and flood risk.

However, quite a lot of the interviewees admitted that their perception of risk might be influenced by the effective coastal protection provided over the past 20 years (especially the nourishment of the beach) and their knowledge of climate change scenarios was limited.

Because of the recognized importance of maintaining the landscape in the area, the existing risk management options (both structural and nontechnical, including land use) were positively assessed by the majority although as many as 30% of interviewees were unsure as to the effectiveness of some measures. Some of the interviewees underlined the controversy regarding the amount of public spending on the nourishment, whereas most of the material vanishes after the winter storms, and it seems that a couple of private firms are benefiting, directly or indirectly, from that process.

The representatives of business sector mentioned coastal risk as an important external source of disruption problems, although with a rather small impact for their activities. In the short term, their main concern is a break in electric energy and water supply, followed by disruption of transport and information technology networks (27.7% and 21.5% of total scores, respectively). Over the longer term, concerns are related to maintaining the "first business class" image of the region and not losing access to the most distinctive tourism attractions (wide, calm beaches).

Focusing on the range of approaches for risk management in the Peninsula, all stakeholders and the competent administration in particular expressed a weak knowledge of nonstructural mitigation measures (e.g., the use of cooperative, innovative and integrated planning, business continuity plans). The concept of system resilience building was not widely known (Table 7.56) and it was not expected to be adopted within the ongoing spatial planning frameworks.

TABLE 7.56 Resilience and the Concept of Spatial Planning: Direct Reflections from Planners on the Role of Planning and Usefulness of Resilience Idea

Questions:	Options of Answer:	Votes
How much is spatial planning important for proper coastal management in terms of flood risks?	Very important	2
	Quite important	6
	Rather of a small role	0
Do you know and use the concept of "resilience" (as known in the management of environment/natural systems)?	Concept known and practiced	2
	Known but not used	2
	Concept not known	4
What are the challenges for broader use of resilience in the spatial planning system?	It is beyond the scope of spatial planning system	0
	The system of spatial planning is not open for new concepts	6
	The planners lack enough knowledge	2

The key points identified in the interviews on spatial planning and spatial management can be synthesized as follows.

1. Flooding and coastal erosion risk management is relevant to spatial planning regulations at many levels, but there is almost no evidence of good practice: the more experienced in strategic decision-making actors denied the consistence of these regulation.
2. The idea of resilience is not known explicitly and/or not understood properly among spatial planners and planning decision makers, even those involved in the environmental management (Natura-2000) at decision-making levels.
3. The spatial planners do not work in partnerships with other flooding and coastal erosion risk management professionals when developing local spatial development plans.
4. There are a few examples of stakeholder engagement in strategic planning process such as Local Agenda 21 or Integrated Coastal Zone Management, but unfortunately not in the framework of spatial planning, including even the quite innovative pilot study of marine spatial planning prepared for the region of Puck Bay.
5. Long-term development issues still need to be incorporated in defense planning, including scenario analysis or envisioning and climate adaptation strategies.

7.8.2.4 Coastal Habitats

The Puck Bay and Hel Peninsula are protected within the Natura-2000 network, as a Special Area of Conservation (Habitats Directive) and as a Special Protection Area (European Union Directive on the Conservation of Wild Birds). Important habitats are listed in Table 7.57.

Main pressures on these nature reserves are due to water pollution and tourism (construction of facilities, e.g., camp sites, excessive traffic). In the past, a serious threat was connected with the exploitation of sand from the bed of Puck Lagoon, used for stabilization and restoration of the Hel Peninsula beaches. At present the beach nourishment with sediments borrowed from the open sea bed moderately transforms the natural landscape and shore habitats (Szymelfenig, Kotwicki, & Graza, 2006).

7.8.3 Application of the SPRC Model

The components of the Sources of flooding on the open coast and in the Gulf of Gdańsk are storm surges, mean SLR, and waves. The SPRC model was used as the conceptual basis for an initial analysis of the flood system on the Hel Peninsula (Narayan, Nicholls, Clarke, & Hanson, 2011). In common with other sites where the flood system is extensive, a nested approach was used. Application of SPRC for the whole Hel Peninsula and consideration of historic flood evidence led to the identification of the root of the spit for a detailed SPRC study. This area (representative for the entire peninsula) was further divided into three subareas indicated as Areas 1, 2, and 3 (Figure 7.99) and a detailed SPRC analysis was carried out to analyze and understand the linkages between main Sources of flooding, the Pathways, and Receptors (THESEUS OD.1.15, 2012). These areas have been identified as the most vulnerable to erosion, despite the presence of the dune system. It is clearly visible that from Puck Bay side terrain is very flat. An in-depth analysis of the cross-sections beginning from root part of peninsula up to Chałupy village helped to identify hot spots of hydraulic vulnerability, where breaching and overflow are more likely to occur.

TABLE 7.57 The Most Important Habitats and Species of the Hel Peninsula

Habitats	Species	Remarks
Sandy beaches	Nitrophilous strandline vegetation, two species of sandhoppers	Found along the both sides of peninsula, heavily used by tourists in the summer period and nourished on the selected areas
Dunes	Dune ecosystems, grassland, heath, forest, particularly rich position of lichens and lichenicolous fungi	Found along the open sea coast, *Hel Dunes Nature Reserve* - a floristic nature reserve
Saltmarshes	*Aster tripolium, Samolus valerandi Juncus ranarius, Centaurium littorale* (Turner/Gilmour), *Blysmus rufus* (Huds.), *Plantago maritima L.*	*Salt Meadows Nature Reserve*, a floristic nature reserve (27.72 ha); also some small areas near Chałupy and Jastarnia
Seagrass Meadows	seagrass, various species pondweeds and algae *Zostera marina, Fucus vesiculosus,* and *Furcelaria fastigata*	In the Puck lagoon
Fish	Shellfish shrimp *Baltic Palaemon adspersus, Nerophis ophidion, Sygnathus typhle, Coregonus lavaretus,* pike *Exos lucius,* bream *Abramis brama,* roach *Rutilus rutilus,* perch *Perca fluviatilis*	Not important species represented presently, degradation of the natural underwater meadows, any commercial fishing
Benthos (open sea)	Crustaceans *Gammarus salinus, Bathyporeia pilosa,* mussels (chert, clams), and polychaete (nereida)	Diversified communities of macroscopic benthic invertebrates (more than 40 species)
Birds	Gulls (*Larus* spp), coots (*Fulica atra*), mute swans (*Cygnus olor*), ducks, dunlin (*Calidris alpina*), common snipe (*Gallinago Gallinago*) and Wood Sandpiper (*Tringa glareola*), and geese	Many species of water birds, an important habitat for migratory birds, including some charadriiformes birds
Marine mammals	Gray seal and harbor porpoise	The Puck Bay is valuable for European migratory marine mammals

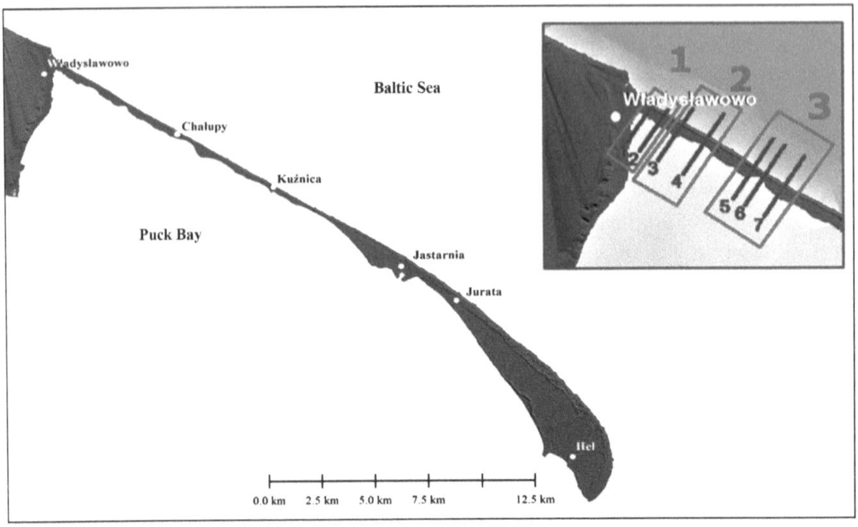

Figure 7.99 Area of general and detailed SPRC model application.

The main sources of flooding, i.e., the open Baltic Sea and waters in Puck Bay, and the main Pathways were identified. The Pathways were: dunes, concrete revetments, groins, nourishment, and seawalls. Because of lower beach elevation, the southern area is more exposed to flood. Key Receptors on the peninsula were identified as the recreational beach, green areas (of different types and including valuable habitats), and as most vulnerable elements the road and rail tracks, the camping sites, small urban land use plots, and very important investment for the analyzed region: the Heat and Power plant. This plant is protected by the concrete seawall and artificial nourishment, regularly carried out along the adjacent coast.

The road and railway track create very good barriers to flooding: before the construction of the existing railway, storm surges, and waves very often caused the breaching of the peninsula during autumn/winter storms. Beaches are periodically destroyed by storm surges and rebuilt by nourishing activities. Beaches are separated from the hinterland by belt of dunes. The area of gray dunes is dominant in the landscape and, since the nineteenth century, has been covered with artificial pine forests planted there with the main function of geomorphologic stabilization of the spit and control of the eolian erosion. The analysis shows that the most threatened is the area along the transport route that runs through the peninsula (railway, roadway).

7.8.4 Meteomarine Climate: Present Conditions and Long-Term Trends

Scenarios of sea-level change were developed using the relationship between global, regional, and local fields and changes in sea level along the Polish coast. Three global model scenarios of greenhouse emissions from Special Report on Emissions Scenarios of the IPCC were taken into account: B1 run1- B1_r1, A1B run1 - A1B_r1, and A2 run1 -A2_r1 (IPCC SRES SPM, 2000). On this basis, ECHAM5 model, run 1 generates the atmospheric circulation at sea level (regional force field) scenarios in the regional scale (Europe and the North Atlantic). ECHAM model was developed by ECMWF and Max Planck Institute for Meteorology to use for climate research, especially to determine the atmospheric pressure field. The local field over the Southern Baltic Sea was calculated by downscaling from the ECHAM5 model (Miętus 2009, Miętus 2010; Jakusik, Wójcik, Biernacik, & Miętus, 2010, p. 59). The local model calibration was done on the basis of data from reanalysis and observation from meteorological and climatologic stations operated by IMGW-PIB (Institute of Meteorology and Water Management - National Research Institute) and located in the coastal zone of Poland (THESEUS OD1.12 2012; Wibig & Jakusik, 2012, pp283; Miętus, 1999, pp157; von Storch & Zwiers, 2001; Miętus & Filipiak, 2002, pp68). Wind is the main driving factor of sea-level changes on tideless seas as Baltic. For this reason the level was calculated as the superposition of the sea level and the wave height effect. Sea-level trends and probability occurrence were calculated based on the yearly maxima for 1955–2009.

Mean sea level around the Gulf of Gdańsk, see Figure 7.100, has risen about 1.5 mm/year in the past 100 years (Kowalska & Sztobryn, 2009; THESEUS OD1.15, 2012). The trend of annual maximal sea level in Wladyslawowo station is also increasing during the observed period, see Figure 7.101. Sea-level return periods for Wladyslawowo were calculated from Gumbel distribution of annual maxima for 1955–2009 (Table 7.58)

Storm surge is defined as the condition when sea level reaches or exceeds 5.70 m, which is exactly 0.70 m above the defined mean sea level (MSL) in Poland. Storm surges along the Southern Baltic Sea coast are most likely to occur after the passage of a low pressure center with a system of atmospheric fronts. The most dangerous storms for the entire southern coast usually occur in correspondence of a stormy depression tracking southeastward from the Norwegian Sea across Scandinavia and the Baltic Sea. Storm surges of this type may last from a few hours to tens of hours. Heavy and long-lasting storms are usually combined with a considerable temporary SLR along the entire coastline (Sztobryn et al., 2005). The highest sea level was observed in 2004 and was equal to 6.44 m (Thesus OD 1.15). Data are synthesized in Table 7.59.

In the Hel Peninsula, storm surges most frequently occurred in 1983 and 2007, with an increasing trend of the number of extreme events (Figure 7.102).

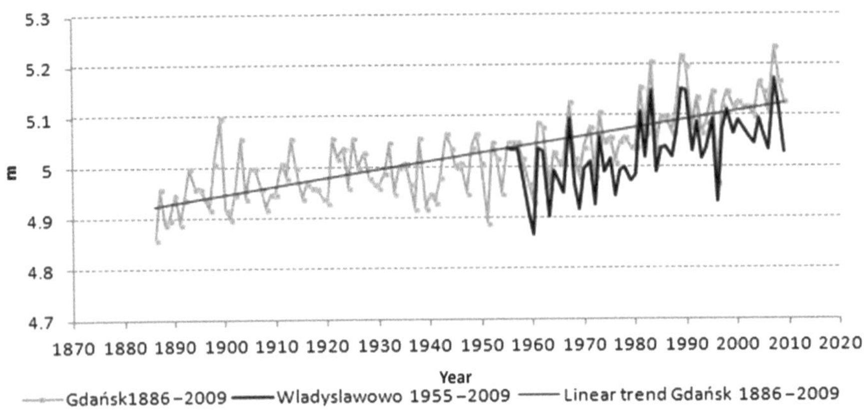

Figure 7.100 Mean annual sea level with trend for Gdańsk 1886–2009 and Wladyslawowo 1955–2009.

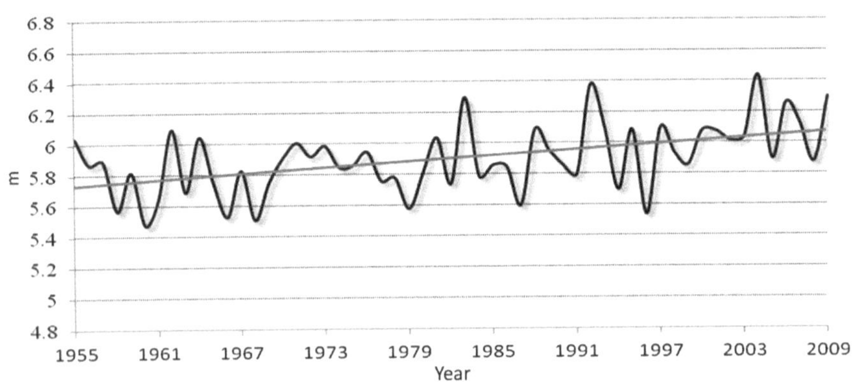

Figure 7.101 Wladyslawowo maximal annual sea levels, 1955–2009.

TABLE 7.58 Sea Level (m) Calculated for Wladyslawowo from the Gumbel Distribution of Annual Maxima for 1955–2009

Wladyslawowo Return Period (Years)	Sea Level (m)	Elevation Above Datum Level (m)
10	6.25	1.17
20	6.40	1.32
30	6.50	1.42
50	6.58	1.50
100	6.72	1.64

TABLE 7.59 Highest Sea-Level Events Around the Gulf of Gdańsk 1955–2009

	Sea Level		
Tide Gauge	Warning Level (m)	Alarm Level (m)	Maximum Observed Level (m)
Gdańsk	5.50	5.70	6.44 (23.11.2004)
Hel	5.50	5.70	6.22 (14.01.1993)
Wladyslawowo	5.50	5.70	6.44 (23.11.2004)

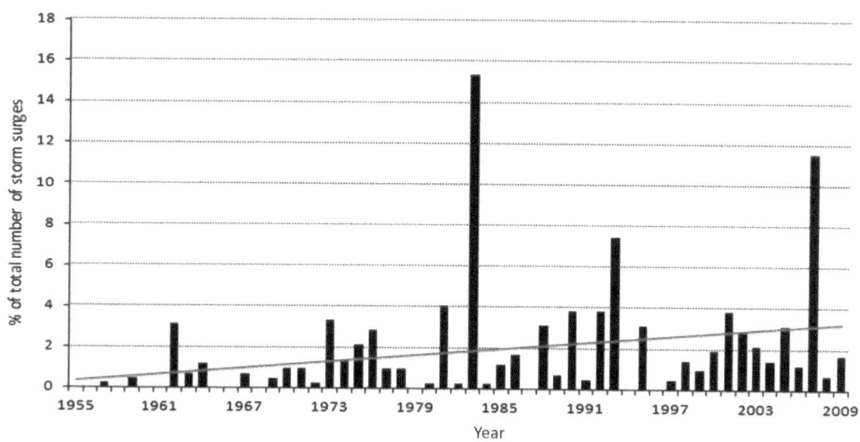

Figure 7.102 Number of the storm surges at Hel for 1955–2009.

The future mean and extreme conditions were calculated as the anomalies (e.g., increase or decrease) from MSL determined for the given reference period. Anomalies of the 99 quintile are reported in Table 7.60. Based on the sea-level results obtained from statistical empirical downscaling for A1B, A2, and B1, the flooding scenario ensembles were prepared for Wladyslawowo, situated on the Hel Peninsula, see Table 7.61.

Risk multipliers (THESEUS OD 1.10, 2012) were analyzed on the basis of SPRC analysis and inundation probability equal to high of sea level and waves with return period of 100 years. On the sandy coast of the Hel Peninsula, there are three main Sources: rise of mean sea level, storm surges, and waves determine the degree of the extend of inundation area, whereas the tides are negligible. The change of the risk probability was calculated on the basis of present probability and results of the modeling from climatic scenarios (Table 7.62). The importance of Sources was established in the rank order and is presented in Table 7.62). The highest increase of the relative importance of flood risk over time was distinguished for wave height, then for the storm surges, and last for the SLR. The level of influence of wave is very high for all analyzed periods, where for storm surges and SLR is medium until the 2020s and 2050s. Only for the 2080s is the influence high.

On the other hand, determination of the offshore wave climate has been based on the numerical prognostic wave model WAM4, in which the input is determined from meteorological (wind and air pressure) fields. The spectral wave model WAM4 was used for reconstruction of the long-term European wave climate from 1958 to 2001 (EU project HIPOCAS). Wind parameters (velocity and direction) at a height of 10 m above the sea and pressure fields were obtained. The results of the wind field were next used in the WAM4 model. Based on this model, wind velocities (mean values and standard deviations) were calculated from Gumbel distribution approximations of annual maxima, see Table 7.63. The representative extreme wind velocities increase systematically in time. This information was used to determine significant wave heights H_S for the same return periods, see Table 7.63.

To derive the joint probability analysis of extreme events due to the combination of wind velocity and water level the Generalized Extreme Value (GEV) analysis was first carried out. The GEV for the wind velocity data and the water level data were separately identified. The logistic model was then applied to calculate the joint extreme value distributions for wind velocity and water level.

The database consisted of the annual maxima of the monthly 99th percentiles of wind velocity and water level, subdivided into three representative subperiods: 2001–2020, 2001–2050, and 2040–2100. The data in Table 7.64 consider both cases of water level as the first variable and wave height as the first variable exceeding the maximum or extreme value. These values were obtained for wind blowing from the azimuth 7° (the direction for which the highest offshore waves nearby Hel Peninsula are generated by wind) and can be helpful in determining the design values (applied in planning of coastal protection against erosion and flooding at the site of Hel Peninsula) in the future.

TABLE 7.60 Anomalies of Maximum Sea Level

Anomalies of Maximum Sea Level	IPCC Scenario	Period	Location: Wladyslawowo	Location: Hel
q99% (m)	B1_r1	2011–2020	0.070	0.067
		2021–2050	0.076	0.077
		2051–2080	0.173	0.176
	A1B_r1	2011–2020	0.045	0.044
		2021–2050	0.108	0.108
		2051–2080	0.215	0.221
	A2_r1	2011–2020	0.056	0.055
		2021–2050	0.099	0.099
		2051–2080	0.214	0.220

TABLE 7.61 Predicted Climate Scenarios with Sea Level and Wave Set Up 100 year Return Period

Period	Sea Level (m)	Wave Set Up (m)	Elevation Above Datum (m)	Sea Level (m)	Wave Set Up (m)	Elevation Above Datum (m)	Sea Level (m)	Wave Set Up (m)	Elevation Above Datum (m)
		A1B			B1			A2	
2012–2020	6.77	0.4	2.09	6.78	0.4	2.10	6.75	0.4	2.07
2021–2050	6.82	0.4	2.14	6.77	0.4	2.09	6.84	0.4	2.16
2051–2080	6.94	0.4	2.26	6.88	0.4	2.20	6.93	0.4	2.25

TABLE 7.62 Relative Importance of Sources of Flood Risk Over Time for the Hel Peninsula (Determined for Scenario A1B)

	Periods			
Rank	2020s	2050s	2080s	Key: Level of Influence
1	Wave	Wave	Wave	Very high
2	Surge	Surge	Surge	High
3	SLR	SLR	SLR	Medium

TABLE 7.63 Wind Velocities (m/s) Calculated from Gumbel Distribution Approximations of Annual Maxima Predicted in the Periods 1960–2020, 1990–2050, and 2040–2100

		1990–2050	
Return Period (Years)	1960–2020	Wind Speed (m/s)	2040–2100
200	27.84 (±1.56)	29.00 (±1.18)	30.36 (±1.25)
100	26.13 (±1.38)	27.58 (±1.02)	28.80 (±1.10)
10	20.37 (±0.79)	22.76 (±0.51)	23.56 (±0.61)
Return period (years)	Significant wave height H_s (m)		
200	11.25	11.71	12.09
100	10.54	11.14	11.63
10	8.51	9.67	10.05

TABLE 7.64 Predicted Water Levels with a 10-Year and 100-Year Return Period and Corresponding Significant Offshore Wave Height H_s

	Water Level is the First Variable		Wave Height is the First Variable	
10 years	Water level (m)	H_s (m)	Water level (m)	H_s (m)
2001–2020	1.07	9.02	0.92	10.22
2001–2050	1.30	9.02	1.01	10.34
2040–2100	1.59	9.24	1.15	11.35
100 years	**Water level (m)**	**H_s (m)**	**Water level (m)**	**H_s (m)**
2001–2020	1.10	9.79	1.03	10.42
2001–2050	1.38	9.96	1.21	11.02
2040–2100	1.74	10.22	1.38	11.53

7.8.5 Flood Mapping

To determine the impact results for the different climate scenarios, the MIKE 3 model was used in conjunction with a GIS program Global Mapper that allowed visualization of flood extents. Flooding maps were based on the results on sea level of return period of 100 years and 200 years calculated for three climate scenarios A1B, B1, and A2 (see Table 7.58). The details are reported in THESEUS OD1.10, 2012 and OD1.15, 2012. If compared with the current risk level, the change is significant for 30–40% of the total area of the spit (Table .7.65).

Extreme storm surge connected with high MSL lead to dune breaching. The root and middle parts of the Peninsula foredunes are subject to destruction by storms almost each year and are reconstructed with artificial nourishment up to the height of 5–6 m above MSL, whereas the hinterland is dominated by lowlands (below 2.5 m MSL). Specific forms of inundation occur along the Puck Bay (because it is only a 1.5–6 m deep lagoon). Oscillation of water level induces overtopping of local defenses and higher ground water levels, combined with salt intrusion in the lowland areas. Furthermore, late winter wind circulation occurs quite often so that the sea ice is moved by waves from sea into the land. These areas, covering the trunk road surface and even demolishing lighter constructions.

The most vulnerable area for 2051–2080 (in case of a 100 year return period water level including wave set up) under scenario A2 is presented in Figure 7.103. The average and extreme event water levels are higher from the open sea side (Hel, Wladyslawowo) than from the Puck Bay.

TABLE 7.65 Area Threatened by Flooding (Including Wave Set up) in Present and in Future Scenarios

Sea Level	*H* (m)	Elevation Above Datum (m)	% of Flooded Area
Present alarm level	6.05	0.97	About 9.5
Scenario A1B-2050	7.22	2.14	About 36
Scenario A1B-2080	7.34	2.26	About 39

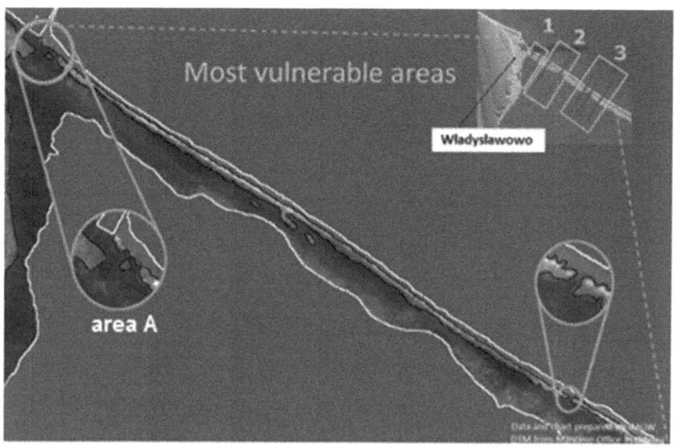

Figure 7.103 Location of possible breaches on Hel Peninsula under the A2 scenario (with return period 100 years) and the most vulnerable area (area A).

Investigations carried out within the project (OD 1.10 THESEUS, 2012) showed that the most vulnerable area was the root of the Peninsula (see area A on the Figures 7.103 and 7.104) not only because of the SLR and waves, but mainly at the location of critical infrastructures in this area (notably the Heat and Power plant). Dune breaching was examined in more details with 3D simulations at the root of the peninsula (see Figure 7.104). The influence of inflow from Puck Bay will not have a significant impact on the present wave regime in the Bay.

7.8.6 Erosion Scenarios

Although wave energy is subject to considerable dissipation on the multibar cross-shore profile (Figure 7.105) of the Hel Peninsula, wave-induced currents can cause severe erosion of the beach, of the dune toe and of the dune itself. which leads to dune breaching and flooding of the hinterland. Direct threat to the dune system (separating the railway and the road from the sea) is significantly amplified by storm surge and wave set-up.

In extreme conditions (100-year return period water level), the wave set-up for the root part of the Hel Peninsula, based on the Battjes and Janssen (1978) approach, reaches up to 0.40 m in 2080 (Table 7.66). When combined with the corresponding extreme storm surge level, the extreme sea level exceeds 2 m in the long-term scenario. Considering that wave run-up on the seaward slope of the dune during the extreme storms is at least 0.5 m, the erosion of the dune foot is unavoidable. The overall hydrological-hydrodynamic effect on a typical cross-shore profile on Hel Peninsula (profile no. 3 in subarea 2, Figure 7.99) is shown in Figure 7.105.

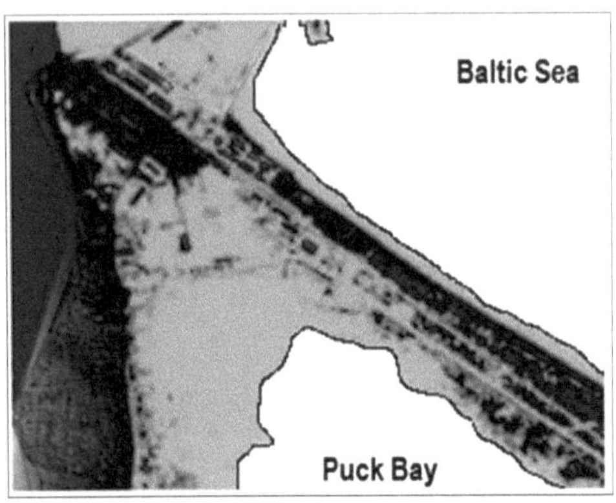

Figure 7.104 Detailed modeling for the most vulnerable area (area A) under scenario A2 for 2051–2080 (gray color) and actual coastline (black line) at and the root of the Peninsula.

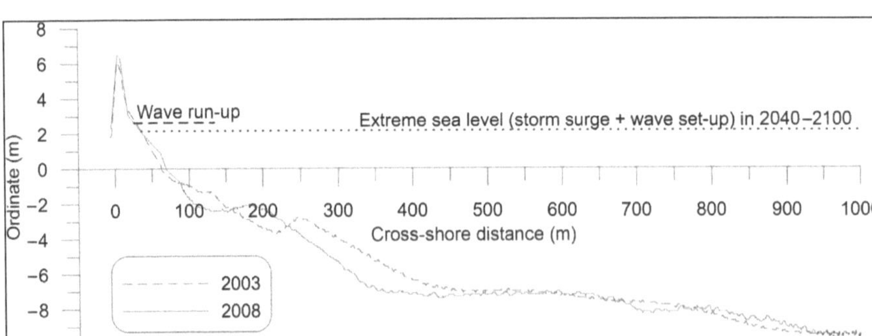

Figure 7.105 Cross-shore profile following nourishment in 2003 and 2008 showing the projected extreme water level in 2080 (Table 7.66).

TABLE 7.66 Water Level and Wave Setup in Extreme Conditions (100 Years Return Period)

Time Slice	Wave Setup (m)	Water Level (m)	Combined Level (m)
1960–2020	0.39	1.10	1.49
1990–2050	0.39	1.38	1.78
2040–2100	0.40	1.74	2.14

The artificial beaches and dunes at Hel Peninsula were designed to satisfy the standard of resistance to a 100 year return-period storm event. This design return period has recently been increased to 200 years by the Maritime Administration to face mid- to long-term scenarios and should be used for the development of coastal management strategies.

Already in recent decades, despite the nourishment of the root of the Peninsula, the cross-shore profiles have shown erosive tendencies and the nourishment activities can barely maintain a "relative" lithodynamic equilibrium. According to the studies carried out by Dubrawski and Zawadzka-Kahlau (2006), the shoreline retreat on a few coastal segments of Hel Peninsula, including its root part, will be about 0.6 m/year (if the nourishment is abandoned and climate severity does not increase). Bearing in mind the anticipated increase of the hydrodynamic and hydrological loads, the material required for beach and dune maintenance should be recalculated.

7.8.7 Mitigation Measures

The selected mitigation options are synthetically described in Table 7.67. Structural solutions (with the exception of submerged breakwaters) correspond to the traditional coastal risk management as described in Section 7.8.2.2.

7.8.7.1 Local Acceptance of Mitigation Options

The acceptance of given mitigation options was analyzed by means of questionnaires and was discussed with local stakeholders (THESEUS OD 1.15, 2012). The preferred option was nourishment because the wide sandy beaches are considered the most important asset to be preserved for tourism and safety of the Hel Peninsula. Stakeholders also indicated that improving coastal monitoring and developing a more operational approach to spatial planning are important institutional measures to minimize consequences of floods. The setup of a combined portfolio of engineering mitigations and nonstructural solutions such as changes of the land use, new built environment codes, reconstruction of resilient habitats, and behavioral changes within the local community was recognized to be a fundamental step toward sustainable development.

TABLE 7.67 Brief Description of the Selected Current and Further Mitigation Options

Mitigation Options	Mitigation for	Brief Description	Management Effort
		Structural/Engineered Measures	
Artificial beach and dune nourishment	Erosion, flood	Shore and beach nourishment have to ensure maintenance of the shoreline position as it was in 2000. Dune toe has to be maintained at an elevation of 1.8 m above the long-term MSL.	High
Groin system	Erosion	The groin system nowadays covers almost 13 km of the Hel Peninsula. Groin crest is about 0.5 m and length is in the range 40–100 m. Repair and renovation is required.	High
Sand accumulation fences	Erosion	Sand accumulation fences are placed in straight line segments, parallel to the dune foot and at a distance 1–3 m. The fascines are about 1.0 m long of which 0.3 m are buried in the sand.	Low
Submerged breakwaters	Erosion	It is envisaged the construction of submerged rubble mound breakwaters, 5 m high and placed at 7.2 m depth. These breakwaters should be constructed in the more vulnerable areas (0–1 km and 11–12 km from the root of Hel Peninsula). Wave height reduction can be up to 53%.	High
		Nonstructural Measures	
Beach monitoring and risk communication	Erosion	Information gaps should be mitigated by more intensive, collaborative efforts of institutions involved in the observatory network to integrate meteomarine and hydrological data with data on dynamic changes of geomorphic structures at the coast. The improved forecasts and warning system can lead to better risk communication to general public and preparedness.	Medium
Integrated coastal zone planning	Flood	Measures include: ■ Strengthening the rail, road, energy, and water supply lines (e.g., by rising elevation, introducing storm culverts) ■ Adaptation of urban environment (e.g., houses basements and higher first floor levels, flood resistant construction) ■ Preservation or restoration of natural beach profile, wetlands habitats, adaptation changes in land reclamation practices. The key challenge for implementation is the involvement of the local community and stakeholders in the strategic visioning and preparation of (spatial) development and action plans.	Medium

7.8.7.2 Strategic Management of the Hel Peninsula

Although all the options synthesized in Table 7.67 offer a certain degree of protection, a combination of measures in a specific strategy could help to achieve a larger effect with lower costs, less disruption to coastal processes and natural landscape, and/or better response to social preference.

The strategy includes as primary measure the maintenance and extension of beach and dunes, by periodic beach nourishment against the 100 years return period storm. As supplementary measures, the modernization of groins system and the broader implementation of sand accumulation fences are proposed. Locally, submerged breakwaters might be also considered for wave attenuation. As an alternative to the construction of new hard defenses, the definition of new safety standards and the managed retreat should be considered in the long term.

The strategy also foresees the enhancement of the community resilience to flood risk by implementing properly designed building codes in new developments or renewal of existing areas, by increasing preparedness and acceptance of certain inevitable processes (temporarily occurred low level inundation) and adaptation to such phenomena by behavioral changes (Chapter 6).

Additionally, the following issues should be addressed for the proper maintenance of the coastal risk mitigation infrastructures at the Hel Peninsula.

Dunes should be properly vegetated and the vegetation maintained. Fencing should be installed to discourage people from walking across dunes and to enhance the effects of eolian accretion of these sandy structures.

Coastal protection structures should be inspected and maintained on a regular basis.

Public awareness has to be improved, with reference to direct consequences of coastal risks (e.g., the quality of life/recreational environment) as well as long-term sustainable development issues.

7.8.7.3 Policy Implications

Because of the decreasing resident population of Hel Peninsula, the timing of extreme events, the ongoing transformation of economic and technical activities (e.g., developing nature friendly tourism, improving observation of natural processes), and the foreseen modification of mitigation options, the vulnerability of society to natural hazards is not expected to increase significantly.

The proposed strategy (in Section 7.8.7.2) focusing on intensive artificial nourishment would maintain the reliance of community on structural measures. This might create a false idea of safety and not encourage the increase of social resilience.

The engineering mitigation options discussed in Section 7.8.7.2 are in agreement with the National Program of Coastal Protection (The 2003 Act), and the preferred nourishment option follows the present trend toward promoting more natural ways to rebalance coastal processes. However, the affordability of the more intense nourishment program—inevitable in terms of future climate scenarios—is disputable within the limited public budget.

7.8.8 Overall Conclusions and Recommendations

The Hel Peninsula coast is subject to increasing erosive and flooding events as a result of climate change according to several IPCC (AR4) scenarios. The existing coastal protection system (groins, revetments, seawalls, supported occasionally with artificial beach nourishment) will need to be upgraded to maintain safety levels for the settlements and basic activities along the root section of the Hel Peninsula. To avoid to the loss of natural ecological habitats, a more extensive, innovative, and sustainable approach to coastal protection is required.

As primary mitigation option in a new defense planning strategy, more intensive artificial beach nourishment is proposed. It should be supported by a system of renovated (or even modernized) groins. Submerged breakwaters can be considered as a supplementary option, to be applied at some locations where the erosive processes are more intense or economic impacts are particularly high.

Alternatively the strategy of limited protection should be considered with a complex of engineered measures, assisted with a range of nonstructural measures in land use and infrastructure. Low-lying areas in the back of the potentially destroyed dunes, whose parameters are to be maintained only by artificial nourishment, should be excluded from further development.

Priority should be given to early warning of the population in areas at risk. Spatial planning and adaptation of buildings to short-term flooding should also be improved. The social dialogue should enable people to realize the inevitability of future climate change and the measures to tackle with them (at a reasonable, socially justified, price). The total socioeconomic benefits and costs of the strategies and multi-criteria assessment methods should be also introduced, e.g. in the framework of Integrated Coastal Zone Management approach proposed by the European Union (Rabski, 2004).

7.9

Flooding and Erosion Risk Analysis in Santander Bay, Spain

Cristina Izaguirre Lasa, Iñigo Losada Rodriguez, Fernando Mendez Incera, Pedro Díaz-Simal, Javier L. Lara, Barbara Ondiviela Eizaguirre, Mirian Jimenez Tobio

Environmental Hydraulics Institute "IH Cantabria", Universidad de Cantabria (UC), Santander, Spain

7.9.1 Description of the Site

The north coast of Spain consists of a series of pocket beaches and small inlets separated by pronounced rocky headlands. Santander Bay is one of the largest inlets on the Cantabrian Coast and is located about 200 km west of the French border (see Figure 7.106). It has a total area of 2270 ha and is 9 km in length and has a maximum width of 5 km. Santander Bay is preceded by El Sardinero inlet, where, on the west side, the tourist beaches of El Sardinero are found. On the east side of the outer bay are the Loredo and Somo beaches and El Puntal Spit, a long narrow stretch of sand that flanks the entrance of the bay together with the Magdalena Peninsula. The bay provides a natural shelter from the waves of the Gulf of Biscay coming from the NNW and is modified mostly by the Magdalena Peninsula that produces wave diffraction and reflection. Most of this wave energy is finally dissipated along the spit.

Santander Bay is a spatial unit that gives home to more than 200,000 people. It shows highly diverse economic activities and human settlement and, hence, a complex framework of conflicting activities can be observed in the area. The area features an urban area specialized in services, an industrial belt with heavy density of infrastructures, cattle farming areas draining into the bay, natural areas (mainly marshes and intertidal areas), and an open water area where maritime activities are present.

Over the years, Santander Bay has undergone many changes because of the development of the city. During past centuries, large marsh areas have been filled in to generate space for economic activities, residential use, and infrastructures, substantially modifying the hydrodynamics of the estuary and the biological framework. The intensive land reclamation (especially in the western edge of the bay) reduced the tidal prism by roughly 40%, producing a deficit in the tidal flat area. The spit experiences a problem of erosion along the seaward side (1 m a year) that results in a narrowing of the spit (now approximately 100 m in width). The westward movement of these eroded sediments has increased the length of the spit at a rate of 15 m per year, resulting in the need to dredge the navigation channel regularly (to a depth of 11 m). This dredged material is replaced at the eastern end of the spit to compensate for erosion losses.

Flood risk is a limited issue in the inner bay, restricted to two to three small locations. Further along the coast in other areas such as Somo, urban development has been located in flood-prone areas and has experienced many flooding events.

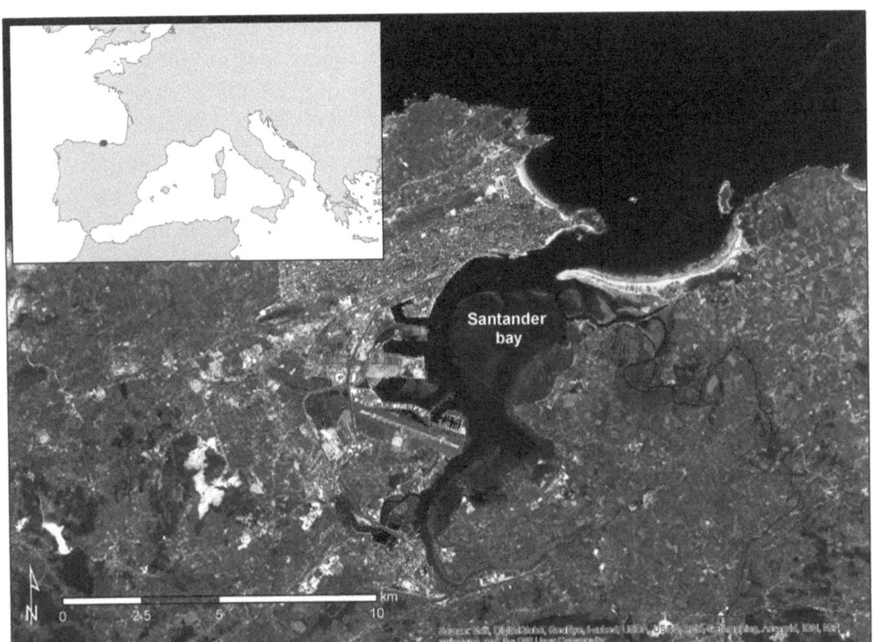

Figure 7.106 Santander Bay.

7.9.2 The Coastal System

7.9.2.1 Existing Policy and Management Strategies

The National Coastal Authority is responsible for the Spanish coastline under the auspices of the Maritime-Terrestrial Public Domain (defined by the Coastal law, 1988). However, in port areas there are site specific arrangements; for Santander Bay, this authority is transferred to the Port Authority of Santander (part of the National Port Authority). The Port Authority is therefore responsible for (1) the port area (western bay coastline) and (2) the use of Santander Bay waters, including the water-quality effects of shipping and the port and activities from fisheries to regattas, including dredging activities. Apart from these two stakeholders, other authorities in the institutional framework include the Navigational Authority of the Santander Bay, Regional Ministry of Environment, North River Basins Authority, Cantabrian Environmental Administration, National Ministry of Environment, and the municipalities.

Spatial planning in Santander Bay is supported by the existing policies based on the National Coastal Law (1988), the Regional Law on Coastal Planning, POL (2003), and the specific spatial planning for each municipality. Local spatial planning has to be adapted to upper level legislation requirements each time new regulations are created, and this process can take a long time to complete. As an example, Santander completed its plan in 2012 and Camargo has restarted its planning process (2013), revising the previous version from 1988.

In summary, the institutional framework in Santander Bay consists of a number of authorities at different scales. There are overlapping/conflicting areas of competence between municipalities, regional government, port authority, and national coastal authority. However, the relatively small area of interest facilitates the transfer of knowledge and cooperative arrangements between authorities.

7.9.2.2 Meteomarine Climate: Present Conditions and Long-term Trends

Flooding levels in Santander Bay are the result of different contributors to extreme water levels: waves, storm surge, astronomical tide, MSL, and river discharge. All of these contributors, mean and extreme, change over time with consequences for the coastal system. Therefore, an analysis of the long-term trends and long-term extreme value distribution of met-hydro-ocean parameters (meteo-ocean climate) is needed.

Wave Modeling

To properly characterize current conditions and long-term trends of wave climate, reanalysis data obtained from the global wave hindcast Global Ocean Wave (GOW) model has been used (Reguero, Menendez, Méndez, Minguez, & Losada, 2012). This reanalysis is obtained using the wave model WaveWatch III (WWIII (Tolman, 2009, pp194) forced by 6-hourly wind fields from the climatic reanalysis NCEP/NCAR (Kalnay et al., 1996). Wave dynamics were transferred into shallow waters using a state-of-the-art methodology combining dynamics (using phase-average model SWAN Booij, Ris, & Holthuijsen, 1999) with statistical downscaling (Camus, Méndez, Medina, Tomas, & Izaguirre, 2013). The propagation process is carried out nesting three computational grids improving the resolution from 1 km in the bigger one, 250 m in the medium one, and 50 m in the detailed grid of Santander area. For the largest grid, the boundary conditions are defined by the real spectrum from GOW along all the sides of the grid. A constant wind field in the computational domain is defined by W_{10} and β_W every hour. To minimize computational resources, several representative sea states have been chosen using the maximum dissimilarity algorithm MaxDiss (Willet, 1996).

The reconstruction of the time series of wave parameters nearshore (H_s, T_p, θ) has been carried out at six representative points of the study area by an interpolation technique based on radial basis functions. For the selection of the representative points we use a methodology based on a classification of areas using the self-organizing maps technique (Kohonen, 2000) taking into account only the source characteristics (similar characteristics of mean significant wave height, H_{s12}, mean energy flux, and mean direction). This methodology can be applied in other study sites, obtaining useful source-related map information. The representative zones and points for Santander are shown in Figure 7.107.

Time-series reconstruction was validated in shallow waters at the available wave buoys (Bilbao buoy from Puertos del Estado, and Virgen del Mar and Santoña buoys from Red Vigia, Gobierno de Cantabria) located in the computational grids obtaining good agreement between both time series (Camus et al. 2013).

Once the coastal wave time series were obtained, the current wave climate conditions were characterized. To assess the present wave climate conditions, a statistical analysis of some representative parameters, such as H_{sm}, H_{s12}, and $H_{90\%}$, is carried out for each objective point, studying different temporal scales.

The statistic correspondent to the significant wave height exceeded 12 h (an indicator of beach erosion) a year was calculated and a regression model was fitted to obtained long-term trends over the past six decades. Results show positive trends for every beach, with the highest increase in Loredo and the lowest in Puntal, inside the Bay (Table 7.68).

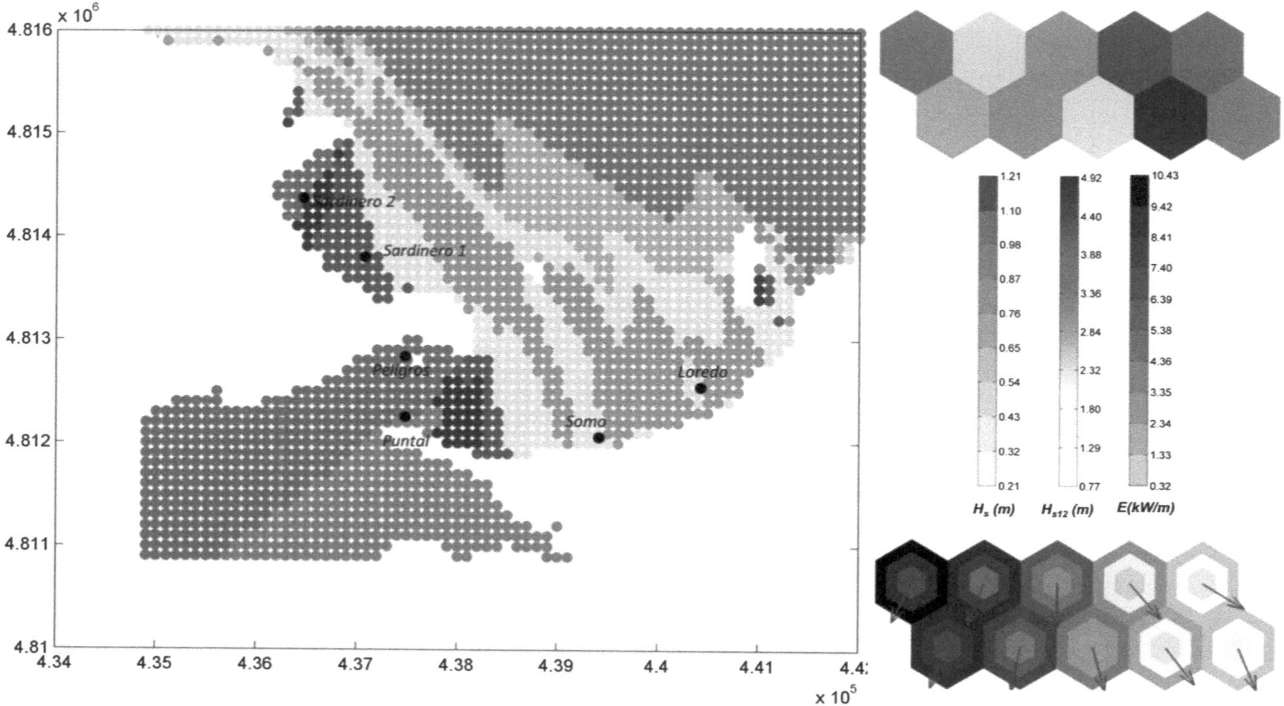

Figure 7.107 Classification of Santander Spit in similar wave characteristic areas and objective points. Each group, defined by a color, has certain conditions of significant wave height (H_s), significant wave height only exceed 12 h a year (H_{s12}), energy (E), and mean direction (represented by blue arrows).

TABLE 7.68 H_{s12} **Mean Trends in the Objective Points**

	Sardinero 2	Sardinero 1	Loredo	Somo	Puntal	Peligros
Mean trend (cm/year)	0.34	0.29	0.67	0.43	0.12	0.07

TABLE 7.69 H_{50} **in the Objective Points**

	Sardinero 2	Sardinero 1	Loredo	Somo	Puntal	Peligros
50-year return level quantile (m)	3.2	3.9	6.8	4.8	2	1.4

The mean energy flux and its long-term trend were also calculated, which is crucial for several coastal processes. Results show a decreasing trend at every point except in Peligros beach, where the waves are completely diffracted.

With regard to the extreme wave climate, extreme events were analyzed by applying the Pareto-Poisson model to obtain the exceedances over a threshold using the POT method (Méndez, Menendez, Luceno, & Losada, 2006). The quantile expected for a 50-year return period is shown in Table 7.69, where one can see that Loredo reaches 6.8 m whereas at El Puntal it does not exceed 2 m.

A nonstationary model was then undertaken, including seasonality and long-term trends (Méndez et al., 2006). Results showed that the most exposed beach is Loredo, which receives the highest waves during the year and presents a strong winter-summer modulation. On the other hand, Peligros is the most sheltered beach; the H_S threshold does not reach 1 m and the variations throughout the year are very smooth.

Seasonality was then combined with the long-term trend for each location. However, the long-term trend is not significant at any objective points (at a 95% confidence interval).

Water Levels

The water level components for coastal flooding are storm surge, astronomical tide, and mean sea level. Storm surge is a random variable that consists of variations in sea level resulting from wind and pressure variations. The largest events of storm surge are associated with storm episodes, so most of the time its occurrence is combined with large waves, although this is not always the case in Santander. The astronomical tide is considered as the movements of up and down of the sea level from the relationship between the Earth and the Moon. It is a deterministic phenomenon with periods close to 12 and 24 h and a maximum range of 5 m in Santander. The information of time series used for each component is summarized in Table 7.70.

TABLE 7.70 Sources of the Information Used for Water-Level Analysis

	Storm Surge	Astronomical Tide	Mean Sea Level
Data source	Santander tide gauge (REDMAR, Puertos del Estado) + global ocean surge, reanalysis (Cid et al. 2014)	Santander tide gauge (REDMAR, Puertos del Estado)	Santander tide gauge (REDMAR, Puertos del Estado)
Temporal resolution	Hourly	Hourly	Monthly

For the calculation of the future flooding and erosion conditions, different relative SLRs are assumed. The selection of the relative SLR is based on the extrapolation of the decadal trends in Santander. The considered scenarios are summarized in Table 7.71.

7.9.2.3 Natural Ecosystems

The main coastal habitats observed in Santander Bay are sandy dunes and beaches, saltmarshes, intertidal mudflats, seagrass beds, and rocky shores. Following is a brief description of each.

Sandy Dunes

The spit at the mouth of Santander Bay is a coastal dune system composed of fluvial (siliceous) and coastal (bioclasts) sediments. The sand dune spit is included in the Natura 2000 site "Dunas del Puntal y Estuario del Miera" and hosts a number of dune habitats of community interest (IH Cantabria, 2010). The site has a significant surface occupied by fixed (3%) and mobile (2%) dunes. The vegetation profile in the dune includes the following species (Loriente, 1996): *Cakile integrifolia* and *Salsola kali* above the tidal zone; *Honckenya peploides* and *Elymus borealis* in the embryonic dunes; *Elymus borealis* and *Euphorbia paralias* in the primary dunes; *Ammophila Australis*, *Calystegia soldanella* and *E. maritimum* in the secondary dunes; and *Helichrysum stoechas* and *Koeleria glauca* in the gray dunes. Some studies indicate the presence of allochthonous species as *Oenothera glazioviana*, *Baccharis halimifolia,* and *Spartina versicolor* (Demarcación de Costas de Cantabria, 2006).

TABLE 7.71 Scenarios of Relative Sea-Level Rise for the Calculation of Future Flooding and Erosion Impact

Scenario	Relative Sea-Level Rise (cm)	Horizon Year
Moderate change	20	2050
High change	40	2080
Pessimistic scenario	100	2080

Sandy Beaches

The main tourist beaches are located at the mouth of the estuary. Sardinero beaches are located in the northeast side of the city, whereas the urban beach of Magdalena is found in the shelter of Magdalena Peninsula, in the east side of the city (see Figure 7.107). This beach is maintained through yearly nourishment. The volume of sand added is between 20,000 and 30,000 m^3. However, new breakwaters are expected to be constructed in the next several years to prevent sand loss. On the right side of the estuary, the sand spit (5.5 km) is included in the Natura 2000 and supports intense sport uses (e.g., surfing and kite surf).

Saltmarshes

Saltmarshes at Santander Bay are located in the inner and sheltered zone of the estuary (mouth of the Miera River). The main communities are formed by *Spartina maritima* and mixed vegetation of *Juncus* spp, *Halimione portulacoides*, and *Sarcocornia* sp. Marshes included in the Site of Community Importance (Habitat Directive) ES1300005 Dunas del Puntal y Estuario del Miera, are managed by the regional authority of Biodiversity (Government of Cantabria). Marshes not included in the Site of Community Importance are managed by the regional authority of hydraulic works and water cycle (Government of Cantabria).

Seagrass Beds

Santander Bay houses important meadows of *Zostera marina* and *Zostera noltii*. The presence of these species is well represented on the tidal flats of the right side of the Bay. In areas with high current velocities and dominated by coarse sediments, *Zostera* sp forms scattered and small meadows (e.g., next to the navigation channel). Meanwhile, in areas with low hydrodynamics and fine sediments, *Zostera* sp grows in dense and wide meadows (e.g., The Barqueria is the largest and best-preserved meadow in the bay of Santander). Although in the past the areas of *Zostera* sp were seriously damaged, today it is present from the mouth to the inner areas of the Bay. In addition, it has been able to settle in areas with levels of human impact formerly incompatible with their presence (Ondiviela et al. 2014).

Intertidal Mudflats

Large areas of Santander Bay (65% of the total area) are characterized by large intertidal mudflats. The importance of these ecosystems is due not only the surface area occupied, but also their ecological and economical value; richness and diversity on the intertidal flats are significantly higher than any other area in the Bay. The intertidal flats play an important role for migrating birds. Moreover, the Bay also contains the *Abra* community (a macroinvertebrate community characterized by *Abra alba*), the richest community identified in the Cantabrian Sea. The macroinvertebrate species present in the estuary of Santander are characteristic of sheltered soft bottoms with a moderate degree of disturbance. The most abundance species are the gastropods *Hydrobia ulvae* and *Rissoa parva*, the bivalves *Abra alba* and *Venus striatula*, the polychaete *Paraonis lyra*, *Notomastus latericius* and *Polydora* sp, and the crustacean *Diogenes pugilator*. It is possible to find species indicative of uncontaminated areas such as the polychaete *Scoloplos armiger*, the sea urchin *Echinocardium* sp, and the crustaceans *Apseudes latrelli* and *Tanais dulongii*.

Rocky Shores

Intertidal rocky shores in Santander Bay are restricted to the cliffs and islands located in the mouth of the estuary. The zoning of the intertidal communities follows the characteristic patterns of the Cantabrian Sea. The middle and lower intertidal levels are occupied by communities of the red algae *Corallina elongata*. In shallow and semihorizontal flats, this community forms a nearly continuous belt, being reduced to a narrow strip on vertical and very exposed walls. The community of the brown algae *Bifurcata bifurcaria* presents a more localized distribution, replacing *C. elongata* in the lower intertidal level of sheltered environments. The presence of *Halopteris scoparia* is scattered, except along horizontal rocky flats, where it becomes codominant with *C. elongata*.

7.9.2.4 Socioeconomic System

Regarding the macroeconomic figures of the area, Cantabria (Santander regional authority) represents around 2% of the overall Spanish economy and has showed a long-term decreasing relative trend. On the other hand, the GDP per person is close to the average in Spain. The main activities are services (with a heavy dependence on tourist activities), industry, and agriculture. The population in the area has followed the global aging trend in Spain and hence has a heavy dependence on social services at present and is expected to increase in the future.

In the city area, the main activities observed are residential with diverse densities and traffic restrictions and a service area specialized both in commercial and administrative activities. This heavily populated area is located on former wetlands that have been reclaimed by the town, the original settlement being located on higher areas. This core urban area is surrounded by a development area where competing demand between residential uses, industrial activities, and infrastructures exists. Because of this, conflicts between urban authorities, port, airport, motorways, railways, and electrical companies and industrial developments exist. A key agent in the area is the Port Authority that competes with urban administration for land area and uses. It develops a subset of industrial and service activities that cover stevedoring, storage, industrial, and maritime maintenance; yachting; and sport activities whose area of influence interact with industrially based activities. These peripheral areas have been located in low areas and coexist with natural and restored natural areas that have not been reclaimed by the cities.

On the eastern area of the bay, numerous cattle farms exist. These have been a traditional source of organic discharge into the marine habitat. This activity has been transformed during the past 20 years, changing from widespread traditional farms into larger concentrated units. This area of the Bay also has a broad set of beautiful beaches that provide the basis for tourism development.

7.9.3 Flood and Erosion Risk Modeling

7.9.3.1 SPRC Approach

The complexity of interactions in the coastal zone suggests that the impacts of climate change can be affecting Santander Bay in many ways. The risk analysis from inundation and coastal erosion is carried out using a methodology based on the SPRC model (see Chapter 2).

Following the SPRC scheme, the main sources are defined (Section 7.9.2.2) and the pathways and receptors characterized to identify the main ecological, social, and economic impacts and their consequences. The coastal system is divided into a number of homogeneous impact units where the hazard statistical distributions, both in intensity and frequency, are examined (Losada, Méndez, Izaguirre, Mínguez, & García, 2012).

7.9.3.2 Statistical Risk Modeling

For every impact unit and hazard, the risk (or expected annual damage) is defined as the sum of the damages that occur throughout the year. For rare events this can be expressed in terms of the PDF of the intensity of the hazard, $f_Z(z)$, and the frequency (Poisson process with parameter λ that models the average number of events per year):

$$R = \lambda \int_{z_u}^{\infty} f_z(z)E(z)v(dz)dz \tag{19}$$

In this equation, $E(z) \cdot V(z)$ represents the exposure and vulnerability functions that are directly linked with the consequences or damages in the inland system.

Depending on the exposure, once the hazard exceeds a threshold (z_u), the damage is generated on the inland system (D). Each pair of exposure level (z) and damage (D) defines a damage function characteristic of each impact unit (Figure 7.118). The intensity of the hazard is modeled as shown in the previous section, whereas the damage follows a Poisson process that can be also modeled through the rate of frequency (λ). The total annual damage in the impact unit is:

$$S = D_1 + D_2 + \ldots + D_N \tag{20}$$

The distribution of the damage is characterized by the mean and the variance:

$$\begin{aligned} \mu_S &= mean(S) = E(N)E(D) \\ \sigma_S^2 &= \text{var}(S) = E(N)\text{var}(D) + \text{var}(N)E^2(N) \end{aligned} \tag{21}$$

Taking into account the assumptions that the intensity of the hazard is Pareto distributed (characterized by parameters σ and ξ), the frequency of the damage follows a Poisson process (with λ parameter) and the damage function is characterized by: $D(Z)=a \cdot Z^b$, the mean and the variance of the damage distribution function (risk) can be obtained by:

$$\mu_S = \lambda u_z \cdot a \left(\frac{\xi}{\sigma}\right)^{-b} \frac{\Gamma(1+b)\cdot\Gamma(-b+1/\xi)}{\Gamma(1/\xi)}$$

$$\sigma_S^2 = \lambda u_z \cdot a^2 \left(\frac{\xi}{\sigma}\right)^{-2b} \left[\frac{\Gamma(1+2b)\cdot\Gamma(-2b+1/\xi)}{\Gamma(1/\xi)} + \frac{[\Gamma(1+b)\cdot\Gamma(-b+1/\xi)]^2}{[\Gamma(1/\xi)]^2}\right] \tag{22}$$

Following this methodology, the current risk in an impact unit associated to a certain hazard (flooding or erosion) can be defined and the future risk associated to climate change conditions estimated.

The risk associated to climate change is a random variable with a normal distribution with mean $\Delta\widehat{R} = \widehat{R}_{2050} - \widehat{R}_{2010}$ and standard deviation $\sigma_{\Delta\widehat{R}} = \sqrt{\widehat{\sigma}^2_{R,2050} + \widehat{\sigma}^2_{R,2010}}$. The statistical characterization of the future damage allows the estimation of the increase of risk.

7.9.3.3 Key Source Drivers of Future Risk Conditions

The flooding and erosion risk assessment at Santander Bay has been developed defining different hazard indicators (shown in parentheses): (1) city flooding (total sea level); (2) beach flooding (flooding level including wave run-up); (3) long-term erosion (shoreline retreat); and (4) episodic erosion (shoreline retreat). Each indicator is obtained combining different marine dynamics parameters such as: annual mean sea level, significant wave height exceeded 12 h every year (H_{s12}), annual mean energy flux direction, and 50-year storm surge level.

The indicator can be associated to any change of the base line (present climate and present socioeconomic and environmental conditions). It is important to highlight that the driver may be associated to deviations of the mean and the variance of the random variable X that defines the hazard.

For future risk conditions, three horizon years, representing short-term (2020s), medium-term (2050s), and long-term (2080s) periods were selected. In Santander Bay, and after a careful examination of the statistical distributions of different marine dynamics parameters, the indicators are ranked in this order: SLR, changes in H_{s12}, and changes in mean wave energy flux direction (see Tables 7.72 and 7.73). SLR is the main driver increasing flood risk in Santander Bay throughout the twenty-first century. No significant changes have been found in storm surge extremes and river discharge (low effect in flooding in Santander).

The anticipated SLR is the most significant indicator of changing risk in the Santander area. As an example, the present total sea-level extreme distribution is bounded tailed (50-year return value is 2.56 m above actual MSL and 200-year return level is 2.76 m, only 20 cm of difference). This means that an increase of sea level of 0.2 m (assuming no changes in storm surge climate) produces a risk multiplier effect of 4 (THESEUS OD1.10, 2012).

TABLE 7.72 Ranking Table for the Hazard Indicators in Santander Bay

	Period		
Rank	2020s	2050s	2080s
1	Sea-level rise	Sea-level rise	Sea-level rise
2	Wave height	Wave height	Wave height
3	Wave direction	Wave direction	Wave direction
4	Surge	Surge	Surge
5	River discharge	River discharge	River discharge

TABLE 7.73 The 50-Year Return Level Quantile of Coastal Flooding (Cl_{50})

	Sardinero 2	Sardinero 1	Loredo	Somo	Puntal	Peligros
50-year return level quantile (m)	8.8	8.6	9.7	9.2	8	7.1

7.9.3.4 Methods for Assessing Flooding and Erosion

The lack of a high-resolution digital terrain model and, therefore, the use of an appropriate flooding model poses limitations in flood and erosion assessments results at a high-resolution scale. To avoid this, flooding and erosion were investigated using the indicators described in the following.

Flooding Levels

Coastal flooding is a random phenomenon, a consequence of the combination of different marine dynamic processes. Flooding level is the result of the addition of four factors: MSL, astronomical tide (AT), storm surge (SS), and run up (RU).

$$\mathrm{FL} = \mathrm{MSL} + \mathrm{AT} + \mathrm{SS} + \mathrm{RU} \tag{23}$$

A pragmatic approach based on the combination of statistical models and different scenarios for SLR is used to assess future flooding levels. The approach consists of extrapolating the long-terms of wave climate, storm surge parameters, PDFs, and assuming the scenarios proposed for mean SLR. Therefore, the calculations of extreme sea level and flooding level PDFs have been obtained perturbing the base scenario (actual climatology) with long-term trends with the different SLR scenarios considered.

Erosion Indicators

Storms induce beach erosion, which results from the removal or redistribution of sediment from the dry beach under the combined action of waves and storm surge. Waves attack the dry beach, entraining sediment into the water column where it is easily transported by the local currents. The storm tide (astronomical tide + storm surge) governs how much of the beach is exposed to direct wave attack, whereas the storm duration determines for how long.

To assess the observed erosion impact, we use the empirical approach developed by Miller and Livermont (2008) that consists of two indices: the Storm Erosion Index (SEI) and the related Peak Erosion Intensity (PEI) index. It differs from traditional measures in that they account for the joint occurrence of storm waves and elevated water levels in a physical way. The PEI index, or equivalently the equilibrium shoreline change for each set of observed forecast conditions is defined as:

$$PEI(t_i) = W_*(t_i)\left[\frac{0.068H_b(t_i) + S(t_i)}{B + 1.28H_b(t_i)}\right] \tag{24}$$

where $W_*(t_i)$ is taken as the distance to the depth limited breakpoint $h_*(t_i) = h_b(t_i) = 0.8H_b(t_i)$. $S(t_i)$ is the water level variation across the surf zone. The berm height B is assumed to be constant over time (see Miller & Livermont, 2008). Note that water levels are relative to MSL and include both the astronomical and surge component. However, the projected erosion impact is calculated considering three factors: MSL, changes in H_{s12}, and changes in the mean direction.

Changes in the beach profile are estimated using Brunn's rule. Beach profile change in a changing climate can be due to SLR or changes in the significant wave height only exceeded 12 hours a year (H_{s12}). SLR induces a sand deficit in the beach active profile, which is compensated by eroding the upper part of the profile, inducing a shoreline retreat of the beach. A commonly applied rule to explain erosion of sandy shores in response to SLR, induced by climate change, is Bruun's rule (Bruun, 1962). This describes the cross-shore response of a beach to SLR ($\Delta\eta$); the profile will retreat landward and upward as the water depth along the beach profile increases. Assuming that the equilibrium profile shape is determined by Dean's formulation (Dean, 1977) and the closure beach is that corresponding to the formulation given by Birkeimer (1985) then the retreat of the beach profile is proportional to the variation on the average level. Changes in the H_{s12} also generate shoreline retreat. This is calculated following the equation:

$$RE = 2.95\frac{\sqrt{H_{s12}}\delta H_{s12}}{\left(0.51w^{0.44}\right)^{3/2}} \tag{25}$$

where δH_{s12} is the driver (difference between future and present H_{s12}). The shoreline retreat is proportional to the variation of significant wave height and the coefficient of proportionality depends on the current significant wave height and grain fall velocity. The formulation obtained indicates that the greater incident significant wave height on the beach and the smaller the size of the sediment, the greater increase of the closure depth in the beach. This action will generate a general retreat in the beach whenever the beach is oversupplied of sand. This retreat will be more important in dissipative beaches.

Concerning the beach plan, its form is governed by the mean energy flux direction associated with incident waves on the beach. If the effect of climate change produces a change in this parameter, the beach will suffer a turn in its plan form. The new orientation will be perpendicular to the new direction of the mean energy flux. This change in orientation will be accompanied by a future advance and retreat of the beach, so that the volume of eroded sand is equivalent to the volume of sand deposited on the front of the beach.

Statistical Analysis

The extreme statistical characterization of flooding and erosion has been carried out using the generalized Pareto distribution (GPD) for characterizing the intensity (with parameters σ and ξ) and Poisson model for characterizing the frequency (with parameter λ) of the events. The sample of maxima is selected using the POT method for independent and identically distributed random variables (Davison & Smith, 1990).

The combination of the Poisson model for frequency and the GPD model for intensity can be expressed (Smith, 2001) in a form compatible with GEV distribution, which characterizes the classical model on extreme value theory. The GEV distribution (characterized by location, μ, scale, ψ, and shape, ξ parameters) includes three distribution families corresponding to the three different types of tail behavior: Gumbel family, with a light tail decaying exponentially; Fréchet distribution, with a heavy tail decaying polynomially; and Weibull family, with a bounded tail.

An extension of the GPD-Poisson model is obtained by allowing its parameters to be time-dependent so that $\nu(t) > 0$, $\sigma(t) > 0$ and $-\infty < \xi(t) < \infty$ may vary throughout the year. Accordingly, the GEV distribution inherits time-dependent parameters $-\infty < \mu(t) < \infty$, $\psi(t) > 0$ and $-\infty < \xi(t) < \infty$. This approach allows modeling the characteristics of the variable that change systematically due to the natural climate variability. The functions $\mu(t)$, $\psi(t)$, and $\xi(t)$ can contain sine waves representing seasonal effects (Katz, Parlange, & Naveau, 2002), linear and/or exponential terms representing long-term trends, and covariates representing environmental processes (e.g., the North Atlantic Oscillation, see Izaguirre, Mendez, Menendez, Luceno, & Losada, 2010) among others. Some examples of this application can be found in Méndez et al. (2006), who used this approach to model the long-term variability of extreme significant wave height in the Pacific (an extended description of the time-dependent model can be found), Menéndez et al. (2008) or Izaguirre et al. (2013).

7.9.3.5 Observed Flooding and Erosion Impacts

Flooding

The time series of the flooding level for the six objective points (see Figure 7.107) were obtained. The flooding level depends on the exposure of the beach and therefore on the energy that it receives. In the case of Loredo Beach, some sea states reach 10 m of flooding level. Puntal and Peligros, which are sheltered by La Magdalena Peninsula and are located in the entrance of the Bay registered the lower flooding levels. To characterize the extreme flooding climate, the statistical model Pareto-Poisson has been applied, obtaining the extreme flooding climate. The 50-year return period quantile of coastal flooding for each beach is summarized in Table 7.71.

A nonstationary model including seasonality (with harmonic functions in the location and scale parameter, see the statistical analysis in Section 7.9.3.4) and long-term trends was carried out, only including them when significant at a 0.05 confidence level (gray values do not satisfy this criterion). Note that SLR is not included in the calculation of flooding level, therefore, although there is no significant long-term trend in the extreme flooding climate it will present higher values because of the SLR. Table 7.74 shows the value of the fitted parameter and its standard errors (in parentheses). The main annual variability is due to the annual cycle, when the high events are registered.

TABLE 7.74 Values of the GEV Parameters Fitted in the Objective Points (Standard Errors are Shown in Parentheses). Values are Given in Meters Except βLT, which is Expressed in Centimeters

Location	β0	β1	β2	β3	β4	βLT	α0	α1	α2	ξ0
Sardinero 2	7.38	0.89	0.16	−0.18	−0.12	0.22	0.31	0.006	0.01	−0.26
	(0.04)	(0.04)	(0.04)	(0.03)	(0.03)	(0.1)	(0.01)	(0.01)	(0.01)	(0.01)
Sardinero 1	7.31	0.85	0.15	−0.17	−0.11	0.21	0.30	0.006	0.01	−0.27
	(0.04)	(0.04)	(0.04)	(0.03)	(0.03)	(0.1)	(0.01)	(0.01)	(0.009)	(0.01)
Loredo	8.10	1.05	0.16	−0.14	−0.10	0.3	0.35	0.02	0.009	−0.28
	(0.05)	(0.04)	(0.04)	(0.04)	(0.04)	(0.1)	(0.01)	(0.01)	(0.01)	(0.02)
Somo	7.62	0.95	0.16	−0.17	−0.11	0.25	0.33	0.01	0.01	−0.26
	(0.05)	(0.04)	(0.04)	(0.03)	(0.03)	(0.1)	(0.01)	(0.01)	(0.01)	(0.01)
Puntal	7.04	0.72	0.12	−0.15	−0.10	0.17	0.27	0.0004	0.008	−0.29
	(0.04)	(0.03)	(0.03)	(0.03)	(0.02)	(0.1)	(0.009)	(0.01)	(0.008)	(0.02)
Peligros	6.34	0.50	0.08	−0.11	−0.08	0.17	0.2	−0.008	0.004	−0.32
	(0.03)	(0.02)	(0.02)	(0.02)	(0.02)	(0.08)	(0.008)	(0.01)	(0.007)	(0.02)

Concerning the frequency of extreme events the biggest rate of flooding events corresponds to Puntal and Peligros beaches during the winter months, which in terms of intensity are the locations with lower flooding level. The annual cycle is the main intraannual variation and is perfectly captured with the fitted model.

Erosion

The PEI index time series was obtained for each point (see Figure 7.107). Loredo is the beach with higher erosion episodes because of its exposure to the most energetic sea states, reaching 50 m of berm erosion in 1965.

A similar analysis of flooding levels has been performed to characterize extreme erosion climate. The quantile of PEI associated to 50-year return period in the present climate correspond to 45 m in Loredo, whereas in Puntal beach it rounds 27 m. Once the stationary extreme erosion climate is characterized, the nonstationary Pareto-Poisson model was fitted to assess variability throughout the year and in the long term. The only significant long-term trend was detected for Peligros Beach. As in the case of flooding, SLR is not considered. Although there is no significant long-term trend from wave climate and storm tide, there will be an increase in erosion from SLR. Consistent with flooding results, Loredo is the most eroded beach, with a higher intensity of PEI index in winter season. In the most exposed beaches, the annual cycle is clear and predominant.

When analyzing the frequency of erosion events, it is clear that, unlike the flooding results that show a perfect annual cycle (maximum flooding events rate in winter), the maximum erosion rate happens in November–December, which means the importance of the semiannual cycle, particularly in the fall (Figure 7.108). The highest erosion rate occurs in Sardinero beaches and Puntal.

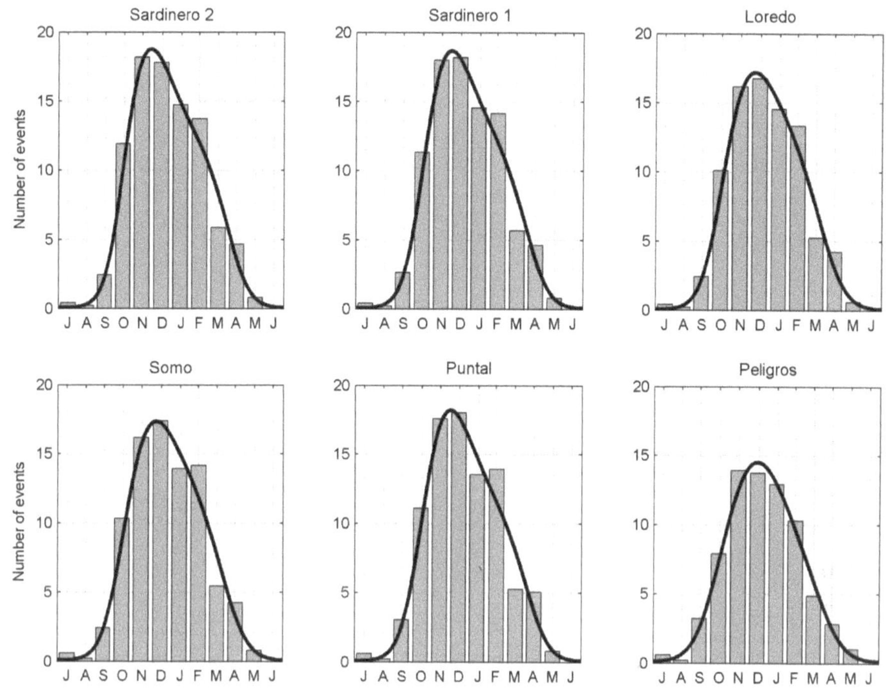

Figure 7.108 Histogram of the number of events of threshold excesses and time-dependent event rate $\widehat{v}(t)$ within a year in an average year (solid line), according to the fitted model.

7.9.3.6 Projected Flooding and Erosion Impacts

Flooding

SLR induces a shift of the PDF such that higher values are more probable. Hence, it can be deduced that logically the maximum flooding elevation will rise in all the beaches. The distortion of the PDFs are associated mainly to changes in mean sea level, increasing the probability of exceedance of a given beach level. As an example, Figure 7.109 shows the PDF of coastal flooding in Sardinero 2 under the present climate conditions, the moderate change (SLR = +0.2 m), the high change (SLR = +0.4 m), and the pessimistic scenario (SLR = +1 m).

Besides, the representative parameters (location, scale, and shape) of the GEV distribution for the six beaches and the scenarios analyzed are summarized in Table 7.75.

The flooding analysis in the inner bay has been carried out using the pragmatic approach described in section 7.9.3.4. The scenarios summarized in Table 7.76 that combine extreme flooding levels with relative SLR projections have been used to assess the flooding in Santander Bay.

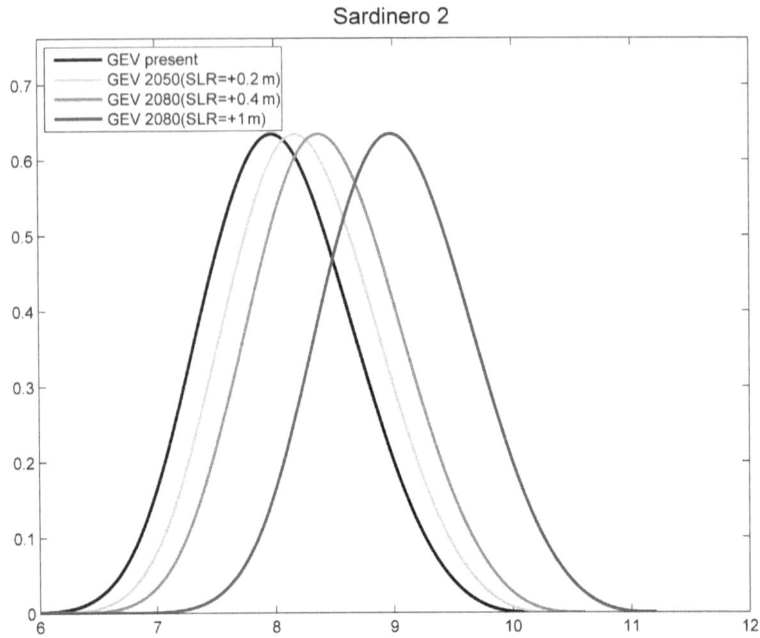

Figure 7.109 PDF of coastal flooding in Sardinero two for the different scenarios of sea-level rise: the present conditions (dark line), moderate change for year 2050 (+0.2 m), high change for year 2080 (+0.4 m), and pessimistic scenario for year 2100 (+1 m).

TABLE 7.75 Summary of the Location, Scale, and Shape Parameters of the GEV Distributions for the Six Beaches Analyzed and Scenarios Proposed

	Location Parameter (μ)				Shape Parameter (ε)				Scale Parameter (Ψ)			
	Present climate	+0.2	+0.4	+1	Present climate	+0.2	+0.4	+1	Present climate	+0.2	+0.4	+1
Sardinero 2	7.81	8.01	8.21	8.81	−0.25	−0.25	−0.25	−0.25	0.6	0.6	0.6	0.6
Sardinero 1	7.72	7.92	8.12	8.72	−0.26	−0.26	−0.26	−0.26	0.35	0.35	0.35	0.35
Loredo	8.62	8.82	9.02	9.62	−0.25	−0.25	−0.25	−0.25	0.43	0.43	0.43	0.43
Somo	8.08	8.28	8.48	9.08	−0.24	−0.24	−0.24	−0.24	0.38	0.38	0.38	0.38
Puntal	7.38	7.58	7.78	8.38	−0.28	−0.28	−0.28	−0.28	0.3	0.3	0.3	0.3
Peligros	6.59	6.79	6.99	7.59	−0.33	−0.33	−0.33	−0.33	0.22	0.22	0.22	0.22

TABLE 7.76 Summary of Flooding Scenarios

Scenario	Year of Return Period	Flooding Level (cm)	Relative Sea-Level Rise (cm)	Total Sea Level	Horizon Year
A	50	256	20	276	2050
B	50	256	40	296	2080
C	50	256	100	356	2080
D	200	276	20	296	2050
E	200	276	40	316	2080
F	200	276	100	376	2080

Figure 7.110 shows the flooding levels associated with a return period of 50 years and an SLR of +1 m (scenario C in Table 7.76). As can be seen in this figure, several areas will be affected by flooding, such as various areas of the Cubas sea inlet (see Figure 7.110(c)). These areas are currently reclaimed, but in the past belonged to the intertidal area of the estuary. Other areas affected belong to Alday marshes (see Figure 7.110(a)). At present, these marshes are regulated hydrodynamically by means of a system of sluices (see Figure 7.110(b)).

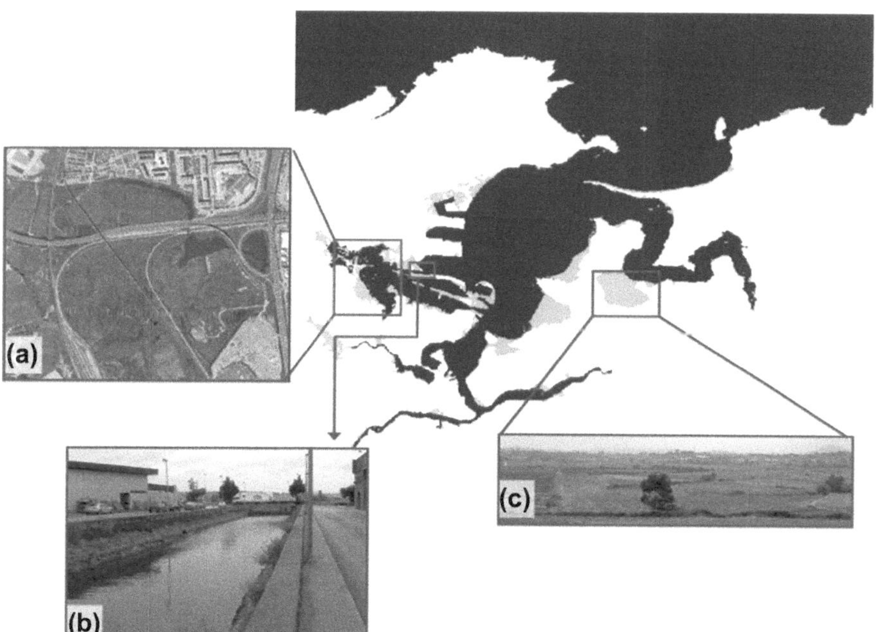

Figure 7.110 Different areas affected by flood level in Santander Bay for scenario C.

TABLE 7.77 Summary of Erosion in Each Beach of Santander Bay for Different Scenarios

	Driver	Horizon Year	Sardinero 2	Sardinero 1	Loredo	Somo	Puntal	Peligros
Shoreline retreat (m/year)	Relative sea-level rise	2080	45.65	39.21	58.95	43.25	16.88	
Shoreline retreat (m)	Changes in H_{s12}	2050	18.4	15.67	52.39	26.66	4.51	1.86
Plan form swing (°)	Changes in the mean energy flux	2080	17.43	20.84	335.84	348.24	19.2	96.26

Erosion: Beach Profile and Plan Form

The shoreline retreat induced by SLR has been obtained based on the Bruun's rule and the SLR projections (Table 7.71). Table 7.77 summarizes the results of the erosion in the beach profile and the plan form. First, the shoreline retreat associated to +1 m of SLR is shown. It is important to remark that these results assume that the only relevant process is the cross-shore sediment transport and that there is no exchange in the longitudinal sections. Second, the total shoreline retreat (in meters) for each beach in 2050, considering the tendencies for H_{s12} can be found. The most affected is Loredo Beach, whereas Peligros Beach has the smallest effects. This is largely because the waves reaching the latter are completely affected by the wave diffraction caused by the Magdalena Peninsula. The results obtained suggest that not only SLR but also long-term changes in H_{s12} can produce important shoreline retreats in the analyzed beaches. Finally, calculation of the mean energy flux direction is crucial for several coastal processes. The long-term trend is negative (i.e., leading to erosion), with the exception of Peligros Beach. These new directions of the mean energy flux will induce changes in the beach orientation and consequently areas of accretion and erosion in the bounds of the beaches.

7.9.4 Mitigation Measures

Several mitigation options were considered, see Table 7.78. These were classified as coastal works, inland intervention, and nonstructural measures, some of which were already implemented in recent years.

TABLE 7.78 Summary of Mitigation Options Including the Type and List of Measures, Objective, and Examples of Application in Santander Bay

Type of Measure	Objective	List of Measure	Example of Application in Santander
Coastal works	Physical protection reducing the run up. Overtopping and velocity.	Beach nourishment Construction of coastal structures: Riprap protection, seawalls, groins Plan retreat Dredging and landfill Increasing the defense crest height Dune rehabilitation	Beach nourishment in the system Magdalen-Peligros beaches, plan retreat developed in the inner bay to recover marsh areas
Inland interventions	To reduce the economic losses and human life at risk by reducing the exposure and, in consequence, the damage function	Construction or upgrading of embankments Landfill Reinforcing of existing infrastructures Protection to drainage systems Relocation of vulnerable activities	Construction of new embankments in El Chiqui area (northwest part of El Sardinero beaches) to reduce the exposure to flooding events.
Nonstructural measures	To reduce economic and social losses acting through policies and management strategic	Land use planning and management Business recovery Evacuation and emergency Insurance Social resilience Dissemination of information on actual risk and self-protection measures Site-specific evacuation plan issues Inventory of critical infrastructures and contingency plans Incentives Networking strengthen Self-aid procedures and public participation on plans	

Specific attention has to be paid to the insurance system as a mitigation option. Obviously the exposed individuals would be better off if they are protected with an insurance coverage and hence individually the instrument is adequate. Nevertheless, the aggregate impact of insurance on the society has to be carefully assessed. On one hand the existence of a collective protection that socially supports individual damages creates an incentive not to take prevention measures, through the well-known concept of moral hazard. On the other hand the redistribution of costs through society is not necessary a global improvement, first because if are not facing uncorrelated risks from a hazard but correlated risks from climate change, and second because rent effects may alter the social welfare balance if redistribution is not cautious made.

Specifically, most of the properties of the state are insured because of financial firm pressures and to the risk memory derived from past experiences on urban fires (mainly the urban fire of 1940). The development of the tourism and the activities in the coastline and the expectation to be protected with public funds and hard infrastructures has resulted in an overexposition to climate change and especially SLR.

To analyze the cost efficiency of the mitigation measures available, proposed schemes for three different areas were investigated. The purpose of the set of proposal was essentially to create a general guide of the available solutions that should be considered when the analysis is replicated in other sites, covering general uses as residential, beach and recreational areas, and important infrastructures. Hence for each hot spot, we have proposed a set of eventual solutions that can be discussed on the decision-maker's desk and show how the climate change science can help in the evaluation process. These proposals have been preliminarily quantified as a draft case that can be exported and cannot be considered a technical proposal; furthermore, some of the proposals would hardly be considered in our specific site, but could be adequate elsewhere. Therefore, not considering them in our case would have imposed our local restrictions to an eventual user that would use our solutions as a catalog.

With this approach in mind, and only with the purpose of scaling the predefined solutions and asset costs, a set of geometric parameters have been introduced in the analysis. The results offered should be considered more a validation test for the methodology than for the solutions.

7.9.4.1 Residential Area: Somo Village

Somo is a small village located in the west side of Santander Bay. It has a permanent population of around 1700 people, growing considerably during summer because of its tourist attraction: long, sandy beaches (Figure 7.111). The problems in this area derive from three main reasons: the village has moved close to the beach, assuming excessive risks; the sand dunes that used to protect the area have been removed; and the increasing storm levels have accelerated the overexposure to risks.

Flooding and erosion impacts that are already identified will be stressed because of changes in the extreme waves and storm events as a consequence of climate change. The current return period for flooding is estimated as 1 year. To protect the population and assets exposed at risk, three mitigation measures are proposed.

The first mitigation option consists of the nourishment of the beach, seeding of the dune with appropriate plants, and the movement of structures. With this option, the objective is to affect the pathway, reducing the coastal flooding, and affect the receptors, reducing the vulnerability. Some design parameters as cross-shore profile, dune elevation, and sediment characteristics (stable sediment distribution, D_{50}) must be defined based on literature formulations and the nearby dunes. Techniques of topographical reconstruction, carried out by supplying sand with machinery, will be used for this purpose. The source of sand may or may be not located within the dune system being rehabilitated. Once the dune ridge has been stabilized, it must be fixed by planting dune species, which in their natural state are responsible for the maintenance of the dune. There is a nursery of dune species near the study area. Taking into account the retreat of socioeconomic assets and the characteristics of this mitigation measure, it has been designed to obtain a return period for coastal flooding of 15 years. This return period design would imply appropriate dimensions of the dune and a long enough period for vegetation to settle and stabilize the dunes.

Second, a submerged dike in front of the beach is proposed to induce wave-breaking (affecting the source), reducing the energy that arrives at the beach and, as a result, erosion rates. Taking into account the cost of the construction works and the high level of socioeconomic activities in the protected area, this measure has been designed to obtain a return period for flood events of 25 years. This implies no requirement of maintenance or failure of the structure over a long enough period to not disturb the assets and socioeconomic activities in the exposed area.

Finally, a riprap wall protecting the beach is proposed to modify the pathway, reducing the coastal flooding. In this case, the mitigation option has been designed to achieve a future flooding return period of 50 years. This high return period design was selected because of the high tourist interest in the area and to avoid frequent disturbance that may be associated with construction, maintenance, and/or the failure of the structure. Note that in all cases the aim is to obtain the same benefit in terms of reducing consequences to the village of Somo.

Figure 7.111 Location of Somo and view of the village.

TABLE 7.79 Summary of the Mitigation Options Proposed for Somo and the Main Design Parameters

Hot Spot	Mitigation Option	Action	Measure	Return Period	Design Parameters
Somo Beach	Do nothing	Wait and pay strategy	0	1 year	NA
	Beach nourishment, plant seeding, and construction retreat	Construction abandonment Dune plant seeding and protecting Beach nourishment	25 houses 50 Ha 1,000,000 m^3	15 years	$D_{50} = 0.025$ (sand obtained from drainage from the sea bed at deep areas) Dune crest = +6 m
	Submerged dike	Dike construction	800 m	25 years	Slope = 2/1 Breakwater crest = 0 m Quarry stones W_{50} = 5 tons (density = 2650 kg/m^3) Crest width = 4 m
	Riprap wall protection	Wall construction Construction abandonment	1000 m 25 houses	50 years	Slope = 2/1 Breakwater crest = + 6 m Quarry stones W_{50} = 5 ton (density = 2650 kg/m^3) Crest width = 4 m

A summary of the mitigation measures described here can be found in Table 7.79, where the action proposed and the design parameters are provided.

7.9.4.2 Beach and Recreational Area: El Chiqui Area

The northwest side of El Sardinero beach is the so-called El Chiqui area (Figure 7.112). This is a recreational area provided with several hotels, a promenade, a scenic overlook, and the tourist attraction of El Sardinero beach (receptors).

The sources affecting this area are extreme waves and storm events. Coastal flooding already regularly causes economic losses affecting the promenade, urban furniture, and the Chiqui Hotel. The existing beach and vertical wall (pathway) provide insufficient physical protection against coastal flooding and will become even less effective with the influence of climate change.

Three mitigation options affecting the pathway are proposed for this site. The first consists of the raising of the sea wall and the elevation of the street, increasing the freeboard level (affecting the pathway). This option is designed with a 10-year return period for flooding events. The second mitigation option is based on a riprap dike protection along the promenade seawall designed for a return period for flooding of 10 years. These two options have been based on a low-medium return period because one of the main valuable assets of this recreational area is the scenery and tourist attractions. A hard and oversized measure has been avoided because it would cause negative side effects to the socioeconomic activities developed in the area. A good complement for these two options would be the design of an early warning system to minimize the results of flooding when the design return period is exceeded.

Finally, a submerged dike in front of the El Chiqui Hotel and its adjacent viewpoint, covering and enlarging the old and obsolete existing structures, is proposed as the third mitigation measure. The aim of this measure is to dissipate wave energy by causing bottom friction and wave-breaking. In this case, the design is for flood events with a 25-year return period because of the costs of construction and maintenance work, and, as a submerged structure, it will avoid any visual impact and associated negative consequences.

7.9.4.3 Infrastructures in the Inner Bay

The natural shelter provided by Santander Bay has caused the development of the city around the inner bay (Figure 7.113). Infrastructure such as the port, industry storage, the airport, or the waste water system have been designed and constructed based on the historical sea levels and storm-surge conditions.

The problem in this area derives from the drainage of wetlands and the development and building of low-elevation facilities. At present, the return period for tidal/meteorological flooding is around 5 years. In a SLR context, the infrastructures (receptors) are assets at risk, being susceptible to losing the designed reliability and operability. To avoid this, two mitigation options have been designed to modify the pathway and the sources.

Figure 7.112 Location of El Chiqui area and view of the hotel and scenic overlook.

Figure 7.113 Location of the inner bay infrastructures and aerial view of the port.

First, the reinforcement of perimeter protection to isolate used areas reaching a new design level is proposed, affecting the pathway to reduce the coastal flooding. This option is designed to achieve a return period for flooding of 20 years, a balance between the reduction of flooding, the visual impact of a hard solution in the socioeconomic activities, and disturbance of construction and maintenance works.

Second, endemic seagrass (*Zortera noltii*) farming in the intertidal flat areas of the bay is proposed in order to modify the wave and prism tide leading to a return period for flooding of 15 years. This return period design is chosen because it allows the settlement and regeneration of this endemic seagrass and its easy maintenance. The coastal protection provided by seagrasses depends on their biological structure and functions within the ecosystem and it is related to their capacity to attenuate waves, reduce current velocities, and promote sediment stabilization. The favorable areas (group 2 in Figure 7.114) represent the zones where the seagrass should be planted as the mitigation option proposed.

7.9.5 Cost and Benefit Analysis for the Mitigation Measures

A quantitative analysis is developed for each of the mitigation measures proposed in the previous section, considering return period for damages, damages avoided, and eventual residual damage after each measure would be eventually implemented.

For the avoided damages, a damage value associated for the affected area and its vulnerability to the phenomena is considered to quantify an expected damage for a catastrophic event. Then the impact of a sequence of catastrophic events were calculated considering (1) a total life of 50 years, (2) a probability of occurrence associated to the return period, (3) the moment of occurrence is determined averaging the discount factor during the return period, (4) 5% and 2% discount factors were applied, and (5) for convenience, the same affection severity was assumed, irrespective of the adopted measure.

To define the damage function we have considered standard percentages of the present estimates for value at risk in the affected area. Hence the residual damage generated is a function of the probability of the event (decreasing with the return period) and the number of similar occurrences in a 50 year period (also decreasing with the return period). The actual flow is adjusted for the extreme distribution previously defined.

The method assumes the convexity of the damage function, although this assumption may be highly unrealistic when low return periods are involved. When no protection measure is taken, the situation of an agent that is flooded every year with high certainty obviously produces a concave behavior when the agent abandons his or her settlement and puts an end to the never-ending theoretical damage flow. This eventual damage would be highly improbable and have been indicated to avoid confusion.

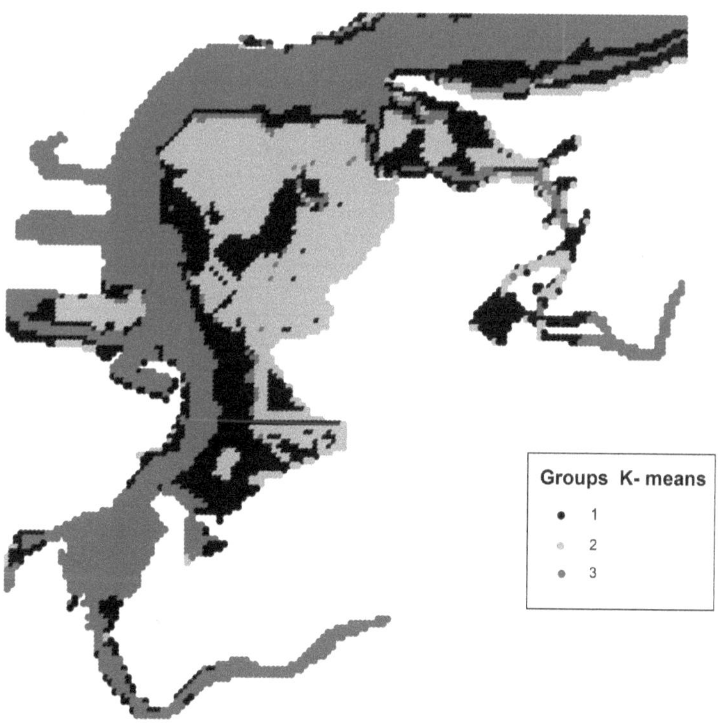

Figure 7.114 Potential habitat distribution (group 2) of *Zostera noltii* in Santander Bay.

7.9.5.1 Residential Area: Somo Village

The costs associated with the mitigation options are shown in Table 7.77 and the tradeoffs between risk reduction, money saved, and environmental impacts are showed in Table 7.78.

7.9.5.2 Beach and Recreational Area: El Chiqui Area

The costs associated with the mitigation options are shown in Table 7.79 and the tradeoffs between risk reduction, money saved, and environmental impacts are showed in Table 7.80.

7.9.5.3 Infrastructure in the Inner Bay

The costs associated with the mitigation options are shown in Table 7.81 and the tradeoffs between risk reduction, money saved, and environmental impacts are showed in Tables 7.82–7.87.

In any of the proposed areas, it can be seen that reaching a higher certainty level for present uses can only be obtained at a relatively high cost with two clear components: financial costs derived from hard measures built and environmental costs derived from avoiding the protection of "quasinatural habitats." Nevertheless, note that a no activity option could be financially attractive in the short term but it does not guarantee the survival of the present situation.

TABLE 7.80 Summary of the Mitigation Options Proposed for the El Chiqui Area and the Main Design Parameters

Hot Spot	Mitigation Option	Action	Measure	Return Period	Design Parameters
El Chiqui area	Do nothing	Wait and pay strategy	NA	5 years	NA
	Elevation of the street and wall	Wall construction	500 m × 50 × 0.5 m = 12,500 m^2	10 years	Promenade crest = +8.5 m
	Riprap dike protection	Dike construction	500 m	10 years	Slope = 2/1 Breakwater crest = +6 m Quarry stones W_{50} = 5 ton (density = 2650 kg/m^3) Crest width = 4 m
	Submerged dike	Dike construction	200 m	25 years	Slope = 2/1 Breakwater crest = 0 m Quarry stones W_{50} = 5 ton (density = 2650 kg/m^3) Crest width = 4 m

TABLE 7.81 Summary of the Mitigation Options Proposed for the Inner Bay Area and the Main Design Parameters

Hot Spot	Mitigation Option	Action	Measure	Return Period	Design Parameters
Inner Bay	Do nothing	Wait and pay strategy	NA	5 years	NA
	Reinforcement of perimeter protection to isolated used areas	Wall construction	5000 m	10 years	Slope = 2/1 Breakwater crest = +2 m Quarry stones W_{50} = 2 ton (density = 2650 kg/m^3) Crest width = 2 m
	Seagrass planting in intertidal flats	Wall construction	1000 m	10 years	Vegetation density = 2000 shoots/m^2
		Seagrass planting	100 Ha	25 years	

TABLE 7.82 Simulated Cost Analysis of Mitigation Measures in Somo Village

Mitigation Option	Action	Measure	Unit Prize	Total (€)
Beach nourishment, plant seeding, and construction retreat	Construction abandonment	25 houses	150,000 €/house	3,750,000
	Dune plant seeding and protecting	50 Ha	10,000 €/Ha	500,000
	Beach nourishment	1,000,000 m^3	4 €/m^3	4,000,000
	Total			8,250,000
Submerge dike	Dike construction	800 m	12,000 €/m^2	9,600,000
Riprap wall protection	Wall construction	1000 m	2500 €/m	2.500.000
	Construction abandonment	25 houses	150,000 €/house	3,750,000
	Total			6,250,000

7.9.6 Policy Implications

To analyze and decide the best appropriate strategic management for the bay, interviews and focus groups were carried out with public and private agents in the area (THESEUS OD1.15, 2012). These invited the participants to discuss their perception of coastal risks and climate change and their opinion on the diverse mitigation options that could be considered. During the summers of 2010 and 2012, two sets of interviews were undertaken. The results are reported in Tables 7.83–7.87. The profiles of the interviewee were both public agents involved in public management and private agents that develop their activities in the surrounding areas. The public group was composed of coastal managers from central and regional governments, Port Authority managers, and local county representatives. In the private group, Chamber of Commerce experts and general firms using coastal areas were questioned about their perceptions of risk exposition, their experiences dealing with past damages, and their conceptual models on the problem when climate change is involved.

For public agents, the main policy measures proposed were hard measures related with the protection of port facilities, coastal settlements, and firms. Institutional style measures based on regulation planning and vulnerability management in general were not considered as the first alternative. Hence institutional awareness and strengthen emerged among the first approaches to be developed.

TABLE 7.83 Summary of Results for Somo Village

Mitigation Option	Environmental Impact	Return Period/ (Prob)	Total Cost (€)	E(RD(residual damage))(*) (×1000 €) $r=5\%$	E(RD(residual damage))(*) (×1000 €) $r=2\%$
Do nothing (annual damage 12,330,000 €)	High	1 year (100%)	0	Not applicable (**)	
Beach nourishment	Marginal	15 years (38.06%)	8,250,000	144.5	240.7
Submerged dike	Medium	25 years (37.54%)	9,600,000	97.5	169.4
Riprap wall protection	High	50 years (37.16%)	6,250,000	46.8	82.9

(*) Expected value of Residual Damage E(RD) flows during 50 years discounted at a 5% and 2% rate.
(**) This total damage value is based on an expected annual damage occurring every year with probability one of 1 and so shows an abnormal value.

TABLE 7.84 Cost Analysis of Mitigation Measures in the El Chiqui area

Mitigation Option	Action	Measure	Unit Prize	Total (€)
Elevation of the street and wall	Wall construction	500 m × 50 × 0.5 m = 12,500 m^2	100 €/m^2	1,250,000
Riprap dike protection	Dike construction	500 m	1000 €/m	500,000
Submerged dike	Dike construction	200 m	12,000 €/m^2	2,400,000

TABLE 7.85 Summary of Results for the El Chiqui Area

Mitigation Option	Environmental Impact	Return Period/ (Prob)	Total Cost (€)	E(PV(residual damage))(*) (×1000 €) $R=5\%$	E(PV(residual damage))(*) (×1000 €) $r=2\%$
Do nothing (annual damage 1,020,000 €/year)	Medium	5 years (40.96%)	0	36.6	63.0
Elevation of the street and wall	Low	10 years (38.74%)	1,250,000	18.8	32.3
Riprap dike protection	Low	10 years (38.74%)	500,000	18.8	32.3
Submerged dike	Medium	25 years (37.16%)	2,400,000	8.3	14.5

TABLE 7.86 Cost Analysis of Mitigation Measures in the Inner Bay

Mitigation Option	Action	Measure	Unit Prize	Total (€)
Reinforcement of perimeter protection to isolated used areas	Wall construction	5000 m	1000 €/m	4,000,000
Seagrass planting in intertidal flats	Wall construction Seagrass planting	1000 m 100 Ha	1000 €/m 5000 €/Ha	1,000,000 500,000 1,500,000

TABLE 7.87 Summary of Results for the Inner Bay

Mitigation Option	Environmental Impact	Return Period	Total Cost (€)	E(PV(residual damage))(*) (×1000 €) $r=5\%$	E(PV(residual damage))(*) (×1000 €) $r=2\%$
Do nothing (annual damage 478,000 €)	Medium	5 years (40.96%)	0	7.1	12.2
Reinforcement of perimeter protection to isolated used areas	High	20 years (37.73%)	4,000,000	4.2	6.6
Seagrass planting in intertidal flats	Low	15 years (38.06%)	1,500,000	5.5	9.2

Insurance policies were proposed as the main mitigation option from affected stakeholders. To clarify the insurance opinion, the private group was specifically questioned about their interaction with insurance and their expectations about insurance coverage and incentives in the future. Two problems were observed: first, the demand of pooling risk from exposed agents should be tied with risk premium prices in insurance to incentive prudent behavior; and, second, insured businesses demand better clauses for their insurance contracts that trade off their financial uncertainty with permanent costs. Again, this is does not produce an adequate incentive to deal with eventual morally hazardous behavior. Transaction costs in the business insurance market are hence extremely relevant and should be focused on in policy instrument definitions.

Interestingly, the following were also key points that arose from analysis of the interviews and focus groups and that might influence future coastal management:

1. The perception of risk associated with climate change is generally not widespread. Only older people show a specific concern about it.
2. Public involvement on climate change issues is not the general rule. Social agents put their confidence in experts and managers to take the needed measures. Nevertheless, interested stakeholders such as nongovernmental organizations are increasingly educating people on the role that they should undertake in such public participation processes.
3. The complexity of the administrative network in coastal areas shows little capacity to produce innovative solutions that help to redefine the situation for coastal areas. Institutional, normative, and administrative frameworks need a critical review.
4. The provision of contingency and business recovery instruments for agents located in the coastal environment is an issue that needs to be encouraged.

7.9.7 Overall Conclusions and Recommendations

A flooding and erosion risk analysis was carried out in Santander Bay using the SPRC approach and state-of-the-art statistical modeling to characterize the damage distribution. The key source drivers identify for future risk conditions are SLR, wave height, and wave direction, which will be the base of the future scenarios considered. The observed flooding levels have been calculated in six beaches of the outer bay, with the most energetic beaches being Somo and Loredo where some sea states reach 10 m of flooding levels. Concerning the frequency of extreme events, the biggest rate of flooding events corresponds to the Puntal and Peligros beaches during the winter months, which in terms of intensity are the locations with lower flooding levels. The observed erosion impact is addressed through the PEI index and its statistical characterization.

The study of projected flooding impacts was based on three scenarios of relative SLR after obtaining flooding maps of Santander Bay. The future erosion impact is study through changes in the wave height and the mean energy flux direction.

Three groups of mitigation options are proposed for reducing flooding and erosion risk: coastal works, inland interventions, and nonstructural measures, where an insurance measure is included. Different examples of these mitigation measures are proposed in three pilot sites of Santander Bay: Somo (a residential area), El Chiqui (beach and recreational area), and the infrastructures of the inner bay. A description of each measure and the cost-impact analysis is provided.

Finally, the complex administrative structure in the Santander site with five national, five regional, and two local government bodies set a complicate framework for strategic management and policy. All of these government institutions participate in the management of the Santander estuary and therefore influence in flood risk and coastal risk management there. However, there is little interest in innovation from the institutions; they were reluctant to see change from traditional practices. The Port Authority in Santander has a crucial role to play in managing the estuary and hence managing the risk of flooding emanating from that source. An important problem is that there is a lack of coordination between organizations and too many parties involved to be able to proceed with innovation. The integrated planning of the estuary is not adequately managed, and although the State Department manages coastal areas, the port of Santander is excluded. Furthermore, the national-level legislative framework was not adequately followed through to the local level (Santander), but this was not seen as a barrier to innovation, rather just an impediment to coherent coastal zone management. Public consultation is resulting in uneven but greater innovation. The necessity is to reinvigorate the process of assessing innovative solutions to coastal risk problems, notwithstanding (or because of) the poor perception of risks that the public has about erosion and flooding into the future. An exogenous process needs to be initiated from central government level to ensure that local communities appraise the risks they face and develop new solutions that both reduce the burden of costs on the public purse and are environmentally sensitive.

The strategic management proposed commits on the reallocation of vulnerable activities (residential) and the modification of urban development procedures, including new exigencies (redundancy of infrastructures, roads, etc.). Furthermore, the institutional strengthen (through existing institutions) and the public coordination of private contingencies seems to be necessary measures. Finally, special importance is given for risk transparency, providing individual communications of exposure risk to property owners and insurance companies (incentives), networking to strengthen institutional agreements on background coverage, and contingency funds availability.

7.10

Teign Estuary, South Devon, UK: Stakeholder-Led Mitigation

David Simmonds [1], Dominic E. Reeve [2], Richard C. Thompson [3], Edmund Penning-Rowsell [4], Robert Nicholls [5], Shunqi Pan [6], Michael Hanley [3], Susan Hanson [5], Andrew Fox [7], Jose M. Horrillo-Caraballo [8], Simon Hoggart [3], Raul Gonzalez [7], Sum Kwan [9], David T. Bilton [3], Simon D. Rundle [3]

[1] *School of Marine Science and Engineering (Faculty of Science & Environment), Plymouth University (UoP), Plymouth, UK*

[2] *School of Marine Science and Engineering, Plymouth University (UoP) (currently: Swn-U College of Engineering, Swansea University) Drake Circus, Plymouth, UK*

[3] *Marine Biology and Ecology Research Centre School of Marine Science and Engineering, Plymouth University (UoP), Plymouth, UK*

[4] *Flood Hazard Research Centre, Middlesex University (MU), Hendon, London*

[5] *Faculty of Engineering and the Environment, University of Southampton (Soton), Southampton, UK*

[6] *School of Marine Science and Engineering, Plymouth University (UoP) (currently: School of Engineering, Cardiff University), Plymouth, UK*

[7] *COAST Research Group, School of Marine Science and Engineering, Plymouth University (UoP), Drake Circus, Plymouth PL4 8AA, UK*

[8] *School of Marine Science and Engineering, University of Plymouth (UoP), Drake Circus, Plymouth, Devon, UK*

[9] *Port Works Division, Civil Engineering and Development Department, HKSAR Government (UoP), 4/F, CEDD Building, 101, Princess Margaret Road, Kowloon, Hong Kong*

7.10.1 Description of the Site

The Teign Estuary is located on the Southwest Peninsula of the UK, on the northern boundary of the English Channel. The Peninsula itself represents one of the most diverse coastal settings in Europe, incorporating a broad range of habitats from exposed rocky and shingle coast to the sheltered mud of flooded valleys, or "rias," alongside densely populated urbanized and industrial zones of Plymouth Sound, Torbay, and Exeter (Horrillo-Caraballo et al., 2012). The Teign Estuary is an example of just such a ria. Running east to west, it is approximately 9 km in extent and less than 1 km in width at its widest point (Figure 7.115).

The River Teign rises on Dartmoor National Park 520 m above UK Ordnance Datum. Its catchment covers an area of 550 km^2 and includes two principal sub-catchments, one containing the Rivers Lemon and Bovey, and another the Aller Brook. The Teign flows in a southeasterly direction, through a diversity of landscapes and habitats before finally meeting its floodplain and discharging into the estuary. These range from open moorland to ancient woodland, improved pasture land, and broad valleys (Reeve et al., 2012).

The landmass in this region of the UK is subject to isostatic rebound following the most recent ice age. This and the slow sea-level rise over the twentieth century continue to have a strong control on the future risk of coastal flooding (Haigh, Nicholls, & Wells, 2009). Over the twenty-first century, an accelerated mean sea-level rise approaching 1 m as well as other anticipated effects of climate change will exacerbate the potential for flooding.

Figure 7.115 To the left, location of the Teign Estuary (UK); to the right, view of the beachfront at Teignmouth.

A major artificial control on the behavior of the estuary and coastal margins is the railway line, which occupies considerable stretches of the coastal frontage to the north of Teignmouth and also along the northern bank of the estuary. The railway runs along a blockwork sea wall beneath cliffs to the north of Teignmouth. It then turns inland and runs through Teignmouth and along the north bank of the estuary to Newton Abbot. Construction of this railway was started in 1846, and since then, the railway sea wall has cut off sediment supply from the red-brown *Permian breccia* cliffs to the littoral. The movement of beach levels against the wall can be large (in excess of 1 m over a single storm), enhancing the possibility of damage to the wall through scour, undermining, block removal, and exacerbating overtopping through larger waves reaching the wall near high tide.

Other manmade pressures are exerted on this system. These include physical disturbance, for example by trampling, dredging, fishing gear, land claim, and adjacent coastal development through the construction of sea defenses and implied potential for changes in the hydrological regime (Reeve et al., 2012).

Another important anthropogenic influence on the estuary is created by the need to maintain access to Teignmouth port. Historically, this used to be important for the export of monumental granite, but now maintains a vital link for the export of fine clay known as "ball clay," used in ceramics and other industries. The mouth of the estuary is marked by a permanent spit, "the Point," on the North bank at Teignmouth extending southwest, and the red-brown cliffs at Shaldon to the south. The shifting sands in the estuary mouth are "plow"-dredged on an almost daily basis.

Potential flooding events in the estuary would have consequences for three significant centers of population, the operations of a small but active port, and the railway, and would disrupt other activities that include small-scale fishing and economically important tourist industries. There is also the possibility of coastal modification due to erosion. The selection of appropriate and acceptable mitigation options is therefore an important issue in the area. Indeed, land use along the Teign Estuary is a complex distribution that includes residential properties, tourism, industry, coastal protection structures, and others.

In the UK, engagement between the authorities charged with managing areas of significant flood risk and the inclusion of stakeholders at risk is now integral to UK flood defense scheme appraisal. Within the estuary communities of Teignmouth and Shaldon, there has been significant levels of active stakeholder interest in recent schemes that affected the choice and design of the flood defense structures. A key aspect of this was the need for a decision balancing the tension between the judgment of expert engineers and scientists against the uncertainty and skepticism of stakeholders whose livelihoods and quality of life depend upon a more conservative approach to the planning and development of their communities. The resulting schemes demonstrate the benefits of engaging in open consultation between flood protection agencies and stakeholders with the clear intention of arriving at mutually acceptable and sustainable solutions. Essential to this is the effective communication of flood risk together with the building of trust through active stakeholder engagement

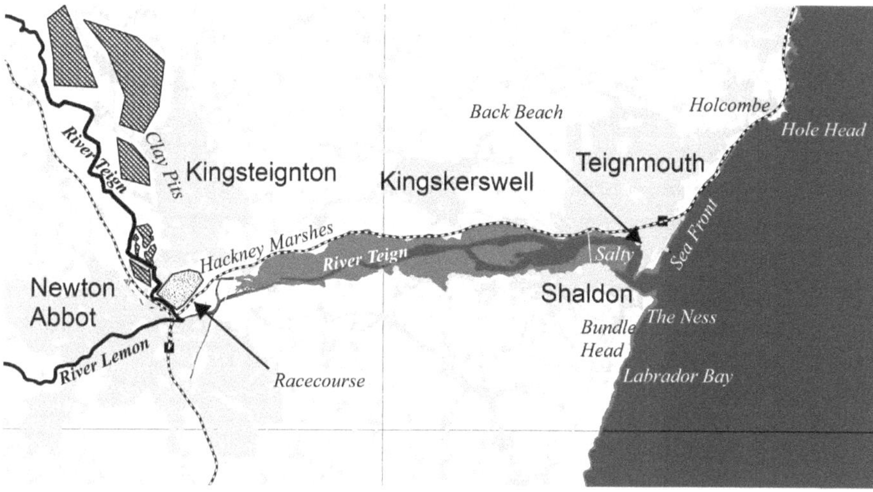

Figure 7.116 Teign Estuary map. *Contains Ordnance Survey Data. © Crown copyright and database. Right 2014.*

and participation. Ultimately this can lead to enhancement of community resilience, economic regeneration, and an increased appreciation of the natural environment.

7.10.2 Settlements

The estuary and the marine environment have a significant physical influence on three of the settlements in this region (locations in Figure 7.116): Newton Abbot and Kingsteignton at the estuary head; the combined settlements of Shaldon and Ringmore on the south side; and Teignmouth on the north side at the estuary mouth. Historically these areas have suffered from "tidal" (extreme sea level) driven flooding. Teignmouth is additionally affected by wave action.

Teignmouth had a recorded population of 20,157 in 2011 (DCC, 2014). The town grew from a fishing port associated with the Newfoundland cod industry to a fashionable resort of some note in Georgian times, with further expansion after the opening of the South Devon Railway in 1846. Today, its port still operates and the town remains a popular seaside holiday location. Teignmouth is linked to the settlement of Shaldon, a village on the Southern shore of the estuary by both a passenger ferry at the river mouth and a road bridge further upstream. Although reduced from its heyday, Teignmouth still receives considerable numbers of holiday makers, in particular day-trippers. The population of Teignmouth is relatively elderly in relation to Devon as a whole and the national average. Some 23% are retired, the Devon average being 18% in 2011. The unemployment rate was 3%, in line with the Devon average. Since then, national and local employment rates have fluctuated, and in 2012–2013 local unemployment was 4.8% as opposed to 7.7% nationally (ONS, 2014). Most employment is in the retail and health sectors (36%, as opposed to 24%

nationally). The area is relatively affluent, with the average house price £226,000 (2011), which is 40% higher than the national average.

Newton Abbot is a market town and civil parish in the Teignbridge District of Devon, on the upper Teign Estuary, with a population of 23,580. The town grew rapidly in the Victorian era as it was home to the South Devon Railway locomotive works. The town has a racecourse nearby and a country park (Decoy). Newton Abbott has a more balanced population than Teignmouth, with proportionately fewer retired people (7.2%) and a more balanced working population, with15% in manufacturing and 18% in the retail sector. Unemployment was low at 2.7% in 2011. House prices are somewhat lower than in Teignmouth, but are still 30% above the national average: this, too, is a relatively affluent area of the UK.

7.10.3 Habitats and Ecology

The Teign features a range of important and sensitive marine habitats including grazing meadows, saltmarshes, mud flats, and rocky and sandy seabeds, together with important biogenic habitats (see Figure 7.117 and Table 7.88).

Intertidal and marsh habitats are important for numerous species of birds, insects, and plants. Collectively these habitats support a diverse flora and fauna including rare and endangered species. The estuary also supports commercially important species including oysters, cockles, and crabs. Despite there being numerous designated sites locally—e.g., the adjacent Exe Estuary is a RAMSAR wetland site of international importance, a Site of Special Scientific Interest, and a Site of Special Protection under the EC Birds directive—there are no such designations operative in the Teign Estuary, probably as a result of local stakeholder opposition.

Figure 7.117 Railway line running along seawall north of Teignmouth.

TABLE 7.88 Important Habitats Found in the Teign Estuary

Habitat	Description
Sandy beaches	Areas of sand beach are found along the front of the town of Teignmouth. This beach is heavily used by tourists and as such represents a poor-quality marine habitat. There are two further stretches of sandy beach: 2 km north of the estuary mouth from Teignmouth to Holcombe and 0.5 km south of the estuary mouth from the Ness to Bundle head.
Saltmarshes:	Approximately 250 m^2 of saltmarsh is found toward the head of the estuary close to Newton Abbot. Small scattered areas of degraded saltmarsh are also found along the northern edge of the estuary adjacent to the railway line.
Sabellaria reefs	Colonies have been found scattered along much of the neighboring coast. With potentially large reefs present at Hole Head at the north end of the open coast section and around Bundle Head and Labrador Bay on the south end of the open coast section
Mussel beds	Small but dense beds of *Mytilus edulis* are found around the area of the Salty, and larger areas of mussels are also found on the intertidal island to the west of the Salty.
Intertidal mudflats	The total area of mudflat within the estuary is 204 ha, and dominates the intertidal in the upper estuary. Species include *Hediste diversicolor, Scrobicularia plana* along with *Fucus* spp., *Macoma balthica,* and *Pygospio elegans.*
Intertidal pebbles and cobbles	This habitat dominates the upper shore all the way along the southern shore. Also the 25 ha area known as the Salty, which comprises an accumulation of pebbles and cobbles, has large aggregations of the edible periwinkle, *Littorina littorea.*
Intertidal rocky shores	An area approximately 1.86 ha is found around the Ness supporting diverse red algae and sponges including *Halisarca dujardini* and *Hymeniacidon perleve;* over-abundant volcano barnacles, *Balanus perforatus;* sea squirts such as *Aplidium punctum*; and dense beds of tiny, <10 mm, edible mussels, *Mytilus edulis.*
Grasslands and woodlands	Areas of grassland fringe much of the coastal area around the estuary within, and approximately 12 Ha are at risk of flooding. The Hackney Marshes Natural Reserve contains a number of habitats under threat: flood meadows, ponds, streams, hedgerows, and woodland. There are two semi-natural meadows. The north meadow is more species-rich because it has not been so intensively farmed. The south meadow has been agriculturally improved.
Freshwater environments	Ponds and streams are found within Hackney marshes. The river Teign that runs through Newton Abbot is tidal for approximately 1.6 km. The freshwater reaches of the river Teign are of high quality.
Artificial structures	Concrete and blockwork seawalls, groynes, and other engineering structures also provide additional habitats. The seawall protecting the railway on the northern bank of the estuary has a very distinct zonation over a vertical expanse of about 2 m. Bands of the brown seaweeds channelled wrack—*Pelvetia canaliculata,* horned wrack (*Fucus ceranoides*), spiral wrack (*Fucus spiralis*), and knotted wrack (*Ascophyllum nodosum*) from the top to the bottom of wall. A short hydraulic structure, a training wall approximately 200 m just off the coast of the Ness, also supports colonies of the same red algae, sponges, and volcano barnacles found on the adjacent rocky shores.

7.10.4 Existing Flood Governance and Management

In order to understand the future management of this estuary, it is necessary to understand the current UK policies for coastal and river management and flood risk assessment. In England, flood protection and coastal defense are the responsibility of the Environment Agency (EA). Coastal management is carried out with reference to the relevant Shoreline Management Plan (SMP; see Nicholls, Townend, Bradbury, Ramsbottom, and Day (2013). This comprises an assessment of the risks of flooding and erosion within a unit of contiguous coastline. In the latest format, the coastlines are treated as systems whose behavior is determined by the interactions of the dominant coastal processes. The SMP strategy addresses the next 100 years of coastal management needs, split into three time horizons or epochs: short term, 2005–2025; medium term, 2025–2055; and long term, 2055–2105.

In order to receive funding from the EA, the coastal protection or defense scheme has to undergo the process of "project appraisal," in which the "benefits" of the scheme have to outweigh the costs measured against the "do-nothing" scenario. The SMP is used to identify strategic approaches to management, appropriate levels of monitoring, and provides a foundation for the appraisal of coastal interventions. The most recent round of SMP development, known for convenience as SMP 2, has resulted from extensive consultation through stakeholder exhibitions and feedback sessions orchestrated by local "coastal groups," which comprise the EA and local government authorities. The impact of this degree of stakeholder involvement is discussed later.

For the Teign Estuary, the SMP (Halcrow, 2011) outlines a policy that is to "hold the line" through the maintenance of existing flood defenses, with a strategy for some "managed realignment" at the head of the estuary over the longer term. Undefended sections of the estuary are to remain as such via "no active intervention." The first sea wall built to protect Teignmouth town was constructed in 1867. The latest seafront scheme was rebuilt and completed in 1991, comprising a re-curved sea wall, secondary defenses, and two floodgates. Wooden groynes are present on the beach.

On the coast, the same "hold-the-line" policy is adopted to ensure continued operation of the railway line. It is envisaged that this might require the building of new coastal structures as the sea-level rise takes place, with steepening, narrowing and possible loss of the beaches adjacent to the seawall and Teignmouth sea defenses. The possibility of managed realignment of the estuary spit has been proposed, though the wider implications for impacts on the estuary and maintaining access to Teignmouth port are under evaluation.

The risk of fluvial-derived flooding is managed through the relevant Catchment Flood Management Plan (CFMP). The main risk of fluvial flooding affects the towns of Newton Abbot and Kingsteignton in addition to the threat from estuarine flooding.

Through development, both settlements have expanded onto the flood plains adjacent to the confluence of the Rivers Teign, Lemon, and the Aller Brook (EA, 2012a). Control of floodwater from the river Lemon entering Newton Abbot is afforded by the Holbeam Dam. However, there is a strong case for finding more space for floodwater storage from both fluvial and estuarine sources. Such space might be found in the Hackney Marsh area and racecourse at the head of the estuary, or further up the Teign valley in the clay pits.

7.10.5 Quantifying Flood Risk Under Climate Change

As mentioned above, appraisal of flooding in the UK has to consider how the threat of flooding will affect a coastal location over the next 100 years and should include a consideration of the current estimates of climate change effects. This section describes an approach used to assess these in the Teign Estuary. The changing risk is assessed over four intervals, or "time slices," that span 1970 to 2100, and consider a standard climate change scenario: the IPCC SRES A1B scenario with greenhouse gas emissions consistent with rapid economic growth, population peak mid-century, and the introduction of new and efficient technologies (Nakicenovic & Swart, 2000; Solomon et al., 2007).

7.10.5.1 Estimation of Flood Sources in the Teign Estuary

To allow an estimate of the likely areas of flooding and erosion, the key forcing parameters should be estimated for the present and, where possible, future time slices. Within an estuary, water levels and hydraulic inputs into the estuary are important. On the coast, water levels and wave height determine the degree of wave damage and overtopping that may lead to both erosion and flooding.

Water Levels

Water levels comprise both deterministic components—tides—and stochastic elements—storm and pressure surges. Over the longer term, the water level in the Teign is also controlled by glacio-isostatic subsidence and projected climate-induced sea-level rise.

Teignmouth has a spring tidal range of approximately 3.8 m, with the maximum tidal current up to 3.0 m/s. Statistical surge levels can be deduced from Weibull and Gumbel regression analysis. The values at Teignmouth were interpolated between the tide-gauge recordings at Plymouth and Weymouth. These indicate 1 in 20-year maximum surges of 1 m for the coastal area of Teignmouth, and 1 in 2000 year surges of up to 1.6–1.8 m.

The tide gauge records also show a steady increase of sea level of 10 cm in the last 20 years, while Haigh et al. (2009) estimate a much smaller rise of 2.1 ± 0.6 mm/year and 1.8 ± 0.3 mm/year since the 1960s at Devonport (Plymouth) and Weymouth, respectively. Recommended allowances for increase in mean water level over the next 100 years are published in the UK (DCLG, 2009) and take a high-end view of likely sea-level rise. The allowance shows an accelerating rate of rise, so that in 2100, there will have been 1 m of sea-level rise. and this is added to the present values calculated above. Note that more flexibility in allowing for future sea-level rise has recently been proposed in UK guidance (Nicholls et al., 2013).

Wave Climate

The wave climate in the area around Teignmouth is known to be predominately driven by swell waves from the Atlantic in the southwest direction, with annual mean significant wave height of 2.0 m (max 4.0 m) and for 1 in 50 year return period, the mean significant wave height is 2.7 m (max 5.3 m) (Reeve et al., 2012).

The future wave climate along the South Devon coast was modeled at Plymouth Sound and the Exe Estuary using the POLCOMMS/ProWam hydrodynamic modeling system reported in Chen, Pan, Hewston, and Cluckie (2010). This uses time series of wind fields and sea-level pressure obtained from meteorological models to hindcast/forecast wave heights in relevant locations. In this case, data were obtained from the Max Planck Institute for Meteorology WDCC CERA database (WDCC, 2012), limited to the IPCC SRES A1B emissions scenario (mentioned earlier). The input data set was spatially extensive, but only coarsely spaced, covering a large proportion of the Atlantic Ocean. Thus, to define wave parameters at a more local scale, a nesting procedure was used to drive increasingly fine wave calculation grids nearer to the site of interest.

The extreme significant wave heights (Table 7.89) at Plymouth and Exmouth were predicted in this way over four intervals: 1970–1999, *present;* 2010–2039, *short-term;* 2040–2069, *mid-term;* 2070–2099, *long-term.* Note that these are different to the standard SMP epochs, but span the same interval in slightly more detail. The calculations indicate a slight, but inconclusive increase in return wave heights in the short term only. This may be due to changes in wind direction as much as wave height and the single realization of future climate. Typically, in deep water the 1 in 20 year return wave height is 11 m near Plymouth and 8 m offshore from Exmouth; the corresponding 1 in 2000 year values are 16.5–17 m and 11.5–11.9 m.

The annual mean and maximum significant wave heights at Plymouth and Exmouth were also modeled. For instance, at Plymouth the maximum annual wave height varied between 6 and 12 m, and at Exmouth, 4–8.5 m. At neither location was any trend indicated with time. Analysis of wave direction indicated the same. Thus,

TABLE 7.89 Predicted (Gumbel) Wave Heights (m)

Return Period (years)	Present (1970–1999)	Short-term (2010–2039)	Mid-term (2040–2069)	Long-term (2070–2099)
	Plymouth			
20	10.9	11.2	10.6	11.0
100	13.0	13.3	12.6	13.0
200	13.9	14.2	13.4	13.9
1000	15.9	17.1	16.3	15.8
	Exmouth			
20	7.9	8.0	7.9	7.5
100	9.3	9.3	9.3	9.4
200	9.9	9.8	9.9	10.0
1000	11.3	11.0	11.4	11.3

areas that are currently at risk of erosion will continue to be eroded to a lesser or greater extent in the future.

River Discharge

An analysis performed using gauging data on the Teign indicate that the yearly average discharge from the river Teign is around 10 m^3/s over the last 51 years. Using a Gumbel analysis, the 1 in 50 year maximum annual discharge is 16.1 m^3/s, and 1 in 2000 year is 250 m^3/s.

Prediction of peak river flows under climate change scenarios has not been performed. However, PPS25 (DCLG, 2009) suggests a precautionary allowance of 10% increase in the time slice 1990–2025 due to a modeled increase in rainfall of 5%, and a 20% increase in peak flow in the time slice 2025–2115 caused by an increase in rainfall of 10–30%. A more local sensitivity testing reported in the South Devon CFMP (EA, 2012a) indicates that in 2100 climate change may account for a 30% increase and land use changes a further 5% increase in the peak values.

7.10.5.2 Flood Mapping and Risk

Having calculated the likelihood and occurrence of the hydraulic forcings, it is possible to link these to corresponding levels of inundation through an understanding of the flood defenses' performance and a suitable hydraulic model of the flood processes.

Flood Processes

Flooding at the coast might be considered to have arisen from one or more processes. These are overflow, where the water level exceeds the defense crest level; overtopping, where waves break on the defenses and run over the crest of the defense; breach, where the combination of wave and water level loading causes a defense to fail, resulting in a lowering of the crest and thus initiating overtopping and/or overflow; and toe failure, where foreshore level is lowered beyond a level that compromises the structural stability of the defense, leading to collapse and then subsequently breaching. Overflow discharges can be estimated using weir discharge formulas or numerical models. Overtopping can be estimated using guidelines, such as EurOtop (Pullen et al., 2008). Existing methods to estimate overtopping of beaches and dunes are much more limited. Breaching and toe failure are more difficult and uncertain to predict.

Flood-Modeling Approach

In order to represent these flood mechanisms effectively, a time-dependant hydraulic inundation model could be used, driven by water levels, surge, waves, and river discharge. However, within the estuary, flood inundation is driven predominantly by water levels and subsidence in the longer term.

For instance, Shaldon and Ringmore, on the south side of the estuary, are low-lying residential areas. Up until recently, minor flooding has occurred (EA, 2012a). Severe damage to properties and businesses would undoubtedly have resulted in the case of a more extreme flood event. Teignmouth also has until recently been at threat of "tidal" or water-level driven flooding from the estuary side.

Because of the steep-sided valleys in which these settlements are sited, the duration of flooding events is almost entirely coupled with the immediate water elevations within the Teign Estuary. Further, the flooded areas are relatively small. Thus, in order to model such events, satisfactory first-order results can be obtained through the use of a "bathtub" model for flooding the measured topography to the water elevation of interest (Table 7.90). This was achieved by Kwan (2011), and some of these results are presented for the Teignmouth Back Beach area for a present and medium term scenario in Figures 7.118 and 7.119.

In the case of flooding that affects a more extensive area, and is driven by more complicated hydraulic inputs, it becomes necessary to employ a more sophisticated, time-dependent flooding model, the data requirements of which also increase in complexity.

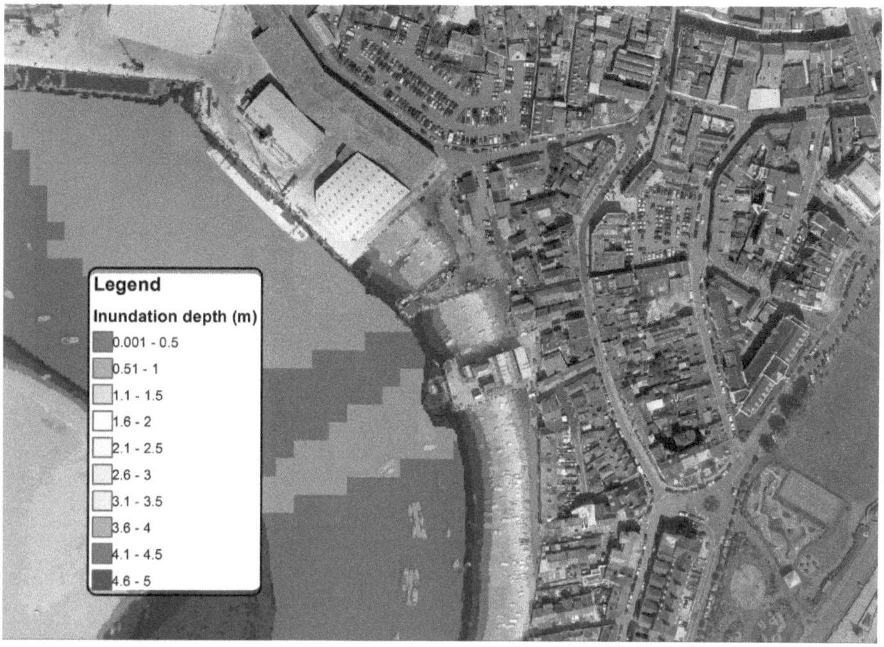

Figure 7.118 Teignmouth flood inundation map: present day, event with 1 in 20 years return period event. *From Kwan (2011).*

TABLE 7.90 Extreme Water-Level Scenarios Within the Teign Estuary; the Scenarios Combine Tide, Surge, and Sea Level Rise

Return Period (years)	Climate Scenarios for Sea-Level Rise			
	Present (1970–1999)	Short-term (2010–2039)	Mid-term (2040–2069)	Long-term (2070–2099)
20	2.96	3.01	3.25	3.59
100	3.16	3.21	3.45	3.80
200	3.24	3.30	3.54	3.88
1000	3.44	3.50	3.74	4.08

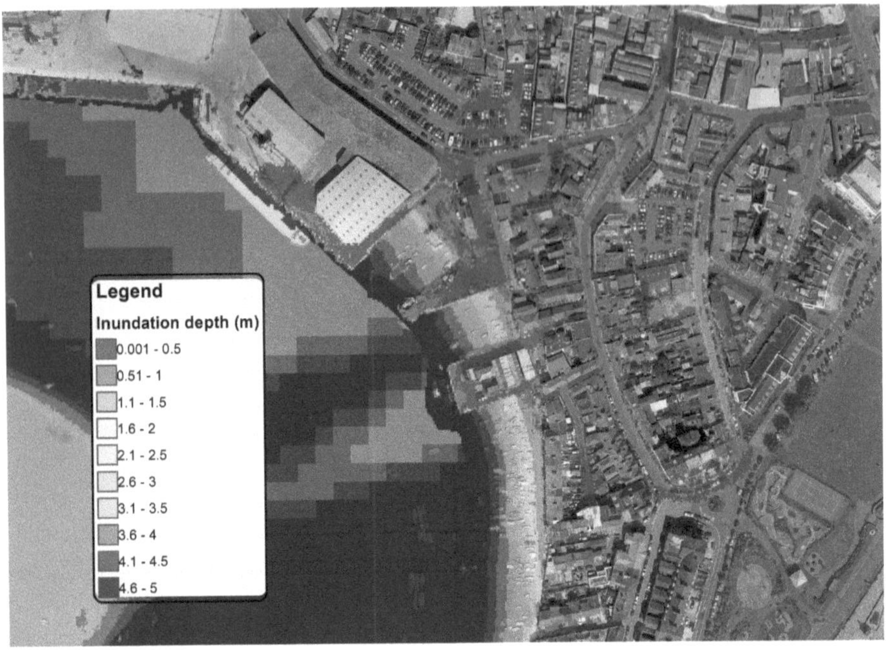

Figure 7.119 Flood inundation map: medium term, with 1/200 years return period event. *From Kwan (2011).*

Flood Risk in the Teign

According to the SMP report (Halcrow, 2011), the long-term vision of the plan for the Teign Estuary is to continue to defend against the risk of flooding to people, property, and infrastructure located around the majority of the estuary. This includes the towns of Teignmouth, Newton Abbot, Kingsteignton, and Shaldon, as well as the Port of Teignmouth and the mainline railway.

The racecourse at Newton Abbot (Fig 7.116) and the adjacent areas were expected to be flooded under most of the scenarios for the instance of a breach of the sluice gate.

In Teignmouth, without the new defenses for the Back Beach area (opened 2012), flooding begins to occur under 1 in 20 year conditions and progresses gradually toward the town under more extreme events. Figure 7.118 shows the effect of a 1 in 20 year event, for the undefended Back Beach in the present day. This clearly shows, that, had the defenses not been built, the town would be at risk of immediate flooding for this short return period scenario. The effect of a 1 in 1000 year event in the medium term, Figure 7.119, shows a considerable increase in the risk of flooding, with much of the lower town shown as flooded. Again, although the simulation does not consider the protection provided by the new flood protection scheme, the increasing threat is graphically demonstrated. Indeed, an increase in the height of the flood walls is scheduled in around 30 years time, as part of this adaptive scheme.

However, the flooding situation is relatively mild in comparison with Shaldon and Ringmore. Examining the future scenarios shows that flooding would not occur until a 1 in 1000 year return period high water in the mid-term scenario, with flooding occurring first in Ringmore. For the long-term scenarios, flooding at Shaldon and Ringmore occurs under extreme events with return periods over 200 years. Water depths in this area could be as high as 2–3 m which will inevitably cause serious economic losses in the area (Kwam, 2011).

The modeling also identified that the railway line along the Teign Estuary is subject to the risk of coastal flooding just downstream of the bridge to Shaldon. Although the flood depth is only around 0.5–1.0 m and would only happen in the long-term scenarios, the disruption of the rail services by the breach of the seawall will inevitably cause economic losses and inconvenience to the passengers. The results do show that the railway station and the railway line at Newton Abbot are flood free under all sea-level rise scenarios.

However, the risk to the coastal sections of railway from overtopping through combined water level and wave activity, of a magnitude sufficient to temporarily close the line, has not been properly considered. Even in the present, single line working and short complete closures of the railway are regular seasonal occurrences, sometimes accompanied by limited damage (breach/failure) of the Brunel blockwork sea wall. In early 2014, the main line was severely damaged just along the coast at Dawlish, forcing completed closure of the line for 8 weeks. An exacerbating feature was the destabilization of the cliffs due to the concurrent high rainfall, unrelated to overtopping.

7.10.6 Stakeholders and Flood Mitigation

The process of selecting appropriate flood management options is determined by the governance structure. i.e., this dictates whether innovation is constrained or enabled (Penning-Rowsell et al., 2014). In the UK (see Section 2.4), there has been an increasing focus on the localization of decision-making. Once experts and agencies have quantified the risks and consequences of flooding in an area, what happens next will depend on the functioning governance structures that relate state to stakeholder and vice versa.

7.10.6.1 Local Risk Perception and Communication

Coastal risk management is centered on stakeholder organizations set up to promote risk reduction at the coast: the members of the risk governance system. These comprise those institutions with concerns in the coast with regard to delivery and maintenance of infrastructure (EA, County and Local Councils), spatial planning, tourism and recreation, nature conservation (Natural England), fishing, etc. In 2009, around the Teign Estuary there were approximately 850 properties at 1% annual risk of fluvial flooding and 0.5% of tidal flooding (EA, 2012a), which has consequences for properties, businesses, habitats, navigation, and the railway .

In Teignmouth there is a history of relatively minor flooding from both the sea—from waves and high tides—and also tidally dominated flooding from the estuary side. Although all new developments in the UK have to consider EA advice, there is no current statutory requirement for this to be heeded. For instance, in Teignmouth, historical houses with ground floor boat houses have had these converted to living rooms over the past decades; although their value has increased, so has their vulnerability.

It is unclear as to the effectiveness of the risk communication process: the EA has published flood risk maps through their website since about 2003 and has engaged in extensive consultation with those end-users directly at risk, but skepticism rises and the level of engagement drops off rapidly from end-users immediately behind the front line. Longer-term residents and those with properties immediately on the river Teign in Teignmouth have greater awareness of historic and potential flooding, but some of those close by or who are new to the coast have little interest in engaging with flood issues (Penning-Rowsell et al., 2014).

As far as emergency response to any natural disaster such as flooding is concerned, the mechanisms in the UK are streamlined and proscribed: consultation is not appropriate. The responsibility lies with the "blue light" emergency services: police, fire brigade, coastguard, emergency elements of the National Health Service, etc. (Penning-Rowsell et al., 2014). Extreme emergencies would be coordinated by COBRA, the UK civil contingencies committee. Flood warnings are provided by the Environment Agency in conjunction with the UK Meteorological Office.

Flood risk planning is a more consultative procedure. In the Teign Estuary, and typical of the UK, flood risk management "grant in aid" is allocated by the Environment Agency directly from central government to schemes, or to local authorities and internal drainage boards. In the Teign case, there is an estuary manager at the county level, as well as local flood and coastal committees, a harbor authority, and an important stakeholder in the form of the Natural England agency. With regard to spatial planning, both the county and the district authorities are involved.

7.10.6.2 Local Response to Recent Flood Mitigation Schemes

Over the last 10 years, three flood mitigation schemes have been proposed for the Teign Estuary and two of these have been accepted. These are discussed in the following, including the reasons for the success and failure of acceptance.

Rejected Scheme: Teignmouth Back Beach (2004)

Although the sea front at Teignmouth has been defended with a coastal scheme built in 1977 and upgraded in 1991 to meet policy requirements at that time, the threat of "tidal," or water-level driven flooding, from the estuary to properties and businesses from the rear of the spit, remained due to a 1 in 20 year event sea level. This would flood the lower town area of Teignmouth.

The EA proposed a flood wall to protect this area, which was in part constituted from historic boat houses converted into waterside private homes and other faculties relating to fishing activity. The scheme put forward comprised a continuous flood wall built to a satisfy a 100-year design life according to shoreline management policy. The incorporation of floodgates was not considered to be viable.

In 2004, the local stakeholders were able to force a rejection of this scheme at the consultation stage on the basis of its visual impact and the restriction of access to the beaches and quays that this would entail. This shows the high value stakeholders place on amenity.

Accepted Scheme: Shaldon and Ringmore (2011)

In Shaldon, a similar risk level for tidal flooding was identified. Following the rejection of the Teignmouth scheme, it became apparent that a more consultative approach had to be taken. This has led to a more adaptive and innovative solution to mitigating the risk of flooding in these settlements in terms of both engineering and social outcomes.

Initially all possible stakeholders were invited to become members of a local Flood Forum, which was instrumental in reviewing prospective flood risk management plans, tailoring them to local circumstances and gaining acceptance and local approval. The Shaldon and Ringmore scheme represents the first application of the EA Building Trust with Communities policy (EA, 2012b). The earlier use of floodgates that had been vetoed by the EA in Teignmouth, was subsequently put forward to remove similar objections from local businesses and waterside residents. This involved a "hand-stitched" approach to fitting flood defense measures into the intricate and historic urban fabric rather than creating a new linear defense feature. An adaptive "wait and see" approach has also been taken to preserve homeowner views of the estuary for as long into the future as possible. This comprises the construction of low walls, with oversized foundations that allow later raising of the defense as relative sea-level effects are observed (Figure 7.120).

As built, the defenses provide a current standard of protection of 1 in 300 years, which decreases with time as global sea level rises and local subsidence occurs. However, as mentioned above, the scheme is planned to be topped up twice over its planned 100-year lifetime to maintain an acceptable standard of protection as the effect of climate change and subsidence are observed. Hence, rather than being precautionary and designed for the worst case, this approach requires monitoring of sea level and recognizing the threshold rise when upgrade is required.

Figure 7.120 Adaptive flood protection scheme in Shaldon. One of the flood warden-operated gates maintains pedestrian access to the foreshore. Elsewhere, access is via attractively designed stepped walkways over the flood wall.

The scheme was opened in May 2011. Continued participation of stakeholders is ensured and reenergized through the participation of some members of the forum as "Flood Wardens" responsible for operating the floodgates when flood warnings are issued under remote direction from the EA. This can be seen to yield extra social benefits allied with "warning and informing" local residents through regular flood drills, empowering stakeholders to take responsibility for their own protection.

Ecological enhancement of these structures was also included. This comprised the inclusion of intertidal niche and rock pool features into the flood wall (Figure 7.121; see also Chapter 4.6.3). Although not suitable for walls under direct wave attack, these features are possible in this location, with little impact or cost to the construction of the flood wall.

Figure 7.121 Engineered intertidal habitats.

There is some weak evidence that the removal of "flood blight" by this choice has initiated an economic regeneration in Shaldon. This is evident in preliminary analysis of property value increases in comparison with national and regional trends.

Accepted Scheme: Teignmouth Back Beach Scheme (2012)

Following the success of the approach taken at Shaldon, a new scheme was proposed to and accepted by the stakeholder forum in Teignmouth for the Back Beach area where the scheme in 2004 was rejected.

The Teignmouth scheme opened in November 2012 and is very similar to that built in Shaldon and Ringmore, comprising the following elements:

- Adaptive concrete flood walls, faced with local stonework, with some incorporation of reinforced glass wall elements to maintain residents sight of the estuary;
- Wooden-faced floodgates operated by neighborhood flood watch volunteers;
- Reinforced flood-proof windows, doors, and property structures (see left hand photo in Figure 7.122).

These defenses raise the existing defenses to a present-day 1 in 1000 year (0.1%) standard of protection. Raising the defense in 30 years time, and possibly a further one or two interventions over the 100-year scheme life, is expected to leave a residual standard of protection at year 100 of 1 in 75 years (1.3%).

As with the Shaldon scheme, this scheme can be seen as a success for stakeholder engagement, which has driven innovation in and acceptance of the need for flood mitigation measures that are both physically and socially acceptable.

Figure 7.122 Teignmouth Back Beach. Reinforcement of the New Quay inn walls and windows (left). Flood Wardens demonstrate operation of gates to university students (right).

7.10.7 Future Mitigation Options for the Teign Estuary

Following the difference in local response to flood mitigation schemes discussed in Section 7.10.6, those responsible for future flood mitigation schemes will need to ensure that the local population is fully engaged with the local decision-making process. If stakeholders are empowered to modify, and possibly compromise the performance of flood defenses, then some mitigation options such as national legislation to enforce the compulsory purchase of flood insurance could be sensible (see Chapter 5). This would mitigate against the danger of "free riding" by those who choose not to take out insurance and would otherwise expect their potential compensation to be born by the state purse. Of course, there are many other strategic issues to consider (Horn & McShane, 2013) and other options can be considered more locally as discussed in the following sections.

7.10.7.1 Ecosystems

Around the Teign Estuary, sandy beaches may function poorly as habitats. Their role at Teignmouth and along the railway line is integral to the performance of the hard structures behind them, helping to dissipate breaking waves and controlling the degree of overtopping. Economically these features are essential to the tourist industry and other users of the foreshore. Just offshore, it has been suggested that the cementatious nature of sabellaria reefs may help to protect the seabed from scour (Section 4.4), though the exact efficacy of this process is not yet established. Similarly, mussel beds are thought to be capable of protecting muddy intertidal substrate within the estuary.

Natural and remediated salt marshes have been suggested as soft engineering options for enhancing flood defences (See Section 4.2) through dissipation of low-energy waves or providing extra resistance to flood inundation episodes. They also might play a role in stabilizing mudflats and channel embankments. Similarly at the head of the Teign, the Hackney Marshes have a flood storage function, retarding and absorbing a proportion of any future flood episode.

7.10.7.2 Managed Realignment

Linked with the creation of new habitats, managed realignment involves the conversion of some regions of the estuary from defended regions (e.g., pasture) to intertidal or marsh environments, through managed realignment (Section 4.2). Existing, possibly uneconomical defenses are degraded in a controlled manner to allow a setback of the shoreline and steady accretion of sediments and habitat transformation to take place.

7.10.7.3 Spatial Planning, Including Time-Limited Planning Approval

Spatial planning can be used to restrict the potential for flood damage into the future by controlling the type and extent of property built in flood-risk areas. Spatial planning could also be used to alter the risk potential of current land uses, by requiring modifications to buildings and other infrastructure, especially during renewal. Within an area of increasing flood risk, the granting of time-limited planning approval can provide extra flexibility in managing the socioeconomic needs of a community.

7.10.7.4 Flood Storage

The aim is to reduce risk of flooding to neighboring high-value, populated areas, by temporarily sacrificing low value, uninhabited areas. In the Teign, Newton Abbot and Kinsteignton have an increasing risk of flooding from the estuary, and also from pluvial input to the rivers feeding the estuary. Locally, the low-lying racecourse and clay pits could be used to temporarily store tidal and fluvial floodwater and this could be further investigated.

7.10.7.5 Flood Warning

Effective flood warnings are now routine in the UK, and they already mitigate the economic impact of a flood event in three ways: (1) saving lives by alerting people and organizations to the flood threat and allowing safe evacuation to risk-free areas, (2) minimizing flood damage by allowing people and organizations to move valuable property from at-risk areas in anticipation of the flooding, and (3) enabling organizations to mobilize in good time and counter the flooding, such as erecting mobile defences, closing floodgates, operating pumps, and providing forecasts of the impending event (Hissel et al, 2014). These could be further improved at the national to local levels.

7.10.7.6 Abandonment

A final alternative to the flooding problem, especially with respect to the railway, could be abandonment of some coastal areas. In the case of the railway, defense of the existing track along the coast is likely to become uneconomical in the future as sea levels rise. Construction of a new railway through the hinterland, joining Plymouth and Exeter, might be preferred and has been discussed since the 1920s. This is currently ruled out by the SMP, but may be more relevant toward the end of this century. It would be hugely expensive, requiring significant tunneling operations, but there may be no alternative as a way of maintaining the West Country main line to London and the north.

7.10.8 Conclusions

The geographic extent of the Teign Estuary, at only 9 km long, is much smaller in comparison to the larger estuarine study sites of the Elbe (Section 7.5), Gironde (Section 7.4), Yangtze (Section 7.3), and Scheldt (Section 7.1). Flooding throughout the estuary is strongly controlled by the tidal and storm surge effects from the sea, and over the longer term, by significant sea-level rise driven by subsidence and the effects of climate change.

However, the scale of the estuary has enabled very local and specialized solutions to the risk of flooding to be considered and initiated. There is clear evidence that stakeholders feel enfranchised with respect to the planning of flood protection and feel involved in the continuing process. Indeed, within and around the Teign Estuary it can be seen that the consultative nature of flood defense management policy in the UK has driven the agencies charged with the safeguarding of communities at risk of flooding to develop innovative new approaches to flood defense that would have been considered too complicated in the past. These comprise novel implementations of flood defense to reflect local policy issues itemized in the SMP—e.g., adaptive defenses, enhancing resilience, and acceptance through stakeholder community engagement.

In the longer term, however, the issues of coastal realignment and abandonment of infrastructure and homes need consideration at some locations. The railway, for instance, is likely to become uneconomic to maintain over the longer term, without major enhancement of sea defenses along the wall. This would impact on other activities such as tourism. Indeed rerouting of the coastal railway is under current consideration in the light of recent line closures caused by storms in 2014. However, finding an inland route will not be straightforward, and the railway is likely to be maintained as is for the immediate future, until the economic argument becomes overwhelming.

Also, further approaches such as reconfiguration of the landscape to accommodate floodwater (making space for water) and softer approaches aimed at changing behavior of inhabitants through insurance and innovative planning laws are already being discussed and may be implemented.

Habitats within the Teign are varied, and often encroached upon by fishing activities or coastal structures. However, there has been sufficient willingness to try out new means for making space for habitats.

In comparison to other sites, the significance of tourism and the value that residents place on being able to enjoy their proximity to the coast and estuary has dictated how flood management has been "hand stitched" for the needs of those living in Shaldon and Teignmouth. This is perhaps to a greater extent than in the larger estuaries, where industry and port activities often dominate the economic landscape, and there are many more stakeholder groups with diverse priorities.

The main lesson to be drawn from the Teign is that when time is taken to engage with and take account of the needs of stakeholders at a very local level, this can yield significant long-term benefits in terms of the resilience and raised awareness of communities at risk of flooding and innovation in the technologies employed to protect them.

References

Abascal, A. J., Castanedo, S., & y Medina, R. (2010). GOS, un reanálisis de marea meteorológica de 60 años de alta resolución para el sur de Europa. *I Encuentro Oceanografía Física Española*. Barcelona (España).

Alcerreca, J. C., Silva, R., & Mendoza, E. (2013). Simple settling velocity formula for calcareous sand. *Journal of Hydraulic Research, 51*(2), 215–219.

Andreeva, N. K., Valchev, N. N., Trifonova, E. V., Eftimova, P. T., Kirilova, D., & Georgieva, M. (2011). Literary review of historical storm events in the western Black Sea. (In Bulgarian). In *Procedings of the Union of scientists. Varna, Bulgaria* (pp. 105–112).

Atkins. (2009). *Shaldon flood defence scheme - Level 1 flood risk assessment [Online]*. Available at: http://docimages.teignbridge.gov.uk/Planning/lg/GFPlanningDocuments.page?Param=lg.Planning&org.apache.shale.dialog.DIALOG_NAME=gfplanningsearch&viewdocs=true&SDescription=09/02217/FUL.

Baeyens, W., van Eck, B., Lambert, C., Wollast, R., & Goeyens, L. (1997). General description of the Scheldt estuary. *Hydrobiologia, 366*(1–3), 1–14. http://dx.doi.org/10.1023/A:1003164009031.

Balderas, O. (2009). *Barómetro Turístico Cancún, quintana roo Diciembre 2009*. México: Asociación de Hoteles de Cancún/ACTI. http://www.acticonsultores.com/acti_indicadores.html. Last accessed 30 September 2013.

Basiński, T. (1995). Protection of the hel spit. *Journal of Coastal Research, SI 22*, 197–201.

Battjes J. A., Janssen J. P. F. M. (1978). Energy loss and set-up due to breaking of random waves, *Proceedings of 16th International Conference on Coastal Engineering*, Vol. I, 569–587.

Bellafiore, D., & Umgiesser, G. (2010). Hydrodynamic coastal processes in the North Adriatic investigated with a 3D finite element model. *Ocean Dynamics, 60*(2), 255–273. http://dx.doi.org/10.1007/s10236-009-0254-x.

Bergemann, M. (1995). Die Lage der oberen Brackwassergrenze im Elbeästuar. *Deutsche Gewässerkundliche Mitteilungen (DGM), 39*(4/5), 134–137.

Bezzi, A., Fontolan, G., Nordstrom, K. F., Carrer, D., & Jackson, N. L. (2009). Beach nourishment and foredune restoration: practices and constraints along the Venetian shoreline, Italy. *Journal of Coastal Research, SI 56*, 287–291.

Birkemeier, W. A. (1985). Field data on seaward limit of profile change. *Journal of Waterway Port Coastal and Ocean Engineering, 111*(3), 598–602.

Boniecka, H., & Zawadzka, E. (2006). Protection of the Polish coast by artificial nourishment. In A. Tubielewicz (Ed.), *Geomorphology and Protection* (pp. 51–60). Gdansk, Poland: EUROCOAST-LITTORAL, University of Technology Gdansk.

Booij, N., Ris, R. C., & Holthuijsen, L. H. (1999). A third-generation wave model for coastal regions - 1. Model description and validation. *Journal of Geophysical Research-oceans, 104*(C4), 7649–7666. http://dx.doi.org/10.1029/98jc02622.

Bouma T.J., van Belzen J., van Dalen J., Klaassen P., Hartog A.M., Callaghan D.P., Temmerman S. & Herman P.M.J. (2014). Short-term mudflats dynamics drive long-term cyclic salt marsh dynamics: identifying key-process determining the tipping points. (submitted for publication).

Bozzeda, F.. (2013) Development and application of fuzzy Bayesian ecological model (FBEM): a general approach for modeling ecosystem changes. Protection and management of natural resources- marine and coastal systems. Phd Thesis. Environmental Sciences, University of Bologna.

Broekx, S., Smets, S., Liekens, I., Bulckaen, D., & De Nocker, L. (2011). Designing a long-term flood risk management plan for the Scheldt estuary using a risk-based approach. *Natural Hazards, 57*(2), 245–266. http://dx.doi.org/10.1007/s11069-010-9610-x.

Bruun, P. (1962). Sea-level rise as a cause of shore erosion. *Journal of the Waterways and Harbors Division, 88*(1–3), 117–130.

Burcharth, H. F., Andersen, T. L., & Lara, J. L. (2014). Upgrade of coastal defence structures against increased Loadings caused by climate change : a first methodological approach. *Coastal Engineering*. ISSN: 0378–3839.

Burzel, A., Dassanayake, D. R., Naulin, M., Kortenhaus, A., Oumeraci, H., Wahl, T., et al. (2010). Integrated flood risk analysis for extreme storm saurges (XtremRisK). In J. McKee Smith, & P. Lynett (Eds.), *Proceedings of the 32nd international conference on coastal engineering (ICCE)* (p. 12). Shanghai, China: Coastal Engineering Research Council.

Calabrese, L., Lorito, S., & Perini, L. (2010). Vulnerability to erosion of the coast close to Ferrara. In L. Perini, & L. Calabrese (Eds.), *The sea-coast system of the Emilia-Romagna Region* (pp. 211–223). Bologna, Italy: Pendragon (in Italian) ed.

Camus, P., Méndez, F. J., Medina, R., Tomas, A., & Izaguirre, C. (2013). High resolution downscaled ocean waves (DOW) reanalysis in coastal areas. *Coastal Engineering, 72*, 56–68. http://dx.doi.org/10.1016/j.coastaleng.2012.09.002.

CCR. (2011). *Compensation scheme for natural disasters*. Paris, France: Caisse Centrale de Réassurance (CCR). http://www.ccr.fr/index.do?aid=2094981368466376057 (in French); http://www.ccr.fr/index.do?aid=2094981368466376508 (in English). Last accessed 11 December 2013.

CFE. (2001). In Q. Roo Cancún, Tercera Etapa, & H. A. B. Juárez (Eds.), *Estudios, Modelos de Simulación y Proyecto Ejecutivo para la Rehabilitación Integral de las playas en el tramo comprendido entre Punta Cancún y Punta Nizuc*. pp.120.

CFE. (2008). *National electricity company database* (Unpublished).

Champion, M. (1862). *Les inondations en France du VIème siècle à nos jours, Tome Quatrieme*. Paris, France: Dunard.

Charles, E., Idier, D., Delecluse, P., Deque, M., & Le Cozannet, G. (2012). Climate change impact on waves in the Bay of Biscay, France. *Ocean Dynamics, 62*(6), 831–848. http://dx.doi.org/10.1007/s10236-012-0534-8.

Chen, Y., Pan, S., Hewston, R., & Cluckie, I. (2010). Ensemble modelling of tides, surge and waves. In J. S. Chung, R. Ayer, S. Prinsenberg, S. W. Hong, & I. Langen (Eds.), *Proceedings of the 20th international offshore and polar engineering conference. June 20–25, Beijing, China*. The International Society of Offshore and Polar Engineers (ISOPE).

Cheng, R. T., Casulli, V., & Gartner, J. W. (1993). Tidal, residual, intertidal mudflat (TRIM) model and its applications to San Francisco Bay, California. *Estuarine Coastal and Shelf Science, 36*(3), 235–280. http://dx.doi.org/10.1006/ecss.1993.1016.

Christianen, M. J. A., van Belzen, J., Herman, P. M. J., van Katwijk, M. M., Lamers, L. P. M., van Leent, P. J. M., et al. (2013). Low-canopy seagrass beds still provide important coastal protection services. *PLoS One, 8*(5), e62413.

Cieślak, A. (2005). Narodowa strategia ZZOP: Zarys polskiego punktu widzenia. In B. Glaeser, A. Sekscinska, & N. Löser (Eds.), *Coastline reports No 6-Integrated coastal zone management at the Szczecin Lagoon: Exchange of experiences in the region* (pp. 11–18). Leiden, The Netherlands: EUCC (Coastal and Marine Union). http://databases.eucc-d.de/files/documents/00000314_02_cieslak.pdf. Last accessed 30 September 2013.

CIMARES. (2009). Política Nacional de Mares y Costas de México - Gestión Integral de las Regiones más Dinámicas del Territorio Nacional. In *Comisión Intersecretarial para el Manejo Sustentable de Mares y Costas* (p. 67).

CLM/SGA. (2010). *CLM/SGA regional climate model dataset.* Hamburg, Germany: Max Planck Institute for Meteorology. http://mud.dkrz.de/projects-at-md/sg-adaptation/clm/clm-info/index.html. Accessed 10 February 2010.

D^{2}SOU project. (2011). *Fiche savoir: Conformité au PPR/PPRI/autres.* Project report http://www.d2sou.com/public_local/documents/savoir/fiche_savoir_ppri.pdf. Last accessed 16 January 2014.

Dachev, V. (2003). Genesis and evolution of Varna Central beach (in Bulgarian). *Proceedings of the Institute of Oceanology, 4*, 74–82.

Davison, A. C., & Smith, R. L. (1990). Models for exceedances over high thresholds. *Journal of the Royal Statistical Society Series B-methodological, 52*(3), 393–442.

DCC. (2014). *Devon county Council. Census profiles 2011.* www.devon.gov.uk. Last accessed 30.01.14.

DCLG. (2009). *Development and flood risk practice guide. Planning Policy Statement 25(PPS25) update.* London, UK: Department for Communities and Local Government (DCLG). https://www.gov.uk/government/publications/development-and-flood-risk-practice-guide-planning-policy-statement-25. Last accessed 17.12.13.

De Nocker, L., Broekx, S., Liekens, I., Bulckaen, D., Smets, S., Gauderis, J., et al. (2006). Cost-benefit analysis to select the optimal flood protection strategy along the Scheldt. In V. Popov, & C. A. Brebbia (Eds.), *Transactions of the Wessex Institute: Ecology and the environment: 91. Risk analysis v: Simulation and hazard mitigation* (pp. 271–277). http://library.witpress.com/pages/paperinfo.asp?PaperID=16394. Last accessed 16.12.13.

de Vries, W. S., Zanuttigh, B., Steendam, G. J., Kloosterboer, H., Van der Nat, A., & Graaff, H. (2011). Integrating science and policy for creating tools for safer European coasts in a changing climate. *Irrigation and Drainage, 60*, 77–83. http://dx.doi.org/10.1002/ird.670.

Dean, R. G. (1977). *Equilibrium beach profiles: US Atlantic and gulf coasts.* Ocean Engineering Technical Report No. 12. Newark, DE, USA: Department of Civil Engineering and College of Marine Studies, University of Delaware.

Dean, R. G. (1987). Coastal sediment processes: toward engineering solutions. In N. Kraus (Ed.), *Proceedings of coastal sediments (1987). May 12-13* (pp. 1–24). New Orleans, LA: American Society of Civil Engineers (ASCE).

Deltacommissie. (2008). *Working together with water; a living land builds for its future. Findings of the Deltacommissie; advice to the Dutch Cabinet.* The Netherlands: Deltacommissie. http://www.deltacommissie.com/en/advies. Last accessed 30.09.13.

Demarcación de Costas de Cantabria. (2006). *Retirada de especies alóctonas en los sistemas dunares del puntal y Berria.* TT.MM. de Ribamontán al Mar y Santoña (Cantabria).

DENUE. (2010). *Directorio Estadístico Nacional de Unidades Economicas* (National Economic Units Statistics Directory).

Dibajnia, M., Sanchez, C., Martinez, M., Lara, A., Nairn, R. B., Marván, F. G., et al. (2004). *Why are Cancun beaches eroding? A question of integrated coastal zone management.* In UPADI (Ed.). Mexico: XIII Congreso Panamericano de Ingeniería Oceánica y Costera.

Diez, J. J., Esteban, M. D., & Paz, R. M. (2009). Cancun-nizuc coastal barrier. *Journal of Coastal Research*, 57–68. http://dx.doi.org/10.2112/06-0822.1.

Dröse, C. (2012). *Untersuchungen zur Seegangsbelastung und Sicherung von künstlichen sandigen Bauwerken im Mündungsbereich der Elbe. Masters degree.* Rostock, Germany: Agrar- und Umweltwissenschaftliche Fakultät, Universität Rostock.

Dubrawski, R., & Zawadzka-Kahlau, E. (2006). *The future of the Polish coastal protection (in Polish).* Gdańsk, Poland: Maritime Institute in Gdańsk.

EA. (2012a). *Summary report. South Devon catchment flood management plan.* Bristol, UK: Environment Agency. http://www.environment-agency.gov.uk/static/documents/Leisure/_CFMP_South_Devon_2012.pdf. Last accessed 17.12.13.

EA. (2012b). *Working with others: Building trust with communities.* A Guide for Staff. Bristol, UK: Environment Agency (EA). http://www.participationcymru.org.uk/working-with-others-building-trust-with-communities. Last accessed 17.12.13.

EDATER. (2001). *Estimation des dégâts après grands événements. Direction de la prévention des pollutions et des risques. Sous-direction de la prévention des risques majeurs.* Ministère de l'aménagement du territoire et de l'environnement. http://catalogue.prim.net/141_estimation-des-degats-apres-grands-evenements-.html. Last accessed 22.01.14

Edmund C. Penning-Rowsell, Wout S. de Vries, Dennis J. Parker, Barbara Zanuttigh, David Simmonds, Ekaterina Trifonova, François Hissel, Jaak Monbaliu, Jacek Lendzion, Nino Ohle, Pedro Diaz, Tjeerd Bouma, Innovation in coastal risk management: An exploratory analysis of risk governance issues at eight THESEUS study sites, Coastal Engineering, Volume 87, May 2014, Pages 210–217, ISSN 0378-3839, http://dx.doi.org/10.1016/j.coastaleng.2013.12.005.

Erdlenbruch, K., Thoyer, S., Grelot, F., Kast, R., & Enjolras, G. (2009). *Risk-sharing policies in the context of the French Flood Prevention Action Programmes.* MPRA Report 20187. Germany: Munich University Library. Munich Personal RePEc Archive. http://mpra.ub.uni-muenchen.de/20187/1/MPRA_paper_20187.pdf. Last accessed 10.12.13.

Ervin, D., Brown, D., Chang, H., Dujon, V., Granek, E., Shandas, V., et al. (2012). Growing cities depend on ecosystem services. *Solutions, 2*(6), 74–86.

Félix Delgado, A., Mendoza, E., Silva, R., Rivera Arriaga, E., & Palacio, A. (2008). *Analysis of the natural and antropogenic evolution of Cancun beach, Mexico, as a contribution to its sustainable management.* The 1st ProCoast Seminar on Coastal Research. Portugal.

Fu, B., Wang, Y. K., Xu, P., & Yan, K. (2013). Mapping the flood mitigation services of ecosystems – a case study in the Upper Yangtze River Basin. *Ecological Engineering, 52*(0), 238–246. http://dx.doi.org/10.1016/j.ecoleng.2012.11.008.

Galiatsatou, P., Prinos, P., Valchev, N., & Trifonova, E. (2012). Analysis of extreme Marine events causing flooding in varna region. *Turkish Journal of Fisheries and Aquatic Sciences, 12*, 523–531. http://dx.doi.org/10.4194/1303-2712-v12_2_44.

Gao, Z. G., & Zhang, L. Q. (2006). Multi-seasonal spectral characteristics analysis of coastal salt marsh vegetation in Shanghai, China. Estuarine. *Coastal and Shelf Science, 69*, 1–2. http://dx.doi.org/10.1016/j.ecss.2006.04.016, 217–224.

Ge, J., Much, D., Kappenberg, J., Ohle, N., Ding, P., & Chen, Z. (2013). Simulating storm flooding maps over HafenCity under present and sea level rise scenarios. *Journal of Flood Risk Management.* http://dx.doi.org/10.1111/jfr3.12054.

Glindemann, H., Thode, K., Dücker, H. P., & Witte, H.-H. (2006). *Concept for a sustainable development of the Tidal Elbe river as an artery of the metropolitan region Hamburg and beyond. A contribution for discussion.* Hamburg, Germany: Hamburg Port Authority and the Federal Administration for Waterways and Navigation. http://www.tideelbe.de/files/strategiepapier_tideelbe_komplett_english_final_020408.pdf. Last accessed 13.11.13.

Goda, Y. (2000). *Random seas and design of maritime structures.* Singapore: World Scientific.

Gonzalez-Leija, M., Mariño-Tapia, I., Silva, R., Enriquez, C., Mendoza, E., Escalante-Mancera, E., et al. (2013). Morphodynamic evolution and sediment transport processes of Cancun beach. *Journal of Coastal Research, 29*(5), 1146–1157. http://dx.doi.org/10.2112/JCOASTRES-D-12-00110.1.

Grozdev, D. (2006). Wind wave climate along the Bulgarian Black Sea coast with international participation. In *Proceedings of the 5th national Geophysical conference "GEOSCIENCES 2006". November 30–December 1, Sofia, Bulgaria* (pp. 296–299).

Guido Aldana, P., Ramírez Camperos, A., Godínez Orta, L., & Cruz León, S. (2009). Estudio de la erosión costera en Cancún y la riviera Maya, México. *Avances en Recursos Hidráulicos, 20*, 41–56.

Haigh, I., Nicholls, R. J., & Wells, N. (2009). Twentieth-century changes in extreme still sea levels in the Engliah Channel. *Continental Shelf Research, 29*, 2083–2098.

Halcrow. (2011). *South Devon & Dorset shoreline management plan review*. Exeter, UK: Halcrow. http://www.sdadcag.org/Enter.htm. Last accessed 25.04.13.

Hanley, M. E., Hoggart, S. P. G., Simmonds, D. J., Bichot, A., Colangelo, M. A., Bozzeda, F., et al. (2014). Shifting sands? Coastal protection by sand banks, beaches and dunes. *Coastal Engineering*. http://dx.doi.org/10.1016/j.coastaleng.2013.10.020.

Herrling, G., & Niemeyer, H. D. (2006). *Hydro- and morphological pressures and impacts*. Harmonised River Basins Strategies North Sea (Interreg3B Project "HARBASINS") project report. Norden, Germany: Niedersachsen Landesbetrieb für Wasserwirtschaft, Küsten- und Naturschutz (NLWKN). http://www.nlwkn.niedersachsen.de/wasserwirtschaft/nordseekueste/FSK/kuesteningenieurwesen/kuesteningenieurwesen-106527.html. Last accessed 24.09.13.

Hissel, F., Morel, G., Pescaroli, G., Graaff, H., Felts, D., & Pietrantoni, L. (2014). Early warning and mass evacuation in coastal cities. *Coastal Engineering*. doi/10.1016/j.coastaleng.2013.11.015.

Hochfeld, B., & Meine, M. (2011). Fünf Jahre Tideelbekonzept. *Wasser und Abfall, Ausgabe 6*, 2–6.

Hoeksema, R. J. (2006). *Dwelling mounds: An escape from the floods. Designed for dry feet: Flood protection and land reclamation in the Netherlands*. Reston, VA, USA: American Society of Civil Engineers (ASCE).

Hori, K., Saito, Y., Zhao, Q. H., & Wang, P. X. (2002). Architecture and evolution of the tide-dominated Changjiang (Yangtze) River delta, China. *Sedimentary Geology, 146*(3–4), 249–264. http://dx.doi.org/10.1016/s0037-0738(01)00122-1.

Horn, D., & McShane, M. (2013). Flooding the market. *Nature Clim. Change, 3*(11), 945–947. http://dx.doi.org/10.1038/nclimate2025.

Horrillo-Caraballo, J. M., Pan, S., Reeve, D., Simmonds, D., Greaves, D., & Fox, A. (2012). Modelling extreme wave events (present and future scenarios) in Southwest England. In P. Lynett, & J. McKee Smith (Eds.), *Proceedings of 33rd conference on coastal engineering. 1–6 july*. Santander, Spain: Coastal Engineering Research Council. http://journals.tdl.org/icce/index.php/icce/article/view/6517.

Huang, H. M.. (2009) Spatial-temporal dynamics of salt marsh vegetation at the intertidal zone in Shanghai based on remote sensing and GIS analysis. PhD Thesis. Eastern China Normal University, Shanghai, China.

IH Cantabria. (2010). *mplementación del plan marco de gestión de los espacios acuáticos de la red Natura 2000 en Cantabria, Gobierno de Cantabria*. Santander, Spain: Consejería de Desarrollo Rural, Ganadería, Pesca y Biodiversidad.

INAH (2013) Instituto Nacional de Antropología (National Institute of Anthropology). www.inah.gob.mx. Last accessed July 2013.

INEGI. (2010). *XII Censo General de Población y Viviendad 2010*. www.inegi.org.mx. Last accessed February 2013.

Instituto de Ingenieria. (1989). *Determinación de las características de oleaje para las playas de Cancún Q. Roo Report for FONATUR*. Universidad Nacional Autónoma de México (UNAM). pp.74.

Instituto de Ingenieria. (1990). *La restitución artificial de playas en Cancún Q.Roo. Report for FONATUR*. Universidad Nacional Autónoma de México (UNAM). pp.56.

IPCC, 2007: Climate Change 2007: Synthesis Report. Contribution of Working Groups I, II and III to the Fourth Assessment Report of the Intergovernmental Panel on Climate Change [Core Writing Team, Pachauri, R.K and Reisinger, A. (eds.)]. IPCC, Geneva, Switzerland, 104 pp.

IPCC SRES SPM. (2000). Summary for Policymakers. In *Emissions scenarios: A special report of IPCC working group iii (PDF)*. IPCC, ISBN 92-9169-113-5.

ISTAT, 2009. http://www3.istat.it/dati/catalogo/20100526_00/Avvio2009.pdf, in Italian.

Izaguirre, C., Méndez, F. J., Espejo, A., Losada, I. J., & Reguero, B. G. (2013). Extreme wave climate changes in Central-South America. *Climatic Change, 119*(2), 277–290.

Izaguirre, C., Mendez, F. J., Menendez, M., Luceno, A., & Losada, I. J. (2010). Extreme wave climate variability in southern Europe using satellite data. *Journal of Geophysical Research-oceans, 115*. http://dx.doi.org/10.1029/2009jc005802.

Izaguirre, C., Méndez, F. J., Espejo, A., Losada, I. J., & Reguero, B. G. (2010). Extreme wave climate changes in Central-South America. *Climatic Change, 119*, 227–290. http://dx.doi.org/10.1007/s10584-013-0712-9.

Jakusik, E., Wójcik, R., Biernacik, D., & Miętus, M. (2010). *Influence of changes of atmospheric circulation above europe and North Atlantic on the mean sea level – Polish coastal zone of Baltic sea.* UJK Kielce. pp. 59–74, (in Polish).

Kalnay, E., Kanamitsu, M., Kistler, R., Collins, W., Deaven, D., Gandin, L., et al. (1996). The NCEP/NCAR 40-Year reanalysis project. *Bulletin of the American Meteorological Society, 77*(3), 437–471. http://dx.doi.org/10.1175/1520-0477 (1996)077<0437:TNYRP>2.0.CO;2.

Katz, R. W., Parlange, M. B., & Naveau, P. (2002). Statistics of extremes in hydrology. *Advances in Water Resources, 25*(8–12), 1287–1304. http://dx.doi.org/10.1016/s0309-1708(02)00056-8.

Keuler, K., Lautenschlager, M., Wunram, C., Keup-Thiel, E., Schubert, M., Will, A., et al. (2009). *Climate simulation with CLM, scenario A1B run no.1, data stream 2: European region MPI-M/MaD.* World Data Center for Climate. http://dx.doi.org/10.1594/WDCC/CLM_A1B_1_D2. http://cera-www.dkrz.de/WDCC/ui/Compact.jsp?acronym=CLM_A1B_2_D2.

Kirwan, M., & Temmerman, S. (2009). Coastal marsh response to historical and future sea-level acceleration. *Quaternary Science Reviews, 28*(17–18), 1801–1808. http://dx.doi.org/10.1016/j.quascirev.2009.02.022.

Kjerfve, B. (1994). Coastal lagoon processes. In B. Kjerfve (Ed.), *Coastal lagoon processes* (pp. 1–8). Amsterdam, The Netherlands: Elsevier Science Publishers B.V.

Kohonen, T. (2000). *Self-organizing maps* (3rd ed.). Berlin, Germany: Springer-Verlag.

Konsulova, T., & Doncheva, V. (2014). Ecological impact assessment of groins in Varna Bay (Black Sea, Bulgaria) - a prerequisite for application of environmentally friendly shore protection structures. *Acta Zoologica Bulgarica, 66*(1).

Kowalska, B., & Sztobryn, M. (2009). Mean sea changes along polish coast. Quaestiones Geographicae, series a. *Physical Geography, 28A/2*, 69–74.

Krestenitis, Y. N., Androulidakis, Y. S., Kontos, Y. N., & Georgakopoulos, G. (2011). Coastal inundation in the north-eastern mediterranean coastal zone due to storm surge events. *Journal of Coastal Conservation, 15*(3), 353–368. http://dx.doi.org/10.1007/s11852-010-0090-7.

Kunkel, K. E., Karl, T. R., Brooks, H., Kossin, J., Lawrimore, J. H., Arndt, D., et al. (2012). Monitoring and understanding trends in extreme storms: state of knowledge. *Bulletin of the American Meteorological Society.*

Kwan, S. H. (2011). GIS modelling of Flood on the South Devon Coast. MSc. thesis. Faculty of Science and Environment, Plymouth University, Plymouth, UK.

Laborie, V., Hissel, F., & Sergent, P. (2012a). Üvaluation des niveaux d'eau extrêmes du futur sur l'estuairede la Gironde. In *Üvénements extrêmes fluviaux et maritimes: Leurs variabilités spatiales et chronologiques dans l'Ouest de l'Europe: Congrès de la Société Hydrotechnique de France 1–2 Fevrier, Siège de l'Unesco, Paris, France.*

Laborie, V., Hissel, F., & Sergent, P. (2012b). Üvolution de l'emprise des zones inondables de l'estuaire de la Gironde sous l'influence du changement climatique. In *XIIièmes journées nationales Génie Côtier - Génie civil. 12–14 June, Cherbourg, France.* http://dx.doi.org/10.5150/jngcgc.2012.103-L. http://www.paralia.fr/jngcgc/12_103_laborie.pdf. Last accessed 11 December 2013.

Laborie, V., Hissel, F., & Sergent, P. (2013). Validation of a 1D simplified model of the Gironde estuary based on a database of TELEMAC simulations. In *Proceedings of the 20th Telemac-Mascaret user conference (TUC 2013). 16–18 october, Karlsruhe, Germany.* http://vzb.baw.de/publikationen/vzb_dokumente_oeffentlich/0/tuc_2013_gesamt_rgb_111013.pdf. Last accessed 22 January 2014.

Li, C. X., Wang, P., Sun, H. P., Zhang, J. Q., Fan, D. D., & Deng, B. (2002). Late Quaternary incised-valley fill of the Yangtze delta (China): its stratigraphic framework and evolution. *Sedimentary Geology, 152*(1–2), 133–158. http://dx.doi.org/10.1016/s0037-0738(02)00066-0.

Li, H., & Zhang, L. (2008). An experimental study on physical controls of an exotic plant *Spartina alterniflora* in Shanghai, China. *Ecological Engineering, 32*(1), 11–21. http://dx.doi.org/10.1016/j.ecoleng.2007.08.005.

Li, J., Dai, Z., Ying, M., Wu, R., Fu, G., & Xu, H. (2007). Analysis on the development and evolution of tidal flats and reclamation of land resource along shore of Shanghai City. *Journal of Natural Resources, 22*(3), 361–371.

Locustella. Anuario de la Naturaleza de Cantabria no 3-2005 "Contribución al conocimiento de las comunidades de zostera spp del cantábrico: La bahía de Santander y las marismas de Santoña" Bárbara Ondiviela Eizaguirre, Manuel A. Alcántara Izquierdo y Lina Fernández Vélez

Loriente, E. (1996). *La vegetación y la flora del gran arenal de Somo-Loredo, Botánica Cántabra* (Santander, Spain).

Losada, I. J., Méndez, F. J., Izaguirre, C., Mínguez, R., & García, E. (2012). Climate change risk from coastal flooding and erosion in Santander. In *33nd international conference on coastal engineering. June 30-July, santander, Spain.*

Loucks, R. G., & Ward, W. C. (2001). Stratification and beach-to-dune transitions in a Holocene carbonate eaolianite complex, Isla Cancun, Quintana Roo, Mexico. In R. Abegg, & M. Harris (Eds.), *Modern and ancient carbonate eaolianites; sedimentology, sequence stratigraphy, and diagenesis* (pp. 57–76). SEPM (Society for Sedimentary Geology) Special Publication No. 71.

LTV 2030. (2001). *Langetermijnvisie Schelde-estuarium (long term vision (LTV) for the scheldt estuary) 2030.* Middelburg, The Netherlands: Ministerie van Verkeer en Waterstaat, Directoraat-Generaal Rijkswaterstaat. Directie Zeeland/Ministerie van de Vlaamse Gemeenschap, Departement Leefmilieu en Infrastructuur. Administratie Waterwegen en Zeewezen. http://www.vnsc.eu/uploads/2012/01/ltv-schelde-estuarium-18-januari-2001.pdf. Last accessed 30.09.13.

Martell, R., Mendoza, E., Mariño, I., Silva, R., & Escalante, E. (2012). Impactos de corto plazo del huracán Dean sobre la morfología de la playa de Cancún, México. *Tecnología y Ciencias del Agua, 3*(4), 89–111.

Martinelli, L., Pezzutto, P., & Ruol, P. (2013). Experimentally based model to size the geometry of a new OWC device, with reference to the Mediterranean Sea wave environment. *Energies, 6*(9), 4696–4720.

Martinelli, L., Zanuttigh, B., & Corbau, C. (2010). Assessment of coastal flooding hazard along the Emilia Romagna littoral, IT. *Coastal Engineering, 57*(11–12), 1042–1058. http://dx.doi.org/10.1016/j.coastaleng.2010.06.007.

McLachlan, A., & Dorvlo, A. (2005). Global patterns in sandy beach macrobenthic communities. *Journal of Coastal Research*, 674–687. http://dx.doi.org/10.2112/03-0114.1.

Meehl, G. A., Stocker, T. F., Collins, W. D., Friedlingstein, P., Gaye, A. T., Gregory, J. M., et al. (2007). Global climate projections. In S. Solomon, D. Qin, M. Manning, Z. Chen, M. Marquis, K. B. Averyt, et al. (Eds.), *Climate Change 2007: The Physical Science Basis. Contribution of Working Group 1 to the Fourth Assessment Report of the Intergovernmental Panel on Climate Change*. Cambridge, UK and New York, NY, USA: Cambridge University Press.

Meire, P., Ysebaert, T., Van Damme, S., Van den Bergh, E., Maris, T., et al. (2005). The Scheldt estuary: a description of a changing ecosystem. *Hydrobiologia, 540,* 1–11. http://dx.doi.org/10.1007/s10750-005-0896-8.

Méndez, F. J., Menendez, M., Luceno, A., & Losada, I. J. (2006). Estimation of the long-term variability of extreme significant wave height using a time-dependent Peak over Threshold (POT) model. *Journal of Geophysical Research-oceans, 111*(C7). http://dx.doi.org/10.1029/2005jc003344.

Menendez, M., Méndez, F. J., Losada, I. J., & Graham, N. E. (2008). Variability of extreme wave heights in the northeast Pacific Ocean based on buoy measurements. *Geophysical Research Letters, 35,* L22607.

Mendoza, E., Silva, R., Enriquez-Ortiz, C., Mariño-Tapia, I. & Felix, A.. (in press) Analysis of the hazards and vulnerability of the cancun Beach system: the case of Hurricane wilma. in Chavez, M., Ghil, M. & Urrutia-Fucugauchi, J. (eds.). Extreme Events: Observations, modelling and economics. AGU Book Series, USA.

Micu, D., Zaharia, T., Todorova, V., & Niţă, V. (2007). *Romanian marine habitats of European interest* (in Romanian), Punct Ochit, Constanţa, Romania.

Miętus, M. (Ed.). (2010). *Report of KLIMAT project.* on http://klimat.imgw.pl.

Miętus, M. (Ed.). (2009). *Report of KLIMAT project.* on http://klimat.imgw.pl.

Miętus, M., & Filipiak, J. (2002). *Wpływ termiki powierzchniowej warstwy wody północnego Atlantyku na wielkoskalową cyrkulację atmosferyczną w rejonie Atlantyku i Europy oraz na warunki termiczne w Polsce w XX w.* In *Materialy Badawcze: Meteorologia, 35.* Warsawa, Poland: Instytut Meteorologii i Gospodarki Wodnej (IMGW). pp.68.

Miętus, M. (1999). *Rola regionalnej cyrkulacji atmosferycznej w kształtowaniu warunków klimatycznych i oceanograficznych w polskiej strefie brzegowej Morza Bałtyckiego.* In *Materialy Badawcze: Meteorologia, 29.* Warsawa, Poland: Instytut Meteorologii i Gospodarki Wodnej (IMGW). pp.157.

Mignot, C. (1969). *Estuaire de la Gironde, synthèse des phénomènes naturels.* Technical report. Chatou, France: Laboratoire central d'hydraulique de France.

Miller, J., & Livermont, E. (2008). An index for predicting storm erosion due to increased waves and water levels. In L. Wallendorf, L. Ewing, C. Jones, & B. Jaffe (Eds.), *Proceedings of solutions to coastal disasters 2008. April 13-16, Turtle Bay, Oahu, Hawaii* (pp. 561–572). American Society of Civil Engineers (ASCE). http://dx.doi.org/10.1061/40968(312) 51.

Miller, J. K., & Dean, R. G. (2004). A simple new shoreline change model. *Coastal Engineering, 51,* 531–556.

Milliman, J. P., & Farnsworth, K. L. (2011). *River discharge to the coastal ocean: A global synthesis.* New York, USA: Cambridge University Press.

Ministerie van Infrastructuur en Milieu, Rijkswaterstaat Zeeland & Uitvoerend secretariaat van de Vlaams-Nederlandse Scheldecommissie. (2013). *T2009-Rapport Schelde estuarium (in dutch).* T2009 Report Scheldt Estuary. pp500 + appendices http://www.vnsc.eu/organisatie/werkgroepen/onderzoek-en-monitoring/voortgang/projectgroep-evaluatie-en-rapportage. Last accessed 9 September 2013.

Mitsch, W. J., & Gosselink, J. G. (2007). *Wetlands* (4th ed.). New York, USA: John Wiley and Sons.

Monbaliu, J., Chen, Z., Felts, D., Ge, J., Hissel, F., Kappenberg, J., et al. (2014). Risk assessment of estuaries under climate change: lessons from Western Europe. *Coastal Engineering.* http://dx.doi.org/10.1016/j.coastaleng.2014.01.001.

Morellato, D., & Benoît, M. (2010). Vagues et changement climatique – simulation des états de mer dans l'Océan Atlantique de 1960 à 2100 pour trois scénarios de changement climatique. *Journées Scientifiques et Techniques du CETMEF,* 8–10 December, Brest France. http://colloques-cetmef.developpement-durable.gouv.fr/IMG/pdf/Resume_Morellato_cle4e1c6d.pdf

Morris, J. T., Sundareshwar, P. V., Nietch, C. T., Kjerfve, B., & Cahoon, D. R. (2002). Responses of coastal wetlands to rising sea level. *Ecology, 83*(10), 2869–2877. http://dx.doi.org/10.2307/3072022.

Municipality of Varna. (2011). *Update of the municipal development plan of the municipality of varna for the period 2007–2013*. Regional Agency for Entrepreneurship and Innovation–Varna. www.varna.bg

Nakicenovic, N., & Swart, R. (Eds.). (2000). *Emissions scenarios. Special report of the intergovernmental panel on climate change*. Cambridge, UK: Cambridge University Press.

Narayan, S., Nicholls, R. J., Clarke, D., Hanson, S., Reeve, D., Horrillo-Caraballo, J., et al. (2014). The SPR systems model as a conceptual foundation for rapid integrated risk appraisals: lessons from Europe. *Coastal Engineering*. http://dx.doi.org/10.1016/j.coastaleng.2013.10.021.

Narayan, S., Nicholls, R., Clarke, D., & Hanson, S. (2011). Investigating the Source – pathway – Receptor – consequence framework for coastal flood system analyses. In *ICE coastal management 2011. 15–16 november, Belfast, northern Ireland*. http://eprints.soton.ac.uk/341048/1/S_Narayan_Investigating-the-SPRC.pdf. Last accessed 30 September 2013.

Nicholls, R. J., Townend, I. H., Bradbury, A. P., Ramsbottom, D., & Day, S. A. (2013). Planning for long-term coastal change: experiences from England and Wales. *Ocean Engineering, 71*, 3–16. http://dx.doi.org/10.1016/j.oceaneng.2013.01.025.

Nicholls, R. J., Wong, P. P., Burkett, V. R., Codignotto, J. O., Hay, J. E., McLean, R. F., et al. (2007). Coastal systems and low-lying areas. In M. L. Parry, O. F. Canziani, J. P. Palutikof, P. van der Linden, & C. E. Hanson (Eds.), *Climate change 2007: Impacts, adaptation and vulnerability. Contribution of working group ii to the Fourth assessment report of the Intergovermental panel on climate change*. Cambridge, UK: Cambridge University Press.

NSI. (2013). *Annual accommodation data*. National Statistical Institute of Bulgaria. http://www.nsi.bg/otrasalen.php?otr=57&a1=2061&a2=2062&a3=2072#cont. Last accessed 14.11.13.

Ntegeka, V., Baguis, P., Boukhris, O., Willems, P., & Roulin, E. (2008). Climate change impact on hydrological extremes along rivers and urban drainage systems: technical report II. In *Study of rainfall and ETo climate change scenarios. Belgian science policy (BELSPO) - SSD research programme* (p. 112). Brussels, Belgium, KU: Leuven-Hydraulics Section and Royal Meteorological Institute of Belgium. http://www.kuleuven.be/hydr/cci/reports/CCI-HYDR_II-ClimateChangeScenarios.pdf. Last accessed 22.01.14.

Ntegeka, V., Decloedt, L. C., Willems, P., & Monbaliu, J. (2012). Quantifying the impact of climate change from inland, coastal and surface conditions. In F. Klijn, & T. Schweckendiek (Eds.), *Proceedings of the 2nd european conference on flood risk management FLOODrisk2012, Rotterdam, the Netherlands, 19–23 november 2012: Comprehensive flood risk management: Research for policy and practice* (pp. 103–111). Leiden, The Netherlands: CRC Press.

Ntegeka, V., Roulin, E., Baguis, P., & Willems, P. (2014). Developing tailored climate change scenarios for hydrological impact assessments. *Journal of Hydrology, 508C*, 307–321. http://dx.doi.org/10.1016/j.jhydrol.2013.11.001.

Ondiviela, B., Losada, I. J., Lara, M., Galván, C., Bouma, T. J., & van Belzen, J. (2014). The role of seagrasses in coastal protection in a changing climate. *Coastal Engineering, 87*, 158–168.

ONERC. (2010). *Recommandations sur l'élévation du niveau de la mer. Synthèse: Prise en compte de l'élévation du niveau de la mer en vue de l'estimation des impacts du changement climatique et des mesures d'adaptation possibles*. La Défense, France: Observatoire National sur les Effets du Réchauffement Climatique (ONERC). http://www.developpement-durable.gouv.fr/IMG/pdf/synth_niveau_mer.pdf. Last accessed 16 January 2014.

Ortiz Pérez, M. A., & de la Lanza Espino, G. (2006). *Diferenciación del espacio costero de méxico: Un inventario regional. Geografía para el siglo XXI*. México: Series de Textos Universitarios, Instituto de Geografía.

Ozer, J., van den Eynde, D., & Ponsar, S. (2008). Trend analysis of the relative mean sea level at Oostende (Southern North Sea-Belgian coast). In D. van den Eynde, R. De Sutter, F. Maes, T. Verwaest, & E. van Bockstaele (Eds.), *Evaluation of climate change impacts and adaptation responses for marine activities: ClLIMAR report CLIMAR/X/JO/200807/EN/TR3*. Brussels, Belgium: Belgium Science Policy.

Pedro Diaz-Simal & Phoebe Koundouri & Benedique Rulleau & Remoundou Kyriaki,. "Valuing Climate Change Mitigation in Coastal Environments Exposed to Extreme Natural Hazards: A choice experiment simulated for different time horizons," DEOS Working Papers 1203, Athens University of Economics and Business.

Penning-Rowsell, E. C., de Vries, W. S., Parker, D. J., Zanuttigh, B., Simmonds, D., Trifonova, E., et al. (2014). Innovation in coastal risk management: an exploratory analysis of risk governance issues at eight THESEUS study sites. *Coastal Engineering*. http://dx.doi.org/10.1016/j.coastaleng.2013.12.005.

Penning-Rowsell, E., Floyd, P., Ramsbottom, D., & Surendran, S. (2005). Estimating injury and loss of life in floods: a deterministic framework. *Natural Hazards, 36*(1–2), 43–64. http://dx.doi.org/10.1007/s11069-004-4538-7.

Plüß, A. D., & Heyer, H. (2007). Morphodynamic multi-model approach for the Elbe estuary. In C. M. Dohmen-Janssen, & S. J. M. H. Hulscher (Eds.), *Proceedings of the 5th IAHR Symposium on river, coastal and estuarine morphodynamics: RCEM 2007. 17–21 september, Enschede, The Netherlands* (pp. 113–117). London, UK: Tylor & Francis Group.

Policy Research Corporation. (2009). *The economics of climate change adaptation in EU coastal areas*. Report for Directorate-General for Maritime Affairs and Fisheries, European Commission. Antwerp, Belgium: Policy Research Corporation http://ec.europa.eu/maritimeaffairs/documentation/studies/documents/report_en.pdf. Last accessed 30 October 2013.

Popov, V., & Mishev, K. (1974). *Geomorphology of the bulgarian black sea coast and shelf. (In bulgarian)*. Sofia, Bulgaria: Publishing House of the Bulgarian Academy of Sciences.

Posada, G., Simmonds, D., Silva, R., & Pedrozo, A. (2008b). A 2D hydrodynamic model with Multi-Quadtree Mesh. In A. L. Prescot (Ed.), *Ocean engineering research advances* (pp. 205–241). New York, USA: Nova Science Publishers.

Preti, M., De Nigris, N., Morelli, M., Monti, M., Bonsignore, F., & Aguzzi, M. (2009). *State of the Emilia-Romagna littoral at 2007 and ten-years management plan* (in Italian, abstract in English). Bologna, Italy: I quaderni dell'ARPA.

ProSes. (2005). *Ontwikkelingsschets 2010 Schelde-estuarium*. Bergen op Zoom, The Netherlands: Projectdirectie ontwikkelingsschets Schelde-estuarium. http://www.vnsc.eu/uploads/2012/01/os2010-samenvatting-engels.pdf. Last accessed 16.12.13.

Pullen, T., Allsop, N. W. H., Bruce, T., Kortenhaus, A., Schuttrumpf, H., & van der Meer, J. W. (2008). *EUROTOP: Wave overtopping of sea defences and related structures: Assessment manual*. Research for UK Environment Agency, German Coastal Research Engineerng Research Council and Riijkswaterstaat, Netherlands. Wallingford, UK: H R Wallingford http://www.overtopping-manual.com. Last accessed 17 December 2013.

RA-IMDC & Vito. (2004). *Conclusies op hoofdlijnen (in Dutch). Maatschappelijke kosten batenanalyse voor de actualisatie van het Sigmaplan* (Social cost-benefit analysis for the actualisation of the Sigma Plan). Antwerp, Belgium: Tijdelijke vereniging RA-IMDC and Vlaamse Instelling voor Technologisch Onderzoek http://www.vnsc.eu/publicaties/vnsc-publicaties/107-maatschappelijke-kosten-batenanalyse-voor-de-actualisatie-van-het-sigmaplan.html. Last accessed 18.09.13.

Reeve, D., Horrillo-Caraballo, J. M., Fox, A., Simmonds, D., Kwan, S., Pan, S., et al. (2012). Coastal defence evaluation; an SPRC approach. In P. Lynett, & J. McKee Smith (Eds.), *Proceedings of 33rd conference on coastal engineering. 1–6 july, santander, Spain*. Coastal Engineering Research Council. http://journals.tdl.org/icce/index.php/icce/article/view/6517.

Reguero, B. G., Menendez, M., Méndez, F. J., Minguez, R., & Losada, I. J. (2012). A Global Ocean Wave (GOW) calibrated reanalysis from 1948 onwards. *Coastal Engineering, 65*, 38–55. http://dx.doi.org/10.1016/j.coastaleng.2012.03.003.

Renn, O. (2008). *Risk governance; coping with uncertainty in a complex world.* London, UK: Earthscan.

Republic of Bulgaria National Assembly. (1998). *Protected areas act.* State Gazette No. 133/11.11.1998 (amended, supplemented and effective State Gazette No. 19/8.03.2011) http://www3.moew.government.bg/files/file/PNOOP/Acts_in_English/Protected_Areas_Act.pdf. Last accessed 30 October 2013.

Republic of Bulgaria National Assembly. (2002). *Biological diversity act.* State Gazette No. 77/9.08.2002 http://europe.bg/upload/docs/CONF_BG_02_03_ad12.pdf. Last accessed 30 October 2013.

Republic of Bulgaria National Assembly. (2010). *Protections against water-related damage and loss.* State Gazette No. 61/2010. Chapter 9 in the Water Act. State Gazette No. 67/27.07.1999 http://www3.moew.government.bg/files/file/PNOOP/Acts_in_English/Water_Act.pdf. Last accessed 30 October 2013.

Rozhdestvenski, A. (1971). *Movements of near-shore sands in Varna Bay (in Bulgarian).* In *Proceedings of the Research Institute of Fisheries and Oceanography* (Vol. 11), 33–41.

Ruíz Martínez, G.. (2009) Determinación del estado morfodinámico de segmentos de playa que poseen obstáculos sumergidos y emergidos. PhD Thesis. Universidad Nacional Autónoma de México (UNAM).

Ruiz, G., Silva, R., & Posada, G. (2013). Comparación morfodinámica de la costa noreste del estado de Quintana Roo, México. *Tecnologiá y Ciencias del Agua, 4*(3), 47–65.

Saito, M., Watanuki, A., Nishigori, W., & Hanzawa, M. (2002). *Stability of a new type of submerged breakwater and its effect on ecosystem,* poster paper no 33. Abstracts of 28th International Conference on Coastal Engineering. Wales, UK: Cardiff.

Savenije, H. (1998). Analytical expression for tidal damping in alluvial estuaries. *Journal of Hydraulic Engineering, 124*(6), 615–618. http://dx.doi.org/10.1061/(ASCE)0733-9429(1998)124:6(615).

Schlichting, H. T., & Gertsen, K. (1999). *Boundary layer theory.* Berlin: Springer-Verlag and Heidelberg GmbH & Co.

Schroevers, M., Huisman, B. J. A., Van der Wal, M., & Terwindt, J. (2011). Measuring ship induced waves and currents on a tidal flat in the Western Scheldt estuary. *Current, Waves and Turbulence Measurements (CWTM)*, 123–129. http://dx.doi.org/10.1109/CWTM.2011.5759539, 2011 IEEE/OES 10th.

Schwarz, J., & Maiwald, H. (2008). Damage and loss prediction model based on the vulnerability of building types. In *4th international Symposium on flood Defence: Managing flood risk, reliability and vulnerability. May 6–8, Toronto, Ontario, Canada.* http://gfzpublic.gfz-potsdam.de/pubman/item/escidoc:6057:6/component/escidoc:6058/74_Schwarz.pdf. Last accessed 9 January 2014.

SEDETUR. (2008). *Tourism statistics.* http://sedetur.qroo.gob.mx/estadisticas/estadisticas.php. Last accessed July 2013.

Shanghai Water Authority. (2010). Shanghai Foreshore Resources Report. *Shanghai Water Authority.* Shanghai (in Chinese).

Shi, B. W., Yang, S. L., Wang, Y. P., Bouma, T. J., & Zhu, Q. (2012). Relating accretion and erosion at an exposed tidal wetland to the bottom shear stress of combined current-wave action. *Geomorphology, 138*(1), 380–389. http://dx.doi.org/10.1016/j.geomorph.2011.10.004.

Silva, R., Borthwick, A. G. L., & Taylor, R. E. (2005). Numerical implementation of the harmonic modified mild-slope equation. *Coastal Engineering, 52*(5), 391–407. http://dx.doi.org/10.1016/j.coastaleng.2004.12.009.

Silva, R., Mariño, I., Enriquez, C., Mendoza, E., Escalante, E., & Ruíz, F. (2006a). Monitoring shoreline changes at Cancún Beach, Mexico: effects of hurricane Wilma. In *30th international conference on coastal engineering.* Singapore: World Scientific.

Silva, R., Mendoza, E., & Losada, M. A. (2006b). Modelling linear wave transformation induced by dissipative structures – regular waves. *Ocean Engineering, 33*(16), 2150–2173. http://dx.doi.org/10.1016/j.oceaneng.2005.11.007.

Silva, R., Ruiz, G., Mariño, I., Posada, G., Mendoza, E., & Escalante, E. (2012). Manmade vulnerability of the cancun Beach system: the case of Hurricane wilma. *CLEAN- Soil, Air, Water, 40*(9), 911–919.

Silva, R., Ruíz, G., Posada, G., Pérez, D., Rivillas, G., Espinal, J. C., et al. (2008). *Atlas de Clima Marítimo para las vertientes mexicanas del Pacífico y Atlántico*. México, Instituto de Ingeniería. Universidad Nacional Autónoma de México.

Slangen, A. B. A., Katsman, C. A., van de Wal, R. S. W., Vermeersen, L. L. A., & Riva, R. E. M. (2012). Towards regional projections of twenty-first century sea-level change based on IPCC SRES scenarios. *Climate Dynamics, 38*(5–6), 1191–1209. http://dx.doi.org/10.1007/s00382-011-1057-6.

Smaal, A. C., Kater, B. J., & Wijsman, J. (2009). Introduction, establishment and expansion of the Pacific oyster *Crassostrea gigas* in the Oosterschelde (SW Netherlands). *Helgoland Marine Research, 63*(1), 75–83. http://dx.doi.org/10.1007/s10152-008-0138-3.

Smaal, A., Stralen, M., & Craeymeersch, J. (2005). Does the introduction of the Pacific Oyster *Crassostrea gigas* lead to species shifts in the Wadden Sea? In R. Dame, & S. Olenin (Eds.), *The comparative roles of suspension-feeders in ecosystems: NATO science series IV: Earth and environmental series* (pp. 277–289) Amsterdam, The Netherlands: Springer Netherlands.

SMIDDEST. (2012). *PAPI d'intention, estuaire de la Gironde*. Blaye, France: Syndicat Mixte pour le Développement Durable de L'Estuarie de la Gironde (SMIDDEST). http://www.sage-estuaire-gironde.org/site/uploads/PAPI_2012.pdf. Last accessed 16 January 2014.

Smith, R. L. (2001). *Environmental statistics*. Chapel Hill, NC, USA: Department of Statistics, University of North Carolina. http://www.stat.unc.edu/postscript/rs/envnotes.pdf. Last accessed 10 December 2013.

SOGREAH. (2010). *Rapport d'étape 4: Définition des événements de référence. Référentiel de protection contre les inondations le long de l'estuaire de la gironde, de la garonne et de la dordogne*. Grenoble, France: Societe Grenobloise d'Etudes et d'Application Hydrauliques (SOGREAH).

Solomon, S., Qin, D., Manning, M., Chen, Z., Marquis, M., Averyt, K. B., et al. (Eds.). (2007). *Climate Change 2007: The Physical Science Basis, Contribution of Working Group I to the Fourth Assessment Report of the Intergovernmental Panel on Climate Change*. Cambridge, UK: Cambridge University Press.

Stakev, M. (1980). About the break-down state of some coastal structures after the storm in February 1979 (in Bulgarian). *Journal of Shipbuilding and Navigation, Bulgaria, 10*, 26–30.

Stanke, C., Murray, V., Amlot, R., Nurse, J., & Williams, R. (2012). The effects of flooding on mental health: outcomes and recommendations from a review of the literature. *PLoS Currents, 4*. http://dx.doi.org/10.1371/4f9f1fa9c3cae.

State Oceanic Administration. (2011). *The sea level bulletin of China in 2010. (in Chinese)*. People's Republic of China. http://www.coi.gov.cn/gongbao/nrhaipingmian/nr2010/201107/t20110729_18924_5.html

State Oceanic Administration. (2012). *The sea level bulletin of China in 2011. (in Chinese)*. People's Republic of China. http://www.coi.gov.cn/gongbao/haipingmian/201303/t20130308_26217.html.

Sterl, A., van den Brink, H., de Vries, H., Haarsma, R., & van Meijgaard, E. (2009). An ensemble study of extreme storm surge related water levels in the North Sea in a changing climate. *Ocean Sci., 5*(3), 369–378. http://dx.doi.org/10.5194/os-5-369-2009.

Stockdon, H. F., Holman, R. A., Howd, P. A., & Sallenger, A. H. (2006). Empirical parameterization of setup, swash, and runup. *Coastal Engineering, 53*(7), 573–588. http://dx.doi.org/10.1016/j.coastaleng.2005.12.005.

Stralberg, D., Brennan, M., Callaway, J. C., Wood, J. K., Schile, L. M., Jongsomjit, D., et al. (2011). Evaluating tidal marsh sustainability in the face of sea-level rise: a hybrid modeling approach applied to San Francisco Bay. *PLoS ONE, 6*(11), e27388. http://dx.doi.org/10.1371/journal.pone.0027388.

Sztobryn, M., Stigge, H. J., Wielbińska, D., Weiding, B., Stanisławczyk, I., Kańska, A., et al. (2005). *Storm surges in the southern Baltic sea (Western and Central parts)*. Hamburg, Germany: Berichte des Bundesamtes für Seeschifffahrt und Hydrographie. http://www.bsh.de/de/Produkte/Buecher/Berichte_/Bericht39/Bericht39E.pdf. Last accessed 20 January 2014.

Szymelfenig, M., Kotwicki, L., & Graza, B. (2006). Benthic recolonisation in post dredging pits in the puck Bay (Southern Balic sea). Estuarine. *Coastal and Shelf Sci., 68*, 489–498.

TELEMAC-MASCARET. (2013). *open TELEMAC-MASCARET.* Artelia, BundesAnstalt für Wasserbau, Centre d'Etudes Techniques Maritimes et Fluviales, Daresbury Laboratory, Electricité de France R&D and HR Wallingford. http://www.opentelemac.org/.

Temmerman, S., Govers, G., Wartel, S., & Meire, P. (2004). Modelling estuarine variations in tidal marsh sedimentation: response to changing sea level and suspended sediment concentrations. *Marine Geology, 212*(1–4), 1–19. http://dx.doi.org/10.1016/j.margeo.2004.10.021.

THESEUS OD1.10. (2012). *Report on quantified scenarios analysis of drivers and impacts of changing flood risk.* http://www.theseusproject.eu/resources/documents/Deliverables/Official-Deliverables-%28OD%29/OD1.10-%28M18%29-Report-on-quantified-scenarios-analysis-of-drivers-and-impacts-of-changing-flood-risk/. Last accessed 9.01.14.

THESEUS OD1.15. (2012). *Integrated report on risk assessment in the study sites.* http://www.theseusproject.eu/resources/documents/Deliverables/Official-Deliverables-%28OD%29/OD1.15-%28M24%29-Integrated-report-on-risk-assessment-in-study-sites./. Last accessed 30.11.13.

THESEUS OD1.17. (2011). *Homogenous methods for the development of local scenarios.* http://www.theseusproject.eu/resources/documents/Deliverables/Official-Deliverables-%28OD%29/OD1.7-%28M16%29-Report-on-consolidated-methologies-for-the-assessment-of-coastal-vulnerability-and-resilience-to-erosion-and-floods/. Last accessed 30.11.13.

Todorova, V. (2012). *Extension of the ecological network Natura 2000 in the Bulgarian Black Sea. (in Bulgarian).* Project report for contract 7976/04.04.2011. Enterprise for Management of Environmental Protection Activities (EMEPA) and Institute Of Oceanology – Bulgarian Academy of Sciences (IO-BAS).

Tolman, H. L. (2009). *User manual and system documentation of Wavewatch-III version 3.14.* NOAA/NWS/NCEP Technical Note (276). Camp Springs, MD, USA: Environmental Modeling Center. http://polar.ncep.noaa.gov/mmab/papers/tn276/MMAB_276.pdf. Last accessed 10 December 2013.

Tomczak, A., & Domachowska, I. (1999). *The shape of the Hel Peninsula in historic times according to cartographic documents.* GTN Gdansk: Peribalticum VII. pp. 99–114.

Toorman, E. (2002). *Modelling of turbulent flow with cohesive sediment.* In J. Winterwerp, & C. Kranenburg (Eds.), *Proceedings in Marine science* (Vol. 5); (pp. 155–169). Elsevier Science.

Toorman, E. A., & Bi, Q. (2012). Improvements to sea-bed boundary conditions. In M. Liste-Muñoz, & J. Monbaliu (Eds.), *Fluxes, interactions and environment at the land-ocean boundary: Downscaling, Assimilation and coupling (FIELD_AC) project report D3.3, section 2.4* (pp. 19–31).

Toorman, E. A., & Bi, Q. (2013a). Hybrid two-phase/mixture modeling of sediment transport. In *Proceedings of the 2nd Symposium on two-phase modelling for sediment dynamics (THESIS-2013). 10–12 June, Chatou, France* (p. 4). Société Hydraulique de Franc.

Toorman, E. A., & Bi, Q. (2013b). A practical model for drag modulation by suspended sediment, with application to the Scheldt estuary. In *12th international conference on Cohesive sediment transport (INTERCOH2013). 21–24 october, Gainesville, Florida.*

Toorman, E. A. (2011). Low-Reynolds modelling of high-concentrated near-bottom suspended sediment transport. In *Proceedings of the Symposium on two-phase modelling for sediment dynamics (THESIS-2011). April 26–28, Paris, France* (p. 4). Société Hydraulique de France. www.vliz.be/imisdocs/publications/220787.pdf. Last accessed 16 December 2013.

Trifonova, E. (2007). Modelling of cross shore profile changes under combination of extreme storm events. In *Proceedings of the 4th international conference "Port development and coastal environment" PDCE'2007. 25–27 september, varna, Bulgaria* (pp. 301–311).

Trifonova, E. V., Valchev, N. N., Andreeva, N. K., & Eftimova, P. T. (2010). Reconstruction of severe storms in the Western Black Sea and assessment of their impact on the coast. In *Proceedings of the 10th international conference "Marine sciences and technologies – Black Sea 2010". 7th–9th October, Varna, Bulgaria* (pp. 254–260).

Trifonova, E. V., Valchev, N. N., Andreeva, N. K., & Eftimova, P. T. (2012). Critical storm thresholds for morphological changes in the western Black Sea coastal zone. *Geomorphology, 143*, 81–94. http://dx.doi.org/10.1016/j.geomorph.2011.07.036.

Trifonova, E., Valchev, N., Galiatsatou, N., Eftimova, P., Andreeva, N., & Kotsev, I. (2014). *Assessment of coastal vulnerability to storm flood for Varna region under conditions of the present climate. (In Bulgarian).* Nikola Vaptsarov Naval Academy Research Papers Accepted for publication.

Truong. H.T. (2013). Climate change impact on floods along River Dender, accounting for the correlation between upstream inland and downstream coastal conditions (Masters thesis). Leuven, Belgium: Water Resources Engineering, KU Leuven and Vrije Universiteit Brussel.

Umgiesser, G., Canu, D. M., Cucco, A., & Solidoro, C. (2004). A finite element model for the Venice Lagoon. Development, set up, calibration and validation. *Journal of Marine Systems, 51*(1–4), 123–145. http://dx.doi.org/10.1016/j.jmarsys.2004.05.009.

Valchev, N. N., Trifonova, E. V., & Andreeva, N. K. (2012). Past and recent trends in the western Black Sea storminess. *Natural Hazards and Earth System Sciences, 12*(4), 961–977. http://dx.doi.org/10.5194/nhess-12-961-2012.

Valchev, N., Davidan, I., Belberov, Z., & Palazov, A. (2007). Practicability of wind waves simulations based on the global reanalysis wind fields in the Black Sea deep and shallow waters. In *Proceedings of the 4th international conference "Port development and coastal environment – 2007". Varna, Bulgaria* (pp. 185–191).

Valchev, N., Davidan, I., Belberov, Z., Palazov, A., & Valcheva, N. (2008). Estimation of wind wave climate of the Western Black Sea during the last 50 years. In *Proceedings of the 9th international conference on "Marine sciences and technologies - Black sea 2008'. Varna, Bulgaria* (pp. 231–239).

Valchev, N., Trifonova, E., Andreeva, N., & Eftimova, P. (2009). Climatic change in the storm occurrence and intensity trends. In *Proceedings of the "International multidisciplinary scientific geo-conference – SGEM'2009". Albena, Bulgaria* (pp. 305–312).

van der Meer, J. W., Briganti, R., Zanuttigh, B., & Wang, B. X. (2005). Wave transmission and reflection at low-crested structures: design formulae, oblique wave attack and spectral change. *Coastal Engineering, 52*(10–11), 915–929. http://dx.doi.org/10.1016/j.coastaleng.2005.09.005.

Vanderlinden, J.-P. (2013). Communicating risk through decision support system design: vulnerability, resilience and the design of cognitive pathways. Workshop presentation at. In *Knowledge, beliefs, representations in understanding climate issues*. Paris Consortium Climate-Environment-Society (GIS Climat- Environnement-Société) and Paragraphe http://www.gisclimat.fr/feedback-workshop-knowledge-beliefs-representations-understanding-climate-issues. Last accessed 22 January 2014.

Vergiev, S., Filipova-Marinova, M., Trifonova, E., Kotsev, I., & Pavlov, D. (2013). The impact of sea water immersion on the viability of psamophilous species *Leymus racemosus* subsp. *sabulosus* and *Ammophila arenaria. Comptes Rendus De L Academie Bulgare Des Sciences, 66*(2), 211–216.

VNSC. (2011). *Indicators of the Long-Term Vision (LTV) 2030 for the Scheldt estuary.* VLIZ Special Publication N0. 50. Oostende, Belgium: Vlaams Instituut voor de Zee (VLIZ) http://www.scheldemonitor.be/indicatoren.php. Last accessed 16 December 2013.

von Storch, H., & Zwiers, F. W. (2001). *Statistical analysis in climate research.* Cambridge, UK: Cambridge University Press.

Wang, J., Gao, W., Xu, S., & Yu, L. (2012). Evaluation of the combined risk of sea level rise, land subsidence, and storm surges on the coastal areas of Shanghai, China. *Climatic Change, 115*(3–4), 537–558. http://dx.doi.org/10.1007/s10584-012-0468-7.

Weerman, E. J., van de Koppel, J., Eppinga, M. B., Montserrat, F., Liu, Q.-X., & Herman, P. M. (2010). Spatial self-organization on intertidal mudflats through biophysical stress divergence. *The American Naturalist, 176*(1), E15–E32. http://dx.doi.org/10.1086/652991.

Weinkle, J., Maue, R., & Pielke, R., Jr (2012). Historical global tropical Cyclone Landfalls*. *Journal of Climate, 25*(13), 4729–4735.

Weisse, R., Bellafiore, D., Menéndez, M., Méndez, F., Nicholls, R. J., Umgiesser, G., et al. (2014). Changing extreme sea levels along European coasts. *Coastal Engineering.* http://dx.doi.org/10.1016/j.coastaleng.2013.10.017.

Wibig, J., & Jakusik, E. (2012). *Climatic and oceanographic conditions in Poland and the south Baltic sea – Important changes and guidelines for the development of adaptation strategies in the domestic economy (in polish).* Warszawa, Poland: Instytut Meteorologii i Gospodarki Wodnej (IMGW PIB). pp. 283.

Willems, P., & Vrac, M. (2011). Statistical precipitation downscaling for small-scale hydrological impact investigations of climate change. *Journal of Hydrology, 402*(3–4), 193–205. http://dx.doi.org/10.1016/j.jhydrol.2011.02.030.

Willems, P. (2013). Revision of urban drainage design rules after assessment of climate change impacts on precipitation extremes at Uccle, Belgium. *Journal of Hydrology, 496*, 166–177.

Willems, P., Olsson, J., Arnbjerg-Nielsen, K., Beecham, S., Pathirana, A., Bülow Gregersen, I., et al. (2012). *Impacts of climate change on rainfall extremes and urban drainage.* IWA Publishing, 252p. Paperback Print ISBN 9781780401256; Ebook ISBN 9781780401263.

Willems, P., Vaes, G., Popa, D., Timbe, L., & Berlamont, J. (2002). Quasi 2D river flood modelling. In D. Bousma, & Y. Zech (Eds.), *River flow 2002: Proceedings of the international conference on fluvial hydraulics* (pp. 1253–1259). Amsterdam, The Netherlands: Taylor and Francis.

Willet, P. (1996). *Molecular diversity techniques for chemical databases.* Information Research, 2, 3. Sweden: Lund University Libraries and University of BorÔs. Last accessed 10 December 2013. http://www.informationr.net/ir/2-3/paper19.html.

Winterwerp, J. C. (2011). Fine sediment transport by tidal asymmetry in the high-concentrated Ems River: indications for a regime shift in response to channel deepening. *Ocean Dynamics, 61*(2–3), 203–215. http://dx.doi.org/10.1007/s10236-010-0332-0.

Woppelmann, G., Miguez, B. M., Bouin, M. N., & Altamimi, Z. (2007). Geocentric sea level trend estimates from GPS analyses at relevant tide gauges world-wide. *Global and Planetary Change, 57*(3–4), 396–406. http://dx.doi.org/10.1016/j.gloplacha.2007.02.002.

Woth, K. (2005). North Sea storm surge statistics based on projections in a warmer climate: how important are the driving GCM and the chosen emission scenario? *Geophysical Research Letters, 32*(22) http://dx.doi.org/10.1029/2005gl023762.

Yang, S. L., Milliman, J. D., Li, P., & Xu, K. (2011). 50,000 dams later: erosion of the Yangtze River and its delta. *Global and Planetary Change, 75*(1–2), 14–20. http://dx.doi.org/10.1016/j.gloplacha.2010.09.006.

Yuan, L., Zhang, L., Xiao, D., & Huang, H. (2011). The application of cutting plus waterlogging to control *Spartina alterniflora* on saltmarshes in the Yangtze Estuary, China. *Estuarine Coastal and Shelf Science, 92*(1), 103–110. http://dx.doi.org/10.1016/j.ecss.2010.12.019.

Zanuttigh, B., Martinelli, L., Lamberti, A., Moschella, P., Hawkins, S., Marzetti, S., et al. (2005). Environmental design of coastal defence in Lido di Dante, Italy. *Coastal Engineering, 52*(10–11), 1089–1125. http://dx.doi.org/10.1016/j.coastaleng.2005.09.015.

Zanuttigh, B., Simcic, D., Bagli, S., Bozzeda, F., Pietrantoni, L., Zagonari, F., et al. (2014). THESEUS decision support system for coastal risk management. *Coastal Engineering*. http://dx.doi.org/10.1016/j.coastaleng.2013.11.013.

Zawadzka, E. (1996). Coastal zone dynamics during artificial nourishment. In *Proceedings of the 25th international conference on coastal engineering (ICCE)* (pp. 2955–2968). Orlando, Fl, USA: American Society of Civil Engineers (ASCE). http://journals.tdl.org/icce/index.php/icce/article/viewFile/5443/5119. Last accessed 30 September 2013.

Zhang, L. Q., & Yong, X. K. (1992). Studies on phenology and spatial distribution of *Scirpus mariqueter* population. *Acta Phytoecologica Sinica, 16*(1), 43–51 (in Chinese with English Abstract).

Zhou, Y. X., & Xie, Y. M. (2012). *Study on wetlands resource investigation and monitoring and evaluation system*. Shanghai, China: Shanghai Science and Technology Press.

Zhu, Z., Zhang, L., Wang, N., Schwarz, C., & Ysebaert, T. (2012). Interactions between the range expansion of saltmarsh vegetation and hydrodynamic regimes in the Yangtze Estuary, China. *Estuarine Coastal and Shelf Science, 96*, 273–279. http://dx.doi.org/10.1016/j.ecss.2011.11.027.

Index

Note: Page numbers followed by "b", "f" and "t" indicate boxes, figures and tables respectively.